Springer Series in Measurement Science and Technology

The Springer Series in Measurement Science and Technology comprehensively covers the science and technology of measurement, addressing all aspects of the subject from the fundamental principles through to the state-of-the-art in applied and industrial metrology, as well as in the social sciences. Volumes published in the series cover theoretical developments, experimental techniques and measurement best practice, devices and technology, data analysis, uncertainty, and standards, with application to physics, chemistry, materials science, engineering and the life and social sciences.

More information about this series at http://www.springer.com/series/13337

Victor Sapritsky · Alexander Prokhorov

Blackbody Radiometry

Volume 1: Fundamentals

Springer

Victor Sapritsky
All-Russian Institute of Optical
and Physical Measurements
Moscow, Russia

Alexander Prokhorov
Virial International, LLC
Gaithersburg, MD, USA

ISSN 2198-7807 ISSN 2198-7815 (electronic)
Springer Series in Measurement Science and Technology
ISBN 978-3-030-57787-2 ISBN 978-3-030-57789-6 (eBook)
https://doi.org/10.1007/978-3-030-57789-6

This Springer imprint is published by the registered company Springer Nature Switzerland AG
The registered company address is: Gewerbestrasse 11, 6330 Cham, Switzerland

Preface

A blackbody is a concept of theoretical physics that is familiar to most of us from courses of thermodynamics, optics, or heat transfer. The blackbodies are artificial devices that reproduce, to some extent, the properties of a homonymous theoretical object. The ability to calculate the characteristics of the radiation of a blackbody from its known temperature and vice versa using the laws of thermal radiation makes blackbodies irreplaceable temperature standards in radiation thermometry and standards of the energy characteristics of optical radiation in optical radiometry.

Over the past decades, blackbody radiometry has undergone revolutionary changes. The emergence and rapid introduction of cryogenic radiometers in the practice of optical radiometry allowed measuring the energy characteristics of optical radiation with previously unattainable small uncertainty. This made it possible to break the previously inextricable connection of blackbody radiometry with the temperature scale. Moreover, the blackbody itself began to be regarded as a measure of thermodynamic temperature, which served as the basis for the development of new methods for establishing the thermodynamic temperature scale.

Over the years, the authors have been deeply involved in development, characterization, and use of blackbodies for both radiation thermometry and blackbody radiometry. Over time, in the life of each researcher (and the authors are no exception), a moment comes when there is a desire to systematize the accumulated knowledge. We believe that the time has come to write a book, in which a consistent presentation of the problem is presented, from the radiation laws of a perfect blackbody to modern advances in blackbody radiometry, especially since no monograph has been published on this subject. The material for this book was compiled from a number of original publications and our own results published in this area. The book is divided into two nearly independent volumes.

The first volume "Fundamentals" provides the introduction to the basic concepts of the optical radiometry on the whole and the blackbody radiometry, in particular, as well as necessary concepts of metrology. We describe materials applicable for manufacturing blackbodies, principles of heating or cooling blackbody radiators, and methods of computational and experimental determination of their characteristics. In the course of presentation of the main content, it was necessary to invoke materials from disciplines laying beyond the mainstream of the book such as material science, heat and mass transfer, semiconductor physics, and so forth. The authors realize that the secondary issues cannot be discussed in sufficient detail without excessive growth of the book's volume, while their superficial description will be hardly useful for readers to whom this book is addressed. Therefore, we are forced to use a palliative solution: we give only brief explanations on such issues, but provide references to the latest review articles on the topic and/or books, in which these problems are explained at an affordable level.

In the second volume "Practical Designs and Applications," which we are continuing to work on, we discuss specific examples of blackbody designs for various applications, as well as the applications themselves, from the realization of radiometric scales in laboratory conditions to the calibration of radiometric remote sensing equipment, both preflight and on-orbit.

The authors hope that both volumes of the book will help a wide audience to familiarize themselves with the general subject, specific topics, and serve as a kind of technical reference.

We express our sincere gratitude to our colleagues, in particular, to

- Dr. Vladimir Khromchenko (NIST, Gaithersburg, USA), discussions with whom at early stages of the manuscript writing helped us to develop its structure;
- Dr. Leonard Hansen (NIST, Gaithersburg, USA) for his kind permission to use the results of his measurements;
- Dr. Vyacheslav Podobedov (NIST, Gaithersburg, USA) for his invaluable assistance in obtaining some hard-to-find literature; and
- Dr. Mikhail S. Matveyev (VNIIM, St. Petersburg, Russia), whose advice was very useful for writing sections related to measuring blackbody temperatures.

We feel indebted to our colleagues from around the world for their help in the discussion and implementation of many of the studies in the field of blackbody radiometry, which constituted the content of this book, namely,

- Dr. Klaus D. Mielenz and Dr. Albert C. Parr (both retired, NIST, Gaithersburg, USA);
- Mr. Robert D. Saunders (deceased, NIST, Gaithersburg, USA);
- Dr. Nigel P. Fox and Dr. Emma R. Woolliams (both NPL, Teddington, UK);
- Dr. Gerhard Ulm (retired, PTB, Braunschweig, Germany);
- Dr. Peter Sperfeld (PTB, Braunschweig, Germany);
- Dr. Jürgen Hartmann (University of Applied Science Würzburg-Schweinfurt, Germany);
- Dr. Leonid N. Samoylov and Dr. Andrey F. Kotyuk (both deceased, VNIIOFI, Moscow, Russia); and
- Dr. Boris B. Khlevnoy and Dr. Svetlana P. Morozova (both VNIIOFI, Moscow, Russia).

We also thank all the people who have kindly provided us with original photographs and graphic materials that allowed us to avoid tedious verbal descriptions and greatly improve the readability of the book. Finally, we wish to express special thanks to our families for their patience and understanding during our work that has occupied a substantial part of our personal time over years.

Moscow, Russia Victor Sapritsky
Gaithersburg, USA Alexander Prokhorov

Contents

Acronyms

AATSR	Advanced along track scanning radiometer
ACR	Absolute cryogenic radiometer
AIRS	Atmospheric infrared sounder
AIST	National Institute of Advanced Industrial Science and Technology (Japan)
ANSI	American National Standards Institute
AOTF	Acousto-optical tunable filter
APRT	Absolute primary radiometric thermometry
ASTER	Advanced spaceborne thermal emission and reflection radiometer
BIPM	Bureau International des Poids et Mesures (Fr.)—International Bureau of Weights and Measures
BRDF	Bidirectional reflectance distribution function
BSSRDF	Bidirectional scattering-surface reflectance distribution function
BSSTDF	Bidirectional scattering-surface transmittance distribution function
BTDF	Bidirectional transmittance distribution function
CCD	Charge-coupled device
CCPR	Consultative Committee for Photometry and Radiometry
CCT	Consultative Committee for Thermometry of the BIPM
CGPM	Conférence Générale des Poids et Mesures (Fr.)—General Conference on Weights and Measures
CIE	Commission Internationale de l'Éclairage (Fr.)—International Commission on Illumination
CIPM	Comité International des Poids et Mesures (Fr.)—International Committee for Weights and Measures
CODATA	Committee on Data for Science and Technology
CTE	Coefficient of thermal expansion
CW	Continuous wave

DE	Distance effect
DHR	Directional-hemispherical reflectance
DI	Dedicated institute
EOS	Earth Observing System
ESA	European Space Agency
ESR	Electrical substitution radiometer
EURAMET	European Association of National Metrology Institutes
FE	Focus effect
FEA	Finite element analysis
FIR	Far infrared
FOV	Field-of-view
FPBB	Fixed-point blackbody
FR	Filter radiometer
FTIR	Fourier transform infrared
FWHM	Full width at half maximum
GUM	Guide to the expression of uncertainty of measurements
HTBB	High-temperature blackbody
HTFP	High-temperature fixed point
IEC	International Electro-technical Commission
IES	Illuminating Engineering Society of North America
IR	Infrared
IS	Integrating sphere
ITS-90	International Temperature Scale of 1990
JCGM	Joint Committee for Guides in Metrology
KRISS	Korea Research Institute of Standards and Science (Republic of Korea)
LDLS	Laser-driven light sources
LHe	Liquid helium
LLG	Liquid light guide
LN	Liquid nitrogen
LNE-INM/Cnam	Laboratoire National de métrologie et d'Essais—L'Institut National de Métrologie/Conservatoire national des arts et métiers (France)
LWIR	Long-wave infrared
MCM	Monte Carlo method
MIR	Middle infrared
MODIS	Moderate resolution imaging spectroradiometer
MSL	Measurement Standards Laboratory (New Zealand)
MWIR	Mid-wave infrared
NASA	National Aeronautics and Space Administration of the United
NBS	National Bureau of Standards (NIST since 1988)
NEDT	Noise equivalent delta temperature
NESDIS	National Environmental Satellite, Data, and Information Service
NIM	National Institute of Metrology, China

NIR	Near infrared
NIST	National Institute of Standards and Technology, USA
NMI	National metrology institute
NMIJ	National Metrology Institute of Japan
NOAA	National Oceanic and Atmospheric Administration of the U.S. Department of Commerce
NPL	National Physical Laboratory, UK
NRC	National Research Council, Canada
OIML	Organisation Internationale de Métrologie Légale (Fr.)—International Organisation of Legal Metrology
PC	Personal computer
PDF	Probability density function
PID	Proportional-integral-differential
ppm	One part per million, 10^{-6}
PRNG	Pseudo-random number generator
PRT	Platinum resistance thermometer
PTB	Physikalisch-Technische Bundesanstalt (Germany)
QTH	Quartz tungsten halogen (lamp)
rms	Root mean square
RPRT	Relative primary radiometric thermometry
RT	Radiation thermometer
SI	Système International d'Unités (Fr.)—International system of units
SIRCUS	Spectral Irradiance and Radiance Calibration with Uniform Sources (facility at NIST)
SNR	Signal-to-noise ratio
SPRT	Standard platinum resistance thermometer
SSE	Size-of-source effect
TD	Trap detector
TIFRI	Technology Innovations For Radiometer Instruments
TIR	Thermal infrared
TPW	Triple point of water
TULIP	Tunable Lasers In Photometry (facility at PTB)
UV	Ultraviolet
VIIRS	Visible/infrared imager/radiometer suite
VIM	International Vocabulary of Metrology (French "Vocabulaire International de Métrologie")
Vis	Visible spectral range
VNIIM	D. I. Mendeleyev Institute for Metrology, St. Petersburg, Russia
VNIIOFI	All-Russian Research Institute for Optical and Physical Measurements, Moscow, Russia
VTBB	Variable-temperature blackbody

Chapter 1
Introduction

Abstract In the introductory chapter, we explain terminological conventions adopted throughout the book. The subject matter of the book is outlined on the base of the definition of blackbody radiometry as a separate part of optical radiometry. The interrelations of radiometric scales based on blackbodies and temperature scale are discussed. The most important applications of blackbody radiometry are briefly overviewed. The expected audience of this book is specified and its chapter-by-chapter synopsis is presented concisely.

Keywords Optical radiometry · Blackbody radiometry · Temperature scale · Primary radiometric thermometry · Metrology

1.1 Terminological Conventions

To avoid erroneous or ambiguous interpretations of the subject of this book, we must first make the necessary terminological clarifications. Historically, the term "blackbody" ("black body" (noun) and "black-body" (adjective) in British English) refers to an idealized object, the product of a brilliant theoretical physics of the early twentieth century. Blackbody, in this sense of the term, although does not exist in the real world being no more than a theoretical abstraction, is an irreplaceable concept of each course of thermodynamics and heat transfer. Almost simultaneously with the deducing the blackbody radiation laws, during their experimental verification, every source of thermal radiation that simulates, with varying degrees of approximation, the characteristics of a blackbody, became also called blackbody. Roughly speaking, we can consider any cavity with a known uniform temperature as an artificial blackbody, to which the radiation laws of a perfect blackbody can be applied with a certain degree of accuracy.

Currently, the blackbodies (the real-world sources of thermal radiation) are among the most widespread instruments used in the experimental physics. To avoid confusion, we shall call "perfect blackbody" an ideal object keeping the term "blackbody" for an artificial source throughout the book.

The term "radiometry" is a less probable source of ambiguity. Fortunately, in the recent years, the noun "radiometry" is closely associated with the electromagnetic

V. Sapritsky and A. Prokhorov, *Blackbody Radiometry*, Springer Series in Measurement Science and Technology, https://doi.org/10.1007/978-3-030-57789-6_1

radiation and much less frequently with radioactivity. In the realm of adjectives, only the polynomial expression "radiometric dating" (determination of the object's age by the content of carbon isotopes) can compete in linguistic usage with such a combination as "radiometric calibration." Any possibility of ambiguous interpretation could be eliminated if we bear in mind the optical radiometry, which can be approximately defined as a science of measuring the energy characteristics of optical radiation. The qualifier "optical" was omitted in the book title for sake of brevity and since the collocation "blackbody radiometry" (sometimes, "blackbody-based radiometry") is already firmly entrenched in the special literature (see, e.g. [6, 38, 43, 72, 75, 76]). Blackbody radiometry is a branch of optical radiometry, in which measurements are performed with the help of blackbodies. More specifically, measurement of radiant energy is made by comparing with the radiation of a blackbody, the characteristics of which can be computed on the basis of fundamental physical laws for a perfect blackbody. In such measurements, the blackbody plays the role of a reference (standard) source of optical radiation.

In the electromagnetic spectrum, optical radiation occupies the wavelength range from about 100 nm to about 100 μm and comprises the ultraviolet (UV), visible (Vis), and infrared (IR) subranges (we compare existing nomenclatures for the spectral subranges in Sect. 2.3.1). In all other cases, the term "radiometry" is usually accompanied by a corresponding qualifier, e.g. X-ray radiometry, microwave radiometry, etc.

Modern radiometric terminology has developed not long ago, circa the mid-1970s. Presently, the basic regulatory documents for the radiometric terms, definitions, and notation is the ILV, International Lighting Vocabulary [40], the standard issued by the CIE.[1] In some countries, their own national standards coexist with the ILV; for instance, the standard ANSI/IES RP-16-10 [3] prepared by the Illuminating Engineering Society of North America is currently in force in the United States. This document entitled "Nomenclature and Definitions for Illuminating Engineering" has some divergences with the ILV, in particular, different boundaries and names for the spectral subranges. Throughout this book, we shall give preference to the international standards, informing the reader about the most significant deviations (if any) in the terminology we use.

Some independent areas of science and technology employ their own terminology, not consistent with that accepted almost ubiquitously. Wherever possible, we will indicate interrelations between professional slang and standardized terminology.

After verbal identifying of the two parts of the composite term "blackbody radiometry," we adopt the working definition of the subject of blackbody radiometry:

> Blackbody radiometry is a branch of optical radiometry that uses radiation sources whose characteristics imitate the characteristics of a perfect blackbody and can be calculated using its radiation laws.

This definition emphasizes the calculability of radiation characteristics of a blackbody radiation source, which means that all characteristics of its radiation can be

[1] International Commission on Illumination (from French: Commission internationale de l'éclairage).

Table 1.1 Base quantities and units used in the SI

	Base quantity	Typical symbol for quantity	Base unit	Symbol for unit
1	Time	t	second	s
2	Length	l, x, r, etc.	meter	m
3	Mass	m	kilogram	kg
4	Electric current	I, i	ampere	A
5	Thermodynamic temperature	T	kelvin	K
6	Amount of substance	n	mole	mol
7	Luminous intensity	I_v	candela	cd

computed from first principles (ab initio) if the temperature of the blackbody is known. This determines the use of blackbodies as the primary standards in optical radiometry. Other sources of radiation (e.g. incandescent lamps) can be calibrated against the blackbody by the method of comparison. The indicating measuring instruments (radiometers or spectroradiometers) can be calibrated against the blackbody by the methods of direct or indirect measurement.

1.2 International System of Units (SI)

Throughout the book, we deal only with the International System of Units (SI[2]), which is used worldwide as the preferred system of units for science, technology, and industry. The SI, adopted in 1960 by the CGPM,[3] contains seven dimensionally independent base units for corresponding base quantities listed in Table 1.1. All other quantities are derived quantities, which can be expressed in terms of base quantities according to the equations of physics. The quantities having the nature of a count (e.g. number of photons) or defined as the ratio of two quantities of the same kind (e.g. the reflectance, which is defined as the ratio of the reflected radiant flux to the incident radiant flux) are the quantities with the associated unit one or dimensionless quantities. The SI includes also 22 derived units with special names and symbols, such as the units of energy and power, the joule (J) and watt (W), respectively ($\mathrm{J} = \mathrm{kg\ m^2\ s^{-2}}$, $\mathrm{W} = \mathrm{kg\ m^2\ s^{-3}} = \mathrm{J/s}$). Since its establishing in 1960, the SI underwent several revisions, mainly related to the definitions of base units and aimed at improving the accuracy and reliability of their realization.

Over the years, the SI has become so familiar to scientists and engineers that it would not be worthwhile to devote a separate section to it if it were not for the recent revision of the definitions of the basic units of SI, which entered into force on May 20, 2019.

---

[2]From French: Système international d'unités.

[3]General Conference on Weights and Measures (from French: Conférence Générale des Poids et Mesures).

Until that date, the structure of the SI base units' definitions reflected its historical evolution. There were three types of these definitions: (i) definitions based on fundamental physical constants, (ii) definitions based on natural objects, and (iii) those based on artifacts.

Among the definitions belonging to the type (i), there was the definition of the length unit, the meter, defined as the length of the path travelled by light in vacuum during a time interval of 1/299 792 458 of a second (it is assumed that the second, the unit of time, was defined previously). The unit of thermodynamic temperature, Kelvin, which, as will be shown below, plays a particularly important role in the blackbody radiometry, was defined as the fraction 1/273.16 of the thermodynamic temperature of the triple point of water (TPW). Thus, the definition of the kelvin belonged to the category (ii). The only and last definition of category (iii) left in SI was the unit of mass, the kilogram, equal to the mass of the international prototype of the kilogram made of platinum-iridium alloy and stored by the BIPM[4] in Sevres, France.

Artifact-based definitions are the oldest, least accurate, and least reproducible. Artifacts are subject to changes in the quantities they reproduce, for example, due to natural aging. Besides, the realization of an artifact-based unit coincides with its definition and is possible only at the artifact location. Currently, artifact-based definitions turned into a kind of relic.

Definitions based on natural objects also have important disadvantages. In particular, the realization of the kelvin based on the TPW is affected by many factors, including impurities, dissolved gases, and the isotopic composition of water.

The SI has been revised several times since its formal adoption in 1960. In the first decade of the 21st century, the main trend in the evolution of the SI was determined: the gradual abandonment of material carriers of the units of measurement and their redefinition in terms of fundamental physical constants. As a practical and dynamic system developed using the latest scientific and technological achievements, the SI allowed a gradual transition from an artifact-based system of units to the system based on fundamental constants. The atomic and quantum phenomena were used in the definitions of second, meter and electrical units. Currently, the accuracy levels in the realization of these units are limited by technical capabilities, rather than the unit definitions. By the first decade of the 21st century, it became clear that the level of development of modern technologies allows realization of the long-standing dream of theoretical physicists about creating a system of units based solely on fundamental physical constants. The philosophy of transition from an artifact-based to a constant-based system of units is discussed in several monographs [30, 62, 71] and numerous articles [10, 21, 24, 80, 83].

In response of Recommendation of the 94th meeting of the CIPM held in October 2005, Mills et al. [57] proposed general principles and a framework of the revision of the SI. The revision of the SI was preceded by extensive preparatory work, mainly

[4]From French: "Bureau international des poids et mesures" (The International Bureau of Weights and Measures). The BIPM is an intergovernmental organization that works together with 59 Member States around the world on all issues related to measurement science and measurement standards.

on refining the values of defining constants, carried out by world leading National Metrology Institutes (NMIs) over the past few decades (see [8, 81], and references therein). The results of years of serious scientific discussion and preparative work (see, e.g. [4, 13, 15, 56, 58–60]) were summarized in the 26th meeting of the CGPM, which took place in November 2018. It was decided to change the definition of the kilogram, the ampere, the kelvin, and the mole on the base of fixed numerical values of the Planck constant h, the elementary charge e, the Boltzmann constant k, and the Avogadro constant N_A, respectively. The adjusted values of these constant were published in the short communication by Newell et al. [63] and in the extended paper by Mohr et al. [61].

The complete system of units is derived in terms of the fixed values of seven defining constants, expressed in the units of the SI. Since May 20, 2019, the SI is defined according to BIPM [9] as follows: "The International System of Units, the SI, is the system of units in which

- the unperturbed ground state hyperfine transition frequency of the cesium 133 atom $\Delta\nu_{Cs}$ is 9 192 631 770 Hz,
- the speed of light in vacuum c is 299 792 458 m/s,
- the Planck constant h is 6.626 070 15 $\times$ 10 $^{-34}$ J s,
- the elementary charge e is 1.602 176 634 $\times$ 10^{-19} C,
- the Boltzmann constant k is 1.380 649 $\times$ 10^{-23} J/K,
- the Avogadro constant N_A is 6.022 140 76 $\times$ 10^{23} mol^{-1},
- the luminous efficacy of monochromatic radiation of frequency 540 $\times$ 10^{12} Hz, K_{cd}, is 683 lm/W,

where the hertz, joule, coulomb, lumen, and watt, with unit symbols Hz, J, C, lm, and W, respectively, are related to the units second, meter, kilogram, ampere, kelvin, mole, and candela, with unit symbols s, m, kg, A, K, mol, and cd, respectively, according to Hz = s^{-1}, J = kg m^2 s^{-2}, C = A s, lm = cd m^2 m^{-2} = cd sr, and W = kg m^2 s^{-3}."

Figure 1.1 illustrates interdependences between seven defining constants and base SI units. The exact values of the defining constants together with values of some other physical constants calculated via the defining constants, which are used in this book, are provided in Appendix A. A more complete list of physical constants (including those known with finite uncertainties) can be found in the NIST Standard Reference Database at https://physics.nist.gov/cuu/Constants/index.html. It is assumed that the maintenance of the SI will be performed through the adjustment of values of the fundamental constants by CODATA, the Committee on Data of the International Science Council (https://www.codata.org/).

The BIPM [9] brochure describing in detail the background, underlying principles, and definition of the new SI can be freely downloaded from the BIPM web site at https://www.bipm.org/en/publications/si-brochure/. A variety of technical means may be used to realize the definitions. Realizations may be revised whenever new experimental means are developed; for this reason, advices on realizing the definitions

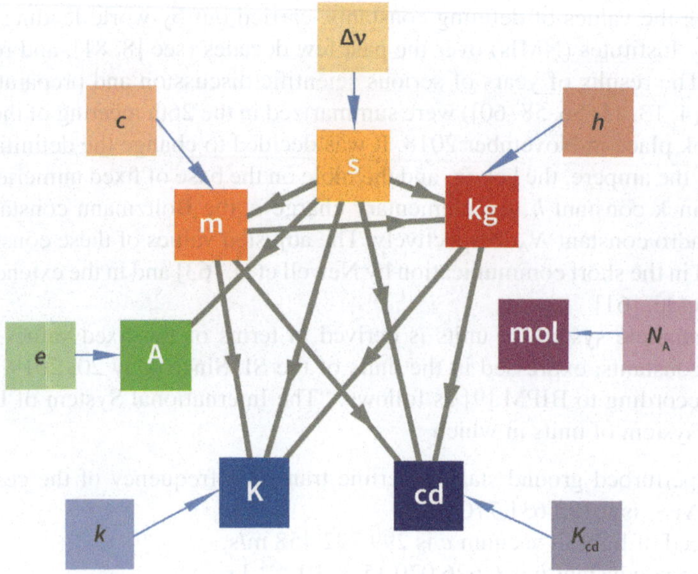

Fig. 1.1 Interdependences between the SI seven base units and corresponding physical constants

are not included in this brochure but is available on the BIPM website as so-called "mises en pratique[5]" for the base units of the SI.

The next regularly scheduled adjustment of defined constants is planned in 2022. The latter revision introduced the most radical change in the SI since its establishing in 1960. However, despite the changes, their impact on everyday measurement practice is expected to be imperceptible. Measurements made using the previous definitions of unit will remain valid within their measurement uncertainty. However, for measurements of the highest accuracy carried out, as a rule, in the NMIs, the corresponding corrections must be made.

1.3 Blackbody Radiation Source: Measurement Principles

According to the International Vocabulary of Metrology (VIM,[6] [41]), blackbodies belong to a special type of measuring instruments—material measures, along with the indicating measuring instruments. The division into material measures and indicating measuring instruments in optical radiometry practically coincides with the traditional division into sources and detectors of optical radiation (the latter have the generic name of radiometers).

[5]French: practical realization (verbatim: put into practice). A collective name for a series of updatable regulatory documents.

[6]From French "Vocabulaire International de Métrologie".

A material measure is used, as a rule, as a measurement standard, which realizes the definition of a given quantity, with an assigned quantity value and associated measurement uncertainty. In this sense, the blackbody differs from the standard resistance of 100 Ω only in that the standard resistance reproduces the only value of a single physical quantity, while the blackbody is a multi-valued measure capable to reproduce different values of different physical quantities (e.g. of radiant flux, spectral irradiance, or radiance temperature) depending of the temperature of the radiating element.

A blackbody is a simulacrum of an ideal object, a perfect blackbody, which is a physical abstraction that does not exist in the real world. The radiation characteristics of the perfect blackbody is completely predictable, if its thermodynamic temperature is known. Thermal radiation emitted by a perfect blackbody is a virtual phenomenon serving as a reference for radiometric measurements involving blackbodies. The classical theory of thermal radiation developed at the turn of the 19th and 20th centuries allows calculating the total and spectral characteristics of thermal radiation of a perfect blackbody using the Stefan-Boltzmann and Planck laws, respectively. So, the radiance L [W m^{-2}sr^{-1}] and spectral radiance L_λ [W m^{-3}sr^{-1}] of the blackbody can be calculated according to the Stefan-Boltzmann and Planck laws, respectively:

$$L(T) = \varepsilon_{eff}(T) \cdot n^2 \sigma T^4, \tag{1.1}$$

$$L_\lambda(\lambda, T) = \frac{c_{1L}\varepsilon_{eff}(\lambda, T)}{n^2\lambda^5\left[\exp\left(\frac{c_2}{n\lambda T}\right) - 1\right]}, \tag{1.2}$$

where T is the thermodynamic temperature; n is the refractive index of the medium, into which the blackbody is radiating; λ is the wavelength in that medium; σ is the Stefan-Boltzmann constant; c_{1L} is the first radiation constant for the spectral radiance; c_2 is the second radiation constant (the numerical values of physical constant are provided in Appendix A); the effective emissivity ε_{eff} expresses the degree of imperfection of a blackbody (by definition, $\varepsilon_{eff} \equiv 1$ for a perfect blackbody).

When geometrical parameters of radiation transfer from the blackbody to the detector are strictly defined, we can use Stefan-Boltzmann's and Planck's laws together with the Lambert law to calculate the irradiance E [W m^{-2}] and the spectral irradiance E_λ [W m^{-3}], respectively.

The VIM [41] defines the measurement principle as a phenomenon serving as a basis of a measurement. The measurement principle of a blackbody radiation source is the phenomenon of emission of electromagnetic radiation by heated bodies. The core of the theory of thermal radiation is the Planck radiation law, which establishes a one-to-one correspondence between the thermodynamic temperature of a perfect blackbody and the energy of its radiation at any wavelength. The Stefan-Boltzmann law can be considered as the result of integration of the Planck law over a semi-infinite wavelength range. All radiometric quantities used in the blackbody radiometry can be derived from Planck's law integrated over spectral and angular domains, finite or semi-infinite.

The one-to-one relationship between the temperature and the radiometric characteristics of a perfect blackbody allows, in principle, to use the same blackbody as a standard for both radiometric quantities and temperature. However, the standards of temperature traditionally belong to the field of thermometry. Blackbodies (mainly operating at high and medium temperatures) have long been used as temperature standards in radiation thermometry (pyrometry is an older name), which deals with remote temperature measurement using thermal radiation emitted by an object. The radiation thermometer (RT), which is the main indicating measuring device in radiation thermometry, differs from the radiometer measuring the radiance only in that the RT is calibrated in terms of temperature.

Although the blackbody radiometry and the radiation thermometry have many intersection points and areas of mutual influence, the methodologies and many concepts of these two areas of measurement technique and science are different. Radiation thermometry is an older and well-developed separate discipline, which is the subject of extensive literature, among which we highlight a two-volume set edited by Zhang et al. [95, 96] and a comprehensive (albeit largely outdated) collective monograph edited by DeWitt and Nutter [22]. A detailed discussion of the problems of radiation thermometry is outside the scope of this book. We will deal with the radiation thermometry only to the extent that it is employed to measure the temperature of blackbodies. Most often, RTs are used for interpolation between blackbodies having known temperatures or extrapolation beyond a known temperature. Other applications of RTs to the blackbody radiometry include evaluation of temperature non-uniformity of radiating surface and tracking changes in radiation of a blackbody over time, in particular, to stabilize its temperature. Despite the limited use of radiation thermometry methods in the blackbody radiometry, we will not neglect the structural elements of blackbodies designed for radiation thermometry if they can also be used for optical radiometry purposes.

1.4 Blackbody Radiometry and Temperature Scale

1.4.1 Preliminaries

The thermodynamic temperature is the critical parameter of the blackbodies by the following reasons:

- radiation characteristics of a blackbody can be calculated using its temperature;
- the operating temperature of the blackbody determines the spectral range of the emitted radiation that can be used for further conditioning and collecting;
- deviation of radiation of the blackbody from that of a perfect blackbody is determined largely by the temperature uniformity over the radiating element;
- in many cases, the uncertainty of the blackbody temperature is the dominant component in the uncertainty budget for realization of radiometric units using

the radiation emitted by the blackbody (this is especially true for the shortwave radiation of high-temperature blackbodies).

The thermodynamic (absolute) temperature, an input parameter in the Planck and Stefan-Boltzmann laws, is one of the principal parameters of classical thermodynamics. There are many texts discussing the concept of temperature in relation to classical thermodynamics and statistical physics, among which the paper by Lindsay [50] and the monograph written by Quinn [69] deserve special mention.

Among seven base quantities of the SI, the temperature is the only non-additive quantity. This means that temperature of a composite body does not equal to the sum of temperatures of parts of it. The value of temperature does not depend on the macroscopic amount of the substance for which the temperature is measured. Such quantities are called intensive; other examples of intensive properties are, e.g., the pressure and the thermal conductivity. Temperature can be determined only using indirect measurements involving some kind of calibrated sensor to convert a measurable quantity (most often, an electric quantity) into a temperature value. The rigorous definition of the thermodynamic temperature must be given from first principles. A straightforward approach based on the second law of thermodynamics and Carnot heat engine is hardly applicable for establishing the thermodynamic temperature scale due to practical impossibility of realizing reversible processes and thermodynamic equilibrium [35]. More appropriate for this purpose is to use the equation of state for an ideal gas, which relate its thermodynamic temperature T with other state properties:

$$PV = nRT, \tag{1.3}$$

where P is the pressure, V is the volume, n is the amount of substance of the gas, and $R \approx 8.314$ J mol^{-1} K^{-1} is the molar gas constant (see Appendix A for more precise value).

Equation 1.3 describes well the behavior of many real gases at moderate temperatures and low pressures. It is supposed that the volume of gas molecules is negligible compared with the volume of container and gas molecules do not interact upon collision. A simple relationship linking the thermodynamic temperature with other measurable parameters gives the opportunity to establish reliably the scale of thermodynamic temperature at least in a limited domain of affecting parameters. Various equations of state are used to realize the kelvin, the SI unit of temperature. In particular, the Planck and Stefan-Boltzmann laws are equations of state for the blackbody radiation; theoretically, they can be employed for establishing the thermodynamic temperature scale.

However, the realization and dissemination of the thermodynamic temperature scale are too complicated and expensive for most practical applications. To make easier routine measurements of temperature in science, industry, and everyday life, the so-called defined temperature scales are used. The defined scales successively adopted by the CIPM[7] are the International Temperature Scale of 1927 (ITS-27), the International Temperature Scale of 1948 (ITS-48), the International Practical Temperature Scale of 1948 (or IPTS-48, the amended editions of 1960), International Practical Temperature Scale of 1968 (IPTS-68, the amended edition of 1975), and, finally, the International Temperature Scale of 1990 (ITS-90), which was adopted by the CIPM in 1989 and still remains in force.

In 1954, the thermodynamic temperature of the TPW was defined to be precisely 273.16 K at 611.657 Pa pressure. Thus, kelvin has is equal numerically to 1/273.16 of the temperature of the TPW. This Celsius scale is defined with two fixed points: 0 °C for the freezing point of water and 100 °C for the boiling point of water at 1 atm pressure. Its unit, the degree Celsius (°C), has the same magnitude as kelvin. Since the freezing point of water is located approximately at 0.01 K below the TPW, the zero point of the Celsius temperature scale is thus redefined at 273.15 K. Both units (K and °C) are currently used on an equal footing in the ITS-90 and the thermodynamic temperature scale.

Although each subsequent practical temperature scale was more accurate approximation of the thermodynamic temperature scale than the previous one, at a certain stage of the temperature metrology development, its accuracy becomes insufficient for some applications, and the ITS-90 is no exception. The current values of the differences between the thermodynamic temperature scale and the defined temperature scales (ITS-48, IPTS-68, and ITS-90) are presented in Fig. 1.2.

1.4.2 International Temperature Scale of 1990 (ITS-90)

The ITS-90 (see [68]) is an internationally agreed best approximation (to the date of the scale adoption) of the thermodynamic temperature scale. The ITS-90 is not a measurement scale sensu stricto, but rather an equipment calibration standard established to perform measurements according to the Kelvin and Celsius temperature scales. The ITS-90 is based on the definition of the unit of thermodynamic temperature, the kelvin, adopted by the 13th CGPM[8] in 1967: "*The kelvin, unit of thermodynamic temperature, is the fraction 1/273.16 of the thermodynamic temperature of the triple point of water*." Presently, this definition is valid only for the ITS-90.

The new definition of the kelvin as the base unit of the SI, adopted by the 26th CGPM in 2018, came into force since May 20, 2019. Later, we discuss this change in definition and its implications for thermometry and blackbody radiometry.

In the definition that is still in effect within the framework of the ITS-90, the triple point of water (TPW) is selected the as the fundamental fixed point. Correspondingly, the thermodynamic temperature of the TPW is exactly $T_{TPW} = 273.16$ K. The TPW is

[7]French: Comité international des poids et mesures (International Committee for Weights and Measures).

[8]French: Conférence Générale des Poids et Mesures (General Conference on Weights and Measures).

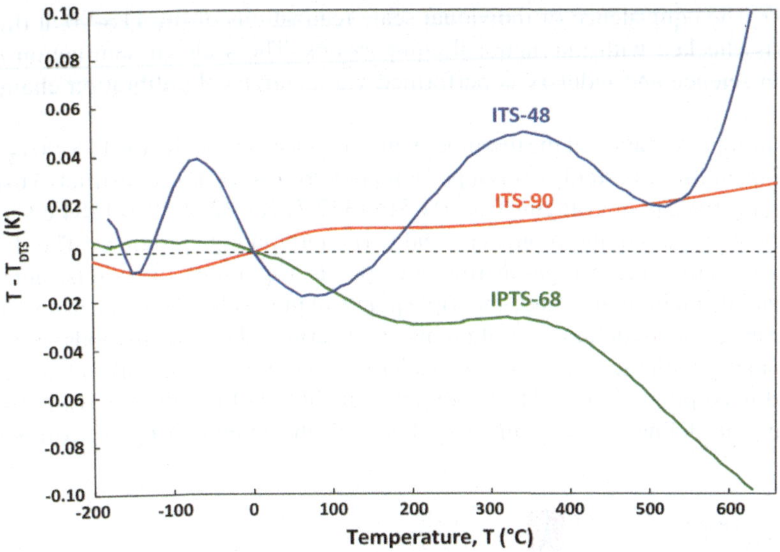

Fig. 1.2 The differences between thermodynamic temperature T and temperature T_{DTS} measured according to the defined temperature scales. Reproduced from Hill and Steele [36] with permission of Taylor & Francis

determined for water with the definite isotopic composition of hydrogen and oxygen. Along with the kelvin, the SI allows the use of the Celsius degree, assuming that the one degree of the Celsius scale has the same magnitude as one kelvin. The difference between the two scales' null points is set to exactly 273.15 K (i.e., 0 K corresponds to – 273.15 °C and 273.16 K corresponds to 0.01 °C). The SI does not specify particular techniques for extrapolation between the absolute zero and the TPW. This, on the one hand, leaves room for different methods of scale realization, but on the other, makes it difficult to ensure the uniformity of these realizations. The ITS-90 defines kelvin and Celsius temperatures, T_{90} and t_{90}, respectively, in the same way as the SI does:

$$t_{90}(^{\circ}\mathrm{C}) = T_{90}(\mathrm{K}) - 273.15. \tag{1.4}$$

The formal text of the ITS-90 gives a very concise definition of temperature scales for T_{90} and t_{90}; the supplementary information concerning the practical realization of the ITS-90 has been published by the BIPM[9] as a separate monograph. The recent updates can be found on the BIPM website at https://www.bipm.org/en/committees/cc/cct/guide-its90.html. Internationally agreed methods for realization of the ITS-90 allow independent realization of the scale by individual authoritative laboratories. The primary realization of the ITS-90 is usually done by national metrology institutes

[9]French: Bureau international des poids et mesures—International Bureau for Weights and Measures.

(NMIs). The equivalence of individual scale realizations of the ITS-90 at different NMIs is checked with international comparisons. The scale dissemination to end users in science and industry is performed via hierarchical calibration chains (see [64]).

As an approximation of the thermodynamic temperature scale, the ITS-90 is subdivided into multiple partially overlapped temperature ranges. It is constructed using (i) 17 defining fixed points ranging from 0.65 to 1357.77 K (−272.50 to 1084.62 °C), (ii) primary interpolating thermometers, and (iii) interpolating equations. The defining fixed points are selected as gas thermometer points, vapor pressure points, and highly reproducible melting, freezing, and triple points of pure substances. Thermodynamic temperatures of the defining fixed points are determined by the methods of primary thermometry with the best accuracy available at the time of the scale adoption. The defined fixed points of the ITS-90 are given in Table B.1 (Appendix B). Figure 1.3 presents the defined fixed points together with the interpolating or extrapolating

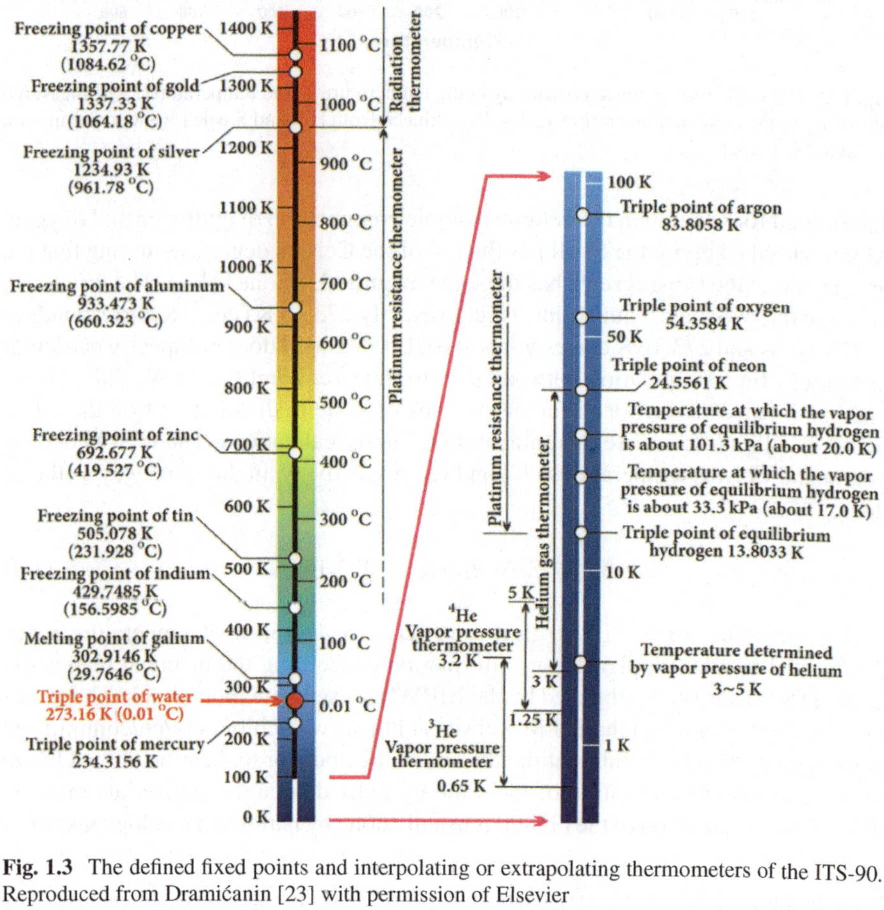

Fig. 1.3 The defined fixed points and interpolating or extrapolating thermometers of the ITS-90. Reproduced from Dramićanin [23] with permission of Elsevier

instruments for realization of the ITS-90 in the temperature ranges between the fixed points. For the blackbody radiometry, the most important instruments for temperature interpolation in the range from 13.8033 to 1234.94 K is the platinum resistance thermometer (PRT). Other types of thermometers (e.g. thermocouples, thermistors, etc.) can be used in the everyday measurement practice, depending on temperature range of interest and requirements to measurement accuracy.

For temperatures above the freezing temperature of silver, the ITS-90 is established by extrapolation using the Planck law and a monochromatic radiation thermometer; the Planckian extrapolation can be started from any of the freezing points of silver, gold or copper. The unknown temperature T_{90} above the freezing temperature of silver is determined from the ratio of the spectral radiances:

$$\frac{L_{\lambda,\mathrm{bb}}(\lambda, T_{90})}{L_{\lambda,\mathrm{bb}}(\lambda, T_{90}(X))} = \frac{\exp\left[c_2/(\lambda T_{90}(X))\right] - 1}{\exp\left[c_2/(\lambda T_{90})\right] - 1}, \tag{1.5}$$

where X denotes any of silver (1234.83 K), gold (1337.33 K), or copper (1357.77 K) freezing points; λ is the wavelength in vacuo, and $c_2 = 0.014388$ m K. This value of the second radiation constant fixed in the original text of the ITS-90 has to be used [94].

Equation 1.5 is an idealized measurement equation written for a perfect blackbody at two temperatures and an ideal RT that registers radiation at an isolated wavelength λ. In practice, the radiation thermometry deals with imperfect blackbodies that play the role of temperature standards and a real narrow-band RT used as a spectral comparator. The so-called fixed-point blackbodies (FPBBs) are the natural standards of temperature in radiation thermometry.

After corrections for the effective emissivities of real-world blackbodies and imperfection of the RT, we obtain the unknown temperature T_{90} of the blackbody, which does not necessarily have to be an FPBB, but can be a variable-temperature blackbody (VTBB), that can operate in a continuous temperature range.

A critical review all the available results obtained with the various techniques of absolute primary thermometry, including constant-volume gas thermometry, acoustic gas thermometry, spectral radiation thermometry, noise thermometry, etc. [28] allowed expressing the difference between T_{90} and T as the polynomial approximation of the fourth order for the temperature range from the TPW (273.1 K) to the copper point (1357.77 K):

$$T - T_{90} = 10^{-3} T_{90} \sum_{i=0\ldots4} c_i \left(273.16/T_{90}\right)^{2i} \tag{1.6}$$

with $c_0 = 0.0497$, $c_1 = -0.3032$, $c_2 = 1.0254$, $c_3 = -1.2895$, and $c_4 = 0.5176$ are the empirical constants. The deviations of ITS-90 from the scale of thermodynamic temperature is illustrated by Fig. 1.4, where the function (1.6) is plotted together with the consensus mean values for the ITS-90 defining fixed points and associated

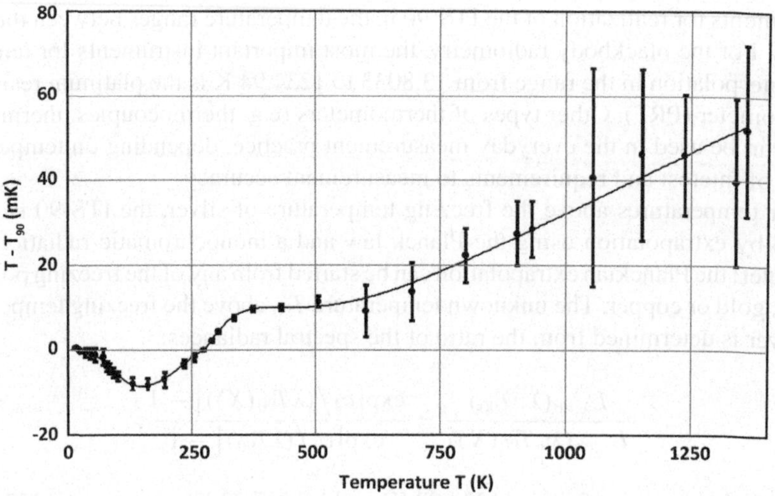

Fig. 1.4 Consensus estimates of $T - T_{90}$ plotted using Eq. 1.6 and data from Fischer et al. [28]. The smooth function (solid line) interpolates the mean values (dots). Error bars represent standard uncertainties

standard uncertainties. As we can see, standard uncertainties are low up to $T = 500$ K but grow rapidly with temperature.

By the time of its adoption, the ITS-90 based on the state-of-the-art determination of temperatures of the fixed points, provided quite satisfactory accuracy for the absolute majority of temperature measurements. Even until now, the uncertainty of fixed point's thermodynamic temperatures can provide quite acceptable uncertainty in reproducing the spectral radiance by means of corresponding FPBBs. However, owing to advances in science and technology, the ITS-90 ceased to meet requirements to temperature metrology of extremely low and extremely high temperatures. To provide accuracy necessary for modern physical research close to the absolute zero, the ITS-90 has been extended in 2000 by the Provisional Low Temperature Scale from 0.9 mK to 1 K (PLTS-2000). The expansion of the ITS-90 above the copper point was postponed, although new technologies urgently required accurate temperature measurements, significantly exceeding the freezing point of copper. In particular, it is very important for the blackbody radiometry to perform radiometric calibrations in the shortwave spectral range, at least down to 200 nm. There are other areas of science, from plasma physics to observational astrophysics, and industry (manufacturing of rocket engine nozzles and re-entry space vehicle fairings, production of nuclear fuel, refractory metals, glass, optical fiber, ceramics and composite materials, safety testing of nuclear reactors etc.), where accurate measurement of high temperatures are vital. At the wavelengths of about 200 nm, even the copper-point (1357.77 K) FPBB has too low spectral radiance. To obtain an acceptable signal-to-noise ratio (SNR) for a signal of a quasi-monochromatic radiation thermometer, the blackbody must have temperature as high as possible, say, 3000 K. However, for temperatures above 1234.93 K (the freezing point of silver), the ITS-90 did not provide other techniques than the Planckian extrapolation using monochromatic radiation thermometer. Extrapolation can be performed starting from silver, gold, or copper point. There is a well-known approximate relationship (see, e.g.[33]), which allows evaluating the uncertainty $u(T_{90})$ of measuring the ITS-90 temperatures T_{90} above the fixed point $T_{90}(X)$:

$$u(T_{90}) = \left[T_{90}/T_{90}(X)\right]^2 u(T_{90}(X)), \tag{1.7}$$

where $u(T_{90}(X))$ is the uncertainty assigned to the FPBB at the freezing temperature of silver, gold, or copper; that is the uncertainty in determining the unknown temperature by extrapolation grows proportional to the square of the ratio of the unknown and known temperatures. Therefore, even relatively small uncertainties $u(T_{90}(X))$ ascribed to the silver, gold, and copper points can lead to unacceptably high uncertainty $u(T_{90})$ and deviation from the thermodynamic temperature scale at the measurement of temperatures $T_{90} >> T_{90}(X)$ using the monochromatic radiation thermometer. In fact, the extrapolatory approach of realization of the ITS-90 limits its applicability to measuring the temperature of HTBBs.

For instance, if to extrapolate from the fixed point of gold (1337.33 K), the temperature uncertainty at the gold point increases fivefold at 3000 K ($(3000/1337.33)^2 \approx 5$). Even if we assume the standard uncertainty $u(T_{Au}) = 15$ mK, then only uncertainty propagation contributes 75 mK to the temperature uncertainty. It should be noted that this is not necessarily the largest component of uncertainty. H. J. Jung, comparing methods of temperature measurement for the high-temperature blackbodies [42], assessed the standard uncertainty in determining temperature of $T_{90} = 3000$ K using the monochromatic (at 650 nm) Planckian extrapolation as 1 K for the "normal" accuracy and 0.3 K for the "best" accuracy. He also predicted possible reducing of the standard uncertainty down to approximately 0.17 K. Comparison of the prognosis made by Jung [42] for 2500 K with the data presented by Fischer et al. [25] shows that the Jung's forecast was even too optimistic. Predicted lowest standard uncertainty of 0.12 K for $T_{90} = 2500$ K was not achieved by 2003 and hardly can be reduced below 0.1 K.

The blackbody radiometry, as it was formed by the 1970s [6, 29, 43, 47, 49, 78, 79], when it stood out as a separate (but not isolated) branch of optical radiometry, was inextricably linked to the current defined temperature scale. While the requirements for the accuracy of radiometric measurements were rather modest, the question of the deviation of the current defined temperature scale from the thermodynamic temperature scale did not arise. Until now, no problems have arisen when the temperatures of low- and medium-temperature blackbodies are measured using contact sensors (for example, platinum resistance thermometers), since the difference between the ITS-90 and the thermodynamic scale is insignificant.

Fig. 1.5 A schematic of measuring the temperature of a HTBB against a FPBB at a fixed point of the ITS-90

For blackbodies operating at temperatures above the freezing point of silver (1234.83 K), the only measuring instrument proposed by the ITS-90 is a monochromatic RT, and the only measurement principle is expressed by Eq. 1.5. A simplified scheme of measuring the temperature of a high-temperature blackbody (HTBB) according Eq. 1.5 is shown in Fig. 1.5. where a tungsten strip lamp is used as a transfer measurement device. Its use is often necessary either because of the inability to place both blackbodies in close proximity to each other, or because of the insufficient stability of the RT, or because its linearity range is insufficient for direct comparison of the spectral radiances of two blackbodies, temperatures of which differ significantly. By the 1990s, this scheme became common for measuring the temperature of the HTBBs, on which the scales of spectral radiance (see, e.g. [85]) and spectral irradiance [86] were based. From the point of view of the blackbody radiometry [72], the biggest inherent drawback of the ITS-90 is manifested in the high-temperature region, where the only way to realize the temperature scale is based on the Planckian extrapolation. Until the mid-1990s, blackbody radiometry was tightly linked to the current defined temperature scale and, consequently, to the radiation thermometry in its traditional form.

By the mid-1990s, the accuracy of the ITS-90 above the freezing point of silver (1234.93 K or 961.78 °C) ceased to satisfy the increased requirements for the accuracy of measuring radiometric quantities.

The only known alternative for measuring high thermodynamic temperature is absolute measurement of some energy characteristics of the blackbody radiation to derive the temperature from the radiation laws of a perfect blackbody. In such a way, the thermodynamic temperature of a blackbody can be determined independently from the ITS-90, when absolute radiometry plays the role of the primary thermometry method.

1.4.3 Radiometric Measurement of Thermodynamic Temperature

The realization of radiometric scales on the base of blackbodies, whose temperature T_{90} is determined with respect to an FPBB at a freezing point of silver, gold, or copper according to Eq. 1.3, got the name source-based, in contrast to detector-based realizations that relays on an absolute radiation detector. This classification of methods for the realization of radiometric scales became especially relevant in the early 1990s, with the emergence of an absolute cryogenic radiometer (ACR),

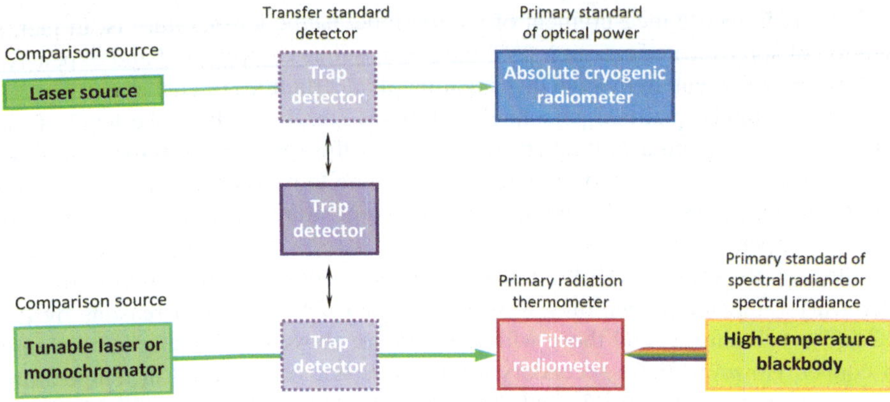

Fig. 1.6 A schematic of the detector-based technique of measuring the temperature of an HTBB

which makes it possible to measure the power of optical radiation (radiant flux) with an unprecedented uncertainty in hundredths of a percent. A generalized scheme for measuring the thermodynamic temperature of an HTBB is depicted in Fig. 1.6.

To date, several independent techniques for the thermodynamic temperature measurements were developed and described in the CCT[10] documents [52, 74] and in a number of publications (e.g. [2, 11, 14, 34, 88, 93]). In general, the process of the thermodynamic temperature measurement consists of four stages:

1. First, the trap detector (TD) is calibrated against an ACR at several wavelengths of laser radiation in terms of the spectral responsivity.
2. Then the spectral responsivity of the TD is interpolated in the continuous spectral range sufficient to calibrate the filter radiometer (FR). The spectral interpolation is performed using the theoretical model of the spectral responsivity of the TD.
3. The spectral responsivity of the FR is determined by comparison of the FR with the TD using a tunable laser or a monochromator as a comparison source.
4. The radiant flux (or another radiometric quantity) of the radiation emitted by the HTBB is measured by the FR within its responsivity band.
5. The thermodynamic temperature of the HTBB is calculated on the base of the Planck law.

Thus, the FR calibrated against the ACR via the TD plays the role of a primary radiation thermometer and makes possible to measure band-limited (within the spectral bandwidth of the FR) radiometric quantities (radiant flux, radiance, or irradiance) with exceptional accuracy and allows to determine thermodynamic temperatures bypassing the ITS-90. The spectral-band radiation thermometry has become a separate branch of primary thermometry. Currently, many countries' national radiometric scales are detector-based and traceable to the ACR. In particular, Parr [66] described the national measurement system for radiometry, photometry, and radiation thermometry of the USA maintained at the NIST that is based on the ACR.

[10]Consultative Committee for Thermometry of the BIPM.

Every radiometric measurement of the thermodynamic temperature is, in fact, a solution of sophisticated research problem and requires complicated and expensive measuring equipment, individual investigation of uncertainties of each measurement facility, and highly qualified personnel. All this is possible only at the level of the NMIs and this situation is unlikely to change in the foreseeable future. In other words, presently, there is a possibility to determine temperatures only for a finite set of reference points for the thermodynamic temperature scale above the freezing point of copper.

Although the original ITS-90, unlike its predecessors, does not provide for any secondary reference points, they were recommended for practical reasons by the CCT [7], including several those with temperature above the freezing temperature of copper. However, these secondary points were not widely used in temperature metrology, partly due to insufficient reproducibility and insufficiently reliable determination of their temperature, partly because of difficulties in their realization. In the 85th Meeting of the CIPM took place in 1996, it was declared "*that a fixed point at temperatures above 2300 K having a reproducibility better than 0.1 K, is highly desirable*" and recommended "*that national laboratories work to develop high-temperature fixed points*" [70]. The introduction of high temperature fixed points (HTFPs) with accurately determined thermodynamic temperature above the highest temperature of the defining fixed point of the ITS-90 promised a great gain in accuracy of high-temperature thermometry and blackbody radiometry providing a highly stable reference source of Planckian radiation.

In response of the CCT recommendation, Yamada et al. [91] of National Metrology Institute of Japan (NMIJ) proposed to employ metal-carbon eutectic alloys as the HTFPs. He radiometrically observed melting and freezing plateaux of some metal–carbon eutectics in the range from 1603 to 2223 K. Sasajima et al. [73] introduced metal carbide–carbon eutectics to raise that the upper reproduces temperatures up to 2750 K. Yamada et al. [92] identified another type of invariant reactions proceeding in binary carbonaceous alloys with intermediate compounds—peritectic reactions, useful for establishing the HTFPs. The HTFP melting/freezing cycles are carried out in radiating cells, similar to those used in FPBBs (although HTFPs can also be realized in cells designed for calibrating contact thermometers). Recognizing the far-reaching potential of the HTFPs, the CCT has launched a series of challenging multi-partner projects whose ultimate goal is to assign reliable thermodynamic temperatures to three metal–carbon eutectic HTFPs: Co–C (1324 °C), Pt-C (1738 °C), and Re-C (2474 °C). The world's leading NMIs have been involved in international cooperation in conducting research in design of HTFP cells and the crucible filling techniques, development of high-temperature furnaces with extended isothermal zones, underlying studies of the ingot microstructure and impurities influencing the HTFP reproducibility, investigations of long-term stability, etc. Research in this area led to significant progress not only in the technique of measuring thermodynamic temperature, but also gave a powerful impulse to the blackbody radiometry as a whole. Currently, the first stage of the introduction of HTFPs in metrological practice can be considered completed: thermometric community received four HTFPs (Co–C, Pd–C, Pt-C, and Re-C) with the assigned values of thermodynamic temperatures and the corresponding uncertainties in their determination. These HTFPs were adopted by the CCT [53] as the reference points of thermodynamic temperature for relative primary radiometric thermometry (a newly introduced term). For Co–C, Pt-C, and Re-C, both points of inflection in the melting curves and equilibrium liquidus thermodynamic temperatures are specified. The point of inflection corresponds to the temperature directly determined from the melting curve. However, it is not the fundamental quantity since it obtained under non-equilibrium conditions. The fundamental quantity obtained under equilibrium conditions is the liquidus temperature, for the calculation method of which there is no consensus (see [51, 89], and references therein). The data for Co–C, Pt-C, and Re-C HTFPs are listed in Table B.2 of Appendix B together with the point of inflection temperature for the Pd–C eutectic. It should be noted that the difference between point of inflection and equilibrium temperature is insignificant for most practical applications. Besides, the CCT plans to update this data on a regular basis.

1.4.4 The Kelvin Redefined and Its Mise En Pratique

The development of methods and equipment for measuring thermodynamic temperature of HTBBs required legislative consolidation in regulatory documents. Such documents called *mise en pratique* for the definition of the kelvin (*MeP*-K) are published by the CCT. The most recent CCT document [16] was timed to the redefinition of the kelvin as part of the SI revision, which entered into force on May 20, 2019. We overviewed briefly the reasons and principles of this revision in Sect. 1.2. Here, we only provide the new definition of the kelvin:

"The kelvin, symbol K, is the SI unit of thermodynamic temperature. It is defined by taking the fixed numerical value of the Boltzmann constant k to be 1.380649 x 10^{-23} when expressed in the unit J K^{-1}, which is equal to kg m^2s^{-2} K^{-1}, where the kilogram, meter and second are defined in terms of h, c and $\Delta\nu_{Cs}$,"

where h is the Planck constant, c is the speed of light in vacuum and $\Delta\nu_{Cs}$ is the cesium frequency corresponding to the transition between the two hyperfine levels of the unperturbed ground state of the ^{133}Cs atom.

This definition implies the exact relation $k = 1.380649 \times 10^{-23}$ J/K. Its effect is that one kelvin is equal to the change of thermodynamic temperature T that results in a change of thermal energy kT by 1.380649×10^{-23} J. The values of physical constants adopted for the new definition [63] are given in Appendix A.

The main purpose of the CCT [16] document ("*Mise en pratique* for the definition of the kelvin in the SI") is to specify how the definition of the kelvin can be realized in practice. The new definition does not provide any specific technique for the practical realization of kelvin.

In principle, one can use any method that allows to derive the thermodynamic temperature value, traceable to a set of seven reference constants that underlie the revised SI [9]. CCT [16] introduces the following important definitions.

Primary thermometry is a method having the highest metrological properties. The primary thermometer is based on a well-understood physical system, for which the equation of state describing the relation between thermodynamic temperature T and other independent quantities can be expressed explicitly without unknown or temperature-dependent constants. Thermodynamic temperature can be obtained by measuring the independent quantities. The acoustic gas thermometry, Johnson noise thermometry, and spectral-band radiometric thermometry can serve as examples of primary methods of thermometry. The primary method of thermometry does not require a reference standard of the temperature.

Absolute primary thermometry allows to measure thermodynamic temperature directly in terms of the definition of the base unit kelvin, i.e. the defined numerical value of the Boltzmann constant. No reference is made to any temperature fixed point. All other parameters specified in the equation of state should be measured or determined by other ways.

Relative primary thermometry allows to measure thermodynamic temperature indirectly using a specified equation of state, with one or more key-parameter values determined from temperature fixed points, for which values for the thermodynamic temperature T and their uncertainties are known a priori from previous application of absolute or relative primary thermometry. In principle, the defining fixed points of the ITS-90 can be used for the relative primary thermometry if the corresponding corrections for the difference $T - T_{90}$ are made.

Defined temperature scales allow to assign temperature values, determined by primary thermometry, to a series of naturally occurring and highly reproducible states (e.g., the freezing, melting, and triple points of pure substances). They also include a specification of the interpolating or extrapolating instruments for particular temperature sub-ranges and necessary interpolating or extrapolating equations. The defined scales are highly prescriptive and define new temperature quantities that provide close approximations to the thermodynamic temperature T. Temperature values assigned to the fixed points of each scale are considered exact (i.e. having zero uncertainty) and are not altered while the scale is in force, even if subsequent research reveals deviation of the assigned values from thermodynamic temperatures. An example of the defined temperature scales is the ITS-90.

Absolute primary radiometric thermometry implies an accurate determination of the optical power, emitted over a known spectral band and known solid angle, by an isothermal cavity of known emissivity. Measurement of the power requires a radiometer, comprising a detector and spectral filter, with known absolute spectral responsivity. The optical system typically includes two co-aligned circular apertures separated by a known distance to define the solid angle, and may additionally include lenses or mirrors. The refractive index of the medium in which the measurement is made must also be known. All measurements of the quantities involved must be traceable to the corresponding units of the SI, namely, the watt and the meter. The most common methods of the absolute primary radiometric thermometry are described by Machin et al. [52] in the corresponding *MeP*-K document, which became the Annex to the CCT [16]. Another important Annex is the *MeP*-K [74] that describes the methods used for determining the uncertainty associated with thermodynamic temperature measured using absolute primary radiometric thermometry. It was concluded that the primary radiometric thermometry makes possible to measure the blackbody temperature at 2800 K with the standard uncertainty of about 0.1 K.

Relative primary radiometric thermometry does not require knowledge of the absolute spectral responsivity of the radiometer and the defining solid angle. Instead, the radiometric quantity (e.g. spectral radiance) is measured relative to homonymous quantity measured for one or more FPBBs, each with known thermodynamic temperature. The three approaches to the relative primary thermometry are possible:

(i) extrapolation from a single fixed point, when only knowledge of the relative spectral responsivity of the radiometer is required;
(ii) interpolation or extrapolation using two fixed points, for which only the bandwidth of the spectral responsivity has to be known;
(iii) interpolation or extrapolation using three or more fixed points, when detailed measurement of the spectral responsivity is not required.

The interpolation and extrapolation are carried out using parametric approximate expression for the radiance of the blackbody averaged over the spectral bandwidth of a radiometer or radiation thermometer, such as the Sakuma–Hattori equation, which is discussed in Chap. 8 of this book. Relative primary radiometric thermometry gives uncertainties that are only slightly higher than absolute primary radiometric thermometry. Guidelines for the realization of the relative primary radiometric thermometry as well as typical uncertainty estimates, can be found in Machin et al. [53] that became the Annex to the CCT [16].

One important consequence of the redefinition of the kelvin is the change in the status of the TPW. The old definition of the kelvin set the temperature of the TPW T_{TPW} to be exactly 273.16 K. The new definition of the kelvin fixes the numerical value of the Boltzmann constant k instead of T_{TPW}, which has to be determined experimentally. The former relative standard uncertainty in the determination of k (3.7×10^{-7}) is transferred to the temperature T_{TPW}. The standard uncertainty of T_{TPW} is hence now $u(T_{\text{TPW}}) = 0.1$ mK.

It is definitely worth noting that CCT [16] does not suspend the action of the ITS-90, but only determines its place as a defined temperature scale. As a defined temperature scale, the ITS-90 remains unchanged throughout the life of the scale [26]. In a wide temperature range around the TPW, most important for practical temperature measurement the differences $T - T_{90}$ (see Fig. 1.3) are small enough to not affect the worldwide uniformity of temperature measurement.

Summing up the recent innovations in the field of thermometry, we can emphasize the provisions that are most important for blackbody radiometry:

1. The most serious changes concern the high-temperature region. Measurements in low- and medium-temperature ranges traceable to the ITS-90 remain valid. Measurements of highest precision carried out previously can be easily corrected for the difference $T - T_{90}$, if necessary.

2. The absolute primary radiometric thermometry becomes a legal mean to measure thermodynamic temperature of HTBBs bypassing the ITS-90. CCT (2019) and the accompanying *MeP*-K documents legalize radiometric realization and dissemination of the thermodynamic temperature scale bypassing the ITS-90. Radiometric measurements of thermodynamic temperature can now be carried out directly in science, industry, and other fields.
3. The methods of the absolute primary radiometric thermometry allow establishing HTFPs, which significantly simplify dissemination of the thermodynamic temperature scale using the relative primary radiometric thermometry.
4. The relative primary radiometric thermometry allows the use of the existing FPBBs at defining fixed points of the ITS-90 (with assigned thermodynamic temperatures) and HTFPs together with the methods for interpolation and extrapolation of temperatures, which are well-developed for the traditional radiation thermometry.

Recent advances in measuring the thermodynamic temperatures of HTBBs have not only led to the emergence of absolute primary radiometric thermometry as a new branch of thermometry, but they have also brought radiation thermometry and black body radiometry closer together. A side effect of the emergence of relative primary radiometric thermometry is that the classification of schemes for the realization of radiometric quantities onto sources-based and detectors-based loses its original meaning. In our opinion, it would be more correct to talk about different approaches to determining the temperature of a blackbody, which is the main source of radiation.

1.5 Applications of Blackbody Radiometry

We could trace the origin of blackbody radiometry to the turn of nineteenth and twentieth centuries, when isothermal cavities began to be widely used in the experimental verification of the laws of thermal radiation and determination of fundamental constants included in them. However, the experimental works carried out by Lummer, Pringsheim, Kurlbaum, Rubens, Gerlach and other prominent scientists of that time (see the review articles by Karoli [43] and Hoffmann [37]) do not allow us to strictly distinguish radiometric applications of blackbodies from thermometric ones.

The earliest known radiometric use of blackbodies was the establishing at the Bureau of Standards (USA) the irradiance scale based on the blackbodies operating at temperatures between 700 and 1200 °C [17, 18]. Since the mid-1950s, the calibration of IR detectors in terms of the total responsivity using temperature-controlled blackbodies becomes a common practice and gradually turns into a commonplace. Bedford [5] described a measurement standard of the total radiant flux and irradiance on the base of blackbody operating over the temperature range from 40 to 150 °C that was used at the NRC[11] to calibrate the thermal detectors of optical radiation with the reproducibility of about 0.1% and the overall uncertainty of 0.3%. In principle,

[11] National Research Center, the NMI of Canada.

the scales of radiance, radiance intensity, radiant flux, and the total radiance temperature can be also realized using blackbody radiation sources and derived from the Stefan-Boltzmann law.

However, growing requirements to measurement accuracy did not allow considering the spectral range of interest as semi-infinite (from $\lambda = 0$ to $\lambda = \infty$), as well as to ignore spectral characteristics of the measuring system. The latter ones can be accounted for the Stefan-Boltzmann law only as a wavelength-independent factor. Presently, the Stefan-Boltzmann law is used only in radiometric applications that deal with very low radiant fluxes (e.g., from cryogenic blackbodies) and do not require extremely high accuracies. If a higher accuracy is required, the band-limited radiance obtained by the integration of Planckian spectral distribution over a finite spectral domain has to be used. At present, the blackbody-based scales of the total radiance are used only in a few NMIs and dedicated institutes (DI), mainly for backward compatibility.

Although blackbodies were widely used in experimental physics and as temperature standards for calibrating optical pyrometers (primarily in metallurgy), their use as radiometric standards has been underestimated for a long time, since the existing electrical substitution radiometers (ESRs) provided a simple and inexpensive way to measure the total radiometric quantities, while measurements of the spectral radiometric quantities still remained little in demand.

This situation has changed in the early 1960s, when the progress in industrial technologies, experimental science, and, especially, the dawn of the space age, led to the need for precise measurements of spectral radiometric quantities. Moreover, exoatmospheric studies expanded the spectral range of radiometric measurements to the areas that are difficult or impossible to access under normal laboratory conditions due to the atmospheric absorption. Artificial sources of blackbody radiation are well suited to spectroradiometric calibrations in a broad spectral range, from the near UV to the far IR, so they have become widely used for pre-launch calibrations of satellite sensors in conditions approaching to those of outer space.

We distinguish the three largest application areas in the modern blackbody radiometry, which were formed in response to current demands of science and industry:

- realization of primary radiometric scales at NMIs, DIs, and specialized facilities,
- pre-launch radiometric calibration of spaceborne sensors performed in cryo-vacuum chambers, and
- in-flight radiometric calibrations of sensors installed on board satellites, spacecrafts, balloons, unmanned aerial vehicles, etc.

Because of many points of intersection, it is often difficult to separate the actual radiometric applications from those belonging traditionally to the radiation thermometry. This is the case, for example, of blackbody-based calibration of thermal imagers for quantitative IR thermography. We still tried not to cross the fuzzy boundary separating these two adjacent disciplines. However, we allowed ourselves to employ the scientific and technical information accumulated in the radiation thermometry, but equally suitable for the blackbody radiometry.

1.5.1 Realization of Radiometric Scales

There are no international standards in the area of optical radiometry. Modern scales of the spectral radiance (from 0.22 to 2.5 μm) and spectral irradiance (from 0.25 to 2.5 μm) are maintained by the world leading NMIs and DIs. The high-temperature blackbodies (see Fig. 1.7) serve as the primary standards and used to calibrate the secondary standard lamps: typically, the tungsten strip lamps for the spectral radiance (and radiance temperature) and tungsten filament lamps for the spectral irradiance.

The CIPM Mutual Recognition Arrangement of the of national measurement standards is supported through participation by NMIs and DIs in the international comparisons of measurement standards (key comparisons and supplementary comparisons). The regional metrology organizations carry out the comparisons of regional scope.

Due to the strong absorption of IR radiation in water vapor, the use of medium- and low-temperature blackbodies in the air is mainly limited to calibrations of broadband radiometers, thermal imagers, and radiation thermometers. Additional problems are associated with background radiation emitted by surrounding objects.

Fig. 1.7 High-temperature blackbodies on a laboratory bench. Courtesy of Dr. Boris Khlevnoy (VNIIOFI)

Fig. 1.8 Photographs of the AIRI facility at NIST: **a** FPBB bench and **b** VTBB/spectral bench (after [55]). Reprinted courtesy of the National Institute of Standards and Technology, U.S. Department of Commerce. Not copyrightable in the United States

A rare example of the realization of the radiance temperature and IR spectral radiance scales at near-ambient temperatures in laboratory environment was given by Mekhontsev et al. [54]. The Advanced Infrared Radiometry and Imaging (AIRI) facility at NIST, whose photographs are presented in Fig. 1.8, includes the FPBBs and precision VTBBs with operating temperatures from −50 to 250 °C. The secondary sources are calibrated against the reference blackbodies using a tunable filter comparator [44]. The typical comparison expanded (at $k = 2$) uncertainty expressed in the radiance temperature units is 25 to 50 mK.

1.5.2 Pre-flight Calibration of Remote Sensing Instrumentation

The required spectral range of blackbody-based calibrations was greatly extended when radiometric measurements of the terrestrial surface were started from satellites in Earth orbit. Optical remote sensing is the generic name of various techniques for measuring optical radiation coming from Earth or other celestial bodies using radiometric instruments installed on board aircrafts (especially, high-altitude ones), unmanned aerial vehicles (UAVs), balloons, satellites, space probes, and manned spacecrafts.

Measurements of radiation coming from objects on Earth, including its surface, oceans, and atmosphere, constitute a significant part of information acquired using various remote sensing techniques.

A sensor[12] mounted on board the satellite (or high-altitude aircraft, balloon, etc.) sees the land or ocean surface through the atmosphere layers, which are opaque to electromagnetic radiation in some spectral ranges and almost transparent in others. The spectral transmittance of the atmosphere is shown in Fig. 1.9 for a nadir's view

[12]In remote sensing, the name of sensors is used traditionally to designate various types of indicating radiometric instruments, including multi-element ones, often together with optics, scanning devices, devices performing spectral selection, etc.

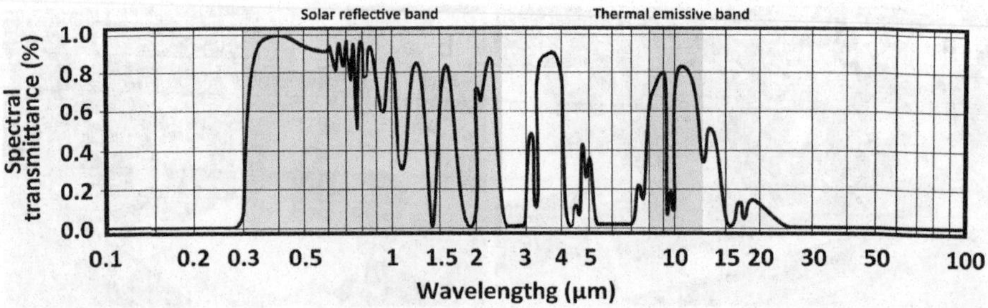

Fig. 1.9 Spectral transmittance of the Earth's atmosphere. Reproduced from [67] with permission of Springer Nature

from the Earth's orbit.

Roughly speaking, the radiation leaving the Earth's atmosphere consists of three parts:

(i) radiation incoming from the Sun (corresponding approximately to the radiation of the perfect blackbody at 6000 K) and reflected by the Earth's surface and scattered by the atmosphere,
(ii) thermal radiation of the Earth's surface that approximately corresponds to the perfect blackbody at 300 K, and
(iii) radiation emitted by the atmosphere that is heterogeneous in temperature, composition and density.

Traditionally, the two spectral ranges that approximately correspond to the so-called atmosphere transmittance windows are distinguished in the Earth's radiation: the solar reflective range of about 0.3 to 2.5 μm and the thermal IR (TIR) range, from 8 to 14 μm.

Modern remote instrumentation for optical sensing covers the spectral range from UV to far IR. Measurements can be carried out with various spectral, spatial, and temporal resolutions. Information delivered by the optical radiation from Earth is not only of scientific, but also of great practical importance: it suffices to mention climate studies, weather forecasts, natural resource management, or ballistic missile defense. Extracting maximum information from observable data requires those measurements be as accurate as possible and be traceable to internationally or nationally recognized measurement standards.

Therefore, calibration of remote sensing instrumentation is required just like calibration of conventional laboratory radiometric instruments. It is distinguished two types of calibrations of remote sensing instrumentation: ground and in-flight calibration.

The ground calibrations are pre-launch *par excellence*, i.e. are carried out before installation of an instrument on board of a satellite and it launch into orbit.

Pre-launch calibration of onboard radiometric sensors must be carried out in the same conditions as they are used in space. This implies a high vacuum and a wide range of temperature variations, from very low to elevated.

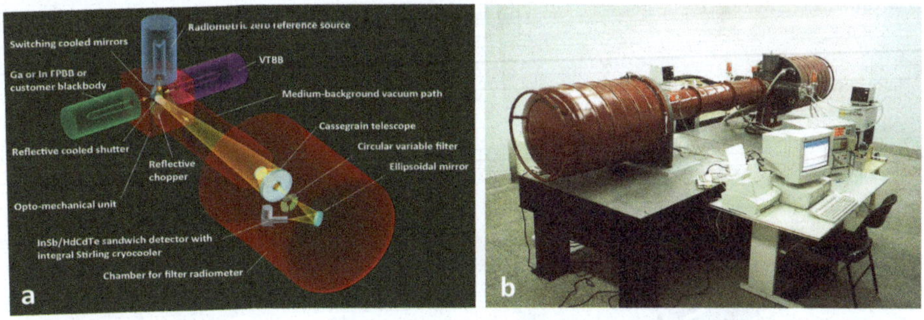

Fig. 1.10 Medium Background Facility (MBF) developed at VNIIOFI: **a** schematic and **b** photograph. Courtesy of Dr. Svetlana Morozova (VNIIOFI)

Calibrations should be carried out in vacuum calibration chambers, which can vary considerably in size and capabilities. Figure 1.10 presents a schematic and photograph of the Medium Background Facility (MBF) developed at VNIIOFI. The MBF contains reference FPBBs (at the melting point of gallium and freezing point of indium) and low-temperature VTBB, as well as helium-cooled blackbody used as radiometric zero reference. The calibration of customers' blackbodies is performed at the background temperature of about 77 K (liquid nitrogen temperature).

Currently, many NMIs and DIs of countries involved in IR remote sensing develop and use similar facilities. Hao et al. [32] described the vacuum radiance-temperature standard facility (VRTSF) at the National Institute of Metrology of China, designed to calibrate IR remote sensors on Chinese meteorological satellites. The VRTSF allows calibrating vacuum blackbodies in terms of radiance temperature, including those used to calibrate IR remote sensors. The layout of the VRTSF is shown in Fig. 1.11. The VRTSF includes the vacuum medium-temperature blackbody (VMTBB), liquid-nitrogen blackbody (LNBB), Fourier transform IR (FTIR) spectrometer, the reduced-background optical system, the vacuum chamber for arranging blackbodies under calibration, the vacuum-pumping system, and the liquid nitrogen-support system. The VRTSF provides the uncertainty of the radiance temperature calibration of 0.026 °C at 30 °C and the wavelength of 10 μm.

The world's leading NMIs, DIs, national space agencies, and specialized laboratories have their own large cryovacuum calibration facilities. Correspondingly, special requirements should apply to blackbodies that are used for pre-launch calibration of satellite sensors. Vacuum blackbodies must combine performance of traditional infrared reference sources with specific features in order to operate in a vacuum chamber at low temperatures.

Figure 1.12 presents a photograph of the LWIRCS (Long Wave Infrared Calibration Source), a blackbody developed at the SDL[13] [48]. It should be noted that low-temperature blackbodies are the most common but not only standard blackbody sources used for pre-launch calibrations.

---

[13]Space Dynamic Laboratory, Logan, UT.

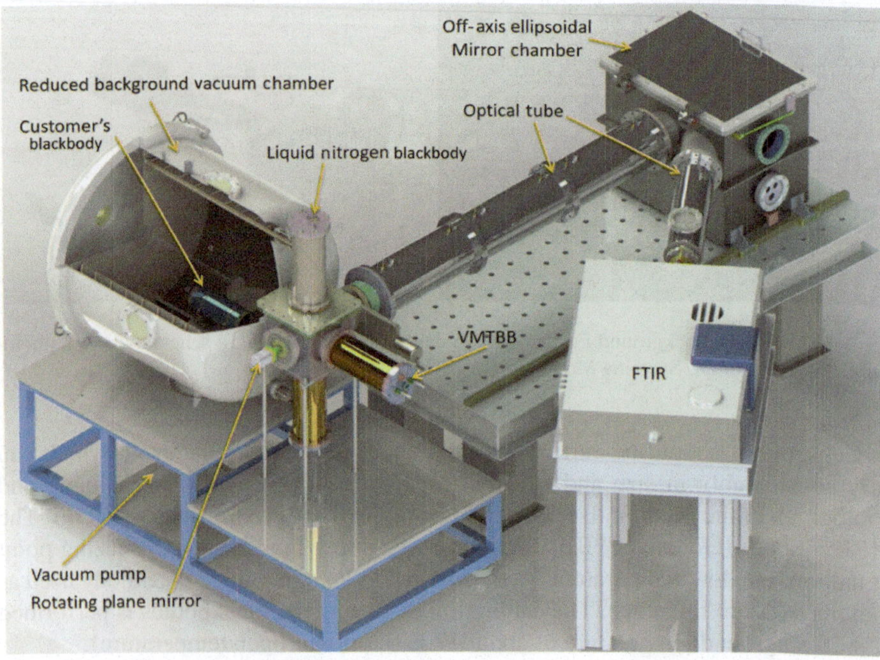

Fig. 1.11 General layout of the VRTSF at NIM (R.P. China). Reproduced from [32] with permission of Springer Nature

The National Institute of Standards and Technology (NIST), the NMI of the USA, elaborated the best practice guideline for pre-launch calibration and radiometric characterization of instruments for passive optical remote sensing [19, 20], which is based on experience gained at NIST, NASA, and NOAA[14] and contains a number of useful examples and recommendations.

1.5.3 In-Flight Radiometric Calibration

On-board satellite sensors, even calibrated before launching into Earth orbit, do not guarantee the accuracy of measurements in orbit, because they are subject to overloads and vibrations at launch, as well as to the effects of cosmic radiation and aging processes in orbit. Monitoring of sensor degradation with time is already a sufficient motivation for post-launch calibration, a significant part of which is carried out using on-board calibration systems.

[14]National Oceanic and Atmospheric Administration (USA).

Fig. 1.12 A photograph of the LWIRCS, a ground calibration blackbody at the SDL. Reproduced from [48] with permission of the author and SPIE

Typically, on-board blackbodies are small or even miniature devices adapted to function autonomously in space environment. As a rule, the accuracy requirements for blackbodies for onboard calibration are less stringent than for blackbodies that play the role of primary standards. The onboard blackbodies are used as secondary standards and should be calibrated against a primary blackbody before launch. However, lower requirements for accuracy are compensated by a variety of specific requirements for onboard blackbodies. To date, dozens of blackbodies have been working or are working on Earth orbit. For instance, the MODIS (Moderate Resolution Imaging Spectroradiometer), one of the most successful imaging sensor launched into Earth orbit by NASA in 1999 on board the Terra (EOS AM) satellite, and in 2002 on board the Aqua (EOS PM) satellite used the flat-plate V-grooved blackbody for in-flight calibration of the MODIS instrument in the thermal infrared (TIR) range.

The VIIRS (Visible Infrared Imaging Radiometer Suite) sensor was launched in October 2011 on board the Suomi National Polar Orbiting Partnership (S-NPP)

Fig. 1.13 The VIIRS on-board calibration blackbody. Reproduced from [45] with permission of Jeremy Kloepfer (Raytheon Company)

satellite. The VIIRS instrument was designed with a strong MODIS heritage [90]. It uses the on-board calibration blackbody (see Fig. 1.13) similar to that of the MODIS.

On-orbit calibrations of the VIIRS sensor is performed by alternating observation in near-parallel rays of the central zone of the blackbody of about 12.5 cm diameter and the cold space. The estimated spectral effective emissivity of the VIIRS onboard blackbody is 0.99820 at 3.39 μm and 0.9975 at 10.6 μm [46]

Another example of simple design of onboard blackbody is given in Fig. 1.14, where the prismatic cavity blackbody developed by ABB Bomem Inc. (Canada) for onboard TIR calibration of the AIRS instrument. The AIRS (Atmospheric Infrared Sounder) was designed to support climate research and improve weather forecasting and also launched on board NASA's Aqua satellite [65].

Depending on type and tasks solved by various IR sensors, the onboard calibration blackbodies differ in size, shape, design, and metrological characteristics. Some IR sensors must be calibrated using high-precision onboard blackbodies, which are not much different in complexity from the best laboratory blackbodies. Figure 1.15 presents the photograph of two blackbody calibration sources of the Advanced Along Track Scanning Radiometer (AATSR).

The AATSR is the instruments launched on board the European Space Agency (ESA)'s Envisat satellite that was active between 2002 and 2012. This instrument is a multi-channel imaging radiometer with the principal goal to provide accurate data

Fig. 1.14 AIRS blackbody calibrator developed by ABB Bomem (after [82]). Reprinted courtesy of the National Institute of Standards and Technology, U.S. Department of Commerce. Not copyrightable in the United States

on global sea surface temperature for monitoring and carrying out research on the Earth's climate.

Presently, the onboard radiometry (including its part related to blackbody-based calibrations) is the rapidly growing field of remote sensing, which deals with long series of observations made with various instruments.

To ensure the uniformity of these observations and make it possible to use them for long-term forecasts in meteorology and climatology, it is necessary to increase the accuracy of radiometric measurements in remote sensing by an order of magnitude and to guarantee their traceability to SI units.

1.5.4 Other Radiometric Applications

Here, we only briefly outlined some application areas of blackbody radiometry. They do not exhaust all applications of blackbodies. We can mention also the use of blackbodies as calculable radiation sources for radiant flux and irradiance calibrations of

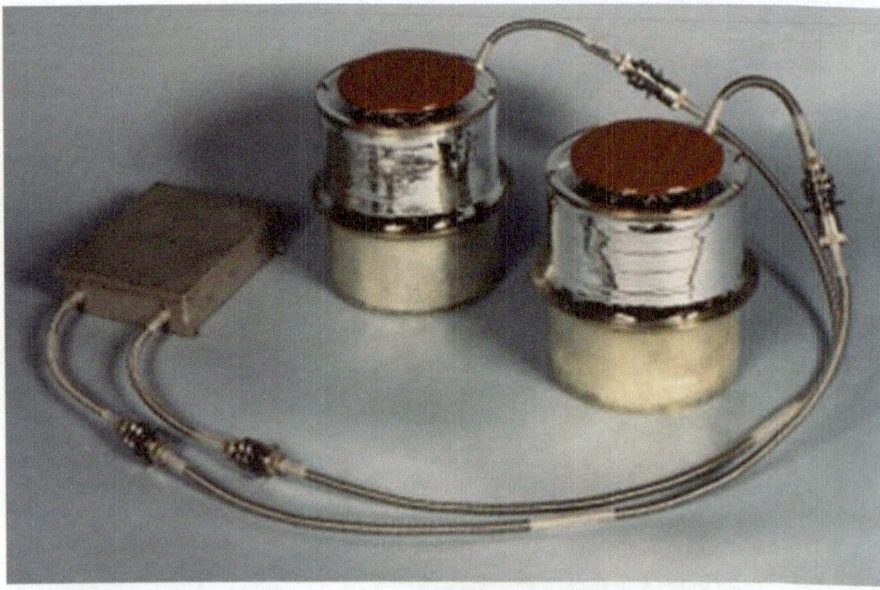

Fig. 1.15 The pair of the AATSR blackbody calibration sources. Reproduced from [77] with permission of Elsevier

heat flux gauges [1, 27, 84] and meteorological instruments (pyranometers, pyrheliometers, net radiometers, etc.) measuring solar radiation [12, 31], for emissivity measurements [39, 87], for determination of absolute responsivity of optical radiation detectors, etc. There are also numerous applications, in which blackbodies are used only as radiation sources of continuous spectra, without need of calculating their magnitudes. Such applications are not relevant to the optical radiometry and, therefore, we will not discuss them.

1.6 Structure of the Book

Like many specialized disciplines, blackbody radiometry is based on a solid foundation of exact sciences, primarily on such fields of physics as optics and thermophysics. As an experimental discipline, the blackbody radiometry uses the results from disciplines common to many branches of scientific instrument designs (e.g. electrical engineering, electronics, signal and data processing, control system engineering, material science, etc.). Finally, the blackbody radiometry has a strong metrological component because it deals with measurements, uncertainties, calibration, etc. Realizing this, the authors did not even try to include all the auxilliary material in this book to make it self-sufficient. However, we could not pass by the fundamental concepts and definitions, without which the uniformity of exposition would not be possible.

For organizational purposes, the book is divided into two volumes. Volume I "Fundamentals" covers the most general theoretical, technological, and metrological ideas and concepts that underlie blackbody radiometry. Volume II "Practical Designs and Applications" is dedicated to practical implementation of general principles presented in the first volume.

1.6.1 Synopsis of Volume I

The first volume of two-volume monograph "Blackbody Radiometry" consists of 9 chapters, which presents general principles of optical and thermal design for blackbodies and the methods commonly used for their characterization, and four Appendices (A through D).

The essential background information is presented, mostly, in Chap. 2. We define the radiometric quantities and their interrelations, review the main concepts, terms, and definition of metrology in connection with the metrological aspects of the blackbody radiometry.

Chapter 3 presents the theoretical basis of blackbody radiometry. We recall briefly the laws of the theory of blackbody radiation, discuss the radiative properties of real bodies (namely, absorptance, reflectance, transmittance, and emissivity), and consider some important problems of radiation heat exchange between diffuse surfaces.

Chapter 4 is dedicated to issues relevant to the effective emissivity, one of the most important characteristics of blackbodies. The isothermal and nonisothermal cavities are discussed with respect to definition, calculation and measurement of their effective emissivities.

Chapter 5 highlights thermophysical and optical aspects of designing blackbodies: methods for heating, cooling, and isothermalization as well as techniques for improving effective emissivities.

Chapter 6 overviews and discusses materials for blackbody radiators, from low-temperature "black" paints and coatings (including modern coatings on the base of carbon nanotubes) to high-temperature carbon materials.

Chapter 7 is dedicated to contact measurements of blackbody temperatures traceable to the ITS-90. The main types of contact temperature sensors (platinum resistance thermometers, thermistors, and thermocouples) applicable to measuring temperature of blackbody radiators as well as some sources of systematic errors in contact temperature measurements of blackbodies are reviewed.

In Chap. 8, the noncontact measurement of blackbody temperature by means of RTs is discussed and the modern radiation thermometers suitable for measuring the temperature of blackbodies are briefly overviewed. Although the presentation is given for conventional radiation thermometry traceable to the ITS-90, the same provisions can be applied to the relative primary radiometric thermometry if to employ thermodynamic temperatures for the fixed points used. We consider the main sources of uncertainty (the nonlinearity, size-of-source effect, etc.) of spectral band RTs.

The techniques of measuring temperature nonuniformities over radiating elements of blackbodies by means of RT are also discussed.

Chapter 9 is the largest chapter, which exposes one of the most important and complicated problem of measuring thermodynamic temperature of high-temperature blackbodies. This problem is crucial not only for blackbody radiometry, but also for the foundations of physical science in general, because its solution allows establishing the thermodynamic temperature scale in the high-temperature range on the base of fundamental physical laws and the most advanced achievements of experimental physics. Chapter 9 covers the principles and instrumental base of the absolute primary radiometric thermometry and describe the four main techniques of measuring the thermodynamic temperature of blackbodies using the ACR, TD, and FR. Design features and metrological characteristics of these measuring instruments are discussed to the extent necessary to consider modern facilities for calibrating the FRs on the base of tunable lasers and monochromators and measurement of blackbody thermodynamic temperatures above the freezing point of silver using these FRs. The sources of uncertainties at calibrations and measurements are considered for different methods of the absolute primary radiometric thermometry.

In this way, the first volume contains the information that can be called definitive for the blackbody radiometry. This volume provides all introductory information needed, explains major radiometric and metrological concepts, gives definition of radiometric values, recall fundamental laws of blackbody radiation, covers general principle of designing blackbodies, and explains major techniques for their characterization. Four Appendices finish Volume I.

Appendix A presents the values of four constant for the revised SI [63] and calculated values of other of fundamental physical constants that are used throughout this book.

Appendix B lists the defining fixed points of the ITS-90 and assigned values for HTFPs.

Appendix C is a glossary of metrology terms used throughout this book and those, through which they are defined. It is composed on the base of the VIM [41] for readers' convenience.

Appendix D is the table of fractional exitance of blackbody radiation that allows evaluation of the fraction of the total exitance of a perfect blackbody at a given temperature, which is contained within some wavelength interval. The guidance on their use is given in Sect. 3.2.5.

The International System of Units (SI) is used throughout the book. Temperatures expressed in °F in the original publications, which we refer to, have been recalculated in kelvins or °C. All linear dimensions in drawings, graphs, and diagrams are expressed in millimeters, unless otherwise indicated.

1.6.2 Volume II: Content at a Glance

The second volume "Practical Designs and Applications" is dedicated to practical implementation of general principles presented in the first volume. We describe in detail the design features of FPBBs operating at temperature of defining fixed points of the ITS-90 and VTBBs of various types: liquid-bath, heat-pipe, high-temperature with the direct resistance heating, etc. Particular attention is paid to HTFPs, their phenomenology, blackbodies on their base, and their use in blackbody radiometry and relative primary radiometric thermometry.

We discuss the most important applications of modern blackbody radiometry. We describe the methods and distinctive features of realization and dissemination of blackbody-based radiometric scales, firstly, the scales of the spectral radiance in the wavelength range from 220 to 2500 nm and spectral irradiance in the wavelength range from 250 to 2500 nm. We discuss the facilities developed at world-leading national metrology institutes blackbody-based calibration facilities, their uncertainty budgets, and methods of calibrating the secondary standards.

Another major area of application of blackbody radiometry is the calibration of radiometric instruments designed for infrared remote sensing of the Earth. This includes various types of instrumentation (staring and scanning, single-element and matrix, multispectral, hyperspectral, and Fourier transform sensors), which must be calibrated before launching into the Earth's orbit and, taking into account the long-term activity of modern satellites, on-orbit with the necessary frequency. We discuss modern ground-based facilities and blackbody standards for prelaunch calibration of remote sensing radiometric equipment and distinctive features of onboard blackbodies.

1.6.3 To Whom is This Book Addressed and How to Use It?

This book is addressed primarily to that part of radiometric community, which consists of professionals involved into design, development, and characterization of blackbodies and radiometric systems on their base, as well as to everyone who uses blackbodies in radiometric calibrations. Typically, these are the staff members (scientists, engineers, and technicians) of research laboratories and government agencies. Radiometrists of other specializations can obtain new information from this book on ways to improve the performance of their instruments, measurement procedures and, as a result, achieve better metrological quality of measurements.

Another type of the intended audience is researchers and engineers who work in areas that are not directly related to the use of blackbodies, but are obligated, in connection with the current project, to shift the focus of their interests to solving blackbody-related problems. For such specialists, this book will make professional reorientation easier due to a lot of useful information gathered in one place. Besides, this book is suitable for self-learning by scientists and engineers who would like to refresh their knowledge in the field.

The third type of potential readers is represented by undergraduate and graduate students whose future work should be related to optical radiometry or neighboring areas. Although our book was not written as a textbook for students, it can be used by students in engineering and physics as an addition to a lecture course.

Finally, many materials presented in the book can be used for tutorials in short courses on optical radiometry and adjacent areas offered by research institutions, universities, professional societies, etc.

Reading of this book requires a basic knowledge in general physics usual for students majoring in engineering or physical sciences. The required mathematical background includes analytic geometry, differential and integral calculi, basics of the probability theory, and elements of mathematical statistics. Some familiarity with metrology and, especially, methods for evaluation of measurement uncertainties is desirable for professionals but not mandatory for general reader. Nevertheless, we refrained from deeply diving into the theory of measurement uncertainty, since the literature on this subject is very extensive. We restricted ourselves to a brief explanation of principal issues providing references to the Joint Committee for Guides in Metrology (JGCM) regulations and the most authoritative sources in the field.

We limited ourselves to a brief explanation of the main issues, providing references to the regulations of the JGCM and the most authoritative sources in this area. Additionally, we complied a thesaurus of metrology-related terms (Appendix C), used in the book.

As a rule, the books devoted to one particular branch of knowledge belong to one of the three categories: (i) textbooks or tutorials, (ii) handbooks or reference guides, and (iii) books containing the original authors' studies. Since the book, offered to the reader, is the first and yet the only monograph in the field, it combines the features of all above-mentioned types. The authors believe that this book will help our readers to gain, along with deeper understanding of general concepts and principles of blackbody radiometry, certain insights on designing of blackbodies and measuring systems on their base, as well as methodology of assessing their uncertainties and possible ways to minimize them. In general, this book can be considered as a compendium of knowledge accumulated on the subject over many years. Besides, we have attempted to describe the state-of-the-art of blackbody radiometry and to outline the ways for its future development.

References

1. A.N. Abdelmessih, T.J. Horn, Experimental and computational characterization of high heat fluxes during transient blackbody calibrations. J. Heat Transfer **132**, 023304 (2010)
2. K. Anhalt, G. Machin, Thermodynamic temperature by primary radiometry. Phil. Trans. R. Soc. A **374**, 20150041 (2016)
3. ANSI/IES RP-16-10, *Nomenclature and Definitions for Illuminating Engineering* (Illuminating Engineering Society of North America, New York, NY, 2010)

4. P. Becker, P. De Bièvre, K. Fujii et al., Considerations on future redefinitions of the kilogram, the mole and of other units. Metrologia **44**, 1–14 (2007)
5. R.E. Bedford, A low temperature standard of total radiation. Can. J. Phys. **38**, 1256–1278 (1960)
6. R.E. Bedford, Blackbodies as absolute radiation standards, in *Precision Radiometry*. Advances in Geophysics, vol. 14, ed. by A.J. Drummond (Acad. Press, New York, 1970), pp. 165–202
7. R.E. Bedford, G. Bonnier, H. Maas, Recommended values of temperature on the international temperature scale of 1990 for a selected set of secondary reference points. Metrologia **33**, 133–154 (1996)
8. W. Bich, The third-millennium International System of Units. Rivista Del Nuovo Cimento **42**(2), 49–102 (2019)
9. BIPM, The International System of Units (SI). 9th edn., vol. 1.06 (2019), https://www.bipm.org/utils/common/pdf/si-brochure/SI-Brochure-9-EN.pdf. Accessed 20 Feb 2020
10. K. Blaum, D. Budker, A. Surzhykov et al., The revised SI: fundamental constants, basic physics and units. Ann. Phys. (Berlin) **531**, 1900148 (2019)
11. L.P. Boivin, C. Bamber, A.A. Gaertner et al., Wideband filter radiometers for blackbody temperature measurements. J. Modern Opt. **57**, 1648–1660 (2010)
12. D. Bolsée, N. Pereira, W. Decuyper et al., Accurate determination of the TOA solar spectral NIR irradiance using a primary standard source and the Bouguer-Langley technique. Solar Phys. **289**, 2433–2457 (2014)
13. C.J. Bordé, Base units of the SI, fundamental constants and modern quantum physics. Phil. Trans. R. Soc. A **363**, 2177–2201 (2005)
14. S. Briaudeau, M. Sadli, F. Bourson et al., Primary radiometry for the *mise-en-pratique*: the laser-based radiance method applied to a pyrometer. Int. J. Thermophys. **32**, 2183–2196 (2011)
15. F. Cabiati, W. Bich, Realization and dissemination of units in a revised SI. Measurement **42**, 1454–1458 (2009)
16. CCT, *Mise en pratique* for the definition of the kelvin in the SI. SI Brochure, 9th edn., Appendix 2 (2019), https://www.bipm.org/utils/en/pdf/si-mep/SI-App2-kelvin.pdf. Accessed 19 Feb 2020
17. W.W. Coblentz, Measurements on standards of radiation in absolute value. Bull. Bureau of Standards **11**, 87–100 (1914)
18. W.W. Coblentz, R. Stair, The present status of the standards of thermal radiation maintained by the Bureau of Standards. Bureau of Standards J. Res. **11**, 79–87 (1933)
19. R.U. Datla, J.P. Rice, K. Lykke et al., *Best Practice Guidelines for Pre-Launch Characterization and Calibration of Instruments for Passive Optical Remote Sensing.* (Report to Global Space-based Inter-Calibration System (GSICS) Executive Panel, NOAA/NESDIS, World Weather Building, Camp Springs, Maryland 20746). NIST Internal Report 7637 (National Institute of Standards and Technology, U.S. Department of Commerce, 2009)
20. R.U. Datla, J.P. Rice, K. Lykke et al., Best practice guidelines for pre-launch characterization and calibration of instruments for passive optical remote sensing. J. Res. Natl. Inst. Stand. Technol. **116**, 621–646 (2011)
21. R. Davis, An introduction to the revised International System of Units (SI). IEEE Instrum. Meas. Mag. **22**(3), 4–8 (2019)
22. D.P. DeWitt, G.D. Nutter (eds.), *Theory and Practice of Radiation Thermometry* (Wiley, New York, 1988)
23. M. Dramićanin, *Luminescence Thermometry: Methods, Materials, and Applications* (Woodhead Publ., Duxford, UK, 2018)
24. J. Fischer, J. Ullrich, The new system of units. Nat. Phys. **12**, 4–7 (2016)
25. J. Fischer, M. Battuello, M. Sadli et al., CCT/03-03. Uncertainty Budgets for Realisation of Scales by Radiation Thermometry (2003), https://www.bipm.org/cc/CCT/Allowed/22/CCT03-03.pdf. Accessed 9 Oct. 2019
26. B. Fellmuth, J. Fischer, G. Machin et al., The kelvin redefinition and its *mise en pratique*. Phil. Trans. R. Soc. A **374**, 20150037 (2016)

27. J.-R. Filtz, T. Valin, J. Hameury et al., New vacuum blackbody cavity for heat flux meter calibration. Int. J. Thermophys. **30**, 236–248 (2009)
28. J. Fischer, M. de Podesta, K.D. Hill et al., Present estimates of the differences between thermodynamic temperatures and the ITS-90. Int. J. Thermophys. **32**, 12–25 (2011)
29. J.C. Flemming, An evaluation of an high temperature blackbody as a working standard of spectral radiance. Appl. Opt. **5**, 195–200 (1966)
30. E.O. Göbel, U. Siegner, *The New International System of Units (SI). Quantum Metrology and Quantum Standards* (Wiley, Weinheim, Germany, 2019)
31. J. Gröbner, Operation and investigation of a tilted bottom cavity for pyrgeometer characterizations. Appl. Opt. **47**, 4441–4447 (2008)
32. X.P. Hao, J. Song, M. Xu et al., Vacuum radiance-temperature standard facility for infrared remote sensing at NIM. Int. J. Thermophys. **39**, 78 (2018)
33. J. Hartmann, High-temperature measurement techniques for the application in photometry, radiometry and thermometry. Phys. Reports **469**, 205–269 (2009)
34. J. Hartmann, K. Anhalt, R.D. Taubert et al., Absolute radiometry for the *MeP*-K: the irradiance measurement method. Int. J. Thermophys. **32**, 1707–1718 (2011)
35 C.M. Herzfeld, The thermodynamic temperature scale, its definition and realization, in *Temperature: Its Measurement and Control in Science and Industry. Part 1. Basic Concepts, Standards and Methods*, vol. 3, ed. by F.G. Brickwedde (Reinhold Publ. Corp., New York, 1962), pp. 41–50
36. K.D. Hill, A.G. Steele, The international temperature scale: past, present, and future. NCSLI Measure **9**, 60–67 (2014)
37. D. Hoffmann, On the experimental context of Planck's foundation of quantum theory. Centaurus **43**, 240–259 (2001). https://doi.org/10.1111/j.1600-0498.2000.cnt430305.x
38. J. Hollandt et al., Primary sources for use in radiometry, in *Optical Radiometry*, ed. by A.C. Parr et al. (Academic Press, 2005), pp. 213–290
39. M. Honner, P. Honnerová, Survey of emissivity measurement by radiometric methods. Appl. Opt. **54**, 669–683 (2015)
40. ILV, *International Lighting Vocabulary*. CIE S 017/E:2011 (CIE Central Bureau, Vienna, 2011)
41. JCGM 200:2012, *International Vocabulary of Metrology—Basic and General Concepts and Associated Terms (VIM)*, 3rd edn., 2008 version with minor corrections (BIPM Joint Committee for Guides in Metrology, Paris, 2012)
42. H.J. Jung, On the determination of the thermodynamic temperature of high temperature blackbodies via ITS-90 or alternative methods, in *Proceedings of the TEMPMEKO '96, 6th International Symposium on Temperature and Thermal Measurements in Industry and Science*, ed. by P. Marcarino (Levrotto & Bella, Torino, Italy, 1997), pp. 235–244
43. A.R. Karoli, Experimental blackbody (absolute) radiometry, in *Precision Radiometry*. Advances in Geophysics, vol. 14, ed. by A.J. Drummond (Acad. Press, New York, 1970), pp. 203–226
44. V.B. Khromchenko, S.N. Mekhontsev, L.M. Hanssen, A tunable filter comparator for the spectral calibration of near-ambient temperature blackbodies. Proc. SPIE **6678**, 66781E (2007)
45. J. Kloepfer, C. Taylor, V. Murgai, Characterization of the VIIRS blackbody emittance, in *Presentation at the Conference on CALCON 2013* (Logan, UT, 2013), https://digitalcommons.usu.edu/cgi/viewcontent.cgi?article=1011&context=calcon. Accessed 17 Oct. 2019
46. D. Kuljis, V. Murgai, J. Kloepfer, Characterization of JPSS J3 and J4 blackbody emissivity. Proc. SPIE **11127**, 111270F (2019)
47. K.C. Lapworth, T.J. Quinn, L.A. Allnutt, A black-body source of radiation covering a wavelength range from the ultraviolet to the infrared. J. Phys. E: Sci. Instrum. **3**, 116–120 (1970)
48. H.M. Latvakoski, M. Watson, S. Topham et al., Accurate blackbodies. Proc. SPIE **7739**, 773919 (2010)
49. A.J. Lichtenberg, S. Sesnic, Absolute radiation standard in the far infrared. J. Opt. Soc. Am. **56**, 75–79 (1966)
50. R.B. Lindsay, The temperature concept for systems in equilibrium, in *Temperature: Its Measurement and Control in Science and Industry*, vol. III, part 1, ed. by C.M. Herzfeld (Reinhold Pub. Corp., New York, 1962), pp. 3–13

51. D.H. Lowe, A.D.W. Todd, R. Van den Bossche et al., The equilibrium liquidus temperatures of rhenium–carbon, platinum–carbon and cobalt–carbon eutectic alloys. Metrologia **54**, 390–398 (2017)
52. G. Machin, K. Anhalt, P. Bloembergen et al., *MeP*-K Absolute Primary Radiometric Thermometry (2012), https://www.bipm.org/utils/en/pdf/si-mep/MeP-K-2018_Absolute_Primary_Radiometry.pdf. Accessed 1 Feb 2020
53. G. Machin, K. Anhalt, P. Bloembergen et al., *MeP*-K Relative Primary Radiometric Thermometry (2017), https://www.bipm.org/utils/en/pdf/si-mep/MeP-K-2018_Relative_Primary_Radiometry.pdf. Accessed 1 Feb 2020
54. S.N. Mekhontsev, V.B. Khromchenko, L.M. Hanssen, NIST radiance temperature and infrared spectral radiance scales at near-ambient temperatures. Int. J. Thermophys. **29**, 1026–1040 (2008)
55. S. Mekhontsev, L. Hanssen, S. Lorentz et al., Primary realization of spectral radiance, emittance and reflectance in the mid- and far-infrared, in *Presentation on the 11th International Conference on New Developments and Applications in Optical Radiometry (NEWRAD 2011)* (Maui, HI, 2011), https://physics.nist.gov/newrad2011/presentations/day1/PM/EAO_OR_009_Mekhontsev_a.pdf. Accessed 5 Feb 2020
56. I.M. Mills, Proposed revisions to the International System of Units. Measurement and Control **47**, 302–307 (2014)
57. I.M. Mills, P.J. Mohr, T.J. Quinn et al., Redefinition of the kilogram, ampere, kelvin and mole: a proposed approach to implementing CIPM recommendation 1 (CI-2005). Metrologia **43**, 227–246 (2006)
58. I.M. Mills, P.J. Mohr, T.J. Quinn et al., Adapting the International System of Units to the twenty-first century. Phil. Trans. R. Soc. A **369**, 3907–3924 (2011)
59. M.J.T. Milton, R. Davis, N. Fletcher, Towards a new SI: a review of progress made since 2011. Metrologia **51**, R21–R30 (2014)
60. P.J. Mohr, Defining units in the quantum based SI. Metrologia **45**, 129–133 (2008)
61. P.J. Mohr, D.B. Newell, B.N. Taylor et al., Data and analysis for the CODATA 2017 special fundamental constants adjustment. Metrologia **55**, 125–146 (2018)
62. W. Nawrocki, *Introduction to Quantum Metrology: The Revised SI System and Quantum Standards*, 2nd edn. (Springer, Cham, Switzerland, 2019)
63. D.B. Newell, F. Cabiati, J. Fischer et al., The CODATA 2017 values of, and for the revision of the SI. Metrologia **55**, L13–L16 (2018)
64. J.V. Nicholas, D.R. White, *Traceable Temperatures: An Introduction to Temperature Measurement and Calibration*, 2nd edn. (Wiley, Chichester, UK, 2001)
65. T.S. Pagano, D.A. Elliot, M.R. Gunson et al., Operational readiness for the atmospheric infrared sounder (AIRS) on the earth observing system aqua spacecraft. Proc. SPIE **4483**, 35–43 (2002)
66. A.C. Parr, A National Measurement System for Radiometry, Photometry, and Pyrometry Based Upon Absolute Detectors. NIST Technical Note 1421. Natl. Inst. Standards. Technol., U.S. Dept. of Commerce (1996), https://nvlpubs.nist.gov/nistpubs/Legacy/TN/nbstechnicalnote1421.pdf. Accessed 1 Feb 2020
67. U. Platt, K. Pfeilsticker, M. Vollmer, Radiation and optics in the atmosphere, in *Springer Handbook of Lasers and Optics*, ed. by Träger (Springer, Dordrecht, Netherlands, 2012), pp. 1475–1517
68. H. Preston-Thomas, The international temperature scale of 1990 (ITS-90). Metrologia **27**, 3–10 (1990)
69. T.J. Quinn, *Temperature*, 2nd edn. (Acad. Press, London, 1990)
70. T.J. Quinn, News from the BIPM. Metrologia **34**, 187–194 (1997)
71. T. Quinn, *From Artefacts to Atoms. The BIPM and the Search for Ultimate Measurement Standards* (Oxford University Press, Oxford, UK, 2012)
72. V.I. Sapritsky, Black-body radiometry. Metrologia **32**, 411–417 (1995/96)
73. N. Sasajima, Y. Yamada, B.M. Zailani et al., Melting and freezing behavior of metal-carbon eutectic fixed-point blackbodies, in *TEMPMEKO 2001. Proceedings of the 8th International Symposium on Temperature and Thermal Measurements in Industry and Science*, vol. 1, ed. by B. Fellmuth, J. Seidel, G. Scholz (VDE Verlag GmbH, Berlin, 2002), pp. 501–506

74. P. Saunders, E. Woolliams, H. Yoon et al., Uncertainty Estimation in Primary Radiometric Temperature Measurement (2018), https://www.bipm.org/utils/en/pdf/si-mep/MeP-K-2018_Absolute_Primary_Radiometry_Uncertainty.pdf. Accessed 1 Feb 2020
75. A.R. Schaefer, R.D. Saunders, Intercomparison between silicon and blackbody-based radiometry using a silicon photodiode/filter radiometer. Appl. Opt. **23**, 2224–2226 (1984)
76. Schaefer et al., Intercomparison between independent irradiance scales based on silicon photodiode physics, gold-point blackbody radiation, and synchrotron radiation. Opt. Eng. **25**, 892–896 (1986)
77. D.L. Smith, Thermal infrared satellite radiometers: design and prelaunch characterization, in *Optical Radiometry for Ocean Climate Measurements*, ed. by G. Zibordi, C.J. Donlon, A.C. Parr (Acad. Press, Amsterdam, Netherlands, 2014), pp. 153–199
78. R. Stair, R.G. Johnston, E.W. Halbach, Standard of spectral radiance for the region of 0.25 to 2.6 microns. J. Res. NBS **64A**, 291–296 (1960)
79. R. Stair, Sources as radiometric standards, in *Precision Radiometry*. Advances in Geophysics, vol. 14, ed. by A.J. Drummond (Acad. Press, New York, 1970), pp. 83–109
80. J. Stenger, J.H. Ullrich, Units based on constants: the redefinition of the International System of Units. Ann. Rev. Condens Matter Phys. **7**, 35–59 (2016)
81. M. Stock, R. Davis, E. de Mirandés et al., The revision of the SI—The result of three decades of progress in metrology. Metrologia **56**, 022001 (2019); Corrigendum—Metrologia **56**, 049502 (2019)
82. J. Tansock, D. Bancroft, J. Butler et al., NIST HB 157. Guidelines for Radiometric Calibration of Electro-Optical Instruments for Remote Sensing. NIST, U.S. Department of Commerce (2015), https://nvlpubs.nist.gov/nistpubs/hb/2015/NIST.HB.157.pdf. Accessed 23 Mar 2020
83. B.N. Taylor, Quantity calculus, fundamental constants, and SI units. J. Res. Natl. Inst. Stan. **123**, 123008 (2018)
84. B.K. Tsai, C.E. Gibson, A.V. Murthy et al., *Heat-Flux Sensor Calibration*. NIST Spec. Publ. 250–65 (U.S. Department of Commerce, Natl. Inst. Stand. Technology, Gaithersburg, MD, 2004)
85. J.H. Walker, R.D. Saunders, A.T. Hattenburg, *NBS Measurement Services: Spectral Radiance Calibrations* (National Bureau of Standards Special Publication 250-1, US. Department of Commerce, Washington, DC, 1987)
86. J.H. Walker, R.D. Saunders, J.K. Jackson et al., *NBS Measurement Services: Spectral Irradiance Calibrations* (National Bureau of Standards Special Publication 250–20, US. Department of Commerce, Washington, DC, 1987)
87. H. Watanabe, J. Ishii, H. Wakabayashi et al., Spectral emissivity measurements, in *Spectrophotometry: Accurate Measurement of Optical Properties of Materials*, ed. by T.A. Germer, J.C. Zwinkels, B.K. Tsai (Acad. Press, Amsterdam, 2014), pp. 333–366
88. E. Woolliams, M. Dury, T. Burnitt et al., Primary radiometry for the mise-en-pratique for the definition of the kelvin: the hybrid method. Int. J. Thermophys. **32**, 1–11 (2011)
89. E.R. Woolliams, K. Anhalt, M. Ballico et al., Thermodynamic temperature assignment to the point of inflection of the melting curve of high-temperature fixed points. Phil. Trans. R. Soc. A **374**, 20150044 (2016)
90. X. Xiong, J. Butler, A. Wu et al., Comparison of MODIS and VIIRS on-board blackbody performance. Proc. SPIE **8533**, 853318 (2012)
91. Y. Yamada, H. Sakate, F. Sakuma et al., Radiometric observation of melting and freezing plateaus for a series of metal-carbon eutectic points in the range 1330 °C to 1950 °C. Metrologia **36**, 207–209 (1999)
92. Y. Yamada, Y. Wang, N. Sasajima, Metal carbide-carbon peritectic systems as high-temperature fixed points in thermometry. Metrologia **43**, L23–L27 (2006)
93. H.W. Yoon, C.E. Gibson, G.P. Eppeldauer et al., Thermodynamic radiation thermometry using radiometers calibrated for radiance responsivity. Int. J. Thermophys. **32**, 2217–2229 (2011)
94. H. Yoon, P. Saunders, G. Machin et al., Guide to the Realization of the ITS-90. Radiation Thermometry. BIPM (2018), https://www.bipm.org/utils/common/pdf/ITS-90/Guide_ITS-90_6_RadiationThermometry_2018.pdf. Accessed 29 Jan 2020

95. Z.M. Zhang et al. (eds.), *Radiometric Temperature Measurements. I. Fundamentals* (Elsevier, Amsterdam, 2010)
96. Z.M. Zhang et al. (eds.), *Radiometric Temperature Measurements. II. Applications* (Elsevier, Amsterdam, 2010)

Chapter 2
Essentials of Optical Radiation Metrology

Abstract The basic concepts of optical radiometry and definitions of radiometric quantities are discussed. The main ideas of metrology are considered in connection with the metrological aspects of optical radiometry. An approach to evaluation of measurement uncertainties described in the Guide to the expression of Uncertainty of Measurements (GUM) is outlined and conditions for its applicability are indicated. The basics of the Monte Carlo simulation of measurement uncertainties (the method of propagation of distributions) is briefly considered and some suitable examples are provided.

Keywords Optical radiometry · Radiometric quantities · Metrology · Uncertainty · Radiometric scale · Monte carlo simulation

2.1 Subject and Foundations of Optical Radiometry

There are two principal points of view on the subject of optical radiometry. The International Lighting Vocabulary (ILV) [62] defines radiometry as the "*measurement of the quantities associated with optical radiation*". Datla and Parr [29] define the subject of optical radiometry as follows: "*Radiometry is the science of measuring electromagnetic radiation in terms of its power, polarization, spectral content, and other parameters relevant to a particular source or detector configuration.*"

We believe that both definitions are overly broad and blur the boundaries with related disciplines. Wavelength is a quantity associated with optical radiation, but measuring wavelengths has never been considered part of radiometry. Measurement of the degree of polarization is the subject of polarimetry [16], but not radiometry. Similarly, the quantitative measurement of the interaction of optical radiation with matter is a traditional subject of spectrophotometry [45].

Our point of view is closer to the definition given by ANSI/IES RP-16-10 [4]: "*Radiometry is the measurement of quantities associated with radiant energy and power*" or the definition proposed by Zalewski [153]: "*Radiometry is the measurement of the energy content of electromagnetic radiation fields and the determination of how this energy is transferred from a source, through a medium, and to a detector.*" These definitions are consistent with the definitions given in the classical books on

V. Sapritsky and A. Prokhorov, *Blackbody Radiometry*, Springer Series in Measurement Science and Technology, https://doi.org/10.1007/978-3-030-57789-6_2

radiometry by Grum and Becherer [51] and Boyd [13], and do not exclude the possible dependence of the energy parameters, say, on the degree of polarization or the degree of coherence of the optical radiation in question.

Radiometry is one of the oldest branches of experimental optics. For a long time, the eye of an observer was the only measuring device sensitive to the energy characteristics of optical radiation, and only within a limited spectral range (the ILV defines the visible radiation by the lower limit taken between 360 and 400 nm and the upper limit between 760 and 830 nm).

We understand classical radiometry as a complex of theoretical concepts and physical laws formulated as early as in eighteenth century (initially in the form of photometry), shortly after the first observation of the diffraction phenomenon (ca. 1660) but before development of the wave theory of light by Thomas Young in 1800s. Classical radiometry adopted the mathematical apparatus and many methods developed in photometry and thus remained within the framework of geometrical (ray) optics.

At the early stage of development of radiometry (in the form of photometry), when no detectors other than the human eye were known, the question of the spectral distribution of the optical radiation energy did not arise. Such measurements became possible only after the development of a monochromator (an optical device that selects a narrow spectral band from the input broadband radiation) and the invention of the first physical sensors of optical radiation (bolometer and radiation thermopile). This can be considered as a starting point for the separation of radiometry and photometry and the transformation of radiometry into an independent branch of optics, although the terms "radiometry" and "spectroradiometry" were coined later, seemingly, by Coblentz [21–23].

At the dawn of radiometry, the validity of its basic concepts and laws, borrowed from photometry, was not questioned. However, after Maxwell [98] developed the theory of electromagnetic radiation, their approximate nature became apparent. At present, when the connection between classical radiometry and the electromagnetic theory of optical radiation is quite understandable, we can formulate the assumptions underlying classical radiometry as follows.

1. The dimensions of the measuring device are much larger than the wavelengths of the radiation, whose characteristics should be measured. This requirement is easily met under the most common measurement conditions; however, for measurements with small apertures and large wavelengths, correction for diffraction effects is usually required.
2. The transfer of electromagnetic energy is considered in terms of a radiant flux carried by the rays propagating along local normals to the wavefront of an electromagnetic wave. The directions of the rays can be easily indicated for a point source emitting spherical waves, and for a collimated beam of rays corresponding to a plane wavefront. The wavefront shape for more complicated cases can be predicted using the Huygens principle, which goes beyond the scope of geometrical optics.

3. The polarization of radiation cannot be taken into account in the framework of geometrical optics. For radiometric systems that include polarization-sensitive components, appropriate corrections must be made using physical (wave) optics.
4. The principle of additivity for radiant fluxes is supported: instead of the amplitudes of electromagnetic fields, the values of radiometric quantities are summed. This means that there is no place for interference in classical radiometry. Again, interference effects that are small enough for most typical radiometric problems can be taken into account as appropriate corrections to the solution obtained in the framework of geometrical optics.
5. The measurements are carried out not for instantaneous and local characteristics of the radiation field, but for those averaged over finite time and space intervals. This assumption allows classical radiometry not to take into account fluctuations of the electromagnetic field, which are subjects of statistical optics [47, 106] and quantum optics (see, e.g. [94] or [52]). Classical radiometry deals with radiation fields that are sufficiently extended in time and space, the amplitudes of stochastic fluctuations of the field should be negligible compared to the random errors of mean values and their variances.
6. Classical radiometry mainly deals with heat sources that generate optical radiation caused by chaotic (uncorrelated) oscillations of particles of matter. Therefore, in classical radiometry, optical radiation is considered as incoherent. The concept of coherence goes beyond geometrical optics. Until the 1960s, when lasers became widely used in scientific research, radiometers completely ignored the coherence of optical radiation. However, an increase in the requirements for radiometric accuracy made a quantitative assessment of the effects associated with the possible partial coherence of the measured optical radiation an urgent problem. Most of the thermal radiation sources used in radiometry are partially coherent in accordance with the theory of partial coherence, which has been developed since the mid-1950s. An interested reader can find the necessary information about the relationship between coherence theory and optical radiometry in a collection of articles edited by Friberg [40] and in selected works by a pioneer in this field, Wolf [144]. Partial coherence theory predicts some phenomena that, in principle, can impose more stringent restrictions on the range of applicability of classical radiometry than the condition of small wavelengths compared to a measuring instrument. In particular, the theory of partial coherence predicts that the image of an incoherent object becomes partially coherent, polychromatic radiation of a point source exhibits time incoherence and the partial space coherence; radiation reflected from a diffuse surface irradiated by a monochromatic beam, on the contrary, exhibits partial temporal coherence and spatial incoherence. The most impressive prediction with far-reaching consequences was the so-called Wolf shift, a correlation-induced modification of the spectral distribution of radiation from a partially coherent non-Lambertian source (a physical source, such as an incandescent lamp, or a secondary source, such as a reflecting surface or a diffracting aperture) when propagating in free space. Fortunately, it has been proven [42, 100, 104] that fundamental changes in the principles of optical radiometry are not required, since the corresponding corrections are

insignificant for typical radiation sources used in radiometry under normal laboratory conditions. The introduction of laser sources into the practice of not only coherent, but also incoherent radiometry still raised some problems associated with the coherence of radiation. For example, irradiating the integrating sphere with a laser source creates a speckle pattern on the inner surface of the integrating sphere. To use it as a source of uniform radiation, this pattern must be eliminated. However, the use of radiation from thermal sources (including blackbodies) does not require correction for partial coherence.

7. The reciprocity principle formulated by von Helmholtz [57] is unconditionally fulfilled in classical radiometry. This principle, sometimes called a theorem, states that the source and detector, separated by a stationary and linear medium, in which any linear optical components such as lenses, mirrors, apertures, etc. can be placed, are interchangeable without changing the measurement results. In an extremely simplified form, the principle of reciprocity is sometimes expressed as "You can see me if I can see you." Unfortunately, it is often forgotten to mention the principle of reciprocity among other fundamentals of optical radiometry; however, it has played and continues to play a very important role in many optical and radiometric applications. Kirchhoff [83] used the principle of reciprocity in deriving the law of thermal radiation, which was later named in his honor. Wien [141] and Planck [116, 117] referred to the Helmholtz reciprocity in their analysis of blackbody radiation. The reciprocity principle is commonly used and strictly observed in reflectometry (see, for example, [17, 20]). The Helmholtz reciprocity is also used in some diffraction problems [25], therefore, its field of applicability is not limited to geometrical optics.

Closing the question of the applicability area of classical radiometry and its relation to wave and quantum optics, we note once again that for most of the problems discussed in this book, satisfactory solutions are obtainable using geometrical optics; wave effects can be taken into account as appropriate corrections. In some cases (for example, to study the photoelectric effect), it is necessary to use the quantum theory of radiation. With the wider use of measuring instruments that employ quantum effects, the role of non-classical radiometry increases, but this has little relevance to the subject of this book.

Although optical radiometry is an old branch of optics, the measurement accuracy of radiometric quantities is relatively modest compared to other physical quantities. For example, the relative uncertainty of the primary standard of frequency maintained by NIST is of the order of 10^{-16} [55]. At the same time, the Primary Optical Watt Radiometer (POWR), which is the basis for all radiometric and photometric units and scales realized at NIST, provides a relative uncertainty of only about 10^{-4} (0.01%) for measuring the optical power of continuous wave (CW) lasers operating within predetermined wavelength and power ranges [60]. The relative uncertainty of the order of 10^{-3} (0.1%) is typical for the realization of spectral irradiance scales in the visible spectral range. Recalling that achieving high accuracy is not an end in itself, but a response to the needs of science, technology, and society, we still try to explain the reasons for such a large difference in the measurement uncertainties of radiometric quantities and, for example, frequency, time, length, etc.

The first reason is the numerous simplifications adopted in classical radiometry and the need to know and/or control many auxiliary parameters and characteristics (spectral and spatial distributions, time and polarization dependences, etc.) when measuring radiometric quantities.

Secondly, each measuring instrument itself is a source of electromagnetic radiation: since each physical body has a nonzero thermodynamic temperature, it emits thermal radiation.

Thirdly, the medium separating the source and the detector, as well as the objects surrounding them, also emit, reflect and scatter optical radiation.

As a result, it is extremely difficult to obtain a detector signal, which depends only on the magnitude of the measured quantity and does not depend on anything else.

2.2 Radiometric Quantities

2.2.1 *Optical Range of Electromagnetic Spectrum*

The International Lighting Vocabulary [62], currently regarded as an international primary authority on terminology in radiometry and adjacent areas, defines the optical radiation as "electromagnetic radiation at wavelengths between the region of transition to X-rays ($\lambda \approx 1$ nm) and the region of transition to radio waves ($\lambda \approx 1$ mm)". The same limits for the optical radiation range are given in the international standard ISO 20473 [63]. The ANSI[1] standard [4] defines the lower bound of the optical range as "approximately 100 nm". This discrepancy in definition of the shortwave boundary of the optical range of the electromagnetic spectrum does not matter to us, since the shortwave limit for blackbody radiometry is about 200 nm.

The most common spectral variable is the wavelength λ equal to the distance in the direction of propagation of an electromagnetic wave between two successive positions, at which the phase of the wave is the same. The wavelength λ in vacuo and wavelength λ' in a medium are related as

$$\lambda' = \lambda / n(\lambda), \tag{2.1}$$

where $n(\lambda)$ is the refractive index of the medium that is equal to the ratio of the velocity of electromagnetic waves in vacuum c to the phase velocity c' of the waves of the monochromatic radiation at wavelength λ' in the medium.

---

[1] American National Standards Institute, a private, non-profit organization dedicated to supporting the voluntary standards for products, services, processes, systems, and personnel in the United States and coordinates U.S. standards with international standards so that American products can be used internationally.

Hereinafter, we will consider optical radiation in vacuo unless otherwise stated. In the optical radiometry, the wavelength is the most frequently used spectral variable. Its most common units are micrometer (μm) and nanometer (nm): 1 μm = 10^{-6} m, 1 nm = 10^{-9} m. In optical spectroscopy, especially when measurements are carried out using Fourier transform spectrometers, the other spectral variable, wavenumber $\tilde{\nu}$, is used. Wavenumber is the reciprocal of the wavelength:

$$\tilde{\nu} = \lambda^{-1}. \tag{2.2}$$

The unit for $\tilde{\nu}$ is cm^{-1}. To convert a wavelength to a wavenumber, simply divide 10,000 by the wavelength expressed in micrometers. For example, a wavelength of 5 μm corresponds to a wavenumber of 10,000/5 = 2000 cm^{-1}.

In works related to the optical radiometry of a terahertz range, the frequency ν [Hz] can be used as a spectral variable.

The angular frequency $\omega = 2\pi\nu$ [rad Hz] is sometimes used in radiometry-related theoretical studies. The spectral variables λ, $\tilde{\nu}$, ν, and ω are related by

$$\lambda = \tilde{\nu}^{-1} = c/(n\nu) = (2\pi c)/(n\omega), \tag{2.3}$$

where c is the velocity of light in vacuo and n is the refractive index of the medium, in which radiation is propagated.

For some applications, instead of the frequency ν, wavelength, λ or other spectral variables, the photon energy E_0 is used in accordance with the Planck relation:

$$E_0 = h\nu = hc/\lambda, \tag{2.4}$$

where h is the Planck constant and c is the velocity of light (see Appendix A for numerical values of physical constants). The photon energy is traditionally expressed in electron-volts (eV), a non-SI unit of energy equal to approximately $1.6602 \cdot 10^{-19}$ J.

The entire range of optical radiation is commonly divided into 3 ranges: ultraviolet (UV) radiation, visible (VIS) radiation, and infrared (IR) radiation. The UV and IR ranges, in turn, are subdivided into smaller subranges.

The three most widely used nomenclatures for the optical diapason according to the ILV [62], the American National Standard ANSI/IES RP-16-10 [4], and the international standard ISO 20473 [63] are presented in Tables 2.1, 2.2, and 2.3, respectively. It should be noted that the boundaries of subranges are rather conditional; even boundaries of the visible spectral range are defined slightly different in these three tables. The subranges of the UV and IR spectral range are often selected for reasons of practical convenience in specific applications, depending on the phenomena of interest.

For instance, the remote sensing community employs the spectral subranges for the UV and IR spectral ranges presented in Table 2.4. As an alternative, optical radiometrists working with remote sensing instruments make extensive use of the informal classification associated with so-called "atmospheric windows".

Table 2.1 The wavelength boundaries of the subranges of optical radiation according to ILV [62] and the corresponding boundary wavenumbers

Subrange name	Wavelengths λ (μm)	Wavenumbers $\tilde{\nu}$ (cm^{-1}), approx
UV-C	0.1–0.28	35,714–100,000
UV-B	0.28–0.315	31,746–35,714
UV-A	0.315–0.4	25,000–31,746
Visible	(0.36…0.4)–(0.76…0.83)	(12,048…13,158)–(25,000…27,778)
IR-A	0.780–1.4	7143–31,746
IR-B	1.4–3	333–7143
IR-C	3–1000	10–333

Table 2.2 Subranges of the optical radiation range according to ANSI/IES RP-16-10 [4]

Subrange name	Wavelengths λ(μm)	Wavenumbers $\tilde{\nu}$(cm^{-1}), approx
Vacuum UV	0.01–0.2	50,000–1,000,000
Extreme UV	0.01–0.1	100,000–1,000,000
Far UV	0.1–0.2	50,000–1,000,000
Middle UV	0.2–0.3	33,333–50,000
Near UV	0.3–0.4	25,000–33,333
Visible	0.38–0.77	12,987–26,316
Near (short wavelength) IR	0.7–1.4	7143–14,286
Intermediate IR	1.5–5	2000–6667
Far (long wavelength) IR	5–1000	10–2000

A sensor[2] mounted on board the satellite (or high-altitude aircraft, balloon, etc.) sees the land or ocean surface through the atmosphere layers, which are opaque to electromagnetic radiation in some spectral ranges and almost transparent in others.

Traditionally, the two spectral ranges that approximately correspond to the so-called atmosphere transmittance windows (see Fig. 1.9 in Chap. 1) are distinguished in the Earth's radiation: the solar reflective range of about 0.3 to 2.5 μm and the thermal IR (TIR) range, from 8 to 14 μm.

In recent decades, a special spectral range has been additionally introduced, the so-called terahertz (THz) band. It includes electromagnetic wave frequencies from 0.3 to 3 THz (less often, from 0.1 or 0.3 to 10 THz). Terahertz radiation is of great interest due to a variety of potential applications in the field of imaging and spectroscopy for medical diagnostics and biology, in defense and security, non-destructive testing and wireless data transmission (see, e.g. [15] or [132]).

---

[2]In remote sensing, the name of sensors is used traditionally to designate various types of indicating radiometric instruments, including multi-element ones, often together with optics, scanning devices, devices performing spectral selection, etc.

Table 2.3 Spectral bands for optics and photonics (after [63])

Designation of the radiation		Short designation			Spectral bands[a]			
					Wavelength λ (nm)	Frequency ν (THz)	Wavenumber $\tilde{\nu}$ (cm^{-1})	Photon energy E_0(eV)
Ultraviolet radiation	extreme UV	UV		EUV	1–100	$3 \cdot 10^5$–3000	10^7–10^6	1240–12.4
	vacuum UV		UV-C	VUV	100–190	3000–1580	10^6–5300	12.4–6.5
	deep UV			DUV	190–280	1580–1070	53,000–36,000	6.5–4.4
	mid UV		UV-B		280–315	1070—950	36,000–32,000	4.4–3.9
	near UV		UV-A		315–380	950–790	32,000–26,000	3.9–3.3
Visible radiation, light		VIS			380–780	790–385	26,000–13,000	3.3–1.6
Infrared radiation	near IR	IR	IR-A[b]	NIR	780–1400	385–215	13,000–7000	1.6–0.9
			IR-B		1400–3000	215–100	7000–3300	0.9–0.4
	mid IR		IR-C	MIR	3000–50,000	100–6	3300–200	0.4–0.025
	far IR			FIR	50,000–10^6	6–0.3	200–10	0.025–0.001

[a]The wavelength values are valid for delimitation of the spectral bands. The values for frequencies, wavenumbers, and photon energies are approximate values given for convenience

[b]For other fields of application, which are excluded from the scope of the ISO 20473 [63], there may be different definitions

Table 2.4 Subranges of the UV and IR spectral ranges used in optical remote sensing (according to various sources)

Subrange name	Wavelengths λ (μm)	Wavenumbers $\tilde{\nu}$ (cm^{-1}) approx
Far ultraviolet	0.01–0.10	100,000–1,000,000
Middle ultraviolet	0.20–0.30	33,333–50,000
Near ultraviolet	0.30–0.38	2632–33,333
Near infrared (NIR)	0.75–1.5	6667–13,333
Short-wave infrared (SWIR)	1.2–3	3333–8333
Mid-wave infrared (MWIR)	3–6	1667–8333
Long-wave infrared (LWIR)	5–15	667–2000
Far infrared (FIR)	14–100	100–714
Submillimeter	100–1000	10–100

In the electromagnetic spectrum, terahertz radiation occupies an intermediate position between microwave and IR radiation and has some common properties with each of them.

The frequency band from 0.3 to 3 THz corresponds to the wavelength from about 100 μm to 1 mm and therefore belongs to the subrange IR-C (FIR, LWIR, or submillimeter) of optical diapason. This allows employing the methodology and measurement equipment developed for very longwave IR radiometry also for at least the shortwave part of the THz diapason [140]. Radiometry of specifically THz radiation is a relatively new direction; it began to develop rapidly in early 2000s.

Figure 2.1 shows an approximate diagram of the optical range of the electromagnetic spectrum with an indication of the most commonly used subranges.

A little running ahead, we point out that the operational spectral region of blackbody radiometry is the wavelength range from 0.2 μm to about 100 μm (sometimes up to 1 mm). Whatever definition for the subject of optical radiometry we accept, we will have to deal with radiometric quantities.

The VIM, the International Vocabulary of Metrology [77], defines the quantity as the property of a phenomenon, body, or substance, where the property has a magnitude that can be expressed as a number and a reference. Thus, a radiometric

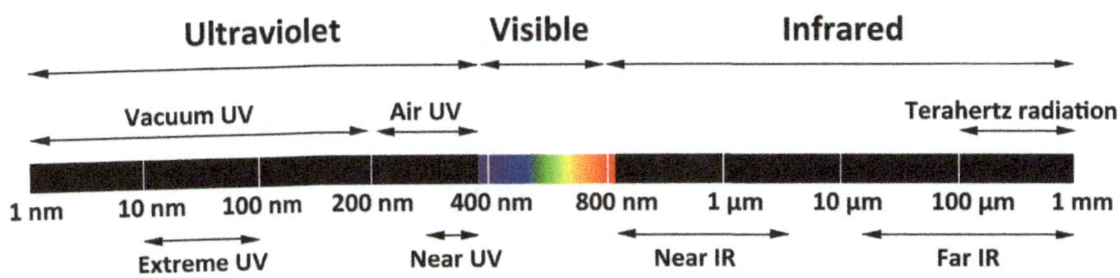

Fig. 2.1 Schematic for the ranges of the optical radiation spectrum

quantity is that related, explicitly or implicitly, to the optical radiation energy. In the following sections, we will discuss radiometric quantities, as well as interrelation between them.

2.2.2 *Total Radiometric Quantities*

The quantities expressing the energy characteristics of polychromatic radiation, we call total (this is not a standardized term) radiometric quantities, referring to the values integrated over the entire spectrum. The most natural way is to start definitions with the optical (radiant) power, P [W] of the energy emitted, transmitted, or received per unit time in the form of radiation. In optical radiometry, this quantity is also called radiant flux and denoted by Φ. The ILV recommends to use symbols P_e and Φ_e; the same subscripts are recommended for all radiometric quantities characterizing the total (integrated over entire spectrum) energy in order to highlight their difference from corresponding spectral, photon, and photometric quantities. In practice, the subscript "e" is usually omitted for sake of brevity. The radiant flux can be arbitrarily distributed over spatial, angular, and spectral domains. All other radiometric quantities are derived from Φ. The ILV gives their definitions verbally using expressions like "the quotient of an element of flux by an element of …" and provides formulae that look like

$$Z = \frac{d\Phi}{dX} \tag{2.5}$$

or

$$Z = \frac{d\Phi}{dXdY}, \tag{2.6}$$

but are not partial derivatives; according to the rules of differential calculus, it would be better to write

$$Z = \frac{\partial^2\Phi}{\partial X \partial Y}. \tag{2.7}$$

However, we deal with not a partial derivative but only symbolism accepted for the quotient of an element of flux Φ by an element of quantity X and an element of quantity Y. More rigorously, Eq. (2.8) can be rewritten as:

$$Z = \lim_{X,Y\to 0} \frac{\Phi}{XY}. \tag{2.8}$$

Often, verbal definitions of radiometric quantities use the phrases "per unit area" and/or "per unit solid angle," but this does not mean 1 m^2 or 1 sr. Actually, the collocation "per unit" means elemental (area or solid angle). A more detailed discussion of this issue was given by Kostkowski [87]. It should be noted that the formalism (2.6) is not commonly accepted. An alternative form recommended by the ANSI standard [4],

$$Z = \frac{d^2\Phi}{dXdY}, \tag{2.9}$$

is supported by some very reputable scientists (see, e.g. [29, 78, 107]).

Using formalism (2.6), we can define the surface (areal) densities of the radiant flux Φ, the radiant exitance and irradiance, the angular (solid angle is meant) density called radiant intensity, and their mix, i.e. "surface-angular" or "angular-surface" density, called radiance.

The surface density of radiant flux at a point on a surface is the quotient of the radiant flux incident on or emitted by an element of the surface area at a given point $\boldsymbol{\xi}$, by the area of this element. The surface density of radiant flux emitted from a surface is called radiant exitance M; the surface density of radiant flux incident on a surface is called irradiance, E:

$$M = \frac{d\Phi}{dA_s} \left[\mathrm{W\,m^{-2}}\right], \tag{2.10}$$

$$E = \frac{d\Phi}{dA_d} \left[\mathrm{W\,m^{-2}}\right], \tag{2.11}$$

where A_s and A_d denote the area elements containing the point $\boldsymbol{\xi}$ on the emitting and the receiving surfaces, respectively.

The radiant intensity I of a source in a given direction $\boldsymbol{\omega}$ is defined as the quotient of the radiant flux dΦ leaving the source and propagated in the element of solid angle dΩ containing the direction $\boldsymbol{\omega}$ by the element of solid angle:

$$I = \frac{d\Phi}{d\Omega} \ [\mathrm{W\,sr^{-1}}]. \tag{2.12}$$

Since the apex of a solid angle must be a point, this definition can only be applied to a point source, or the surface element of a radiation source, or the element of a surface reflecting the incident radiation. A point source is a source, the dimensions of which are negligibly small as compared with the distance between the source and the irradiated surface. A point source is an idealization that does not exist in reality, because it should have an infinite density of energy that is physically impossible. However, the concept of a point source simplifies calculations in many circumstances, for instance, the radiant intensity of a point source is convenient for describing the angular distribution of the radiant flux of real-world sources (lamps, LEDs, etc.) In

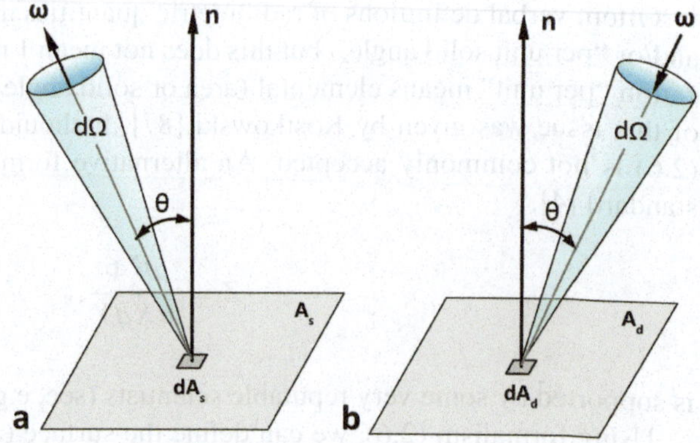

Fig. 2.2 Schematics to the definition of radiance for: **a** outgoing radiation and **b** incident radiation

this case, it is important to make sure that the basic assumption is correct: the source should be sufficiently distant from the detector or the point of observation.

The radiance L in a given direction $\boldsymbol{\omega}$ at a given point $\boldsymbol{\xi}$ of the surface of a source, of a detector, or of any other real or virtual surface is the quotient of the radiant flux leaving, passing through, or arriving at an element of the surface surrounding the point $\boldsymbol{\xi}$, and propagated in directions defined by an elementary cone containing the direction $\boldsymbol{\omega}$, by the product of the solid angle dΩ of the cone and the area of the orthogonal projection of the element dA of the surface on a plane perpendicular to the given direction:

$$L = \frac{d\Phi}{d\Omega\, dA_p} = \frac{d\Phi}{d\Omega\, dA \cos\theta}, \tag{2.13}$$

where $dA_p = dA\cos\theta$ is the elementary projected area, θ is the acute angle between the direction of the radiant flux propagation and the normal to the surface at the point of beam-surface interaction (see Fig. 2.2).

The dimension of the radiance is the same as of exitance and irradiance but the commonly used unit is [W m^{-2} sr^{-1}]. The equivalent and useful formulae for L can be easily obtained by expressing it via radiant intensity I, radiant exitance M, or irradiance E. For a source:

$$L = \frac{dI}{dA_s \cos\theta}, \tag{2.14}$$

$$L = \frac{dM}{d\Omega \cos\theta}. \tag{2.15}$$

For an irradiated surface:

$$L = \frac{dE}{d\Omega \cos\theta}. \tag{2.16}$$

Above, the definitions of radiometric quantities have been written down by means of derivative-like expressions ascending to the radiant flux Φ. Useful equations are obtained if we move in the opposite direction and use integration over appropriate spatial and angular domains. In such a way, equivalent definitions of the radiant exitance and irradiance can be written:

$$M = \int_{2\pi} L \cos\theta \, d\Omega, \tag{2.17}$$

$$E = \int_{2\pi} L \cos\theta \, d\Omega, \tag{2.18}$$

where the integrals are taken over the hemisphere visible from the given point.

For a point source, the radiant flux can be expressed via the radiant intensity I:

$$\Phi = \int_{\Omega} I \, d\Omega, \tag{2.19}$$

where Ω is the solid angle (2π for the point source on a surface or 4π for the point source in the 3D space).

The radiant flux can be expressed via the irradiance and radiant exitance:

$$\Phi = \int_{A_d} E \, d\mathrm{A_d}, \tag{2.20}$$

$$\Phi = \int_{A_s} M \, d\mathrm{A_s}. \tag{2.21}$$

The radiance is the most versatile radiometric quantity, equally suitable for characterization of various radiation fields and for performing complicated radiometric calculations. Properties of radiance are discussed in detail by Nicodemus and Kostkowski [103], Datla and Parr [29], and Johnson et al. [78].

2.2.3 Spectral Radiometric Quantities

In order to quantify the contribution of different spectral components into the spectrum of polychromatic radiation, the spectral quantities (a.k.a. spectral densities or spectral concentrations of radiometric quantities) must be introduced. In general, a spectral radiometric quantity Q_x is a spectral density for the spectral variable x (wavelength λ, wavenumber $\tilde{\nu}$, frequency ν, or angular frequency ω). It is defined by the ILV as the quotient of the radiant quantity $\mathrm{d}Q(x)$ contained in an elementary range of $\mathrm{d}x$ at the spectral variable x by that range:

$$Q_x = dQ(x)/dx. \tag{2.22}$$

A spectral radiometric quantity Q_x, considered as a function of the spectral variable x over a range of x, represents the spectral distribution of Q. Below, we give definitions of the most commonly used spectral radiometric quantities (the spectral radiant flux Φ_λ, spectral radiant exitance M_λ, spectral irradiance E_λ, spectral radiance intensity I_λ, and spectral radiance L_λ) in the wavelength scale following a common approach expressed by Eq. 2.22:

$$\Phi_\lambda = d\Phi/d\lambda\ \left[\mathrm{W\,m^{-1}}\right], \tag{2.23}$$

$$M_\lambda = dM/d\lambda\ \left[\mathrm{W\,m^{-3}}\right], \tag{2.24}$$

$$E_\lambda = dE/d\lambda\ \left[\mathrm{W\,m^{-3}}\right], \tag{2.25}$$

$$I_\lambda = dI/d\lambda\ [\mathrm{W\,sr^{-1}\,m^{-2}}]. \tag{2.26}$$

$$L_\lambda = dL/d\lambda\ \left[\mathrm{W\,m^{-3}\,sr^{-1}}\right]. \tag{2.27}$$

Alternative definitions for spectral irradiance and spectral radiance can be also given through the radiant flux and the spectral radiant flux, respectively:

$$E_\lambda = \frac{d^2\Phi(\lambda)}{dAd\lambda}, \tag{2.28}$$

$$L_\lambda = \frac{d^2\Phi_\lambda(\lambda)}{dA\cos\theta d\Omega d\lambda}. \tag{2.29}$$

A spectral radiometric quantity determined for the spectral variable x can be converted into the homonymous spectral quantity determined for the spectral variable y by applying the following relationship:

$$Q_y = Q_x\left|dx/dy\right|. \tag{2.30}$$

When operations with another spectral variable are necessary, conversion can be done using Eq. 2.30. The most commonly used conversion is the transformation between wavelengths and wavenumbers:

$$d\tilde{\nu} = \frac{1}{\lambda^2}d\lambda, \tag{2.31}$$

$$d\lambda = \frac{1}{\tilde{\nu}^2}d\tilde{\nu}. \tag{2.32}$$

For instance, for the spectral radiance flux in wavelength and wavenumber scales:

$$\Phi_\lambda = \tilde{\nu}^2 \, \Phi_\nu, \tag{2.33}$$

$$\Phi_{\tilde{\nu}} = \lambda^2 \, \Phi_\nu. \tag{2.34}$$

Other examples of such conversions we provide in Sect. 3.2.2, where the Planck law is written for different spectral variables.

The (total) radiant flux is computed via integrating the spectral radiant flux by an appropriate spectral variable, e.g.:

$$\Phi = \int_0^\infty \Phi_\lambda d\lambda, \tag{2.35}$$

$$\Phi = \int_0^\infty \Phi_{\tilde{\nu}} d\tilde{\nu}. \tag{2.36}$$

2.2.4 Responsivity of Radiation Detector

The radiant flux falling onto an optical radiation detector generates an output signal, usually electrical current or voltage, whose magnitude depends on the incident flux. Responsivity s of a detector for a general case of non-zero output Y_d in the absence of input (so-called "dark" signal) is defined by the ILV as the quotient

$$s = (Y_c - Y_d)/X, \tag{2.37}$$

where X is the detector input (usually, radiant flux Φ) and Y_c is the detector output for the current value of the input.

If the detector input is the radiant flux and the output is the electric current or voltage, the units of responsivity are A W^{-1} or V W^{-1}, respectively. Relative responsivity of a detector s_r is defined as the ratio of the responsivity s, when the detector input equals X, to the responsivity s_0 at the reference input X_0:

$$s_r = s/s_0. \tag{2.38}$$

If the detector's output depends linearly on the detector's input (i.e. the responsivity does not depend on the input's level), this detector is considered linear. Otherwise, we deal with a non-linear detector.

Sometimes, the above-defined quantities are referred to as total or integrated responsivities because they are defined regardless of the spectral composition of the incident radiation. Therefore, it makes sense to compare responsivities of different detectors only if they are irradiated with radiant fluxes having identical spectral compositions or if detectors are non-selective with respect to the spectral variables. The thermal detectors of optical radiation provide the lowest spectral selectivity within certain spectral range; the spectral selectivity of a planar thermal detector is determined by the spectral selectivity of its absorbing coating; the selectivity of a cavity detector depends also on the receiving element geometry.

A real-world detector is always spectrally selective. It reacts differently to the monochromatic radiation at different wavelengths even if incident radiant fluxes are equal. In order to establish relationship between the spectral composition of radiation falling onto a detector and its output, spectral responsivity of a detector $s(\lambda)$ is defined as the wavelength-dependent quotient

$$s(\lambda) = dY(\lambda)/dQ(\lambda), \tag{2.39}$$

where $dY(\lambda)$ and $dQ(\lambda) = Q_\lambda(\lambda)d\lambda$ are the detector output and monochromatic input in the wavelength interval $d\lambda$, respectively; Q and Q_λ are a radiometric and a corresponding spectral radiometric quantity.

As a rule, spectral responsivity is determined with respect to radiant flux. This quantity is designated as $s_\Phi(\lambda)$ or simply $s(\lambda)$ and has the unit [W A^{-1}] or [W V^{-1}]. However, depending on the radiometric quantity that is considered as the input, we can define, for convenience, spectral irradiance responsivity, $s_E(\lambda)$ [W m^{-2}A^{-1}] or [W m^{-2} V^{-1}], and spectral radiance responsivity, $s_L(\lambda)$ [W m^{-2}sr^{-1}A^{-1}] or [W m^{-2}sr^{-1} V^{-1}].

Dimensionless relative spectral responsivity of a detector $s_r(\lambda)$ is the ratio of the spectral responsivity $s(\lambda)$ of the detector at wavelength λ to a given reference value s_m:

$$s_r(\lambda) = s(\lambda)/s_m. \tag{2.40}$$

The reference value s_m can be defined as an average, a maximum, or an arbitrarily chosen value of $s(\lambda)$. The signal Y of a detector at a polychromatic input Q can be expressed via the spectral responsivity $s(\lambda)$ as

$$Y = \int_0^\infty s(\lambda) Q_\lambda(\lambda) d\lambda \tag{2.41}$$

or via the relative spectral responsivity $s_r(\lambda)$ as

$$Y = s_m \int_0^\infty s_r(\lambda)\, Q_\lambda(\lambda) d\lambda. \tag{2.42}$$

In practice, instead of integration from $\lambda = 0$ to $\lambda = \infty$, it is enough to compute integrals in Eqs. 2.41 and 2.42 only over a finite interval $[\lambda_{min}, \lambda_{max}]$, where integrands are different from zero.

Semiconductor photodetectors are widely used in the modern radiometry. Their ability of converting optical radiation to electric signal is often characterized by so-called quantum efficiency, along with spectral responsivity. According to ILV [62], the quantum efficiency η of a detector is the ratio of the number of elementary events (such as release of an electron) contributing to the detector output, to the number of incident photons. For convenience, the two kind of quantum efficiency are considered: the external quantum efficiency (EQE) and the internal quantum efficiency (IQE). The EQE η_e is the ratio of the number of elementary events contributing to the detector output, to the number of incident photons, including those reflected by the detector. The IQE η_i is the ratio of the number of elementary events contributing to the detector output, to the number of photons absorbed by the detector. In this book, we will consider only windowless front-illuminated photodiodes, which occupy an exceptionally important place in the radiometry of UV, VIS, and NIR spectral range. For photodiodes of this type, such an elementary event is absorption of a photon in a semiconductor resulting in the excitation of an electron from the valence band to the conduction band (in other words, the generation of photon-hole pair due to internal photoelectric effect). Only a part of photons incident onto a photodiode is absorbed and generates free electron-hole pairs that can be collected. Some pairs may be trapped or recombined without contribution to the photocurrent. The EQE is a figure of merit of the entire photodiode, while the IQE can be considered as a quality score for the semiconductor structure that forms the photodiode. The EQE and IQE of a windowless front-illuminated photodiode are related by the simple and obvious formula:

$$\eta_i(\lambda) = \frac{\eta_e(\lambda)}{1 - \rho(\lambda)}, \tag{2.43}$$

where $\rho(\lambda)$ is the spectral reflectance of the photodiode's receiving surface and λ is the wavelength of incident radiation.

The EQE of a photodiode is always less than unity and depends upon photodiode's structure. As can be seen from Eq. 2.44, $\eta_e < \eta_i$. The EQE of a photodiode can be increased by reducing the reflectance $\rho(\lambda)$ of the receiving surface, increasing absorption within the depletion layer[3] and preventing the recombination or trapping of charge carriers prior to their collection.

The spectral responsivity $s(\lambda)$ of a such photodiode can be easily expressed through its spectral IQE $\eta_i(\lambda)$. The measured photocurrent I_{ph} in an external circuit occurs due to the flow of charge carriers to the photodiode's terminals. Number of electrons collected per second equals I_{ph}/e, where e is the elementary charge. If Φ is the monochromatic incident radiant flux then the number of photons arriving

[3]Depletion layer is a region around the metal–semiconductor junction, where recombination of electrons and holes has reduced substantially the number of equilibrium majority carriers (for details, see any textbook on semiconductor physics).

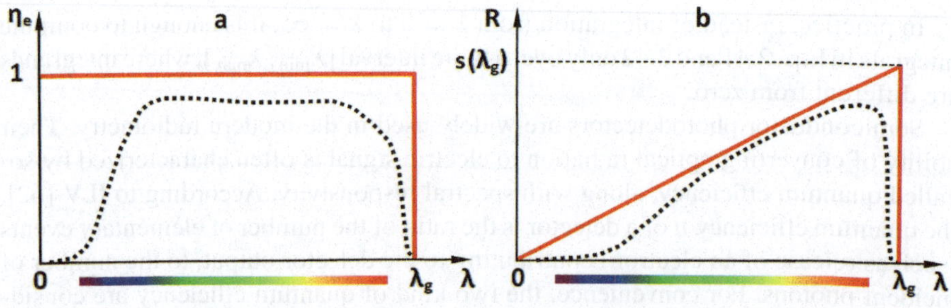

Fig. 2.3 Characteristics of an ideal photodiode (solid lines) and a real photodiode (dotted lines) plotted against wavelength λ: **a** external quantum efficiency η_e and **b** spectral responsivity $s(\lambda)$

per second equals $\Phi/(h\nu)$, where h is the Planck constant and ν is the frequency of incident radiation. Therefore, defining the photodiode's responsivity as $s = I_{ph}/\Phi$ and going from frequencies to wavelengths, we can write down the expression for s as

$$s(\lambda) = \frac{\lambda e \eta_e(\lambda)}{hc} = \frac{[1 - \rho(\lambda)] \cdot \lambda e \eta_i(\lambda)}{hc}. \quad (2.44)$$

For an ideal photodiode, $\rho(\lambda) \equiv 0$, both EQE and IQE do not depend on wavelength up to $\lambda_g = (hc)/E_g$, where E_g is the semiconductor's energy band gap and c is the speed of light in vacuum. Therefore, the spectral responsivity $s(\lambda)$ of an ideal photodiode must depend linearly on the wavelength λ. For any real-world photodiode, $\eta_i(\lambda) < 1$ and the spectral responsivity curve lies below the line for an ideal photodiode (see Fig. 2.3).

Quantum efficiencies for polychromatic incident radiation can be defined following the common rules given in the next Section.

2.2.5 *Band-Limited and Spectrally Weighted Quantities*

A total radiometric quantity Q can be expressed through the corresponding spectral radiometric quantity Q_λ (more precisely, through the spectral distribution $Q_\lambda(\lambda)$) using an obvious relationship:

$$Q = \int_0^\infty Q_\lambda(\lambda) d\lambda. \quad (2.45)$$

If integration is performed only over a finite wavelength range $[\lambda_a, \lambda_b]$, we obtain so-called band-limited radiometric quantity $Q(\lambda_a, \lambda_b)$:

$$Q(\lambda_a, \lambda_b) = \int_{\lambda_a}^{\lambda_b} Q_\lambda(\lambda) d\lambda, \quad (2.46)$$

Although the term "band-limited" is not standardized, band-limited quantities are frequently used in optical radiometry to evaluate the fraction of a radiometry quantity contained in certain spectral interval. For instance, in optical remote sensing, the values of band-limited quantities are used for rough estimates of radiation to be registered by channels of multispectral or hyperspectral sensors. More results that are accurate can be obtained if to employ spectral responsivity $s(\lambda)$ as a weighting function.

Corresponding quantities are usually called spectrally weighted (this is also non-standardized term) and can be defines as:

$$\overline{Q} = \int_0^\infty s(\lambda) Q_\lambda(\lambda) d\lambda \Big/ \int_0^\infty s(\lambda) d\lambda. \quad (2.47)$$

Spectrally weighted quantities are especially popular in biochemistry, biophysics and medicine. For instance, it is well known that radiation of UV-B spectral range contained in radiation from the sun and some artificial sources (UV lamps, electric arc used in welding), has an erythemal (sunburn, i.e. relating to abnormal redness) effect on the human skin. The UV radiation is widely used in phototherapy but excessive UV irradiation may cause skin cancer. Therefore, measuring of the spectral irradiance weighted with erythemal sensitivity of human skin (called the erythema action spectrum $s_{er}(\lambda)$) is very important for public health. The international standard ISO/CIE [72] provides the reference erythema action spectrum and regulates permissible erythema doses.

The most known spectrally weighted quantities are photometric ones. In their definitions, the CIE spectral responsivity of the human eye acknowledged by an international standard [64] plays the role of the weighting function. The relative spectral responsivity of the human eye with the maximum accepted as the reference value is called the spectral luminous efficiency. For photopic (day) and scotopic (night) visions, spectral luminous efficiencies are denoted by $V(\lambda)$ and $V'(\lambda)$, respectively. Both functions are depicted in Fig. 2.4. As a rule, photometric quantities are expressed in practice via $V(\lambda)$; quantities for scotopic vision are typically used only in dedicated research.

A photometric quantity Q_v is defined in relation to the corresponding spectral radiometric quantities Q_λ by the following equation

$$Q_v = K_{cd} \int_{\lambda_1}^{\lambda_2} Q_\lambda(\lambda) V(\lambda) d\lambda, \quad (2.48)$$

where $\lambda_1 = 360$ nm and $\lambda_2 = 830$ nm.

The units of photometric and corresponding radiometric quantities do not match; therefore, the constant K_{cd}called, according to SI [11] the luminous efficacy of

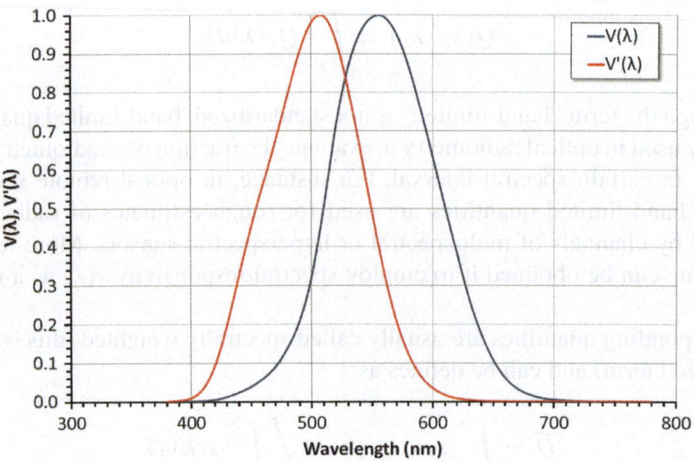

Fig. 2.4 The spectral luminous efficiencies for photopic vision $V(\lambda)$ and for scotopic vision $V'(\lambda)$

Table 2.5 Major radiometric and photometric quantities and units

Radiometry			Photometry		
Quantity	Symbol	Unit	Quantity	Symbol	Unit
Radiant flux	Φ	W	Luminous flux	Φ_v	lm
Radiant exitance	M	W m^{-2}	Luminous exitance	M_v	lm m^{-2}
Irradiance	E	W m^{-2}	Illuminance	E_v	lx = lm m^{-2}
Radiant intensity	I	W sr	Luminous intensity	I_v	lm sr^{-1} = cd
Radiance	L	W m^{-2} sr^{-1}	Luminance	L_v	lm m^{-2} sr^{-1}

monochromatic radiation of frequency 540 × 1012 Hz, is not dimensionless. Its exact value is 683 lm/W.

We do not consider in this book the photometric quantities and photometry issues, thence we just mention the correspondence between photometric and radiometric quantities presented in Table 2.5, where lm, lx, and cd denote the photometric units: lumen, lux, and candela, respectively.

We refer the readers interested in deeper knowledge in photometry to the BIPM monograph [148] and the handbook edited by DeCusatis [32]. The overview of modern photometry is given by Ohno [108].

2.2.6 *Photon Counterparts of Radiometric Quantities*

Photon quantities characterize electromagnetic radiation in terms of the number of photons. Due to the dual nature of optical radiation, each radiometric quantity Q has its own photon analog Q_p and each spectral radiometric quantity Q_λ has a photon analog $Q_{p,\lambda}$. The energy of a quantum of electromagnetic radiation in vacuum equals

$$q = h\nu = hc/\lambda, \tag{2.49}$$

where h is the Planck constant, c is the speed of light, ν and λ are the frequency and the wavelength of radiation. The values of the constants h and c are provided in Appendix A.

The base photon quantity is the photon flux, Φ_p that is measured in s^{-1} and equal to the quotient of the number of photons, dN_p, emitted, transmitted, or received in an element of time, dt, by that element of time:

$$\Phi_p = dN_p/dt. \tag{2.50}$$

Since N_p is the discrete variable, one must treat Eq. 2.50 only as the limiting value of the quotient $\Delta N_p/\Delta t$ at a sufficiently small time interval Δt. The definitions of photon quantities are completely analogous to the definitions of radiometric quantities. For instance, the photon radiance L_p [s^{-1} m^{-2} sr^{-1}] in a given direction, at a given point of a real or imaginary surface is defined by the equation

$$L_p = \frac{d\Phi_p}{dA\cos\theta d\Omega} \tag{2.51}$$

is the analogue of Eq. 2.14 for the radiance. The spectral photon flux $\Phi_p(\lambda)$ and the spectral radiant flux $\Phi_\lambda(\lambda)$ are related as

$$\Phi_p = \frac{1}{hc}\int\limits_0^\infty \Phi_\lambda(\lambda)\lambda d\lambda. \tag{2.52}$$

The photon quantities and units are presented in Table 2.6 with the corresponding radiometric quantities.

Table 2.6 Most important photon quantities and units

Photon quantity	Symbol	Units	Radiometric analogue
Photon flux	Φ_p	s^{-1}	Radiant flux
Photon exitance	M_p	s^{-1} m^{-2}	Radiant exitance
Photon irradiance	E_p	s^{-1} m^{-2}	Irradiance
Photon intensity	I_p	$s^{-1}sr^{-1}$	Radiant intensity
Photon radiance	L_p	$s^{-1}sr^{-1}$ m^{-2}	Radiance

Spectral photon quantities $Q_{p,\lambda}$ for broadband radiation can be also obtained from the corresponding photon quantities Q_p by analogy with radiometric quantities:

$$Q_{p,\lambda} = dQ_p/d\lambda. \tag{2.53}$$

Conversion of a spectral photon quantity to the corresponding spectral radiometric quantity can be done using the following formula:

$$Q_\lambda(\lambda) = \frac{hc}{\lambda} Q_{p,\lambda}(\lambda). \tag{2.54}$$

Photon quantities are closely related to measurement of ultra-low levels of optical radiation (photon fluxes of order of 10^{-12} down to 10^{-18} W) by means of photon counting technique, in which individual photons are counted using a single-photon detector. For a long time after 1940s, the photomultiplier tubes were the only detectors capable to operate in the photon-counting mode. Due to the photoemission threshold of the photocathode materials, the use of photomultipliers was limited by wavelengths not exceeding, at best, 1.1 μm. The shortwave limit was defined by the spectral transmittance of the vacuum tube window. The silicon avalanche photodiodes developed in the early 1960s were the first solid-state detectors with the sensitivity sufficient to register single photons.

After significant improvement of their design and manufacturing technology, including the use of novel semiconductor materials, the avalanche photodiodes become commonly used as receivers in fiber-optics communications.

Presently, there are many photodetectors, including solid-state photomultipliers, some charge-coupled devices, avalanche photodiodes, superconducting nanowire and transition edge detectors, quantum well IR detectors, etc. that be configured to detect individual photons. An overview of modern single-photon detectors can be found in Chaps. 4 through 7 of the collective monograph edited by Migdall et al. [101].

To express extremely low-level optical radiation employed in quantum information processing and technology, quantum cryptography and communication, to characterize new photonics devices such as photonic crystal fibers, computer generated holograms, and the detectors mentioned above, photon quantities are more convenient than usual radiometric quantities. Photon quantities are often used for characterization of remote sensing radiometric instruments such as focal-plane arrays (FPAs) and highly sensitive detectors with digital output; the use of photon quantities in blackbody radiometry is limited mainly by measuring systems with cryogenic blackbodies.

2.3 An Overview of Basic Concepts of Metrology

2.3.1 *Measurement*

Metrology is the science of measurement, covering both experimental and theoretical issues of measurements in any field of science, engineering, and technology. Metrology is subdivided into scientific metrology, applied (technical or industrial) metrology, and legal metrology.

Scientific metrology deals with the establishing the units of measurement, physical constants, development of new measurement methods, creation and maintaining the measurement standards, and providing the uniformity of measurements. Fundamental metrology, a frequently used term that does not have an internationally accepted definition, is considered either as a synonym for scientific metrology or as its branch of the highest level of accuracy. The BIPM subdivides the scientific metrology into nine technical subject fields of measuring (i) mass, (ii) electricity, (iii) length, (iv) time and frequency, (v) temperature, (vi) ionizing radiation and radioactivity, (vii) photometry and radiometry, (viii) acoustics, and (ix) amount of substance. From this point of view, optical radiometry is itself a subdivision of the scientific metrology.

Applied metrology concerns the application of measurements to industrial processes, and their use in society, ensuring the suitability and adequate functioning of measuring instruments, their calibration and quality control. Legal metrology is the application of legal requirements to measurements and measuring devices. It deals with measurements that affect the transparency of economic transactions, particularly where there is a need for legal verification of the measuring instrument.

Metrology (perhaps with the exception of legal metrology or some part of it) is an exact science that deals with the values of physical quantities, measurement errors, and their statistical characteristics. Like any exact science, it needs strict definitions of basic concepts. Currently, the main regulatory document in this area is the 3rd edition of the VIM [77].

Below, we briefly discuss and comment on some of the concepts that are often used in this book. The verbatim definitions from the VIM for the terms in bold in this and the next section are given in the Glossary of Metrology Terms (Appendix B) for readers' convenience.

Measurement is the main source of knowledge in the exact sciences. Measurement of a physical (including radiometric) quantity is the process of experimental determination of a **quantity value** to assign a product of a number and a measurement unit to that quantity value. The quantity that has to be measured is called **measurand**.

To obtain a **measurement result**, measurement should be conducted using a calibrated **measuring instrument** that operates in specified measurement condition according to the specified **measurement procedure**.

Measurement procedure is a detailed description of a measurement according to one or more measurement principles and to a given measurement method. Measurement procedure is based on a **measurement model.** It may include calculations for obtaining a measurement result.

Measurement principle is a phenomenon serving as a basis of a measurement. For instance, dependence of electrical resistance upon temperature is the principle of temperature measurements using resistance thermometers; the thermoelectric (Seebeck) effect is the basis of measuring temperature by means of thermocouples; the photovoltaic effect in semiconductors is the underlying measurement principle of photodiode detectors of optical radiation.

Measurement method is a description of a logical structure and sequence of operations performed during a measurement. There is no generally accepted classification of measurement methods except for their subdividing onto direct and indirect ones. Some methods are equally applicable to measurements of various physical quantities, such as the comparison method, the substitution method, differential measurement method, etc. Other methods can be applied to certain measurands only. In the optical radiometry, the best-known methods are the method of direct comparison and so-called electrical substitution method.

The method of direct comparison can be employed to compare both radiation sources and radiation detectors. The electrical substitution method is applied to measuring radiant flux using a thermal detector, whose receiving element are heated alternately up to the same temperature by the optical radiation and the electric current. Such a detector is commonly called electrical substitution radiometer (ESR); we discuss the electrical substitution method in Sect. 9.2.1.

The measurement model as the mathematical relation among all quantities known to be involved in a measurement. A general form of a measurement model is the equation

$$h(Y, X_1, X_2, \ldots, X_N) = 0, \tag{2.55}$$

where Y is the measurand (**output quantity**) of the measurement model, for which the quantity value has to be inferred from information about **input quantities** $X_1, X_2, \ldots, X_N$ of the measurement model. In the cases of simultaneous measurements of more than one quantity of the same or different kind, the measurement model also contains more than one equation.

Measurement function is considered as a special case of a measurement model, when Eq. 2.55 can be explicitly resolved for the measurand Y:

$$Y = f(X_1, X_2, \ldots, X_N), \tag{2.56}$$

where f to symbolize an algorithm, yielding a unique output quantity value $y = f(x_1, x_2 \ldots, x_N)$ for corresponding input quantity values $x_1, x_2, \ldots, x_N$. Sophisticated readers, of course, recognized in Eq. 2.56 an expression for indirect measurements. We recommend less experienced readers to turn to Rabinovich [123], where the Chap. 5 is entirely dedicated to indirect measurements.

Here, it should be noted that the concept of measurement function is similar, if not equivalent, to the concept introduced by Kostkowski and Nicodemus [86] of NBS as the central concept for all optical radiation measurements. They defined the measurement equation that "*relates the output signal from a radiometric instrument or measuring device to the distribution of radiation incident on it.*" The concept of measurement equation became widespread not only in official publications of the NBS/NIST (e.g. [46, 139]) and research papers on optical radiometry written by its staff (see, e.g. [99, 113, 150]), but also among radiometric and pyrometric communities around the world [50, 105, 145]. Examples of measurement equations for typical radiometric measurements were provided by Tsai and Johnson [137], Datla and Parr [29], Johnson et al. [78].

2.3.2 Performance Characteristics of Measurement

In contrast to qualitative observation, measurement allows us to express empirical knowledge in a mathematical form and makes possible a quantitative description of objects and processes. A measurement is not complete until the **measurement uncertainties** are evaluated and an **uncertainty budget** is prepared.

Many generations of scientists and engineers operated within the following paradigm. The aim of measurement is experimental determination of the **true quantity value** of the measurand. None measurement can be perfectly accurate. Every measurement is subject to error, due to which the result of a measurement differs from the true value of the measurand. The measurement error is defined as the difference between the measured value and the true value of the measurand. Since the measurement errors are random variables, they can be analyzed using the mathematical statistics methods. Measurement errors are subdivided into **random measurement errors** (relative to the expected value of sampling distribution) and **systematic measurement errors** (deviation of the expected value from the true value). The imprecision is quantified by the estimated standard deviation (termed **standard measurement uncertainty**) of the sampling distribution of measured values or by a confidence (coverage) interval for the unknown expected value.

This approach is referred to as *Error Analysis* and in somewhat different variations is presented by many authors, among which the most authoritative are Mandel [93], Grabe [49], and Rabinovich [123]. The mathematical apparatus developed within the framework of the *Error Analysis* is not changed so far and is not questioned. However, a poorly developed conceptual base of the *Error Analysis* may impede, in some important cases, unambiguous identification of the input values as well as interpretation of the results. The most criticized concept is **true quantity value**.

We can never know the measurand true value that is unknown and unknowable. Correspondingly, measurement uncertainty can be determined in different ways. Recognizing the lack of international consensus on the expression of uncertainty in measurement, the International Organization for Standardization (ISO) published the first version of the Guide to the expression of Uncertainty of Measurements (GUM) in 1993, the latest version of which [73] was published in 2008 by the Joint Committee for Guides in Metrology. The GUM introduces the *Uncertainty Analysis* approach aimed to overcoming the drawbacks of the *Error Analysis*. Its philosophy is explained by Howarth and Redgrave [61]. Taylor and Kuyatt [134] briefly outlined the GUM approach in a brochure originally intended for the NIST internal use, but becoming very popular in the metrology community. A good introduction to measurement uncertainty has been issued by the NPL [8].

While the GUM was able to provide a uniformity in evaluation and expression of measurement uncertainties, it could not complete the transition from the *Error Analysis* to the *Uncertainty Analysis*. Despite the fact that the concepts of true value and error were excluded from the GUM, practitioners continued to use these concepts in the traditional sense.

Kacker et al. [80] discussed the logic of development of the concept of measurement uncertainty and the methods for its quantification from *Error Analysis* to the modern approach (*Uncertainty Analysis*) based on the GUM. The latter edition of the VIM [77] redefined the concepts of the true quantity value, measurement error, and measurement uncertainty, aligning them with both the GUM and the needs of practitioners. In addition to redefinition of the **true quantity value** (now, it is simply a quantity value consistent with the definition of a quantity, i.e. does not pretend to be a single value), the VIM introduces the **reference quantity value** as a quantity value used as a basis for comparison with values of quantities of the same kind. Correspondingly, the measurement error is defined as a measured quantity value minus a reference quantity value. Such a palliative solution allows to achieve an additional goal to maintain compatibility with existing ISO group of standards under the common name "Accuracy (trueness and precision) of measurement methods and results": ISO 5725-1:1994, ISO 5725-2:1994, ISO 5725-3:1994, ISO 5725-4:1994, ISO 5725-6:1994, and ISO 5725-5:1998 [66–71].

Measurement accuracy, **measurement trueness**, and **measurement precision** are the three different performance characteristics of measurements. These terms should not be used one instead of another, especially since only precision is a quantitative characteristic, whereas accuracy and trueness are not quantities and, therefore, can only be used in comparative statements.

but always has the status**Measurement accuracy** indicates the closeness of agreement between a measured quantity value and a true quantity value of a measurand. As we already mentioned, the true value is unknown. Therefore, the accuracy of the measurement cannot have a numerical value. It can only be used as a qualitative characteristic of the measurement. **Measurement trueness** is defined as the closeness of agreement between the average of an infinite number of replicate measured quantity values and a reference quantity value. Measurement trueness is inversely related to systematic measurement error, but not related to random error. This concept is used in the definition of reference measurement procedure, which is used, in turn, to define the concept of calibration. It should be noted that the concept of **measurement trueness** is not often used in works dedicated to solution of practical metrological problems.

Precision is the most useful performance characteristic of a measurement that indicates the closeness of agreement between indications or measured quantity values obtained by repetitive measurements of the same measurand under specified measurement conditions. The measurement precision can be expressed numerically by standard deviation, variance, or coefficient of variation under the specified conditions of measurement. The VIM defines three category of measurement precision: **measurement repeatability**, **intermediate measurement precision**, and **measurement reproducibility**. They are defined for three types of relevant measurement conditions:

- **repeatability conditions of measurement** are the condition that assumes the same measurement procedure, same operators, same measuring system, same operating conditions, same location, and replicate measurements on the same or similar objects over a short period of time;
- **intermediate precision conditions of measurement** are the condition that includes the same measurement procedure, same location, and replicate measurements on the same or similar objects over an extended period of time, but may include other conditions involving changes;
- **reproducibility conditions of measurement** are the condition that includes different locations, operators, measuring systems, and replicate measurements on the same or similar objects.

2.3.3 Calibration and Traceability of Measuring Instruments

Measuring instrument is a device used for measuring. There are two kinds of measuring instrument: (i) an **indicating measuring instrument** and (ii) a **material measure**. A common name of the indicating measuring instrument in radiometry is a *radiometer*. Often, the term "detector" is used instead, following the historical tradition. We also will use it when we talk about measuring devices with long-established names (e.g., trap detectors). In optical remote sensing, the term "**sensor**" is applied to any type of radiometers including the imaging ones, i.e., the FPA together with optical elements, cooling system, etc., despite the fact that, according to the VIM, a sensor should be considered as the element of a radiometer that is directly affected by radiation, whose characteristics have to be measured.

Depending on the design of a radiometer and/or way of using it, a radiometer allows measuring a specific characteristic of optical radiation energy, e.g. radiant flux, radiance or spectral radiance, irradiance or spectral irradiance. A radiometer designed to measure the spectral quantities is often (but not always) called a spectroradiometer.

A blackbody represents another type of measuring instruments—a material measure, which reproduces radiometric quantities (e.g. spectral radiance or irradiance) or radiance temperature, each with an assigned quantity value.

To explain the expression "calibrated measuring instrument" it makes sense to undertake a brief historical excursus. At the dawn of metrology, the indicating measuring instruments were divided into absolute and relative. An instrument was considered as relative if it was not capable to indicate measurement results in the units of a measured quantity. Since the relationship between the indications of a relative instrument and the measured quantity value may be non-linear, the appropriateness of using such instruments has been questioned. The absolute measuring instruments (sometimes called self-calibrating) were able to indicate the measurement results in commonly accepted units or allowed such results to be derived by means of simple calculations. The so-called absolute pyrheliometers (see Vignola et al. [138]) developed at the turn of the 19th and 20th centuries to measure the direct solar irradiance were the first absolute radiometers allowing to conduct measurements in commonly accepted units (in calories per minute per square centimeter; the SI unit is W/m^2).

The fact that a measuring instrument is "absolute" does not mean that it is capable to carry out perfectly accurate measurements, i.e. to obtain the true value of a measurand, in modern terminology. For instance, the values of solar irradiance measured by the first pyrheliometers had to be corrected for various systematic effects. In fact, every instance of absolute pyrheliometer required individual characterization and prescribed storage conditions to achieve a modest accuracy of several percents [41]. This makes absolute instruments too expensive for everyday use. It is more reasonable to employ for routine measurements less expensive non-absolute instruments, which are regularly compared with (calibrated against) an absolute instrument. The latter plays the role of a reference or, in modern terms, a **primary measurement standard**.

Calibration can be carried out not only directly, but also through a **metrological traceability chain**, a hierarchical structure, in the nodes of which the **measurement standards** are placed. Within the framework of the existing measurement paradigm, the concept of an absolute measuring instrument has been superseded by the concept of a calibrated measuring instrument. Only a primary measurement standard can be considered as an absolute measuring instrument.

Although the VIM does not bring under regulations the usage of the qualifier "absolute" with the terms "measurement," "measuring instrument," etc., this usage still occurs in optical radiometry in such long-established collocations as "absolute cryogenic radiometer" or "absolute radiometry," and even "absolute radiometric calibration." It should be understood that the adjective "absolute" in the following examples has different meanings:

- absolute cryogenic radiometer is the measuring instrument that measures optical power (radiant flux) in watts and does not require calibration as a primary standard;
- absolute radiometry means that measurements of radiometric quantities are conducted in corresponding units of the SI (as opposed to relative radiometry

that deals with measurement of the ratios of radiometric quantities of the same kind);

- absolute radiometric calibration is a kind of an oxymoron used, perhaps, in contrast to nonexistent "relative calibration," for which it would be better to use the term "characterization." It should be noted that the collocation "absolute calibration" has recently been used for radiation thermometers [3; 81; 95; 149] to distinguish calibration based on measuring their absolute responsivity from calibration against blackbodies at the fixed points on the ITS-90.

Such usage of the qualifier "absolute" must be distinguished from its use in the word groups, in which both adjectives "absolute" and "relative" are admissible (e.g. "absolute responsivity" and "relative responsivity").

Calibration is a two-stage process carried out under specified conditions. First, a relation between the quantity values provided by a measurement standard (**calibrator**) and corresponding indications of a calibrated instrument is established. Then this information is used to establish a relation for converting an indication of a calibrated instrument into a measurement result. Sometimes, the first stage alone is treated as a calibration. The calibration standard, in turn, can be calibrated against a standard of higher rank and so forth. In such a way, the calibration chain forms a hierarchical structure. Only a measuring instrument that does not require calibration (i.e. absolute instrument) must be at the top of the calibration chain. All quantity values involved in calibration are considered together with associated **measurement uncertainties** (see Sect. 2.4.2.) The **measurement accuracy** of the calibrated instrument is always lower than that of calibrator, which is not necessarily an absolute measuring instrument but always has the status of measurement standard.

Calibration is carried out in accordance with the special kind of measurement procedure, so-called **reference measurement procedure**. The VIM distinguishes among reference measurement procedures the **primary reference measurement procedures** as the reference procedures used to obtain a measurement result without relation to a measurement standard for a quantity of the same kind (but with possible referencing to measurement standards for other quantities). Therefore, we can redefine absolute measurements as measurements conducted according to the primary reference measurement procedure.

Calibration should not be confused with **verification** or **validation**.

Verification is provision of objective evidence that a given item fulfils specified requirements, for example:

(i) confirmation that a given FEL lamp meets the requirements of the calibration standard of spectral irradiance in the prescribed wavelength range;
(ii) confirmation that the temperature uniformity of a given flat-plate blackbody is within limits claimed by the manufacturer;
(iii) confirmation that a target measurement uncertainty of a radiometer is met.

Validation is a verification, where the specified requirements are adequate for an intended use, for example, a spectroradiometer can be validated also for operating in space to measure the thermal infrared (TIR) radiance of the Earth.

Now, we approached closely to the definition of a measurement standard, the concept playing one of the key roles in metrology.

Measurement standard is the realization of the definition of a given quantity, with the stated quantity value and associated measurement uncertainty. This realization is intended to serve as a reference and can be implemented in a material measure, measuring instrument or measuring system, or a reference material. Several quantities of the same or of different kinds may be realized in one measurement standard.

The term "realization" is frequently used in metrology in various contexts denoting the three different procedures:

1. Realization in the narrow sense implies the physical realization of the measurement unit according to its definition.
2. Realization as a reproduction denotes establishing a measurement standard on the base of a highly reproducible measuring instrument that operates according a certain measuring principle (using a certain physical phenomena).
3. Realization as adopting a material measure or a set of material measures, each reproducing one value of a physical quantity, as a measurement standard.

The VIM distinguishes measurement standards by (a) their metrological level (primary, secondary, working) and (b) by the place of their permanent storage and the area of their application (international, national, reference). International measurement standards are established by convention, following resolutions of the GCPM; they are recognized and served worldwide according to international agreements. National measurement standards are recognized by national authorities as the basis for assigning quantity values to other standards for the quantity of the same kind.

Primary standard heads the calibration chain and does not require calibration in terms of a quantity it reproduces. Primary standards are the most accurate measuring instrument in their area of use established using the primary reference measurement procedure, i.e. the procedure used to obtain a measurement result without referencing to another measurement standards for a quantity of the same kind.

There are no international standards in the field of optical radiometry. The primary national standards, as a rule, are stored in and maintained by National Metrology Institutes (NMIs). The prime responsibility of the NMIs is to be the source of traceability of the highest metrological level (to the SI or to other internationally agreed references) for users in their countries. The three world's leading NMIs are the oldest ones. These are the PTB (Physikalisch-Technische Bundesanstalt, established in 1887 as Physikalisch-Technische Reichsanstalt,[4] PTR) in Germany, the NPL (National Physical Laboratory, established in 1900) in the UK, and the NIST (National Institute of Standards and Technology, established as Bureau of Standards in 1901, and known as the NBS, National Bureau of Standards, between 1904 and 1988) in the USA. Readers interested in history of science can find many useful materials and references on the early years of these NMIs in Johnston [79].

In some countries, there are several metrological organizations operating at the top of the national metrology system, each in their own field of activity, and sharing the

[4] Imperial Physical Technical Institute (German).

responsibility of the NMI. Such organizations are classified as Designated Institutes (DI) although, in most cases, they differ from the NMIs only nominally. For instance, the NMI of the Russian Federation (and of the former Soviet Union) is the VNIIM, D. I. Mendeleyev Institute for Metrology (St. Petersburg, Russia, https://www.vniim.ru/) but photometry and spectral radiometry are the responsibility areas of the VNIIOFI, the All-Russian Research Institute for Optical and Physical Measurements (Moscow, Russia, https://www.vniiofi.com/).

A measurement standard established by means of calibration against a primary standard for a quantity of the same kind is qualified as a **secondary measurement standard**.

Measurement standard used for routine calibrations of measuring instruments is called **working measurement standard**. Working standards are typically calibrated against a reference measurement standard designated for calibration of other standards for quantities of the same kind in a given organization or at a given location. **Travelling measurement standard** is designed to be transportable between different locations; the travelling (portable) standard may have a special construction.

One more terminological remark: the term "transfer standard" widely used in many areas of measurement, including radiometry, denotes a measuring instrument used as an intermediary (e.g. at comparison of measurement standards) and is not necessarily a measurement standard. The VIM recommends the term "**transfer measurement device**" instead; presently, both terms are in circulation, along with "comparison artifact" ("artefact" in British English) and more specific "transfer detector," "transfer radiometer," "transfer lamp," etc.

A property of a measurement result, through which it can be related to a reference (a definition of a measurement unit through its practical realization, a reference measurement procedure, or a measurement standard) using a documented unbroken chain of calibrations, each contributing to the measurement uncertainty, is called **metrological traceability**. Metrological traceability requires an established calibration hierarchy, along with some other relevant metrological information about the reference. For measurements with more than one input quantity in the measurement model, each of the input quantity value should itself be metrologically traceable via its own calibration hierarchy, so the entire metrological traceability chain may have a branched structure or form a network. If the reference is the definition of a measurement unit through its practical realization, one can talk about the metrological traceability to a measurement unit. The expression "traceability to the SI" means metrological traceability to a measurement unit of the International System of Units. An example of the branched traceability chain to the SI unit is presented schematically in Fig. 2.5. The SI derived unit [Q] for the quantity Q is supposed to be defined via the SI base units $[Q_1]$, $[Q_2]$, and $[Q_3]$.

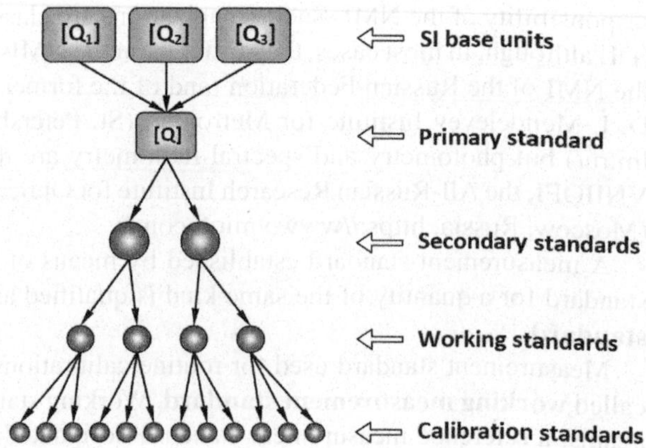

Fig. 2.5 Schematic of the metrological traceability chain to the SI derived unit [Q] of the measured quantity Q; [Q_1], [Q_2], and [Q_3] are the SI base units

2.3.4 *Radiometric Scales*

The concept of measurement scale is one of the key concepts of fundamental metrology. The VIM provides an over-formalized definition of the measurement (quantity-value) scale (not to be confused with the scale of a displaying measuring instrument!) as "*ordered set of quantity values of quantities of a given kind of quantity used in ranking, according to magnitude, quantities of that kind,*" which gives little for this concept understanding. Relying on an intuitive comprehension of the measurement scale of a continuously varying quantity and referring to the practice of using the term "scale" in, e.g. thermometry, we can try to translate the VIM's definition into plain language as follows:

1. To establish a scale of some physical quantity, we must have a possibility to reproduce several (at least two) reference values of that quantity. This can be done with the help of a primary measurement standard. For radiometric scales, one value is enough because the second value—so-called "radiometric zero" (no optical radiation input) can be always easily obtained.
2. To realize the scale between the values, reproduced by the standard and, perhaps, out of their range, we must have a measuring instrument with the known measurement equation to perform interpolation (and, perhaps, extrapolation) using the values of the quantity reproduced by the primary standard.
3. The scale established in such a way should not remain the "thing-in-itself" but should be available to end users in their measurement practice. Therefore, we must have a possibility to disseminate the scale using secondary standards traceable to the primary standard.

The collocations like "realization and dissemination of the (…) scale," where (…) denote the name of a radiometric quantity (e.g. spectral radiance, spectral irradiance, etc.) are very often used in the works completely or partially dedicated to metrological aspects of optical radiometry. We should figure out what the term "scale" means in such a context. Going back to the early years of radiometry, we encounter for the first time the concept of measurement scale at the beginning of the 20th century. At that time, two competing scales for measurement of direct solar irradiance coexisted: the pyrheliometric scale of Ångström based on the so-called Compensation Pyrheliometer [7] in Europe and the Smithsonian pyrheliometric scale based on the Continuous Flow Standard Pyrheliometer [1] in North America. Throughout the first half of the twentieth century, these scales have repeatedly been compared with each other and the results of comparisons were expressed numerically [41]. Therefore, the term "scale" in radiometry was coined much earlier than the first definition of the measurement scale was given. During a long time, radiometrists employed the term "scale" considering it as obvious without having a strict definition.

An attempt to formalize the concept of measurement scale (again, without an explicit definition) was undertaken by [133], the prominent American psychologist worked at Harvard University. He proposed the classification of measurement scales (nominal, ordinal, interval, and ratio) on the base of (a) variety of rules for the assignment of numerals to the measurement results, (b) mathematical properties (group structure) of the resulting scales, and (c) statistical operations allowable for measurements made with the scales of different types. Although Stevens' classification has been severely criticized afterwards, its principles, in somewhat modified form, remains in use in many fields up to date. In our view, the main drawback of the approach accepted by Stevens and his followers is the complete detachment of measurement scales from measuring instruments. Such an approach may seem reasonable in psychology, sociology, and some other non-exact sciences but creates many unsolvable problems e.g., in physics.

It should be noted that the Stevens' understanding of measurement scale is not the only one, although it is the most widespread. The concept of measurement scale was developed by many authors (see, e.g. [9, 38, 115]). Most of them linked definition of a scale to a specific measuring instrument.

An evidence of doubts in applicability of usual understanding of the concept of measurement scale to the radiometric scales even of renowned experts in the field is the letter to the editor of the journal "Metrologia" [43]. In response to this inquiry, Quinn (2000) wrote: "I would like to emphasize the problems that can arise if a clear distinction is not made between a scale, which I understand to be *a recipe or a procedure for realizing a physical quantity*, and the physical quantity itself" and further: "My recommendation, therefore, is to be prudent in the use of the term 'scale' and make sure that it is not taken to refer to a physical quantity. *In any quantitative statement, the term should be avoided*" (italics supplied). Analysis of the usage of the term "scale" in the radiometric context shows that the word groups "realization of a scale" and "dissemination of a scale" have the following meanings.

To realize a radiometric quantity means to establish a primary standard realizing the unit of this quantity.

To disseminate a radiometric scale means to establish the calibration procedures to compare the secondary and working standards with a primary standard, and so forth, downwards along the traceability chain.

It should also be noted that radiometrists still do not follow Quinn's suggestion to avoid the term "scale" in quantitative expressions. The expressions like "scale accuracy," "scale reproducibility," "difference between radiometric scales," "uncertainty of the scale," etc., all with the indicated numerical values can be found in many works until now (see, e.g. [48, 112, 151]). However, it is also clear, that such a usage is no more than a tribute to tradition or abbreviated expressions for "accuracy (reproducibility) of values obtained for the scale," "difference between values obtained for these scales," "uncertainty of a value obtained according to a scale" and so forth.

2.4 GUM Approach for Measurement Uncertainty Evaluation

2.4.1 *Evaluation of Measurement Uncertainties*

The Guide to the Expression of Uncertainty in Measurement, known as "GUM," provides guidance on how to determine, combine and express uncertainty. It was developed by the Joint Committee for Guides in Metrology (JCGM), a joint committee of the BIPM and other relevant standards organizations such as ISO (International Organization for Standardization). The latter edition of the GUM is JCGM 100:2008 [73]. We cannot and do not consider it necessary to describe in detail all the features of the GUM's approach to the analysis, calculation and expression of measurement uncertainties, especially because a huge amount of literature on the evaluation of measurement data has been published since 1993, when the first GUM edition was released. Therefore, we highlight only the main points that we will use throughout this book.

The standard uncertainty of the measurement result (i.e. uncertainty expressed as a standard deviation) is a central concept of the GUM [73], which explains the most common features the *Uncertainty Analysis* approach. Instead of dividing the errors into random and systematic, the GUM suggests using two methods to evaluate the different components of standard uncertainty. The type A evaluation employs the statistical analysis of series of observations. Let x_k, $k = 1, 2, \ldots, n$ are n independent observations obtained under the same measurement conditions for the measured quantity (measurand) X that varies randomly. Most often, the best estimate of the expectation or expected value of the random variable x is the arithmetic mean or average:

$$\overline{x} = \frac{1}{n}\sum_{k=1}^{n} x_k. \quad (2.57)$$

The individual observations x_k differ in value because of random variations in the influence quantities, or random effects. The experimental variance of the observations (the estimate of the variance for the probability distribution of x), is calculated as

$$s^2(x_k) = \frac{1}{n-1}\sum_{k=1}^{n}(x_k - \overline{x})^2 = \frac{1}{n-1}\left[\sum_{k=1}^{n} x_k^2 - \frac{1}{n}\left(\sum_{k=1}^{n} x_k\right)^2\right]. \tag{2.58}$$

The best estimate of the variance of the mean, is given by

$$s^2(\overline{x}) = \frac{s^2(x_k)}{n}. \tag{2.59}$$

The experimental standard deviation of the mean $s(\overline{x})$, equal to the positive square root of $s^2(\overline{x})$ can be used as a measure of the uncertainty of $\overline{x}$, quantifying how well $\overline{x}$ estimates the expectation of x.

Most often, it is assumed that the distribution, which best describes the realization of a random variable in multiple independent measurements, is a Gaussian distribution. When the uncertainty is evaluated from a small-volume sampling of a measurand characterized by a Gaussian distribution, the corresponding distribution can be taken as the Student t-distribution.

All other (non-statistical) methods of uncertainty evaluation belong to the type B. The Type B evaluation typically assumes the use of a priori knowledge or information adopted from external sources. This information may incorporate uncertainties determined in manufacturer's specifications, calibration certificates, borrowed from handbooks and recognized databases, or simply experience with, or general facts on the behavior of materials or instruments.

There is no direct correspondence between the classification of the methods of evaluation as Type A or Type B and the identification of uncertainties as arising from random effects or from the corrections for systematic effects. However, the components of uncertainty arising from random effects are usually evaluated by statistical methods (Type A) while the components of uncertainty arising from the corrections for recognized systematic effects are most often evaluated by other methods (Type B).

2.4.2 *Combined Uncertainty*

In most practical cases, the value of a measurand Y is obtained from indirect measurements, when directly measured quantities $X_1, X_2, \ldots, X_N$ are related with Y through the explicit function f (Eq. 2.59). Individual uncertainties $u(x_1), u(x_2), \ldots, u(x_N)$, whether they are of Type A or Type B, are combined by applying the so-called law of propagation of uncertainty to obtain a combined standard uncertainty $u_c(y)$ using a Taylor series expansion limited to first-order terms:

$$u_c(y) = \sqrt{\sum_{i=1}^{N} c_i^2 u^2(x_i) + 2\sum_{i=1}^{N-j}\sum_{j=i+1}^{N} c_i c_j u(x_i, x_j)}$$
$$= \sqrt{\sum_{i=1}^{N} c_i^2 u^2(x_i) + 2\sum_{i=1}^{N-j}\sum_{j=i+1}^{N} c_i c_j r(x_i, x_j) u(x_i) u(x_j)}, \quad (2.60)$$

where

$$c_i = \frac{\partial f}{\partial x_i} = \left.\frac{\partial f}{\partial X_i}\right|_{x_1, x_2, \ldots, x_N} \quad (2.61)$$

is the sensitivity coefficient obtained through the partial derivative of the function f with respect to each of the input quantities X_i taken at the expectation values x_i, $i = 1, 2, \ldots, N$ that converts the input unit of x_i to the unit of the measurand;

$$u(x_i, x_j) = u(x_j, x_i) = \frac{1}{n(n-1)}\sum_{k=1}^{n}(x_{i,k} - \overline{x}_i)(x_{j,k} - \overline{x}_j) \quad (2.62)$$

is the estimated covariance associated with x_i and x_j;

$$r(x_i, x_j) = \frac{u(x_i, x_j)}{u(x_i)u(x_j)} = \frac{\sum_{k=1}^{n}(x_{i,k} - \overline{x}_i)(x_{j,k} - x_j)}{\sqrt{\sum_{k=1}^{n}(x_{i,k} - \overline{x}_i)^2 \sum_{k=1}^{n}(x_{j,k} - \overline{x}_j)^2}} \quad (2.63)$$

is the correlation coefficient between the input quantities X_i and X_j; $-1 \leq r(x_i, x_j) \leq 1$. Often, for faster single-pass calculation (without preliminary calculation of $\overline{x}_i$ and $\overline{x}_j$), an equivalent formula is more convenient for calculating sample correlations:

$$r(x_i, x_j) = \frac{n\sum_{k=1}^{n} x_{i,k}x_{jk} - \sum_{k=1}^{n} x_{i,k}\sum_{k=1}^{n} x_{j,k}}{\sqrt{n\sum_{k=1}^{n} x_{i,k}^2 - \left(\sum_{k=1}^{n} x_{i,k}\right)^2}\sqrt{n\sum_{k=1}^{n} x_{j,k}^2 - \left(\sum_{k=1}^{n} x_{j,k}\right)^2}}, \quad (2.64)$$

where $x_{l,k}$ and $x_{j,k}$ are realization of the random variates x_i and x_j. If x_i and x_j are independent, $r(x_i, x_j) = 0$, and Eq. 2.60 is reduced to

$$u_c(y) = \sqrt{\sum_{i=1}^{N} c_i^2 u^2(x_i)}. \quad (2.65)$$

When $r(x_i, x_j) = \pm 1$, the estimates x_i and x_j are linearly correlated.

2.4.3 Expanded Uncertainty

The combined standard uncertainty $u_c(y)$ is a point estimation for the uncertainty of the measurement result Y. However, in some applications of highest accuracy and responsibility, it is desired to have an interval estimation, i.e. to determine the interval of values, in which the measurement result is expected to fall with a reasonably chosen probability. An expanded measurement uncertainty U is the product of a combined standard measurement uncertainty $u_c(y)$ and a coverage factor k:

$$U = k \cdot u_c(y), \tag{2.66}$$

where k is defined for a given coverage probability (level of confidence). In Eq. 2.66, U defines an interval about the measurement result encompassing a large fraction p of the probability density function (PDF) that characterizes this result and its combined standard uncertainty $u_c(y)$. For instance, for a Gaussian distribution with a 95.45% coverage probability, $k = 2$. This means that the measured value y fall in the interval $[y - 2u_c(y), y + 2u_c(y)]$ with the probability of 0.9545. The coverage factor values of 2 and 3 are the most typical in metrological practice. If we can assume the normal (Gaussian) probability distribution for u_c, the most popular values of the coverage factor k for the coverage probability p can be taken from Table 2.7.

The assumption of a normal distribution is easily fulfilled if there are several (usually, $N \geq 3$) independent uncertainty components having well-behaved probability distributions, e.g. normal or rectangular (uniform) and, at the same time, none of the components has a clearly dominant magnitude. In such a case, the conditions of the Central Limit Theorem are met and we can confidently assume that the distribution of the output quantity is very close to normal.

Degrees of Freedom In practice, the number of repeated observations is often small, say, fewer than 10. When the mean of a normally distributed population is estimated using a small sample size, the Gaussian (normal) probability distribution may not be applicable to obtaining the coverage factor for evaluating the expanded uncertainty.

In such a case, the coverage factor must be computed on the base of Student's t-distribution that is the probability distribution of a continuous random variable $-\infty < t < \infty$, the PDF of which is

Table 2.7 Values of the coverage factor k that produces an interval having coverage probability p assuming a normal probability distribution (after [73])

Coverage probability p (%)	Coverage factor$k$
68.27	1
90	1.645
95	1.960
95.45	2
99	2.576
99.73	3

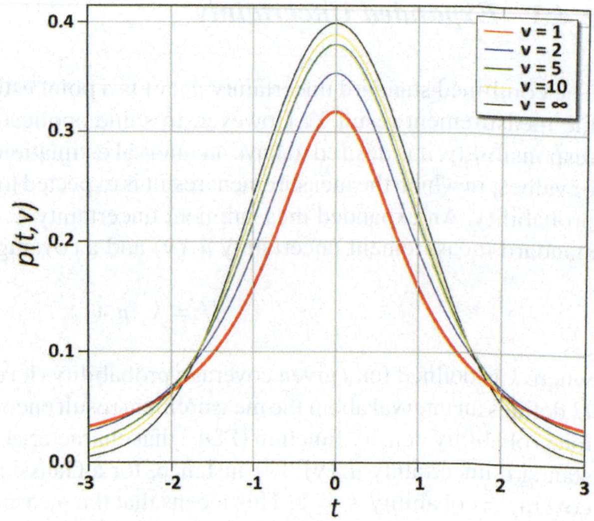

Fig. 2.6 The Student t-distribution for several values of degrees of freedom ν; $\nu = \infty$ corresponds to the standard normal probability distribution

$$p(t, \nu) = \frac{\Gamma[(\nu + 1)/2]}{\sqrt{\pi \nu}\Gamma(\nu/2)[1 + t^2/\nu]^{(\nu+1)/2}}, \tag{2.67}$$

where Γ is the gamma function and $\nu > 0$ is the so-called degree of freedom.

The expectation of the t-distribution is zero and its variance is $\nu/(\nu - 2)$ for $\nu > 2$. If $\nu \to \infty$, the t-distribution approaches a standard normal distribution (the Gaussian distribution with the zero mean value and the standard deviation equal to unity).

As can be seen in Fig. 2.6, the lower the value of ν the more noticeable the difference between the t-distributions and the standard normal distribution. Estimating the variance using the sample from a normally distributed population results in greater uncertainty (the "tails" of the t-distribution are higher than those of the normal distribution).

For a combined standard uncertainty $u_c(y)$, the "effective degrees of freedom" ν_{eff} of $u_c(y)$, which can be calculated using from the Welch-Satterthwaite formula is

$$\nu_{eff} = \frac{u_c^4(y)}{\sum_{i=1}^{N} \frac{c_i^4 u^4(x_i)}{\nu_i}}, \tag{2.68}$$

where $c_i = \partial f/\partial x_i$, all of $u(x_i)$ are mutually statistically independent, N is the number of input variables, and ν_i is the degrees of freedom of $u(x_i)$. If it is known that x_i is sampled from the Gaussian distribution, the corresponding degrees of freedom $\nu_i = n - 1$, where n is the number of independent repeated observations. If a standard uncertainty $u(x_i)$ is obtained via the Type B evaluation, it is commonly accepted to set its degrees of freedom $\nu_i = \infty$ to avoid underestimation.

If the effective degrees of freedom ν_{eff} computed via Eq. 2.68 is not integer, it should be truncated to the next lower integer. By choosing the coverage probability (confidence levels)p, we can calculate using (2.68) the value $t_p(\nu)$ of coverage factor for a given degrees of freedom ν or an effective degrees of freedom ν_{eff}. The expanded uncertainty

$$U = t_p(\nu)u_c(y) \tag{2.69}$$

defines an interval $y - U$ to $y - U$ (conveniently written as $Y = y \pm U$) that could reasonably be attributed to Y and that is expected to encompass a fraction p of the area under the distribution curve between $t = -t_p(\nu)$ and $t = t_p(\nu)$.

The values of coverage factor $t_p(\nu)$ for various degrees of freedom ν and coverage probabilities p of 68.27, 90, 95.45, 99, and 99.73% are plotted in Fig. 2.7.

Asymptotic behavior of curves $t_p(\nu)$ for the degrees of freedom $\nu \to \infty$ and different values of the coverage probabilities p shows that the limiting values of $t_p(\nu)$ coincide with the coverage factors k for the corresponding values of p from Table 2.7. Sufficiently detailed tables of the values of $t_p(\nu)$can be found in the GUM [73], in [92, 102], or [53].

There are also many easily accessible programs, from free to commercial, from online applets to desktop applications, from spreadsheets (such as MS Excel) to

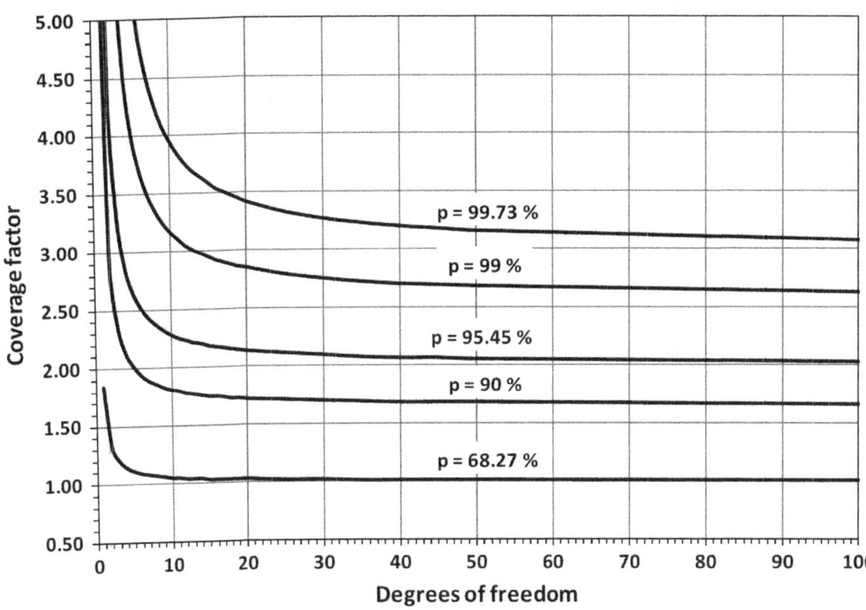

Fig. 2.7 Coverage factors derived from the Student t-distributions and plotted against the degrees of freedom for several values of the coverage probability p

statistical software packages, which allow, among other things, to calculate Student t-distribution values.

The data processing flow of the GUM uncertainty framework is depicted schematically in Fig. 2.8. The upper and lower parts of the diagram illustrate the point estimation (combined standard uncertainty) and interval estimations (coverage interval), respectively.

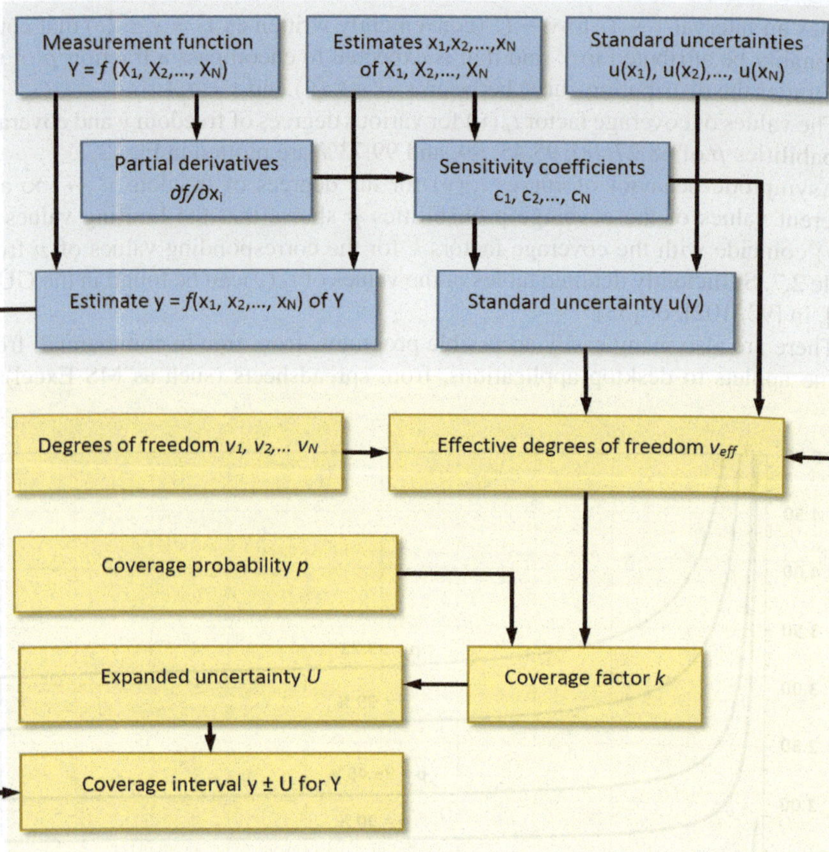

Fig. 2.8 Schematic of the data processing flow of the GUM framework. The upper and lower parts of the diagram illustrate the point and interval estimations, respectively

2.4.4 Reporting Measurement Results

The GUM suggests stating the numerical result of the measurement with the combined standard uncertainty (this is typical for reporting fundamental measurement research and comparisons of measurement standards) in one of four formats. Below, we illustrate them with two examples.

A. Let the quantity, whose value is being reported, is the melting temperature of gallium (a nominal value is 302.9146 K). Let the estimates of the result and the combined standard uncertainty are 302.9231 K and 1.7 mK, respectively. The four possible formats to report this temperature measurement are (the words in parentheses may be omitted for brevity if u_c is already defined in the document reporting the result):

   (i) $T_{Ga} = 302.9231$ K with (a combined standard uncertainty) $u_c = 1.7$ mK.
   (ii) $T_{Ga} = 302.9231(17)$ K, where the number in parentheses is the numerical value of (the combined standard uncertainty) u_c referred to the corresponding last digits of the quoted result.
   (iii) $T_{Ga} = 302.9231(0.0017)$ K, where the number in parentheses is the numerical value of (the combined standard uncertainty) u_c expressed in the unit of the quoted result.
   (iv) $T_{Ga} = (302.9231 \pm 0.0017)$ K, where the number following the symbol $\pm$ is the numerical value of (the combined standard uncertainty) u_c.

B. Let the quantity, whose value is being reported, is the spectral radiance L_λ at 656.3 nm of a tungsten strip lamp. The result of measurement can be reported as:

   (i) $L_\lambda = 4.651 \times 10^{10}$ W m^{-3} sr^{-1} with (a combined standard uncertainty) $u_c = 0.025 \times 10^{10}$ W m^{-3} sr^{-1}.
   (ii) $L_\lambda = 4.651(25) \times 10^{10}$ W m^{-3} sr^{-1}.
   (iii) $L_\lambda = 4.651(0.025) \times 10^{10}$ W m^{-3} sr^{-1}.
   (iv) $(4.651 \pm 0.025) \times 10^{10}$ W m^{-3} sr^{-1}.

In both examples, the most common form is (i), while form (iv) is recommended to be avoided whenever possible, since it is traditionally used to indicate the coverage interval corresponding to a high level of confidence, and therefore it can be confused with expanded standard uncertainty. As a rule, it is sufficient to quote the combined standard uncertainty to at most two significant digits (although, in some circumstances, additional digits may be necessary to avoid round-off errors in subsequent calculations).

When the result of measurement is reported together the expanded uncertainty $U = ku_c(y)$ or $U = t_p(\nu)u_c(y)$, one should

(i) give a full description of how the measurand Y is defined;
(ii) state the result of the measurement as $Y = y \pm U$ and give the units of y and U;

(iii) include the relative expanded uncertainty $U/|y|$, $y \neq 0$, when appropriate;
(iv) give the value of k used to obtain U (or, for the convenience of the user of the result, give both k and $u_c(y)$);
(v) give an approximate level of confidence associated with the interval $y \pm U$ and state how it was determined;
(vi) give the detailed information on each estimate and method used to obtain it, on degrees of freedom for the standard uncertainty of each input estimate and how it was obtained, the measurement function, its partial derivatives or sensitivity coefficients, etc. Finally, one should refer to published documents containing any useful information.

Uncertainty Budget Every responsible metrological investigation is accompanied by the uncertainty budget. According to VIM [77], an uncertainty budget is a statement of a measurement uncertainty, of the components of that measurement uncertainty, and of their calculation and combination. The uncertainty budget should describe the measurement model, types of evaluation of each component of measurement uncertainty, estimates, measurement uncertainties associated with each quantity in the measurement model, type of applied probability density functions, covariances and/or correlation coefficients (matrix), degrees of freedom, and a coverage factor. The uncertainty budget, as a rule, is presented in the tabular form. Good examples of uncertainty budgeting in optical radiation measurements based on the GUM framework can be found in e.g., Khlevnoy [82], Yoon and Gibson [151], Goodman et al. [48].

2.4.5 *Conditions for the Application of the GUM Framework*

GUM relies on a rigorous foundation of the probability theory and mathematical statistics. It covers the most common cases in practical uncertainty evaluation. However, there are measurement scenarios, for which application of the GUM framework becomes too laborious or cannot guarantee an acceptable quality of estimates. The document [76] describes some situations, where the GUM uncertainty framework might not be satisfactory:

- The GUM framework is based on the first-order Taylor series expansion of the measurement function. This framework may result in too crude estimate of uncertainty if the measurement function f has significant nonlinearities. For measurement scenarios, where nonlinearities cannot be neglected, the GUM recommends to include the higher-order terms in the Taylor series expansion in Eq. 2.60. When the distribution of each X_i is normal and the input quantities are uncorrelated, the most important terms of next highest order have to be added to the terms of Eq. 2.65:

$$u_c(y) = \sqrt{\sum_{i=1}^{N}\left(\frac{\partial f}{\partial x_i}\right)^2 u^2(x_i) + \sum_{i=1}^{N}\sum_{j=1}^{N}\left[\frac{1}{2}\left(\frac{\partial^2 f}{\partial x_i \partial x_j}\right)^2 + \frac{\partial f}{\partial x_i}\frac{\partial^3 f}{\partial x_i \partial x_j^2}\right] u^2(x_i)u^2(x_j)}. \tag{2.70}$$

Partial derivatives in Eq. 2.70 can be easily calculated analytically if the measurement function f is given in a closed analytical form, as an explicit function of $X_1, X_2, \ldots, X_N$. Although the case of an implicit measurement function (measurement equation)

$$h(Y, X_1, X_2, \ldots, X_N) = 0 \tag{2.71}$$

is not directly considered in the GUM [73], one can apply the rules for differentiating implicit functions and reduce the case of an implicit measurement function to uncertainty evaluation using the GUM framework. However, such an approach is rather speculative, since the nonlinearity of the measuring function leads to the need for numerical methods, including numerical differentiation using finite differences. This operation is inherently ill-conditioned, which means that even small uncertainty in function may result in large uncertainties in its derivatives. If the measurement model cannot be written in closed form (e.g. when the measurement model is described by an algorithm), finite differences are the only way to calculate partial derivatives. The reliability of numerical calculation of derivatives of higher orders is doubtful. In practice, this limits the applicability of the GUM framework to non-linear measurement functions.

- The GUM framework is not capable of handling the input quantities associated with asymmetric PDFs. It should be noted that asymmetric PDFs are very common in the blackbody radiometry. For example, the measurement uncertainty of the emissivity, which is an input quantity in many measurement models, must have an asymmetric distribution, being bounded both above and below. This is especially true for very low and very high emissivity values.
- One of the corollaries of the central limit theorem in the probability theory states that the PDF of the random variate composed from a sufficiently large number of independent random components tends to a Gaussian (normal) distribution or Student's t-distribution (when small sample from the Gaussian population is made). The central limit theorem holds even if individual components are not normally distributed. The conditions for its applicability are the "good" behavior of the measurement function and the absence of dominant components of uncertainty. However, the uncertainty contributions $|c_1|u(x_1), |c_2|u(x_2), \ldots, |c_N|u(x_N)$ have often dramatically different magnitudes, which can lead to erroneous results for obtaining the coverage interval for Y.
- The Welch–Satterthwaite formula (2.68) recommended in the GUM can produce incorrect results when the input quantities are correlated. A generalization of the Welch–Satterthwaite formula proposed by Willink [143] can be used.

However, there is another serious drawback of the GUM approach, which, although is not fundamental, can make its use unacceptably difficult. There are many situations in modern metrology, when the measurand can be obtained only from combined measurements (if several quantities of the same kind are measured) or, if quantities of different kinds are measured, from simultaneous measurements (see Rabinovich [123]). Modern measurement facilities are often complicated systems, which include a variety of measuring instruments and control information flows that are processed in digital form using a computer or microprocessor devices. In many cases, the input and output values cannot be interconnected by a simple expression suitable for analytical differentiation. For example, this relationship can only be expressed by a numerical algorithm.

To overcome limitations of the GUM framework, the JCGM has issued several Supplements to the GUM among which the most influential is the Supplement 1 to the GUM "Propagation of Distributions Using a Monte Carlo Method" [74]. We will call it GUM-S1 and discuss it briefly in Sect. 2.5.

Another important Supplement [75] is devoted to extension of GUM and GUM-S1 to an arbitrary number of output quantities by replacing the scalar measurand and its variance by a vector measurand and a covariance matrix. JCGM 102:2011 [74] covers the fitting, interpolation, and non-linear regression problems that are usually solved using method of least squares or multi-dimensional optimization. We will not consider especially the problems of combined and simultaneous measurements; the interested readers can find a few examples for the areas relevant to the subject of this book in papers by Pearce et al. [114], Rosenkranz [125], Saunders [127], and Woolliams [147].

Concluding this section, we should note that we presented just a general outline of the GUM framework. Readers who need in-depth study of the GUM uncertainty framework may refer to the books written by Kirkup and Frenkel [34], Grabe [49], Dieck [84] or Ratcliffe and Ratcliffe [124]. Of course, the original JCGM documents, freely available at https://www.bipm.org/en/publications/guides/gum.html, remain the primary source of information on this subject. These documents are also published by the member organizations of the JCGM: the BIPM,[5] IEC,[6] OIML,[7] et al.

[5]Bureau International des Poids et Mesures (Fr.)—International Bureau of Weights and Measures (https://www.bipm.org/en/about-us/).

[6]International Electro-technical Commission (https://www.iec.ch/).

[7]Organisation Internationale de Métrologie Légale (Fr.)—International Organisation of Legal Metrology (https://www.oiml.org/en).

2.5 Monte Carlo Modeling of Measurements for Uncertainty Analysis

2.5.1 The Main Ideas of Stochastic Simulation of Measurement Errors

The concept adopted by the GUM-S1 [74] is the propagation of distributions unlike propagation of uncertainties proclaimed in GUM [73]. The propagation of distributions involves calculations of the probability distribution for the measurand from the distributions associated with the input quantities of the measurement model. These calculations are carried out using the Monte Carlo method (MCM), the powerful tool successfully employed in various fields of mathematics, physics, and engineering. The MCM is a common name of the family of numerical algorithms based on repeated random sampling to obtain the expectation value as an approximation of the exact result. The mathematical foundations of the MCM are exposed in numerous books (e.g. [35, 39, 88]).

The Monte Carlo modeling of the measurement uncertainties is based on the following ideas:

- Replacement of a measurement (full-scale, nature) experiment with a computational (numerical, virtual) experiment that is performed on a measurement model using special simulation software and a digital computer.
- Instead of uncertainty propagation law, the principle of propagation of distributions is used. It determines the probability density functions (PDFs) for an output quantity from the PDFs assigned to the input quantities, on which the output quantity depends.
- In contrast to the GUM (the first-order approximation of Taylor series) approach, the MCM is a purely numerical technique. Principle of propagation of probability distributions is implemented numerically using stochastic modeling. It requires the use of a computer.
- Numerical implementation is based on repeated sampling from the PDFs for the input quantities and evaluation of the PDF for the output quantity.
- The joint N-dimensional PDF $g_X(\boldsymbol{\xi})$ should be used for the input quantities $\mathbf{X}$.
- Instead of continuous distribution, their discrete representations are propagated.
- The results of M repeated numerical experiments (trials) are subjected to the same statistical analysis as the results of repeated measurements. The primary output of the MC method is the model quantities Y, from which the expectation and variance are directly determined.
- The coverage interval for Y is calculated as the boundaries y_{low} and y_{high} of the interval into which the output value falls with a given coverage probability p.

With the proliferation of personal computers, the MCM is increasingly used to solve mathematical problems and model complex systems. The first attempts of applying stochastic modeling to assessment of measurement uncertainties in specific

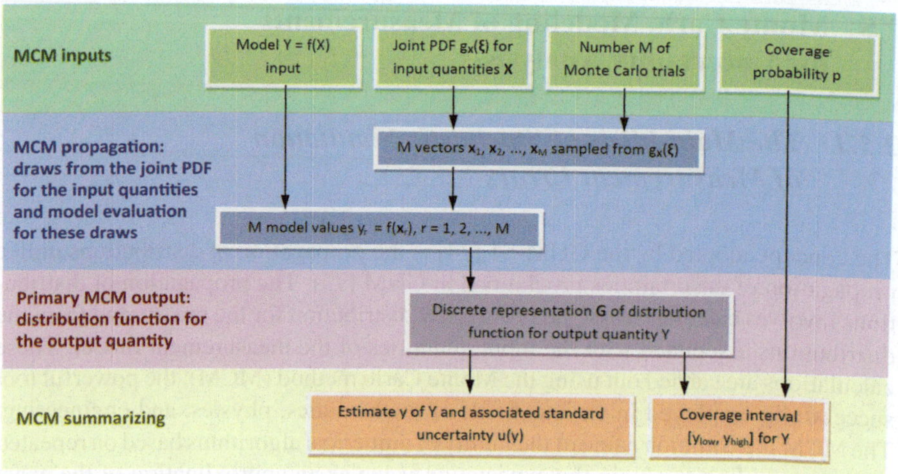

Fig. 2.9 The propagation and summarizing stages of uncertainty evaluation using the Monte Carlo modeling to implement the propagation of distributions

measuring tasks were made in the last decade of twentieth century. The collected experience has been systematized in a series of papers by Cox et al. [26], Cox and Siebert [27], Sommer and Siebert [131].

In response to the requests of practicing metrologists, the JCGM has adapted the GUM-S1 [74], a Supplement 1 to the GUM, in which a uniform treatment of the Monte Carlo simulation for models having any number of input quantities and a single output quantity. The stages of uncertainty evaluation using the propagation of distributions by the Monte Carlo modeling are depicted schematically in Fig. 2.9.

2.5.2 *Algorithm of the Monte Carlo Simulation*

Although the joint N-dimensional PDF $g_X(\boldsymbol{\xi})$ is required at the first stage of uncertainty evaluation using the Monte Carlo simulation (Fig. 2.9), most often, it is unknown. In practice, it is sufficient to attribute a suitable one-dimensional PDF $g_{X_i}(\xi_i)$ to each input quantity X_i. The correlations between some input quantities can be defined by a covariance matrix, which, as a rule, is a sparse matrix, i.e. has only few nonzero elements. A diagram of the Monte Carlo modeling of uncertainty for uncorrelated input quantities $\mathbf{X} = X_1, \ldots, X_N$ and a single output quantity Y is presented in Fig. 2.10.

A necessary element of any Monte Carlo modeling procedure is the program generator of pseudorandom number (PRNG) with the uniform PDF on the interval [0, 1]. Such properties of a PRNG as extremely long period, uniformity of the PDF, no correlation between successive pairs, triples, etc. of random numbers, and high speed [36, 56] are critically important for simulation-based uncertainty evaluation.

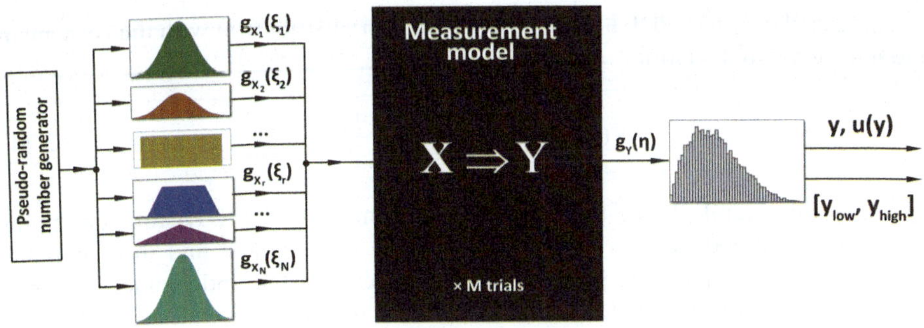

Fig. 2.10 Diagram of the MC modeling of uncertainty for uncorrelated input quantities $\mathbf{X} = X_1, \ldots, X_N$ and a single output quantity Y

Since the pseudorandom numbers are generated by deterministic algorithms (44), they are as "random" as they pass statistical tests and test suites developed to assess the "randomness" of PRNGs [89, 96, 126]. The ISO standard [65] recommends some popular algorithms of PRNG.

L'Ecuyer [90] identified absolute leaders among the PRNGs in the Monte Carlo modeling applications. These are the families of various program implementations of the Mersenne-Twister PRNG [97] and based on similar algorithm the WELL (Well Equidistributed Long-period Linear) PRNG [111], which have become the default PRNGs in many modern software packages.

The uniformly distributed random numbers are used to generate random variates with non-uniform PDFs. There is an extensive literature on general methods for generating non-uniform distributions of one-dimensional and multidimensional random variables [33, 44, 58] and particular cases for the most common simple distributions (for metrology application, these are, typically, the normal, rectangular, triangular, or trapezoidal distribution).

The next step of the modeling algorithm is generating a set of N input parameters $x_1, x_2, \ldots, x_N$, which are random variables distributed according to a PDF assigned to each input parameter. These values are fed to the measurement model represented in Fig. 2.10 as the black box with the N inputs and one output.

The source material for sampling from any continuous PDFs is the pseudo-random numbers η sampled from the uniform PDF $U(\eta)$:

$$U(\eta) = \begin{cases} 1 \text{ if } 0 \leq \eta \leq 1 \\ 0 \text{ otherwise} \end{cases}. \tag{2.72}$$

When we need to generate random variates uniformly distributed on the interval $[a, b]$, it is enough to make a linear transformation:

$$\eta_{[a,b]} = (b - a)\eta + a. \tag{2.73}$$

Another elementary PDF is the Gaussian (normal) distribution with the zero mean and unit standard deviation:

$$N(\eta_N) = \frac{1}{\sqrt{2\pi}} \exp\left(-\frac{\eta_N^2}{2}\right). \tag{2.74}$$

There are several simple algorithms (see Thomas et al. [135]) for generating normally distributed random variates η_N. One of the simplest and most popular is the Box-Muller algorithm [12]. It relies on generating two independent random variates η_1 and η_2 from $U(\eta)$. The following two variates

$$\eta_{N,1} = \sqrt{-2\ln\eta_1}\,\sin(2\pi\eta_2), \tag{2.75}$$

$$\eta_{N,1} = \sqrt{-2\ln\eta_1}\,\cos(2\pi\eta_2) \tag{2.76}$$

are independent random variates with distribution $N(\eta_N)$.

To obtain the random variate $\eta_{N(\mu,\sigma)}$ normally distributed with the mean value μ and the standard deviation σ, i.e.

$$N(\eta_N, \mu, \sigma) = \frac{1}{\sigma\sqrt{2\pi}} \exp\left[-\frac{1}{2}\left(\frac{\eta_N - \mu}{\sigma}\right)^2\right], \tag{2.77}$$

the random variate $\eta_{N,\mu,\sigma} \sim N(\eta_N, \mu, \sigma)$ must be linearly transformed:

$$\eta_{N,\mu,\sigma} = \sigma\eta_N + \mu. \tag{2.78}$$

The black box in Fig. 2.10 symbolizes a measurement model, which can be a measurement equation, a measurement function, or some algorithm mapping the set of the input parameters $\mathbf{X}$ to a set of the output values Y. This process should be repeated M times for every $\mathbf{X}$ to obtain M output values of Y. The resulting sample can be used to estimate the PDF of the random variate y. If necessary, the histogram (scaled frequency distribution) approximating the PDF can be plotted and the statistical hypotheses about its shape can be checked.

The mean value and the standard deviation can be calculated from output values $y_1, y_2, \ldots, y_M$ obtained from M trials.

The GUM-S1 recommend to use $M = 10^6$ to estimate a 95% coverage interval for the output quantity. Alternatively, the minimum necessary number of trials M, which ensured the coverage interval $[y_{low}, y_{high}]$ at a coverage probability p, can be found as described by Cox and Siebert [27] and Willink [142]. A less complicated and more illustrative method is based on the visual control of the histogram for the PDF of the output value plotted at growing values of M as it is shown in Fig. 2.11. With the reasonable choice of the width of the histogram bins [10, 85, 129, 130], a sufficient smoothness of the histogram can serve as an intuitive criterion for the end of modeling.

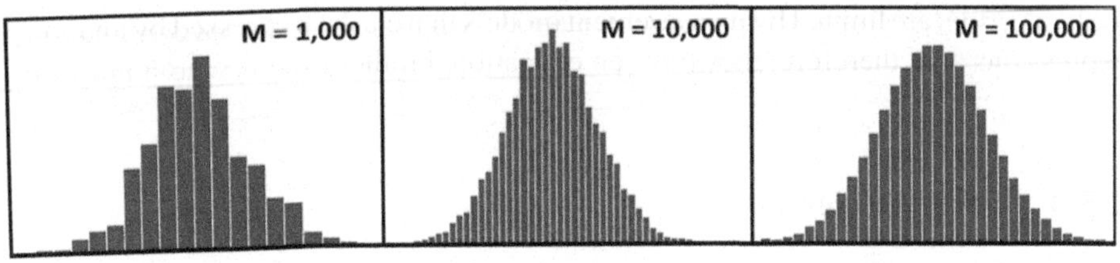

Fig. 2.11 Histograms of the output value with a near-normal PDF plotted successively for three numbers of trials M

Many modern researchers consider the Monte Carlo modeling as an alternative to evaluating measurement uncertainty in accordance with the GUM approach. Comparison of these techniques for indirect measurements shows that they lead to equivalent results, if both approaches are equally applicable [80, 114, 121]. Moreover, Cox and Harris [28] regard "Monte Carlo as the primary method for providing a benchmark for general uncertainty evaluation problems."

Some authors believe that the greatest drawback of the Monte Carlo modeling is a large duration of calculations. This was true several decades ago, but the performance of modern PCs makes this drawback irrelevant.

Impossibility to analyze contributions of individual sources of uncertainty is often considered another disadvantage of the Monte Carlo approach, but this is only a seeming drawback. However, this is rather a question of the proper organization of calculations. Allard and Fisher [2] showed that sensitivity analysis can be effectively performed within the Monte Carlo modeling by holding all input quantities but one fixed at their best estimate and performing a separate Monte Carlo simulation for each input quantity.

Certainly, need of writing a simulation code for each specific task to implement the algorithms of the MCM can be regarded as a principal drawback. This increases the requirements for the user's qualifications. However, sophisticated problems in the uncertainty evaluation using the GUM approach, as a rule, also require of some kind of programming work.

Requirements to the performance of computing systems and software for Monte Carlo simulation of the distributions propagation depends, first and foremost, on the complexity of the measurement model. If the measurement model includes, say, numerical integration of a complicated function, we must perform this procedure M (usually, 10^4–10^6) times, and this may require compilation of the code, parallel computing, fastest processor, or even multiprocessor system. Esward et al. [37] and Woolliams et al. [146] described application of distributed computing system for the Monte Carlo simulation to the analysis of uncertainties in a measurement system to estimate the absolute thermodynamic temperatures of two high-temperature blackbodies (HTBBs) by measuring the ratios of their spectral radiances.

For simple and/or typical metrology problems, different manufacturers offer software tools of various degrees of complexity, from Excel add-ins to standalone executables and Web-based software applications. However, the possibility of these tools, as a rule, are limited by measurement models that can be expressed by analytical expressions and, therefore, they have an educational rather than research purpose.

2.5.3 *Case Studies*

Propagation of distributions is a relatively new approach. GUM-S1 explains most important concepts in the Monte Carlo evaluation of measurement uncertainties and gives general recommendations on its software implementation. Besides, the GUM-S1 illustrates some specific issues by application examples. Many works on the Monte Carlo uncertainty evaluation applied to particular problems related to materials of this book or based on similar mathematical models have been published in two recent decades. We will mention only a few such works to illustrate the circle of problems, to which the MCM can be applied. We will not analyze their results, since these are solutions of particular problems. However, some details, which seem important to us and can serve as guidelines for the Monte Carlo solutions, will be highlighted.

Poikonen et al. [118] applied the MCM to the analysis of the photometer quality factor f_1' that characterizes the spectral mismatch with respect to a human eye and is defined [18] as

$$f_1' = \int_0^\infty \left|s_{rel}^*(\lambda) - V(\lambda)\right| d\lambda \Bigg/ \int_0^\infty V(\lambda) d\lambda, \tag{2.79}$$

where $V(\lambda)$ is the spectral luminous efficiencies for photopic vision (see Fig. 2.4); $s_{rel}^*(\lambda)$ is the normalized relative spectral responsivity of the photometer defined, in turn, as

$$s_{rel}^*(\lambda) = s_{rel}(\lambda) \cdot \int_0^\infty S_A(\lambda) V(\lambda) d\lambda \Bigg/ \int_0^\infty S_A(\lambda) s_{rel}(\lambda) d\lambda, \tag{2.80}$$

where $S_A(\lambda)$ is the relative spectral distribution of the CIE standard illuminant A[8] and $s_{rel}(\lambda)$ is the relative spectral responsivity of the photometer.

In the ideal case, $s_{rel}^*(\lambda) = V(\lambda)$, which results in $f_1' = 0$. Poikonen et al. [118] described the algorithm for determining the uncertainty of photometer quality factor f_1' using Monte Carlo simulations in presence of systematic and random errors in the relative spectral responsivity of the photometer. The simulation results supported by analytical considerations showed that the neglecting the systematic errors (use of random errors alone) leads to underestimating the uncertainty of f_1'. It was also demonstrated a high sensitivity of the uncertainty of the quality factor f_1' to fine

[8]CIE standard illuminant A has the relative spectral power distribution is that of a Planckian radiator at a temperature of approximately 2856 K and mimics typical indoor tungsten-filament lighting.

details of the photometer spectral responsivity when the both systematic and random errors are modeled together.

Later, Poikonen et al. [119] applied similar Monte Carlo technique to analysis of photometer directional response factor f_2 defined [19] as

$$f_2 = \int_0^{1.484} |f_2(\varepsilon, \phi)| \sin 2\varepsilon d\varepsilon, \tag{2.81}$$

where the upper integration limit of 1.484 rad corresponds to the angle of 85°; $f_2(\varepsilon, \phi)$ is the deviation of the photometer directional response to the incident radiation is expressed as

$$f_2(\varepsilon, \phi) = \frac{R(\varepsilon, \phi)}{R(0, \phi) \cos \varepsilon} - 1, \tag{2.82}$$

where $R(\varepsilon, \phi)$ is the signal output of the photometer as a function of the incidence and azimuth angles ε and ϕ, respectively.

The uncertainty of the photometer directional response index f_2 was investigated using Monte Carlo simulations and error models derived for the characterization measurements of three photometers. Both random and systematic errors are modeled simultaneously to mimic the measurement noise, drift of the light source, alignment of the photometer, and the errors of the rotary stage. The values of f_2 obtained for different azimuth angle of the photometer are found to be very sensitive to the asymmetry of the directional response.

De Lucas and Segovia [30] applied the MCM to analyze uncertainty of the effective emissivity ε_{eff} of isothermal diffuse cylindro-conical cavities of variable temperature blackbodies (VTBB) of a high metrological quality, such as heat-pipe blackbodies. A distinctive feature of this work is the absence of a closed-form equation for the effective emissivity, which is calculated using the Monte Carlo ray tracing (see Sect. 4.4), a time-consuming algorithm itself.

The effect of five variables (cavity wall emissivity ε, length L of the cylindrical part of the cavity, cavity diameter D, aperture diameter A, and cone apex angle Ω) with the mean values ε_0, L_0, D_0, A_0, and Ω_0, respectively, and widths of corresponding uniform distributions $\Delta\varepsilon$, ΔL, ΔD, ΔA, and $\Delta\Omega$. For each calculation of the effective emissivity, the following set of initial data has been used:

$$\begin{cases} \varepsilon = \varepsilon_0 + (2\eta_1 - 1)\Delta\varepsilon \\ L = L_0 + (2\eta_2 - 1)\Delta L \\ D = D_0 + (2\eta_3 - 1)\Delta D, \\ A = A_0 + (2\eta_4 - 1)\Delta A \\ \Omega = \Omega_0 + (2\eta_5 - 1)\Delta\Omega \end{cases} \tag{2.83}$$

where $\eta_1, \ldots, \eta_5$ are the independent random numbers uniformly distributed on the interval [0, 1].

The ray tracing algorithm was optimized, which allowed to write the modeling code using the Visual Basic for Applications programming language for the MS Excel and perform calculations using the PC with 3.2 GHz processor and 4 GB of RAM. For each calculation of the effective emissivity, 10^6 ray trajectories were traced; $M = 10^4$ simulations were performed to build the probability distribution of ε_{eff} from the affecting quantities (2.83).

Later, De Lucas and Segovia [31] expanded the above analysis to nonisothermal cavities. Usually, when the effect of nonuniform temperature distribution on the effective emissivity is investigated, the modeled distributions, at best, are represented by a polygonal line. The author proposed to model the axial temperature distribution $T(x) = T_0 + \delta T(x)$, where T_0 is the reference temperature and $\delta T(x)$ is given by polynomial families up to order 3 with coefficients chosen randomly from uniform PDFs taking into account physical plausibility of resulting temperature distributions. However, it remains unclear whether such a generation method leads to an equal probability of significant and insignificant deviations from the uniform temperature distribution and whether this affects the uncertainty of the effective emissivity.

The number of publications on the use of MCM for uncertainty analysis is steadily growing. Only in the last decade, the interesting studies were appeared in the areas of radiometry, radiation thermometry, IR thermography [5, 6, 24, 59, 91, 109], contact thermometry [110, 136], and other fields. In all of these works, the MCM was applied to solving problems that cannot be solved using the GUM uncertainty framework without significant simplifications and approximations.

We restrict ourselves to a slightly more detailed consideration of the method for determination of the blackbody thermodynamic temperatures based on the measuring the signal ratios for two filter radiometers (FRs), or a single FR with two exchangeable optical filters [122]. The idea of the method of ratios (also referred to as "the double-wavelength technique") is measuring the ratios the spectral radiances of the blackbody at two different temperatures to determine them. This idea is not new and based in believe in self-sufficiency of Planck's law for determining the temperature of a blackbody, i.e. the possibility to establish the temperature scale on the base of the scale of wavelengths and only relative measurements of blackbody radiances. Mathematically, this problem is described by the system of equations:

$$\begin{cases} X_1(T_1, T_2) = \dfrac{S_1(T_1)}{S_1(T_2)} = \dfrac{\int_0^\infty s_{rel,1}(\lambda) L_{\lambda,bb}(\lambda, T_1) d\lambda}{\int_0^\infty s_{rel,1}(\lambda) L_{\lambda,bb}(\lambda, T_2) d\lambda} \\ X_2(T_1, T_2) = \dfrac{S_2(T_1)}{S_2(T_2)} = \dfrac{\int_0^\infty s_{rel,2}(\lambda) L_{\lambda,bb}(\lambda, T_1) d\lambda}{\int_0^\infty s_{rel,2}(\lambda) L_{\lambda,bb}(\lambda, T_2) d\lambda} \end{cases}, \tag{2.84}$$

where $S_1(T_i)$ and $S_2(T_i)$ are the signals of the first and second FRs, respectively, when the temperatures T_i $(i = 1, 2)$ are measured; $s_{rel,1}(\lambda)$ and $s_{rel,2}(\lambda)$ are the relative spectral responsivities of the first and second FRs, respectively; $L_{\lambda,bb}(\lambda, T)$ is the spectral radiance of the perfect blackbody at the wavelength λ and temperature T expressed by the Planck law:

$$L_{\lambda,bb}(\lambda, T) = \frac{c_{1L}}{\lambda^5\left[\exp\left(\frac{c_2}{\lambda T}\right) - 1\right]}, \tag{2.85}$$

where c_{1L} and c_2 are the first radiation constant for the spectral radiance and the second radiation constant, respectively (see Appendix A). After measuring the FR signals and forming the ratios X_1 and X_2, the system of nonlinear Eq. (2.84) must be solved with respect to T_1 and T_2, one of which is the desired value and the other is auxiliary value, the knowledge is not required.

A common way to solve the system (2.84) for T_1 and T_2 is to convert it to an equivalent problem of minimization for the Euclidean distance between an initial guess and the true solution. This approach leads to the minimization problem for the objective function,

$$F(T_1, T_2) = \left[X_1\left(T_{1,true}, T_{2,true}\right) - X_1(T_1, T_2)\right]^2 + \left[X_2\left(T_{1,true}, T_{2,true}\right) - X_2(T_1, T_2)\right]^2, \tag{2.86}$$

where $(T_{1,true}, T_{2,true})$ is the solution of the system (2.84).

There were several previous works dedicated to determination of the uncertainties in T_1 and T_2. Woolliams et al. [146] used the MCM to compute $u(T_1)$, $u(T_2)$, and $r(T_1, T_2)$ for the Si photodiode with the center wavelength of about 700 nm and the InSb detector with the center wavelength around 4.55 μm. For these hypothetic FRs, the complex model of errors consisting of both multiplicative (expressed by relative uncertainties) and additive (expressed by absolute uncertainties) terms was built. The uncertainties in determination of signal ratios were considered as the greatest components; the Gaussian PDFs were ascribed to the corresponding errors.

An attempt to applying the GUM approach to the method of ratios was undertaken by Saunders [127]. He substituted the integrals in (2.84) with the approximate Sakuma-Hattori equations (see Sect. 8.4.1) and performed the implicit differentiation to derive the sensitivity coefficients. It was proposed that measurements be carry out using a two-color Si/InGaAs photodiode with and without a narrowband filter to obtain detector wavebands that differ as much as possible in bandwidth.

A similar problem was examined by Prokhorov et al. [122], but the well-studied Hamamatsu S1337 ([54]) silicon photodiode was proposed to be used as a detector with two different filters: the first, a narrow-band interference filter with the Gaussian spectral transmittance centered at 650 nm and an the full width half maximum (FWHM) of 10 nm, and the second, a wide-band Schott KG5 glass filter ([128]). Relative spectral responsivities of the FR with two these filters are shown in Fig. 2.12.

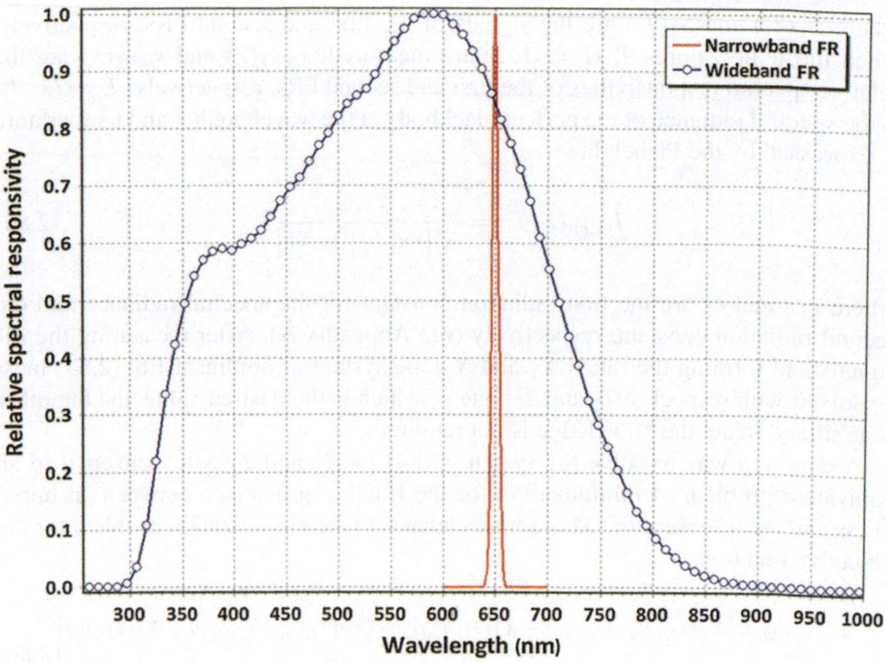

Fig. 2.12 Relative spectral responsivities of the silicon photodiodes with two filters. Reproduced from [122] with permission of Springer nature

The system (2.77) is the system of two non-linear equations, implicit with respect to unknown variables T_1 and T_2, placed under the integral signs. The unknown variables cannot be expressed from (2.77) in a closed form. Therefore, we cannot use the GUM approach directly, without substantial simplification and approximations.

The Levenberg–Marquardt (LM) algorithm [120] was used to solve non-linear least squares problems (2.84). Figure 2.13a presents the surface of the function $F(T_1, T_2)$ for $T_{1,true} = 1357.77$ K (the freezing temperature of copper) and $T_{2,true} = 2747$ K (the melting temperature of Re–C eutectic alloy) plotted in the logarithmic scale within the square 100 K × 100 K using a regular grid with 2001 × 2001 nodes. Figure 2.13b shows the three-dimensional (3D) plot of the same function for $T_{1,true} = 2500$ K and $T_{2,true} = 3500$ K. Both surfaces shown in Fig. 2.13 have a similar shape: the very narrow deep minimum ($F = 0$; $\lg(F) = -\infty$) corresponds to $T_1 = T_{1,true}$ and $T_2 = T_{2,true}$ lies at the bottom of a narrowing down valley.

The convergence domain of the LM algorithm was found to be large enough although it depends on initial settings (conditions of the iterative process termination, initial step bound, and the mode of variables scaling). No problems with convergence arise if an initial guess is chosen in such a way that the first (smaller) of the starting temperatures is less than $\min(T_{1,true}, T_{2,true})$, while the second (larger) starting temperature is greater than $\max(T_{1,true}, T_{2,true})$. For most cases, the solution is obtained in modern PCs almost instantaneously, after several iterations.

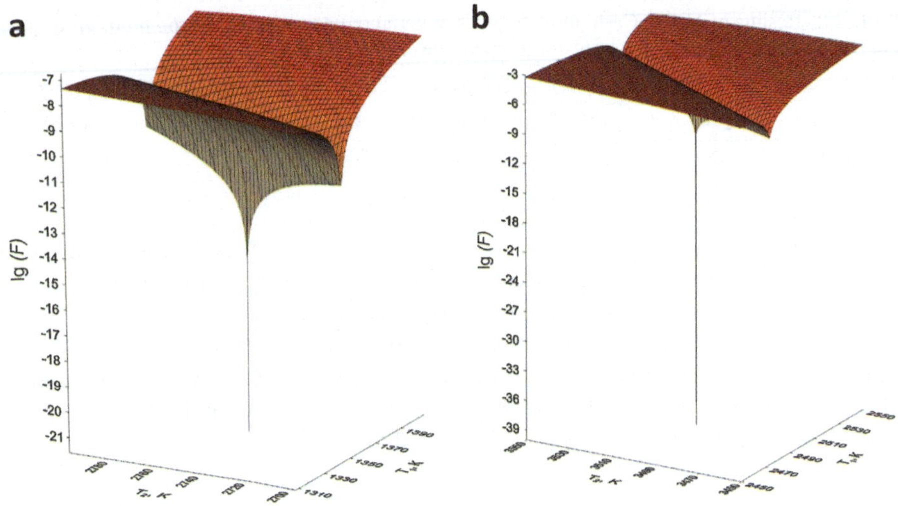

Fig. 2.13 Three-dimensional plots of the decimal logarithm of the objective function $\lg[F(T_1, T_2)]$ for two pairs of the true temperatures: **a** $T_{1,true} = 1357.77$ K and $T_{2,true} = 2747$ K and **b** $T_{1,true} = 2500$ K and $T_{2,true} = 3500$ K. Reproduced from [122] with permission of Springer nature

The Monte Carlo modeling was performed to assess the stability of solutions to the random errors added to measured signals.

For numerical experiments, $X_{1,true}$ and $X_{2,true}$ were pre-computed for the known (true) values of $T_{1,true}$ and $T_{2,true}$; the same $s_{rel,1}(\lambda)$, and $s_{rel,2}(\lambda)$ depicted in Fig. 2.12 were used for all cases.

It was assumed that the output signal of the FR can be measured with the random error characterized by the Gaussian probability distribution having a zero mean and a relative standard deviation (relative standard uncertainty) $u_{rel}(S)$. The essence of the modeling algorithm is the application of the LM algorithm to find solutions of (2.77) for multiple realization of the random values of the FR signals:

$$S = [1 + u_{rel}(S)\eta_N(0, 1)]S_{true}, \tag{2.87}$$

where S_{true} is any of $S_1(T_{1,true})$, $S_2(T_{1,true})$, $S_1(T_{2,true})$, and $S_2(T_{2,true})$; $\eta_N(0, 1)$ is the normally distributed random variable with the zero mean and unit standard deviation.

To generate random normal variates, the Box–Muller method (see Eqs. 2.75 and 2.76) was used. The resulting random variates $\eta_{N,1}$ and $\eta_{N,2}$ are independent and normally distributed with the zero mean and unit standard deviation. Therefore, to generate four normally distributed random errors, it if enough to generate four variates uniformly distributed on [0, 1], employ the Box–Muller method and linearly transform them according Eq. 2.87.

Table 2.8 Results of Monte Carlo modeling of temperature uncertainties for the method of ratios and two variants (**a** and **b**) of sought-for temperatures (after [122])

	$T_{1,true}$(K)	$T_{2,true}$(K)	$\overline{T}_1$(K)	$\overline{T}_2$(K)	$u(T_1)$(mK)	$u(T_2)$(mK)	$r(T_1, T_2)$
a	1357.77	2747	1357.768	2746.994	130	497	0.9978
b	2500	3500	2500.002	3500.004	781	1472	0.9994

The signals $S_1(T_{1,true})$, $S_2(T_{1,true})$, $S_1(T_{2,true})$, and $S_2(T_{2,true})$ were computed by numerical integration of appropriate Planckian functions multiplied by relative spectral responsivities; then the multiplicative random Gaussian error with a given standard deviation was added to each signal, and the ratios were computed. For each pair of X_1 and X_2 obtained in such a way, T_1 and T_2 were found by the LM method. Their mean values $\overline{T}_1$ and $\overline{T}_2$, standard uncertainties $u(\overline{T}_1)$ and $u(\overline{T}_2)$, and the correlation coefficient $r(T_1, T_2)$ were computed as

$$\overline{T}_{1,2} = \frac{1}{N}\sum_{i=1}^{N} T_{i,2,i}, \tag{2.88}$$

$$u(\overline{T}_{1,2}) = \sqrt{\frac{1}{N(N-1)}\sum_{i=1}^{N}\left(T_{1,2,i} - \overline{T}_{1,2}\right)^2}, \tag{2.89}$$

$$r(T_1, T_2) = \frac{\sum_{i=1}^{N}\left(T_{1,i} - \overline{T}_1\right)\left(T_{1,i} - \overline{T}_1\right)}{\sqrt{\sum_{i=1}^{N}\left(T_{1,i} - \overline{T}_1\right)^2}\sqrt{\sum_{i=1}^{N}\left(T_{2,i} - \overline{T}_2\right)^2}}, \tag{2.90}$$

where $T_{1,i}$ and $T_{2,i}$ are results for i-th trial, N is the number of trials. The results of statistical processing are collected in Table 2.8. Figure 2.14 presents the scatter plots of errors $T_1 - T_{1,true}$ and $T_2 - T_{2,true}$.

These graphs were obtained using 10,000 trials for ($T_{1,true}$ = 1357.77 K, $T_{2,true}$ = 2747 K) and ($T_{1,true}$ = 2500 K, $T_{2,true}$ = 3500 K), respectively. In both cases, the correlation coefficients $r(T_1, T_2)$ are very close to unity, which reflects strong mutual interdependence of T_1 and T_2 through (2.84).

If to compare the data from Table 2.8 with those provided in Table B.2 (see Appendix B), one can see that the uncertainties obtained by the method of ratios are only two times greater than the best uncertainties achieved by world leading NMIs using the absolute primary radiometric thermometry (see Chap. 9).

We believe that the method of ratios, as well as other relative methods, which are not connected with the standard techniques of the radiation thermometry, nor with absolute measurements of the radiation characteristics of blackbodies, can be used to verify the results obtained by the above methods and, thus, increase the reliability of such results, or, conversely, detect unaccounted for systematic errors, the sources of which differ significantly.

It is too early to draw any conclusions regarding the method of ratios, especially since research in this area is ongoing [14, 152].

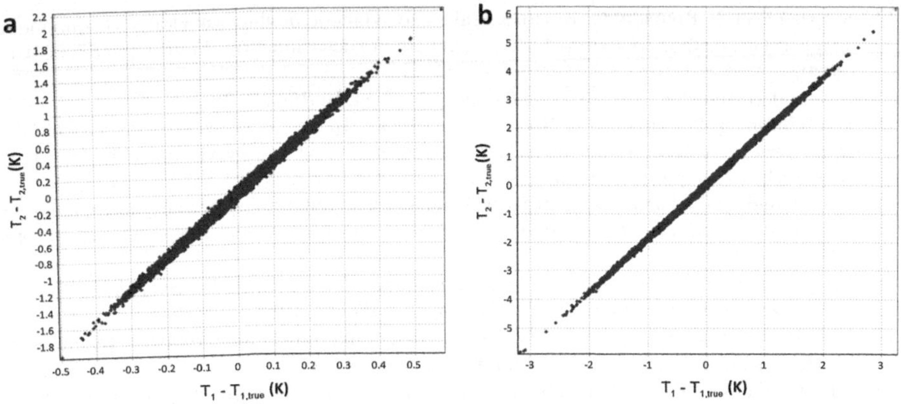

Fig. 2.14 Scatter plots of errors for determination of T_1 and T_2 at the relative standard uncertainties of signal measurements $u_{rel}(S_1) = u_{rel}(S_2) = 0.01\%$ for: **a** $T_{1,true} = 1357.77$ K and $T_{2,true} = 2747$ K and **b** $T_{1,true} = 2500$ K and $T_{2,true} = 3500$ K. Reproduced from [122] with permission of Springer nature

References

1. C.G. Abbot, F.E. Fowle, Part I. Determination of the intensity of the solar radiation outside the Earth's atmosphere, otherwise termed "the solar constant of radiation." Ann. Astrophys. Observ. Smith. Inst. **II**, 11–158 (1908)
2. A. Allard, N. Fischer, Sensitivity analysis in practice: providing an uncertainty budget when applying supplement 1 to the GUM. Metrologia **55**, 414–426 (2018)
3. K. Anhalt, A. Zelenjuk, D.R. Taubert et al., New PTB setup for the absolute calibration of the spectral responsivity of radiation thermometers. Int. J. Thermophys. **30**, 192–202 (2009)
4. ANSI/IES RP-16-10, *Nomenclature and Definitions for Illuminating Engineering* (Illuminating Engineering Society of North America, New York, 2010)
5. A. Araújo, S. Silvano, N. Martins, Monte Carlo uncertainty simulation of surface emissivity at ambient temperature obtained by dual spectral infrared radiometry. Infrared Phys. Technol. **67**, 131–137 (2014)
6. A. Araújo, N. Martins, Monte Carlo simulations of ambient temperature uncertainty determined by dual-band pyrometry. Meas. Sci. Technol. **26**, 085016 (2015)
7. K. Ångström, Ueber absolute Bestimmungen der Wärmestrahlung mit dem elektrischen Compensationspyrheliometer, nebst einigen Beispielen der Anwendung dieses Instrumentes. Ann. Phys. **303**, 633–648 (1899)
8. S. Bell, *A beginner's guide to uncertainty of measurement.* Measurement Good Practice Guide No. 11 (Issue 2) (NPL, UK, 2001), https://www.npl.co.uk/publications/a-beginners-guide-tou ncertaintyin-measurement. Accessed 21 Jan 2019
9. K. Berka, Scales of measurement, in *Language, Logic, and Method*, ed. by R.S. Cohen, M.W. Wartofsky (D. Reidel Publ. Co., Dordrecht, Holland, 1983), pp. 1–73
10. L. Birgé, Y. Rozenholc, How many bins should be put in a regular histogram. ESAIM: P&S **10**, 24–45 (2006). https://doi.org/10.1051/ps:2006001
11. BIPM, *The International System of Units (SI)*, 9th ed. (2019), https://www.bipm.org/utils/common/pdf/si-brochure/SI-Brochure-9-EN.pdf. Accessed 5 Jan 2020
12. G.E.P. Box, M.E. Muller, A note on the generation of random normal deviates. Ann. Math. Statist. **29**, 610–611 (1958)
13. R.W. Boyd, *Radiometry and the Detection of Optical Radiation* (Wiley, New York, 1983)

14. S. Briaudeau, F. Bourson, O. Kozlova et al., The synthetic double wavelength technique: a simple robust method for thermodynamic temperature determination. Metrologia **57**, 025014 (2020)
15. E. Bründermann et al., *Terahertz Techniques* (Springer, Berlin, 2012)
16. R.A. Chipman, Polarimetry, in *Handbook of Optics, Volume I: Geometrical and Physical Optics, Polarized Light, Components and Instruments*, 3rd ed., ed. by M. Bass (McGraw-Hill, New York, 2010), pp. 15.1–15.46
17. C.J. Chunnilall, E. Theocharous, Infrared hemispherical reflectance measurements in the 2.5 μm to 50 μm wavelength region using a Fourier transform spectrometer. Metrologia **49**, S73–S80 (2012)
18. CIE, *Methods of Characterizing the Performance of Radiometers and Photometers*. CIE Publ. No 53. (International Commission on Illumination, Vienna, 1982)
19. CIE, *Methods of Characterizing Illuminance Meters and Luminance Meters*. CIE Publ. No 69. (International Commission on Illumination, Vienna, 1987)
20. F.J.J. Clarke, D.J. Parry, Helmholtz reciprocity: its validity and application to reflectometry. Light. Res. Technol. **17**, 1–11 (1985)
21. W.W. Coblentz, Instruments and methods used in radiometry. Bull. Bur. Stand. **4**, 391–460 (1907–1908)
22. W.W. Coblentz, Instruments and methods used in radiometry II. Bull. Bur. Stand. **9**, 7–63 (1913)
23. W.W. Coblentz, Methods and apparatus in spectroradiometry. J. Opt. Soc. Amer. **7**, 439–454 (1923)
24. R.R. Cordero, G. Seckmeyer, S. Riechelmann et al., Monte Carlo-based uncertainty analysis of UV array spectroradiometers. Metrologia **49**, 745–755 (2012)
25. J.M. Cowley, *Diffraction Physics*, 3rd edn. (Elsevier, Amsterdam, 1995)
26. M.G. Cox, M.P. Dainton, A.B. Forbes et al., Use of Monte Carlo simulation for uncertainty evaluation in metrology, in *Advanced Mathematical and Computational Tools in Metrology V*, ed. by P. Ciarlini, M.G. Cox, E. Filipe et al. (World Scientific Publishing Co., River Edge, NJ, 2001), pp. 93–105
27. M.G. Cox, B.R.L. Siebert, The use of a Monte Carlo method for evaluating uncertainty and expanded uncertainty. Metrologia **43**, S178–S188 (2006)
28. M.G. Cox, P.M. Harris, Validating the applicability of the GUM procedure. Metrologia **51**, S167–S175 (2014)
29. R.U. Datla, A.C. Parr, Introduction to optical radiometry, in *Optical Radiometry*, ed. by A.C. Parr et al. (Academic Press, Amsterdam, Netherlands, 2005), pp. 1–34
30. J. De Lucas, J.J. Segovia, Uncertainty calculation of the effective emissivity of cylinder-conical blackbody cavities. Metrologia **53**, 61–75 (2016)
31. J. De Lucas, J.J. Segovia, Measurement and analysis of the temperature gradient of blackbody cavities, for use in radiation thermometry. Int. J. Thermophys. **39**, 57 (2018)
32. C. DeCusatis (ed.), *Handbook of Applied Photometry* (AIP Press, Woodbury, NY, 1998)
33. L. Devroye, *Non-Uniform Random Variate Generation* (Springer, New York, 1986)
34. R.H. Dieck, *Measurement Uncertainty: Methods and Applications*, 5th ed. (ISA—The Instrumentation, Systems, and Automation Society, Research Triangle Park, NC, 2017)
35. W.L. Dunn, J.K. Shultis, *Exploring Monte Carlo Methods* (Academic Press, Amsterdam, 2012)
36. K. Entacher, Bad subsequences of well-known linear congruential pseudorandom number generators. ACM Trans. Model. Comput. Simul. **8**, 61–70 (1998)
37. T.J. Esward, A. de Ginestous, P.M. Harris et al., A Monte Carlo method for uncertainty evaluation implemented on a distributed computing system. Metrologia **44**, 319–326 (2007)
38. L. Finkelstein, Fundamental concepts of measurement: definition and scales. Meas. Control **8**, 105–111 (1975)
39. G.S. Fishman, *Monte Carlo. Concepts, Algorithms, and Applications* (Springer, New York, 1996)

40. A.T. Friberg (ed.), *Selected Papers on Coherence and Radiometry*. SPIE Milestone Series, MS 69 (SPIE Press, Bellingham, WA, 1993)
41. C. Fröhlich, History of solar radiometry and the world radiometric reference. Metrologia **28**, 111–115 (1991)
42. J.L. Gardner, Partial coherence and practical radiometry. Metrologia **30**, 419–423 (1993)
43. J.L. Gardner, On the use of the term "scale(s)" in radiometric and photometric metrology. Metrologia **37**, 547 (2000)
44. J.E. Gentle, *Random Number Generation and Monte Carlo Methods*, 2nd edn. (Springer, New York, 2003)
45. T.A. Germer et al. (eds.), *Spectrophotometry: Accurate Measurement of Optical Properties of Materials* (Academic Press, Amsterdam, 2014)
46. C.E. Gibson, B.K. Tsai, A.C. Parr, *NIST Measurement Services: Radiance Temperature Calibrations* (NIST Spec. Publ. 250-43. NIST, U. S. Dept. of Commerce, Washington, DC, 1998)
47. J.W. Goodman, *Statistical Optics*, 2nd edn. (Wiley, Hoboken, NJ, 2015)
48. T. Goodman, W. Servantes, E. Woolliams et al., *Final Report on the EURAMET.PR-K1.a-2009 Comparison of Spectral Irradiance 250–2500 nm*. NPL Report OP 12 (National Physical Laboratory, Teddington, UK, 2015), https://www.bipm.org/utils/common/pdf/final_reports/PR/K1/EURAMET-PR-K1.a-Final-Report.pdf. Accessed 16 Jan 2020
49. M. Grabe, *Measurement Uncertainties in Science and Technology*, 2nd edn. (Springer, Berlin, 2014)
50. B.G. Grant, *Field Guide to Radiometry* (SPIE Press, Bellingham, WA, 2011)
51. F. Grum, R.J. Becherer, *Radiometry* (Academic Press, Orlando, FL, 1979)
52. G. Grynberg, A. Aspect, C. Fabre, *Introduction to Quantum Optics: From the Semi-classical Approach to Quantized Light* (Cambridge University Press, Cambridge, UK, 2010)
53. S.V. Gupta, *Measurement Uncertainties: Physical Parameters and Calibration of Instruments* (Springer, Berlin, 2012)
54. Hamamatsu: Si photodiodes. Selection guide. Hamamatsu Photonics K.K. (2020), https://www.hamamatsu.com/resources/pdf/ssd/si_pd_kspd0001e.pdf. Accessed 11 Jun 2020
55. T.P. Heavner, E.A. Donley, F. Levi et al., First accuracy evaluation of NIST-F2. Metrologia **51**, 174–182 (2014)
56. P. Hellekalek, Good random number generators are (not so) easy to find. Math. Comput. Simul. **46**, 485–505 (1998)
57. H. von Helmholtz, *Handbuch der physiologischen Optik*, Band 1, S. 169 (Verlag von Leopold Voss, Leipzig, 1856)
58. W. Hormann, J. Leydold, G. Derflinger, *Automatic Nonuniform Random Variate Generation* (Springer, Berelin, 2004)
59. K. Horne, A. Fleming, B. Timmins et al., Monte Carlo uncertainty analysis for photothermal radiometry measurements using a curve fit process. Metrologia **52**, 83–792 (2015)
60. J.M. Houston, J.P. Rice, NIST reference cryogenic radiometer designed for versatile performance. Metrologia **43**, S31–S35 (2006)
61. P. Howarth, F. Redgrave, *Metrology—In Short*, 3rd ed. (EURAMET, 2008), https://www.euramet.org/publications-media-centre/documents/metrology-in-short/. Accessed 21 Jan 2019
62. ILV, *International Lighting Vocabulary*. CIE S017/E:2011 (CIE Central Bureau, Vienna, 2011)
63. ISO 20473:2007, *Optics and Photonics—Spectral Bands* (International Standard, ISO, Geneva, 2007)
64. ISO 23539:2005 (CIE S010/E:2004), *Photometry—The CIE System of Physical Photometry* (International Standard, ISO, Geneva, 2005)
65. ISO 28640:2010, *Random Variate Generation Methods—Specifies Methods for Generating Uniform and Non-uniform Random Variates for Monte Carlo Simulation Purposes* (International Organization for Standardization, Geneva, 2010)
66. ISO 5725-1:1994, *International Standard. Accuracy (Trueness and Precision) of Measurement Methods and Results—Part 1: General Principles and Definitions* (International Organization for Standardization (ISO), Geneva, 1994)

67. ISO 5725-2:1994, *International Standard. Accuracy (Trueness and Precision) of Measurement Methods and Results—Part 2: Basic Method for the Determination of Repeatability and Reproducibility of a Standard Measurement Method* (International Organization for Standardization (ISO), Geneva, 1994)
68. ISO 5725-3:1994, *International Standard. Accuracy (Trueness and Precision) of Measurement Methods and Results—Part 3: Intermediate Measures of The Precision of a Standard Measurement Method* (International Organization for Standardization (ISO), Geneva, 1994)
69. ISO 5725-4:1994, *International Standard. Accuracy (Trueness and Precision) of Measurement Methods and Results—Part 4: Basic Methods for the Determination of The Trueness of a Standard Measurement Method* (International Organization for Standardization (ISO), Geneva, 1994)
70. ISO 5725-6:1994, *International Standard. Accuracy (Trueness and Precision) of Measurement Methods and Results—Part 6: Use in Practice of Accuracy Values* (International Organization for Standardization (ISO), Geneva, 1994)
71. ISO 5725-5:1998, *International Standard. Accuracy (Trueness and Precision) of Measurement Methods and Results—Part 5: Alternative Methods for the Determination of the Precision of a Standard Measurement Method* (International Organization for Standardization (ISO), Geneva, 1998)
72. ISO/CIE 17166:2019, *Erythema Reference Action Spectrum and Standard Erythema Dose* (International Organization for Standardization (ISO), Geneva, Switzerland, 2019)
73. JCGM 100:2008, GUM, *Evaluation of Measurement Data. Guide to the Expression of Uncertainty in Measurement.* (GUM 1995 with Minor Corrections) (BIPM Joint Committee for Guides in Metrology, Paris, 2008)
74. JCGM 101:2008. GUM, *Evaluation of Measurement Data—Supplement 1 to the "Guide to the Expression of Uncertainty in Measurement"—Propagation of Distributions Using a Monte Carlo Method* (BIPM Joint Committee for Guides in Metrology, Paris, 2008)
75. JCGM 102:2011. *Evaluation of Measurement Data—Supplement 2 to the "Guide to the Expression of Uncertainty in Measurement"—Extension to any Number of Output Quantities* (BIPM Joint Committee for Guides in Metrology, Paris, 2011)
76. JCGM 104:2009, *Evaluation of Measurement Data—An Introduction to the "Guide to the Expression of Uncertainty in Measurement" and Related Documents* (BIPM Joint Committee for Guides in Metrology, Paris, 2009)
77. JCGM 200:2012, *International Vocabulary of Metrology—Basic and General Concepts and Associated Terms (VIM)*, 3rd ed.—2008 Version with Minor Corrections. (BIPM Joint Committee for Guides in Metrology, Paris, 2012)
78. B.C. Johnson, H. Yoon, J.P. Rice, A.C. Parr, Principles of optical radiometry and measurement uncertainty, in *Optical Radiometry for Ocean Climate Measurements*, ed. by G. Zibordi, C.J. Donlon, A.C. Parr (Academic Press, Amsterdam, 2014), pp. 13–67
79. S.F. Johnston, *A History of Light and Colour. Measurement Science in the Shadows* (Institute of Physics Publishing, Bristol, UK, 2001)
80. R. Kacker, K.-D. Sommer, R. Kessel et al., Evolution of modern approaches to express uncertainty in measurement. Metrologia **44**, 513–529 (2007)
81. T. Keawprasert, K. Anhalt, D.R. Taubert, A comparison of absolute calibrations of a radiation thermometer based on a monochromator and a tunable source. AIP Conf. Proc. **1552**, 682–687 (2013)
82. B. Khlevnoy, CCPR S1 supplementary comparison. Final Report on CCPR-S1: Spectral Radiance 220 nm to 2500 nm. Metrologia **45**, 02001 (2008), https://www.bipm.org/utils/common/pdf/final_reports/PR/S1/CCPR-S1.pdf. Accessed 16 Jan 2020
83. G.R. Kirchhoff, Über die Verhältnis zwischen dem Emissionsvermögen und dem Absorptionsvermögen der Körper für Wärme und Licht. Ann. Phys. **109**, 275–301 (1860)
84. L. Kirkup, R.B. Frenkel, *An Introduction to Uncertainty in Measurement Using the GUM (Guide to the Expression of Uncertainty in Measurement)* (Cambridge University Press, Cambridge, UK, 2006)

85. K.H. Knuth, Optimal data-based binning for histograms and histogram-based probability density models. Digital Signal Processing **95**, 102581 (2019)
86. H.J. Kostkowski, F.E. Nicodemus, An introduction to the measurement equation, in *Self-Study Manual on Optical Radiation Measurements: Part I—Concepts*, Chapters 4 and 5. NBS Technical Note 910-2, ed. by F.E. Nicodemus (NBS, U.S. Dept. of Commerce, Washington, DC, 1978), pp. 58–92
87. H.J. Kostkowski, *Reliable Spectroradiometry* (Spectroradiometry Consulting, La Plata, MD, 1997)
88. D.P. Kroese, T. Taimre, Z.I. Botev, *Handbook of Monte Carlo Methods* (Wiley, Hoboken, NJ, 2011)
89. P. L'Ecuyer, R. Simard, TestU01: A C library for empirical testing of random number generators. ACM Trans. Math. Softw. **33**, article 22, 1–40 (2007)
90. P. L'Ecuyer, Uniform random number generators, in *International Encyclopedia of Statistical Science*, ed. by M. Lovric (Springer, Heidelberg, 2011), pp. 1625–1630
91. B. Lane, E. Whitenton, V. Madhavan et al., Uncertainty of temperature measurements by infrared thermography for metal cutting applications. Metrologia **50**, 637–653 (2013)
92. L. Laurencelle, F.-A. Dupuis, *Statistical Tables. Explained and Applied* (World Scientific Publ, River Edge, NJ, 2000)
93. J. Mandel, *The Statistical Analysis of Experimental Data* (Interscience Publ, New York, 1964)
94. L. Mandel, E. Wolf, *Optical Coherence and Quantum Optics* (Cambridge University Press, Cambridge, UK, 1995)
95. J.M. Mantilla, M.L. Hernanz, J. Campos et al., Monochromator-based absolute calibration of a standard radiation thermometer. Int. J. Thermophys. **35**, 493–503 (2014)
96. G. Marsaglia, W.W. Tsang, Some difficult-to-pass tests of randomness. J. Stat. Software **7**, 1–9 (2002)
97. M. Matsumoto, T. Nishimura, Mersenne Twister: a 623-dimensionally equidistributed uniform pseudo-random number generator. ACM Trans. Model. Comput. Simul. **8**, 3–30 (1998)
98. J.C. Maxwell, A dynamical theory of the electromagnetic field. Phil. Trans. R. Soc. Lond. **155**, 459–512 (1865)
99. K.D. Mielenz, R.D. Saunders, J.B. Shumaker, Spectroradiometric determination of the freezing temperature of gold. J. Res. Natl. Inst. Stand. Technol. **95**, 49–67 (1990)
100. K.D. Mielenz, "Wolf shifts" and their physical interpretation under laboratory conditions. J. Res. Natl. Inst. Stand. Technol. **98**, 231–240 (1993)
101. A. Migdall, S. Polyakov, J. Fan, J. Bienfang (eds.), *Single-Photon Generation and Detection* (Academic Press, Amsterdam, 2013)
102. G.J. Myatt, *Making Sense of Data. A Practical Guide to Exploratory Data Analysis and Data Mining* (Wiley, Hoboken, NJ, 2007)
103. F.E. Nicodemus, H.J. Kostkowski, Distribution of optical radiation with respect to position and direction—radiance, in *Self-study Manual on Optical Radiation Measurements: Part I—Concepts*, Chapters 1 to 3. NBS Tech. Note 910-1, ed. by F.E. Nicodemus (NBS, U.S. Dept. of Commerce, Washington, DC, 1976), pp. 10–44
104. K.A. Nugent, J.L. Gardner, Radiometric measurements and correlation-induced spectral changes. Metrologia **29**, 319–324 (1992)
105. G.D. Nutter, Radiation thermometers: design principles and operating characteristics, in *Theory and Practice of Radiation Thermometry*, ed. by D.P. DeWitt, G.D. Nutter (Wiley, New York, 1988), pp. 231–337
106. E.L. O'Neill, *Introduction to Statistical Optics* (Dover, Mineola, NY, 1992)
107. Y. Ohno, Basic concepts in photometry, radiometry, and colorimetry, in *Handbook of Optoelectronics, Volume 1: Concepts, Devices, and Techniques*, 2nd ed., ed. by J.P. Dakin, R.G.W. Brown (CRC Press, Boca Raton, FL, 2018), pp. 265–279
108. Y. Ohno, Photometry, in *Optical Radiometry*, ed. by A.C. Parr et al. (Academic Press, Amsterdam, 2005), pp. 327–366
109. ÅA.F. Olsen, R.A. Bergerud, Traceable calibration of a radiation thermometer in the range 100 °C to 300 °C by model fitting. Int. J. Thermophys. **36**, 1803–1812 (2015)

110. R. Palenčár, P. Sopkuliak, J. Palenčár et al., Application of Monte Carlo method for evaluation of uncertainties of ITS-90 by standard platinum resistance thermometer. Meas. Sci. Rev. **17**, 108–116 (2017)
111. F.O. Panneton, P. L'Ecuyer, P. Matsumoto, Improved long-period generators based on linear recurrences modulo 2. ACM Trans. Math. Softw. **32**, 1–16 (2006)
112. A.C. Parr, The candela and photometric and radiometric measurements. J. Res. NIST **106**, 151–186 (2000)
113. A.C. Parr, B.C. Johnson, The use of filtered radiometers for radiance measurement. J. Res. Natl. Inst. Stand. Technol. **116**, 751–760 (2011)
114. J.V. Pearce, P.M. Harris, J.C. Greenwood, Evaluating uncertainties in interpolations between calibration data for thermocouples. Int. J. Thermophys. **31**, 1517–1526 (2010)
115. J. Piotrowski, *Theory of Physical and Technical Measurement* (Elsevier, Amsterdam, 1992)
116. M. Planck, *The Theory of Heat Radiation* (P. Blakiston's Son & Co., Philadelphia, PA, 1914)
117. M. Planck, *Theory of Light* (Macmillan and Co., London, 1932)
118. T. Poikonen, P. Kärhä, P. Manninen et al., Uncertainty analysis of photometer quality factor. Metrologia **46**, 75–80 (2009)
119. T. Poikonen, P. Blattner, P. Kärhä et al., Uncertainty analysis of photometer directional response index using Monte Carlo simulation. Metrologia **49**, 727–736 (2012)
120. W.H. Press, S.A. Teukolsky, H.A. Bethe et al., *Numerical Recipes: The Art of Scientific Computing*, 3rd ed. (Cambridge University Press, Cambridge, UK, 2007)
121. M. Priel, From GUM to alternative methods for measurement uncertainty evaluation. Accred. Qual. Assur. **14**, 235–241 (2009)
122. A. Prokhorov, V. Sapritsky, B. Khlevnoy, Alternative methods of blackbody thermodynamic temperature measurement above silver point. Int. J. Thermophys. **36**, 252–266 (2015)
123. S.G. Rabinovich, *Evaluating Measurement Accuracy. A Practical Approach*, 3rd ed. (Springer, Cham, Switzerland, 2017)
124. C. Ratcliffe, B. Ratcliffe, *Doubt-Free Uncertainty in Measurement. An Introduction for Engineers and Students* (Springer, Cham, Switzerland, 2015)
125. P. Rosenkranz, Uncertainty propagation for platinum resistance thermometers calibrated according to ITS-90. Int. J. Thermophys. **32**, 106–119 (2011)
126. A. Rukhin, J. Soto, J. Nechvatal et al., *A Statistical Test Suite for Random and Pseudorandom Number Generators for Cryptographic Applications*. NIST Spec. Publ. 800-22 Revision 1a (U.S. Dept. of Commerce, NIST, 2010)
127. P. Saunders, Analysis of the potential accuracy of thermodynamic measurement using the double-wavelength technique. Int. J. Thermophys. **35**, 417–437 (2014)
128. SCHOTT, KG5 Data Sheet. SCHOTT North America, Inc. (2014), https://www.sydor.com/wp-content/uploads/2019/06/SCHOTT-KG5-Shortpass-Filter.pdf. Accessed 11 Jun 2020
129. D.W. Scott, Sturges' rule. WIREs. Comput. Stat. **1**, 44–48 (2009)
130. D.W. Scott, Histogram. WIREs. Comput. Stat. **2**, 203–306 (2010)
131. K.D. Sommer, B.R.L. Siebert, Systematic approach to the modelling of measurements for uncertainty evaluation. Metrologia **43**, S200–S210 (2006)
132. H.-J. Song, T. Nagatsuma (eds.), *Handbook of Terahertz Technologies. Devices and Applications* (CRC Press, Boca Raton, FL, 2015)
133. S.S. Stevens, On the theory of scales of measurement. Science **103**, 677–680 (1946)
134. B.N. Taylor, C.E. Kuyatt, *Guidelines for Evaluating and Expressing the Uncertainty of NIST Measurement Results*. NIST Technical Note 1297 (1994 Edition.) (NIST, U.S. Dept. of Commerce, Washington, DC, 1994)
135. D.B. Thomas, W. Luk, P.H.W. Leong et al., Gaussian random number generators. ACM Comput. Surv. **39**, 11 (2007)
136. A.S. Tistomo, D. Larassati, A. Achmadi et al., Estimation of uncertainty in the calibration of industrial platinum resistance thermometers (IPRT) using Monte Carlo method. MĀPAN—J. Metrol. Soc. India **32**, 273–278 (2017)
137. B.K. Tsai, B.C. Johnson, Evaluation of uncertainties in fundamental radiometric measurements. Metrologia **35**, 587–593 (1998)

138. F. Vignola, J. Michalsky, T. Stoffel, *Solar and Infrared Radiation Measurements* (CRC Press, Boca Raton, FL, 2012)
139. J.H. Walker, R.D. Saunders, A.T. Hattenburg, The NBS scale of spectral radiance. Metrologia **24**, 79–88 (1987)
140. L. Werner, H.-W. Hübers, P. Meindl et al., Towards traceable radiometry in the terahertz region. Metrologia **46**, S160–S164 (2009)
141. W. Wien, Temperatur und Entropie der Strahlung. Ann. Phys. **288**, 133–165 (1894)
142. R. Willink, On using the Monte Carlo method to calculate uncertainty intervals. Metrologia **43**, L39–L42 (2006)
143. R. Willink, A generalization of the Welch-Satterthwaite formula for use with correlated uncertainty components. Metrologia **44**, 340–349 (2007)
144. E. Wolf, *Selected Works of Emil Wolf With Commentary* (World Scientific Publishing, Singapore, 2001)
145. W.L. Wolfe, *Introduction to Radiometry* (SPIE Press, Bellingham, WA, 1998)
146. E.R. Woolliams, R. Winkler, S.G.R. Salim et al., The double-wavelength technique—an alternative technique to determine thermodynamic temperature. Int. J. Thermophys. **30**, 144–154 (2009)
147. E.R. Woolliams, Uncertainty analysis for filter radiometry based on the uncertainty associated with integrated quantities. Int. J. Thermophys. **35**, 1353–1365 (2014)
148. G. Wyszecki, W.R. Blevin, K.G. Kessler, K.D. Mielenz, *Principes régissant la photométrie/Principles Governing Photometry* (BIPM, Sèvres, France, 1983), https://www.bipm.org/utils/common/pdf/monographies-misc/Monographie1983-1.pdf. Accessed 28 Mar 2017
149. Y. Yamaguchi, Y. Yamada, J. Ishii, Supercontinuum-source-based facility for absolute calibration of radiation thermometers. Int. J. Thermophys. **36**, 1825–1833 (2015)
150. H.W. Yoon, C.E. Gibson, P.Y. Barnes, Realization of the National Institute of Standards and Technology detector-based spectral irradiance scale. Appl. Opt. **41**, 5879–5890 (2002)
151. H.W. Yoon, C.E. Gibson, *NIST Measurement Services: Spectral Irradiance Calibrations*. NIST Spec. Publ. 250-89 (NIST, Gaithersburg, MD, 2011)
152. S. Yurtseven, O. Pehlivan, H. Nasibov, Size-of-source (SSE) effect of broadband and interference filters for double wavelength thermometers. J. Phys. Conf. Ser. **1065**, 122015 (2018)
153. E.F. Zalewski, Radiometry and photometry, in *Handbook of Optics. Volume II: Design, Fabrication, and Testing; Sources and Detectors; Radiometry and Photometry*, 3rd ed., ed. by M. Bass (McGraw-Hill, New York, 2010), pp. 34.3–34.47

Chapter 3
Theoretical Basis of Blackbody Radiometry

Abstract The theoretical basis of blackbody radiometry includes the radiation laws of a perfect blackbody discovered at the turn of the nineteenth and twentieth centuries and the theory of the radiation heat transfer developed by the mid-1960s. To ensure uniformity of presentation throughout the book, the synopsis of fundamental ideas, definitions, and laws are given in this chapter in the modern form. The concept of a perfect blackbody and its fundamental radiation laws are introduced and explained. The nomenclature of radiative properties of surfaces are considered in connection with the reciprocity principle, Kirchhoff's law, and the energy conservation law. The radiance temperature concept is discussed. The simplest methods for calculation of radiative heat transfer in geometrical configurations, typical for radiometric systems, are outlined. Thus, the chapter allows readers to update their knowledge to proceed to the next chapters.

Keyword Perfect blackbody · Blackbody radiation laws · Radiative properties · Radiance temperature · Diffuse enclosure

3.1 Introductory Notes

This chapter focuses on relationships, which are ubiquitously used in blackbody radiometry and will be used throughout this book. First, these are the radiation laws of a perfect blackbody. Our goal is not to derive these laws anew or to describe the history of scientific thought in this area. This is not necessary because there are many excellent books devoted, in whole or in part, to the historical context of theoretical studies in the field of thermal radiation with the usual emphasis on the origin of quantum theory (see, for example, [1, 8, 38, 42, 80, 83]). The foundations of the theory of thermal radiation were laid in the late 19th and early twentieth centuries, at the dawn of an era leading to the creation of quantum theory and is known as the golden age of theoretical physics. The theory of thermal radiation is a well-studied, developed, and self-consistent system of concepts and laws, which was completed mainly when Planck [59] derived the blackbody radiation law named

---

The original version of this chapter was revised: The equation (3.143) has been corrected. The correction to this chapter is available at https://doi.org/10.1007/978-3-030-57789-6_10

V. Sapritsky and A. Prokhorov, *Blackbody Radiometry*, Springer Series in Measurement Science and Technology, https://doi.org/10.1007/978-3-030-57789-6_3

Table 3.1 Chronology of milestone works related to the theory of blackbody radiation

Year	Event	Reference
1760	J. H. Lambert proposes cosine law for diffuse surfaces	[43]
1859	G. Kirchhoff, by means of a thought experiment, derives his law for discrete wavelengths	[39]
1860	Kirchhoff generalizes his law to the continuous spectrum and introduces the concept of a perfect blackbody	[40]
1879	J. Stefan finds experimentally the "fourth-power law" for blackbody radiation	[79]
1884	From thermodynamic considerations, L. Boltzmann derives the dependence found by Stefan in 1879	[15]
1893	W. Wien establishes the displacement law using thermodynamic considerations	[86]
1896	Wien theoretically obtains the distribution of energy in the emission spectrum of a perfect blackbody	[87]
1900	M. Planck "enhances" Wien's law introducing an equation for the distribution of energy in the radiation spectrum of a perfect blackbody	[58]
1900	Planck proposes the hypothesis of energy quanta and derives the radiation law named after him	[59]
1900	Lord Rayleigh derives the T/λ^4 spectral distribution on the base of classical thermodynamics	[62]
1909	J. Jeans derives Rayleigh's distribution from Planck's law	[37]

after him. Currently, having an accomplished theory of thermal radiation, we are not obliged to follow the timeline of theoretical and experimental achievements in this area presented in Table 3.1 but can safely build a convenient and reasoned exposition.

All works collected in Table 3.1 are theoretical except for that by Stefan [79], which was based on the results of experimental studies. Of course, along with theoretical studies, a number of measurement experiments were carried out to prove or disprove the corresponding working hypotheses. Although these experiments are currently of historical interest only, we would like to draw the attention of interested readers to the article by Hoffmann [27], which presents a fascinating history of experiments conducted by German physicists in support of the development of the blackbody radiation theory.

For the sake of convenience of further presentation, we consider, for instance, the radiation laws of Wien and Rayleigh-Jeans as approximations of Planck's law, although they were its forerunners. Similarly, the Stefan-Boltzmann law and the Wien displacement law were obtained before Planck's law, but it is more convenient to derive them directly from the Planck law.

By introducing the emissivity, we can apply, with some reservations, the radiation laws of a perfect blackbody to describe thermal radiation of real bodies. The concept of emissivity allows, in turn, introducing so-called radiometric temperatures, which play an important role in radiation thermometry and provide its connection with blackbody radiometry.

Unlike a perfect blackbody that is an idealized virtual object, the blackbody radiator standards are artifacts made of real-world materials. The metrological application of blackbodies requires a quantitative assessment of the difference between their radiation characteristics and the characteristics of a perfect blackbody. Therefore, it is necessary to be able to calculate the radiative heat exchange in the system of real bodies, for which a system of radiative characteristics (first of all, reflection) should be established. Calculation of the radiation heat transfer is an integral part of the optical radiometry and the blackbody radiometry in particular. We need in appropriate analytical and/or numerical tools to compute the radiation heat exchange inside a blackbody cavity, to calculate the radiation transfer from a blackbody to a detector, to evaluate the stray radiation in the entire radiometric system, etc. It is impossible to cover all aspects of radiation heat exchange here. However, this is not necessary, since there are excellent books on this branch of thermal physics, for example, the classic monographs [30, 49, 77], to which we will repeatedly refer.

Ensuring uniformity of terms, definitions, and notation throughout the book is one of the objectives of this chapter, which is a concise compendium of basic laws and relationships regularly used in blackbody radiometry rather than a comprehensive exposition of the problem.

3.2 Blackbody Radiation

3.2.1 Concept of a Perfect Blackbody

The central concept of theory of thermal radiation, a perfect blackbody, was introduced by Kirchhoff [39] to establish, by means of a thought experiment, the connection between the absorption of an object and its thermal emission. In a subsequent article, where Kirchhoff [40] generalized his argumentation from discrete wavelengths to the continuous spectrum, he wrote, "*The proof I am about to give of the law above stated, rests on the supposition that bodies can be imagined which, for infinitely small thicknesses, completely absorb all incident rays, and neither reflect nor transmit any. I shall call such bodies perfectly black, or, more briefly, black bodies. It is necessary in the first place to investigate the radiating power of bodies of this description.*"[1]

For a perfect blackbody, the absorptance (the ratio of absorbed and incident radiant fluxes) is equal to unity for all wavelengths, regardless of the temperature. The Kirchhoff law, written in modern notation, states that for a real body with the absorptance α in thermodynamic equilibrium,

$$M_\lambda(\lambda, T)\big/\alpha(\lambda, T) = M_{\lambda,bb}(\lambda, T), \tag{3.1}$$

[1] Cited in English translation [41] of the original article [40].

and

$$M(T)\big/\alpha(T) = M_{bb}(T) \tag{3.2}$$

where M_λ and $M_{\lambda,bb}$ are the spectral exitance (the surface density of the emitted spectral radiant flux) of a real body and a perfect blackbody, respectively; M and M_{bb} are the corresponding exitances.

The Kirchhoff law is valid for any surface or body, for any wavelength λ and temperature T, and for arbitrarily polarized radiation. Since the absorptance of a perfect blackbody is equal to unity for any wavelength, any direction, and any polarization, its emitted radiation is unpolarized. In such a way, Kirchhoff's law establishes the relationship between the emissive and absorptive characteristics of any body through the spectral exitance of a perfect blackbody, which is the reason for a thorough study of its radiation.

Besides, the Kirchhoff law states that objects are good emitters of radiation then and only then, if they are good absorbers of radiation. The definition of a perfect blackbody has not changed significantly over time. Richmond and Nicodemus [64] defined it as "*a surface (material or geometrical) that absorbs all radiant flux of all wavelengths and polarizations incident upon it from all possible directions (absorptance α equal to one, and reflectance ρ and transmittance τ both equal to zero, regardless of the geometry, spectrum, or polarization of the incident rays).*" They emphasized that this definition is given for the surface, because in radiometric analyzes and calculations it is more convenient to assume that absorption occurs on a surface, real or virtual.

Some important properties of a perfect blackbody can be derived by simple thought experiments on the base of Kirchhoff's definition of a perfect blackbody.

Let us consider a perfect blackbody with a constant temperature, placed inside a completely isolated cavity of arbitrary shape, the walls of which are also perfectly black and have a constant temperature that differs at the initial moment from the temperature of the perfect blackbody inside the cavity. After some time, the internal body and the closed cavity will have a common equilibrium temperature. In equilibrium conditions, a perfect blackbody must emit exactly the same amount of radiation as it absorbs. The proof of this fact is by contradiction. Let us consider what happens if the energy of the incoming and outgoing radiation is not equal. In this case, we get the heat transfer from a colder body to a hotter one, which contradicts the second law of thermodynamics. Since by its definition, a perfect blackbody absorbs the maximum possible amount of radiation coming in any direction at any wavelength, it must also emit the maximum possible amount of radiation. In other words, a perfect absorber must be also a perfect emitter, as we have already noticed.

Let us consider again an isothermal closed cavity of arbitrary shape with perfectly black walls and move the perfect blackbody inside the cavity to another position and/or change its orientation. The inner body must maintain the same temperature, since the entire closed system remains isothermal. Consequently, an internal body must emit the same amount of radiation as before. Being in equilibrium, it should receive the same amount of radiation from the walls of the cavity. Thus, the radiation energy received by the internal body should not depend on its position or on its orientation in the cavity. Therefore, radiation passing through any point inside the cavity should not depend on its position or direction. This means that the radiation of a perfect blackbody is isotropic, i.e. its radiance should not depend on the viewing direction. This statement is expressed by the Lambert law of emission that can be expressed in terms of radiance L_{bb} and spectral $L_{\lambda,bb}$ radiance as:

$$L_{bb}(\theta) = Const, \tag{3.3}$$

$$L_{\lambda,bb}(\theta) = Const, \tag{3.4}$$

or in terms of radiant intensity I_{bb} and spectral radiant intensity $I_{\lambda,bb}$:

$$I_{bb}(\theta) = I_{bb}(0)\cos\theta, \tag{3.5}$$

$$I_{\lambda,bb}(\theta) = I_{\lambda,bb}(0)\cos\theta, \tag{3.6}$$

where θ is the angle between the viewing direction and the normal to the surface. The Lambert law makes it possible to find simple relationships between the exitance M_{bb} or spectral exitance $M_{\lambda,bb}$ of a perfect blackbody and its radiance L_{bb} or spectral radiance $L_{\lambda,bb}$:

$$M_{bb} = L_{bb}\int_{\varphi=0}^{2\pi} d\varphi \int_{\theta=0}^{\frac{\pi}{2}} \cos\theta\,\sin\theta\,d\theta = \pi\,L_{bb}, \tag{3.7}$$

$$M_{\lambda,bb} = L_{\lambda,bb}\int_{\varphi=0}^{2\pi} d\varphi \int_{\theta=0}^{\frac{\pi}{2}} \cos\theta\,\sin\theta d\theta = \pi\,L_{\lambda,bb}. \tag{3.8}$$

A perfect blackbody is a virtual object that is no more than a physical abstraction. However, although a perfect blackbody does not exist in the real world, it is possible to create an artificial object behaving largely like a perfect blackbody. Kirchhoff suggested using an isothermal enclosure with a tiny hole in the wall as such an artifact. Any ray falling into this cavity undergoes multiple reflections and loses the energy absorbed by the cavity wall at each reflection. As a result, only a small fraction of the radiant flux incident into the cavity, can escape it through the hole. The smaller the ratio of the area of the hole to the area of the enclosure, the closer to unity the absorptance of the cavity. Since such a cavity is a good absorber, it must be also a good thermal emitter in accordance with the Kirchhoff law. To experimentally verify Kirchhoff's theory, Wien and Lummer [85] proposed to investigate the spectrum of radiation exiting through a narrow opening in the wall of the hollow sphere, the diameter of which is much large than that of the opening.

In summary, we can list, in qualitative terms, the properties of thermal radiation of a perfect blackbody as follows:

- Radiation emitted by a perfect blackbody is unpolarized and isotropic;
- Radiation emitted at a given wavelength by a perfect blackbody depends only on its thermodynamic temperature;
- A perfect blackbody emits more radiation than any other thermal radiator at the same temperature;
- Any perfect blackbodies at the same temperature are indistinguishable, i.e. they emit exactly the same radiation.

3.2.2 *Planck's Distribution and Wien's Displacement Law*

The Planck law, or Planck's distribution, which is the main relation of the blackbody radiometry, can be written using the wavelength λ as a spectral variable for the spectral exitance (the radiant flux emitted per unit of area and per unit of wavelength), $M_{\lambda,bb}$, of a perfect blackbody as

$$M_{\lambda,bb}(\lambda, T) = \frac{c_1}{\lambda^5\left[\exp\left(\frac{c_2}{\lambda T}\right) - 1\right]}, \tag{3.9}$$

where T is the thermodynamic temperature, $c_1 = 2\pi h c^2$ is the first radiation constant, $c_2 = hc/k$ is the second radiation constant, h is the Planck constant, c is the speed of light in vacuum, and k is the Boltzmann constant (the numerical values of physical constant are provided in Appendix A). In the blackbody radiometry, the Planck law is more often written for the spectral radiance $L_{\lambda,bb}$ (the radiant flux emitted per unit of area, per unit of solid angle, and per unit of wavelength) using the constant $c_{1L} = c_1/\pi = 2h c^2$ as

$$L_{\lambda,bb}(\lambda, T) = \frac{c_{1L}}{\lambda^5\left[\exp\left(\frac{c_2}{\lambda T}\right) - 1\right]}. \tag{3.10}$$

The relation between the first radiation constant c_1 and the first radiation constant for spectral radiance c_{1L} follows directly from Eq. 3.8 that expresses the Lambert law. The spectral radiance can be expressed in W·m^{-3}·sr^{-1}, W·m^{-2}·µm^{-1}·sr^{-1}, or

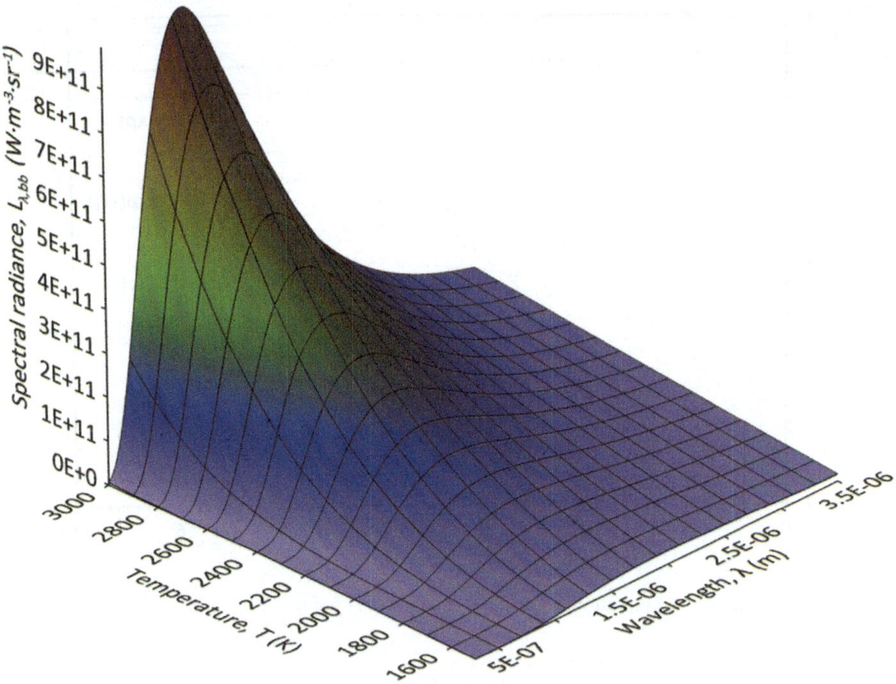

Fig. 3.1 The 3D representation of the Planck distribution

W·m^{-2}·nm^{-1}·sr^{-1}. The three-dimensional (3D) graph of the Planck distribution is presented in Fig. 3.1.

The graph of the Planckian function has only one maximum. Its position in the blackbody radiation spectrum depends on the temperature of the blackbody and moves to the shortwave region of the spectrum with increasing temperature. This phenomenon is described by the Wien displacement law. Its expression depends on spectral representation: the Wien displacement law is written differently in the wavelength, wavenumber, frequency, etc. domains.

Let us find the position of the maximum $\lambda_{\max}$ of the Planckian distribution and the spectral radiance $L_{\lambda,bb,\max}$ corresponding to that wavelength. We have to find the first derivative of the spectral radiance expressed by Eq. 3.10 with respect to the wavelength and set it to zero. After simple transformations, we get the transcendental equation for the unknown value of $x_\lambda = hc/(\lambda kT)$:

$$x_\lambda = 5\left(1 - e^{-x_\lambda}\right). \tag{3.11}$$

Equation 3.11 has two roots. In Fig. 3.2, they are presented by the intersection points of the curve $y = 5\left(1 - e^{-x}\right)$ and the straight line $y = x$. The first root of

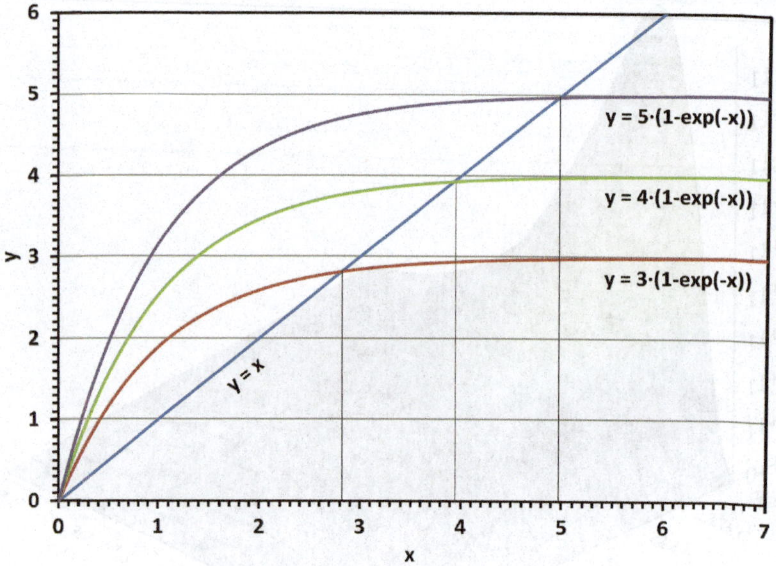

Fig. 3.2 To the solution of equations $x = n \cdot (1 - \exp(-x))$ for $n = 3$, 4, and 5

Eq. 3.11, $x_{\lambda,1} = 0$ corresponds to the trivial case of $\lambda = \infty$; the second root $x_{\lambda,2} \approx 4.96511423174$ can be easily found numerically.

The modern form of the Wien displacement law is the wavelength domain is

$$\lambda_{\max} T = b = c_2 / x_{\lambda,2}, \tag{3.12}$$

where b is the Wien displacement law constants (see Appendix A).

The maximum value $L_{\lambda,bb,\max}$, found by substituting (3.12) into (3.10), is

$$L_{\lambda,bb,\max} = \frac{2x_{\lambda,2}^5 k^5}{h^4 c^3 (e^{x_{\lambda,2}} - 1)} T^5 [\mathrm{W} \cdot \mathrm{m}^{-3} \cdot \mathrm{sr}^{-1}], \tag{3.13}$$

that is, the peak spectral radiance is proportional to the fifth power of the temperature.

The distributions of the spectral radiance of a perfect blackbody in the wavelength domain are presented for several temperatures, for the vertical axis linear scale (Fig. 3.3a) and the logarithmic scale (Fig. 3.3b), together with the curves for the Wien displacement law (loci of $L_{\lambda,bb,\max}$ values) expressed by Eqs. 3.12 and 3.13.

In the wavenumber $\tilde{\nu}$ [cm^{-1}] representation, which is often used in the IR spectrometry and is typical for the use of Fourier transform spectrometers, the Planck law is written as

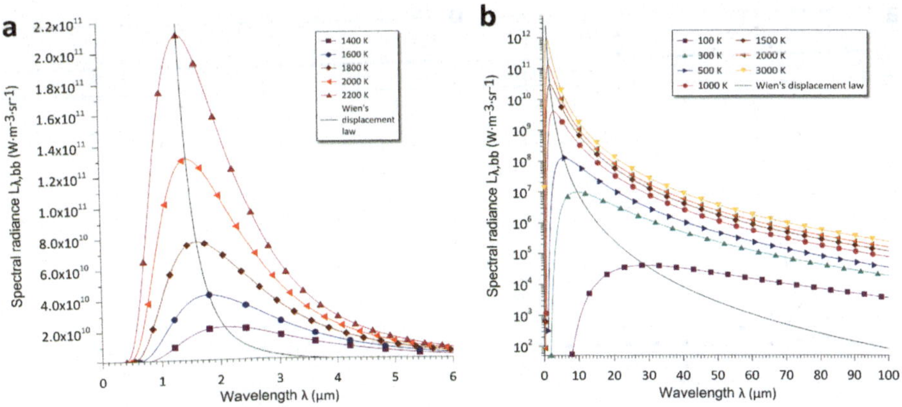

Fig. 3.3 Distributions of spectral radiance of a perfect blackbody in the wavelength domain: **a** with a linear scale along the ordinate, at five temperatures and **b** with a logarithmic scale along the ordinate, at seven temperatures shown the legends. The curves expressing the Wien displacement law are plotted together with the Planckian curves

$$L_{\tilde{\nu}}(\tilde{\nu}, T) = \left|\frac{d\lambda}{d\tilde{\nu}}\right| L_\lambda(\lambda, T) = \frac{2 \cdot 10^8 hc^2 \tilde{\nu}^3}{\exp\left(\frac{100 \cdot hc\tilde{\nu}}{kT}\right) - 1} \ [\mathrm{W \cdot m^{-2} \cdot sr^{-1} \cdot (cm^{-1})^{-1}}]. \tag{3.14}$$

The position of maximum of the Planckian distribution in the wavenumber representation is found through the numerical solution of the transcendental equation for the variable $x_{\tilde{\nu}} = 100hc\tilde{\nu}/(kT)$:

$$x_{\tilde{\nu}} = 3\left(1 - e^{-x_{\tilde{\nu}}}\right). \tag{3.15}$$

The non-trivial solution of Eq. 3.15 is $x_{\tilde{\nu},2} \approx 2.82143937212$.
The wavenumber of the maximum for $L_{\tilde{\nu}}(\tilde{\nu}, T)$ and its maximum value are

$$\tilde{\nu}_{\max} = x_{\tilde{\nu},2} kT/(100hc) \ [\mathrm{cm^{-1}}], \tag{3.16}$$

$$L_{\tilde{\nu},bb,\max} = \frac{200 x_{\tilde{\nu},2}^3 k^3}{h^2 c (e^{x_{\tilde{\nu},2}} - 1)} T^3 \left[\mathrm{W \cdot m^{-2} \cdot sr^{-1} (cm^{-1})^{-1}}\right]. \tag{3.17}$$

The dependences calculated via Eq. 3.14 for temperatures of 300 K, 500 K, 750 K, and 100 K are plotted in Fig. 3.4a together with the dependence that expresses the Wien displacement law in the wavenumber domain. For the frequency ν [Hz] as a spectral variable, the Planck law is written as:

$$L_\nu(\nu, T) = \frac{2h}{c^2} \cdot \frac{\nu^3}{\exp\left(\frac{h\nu}{kT}\right) - 1} \ [\mathrm{W \cdot m^{-2} \cdot sr^{-1} \cdot Hz^{-1}}]. \tag{3.18}$$

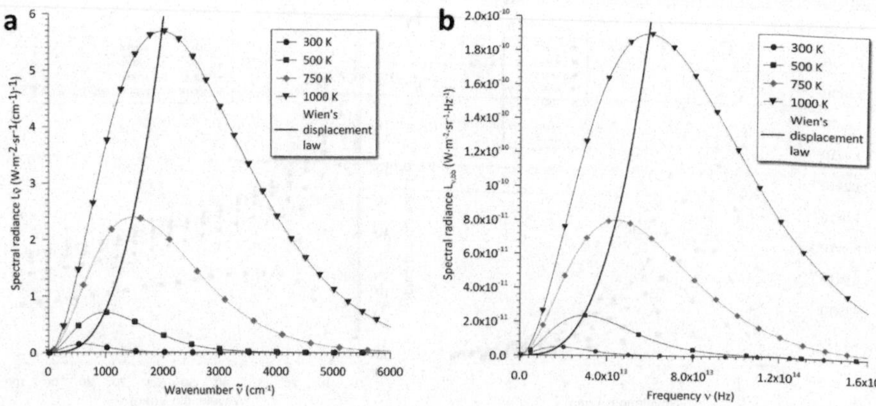

Fig. 3.4 The Planckian curves for temperatures of 300, 500, 750, and 1000 K and the dependence expressing Wien's law: **a** in the wavenumber domain and **b** in the frequency domain

Performing the same operations as in the two previous cases, we come to the transcendental equation for $x_\nu = h\nu/(kT)$:

$$x_\nu = 3\left(1 - e^{-x_\nu}\right). \tag{3.19}$$

Its non-trivial numerical solution is the same as for Eq. 3.15: $x_{\nu,2} \approx$ 2.82143937212. The Wien displacement law in the frequency domain is written as:

$$\nu_{\max} = \frac{x_{\nu,2}k}{h} T\,[\mathrm{Hz}]. \tag{3.20}$$

The maximum value of the Planckian curve in the wavenumber domain is

$$L_{\nu,bb,\max} = \frac{2x_{\nu,2}^3 k^3}{h^2c^2\left(e^{x_{\nu,2}} - 1\right)} T^3\ [\mathrm{W \cdot m^{-2} \cdot sr^{-1} \cdot Hz^{-1}}]. \tag{3.21}$$

The dependences calculated using Eq. 3.18 for temperatures of 300, 500, 750, and 1000 K are plotted in Fig. 3.4b along with the dependence expressing the Wien displacement law in the frequency domain. The similarity of graphs presented in Fig. 3.4a and 3.4b is obvious: it is based on the similarity of Eqs. 3.15 and 3.19; in both wavenumber and frequency domains, the maxima of Planckian curves are proportional to T^3 (sf. Equations 3.17 and 3.21).

An important consequence from the Wien displacement law is the scaling equation that establishes the similarity of shapes for Planckian curves plotted for any temperature T_1 and T_2:

$$L_{\lambda,bb}(\lambda, T_1) = \left(\frac{T_1}{T_2}\right)^5 L_{\lambda,bb}\left(\lambda \cdot \frac{T_1}{T_2}, T_2\right). \tag{3.22}$$

Similar scale relations can be obtained for other spectral variables.

It should be borne in mind the well-known "paradox" (see, e.g. [26, 67]) associated with the presentation of the Wien displacement law with different spectral variables: the maximum of the Planckian spectral curve obtained for one spectral variable at a given temperature is not identical to the maximum obtained for another spectral variable at the same temperature. For instance, if $\lambda_{\max}$ is the wavelength of the maximum found for the Planck law expressed for the wavelength domain and $\tilde{\nu}_{\max}$ is the wavenumber of the maximum found for the Planck law at the same temperature expressed for the wavenumber domain, then generally $\lambda_{\max} \neq 1/\tilde{\nu}_{\max}$ (for more discussion, see Chap. 2 in [17]).

3.2.3 *Approximations of Planck's Law*

The Planck law expressed by Eq. 3.10 (in wavelength representation) does not look too complicated from a mathematical point of view. Now, using a personal computer, we can easily make any calculations according to Planck's law as accurately as necessary. However, the equation expressing Planck's law remains not very convenient for analytical transformations because it (i) cannot be completely converted to the sum by taking the logarithm of both sides and (ii) is not integrable in elementary functions. A number of various approximations were developed in order to simplify calculating the definite integrals of the Planckian distribution over the finite spectral range. With the ubiquitous use of computers, the need for most of the developed approximations has become questionable. In fact, only equations of Wien and Rayleigh–Jeans have retained their significance to the present day. They are currently regarded as approximations of the Planck law, although they have historically been the earliest attempts to obtain the energy distribution in the spectrum of a perfect blackbody within the framework of classical physics.

These approximations, as special cases of Planck's law for small and large values of the product λT, are still present in the literature since they can be useful for studying the asymptotic behavior of Planck's equation and for obtaining analytic expressions for measurement uncertainties. Representing the exponential term in Eq. (3.10) as a power series

$$\exp\left(\frac{c_2}{\lambda T}\right) = 1 + \frac{c_2}{\lambda T} + \frac{1}{2} \cdot \left(\frac{c_2}{\lambda T}\right)^2 + \frac{1}{6} \cdot \left(\frac{c_2}{\lambda T}\right)^3 + \dots, \tag{3.23}$$

we can limit ourselves by two first terms if $\lambda T >> c_2$. This is the case of very long wavelengths and/or very high temperatures, known as the Rayleigh-Jeans law (the Rayleigh-Jeans approximation):

$$L_{\lambda,RJ}(\lambda, T) = \frac{c_{1L} T}{c_2 \lambda^4}, \tag{3.24}$$

In many old reference books and textbooks, it is indicated that the error of approximation does not exceed 1% if $\lambda T > 7.7 \cdot 10^6$ μm·K.

Another approximation of Planck's law is obtained by neglecting the −1 summand in the denominator of the right-hand part of Eq. 3.10. This is possible if the exponential term is much larger than one, i.e. if $\lambda T << c_2$. The resulting equation for the blackbody spectral radiance is known as Wien's law (the Wien approximation) and is expressed as

$$L_{\lambda,\mathrm{Wien}}(\lambda, T) = \frac{c_{1L} e^{-\frac{c_2}{\lambda T}}}{\lambda^5}. \tag{3.25}$$

The relative error of the Wien approximation with respect to Planck's exact formula is

$$\frac{L_{\lambda,\mathrm{Wien}} - L_{\lambda,bb}}{L_{\lambda,bb}} = -e^{-\frac{c_2}{\lambda T}}, \tag{3.26}$$

i.e. Wien's formula always underestimates the spectral radiance. The Wien approximation holds to better than 1% if $\lambda T < 3000$ μm·K, which is suitable for many industrial applications. The relative error $\Delta L_{\lambda,bb} / L_{\lambda,bb} = \left(L_{\lambda,approx} - L_{\lambda,bb}\right) / L_{\lambda,bb}$ introduced by using an approximate value $L_{\lambda,\mathrm{approx}}$ of the spectral radiance instead of the value $L_{\lambda,bb}$ computed according to the Planck equation is presented in Fig. 3.5 for the Wien and Rayleigh-Jeans laws as functions of the product λT. With the advent and widespread use of computers, approximate expressions for the Planck function have become much less common. Currently, they can be used if a compact form of analytical expressions is desired, but the accuracy of the calculations is not a priority. This is the case of the uncertainty propagation, where analytical expressions for first derivatives are used (see the next Sect. for details). The Wien approximation allows obtaining slightly less complex expressions

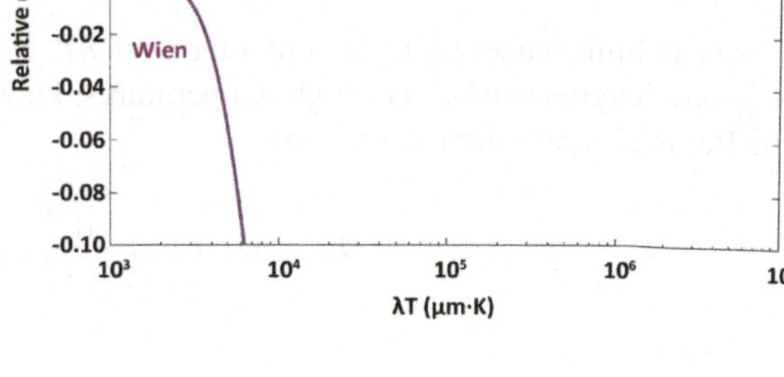

Fig. 3.5 Relative errors of the Wien and Rayleigh–Jeans approximations from the Planck equation plotted against the product λT

for derivatives than those obtained using the Planck law:

$$\frac{\partial L_{\lambda,Wien}}{\partial \lambda} = \frac{c_{1L}}{\lambda^6} e^{-\frac{c_2}{\lambda T}} \left(\frac{c_2}{\lambda T} - 5 \right), \tag{3.27}$$

$$\frac{\partial L_{\lambda,Wien}}{\partial T} = \frac{c_{1L} c_2}{\lambda^6 T^2} e^{-\frac{c_2}{\lambda T}}. \tag{3.28}$$

However, the closeness of functions does not necessarily entail the proximity of their derivatives. If we consider the relative errors of the first derivatives:

$$\frac{\frac{\partial L_{\lambda,Wien}}{\partial \lambda} - \frac{\partial L_{\lambda,bb}}{\partial \lambda}}{\frac{\partial L_{\lambda,bb}}{\partial \lambda}} = -\frac{e^{-\frac{c_2}{\lambda T}} \left[e^{\frac{c_2}{\lambda T}} \left(\frac{2c_2}{\lambda T} - 5 \right) - \frac{c_2}{\lambda T} + 5 \right]}{e^{\frac{c_2}{\lambda T}} \left(\frac{2c_2}{\lambda T} - 5 \right) + 5}, \tag{3.29}$$

$$\frac{\frac{\partial L_{\lambda,Wien}}{\partial T} - \frac{\partial L_{\lambda,bb}}{\partial T}}{\frac{\partial L_{\lambda,bb}}{\partial T}} = e^{-\frac{2c_2}{\lambda T}} \left(1 - 2e^{\frac{c_2}{\lambda T}} \right), \tag{3.30}$$

and their graphs plotted against the product λT in Fig. 3.6, we will see that their behavior is different. While the derivative (3.28) with respect to temperature approximates the true value more than satisfactory throughout the full range of the Wien approximation validity, this cannot be said about the derivative with respect to the wavelength.

The denominator in the right-hand side of Eq. 3.29 vanishes if $\lambda T = b$, where $b \approx 2897.77$ μm·K is the constant in the Wien displacement law (depicted by

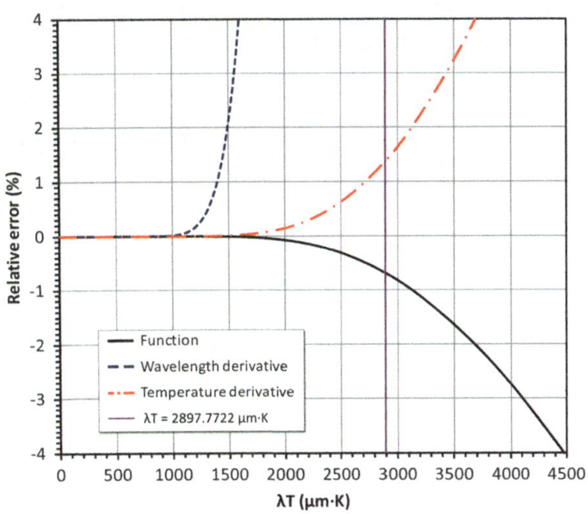

Fig. 3.6 Relative errors of Wien's approximation and its first derivatives with respect to wavelength and temperature relative to the Planck function (according to Eqs. 3.26, 3.29, and 3.30). The vertical line corresponds to $\lambda T = b$

the vertical line in Fig. 3.6. When approaching this value, the relative error of the derivative with respect to wavelength (3.29) increases sharply and already at $\lambda T \approx$ 1500 μm·K reaches 2%. Figure 3.6 show that we can use the Wien approximation when computing the derivative with respect to the temperature but not with respect to the wavelength.

3.2.4 *Sensitivity Coefficients for Planck's Equation*

The equation expressing Planck's law is the central equation of the blackbody radiometry. It is the base of the measurement equations for reproducing the spectral radiance and spectral irradiance, principal spectral quantities in the blackbody radiometry. Hereinafter, we will use the wavelength representation for the spectral radiance unless otherwise stated.

In order to calculate the spectral radiance of a perfect blackbody, we need to measure only two values: the wavelength λ and the thermodynamic temperature T. If we assume that λ and T are independent (i.e. their measurement uncertainties are uncorrelated) and the constants c_{1L} and c_2 are known exactly, than the uncertainty of $L_{\lambda,bb}$ is expressed as

$$u\left(L_{\lambda,bb}\right) = \sqrt{(c_\lambda u(\lambda))^2 + (c_T u(T))^2}, \tag{3.31}$$

where $c_\lambda = \left.\frac{\partial L_{\lambda,bb}}{\partial \lambda}\right|_{\substack{\lambda=\lambda^* \\ T=T^*}}$ and $c_T = \left.\frac{\partial L_{\lambda,bb}}{\partial T}\right|_{\substack{\lambda=\lambda^* \\ T=T^*}}$ are the sensitivity coefficients for λ and T, respectively, evaluated at the expectation values λ^* and T^* of the given wavelength and temperature. The partial derivatives in (3.31) are easily obtainable analytically:

$$\frac{\partial L_{\lambda,bb}}{\partial \lambda} = \frac{c_{1L}}{\lambda^7 T} \cdot \frac{(c_2 - 5\lambda T)\exp\left(\frac{c_2}{\lambda T}\right) + 5\lambda T}{\left[\exp\left(\frac{c_2}{\lambda T}\right) - 1\right]^2}, \tag{3.32}$$

$$\frac{\partial L_{\lambda,bb}}{\partial T} = \frac{c_{1L} c_2}{\lambda^6 T^2} \cdot \frac{\exp\left(\frac{c_2}{\lambda T}\right)}{\left[\exp\left(\frac{c_2}{\lambda T}\right) - 1\right]^2}. \tag{3.33}$$

It is convenient to transform Eq. 3.31 to a form with the relative uncertainties:

$$\frac{u\left(L_{\lambda,bb}\right)}{L_{\lambda,bb}} = \sqrt{\left(C_\lambda \frac{u(\lambda)}{\lambda}\right)^2 + \left(C_T \frac{u(T)}{T}\right)^2}, \tag{3.34}$$

where

$$C_\lambda = \left(\frac{\lambda}{L_{\lambda,bb}} \cdot \frac{\partial L_{\lambda,bb}}{\partial \lambda}\right)_{\substack{\lambda=\lambda^* \\ T=T^*}} \tag{3.35}$$

and

$$C_T = \left(\frac{T}{L_{\lambda,bb}} \cdot \frac{\partial L_{\lambda,bb}}{\partial T}\right)_{\substack{\lambda=\lambda^* \\ T=T^*}}. \tag{3.36}$$

It is not difficult to obtain analytical expressions for C_λ and C_T:

$$C_\lambda = \frac{\lambda}{L_{\lambda,bb}} \frac{\partial L_{\lambda,bb}}{\partial \lambda} = \frac{\exp\left(\frac{c_2}{\lambda T}\right)(c_2 - 5\lambda T) + 5\lambda T}{\lambda T\left[\exp\left(\frac{c_2}{\lambda T}\right) - 1\right]}. \tag{3.37}$$

$$C_T = \frac{T}{L_{\lambda,bb}} \frac{\partial L_{\lambda,bb}}{\partial T} = \frac{c_2 \exp\left(\frac{c_2}{\lambda T}\right)}{\lambda T\left[\exp\left(\frac{c_2}{\lambda T}\right) - 1\right]}. \tag{3.38}$$

As can be easily seen, expressions for C_λ and C_T are not only similar, they differ by a constant:

$$C_T - C_\lambda = 5. \tag{3.39}$$

The relative sensitivity coefficients C_λ and C_T as functions of the product λT are plotted in Fig. 3.7. They asymptotically approach 1 and -4, respectively, when λT increases. Both coefficients, C_λ and C_T, grow indefinitely if their arguments tend to zero.

The dependence of the uncertainty in the spectral radiance on the uncertainty in the temperature is dominant since the relative uncertainty of λ negligible as compared with the uncertainty of temperature T for most practical cases, so

$$\frac{u\left(L_{\lambda,bb}\right)}{L_{\lambda,bb}} = |C_T|\frac{u(T)}{T}. \tag{3.40}$$

Often, it is more convenient to rewrite Eq. 3.38 for the dimensionless variable $x = \lambda / \lambda_{\max}$, which can be easily done using the Wien displacement law (3.12):

$$C_T(x) = \frac{c_2}{bx} \exp\left(\frac{c_2}{bx}\right)\left[\exp\left(\frac{c_2}{bx}\right) - 1\right]^{-1}. \tag{3.41}$$

The function $C_T(x)$ is plotted in Fig. 3.8.

At the wavelength $\lambda_{\max}$ ($x = 1$), the relative uncertainty of the spectral radiance is exactly five times greater than that of temperature whatever temperature is. For

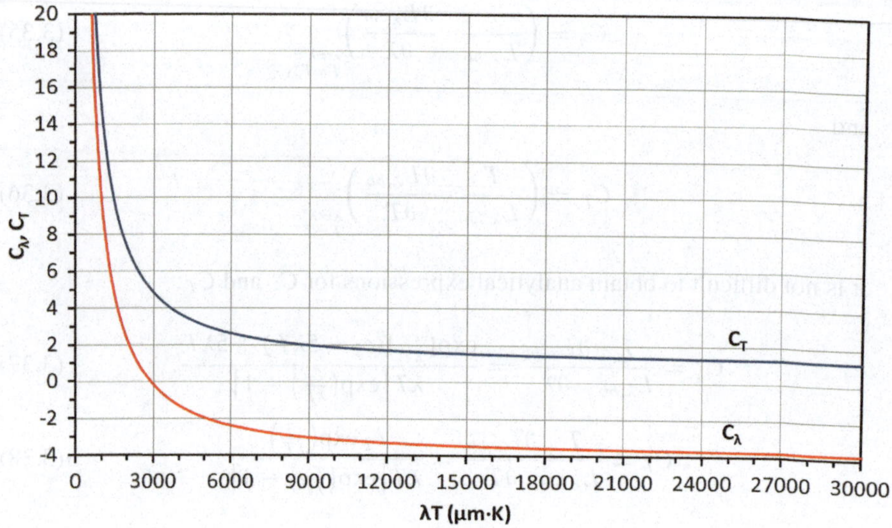

Fig. 3.7 Relative sensitivity coefficients C_λ and C_T

Fig. 3.8 Semi-log plot of the function $C_T(x)$ according to Eq. 3.41

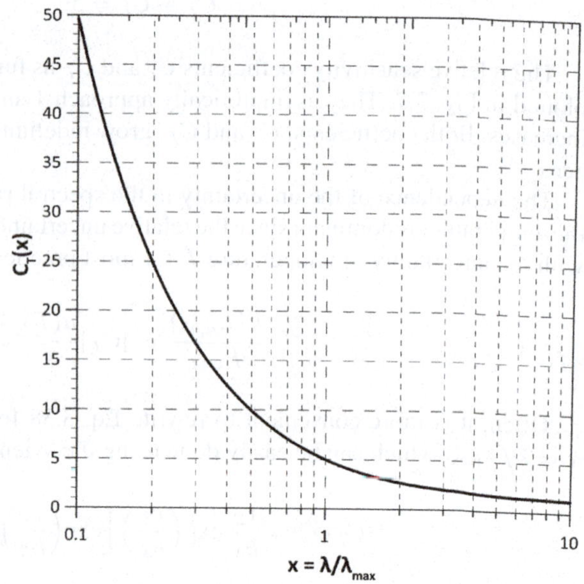

wavelengths greater than λ_{max}, $C_T(x)$ decreases and asymptotically approaches to unity when x increases. At $\lambda < \lambda_{max}$, $C_T(x)$ rises steeply; this reflects a steeper rise of the ascending branch of the Planckian curve. The choice of the operational spectral range should be made on the base of allowable ratio of the desired relative uncertainty in the spectral radiance determination and achievable uncertainty of the blackbody temperature measurement. For instance, a blackbody operating at the freezing temperature of silver (1234.93 K) has $\lambda_{max} \approx 2.35$ μm. For a monochromatic radiation thermometer operating at wavelengths of 0.65 μm ($x \approx 0.28$), 0.9 μm ($x \approx 0.38$), and 1.5 μm ($x \approx 0.64$), the relative uncertainty in determination of the spectral radiance is greater than the relative uncertainty in temperature measurements of about 18, 13, and 7.8 times, respectively. Equations 3.39 and 3.40 tightens significantly the requirements to accuracy of determining the temperature of high-temperature blackbodies used for establishing radiometric scales in the shortwave spectral range.

3.2.5 *Integration of the Planck Equation. The Stefan-Boltzmann Law*

To find the radiant exitance M_{bb} or the radiance L_{bb} of a perfect blackbody at the temperature T, we need to compute an improper integral of the corresponding spectral quantity over the full range of respective spectral variable:

$$M_{bb}(T) = \int_0^\infty M_{\lambda,bb}(\lambda, T)\,d\lambda = \int_0^\infty M_{\tilde{\nu},bb}(\lambda, T)\,d\tilde{\nu} = \int_0^\infty M_{\nu,bb}(\lambda, T)\,d\nu, \quad (3.42)$$

$$L_{bb}(T) = \int_0^\infty L_{\lambda,bb}(\lambda, T)\,d\lambda = \int_0^\infty L_{\tilde{\nu},bb}(\lambda, T)\,d\tilde{\nu} = \int_0^\infty L_{\nu,bb}(\lambda, T)\,d\nu. \quad (3.43)$$

Let us compute the radiant exitance of a perfect blackbody by integrating the spectral radiant exitance over all wavelengths λ from 0 to ∞ of:

$$M_{bb} = \int_0^\infty M_{\lambda,bb}(\lambda, T)\,d\lambda = c_1 \int_0^\infty \frac{d\lambda}{\lambda^5\left[\exp\left(\frac{c_2}{\lambda T}\right) - 1\right]}. \quad (3.44)$$

The latter integral can be rewritten after the change of variable $x = c_2 \big/ (\lambda T)$ as

$$M_{bb} = \frac{2\pi k^4}{h^3 c^2} T^4 \int_0^\infty \frac{x^3 dx}{\exp(x) - 1}. \quad (3.45)$$

The integral in the right-hand part of Eq. 3.45 cannot be expressed in terms of elementary functions but easily written using the gamma function $\Gamma(s)$ and Riemann's zeta function $\zeta(s)$. According to [5],

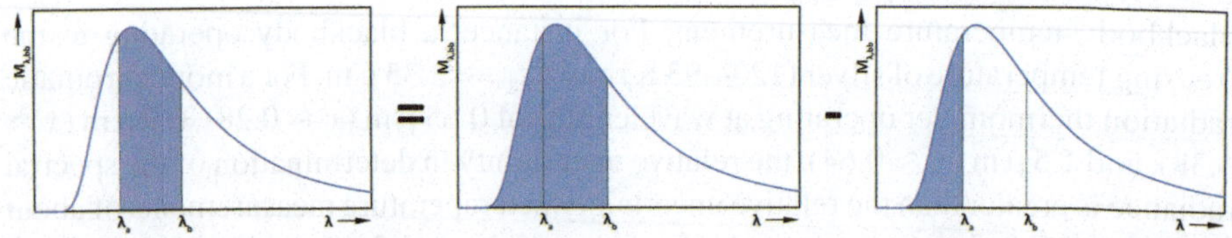

Fig. 3.9 A diagram illustrating calculation of the integral of Planck's function over a finite wavelength interval $[\lambda_a, \lambda_b]$

$$\zeta(s) = \frac{1}{\Gamma(s)} \int_0^\infty \frac{x^{s-1}dx}{e^x - 1}, \quad \text{Re}s > 1. \tag{3.46}$$

Using known particular values $\zeta(4) = \pi^4/90$ and $\Gamma(4) = 3! = 6$, we obtain the simple expression known as the Stefan-Boltzmann law:

$$M_{bb} = \sigma T^4, \tag{3.47}$$

where $\sigma = \frac{\pi^4 c_1}{15 c_2^4} \approx 5.67 \cdot 10^{-8}\ \text{W} \cdot \text{m}^{-2} \cdot \text{K}^{-4}$ is the Stefan-Boltzmann constant (its value can be found in Appendix A).

For various applications, it is often necessary to find the band-limited radiant exitance $M_{bb}(\lambda_a, \lambda_b, T)$ (or other band-limited quantity) of a perfect blackbody in a certain wavelength interval $[\lambda_a, \lambda_b]$ or the ratio $M_{bb}(\lambda_a, \lambda_b, T)/M_{bb}(T)$, i.e. fraction of the radiant exitance of a perfect blackbody at a temperature T falling into the wavelength interval $[\lambda_a, \lambda_b]$. This fraction can be expressed via the difference of two integrals with the finite limits as illustrated by Fig. 3.9:

$$\frac{M_{bb}(\lambda_a, \lambda_b, T)}{M_{bb}(T)} = \frac{1}{\sigma T^4} \int_{\lambda_a}^{\lambda_b} M_{\lambda,bb}(\lambda, T) d\lambda = \frac{1}{\sigma T^4} \left[\int_0^{\lambda_b} M_{\lambda,bb}(\lambda, T) d\lambda - \int_0^{\lambda_a} M_{\lambda,bb}(\lambda, T) d\lambda \right], \tag{3.48}$$

The integrals in the right-hand part of (3.48) can be calculated using adaptive numerical integration technique [65], however, currently, the most common is the method proposed in [18]. According to this method, the cumulative fractional blackbody function, $F(\lambda T)$, is presented in the form of rapidly converging series:

$$F(\lambda T) = \frac{M_{bb}(0, \lambda, T)}{M_{bb}(T)} = \frac{L_{bb}(0, \lambda, T)}{L_{bb}(T)} = \frac{15}{\pi^4} \sum_{n=1}^{\infty} \left[\frac{e^{-nx}}{n} \left(x^3 + \frac{3x^2}{n} + \frac{6x}{n^2} + \frac{6}{n^3} \right) \right], \tag{3.49}$$

where $x = \frac{c_2}{\lambda T}$. Alternatively, the cumulative fractional blackbody function can be presented with the wavelength ratio λ/λ_{max} as the argument, via the Wien displacement law. The graph of the cumulative fractional blackbody function F is presented in Fig. 3.10, the table of its values is given in Appendix B.

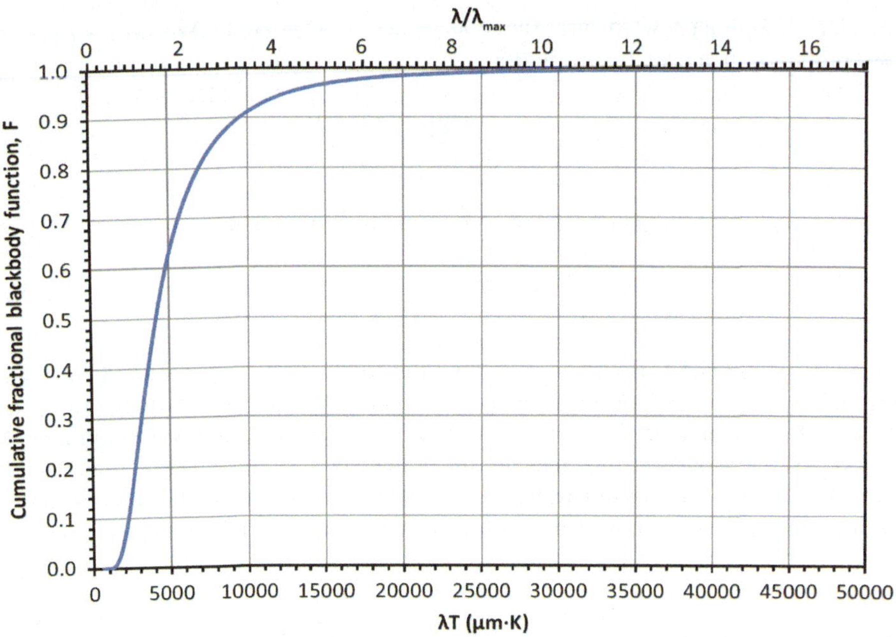

Fig. 3.10 Cumulative fractional blackbody function F of the arguments λT and λ/λ_{max}

The λ/λ_{max} representation allows to see that about 25% of the thermal radiation energy is emitted by a perfect blackbody at $\lambda < \lambda_{max}$; greater than 95% is emitted in the spectral range between 0.5 λ_{max} and 50 λ_{max}, and about 99.9% lies in the wavelength interval from 0.36 λ_{max} to 23 λ_{max}. Of particular interest is the question posed and successfully solved by Zhigang [89]: at what temperature T, for given wavelengths λ_a and λ_b, the maximum of the function

$M_{bb}(\lambda_a, \lambda_b, T)/M_{bb}(T)$ is reached? Mathematically, this problem is formulated as

$$\frac{\partial}{\partial T}\left[\frac{1}{\sigma T^4}\int\limits_{\lambda_a}^{\lambda_b} M_{\lambda,bb}(\lambda, T)d\lambda\right] = 0. \tag{3.50}$$

After some transformations, Zhigang [89] derived the transcendental equation

$$\int\limits_{\lambda_a}^{\lambda_b}\left\{1 - \frac{c_2\exp\left(\frac{c_2}{\lambda T}\right)}{4\lambda T\left[\exp\left(\frac{c_2}{\lambda T}\right) - 1\right]}\right\}\cdot\frac{d\lambda}{\lambda^5\left[\exp\left(\frac{c_2}{\lambda T}\right) - 1\right]} = 0, \tag{3.51}$$

which, after integration, will have the form of $f(\lambda_a, \lambda_b, T) = 0$. Its single root can be found numerically for any given λ_a and λ_b. For a symmetrical with respect to a central wavelength λ_0 a narrow spectral range $\lambda_0 - \Delta\lambda \leq \lambda \leq \lambda_0 + \Delta\lambda$, $\Delta\lambda/\lambda_0 << 1$, [89] proposes an approximate solution:

$$\lambda_0 T = \frac{c_2}{4} \cdot \frac{\exp\left[c_2/(\lambda_0 T)\right]}{\exp\left[c_2/(\lambda_0 T)\right] - 1}. \tag{3.52}$$

Equation 3.52 does not depend on $\Delta\lambda$; it can be rewritten as

$$x = 4\left(1 - e^{-x}\right), \tag{3.53}$$

where $x = c_2/(\lambda T)$. Equation 3.53 has non-trivial root $x \approx 3.920690394$ (see Fig. 3.2 again). The relationships 3.51 and 3.52 can be useful when the temperature of the blackbody should be chosen on the base of the given operating wavelength range and vice versa. However, this does not mean that such a choice can be made using the only criterion of maximum in-band spectral sensitivity of the radiometric system: out-of-band sensitivity (out-of-band leakage) should also be taken into account.

3.2.6 Blackbody Radiation in Refractive Media

In all the above equations, we assumed that the radiation of a perfect blackbody occurs in a vacuum, since all original classical works of the founding fathers of the theory of thermal radiation were based on the assumption that a perfect blackbody emits in a vacuum. However, in many practical cases we are forced to take into account the propagation of optical radiation in media other than vacuum (usually, gases such as air or argon). The case of blackbody radiation in a non-dissipative passive medium, whose refractive index $n > 1$, has been considered by Planck [57]. Since the frequency is the basic spectral variable, which remains unchanged as a ray passes through different media, he obtained the energy distribution over frequencies of electromagnetic waves. In the optical radiometry, wavelengths are used more often than frequencies, so we can simply rewrite Planck's formula in modern notation for the spectral radiant exitance $M_{\lambda,bb}$ and spectral radiance $L_{\lambda,bb}$ in a lossless medium with a spectral refractive index n and a wavelength λ in that medium as the spectral variable:

$$M_{\lambda,bb}(\lambda, T) = \frac{c_1}{n^2\lambda^5\left[\exp\left(\frac{c_2}{n\lambda T}\right) - 1\right]}, \tag{3.54}$$

$$L_{\lambda,bb}(\lambda, T) = \frac{c_{1L}}{n^2\lambda^5\left[\exp\left(\frac{c_2}{n\lambda T}\right) - 1\right]}. \tag{3.55}$$

Relative error introduced by neglecting the refractive index is

$$e = \frac{M_{\lambda,bb}^{(vacuum)} - M_{\lambda,bb}^{(\text{medium})}}{M_{\lambda,bb}^{(\text{medium})}} = \frac{L_{\lambda,bb}^{(vacuum)} - L_{\lambda,bb}^{(\text{medium})}}{L_{\lambda,bb}^{(medium)}} = n^2 \frac{\exp\left(\frac{c_2}{n\lambda T}\right) - 1}{\exp\left(\frac{c_2}{\lambda T}\right) - 1} - 1. \quad (3.56)$$

The graph of e plotted against λT and $\lambda / \lambda_{\max}$ in Fig. 3.11 gives an idea of the order of this error for the constant value of refractive index $n = 1.0003$. This error tends to $-\infty$ when $\lambda T \to 0$ and asymptotically approaches to $n - 1$ when $\lambda T \to \infty$.

The corrections for the refractive indices of gaseous media are small (about a few hundredths of a percent). Therefore, they are necessary only for radiometric measurements of the highest accuracy. The Stefan-Boltzmann law should be rewritten for a refractive medium as

$$L_{bb,m}(T) = n^2 L_{bb}(T) = n^2 \sigma T^4 / \pi, \quad (3.57)$$

$$M_{bb,m} = n^2 \sigma T^4. \quad (3.58)$$

Finally, the Wien displacement law takes the form:

$$\lambda_{\max} = b / (nT). \quad (3.59)$$

The spectral refractive index or air has been intensively studied by experimental methods. As a rule, experimental data are fitted to an empirical equation such as the Sellmeier equation [9]. Since the refractive index depends on temperature, pressure, humidity and air composition, measurements are usually made for so-called standard air, which is dry air at a temperature of 15 °C, a pressure of 101.325 kPa (1 atm) and

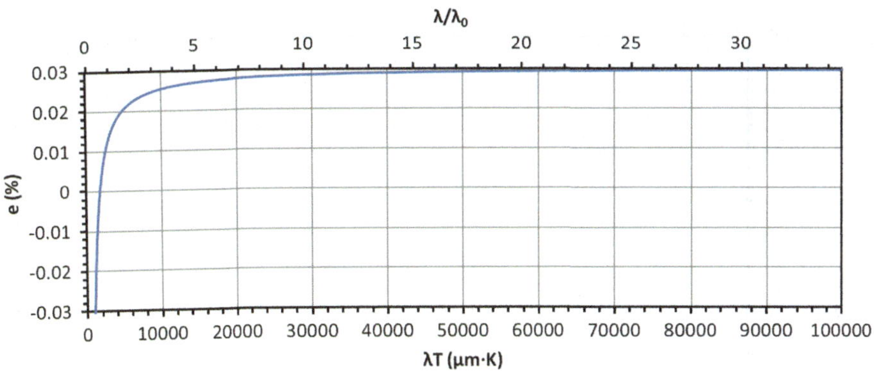

Fig. 3.11 Relative error due to neglecting the refractive index of air calculated according to Eq. 3.56 for $n = 1.0003$

at a CO_2 content of 450 ppm. One of the first equations for the refractive index of the standard air for the near UV to near IR spectral diapason was proposed by Edlén [23]:

$$n = 1 + 0.00008341 + \frac{0.02406030}{130 - \lambda^{-2}} + \frac{0.00015997}{38.9 - \lambda^{-2}}, \tag{3.60}$$

where λ is the vacuum wavelength in micrometers. Another dispersion formula for the standard air for the VIS and NIR spectral ranges was presented by Ciddor [19]:

$$n = 1 + \frac{0.05792105}{238.0185 - \lambda^{-2}} + \frac{0.00167917}{57.362 - \lambda^{-2}}. \tag{3.61}$$

The graph for $n(\lambda)$ for Eq. 3.61 is plotted in Fig. 3.12. It should be noted that the values of $n(\lambda)$ calculated via Edlén's and Ciddor's formulae differ in the spectral range of 0.25 to 1.6 μm by less than 10^{-6}, so both equations are in use until now.

To introduce correction for deviations of the temperature, pressure, and CO_2 content from the standard air, the fitting constants are assumed independent of the above parameters; their influence is taken into account using the appropriate correction factors. One of the first such corrections was proposed by Owens [55]. His formulae introduce multiplicative correction in the Edlén equation for a wide range of pressure, temperature, and composition.

There are online applications that allow computing the refractive index of air for one wavelength [81] or several wavelengths [60]. Alternatively, the refractive indices of air can be computed in the Matlab environment from MathWorks® using free m-files [50, 72]. In many cases, high accuracy in calculating the refractive index of air is not needed. If measurements of the refractive index of air were not carried

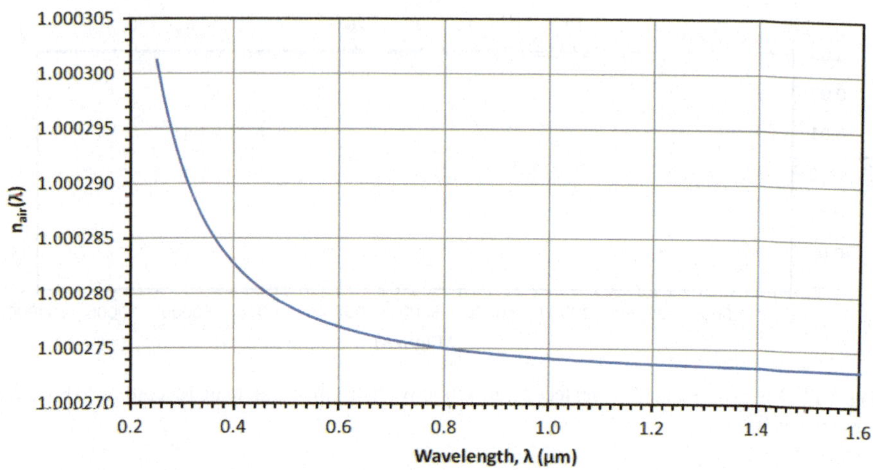

Fig. 3.12 Dispersion curve for the standard air calculated according to [19]

out specifically for the given measurement conditions, it is usually enough within the spectral range 0.25 μm ≤ λ ≤ 2.5 μm to set $n \equiv 1.00029$ with the expanded uncertainty at the coverage factor $k = 2$ of 0.00005 [88].

The atmospheric air is less transparent in the mid- and far-IR spectral ranges, so the high-accuracy low-temperature blackbodies operate, as a rule, inside evacuated chambers. Therefore, the longwave refractive indices of air are less demanded in the blackbody radiometry. They are affected mostly by the content of water vapor. Mathar [48] provided the fitting coefficients for the multivariate Taylor expansion for the refractive index of air in the spectral range from 1.3 μm to 24 μm as a function of temperature (from 10 to 25 °C), pressure (from 500 to 1023 hPa), and relative humidity (from 5 to 60%). These data are available in the tabular and graphical form at https://refractiveindex.info.

3.3 Definition of Radiative Properties of Real Bodies

3.3.1 Terminology Notes

Radiative properties describe three kind of radiation-matter interaction (namely, the absorption, reflection, transmission, and thermal emission) at the phenomenological level. The ILV [35] defines the absorptance, reflectance, and transmittance as the ratios of the absorbed radiant flux Φ_a, reflected radiant flux Φ_r, and transmitted radiant flux Φ_t, respectively, to the incident flux Φ_i under specified conditions:

$$\rho = \frac{\Phi_\mathrm{r}}{\Phi_\mathrm{i}}, \tag{3.62}$$

$$\alpha = \frac{\Phi_\mathrm{a}}{\Phi_\mathrm{i}}, \tag{3.63}$$

$$\tau = \frac{\Phi_\mathrm{t}}{\Phi_\mathrm{i}}. \tag{3.64}$$

Compliance with the energy conservation law is expressed by the following two equations:

$$\Phi_i = \Phi_r + \Phi_a + \Phi_t, \tag{3.65}$$

$$\rho + \alpha + \tau = 1. \tag{3.66}$$

These three properties (ρ, α, and τ) are dimensionless and may take values from 0 to 1. They are considered as traditional subjects of spectrophotometry [25], but play a significant role in radiative heat transfer, radiometry, and many other fields of science and engineering.

The most of components used in the blackbody-based radiometric systems are opaque bodies (exceptions are such optical components as lenses and filters), for which $\tau = 0$ and $\rho + \alpha = 1$. A perfect blackbody absorbs all incident radiation; therefore, for a perfectly black surface, $\alpha = 1$ and $\rho = \tau = 0$.

All bodies at a nonzero temperature emit thermal radiation. To characterize the thermal emissive property, the ILV [35] provides separately two definitions:

1. The hemispherical emissivity is the ratio of the radiant exitance $M(T)$ of a radiator to that of a Planckian radiator $M_{bb}(T)$ at the same temperature T:

$$\varepsilon_h = \frac{M(T)}{M_{bb}(T)}. \tag{3.67}$$

2. The directional emissivity of a thermal radiator, in a given direction, is the ratio of the radiance $L(\boldsymbol{\omega}, T)$ of the radiator in the given direction $\boldsymbol{\omega}$ to the radiance $L_{bb}(T)$ of a Planckian radiator at the same temperature T:

$$\varepsilon = \frac{L(\boldsymbol{\omega}, T)}{L_{bb}(T)}. \tag{3.68}$$

According to the Kichhoff law, the maximum possible energy at a given temperature is emitted by a perfect blackbody; therefore, both hemispherical and directional emissivities are also dimensionless and can take values from 0 to 1. For a perfect blackbody, $\varepsilon_h = \varepsilon = 1$.

These terms are not ubiquitously accepted; some of them have synonyms in adjacent areas of physics and engineering. Some authors use the ending "*ivity*" (absorptivity, reflectivity, transmissivity, and emissivity) for radiative characteristics of an idealized material (a pure, homogeneous material with an optically smooth surface or a layer thick enough to neglect the effect of its boundaries) and "*ance*" (absorptance, reflectance, transmittance, and emittance) for a real material with a rough and contaminated surface. Despite the fact that this discussion has been going on since the early 1960s, and the logic of introducing two terms for the same physical quantity now looks factitious, this problem is still not completely resolved. There are disagreements between two most authoritative regulations, the standard ANSI/IES RP-16–10 [4] and the International Lighting Vocabulary (ILV) [35]. They concertedly define only the terms "reflectance" and "reflectivity" in the very sense as mentioned above. In the same sense, the ILV [35] distinguishes "transmittance" and "transmissivity," "absorptance" and "absorptivity," but the ANSI standard uses the term "transmissivity" only in the collocation "atmospheric transmissivity" and does not use the term "absorptivity" at all. Finally, ANSI/IES RP-16–10 [4] alternately uses the terms "emittance" and "emissivity" without obvious logical grounds, while ILV [35] employs the term "emissivity" and does not employ "emittance."

In order to eliminate the lack of uniformity in terminology and avoid terminological confusion, we will use the terms "reflectance," "transmittance," "absorptance," and "emissivity" throughout the book, with modifiers specifying the spectral and directional domains for incoming and/or outgoing radiation beams, to which a particular property corresponds. The following modifiers are used in this book to refer radiative characteristics to the common spectral domains.

Spectral (monochromatic) radiative property is defined for a very narrow spectral band. For definiteness, we will operate with a wavelength λ as a spectral variable. Symbolically, the spectral radiative characteristics are denoted by $\rho(\lambda)$, $\alpha(\lambda)$, $\tau(\lambda)$, and $\varepsilon(\lambda)$. Unlike a radiometric quantity, in the name of which the modifier "spectral" means the spectral density (denoted by the corresponding spectral variable in the subscript) as well as spectral dependence (denoted by the argument), the modifier "spectral" in the name of a radiative property means only dependence on the corresponding spectral variable.

Total radiative property is defined for the entire spectrum, for $0 < \lambda < \infty$. The modifier "total" can be omitted, if this does not lead to ambiguity. All total radiative properties can be obtained from the spectral ones by weighted averaging over the entire wavelength range, from $\lambda = 0$ to $\lambda = \infty$. For the absorptance, reflectance, and transmittance, the spectral distribution of the incident radiation is used as the weighting function. For the emissivity, the weighting function is the Planckian distribution.

The band-limited radiative properties are obtained by the averaging over a finite spectral interval, say, $\lambda_a \leq \lambda \leq \lambda_b$. This modifier is not widely used, but may be convenient for some applications. The band-limited radiative properties are defined in the same manner as the total radiative properties.

An important idealization of the spectral dependence for radiative properties is the gray body approximation, according to which the radiative properties are assumed wavelength-independent. For a gray body, each spectral radiative property is equal to the homonymous total and band-limited values. The gray body approximation simplifies calculations of radiation heat exchange.

We will assume that all radiative properties can be attributed to a point of a surface, or the surface is supposed to be homogeneous. We will use two most common geometric descriptors for incoming and outgoing radiation beams: directional (and normal as a special case) and hemispherical.

3.3.2 Absorptance

By analogy with the formal definition of ILV [35], the spectral directional absorptance can be defined as the absorbed fraction $\Phi_{\lambda,a}$ of spectral radiant flux $\Phi_{\lambda,i}$ falling on a surface within an elementary solid angle $d\Omega_i$ around the direction $\boldsymbol{\omega}_i$ of incidence (see Fig. 3.13a):

$$\alpha(\lambda, T, \boldsymbol{\omega}_i) = \frac{\Phi_{\lambda,a}(\lambda, T, \boldsymbol{\omega}_i)}{\Phi_{\lambda,i}(\lambda, \boldsymbol{\omega}_i)}. \tag{3.69}$$

The spectral hemispherical absorptance $\alpha(\lambda, T)$ is the absorbed fraction $\Phi_{\lambda,a}$ of the spectral radiant flux Φ_λ incident from all directions in a hemispherical solid angle Fig. 3.15b. If the hemispherical distribution of the incident flux is given by the spectral radiance $L_{\lambda,i}(\lambda, \theta_i, \varphi_i)$ in the spherical coordinate system, the spectral hemispherical absorptance can be obtained from the spectral directional absorptance as follows:

$$\alpha(\lambda, T) = \frac{\int_{\phi_i=0}^{2\pi} \int_{\theta_i=0}^{\frac{\pi}{2}} \alpha(\lambda, T, \theta_i, \phi_i) L_{\lambda,i}(\lambda, \theta_i, \phi_i) \cos\theta_i \sin\theta_i d\theta_i d\phi_i}{\int_{\phi_i=0}^{2\pi} \int_{\theta_i=0}^{\frac{\phi}{2}} L_{\lambda,i}(\lambda, \theta_i, \phi_i) \cos\theta_i \sin\theta_i d\theta_i d\phi_i}. \tag{3.70}$$

The total directional absorptance $\alpha(T, \boldsymbol{\omega}_i)$ and the total hemispherical absorptance $\alpha(T)$ can be obtained by averaging over entire spectral domain:

$$\alpha(T, \boldsymbol{\omega}_i) = \frac{\int_0^\infty \alpha(\lambda, T, \boldsymbol{\omega}_i) \Phi_{\lambda,i}(\lambda, \boldsymbol{\omega}_i) d\lambda}{\int_0^\infty \Phi_{\lambda,i}(\lambda, \boldsymbol{\omega}_i) d\lambda} = \frac{\int_0^\infty \alpha(\lambda, T, \boldsymbol{\omega}_i) L_{\lambda,i}(\lambda, \boldsymbol{\omega}_i) d\lambda}{\int_0^\infty L_{\lambda,i}(\lambda, \boldsymbol{\omega}_i) d\lambda}, \tag{3.71}$$

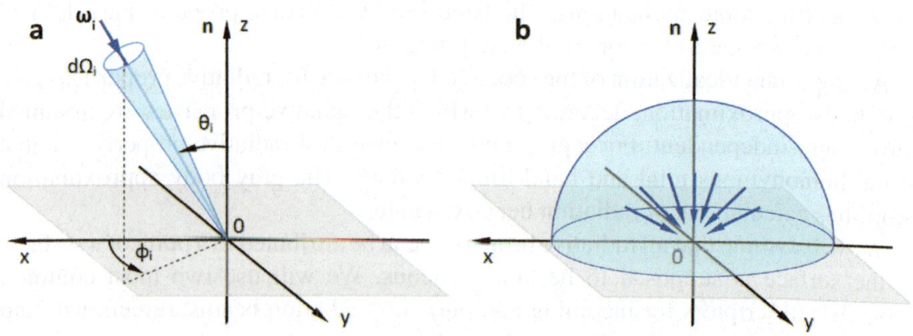

Fig. 3.13 Schematics for definitions of absorptances: **a** directional and **b** hemispherical

$$\alpha(T) = \frac{\int_0^\infty \alpha(\lambda, T)\Phi_{\lambda,i}(\lambda)d\lambda}{\int_0^\infty \Phi_{\lambda,i}(\lambda)d\lambda} = \frac{\int_0^\infty \alpha(\lambda, T)L_{\lambda,i}(\lambda)d\lambda}{\int_0^\infty L_{\lambda,i}(\lambda)d\lambda}. \tag{3.72}$$

3.3.3 Reflectance

The reflectance of a surface is easier (in a certain sense) to measure but more complicated to specify than the absorptance. The reflected radiant flux depends not only on the incidence direction but also on the direction, in which the reflected flux is observed. Although the reflectance depends of the temperature of the reflecting surface, we will omit this dependence for brief but it should be remembered that this dependence is implicitly present. The nomenclature for reflection properties was established by Nicodemus et al. [52]. Presently, it is used ubiquitously, at least in radiometry-related applications. To characterize the directional distribution of the reflected radiation for each wavelength, the most complete characteristic of a reflecting surface, the spectral bidirectional reflectance distribution function $f_r(\lambda, \boldsymbol{\omega}_i, \boldsymbol{\omega}_v)$, is used:

$$f_r(\lambda, \boldsymbol{\omega}_i, \boldsymbol{\omega}_v) = \frac{dL_{\lambda,r}(\lambda, \boldsymbol{\omega}_i, \boldsymbol{\omega}_v)}{dE_{\lambda,i}(\lambda, \boldsymbol{\omega}_i)} = \frac{dL_{\lambda,r}(\lambda, \boldsymbol{\omega}_i, \boldsymbol{\omega}_v)}{L_{\lambda,i}(\lambda, \boldsymbol{\omega}_i)\cos\theta_i d\Omega_i}, \tag{3.73}$$

where $L_{\lambda,i}$ and $L_{\lambda,r}$ are the spectral radiance of the incident and reflected radiation, respectively; $E_{\lambda,i}$ is the spectral irradiance produced by the incident radiation at the point of reflection; $d\Omega_i$ is the elementary solid angle around the direction of incidence $\boldsymbol{\omega}_i$; $\boldsymbol{\omega}_v$ is the viewing direction; θ_i is the incidence angle (see Fig. 3.14).

Unlike other radiative properties, the bidirectional reflectance distribution function (BRDF) has dimensionality of the reciprocal steradian (sr^{-1}). Such an unusual

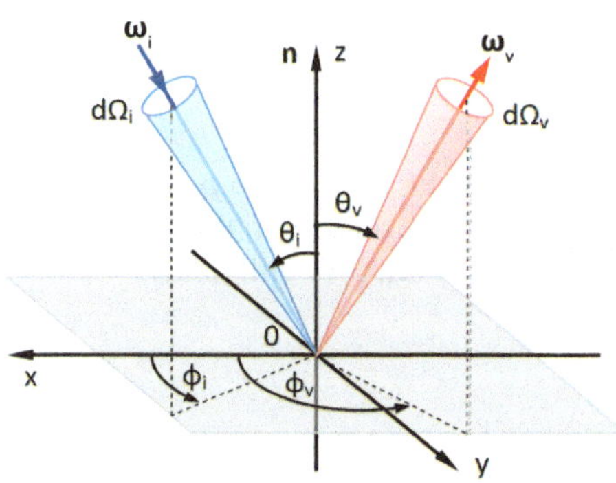

Fig. 3.14 To the definition of the bidirectional reflectance distribution function

definition of the BRDF enforces the reciprocity principle in the simplest form:

$$f_r(\lambda, \omega_i, \omega_v) = f_r(\lambda, -\omega_i, -\omega_v). \tag{3.74}$$

Using spherical coordinates instead of a vector form, we can rewrite the reciprocity principle for BRDF as

$$f_r(\lambda, \theta_i, \varphi_i, \theta_v, \varphi_v) = f_r(\lambda, \theta_v \varphi_v, \theta_i, \varphi_i). \tag{3.75}$$

As examples of simple BRDFs, we can write expressions of two reflection models that are widely used idealizations in calculation of radiative heat transfer. The Lambertian (perfectly diffuse) reflection is characterized by the following BRDF:

$$f_{r,d}(\lambda, \theta_i, \varphi_i, \theta_v, \varphi_v) = \rho_d(\lambda)/\pi, \tag{3.76}$$

where ρ_d is the diffuse reflectance that does not depend on the incidence and viewing directions.

According to [52], the BRDF expressing the perfectly specular (mirror-like) reflectance can be written in form

$$f_{r,sp}(\lambda, \theta_i, \varphi_i, \theta_v, \varphi_v) = 2\rho_{sp}(\lambda) \cdot \delta\big(\sin^2\theta_v - \sin^2\theta_i\big) \cdot \delta(\varphi_v - \varphi_i \mp \pi), \tag{3.77}$$

where ρ_{sp} is the specular reflectance,$\delta(\cdot)$ is the Dirac delta function [6], for which

$$\delta(x) = 0, \quad x \neq 0, \tag{3.78}$$

$$g(0) = \int_a^b g(x)\delta(x)dx, \tag{3.79}$$

where $g(x)$ is any well-behaved function and the integration includes $x = 0$. As a special case of (3.79) with $g(x) \equiv 1$, $a = -\infty$, and $b = \infty$:

$$\int_{-\infty}^{\infty} \delta(x)dx = 1. \tag{3.80}$$

The specular reflectance ρ_{sp} in (3.76) may depend on the incidence angle. Since for the specular reflection, the angle of incidence equals the angle of reflection, the reciprocity in Eq. 3.77 still holds.

The BRDF can be considered as the primary reflection property. All others can be expressed through it using the general scheme, in which the uniformity and isotropy of the incidence beam is assumed. Nicodemus et al. [52] introduced nine kinds of reflectance for nine combinations of the base configurations (directional, conical, and hemispherical) of the incidence and viewing beams as shown in Fig. 3.15. All types of reflectance with the directional incidence and/or viewing angular domains are only conceptual properties because it is impossible to realize irradiation or reflected radiation registration within infinitely small solid angle. Besides, if the reflected radiation is collected in an infinitesimal solid angle, the corresponding types of reflectance must be designated as differentials. The measurable quantities are shaded in gray in Fig. 3.15, the reflective properties with directional incidence and/or viewing conditions can be approximated by the corresponding conical conditions with narrow but finite conical solid angles. For our purposes, it is important to specify the relations between the BRDF and the directional-hemispherical reflectance (DHR):

$$\rho(\lambda, \theta_i, \varphi_i, 2\pi) = \rho(\lambda, \theta_i, \varphi_i) = \int_{2\pi} f_r(\lambda, \theta_i, \varphi_i, \theta_v, \varphi_v) d\Omega_v. \tag{3.81}$$

Integration in (3.81) is performed over the upper hemisphere. In the explicit form, this integral can be written as

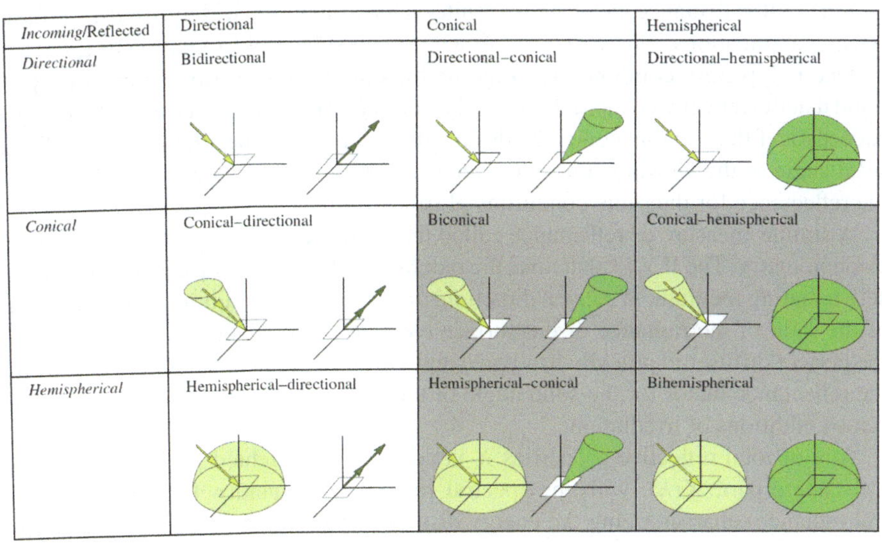

Fig. 3.15 Geometrical-optical concepts for the reflectance terminology. Grey cells correspond to measurable quantities; the others denote conceptual quantities. Reproduced from [71] with permission of Elsevier

$$\rho(\lambda, \theta_i, \varphi_i) = \int_{\varphi_v=0}^{2\pi} \int_{\theta_v=0}^{\frac{\pi}{2}} f_r(\lambda, \theta_i, \varphi_i, \theta_v, \varphi_v) \sin\theta_v \cos\theta_v d\theta_v d\varphi_v. \quad (3.82)$$

In addition to nine types of reflectance, the nine type of so-called reflectance factors were introduced in [52]. The ILV [35] defines the reflectance factor R (at a surface element, for the part of the reflected radiation contained in a given cone with apex at the surface element, and for incident radiation of given spectral composition, polarization and geometric distribution) as the ratio of the radiant flux reflected in the directions delimited by the given cone to that reflected in the same directions by a perfect reflecting diffuser identically irradiated. Therefore, the reflectance factor is the dimensionless property relied on an ideal (lossless) perfectly diffuse (Lambertian) surface. The reflectance factors can be easily expressed through the BRDF. For instance, the bidirectional reflectance factor (BRF) is equal

$$R(\lambda, \theta_i, \varphi_i, \theta_v, \varphi_v) = \pi f_r(\lambda, \theta_i, \varphi_i, \theta_v, \varphi_v). \quad (3.83)$$

For the most general case of biconical reflectance factor:

$$R(\lambda, \omega_i, \omega_v) = \frac{\pi}{\Omega_i \Omega_v} \int_{\Omega_i} \int_{\Omega_v} f_r(\lambda, \theta_i, \varphi_i, \theta_v, \varphi_v) d\Omega_i d\Omega_v. \quad (3.84)$$

Since all reflectance factors take finite values, they are widely used in remote sensing applications, where the directional or hemispherical irradiation of the Earth surface is typically combined with the unidirectional observation. It must keep in mind that the reflectance factor for specular surfaces irradiated by a narrow beam may exceeds 1, if the viewing cone includes the direction of the specular reflection. If the solid angle of the viewing cone approaches 2π sr, the reflectance factor approaches the reflectance for the same conditions of irradiation.

A similar measure of reflectance called the radiance factor β is used in the spectrophotometry. The ILV [35] defines the radiance factor (at a surface element of a non self-radiating medium, in a given direction, under specified conditions of irradiation) as the ratio of the radiance of the surface element in the given direction to that of the perfect diffuser identically irradiated and viewed. The radiance factor is equal to the reflectance factor for the solid angle of the viewing cone approaching 0 and the same conditions of irradiation.

Any of total reflective properties X is defined via weighted averaging its spectral counterpart, $X(\lambda)$, with the spectral radiometric quantity Y_λ (e.g., the spectral radiance) characterizing the energy distribution in the spectrum of the incident radiation:

$$X = \frac{\int_0^\infty X(\lambda) Y_\lambda(\lambda) d\lambda}{\int_0^\infty Y_\lambda(\lambda) d\lambda}. \quad (3.85)$$

In the same way we can define the band-limited properties for any wavelength band $\lambda_a \le \lambda \le \lambda_b$:

$$X(\lambda_a, \lambda_b) = \frac{\int_{\lambda_a}^{\lambda_b} X(\lambda) Y_\lambda(\lambda) d\lambda}{\int_{\lambda_a}^{\lambda_b} Y_\lambda(\lambda) d\lambda}. \tag{3.86}$$

The most widely used reflection characteristics are the BRDF and DHR (as a limiting case of biconical reflectance); we will use them throughout the book.

3.3.4 *Transmittance*

A description of the transmittance of a semitransparent material is as complex as of reflection. Fortunately, there is no need to repeat all of the above considerations for transmittance because they are almost identical to those for reflectance. The expressions for the reflectance ρ can be rewritten for the transmittance τ with replacing the upper hemisphere with the lower Fig. 3.16 and substitution of subscript "r" with "t". The spectral bidirectional transmittance distribution function (BTDF) is defined by analogy with the BRDF:

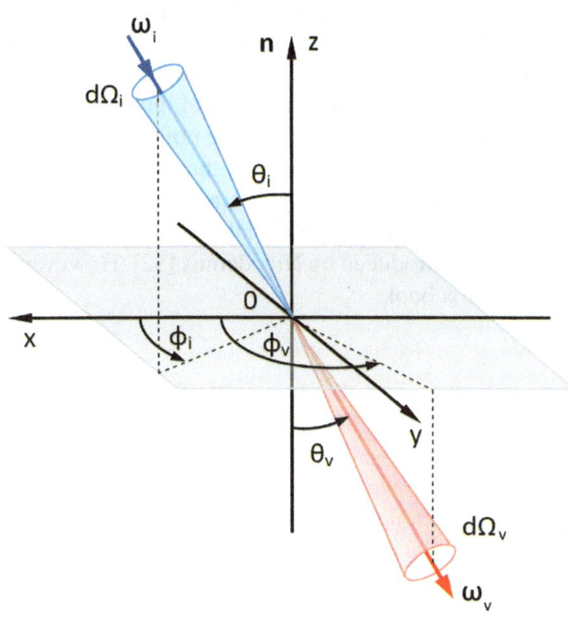

Fig. 3.16 A diagram for definition of the BTDF

$$f_t(\lambda, \omega_i, \omega_v) = \frac{dL_{\lambda,t}(\lambda, \omega_i, \omega_v)}{dE_{\lambda,i}(\lambda, \omega_i)} = \frac{dL_{\lambda,t}(\lambda, \omega_i, \omega_v)}{L_{\lambda,i}(\lambda, \omega_i)\cos\theta_i d\Omega_i}, \tag{3.87}$$

where $L_{\lambda,t}$ is the spectral radiance of the transmitted radiation. In the case of regular transmittance (without internal scattering), the BTDF is expressed by the delta function, similar to the specular reflectance for the BRDF. In the case of a perfectly diffuse transmission, the radiance of the transmitted radiation does not depend on the viewing direction, i.e. the sample is a Lambertian transmitter and its BTDF is

$$f_{t,d}(\lambda, \theta_i, \varphi_i, \theta_v, \varphi_v) = \tau_d(\lambda)/\pi, \tag{3.88}$$

where $\tau_d(\lambda)$ is the spectral diffuse transmittance.

The spectral directional-hemispherical transmittance is then defined as

$$\tau(\lambda, \theta_i, \varphi_i) = \int_{2\pi} f_t(\lambda, \theta_i, \varphi_i, \theta_v, \varphi_v)\cos\theta_i d\Omega_t \tag{3.89}$$

and is an analog of the spectral DHR for the reflection.

The spectral BTDF is the basic parameter for spectral and geometric description of the transmission properties of thin samples with negligible internal scattering such as thin scattering film, when the transmitted radiation emerges from a point not too far from the point of the beam incidence. The polarization aspects must be defined for complete specification of transmission. For a thick volumetrically scattering semi-transparent sample, when the point of incidence on the one side of the sample can be at a considerable distance from the point/region of the radiation output. In this case, it is more appropriate to use of the bidirectional scattering-surface transmittance distribution function (BSSTDF), a non-standardized quantity sometimes employed by analogy with the bidirectional scattering-surface reflectance distribution function (BSSRDF) introduced by Nicodemus [52]. However, these aspects go far beyond the scope of this book.

3.3.5 *Emissivity and Kirchhoff's Law*

The basic radiative property for thermal emission from an opaque surface is the spectral directional emissivity defined as the ratio of the spectral radiance of the specimen to that of the perfect blackbody at the same temperature T and wavelength λ:

$$\varepsilon(\lambda, T, \boldsymbol{\omega}_v) = \frac{L_\lambda(\lambda, T, \boldsymbol{\omega}_v)}{L_{\lambda,bb}(\lambda, T)}, \tag{3.90}$$

where L_λ and $L_{\lambda,bb}$ are the spectral radiance of the sample under study and a perfect blackbody, respectively, $\boldsymbol{\omega}_v$ is the vector specifying the viewing direction (see Fig. 3.17a). Since the radiation of a perfect blackbody obeys Lambert's law, its radiance and spectral radiance do not depend on direction. Usually, a Cartesian coordinate system with the associated spherical system is mounted in such a way to superpose the origin of coordinate system with the vertex of elementary viewing cone $d\Omega_v$ around the viewing direction $\boldsymbol{\omega}_v$, the axis z coincides with the normal **n** to the surface, and the axes y and z lie in the tangent plane to the surface xy.

The direction $\boldsymbol{\omega}_v$ is specified by the polar angle θ_v from the axis z to $\boldsymbol{\xi}_v$ and the azimuthal angle φ from the axis x with arbitrary chosen direction and the projection of $\boldsymbol{\omega}_v$ onto the plane xy.

In the spherical coordinate system, the spectral directional emissivity can be written as

$$\varepsilon(\lambda, T, \theta_v, \varphi_v) = \frac{L_\lambda(\lambda, T, \theta_v, \varphi_v)}{L_{\lambda,bb}(\lambda, T)}. \tag{3.91}$$

The spectral normal emissivity $\varepsilon_n(\lambda, T)$ is an important special case of the spectral directional emissivity. The spectral hemispherical emissivity (see Fig. 3.17b) can be obtained from the spectral directional emissivity by integration of the numerator and denominator in the right-hand part of Eq. 3.91:

$$\varepsilon_h(\lambda, T) = \varepsilon(\lambda, T) = \frac{M_\lambda}{M_{\lambda,bb}} = \frac{1}{\pi}\int_{\varphi_v=0}^{2\pi}\int_{\theta_v=0}^{\frac{\pi}{2}} \varepsilon(\lambda, T, \theta_v, \varphi_v)\cos\theta_v \sin\theta_v d\theta_v d\varphi_v. \tag{3.92}$$

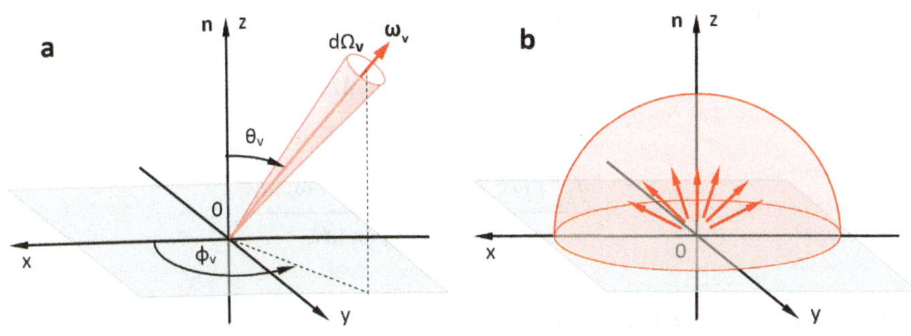

Fig. 3.17 Schematics for definitions of emissivities: **a** directional and **b** hemispherical

For an isotropic surface, i.e., a surface, for which the composition, structure, relief, etc. do not depend on azimuthal angle φ_v, Eq. 3.92 reduces to

$$\varepsilon_h(\lambda, T) = \varepsilon(\lambda, T) = 2\int_0^{\frac{\pi}{2}} \varepsilon(\lambda, T, \theta_v)\cos\theta_v \sin\theta_v d\theta_v. \quad (3.93)$$

Thermal emission of a perfectly diffuse surface obeys the Lambert law; for a Lambertian radiator, $\varepsilon(\lambda, T, \boldsymbol{\omega}_i) = \varepsilon_h(\lambda, T)$ for any $\boldsymbol{\omega}_v$. The total directional emissivity is defined as the ratio of spectral radiances:

$$\begin{aligned}\varepsilon(T, \boldsymbol{\omega}_v) &= \frac{L(T, \boldsymbol{\omega}_i)}{L_{bb}(T)} = \frac{1}{L_{bb}(T)}\int_0^{\infty} L_\lambda(\lambda, T, \boldsymbol{\omega}_i)d\lambda \\ &= \frac{\pi}{\sigma T^4}\int_0^{\infty} \varepsilon(\lambda, T, \boldsymbol{\omega}_i)L_{\lambda,bb}(\lambda, T)d\lambda \\ &= \frac{1}{\sigma T^4}\int_0^{\infty} \varepsilon(\lambda, T, \boldsymbol{\omega}_i)M_{\lambda,bb}(\lambda, T)d\lambda.\end{aligned} \quad (3.94)$$

The total hemispherical emissivity, defined as the ratio of exitances, can be expressed by averaging of the spectral emissivity with the spectral exitance of a perfect blackbody as the weighting function:

$$\varepsilon_h(T) = \varepsilon(T) = \frac{M(T)}{M_{bb}(T)} = \frac{1}{\sigma T^4}\int_0^{\infty} \varepsilon_h(\lambda, T)M_{\lambda,bb}(\lambda, T)d\lambda. \quad (3.95)$$

The band-limited directional and hemispherical emissivities are defined using the similar averaging over the finite wavelength range $[\lambda_a, \lambda_b]$:

$$\begin{aligned}\varepsilon(\lambda_a, \lambda_b, T, \boldsymbol{\omega}_i) &= \frac{\int_{\lambda_a}^{\lambda_b} \varepsilon(\lambda, T, \boldsymbol{\omega}_i)M_\lambda(\lambda, T, \boldsymbol{\omega}_i)d\lambda}{\int_{\lambda_a}^{\lambda_b} M_\lambda(\lambda, T, \boldsymbol{\omega}_i)d\lambda} \\ &= \frac{\int_{\lambda_a}^{\lambda_b} \varepsilon(\lambda, T, \boldsymbol{\omega}_i)L_\lambda(\lambda, T, \boldsymbol{\omega}_i)d\lambda}{\int_{\lambda_a}^{\lambda_b} L_\lambda(\lambda, T, \boldsymbol{\omega}_i)d\lambda},\end{aligned} \quad (3.96)$$

where either spectral exitance or spectral radiance can be used as weighting function.

For a gray surface, $\varepsilon(\lambda, T, \boldsymbol{\omega}_i) = \varepsilon(\lambda_a, \lambda_b, T, \boldsymbol{\omega}_i) = \varepsilon(T, \boldsymbol{\omega}_i)$ and $\varepsilon_h(\lambda, T) = \varepsilon_h(\lambda_a, \lambda_b, T) = \varepsilon_h(T)$ for any λ_a and λ_b. If, additionally, this surface is Lambertian, it can be characterized by a single constant value of emissivity ε, which simplify significantly engineering calculations.

When the absorptance and emissivity are defined strictly, we can return to Kirchhoff's law written in form of Eqs. 3.1 and rewrite it in the following simpler and more convenient form:

$$\alpha(\lambda, T, \boldsymbol{\omega}_i) = \varepsilon(\lambda, T, -\boldsymbol{\omega}_i), \tag{3.97}$$

This most general form of Kirchhoff's law holds for the bodies in thermodynamic equilibrium, for which the Kirchhoff law was derived. The thermodynamic equilibrium is observed strictly only in an isothermal enclosure, where the algebraic sum of incoming and outgoing heat fluxes is equal to zero. If heat transfer is present, Eq. 3.94 holds only approximately. Experimental evidences and detailed analysis indicate that, in most practical applications, the surrounding radiation field weakly affects the spectral directional absorptance and emissivity of solid and liquid bodies, which maintain themselves in a local thermodynamic equilibrium. This means that the intensive parameters in these bodies are varying in space and time so slowly that we can assume thermodynamic equilibrium in some volume around each point.

Since the thermal radiation of many real bodies is polarized (see, e.g. [12, 63, 68]), Eq. 3.94 is valid only for each component of polarization:

$$\alpha_s(\lambda, T, \boldsymbol{\omega}_i) = \varepsilon_s(\lambda, T, -\boldsymbol{\omega}_i), \tag{3.98}$$

$$\alpha_p(\lambda, T, \boldsymbol{\omega}_i) = \varepsilon_p(\lambda, T, -\boldsymbol{\omega}_i), \tag{3.99}$$

where the subscripts "*s*" and "*p*" denote components with the electric field oscillating in the plane perpendicular and parallel to the plane of incidence (emission), respectively, and

$$\alpha(\lambda, T, \boldsymbol{\omega}_i) = \tfrac{1}{2}\left[\alpha_p(\lambda, T, \boldsymbol{\omega}_i) + \alpha_s(\lambda, T, \boldsymbol{\omega}_i)\right], \tag{3.100}$$

$$\varepsilon(\lambda, T, \boldsymbol{\omega}_i) = \tfrac{1}{2}\left[\varepsilon_p(\lambda, T, \boldsymbol{\omega}_i) + \varepsilon_s(\lambda, T, \boldsymbol{\omega}_i)\right]. \tag{3.101}$$

Equation 3.97 is valid for all incident flux only if the incident radiation has equal components of polarization. The Kirchhoff law is often written out for the total directional absorptance and emissivity:

$$\alpha(T, \boldsymbol{\omega}_i) = \varepsilon(T, -\boldsymbol{\omega}_i). \tag{3.102}$$

However, this equation is accomplished only for the special case, when the spectral distribution of the incident radiation is proportional to that of a perfect blackbody, i.e.

$$L_{\lambda,i}(\lambda, \omega_i) = C(\omega_i)L_{\lambda,bb}(\lambda, T). \tag{3.103}$$

The most often case, when Eq. 3.103 fulfills, is the irradiation of a specimen by the thermal radiator, which can be considered as a gray body. Another important case, when Eq. 3.102 is valid, is the gray specimen. We can write the Kirchhoff law also for hemispherical spectral and total absorptances and emissivities, although special conditions must be met to fulfill these equalities:

$$\alpha_h(\lambda, T) = \varepsilon_h(\lambda, T), \tag{3.104}$$

$$\alpha_h(T) = \varepsilon_h(T). \tag{3.105}$$

The restrictions on the applicability of the Kirchhoff law for various modifications of the absorptance and emissivity are summarized in Table 3.2 adopted from [30], where these restrictions are discussed in detail.

Approximations of a perfectly diffuse (Lambertian) surface, whose radiative properties do not depend on direction, of a gray surface, whose radiative properties do not depend on a spectral variable, and of a diffuse gray surface, whose radiative properties are direction- and wavelength-independent allow expanding the area of applicability of Kirchhoff's law and simplifying significantly calculations of radiative heat exchange.

3.3.6 Radiometric Temperatures

Radiometric temperatures are the collective name of apparent (fictitious) temperatures determined by the radiation of real or virtual radiation sources. In general, a radiometric temperature of a real body does not coincide with its thermodynamic temperature. The measurement of temperature by the thermal radiation is the subject of radiation thermometry (pyrometry). The radiation thermometry has many intersections with the optical radiometry and, especially, with the blackbody radiometry. Many related measuring instruments, terms, definitions, concepts, and quantities are used in both disciplines. The radiance temperature is the most known and most widely used among these quantities. In some areas such as remote sensing of the Earth, the term "brightness temperature" is common instead of "radiance temperature" recommended by the ILV [35].

Let us consider the radiating body at the true (thermodynamic) temperature T, having the spectral emissivity $\varepsilon(\lambda, \theta_v)$. Its radiance temperature T_S at a wavelength λ is the temperature of the perfect blackbody, for which the spectral radiance

Table 3.2 The Kirchhoff law rules of applicability to absorptance and emissivity (after [30])

Absorptance and emissivity modifiers	Equation	Restrictions
Spectral directional	$\alpha(\lambda, T, \boldsymbol{\omega}_i) = \varepsilon(\lambda, T, -\boldsymbol{\omega}_i)$	None
Total directional	$\alpha(T, \boldsymbol{\omega}_i) = \varepsilon(T, -\boldsymbol{\omega}_i)$	• The spectral distribution of incident radiation is proportional to that of a perfect blackbody, or • Absorptance and emissivity are wavelength-independent (gray surface)
Spectral hemispherical	$\alpha_h(\lambda, T) = \varepsilon_h(\lambda, T)$	• Incident radiation is direction-independent, or • Absorptance and emissivity are direction-independent (diffuse surface)
Total hemispherical	$\alpha_h(T) = \varepsilon_h(T)$	• Incident radiation is direction-independent and have a spectral distribution proportional to that of a perfect blackbody, or • Incident radiation is direction-independent and absorptance and emissivity are wavelength-independent (gray surface), or • Incident radiation from each direction has spectral distribution proportional to that of a perfect blackbody and absorptance and emissivity are wavelength-independent (gray surface), or • Absorptance and emissivity are direction- and wavelength-independent (diffuse gray surface)

$L_{\lambda,bb}(\lambda, T_S)$ is equal to the spectral radiance of the thermal radiator considered. The radiance temperature $T_S(\lambda)$ can be inferred from the equation

$$\varepsilon(\lambda, \theta_v) L_{\lambda,bb}(\lambda, T) = L_{\lambda,bb}(\lambda, T_S). \tag{3.106}$$

After applying the Planck law, we obtain:

$$T_S(\lambda, \theta_v) = \frac{c_2}{\lambda \ln\left[1 + \frac{c_{1L}}{\lambda^5 \varepsilon(\lambda, \theta_v) L_{\lambda,bb}(\lambda, T)}\right]}. \tag{3.107}$$

As can be seen, the radiance temperature is the function of the wavelength and the viewing direction, as well as the thermodynamic temperature T and the emissivity ε of the radiating surface. Equation 3.107 becomes simpler if Wien's approximation can be used:

$$\frac{1}{T_S} = \frac{1}{T} - \frac{\lambda}{c_2} \ln \varepsilon(\lambda, \theta_v). \tag{3.108}$$

Hence, the difference between the true temperature T and the radiance temperature T_S is

$$\Delta T = T - T_S = -\frac{\lambda T_S}{c_2} T \cdot \ln \varepsilon(\lambda, \theta_v) \tag{3.109}$$

The relative uncertainty in the measured temperature due to the uncertainty in the emissivity can be evaluated by differentiating Eq. 3.108 with respect to ε at $T_S = Const$. After substituting the differentials with the finite differences, we obtain:

$$\frac{\Delta T}{T} = -\frac{\lambda T}{c_2} \frac{\Delta \varepsilon(\lambda, \theta_v)}{\varepsilon(\lambda, \theta_v)}, \tag{3.110}$$

where $\Delta\varepsilon$ is the approximation of the uncertainty $u(\varepsilon)$ in the determination of the uncertainty. By applying the Wien displacement law to Eq. 3.110 and using the approximate relationship $c_2/b \approx 5$, we can write

$$\frac{\Delta T}{T} \approx -\frac{\lambda}{5\lambda_{\max}} \cdot \frac{\Delta \varepsilon(\lambda, \theta_v)}{\varepsilon(\lambda, \theta_v)}. \tag{3.111}$$

Equations 3.107 and 3.108 are the basis of the narrow-band (spectral) radiation thermometry. For the total radiation thermometry, which retains its importance for cold targets such as cryogenic blackbodies, the total radiance temperature T_R (sometimes called radiation temperature) is defined via the Stefan-Boltzmann law:

$$T_R(\theta_v) = T \sqrt[4]{\varepsilon(\theta_v)}, \tag{3.112}$$

where $\varepsilon(\theta_v)$ is the directional total emissivity of the target. The concept of the total radiance temperature has limited practical usefulness because of impossibility of measuring the radiant power over the entire spectrum. For the wideband radiation thermometers, it is more useful to introduce the band-limited radiance temperature $\overline{T}_S$ as the temperature of a perfect blackbody having the same band-limited radiance over the prescribed spectral band $[\lambda_a, \lambda_b]$ as the surface under consideration:

$$\int_{\lambda_a}^{\lambda_b} \varepsilon(\lambda, \theta_v) L_{\lambda,bb}(\lambda, T) d\lambda = \int_{\lambda_a}^{\lambda_b} L_{\lambda,bb}(\lambda, \overline{T}_S) d\lambda. \tag{3.113}$$

This definition is not commonly accepted. In the remote sensing applications, where the measured quantity is the land or sea temperature, the so-called "spectral band brightness temperature" is often defined with respect to the spectral responsivity of a given sensor and/or the instrument slit function (see, e.g. [11]). Additionally, the spectral transmittance of the atmosphere along the line of sight can be added under the integral sign.

Equation 3.113 or similar cannot be resolved for the band-limited radiance temperature $\overline{T}_S$ in the closed form; however, there are standard methods that allow evaluating its value. The simplest algorithm implies introducing an effective wavelength, rewriting Eq. 3.113 in the form similar to (3.107) and inverting the Planck function for that wavelength. Alternatively, the iterative approach, despite its computational complexity, can be used to obtain $\overline{T}_S$ with sufficient accuracy. Finally, the empirical fit of $\overline{T}_S$ to the spectral channel-dependent band-limited radiance is typical at the processing of satellite multispectral sensor data [2]. A parameterization performed during pre-launch calibration provides sufficient accuracy and efficient computation needed for the real-time processing of a large volume of measurement information.

Incomplete information on emissivity (spectral and directional) is the main problem of radiation thermometry, to solve which, at least partially, many measurement methods were developed. Accordingly, for these purposes, specific types of radiometric temperature (such as the ratio temperature—see [82] were introduced. Fortunately, the radiation emitted by artificial blackbodies is very close to that of a perfect blackbody. This means not only that we deal with the radiation source with an extremely predictable angular and spectral distribution, but also with significantly reduced, due to the "cavity effect" (see Sect. 4.1.2), uncertainty in determining and spectral variations of the emissivity. For instance, while the typical values of the relative measurement uncertainty for the emissivity of flat samples are from 0.5 to 5%, the effective emissivities of precision blackbodies can be determined with the relative uncertainty of 0.05% or even less. This allows, for example, measuring the temperature of high-temperature blackbodies with an accuracy comparable to or higher than the accuracy achievable using contact sensors. Exceptionally simple relationship between spectral radiance and radiance temperature of a blackbody leads to similarity of blackbody-based measurement facilities designed for the spectral radiance and radiance temperature. However, by tradition, these facilities are traceable to different primary standards. Just recently, Jablonski et al. [36] outlined a possibility of traceable measurements of both radiance temperature and spectral radiance in the range of 0.3–14 μm using the same facility.

Radiance temperature is the most widespread but not the only fictitious temperature that can be derived from the optical radiation of the real bodies. The ILV [35] defines several kinds of such temperatures via the concept of a perfect blackbody (Planckian radiator), among which the color temperature T_c, the correlated color temperature T_{cp}, and the distribution temperature T_D. We define strictly only the latter one, which requires no auxiliary entities to be introduced.

The distribution temperature T_D of a source is the temperature of the Planckian radiator whose relative spectral distribution $S_{bb}(\lambda, T)$ is the same or nearly the same as $S(\lambda)$ of the radiation considered in the spectral range $[\lambda_a, \lambda_b]$ of interest, for which the following integral is minimized by adjustment of the scaling factor a and T:

$$\int_{\lambda_a}^{\lambda_b}\left[1-\frac{S(\lambda)}{aS_{bb}(\lambda,T)}\right]^2 d\lambda. \tag{3.114}$$

where $S_t(\lambda)$ is the relative spectral distribution of the radiation being considered, a is a, and

$$S_{bb}(\lambda,T)=\lambda^{-5}\left(e^{\frac{c_2}{\lambda T}}-1\right)^{-1} \tag{3.115}$$

is the relative spectral distribution of the Planckian radiator at temperature T. The distribution temperature is very convenient characteristics of radiation sources with continuous spectrum similar to the spectrum of a perfect blackbody, e.g. incandescent lamps (e.g. tungsten filament lamps). For a gray radiator, the distribution temperature coincides with the thermodynamic temperature since the functional (3.114) reach the absolute minimum, equal to zero, at $S(\lambda)=\frac{\varepsilon}{c_{1L}}L_{\lambda,bb}(\lambda,T)$. Computational aspects of the distribution temperature determination are discussed in [3, 24, and 66].

The color temperature T_c and the correlated color temperature T_{cp} are defined as the temperatures of a perfect blackbody that emits light of colors comparable (in certain sense) to that of the sources under characterization. The color and correlated color temperatures have important applications in photometry, illumination engineering, photodetection, astrophotometry, and other fields. Since their rigorous definitions require diving into the color science, we refer interested readers to the books on colorimetry (e.g. [34, 53], or [54]).

3.4 Radiation Heat Transfer in Nonparticipating Media

3.4.1 Preliminary Matters

Radiation heat transfer is one of three modes of the heat transfer, along with heat conduction and convection. Unlike the latter two, radiative energy is transmitted between the distant elements without requiring an intermediate medium. While heat conduction and convection depend linearly on the temperature difference, radiative heat transfer is inherently non-linear: according to the Stefan-Boltzmann law, the thermal radiation emitted by a heated body is proportional to the fourth power of the temperature.

The radiation heat transfer is the well-developed section of the thermal physics. Presently, it includes problems of radiative heat exchange among surfaces with arbitrary spectral and angular properties, radiation combined with conduction and

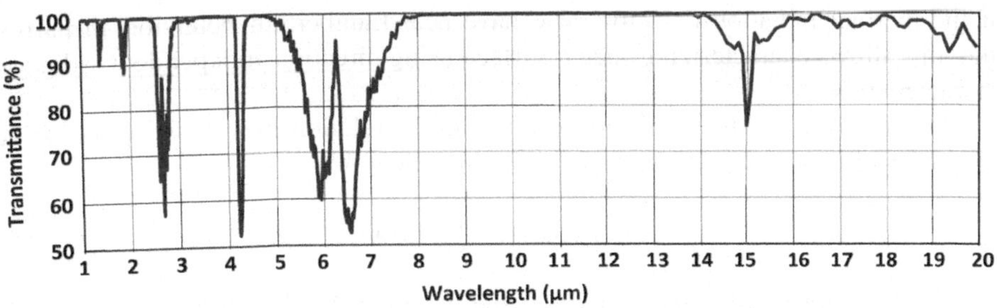

Fig. 3.18 Spectral transmittance of an air path 1 m long with 300 ppm CO_2 concentration, an air temperature of 25 °C and a relative air humidity of 50%. Reproduced from [28] with permission of Elsevier

convection, radiative heat transfer in participating (absorbing, scattering, and emitting) media, etc. However, for many engineering applications, not too strict requirements for calculation accuracy allow us to adopt some limitations and assumptions, significantly simplifying calculations of radiation heat transfer.

Such a simplified interpretation of the processes of radiative heat transfer is perfectly suited for describing the transfer of thermal radiation in blackbody-based radiometric systems.

For many radiometric measurements, carried out in laboratory conditions, we can neglect the influence of the medium separating the surfaces on the radiative heat transfer between them. The low-resolution transmittance spectrum for an air path of 1 m long is shown in Fig. 3.18. The temperature of air is 25 °C, relative humidity of 50%, and the concentration of CO_2 is 300 ppm. The spectral transmission of atmospheric air, which is a composition of nitrogen, oxygen, and other gaseous species, is most affected by water vapor and carbon dioxide. However, in laboratory conditions, the length of the path traveled by optical radiation in the air is relatively small (usually not more than 1 m),therefore, absorption corrections are usually not made; however, it is desired to choose the spectral ranges of radiometric measurements so that to avoid the molecular absorption bands that separate the "transparency windows" in Fig. 3.18. The typical measuring windows are 1.5 to 1.8 μm, 2 to 2.5 μm, 3 to 5 μm and 8 to 14 μm. The atmospheric air in laboratory conditions can be considered as a nonparticipating medium in many respects. We can neglect the own thermal radiation of air at room temperatures and consider the scattering negligible if the air is free of dust and aerosols. Although the true nonparticipating medium is vacuum, most monatomic and diatomic gases can be considered as such. The second assumption that we can confidently accept is the Lambertian behavior of emission and reflection. The Lambert cosine law holds with a high precision for the radiant intensity of the radiation of apertures of blackbody sources. The integrating spheres, often used to compare the radiation sources in VIS and NIR spectral ranges, have the walls made of materials that reflect in accordance with Lambert's law. Many absorptive ("black") paints and coatings used to suppress the stray radiation and as receiving surfaces of the thermal radiation detectors also have near-Lambertian behavior. In particular, this allows characterizing such a surface using direction-independent radiation characteristics.

This, in turn, leads to exceptionally simple relationship between radiative properties of opaque diffuse surfaces at any wavelength:

$$\varepsilon(\lambda) = \alpha(\lambda) = 1 - \rho(\lambda). \tag{3.116}$$

Therefore, for Lambertian surfaces (usually called simply "diffuse" in the radiation heat transfer), we can forget about division of emissivity, absorptance, and reflectance into directional, hemispherical, etc. This makes it possible to use for calculations of the thermal energy exchange in a system of diffuse surfaces, the formalism of diffuse view factors, purely geometric quantities, which depend only on the shapes of the surfaces included in the system and their mutual arrangement.

The third simplification, usually adopted in engineering calculations of radiative heat transfer, is a model of a gray surface, which means that the radiation properties of the surface are independent of the wavelength (or other spectral variables). This is not a principal limitation, but it allows us to forget about the difference between the spectral, total, and band-limited radiation properties. The diffuse gray surfaces are the basic and simplest components of a purely radiative system (i.e., a system, in which the only mode of the heat transfer is thermal radiation).

There are many excellent books on these and more complex problems of radiation heat transfer, among which we recommend the classic monographs [30, 49, 77].

3.4.2 *Diffuse View Factors*

The diffuse view factor (alternatively called configuration factor, angle factor, shape factor, or form factor) is an important concept for analyzing thermal radiation exchange among diffuse surfaces. A view factor is the fraction of diffuse radiation leaving one surface that directly reaches another surface. In this Section, we will assume that all surfaces considered are perfectly black. This will allow us to ignore the increase in the radiant exitance of the emitting surface due to radiant fluxes reflected by neighboring surfaces and returned back. We consider application of diffuse view factors to analyze the systems of reflecting surfaces in the next section.

Let us consider two arbitrarily oriented perfectly black surface elements dA_1 at a temperature T_1 and dA_2 at T_2 shown in Fig. 3.19.

Let $\mathbf{n}_1$ and $\mathbf{n}_2$ are the normals to those surface elements, the line of the length d_{12} connects these elements, and θ_1 and θ_2 are the angles formed by this line with the surface normals. The radiant flux leaving the element dA_1 and incident on dA_2 is

$$d^2\Phi_{dA_1-dA_2} = dI_1 \cos\theta_1 d\Omega_1 = L_1 dA_1 \cos\theta_1 d\Omega_1, \tag{3.117}$$

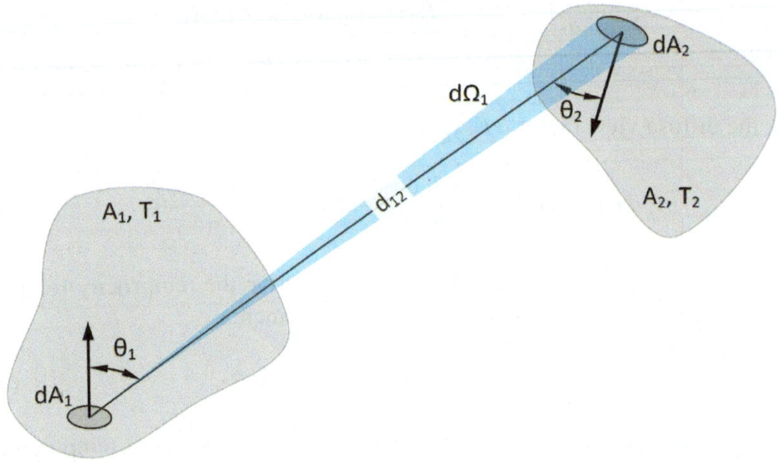

Fig. 3.19 Geometry for definitions of view factors

where dI_1 is the radiant intensity, L_1 is the radiance of the surface element dA_1, and $d\Omega_1$ is the solid angle subtended by dA_2, when viewed from dA_1. Combining the expression for the solid angle $d\Omega_1$

$$d\Omega_1 = \frac{dA_2 \cos\theta_2}{d_{12}^2} \tag{3.118}$$

with Eq. 3.117, we obtain

$$d^2\Phi_{dA_1-dA_2} = \frac{L_1 \cos\theta_1 \cos\theta_2 dA_1 dA_2}{d_{12}^2}. \tag{3.119}$$

For the Lambertian surface element dA_1, the total radiant flux leaving it within the hemispherical solid angle over dA_1 is equal to

$$d\Phi_1 = \pi L_1 dA_1. \tag{3.120}$$

The fraction of the radiant flux leaving dA_1 that arrives at element dA_2 is defined as the diffuse view factor $dF_{dA_1-dA_2}$:

$$dF_{dA_1-dA_2} = \frac{\cos\theta_1 \cos\theta_2}{\pi d_{12}^2} dA_2 = \frac{\cos\theta_1 d\Omega_1}{\pi}. \tag{3.121}$$

An analogous derivation for the radiation leaving a diffuse element dA_2 that reaches dA_1 results in

$$d^2\Phi_{dA_2-dA_1} = \frac{L_2 \cos\theta_1 \cos\theta_2 dA_1 dA_2}{d_{12}^2} \tag{3.122}$$

and for the diffuse view factor dF_{dA_2-dA1}

$$dF_{dA_2-dA_1} = \frac{\cos\theta_1 \cos\theta_2}{\pi d_{12}^2} dA_1 = \frac{\cos\theta_2 d\Omega_2}{\pi}. \tag{3.123}$$

After comparing (3.121) and (3.123), we can derive the reciprocity relationship for the diffuse view factors between two surface elements:

$$dF_{dA_1-dA_2} dA_1 = dF_{dA_2-dA_1} dA_2. \tag{3.124}$$

Consider two perfectly black isothermal surfaces A_1 and A_2 at temperatures T_1 and T_2, respectively. Let us consider also elementary areas dA_1 and dA_2 as shown in Fig. 3.19. The view factor between the element dA_1 and the finite area A_2 can be obtained by integration over A_2:

$$F_{dA_1-A_2} = \int_{A_2} \frac{\cos\theta_1 \cos\theta_2}{\pi d_{12}^2} dA_2 = \int_{A_2} dF_{dA_1-dA_2}. \tag{3.125}$$

By analogy, the view factor between the element dA_2 and the finite area A_1 can be obtained by integration over A_1:

$$F_{dA_2-A_1} = \int_{A_1} \frac{\cos\theta_1 \cos\theta_2}{\pi d_{12}^2} dA_1 = \int_{A_1} dF_{dA_2-dA_1}. \tag{3.126}$$

Further, we can introduce the view factors from the finite areas to the area elements:

$$dF_{A_1-dA_2} = \frac{dA_2}{A_1} \int_{A_1} \frac{\cos\theta_1 \cos\theta_2}{\pi d_{12}^2} dA_1, \tag{3.127}$$

$$dF_{A_2-dA_1} = \frac{dA_1}{A_2} \int_{A_2} \frac{\cos\theta_1 \cos\theta_2}{\pi d_{12}^2} dA_2, \tag{3.128}$$

The reciprocity relationships can be written for the view factors $F_{dA_1-A_2}$ and $dF_{A_2-dA_1}$, $F_{dA_2-A_1}$ and $dF_{A_1-dA_2}$:

$$F_{dA_1-A_2} dA_1 = dF_{A_2-dA_1} A_2, \tag{3.129}$$

$$F_{dA_2-A_1} dA_2 = dF_{A_1-dA_2} A_1. \tag{3.130}$$

Let us provide a practically important example of the use of view factor from area element to the finite-size surface. Figure 3.20 presents a planar element dA_1

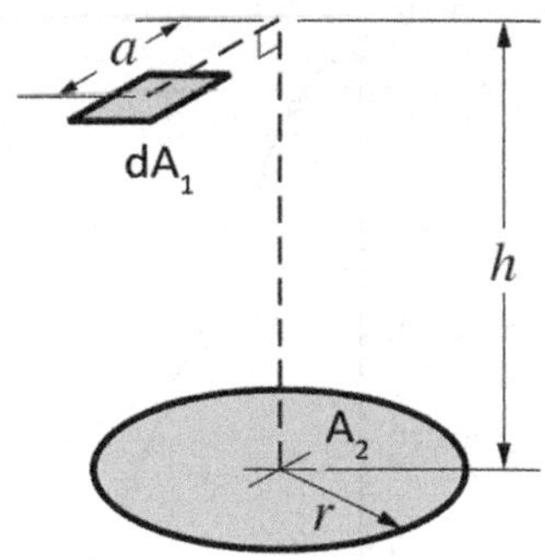

Fig. 3.20 Geometry for the view factor from differential planar element to circular disk lying in parallel planes

and a circular disk A_2, lying in parallel planes, the distance between which is h. The element is offset from the center of the disk by the distance a. According to many source (see, e.g. [30]),

$$F_{dA_1-A_2} = \frac{1}{2}\left[1 - \frac{X - 2R^2}{\sqrt{X^2 - 4R^2}}\right], \tag{3.131}$$

where $X = 1 + H^2 + R^2$, $H = h/a$, and $R = r/a$. Using (3.131), we can obtain the distribution of irradiance $E(a)$ created by an isothermal disk with uniform exitance M in the plane parallel to the disk:

$$E(a) = \frac{MA_2}{dA_1} dF_{A_2-dA_1} = MF_{dA_1-A_2}. \tag{3.132}$$

The relative distributions of the irradiance

$$E_{rel} = E/M = F_{dA_1-A_2} \tag{3.133}$$

are plotted in Fig. 3.21 against the relative distance from the disk axis a/r for three values of h/r. Another possible application of the view factor for the system depicted in Fig. 3.21 is the calculating the distribution of radiation heat loss across the bottom of cylindrical cavity through its aperture.

Next, we can introduce the view factors for radiation between two diffuse surfaces of finite size, A_1 and A_2:

$$F_{A_1-A_2} = \frac{1}{A_1}\int_{A_1}\int_{A_2} \frac{\cos\theta_1 \cos\theta_2}{\pi d_{12}^2} dA_1 dA_2, \tag{3.134}$$

$$F_{A_2-A_1} = \frac{1}{A_2}\int_{A_1}\int_{A_2} \frac{\cos\theta_1 \cos\theta_2}{\pi d_{12}^2} dA_1 dA_2. \tag{3.135}$$

The reciprocity relation for the view factors between finite areas is

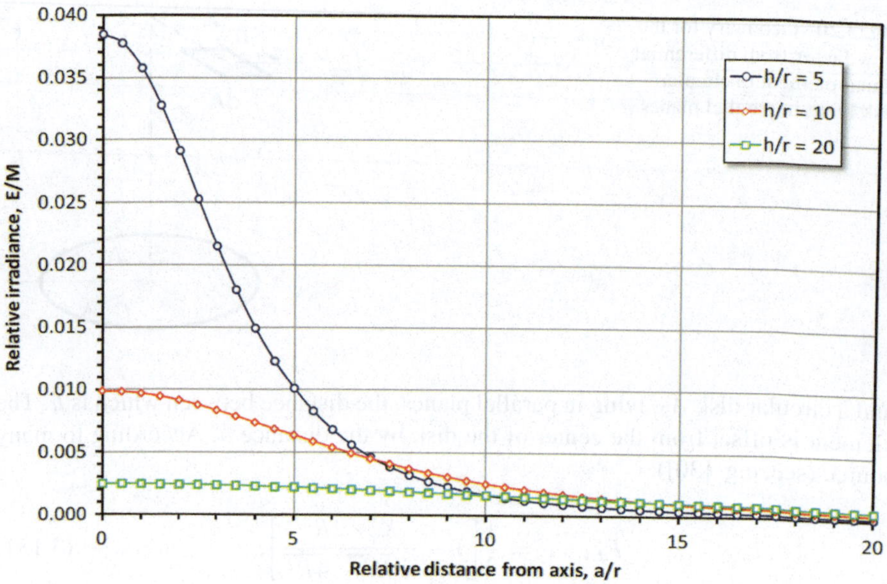

Fig. 3.21 Relative irradiance E/M produced by a disk of uniform exitance M in the plane parallel to that disk at a relative distance a/r from the disk axis for three values of h/r

$$A_1 F_{A_1-A_2} = A_2 F_{A_2-A_1}. \tag{3.136}$$

As an example of a view factor between two finite-area surfaces, we consider the system of two coaxial disks lying in parallel planes Fig. 3.22. This configuration is frequently employed in the blackbody radiometry when computed radiation transfer between the pair of precision apertures defining the geometry of the measured radiant flux of a blackbody.

The view factor from the first to the second disk is equal

$$F_{A_1-A_2} = \frac{1}{2}\left[X - \sqrt{X^2 - 4\left(\frac{R_2}{R_1}\right)^2}\right], \tag{3.137}$$

Fig. 3.22 Geometry for the view factor between two parallel coaxial disks

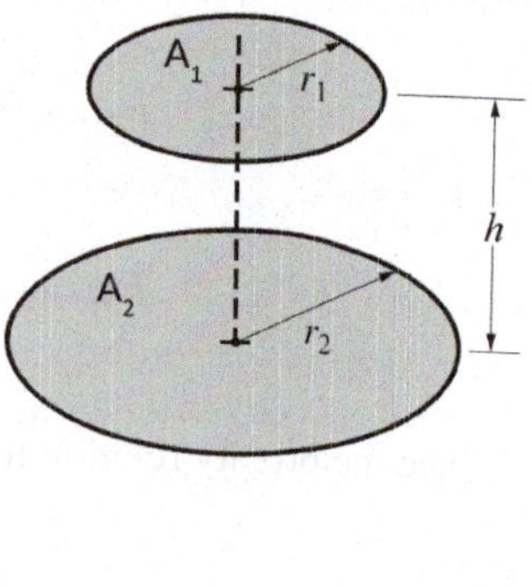

where $X = 1 + \frac{1+R_2^2}{R_1^2}$, $R_1 = r_1/h$, and $R_2 = r_2/h$. If M is the radiant exitance of the disk A_1 then the radiant flux transferred from A_1 to A_2 equals

$$\Phi_{A_1-A_2} = M F_{A_1-A_2} A_1. \tag{3.138}$$

We compare the fractions of the radiant flux transferred between two diffuse disks computed using Eq. 3.137 with the fraction $F'_{A_1-A_2}$ computed using the inverse square law:

$$F'_{A_1-A_2} = \frac{A_2}{h^2} \tag{3.139}$$

and with the help of more accurate approximation

$$F''_{A_1-A_2} = \frac{A_2}{r_1^2 + r_2^2 + h^2}. \tag{3.140}$$

Let us assume, for simplicity, that both disks are identical ($A_1 = A_2 = A$, $r_1 = r_2 = r$) to assess the errors e' and e'' introduced by the use of Eqs. 3.139 and 3.140 instead of Eq. 3.13:

$$e' = 100\% \cdot \left(1 - F'_{A_1-A_2} / F_{A_1-A_2}\right), \tag{3.141}$$

$$e'' = 100\% \cdot \left(1 - F''_{A_1-A_2} / F_{A_1-A_2}\right). \tag{3.142}$$

The fractions $F_{A_1-A_2}$, $F'_{A_1-A_2}$, and $F''_{A_1-A_2}$ of the radiant flux transferred between two identical diffuse disks are plotted in Fig. 3.23 against the relative distance h/r between disks. The corresponding relative errors of approximations, e' and e'' are presented in Fig. 3.24 also as functions of h/r.

In practice of radiometric measurements of thermodynamic temperature of blackbody sources (see details in Chap. 9), a geometric factor g defined as the product

$$g = A_1 F_{A_1-A_2} = \frac{2\pi r_1^2 r_2^2}{r_1^2 + r_2^2 + d^2 + \sqrt{\left(r_1^2 + r_2^2 + d^2\right)^2 - 4 r_1^2 r_2^2}} \tag{3.143}$$

is often used instead of a view factor [70].

Due to the reciprocity of view factors, the geometric factor is symmetrical with respect to the two apertures, so it does not matter which aperture is considered the first, and which second. The use of the geometric factor allows expressing the radiant flux (or its spectral density) at the output of the aperture pair via the radiance (or spectral radiance) of the blackbody as

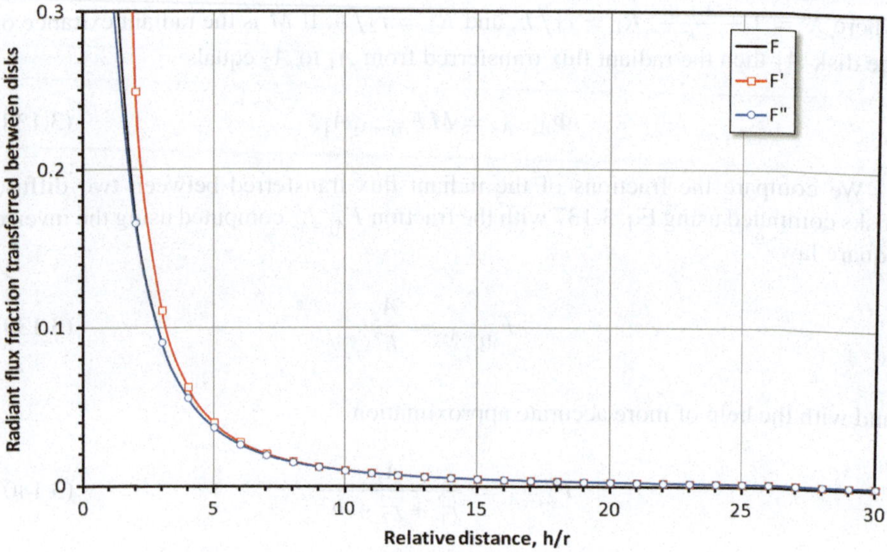

Fig. 3.23 Fractions of the radiant flux transferred between two coaxial parallel diffuse disks of identical radii r plotted against the relative distance h/r between disks

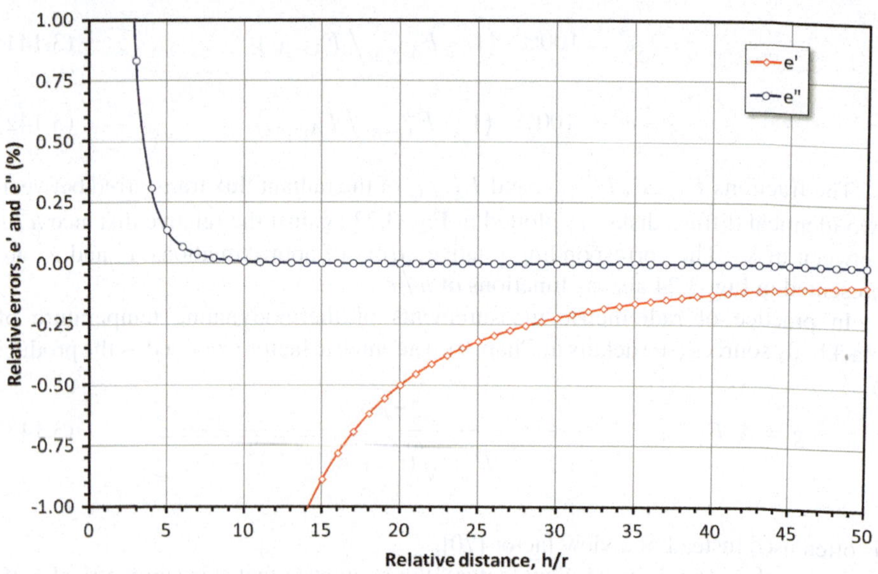

Fig. 3.24 Relative errors of approximations, e' and e'' (Eqs. 3.141 and 3.142) as functions of the relative distance h/r between two disks of the same radii r

$$\Phi = g\pi L_{bb}. \tag{3.144}$$

The view factors allow calculation of radiation heat transfer in various systems by referring to formulae, graphs, and tabulated values for the typical geometric configuration between two surfaces. The collections of formulae for diffuse view factors are provided in Appendices to monographs [32, 49, 77]. An extensive online catalogue of view (configuration) factors is maintained by the University of Texas at Austin [31].

There are many techniques that allow avoiding calculations of integrals over a surface in (3.127) and (3.128) or double surface integrals in accordance with (3.135) and (3.136) For instance, the replacement of area integrals by contour integrals (see, e.g. [16, 74]) simplifies the most time-consuming part of the analysis.

Various numerical methods for calculating view factors were designed at early stages of computer graphics development [20], when the light transfer among Lambertian surfaces were considered its most challenging subject. The so-called view factor algebra provides an opportunity of various manipulations among already known (catalogued) view factors to derive new ones. It is rather a set of practical rules than a strictly formalized system. Modest [49] defines view factor algebra as "*repeated application of the rules of reciprocity and the summation relationship.*"

We can formulate the additivity property (or summation relationship) for the view factors as follows. Situation depicted in Fig. 3.25a for the surface A_2 that is the union of subsurfaces A_{21} and A_{22} ($A_2 = A_{21} \cup A_{22}$ entails $F_{A_1-A_2} = F_{A_1-A_{21}} + F_{A_1-A_{22}}$) can be extended to the union of an arbitrary number of subsurfaces (see Fig. 3.25b):

$$\sum_{j=1}^{n} F_{A_i-A_j} = 1, \quad i = 1,\ 2,\ \ldots,\ n, \tag{3.145}$$

where $F_{A_j-A_j} = 0$, if A_j is a flat or convex surface. Equation 3.145 expresses the energy conservation for view factors in a closed system of diffuse surfaces. The additivity property allows not only summarizing view factors but also subtracting

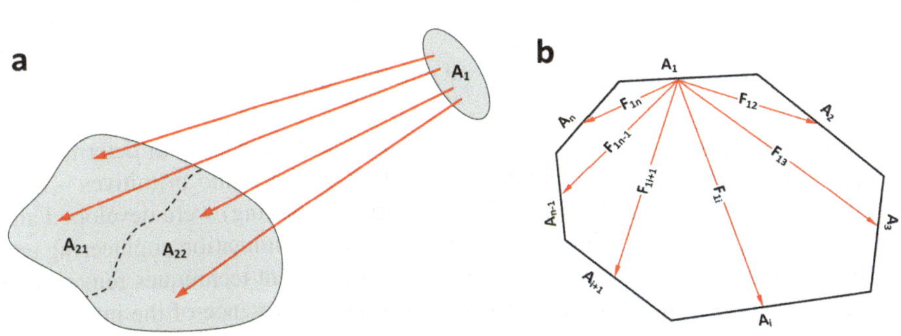

Fig. 3.25 Illustration to the summarizing of view factors for: **a** two surfaces and **b** many surfaces integrated into an enclosure

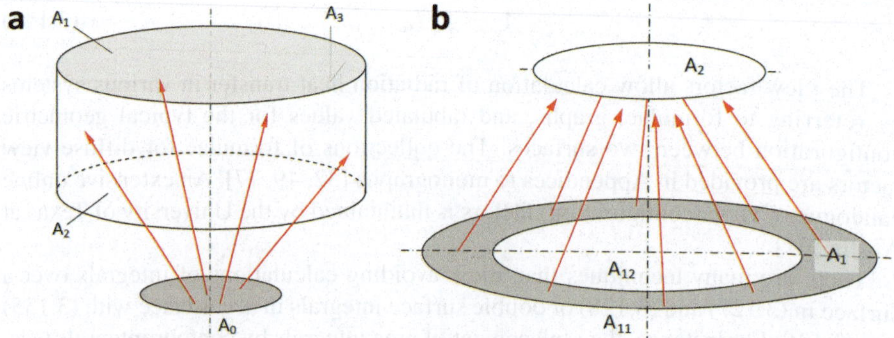

Fig. 3.26 Illustrations to the view factors subtraction

them. In Fig. 3.26a, the inner side of the cylindrical surface irradiates the disk lying in the plane perpendicular to their common axis.

After introducing two virtual disks, A_1 and A_2, closing the cylindrical surface from both ends, we can write

$$F_{A_0-A_3} = F_{A_0-A_2} - F_{A_0-A_2}. \tag{3.146}$$

In Fig. 3.26b, the concentric ring A_1 irradiates the coaxial disk A_2 lying in the parallel plane. The ring A_1 can be represented in terms of the set theory as the difference of a larger disk A_{11} and a smaller disk A_{12}, i.e. as the locus of points that belongs to A_{11} but does not belong to A_{12}.

The view factor from the disk A_2 to the ring A_1 is equal

$$F_{A_2-A_1} = F_{A_2-A_{11}} - F_{A_2-A_{12}}. \tag{3.147}$$

After applying the reciprocity relation (3.136), we obtain

$$F_{A_1-A_2} = \frac{A_2}{A_1} F_{A_2-A_{11}} - F_{A_2-A_{12}}. \tag{3.148}$$

More examples of the practical use of the view factor algebra can be found in [30, 49, 77].

For calculation of view factors, various numerical methods, both deterministic (adaptive integration, contour integration, splitting surfaces into primitives – rectangular or triangular) and stochastic (Monte Carlo ray tracing) were developed and realized in specialized software for thermal modeling, illumination engineering and lightning, image synthesis, etc. A combination of different techniques removes all restrictions on the complexity of the system geometry, presence of the intervening surfaces that partially obscure one surface from another, etc., but allows to obtain only the numerical values (instead of analytical expressions) for view factors, which are difficult to analyze. Therefore, an analytical approach for determining the view factors retains its importance for the study of not too complex systems, for which, in addition, the diffuse approximation cannot introduce significant errors.

3.4.3 Radiation Exchange in Diffuse Gray Enclosures

Let us consider a closed system, the cross section of which is shown in Fig. 3.27, consisting of n finite-size surfaces.

We will assume that:

1. All surfaces are opaque, spatially uniform, and gray (i.e. each surface can be are characterized by a wavelength-independent emissivity ε_i or reflectance $\rho_i = 1 - \varepsilon_i$). It should be noted that the assumption that the radiation properties are independent of the wavelength is not mandatory, but rather adopted to simplify the analysis.
2. All surfaces are diffuse emitters, absorbers, and reflectors, i.e. Lambert's cosine law is obeyed for emission, absorption, and reflection. This assumption is central to the method we are describing; it will not work if surfaces are not diffuse.
3. The medium inside the enclosure is radiatively non-participating (non-emitting, non-scattering, and non-absorbing).
4. All surfaces are isothermal (each at temperature T_i).
5. The incident, emitted, and reflected radiant fluxes are distributed uniformly over each surface. The analysis of this assumption will be given later.

The final objective of the problem we are considering is the calculation of spatial, angular, and spectral distributions of radiant fluxes. Since we adopted the diffuse gray approximation, we are not required to consider spectral radiance; it is sufficient to find the surface density of corresponding radiant fluxes for each surface. The traditional approach known since 1930s, presented in an almost modern form in [73], and has not undergone significant changes to date, is to determine the radiosity B_i defined as the effective radiant exitance (i.e. the sum of the own thermal and reflected radiation)

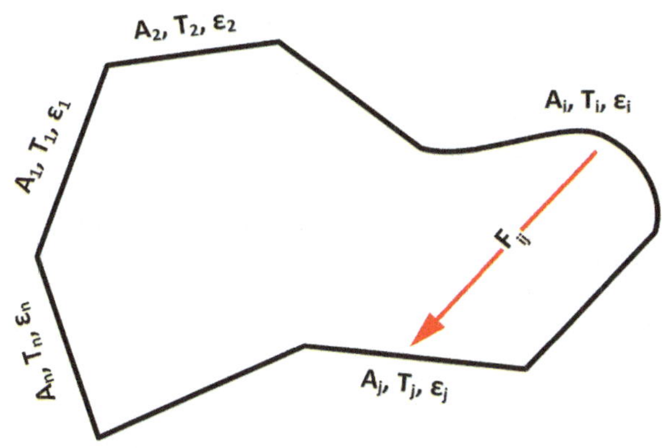

Fig. 3.27 A cross section of a hollow enclosure bounded by isothermal diffuse surfaces

of each surface A_i:

$$B_i = \varepsilon_i \sigma T_i^4 + \rho_i E_i, \tag{3.149}$$

where σ is the Stefan-Boltzmann constant and E_i is the irradiance of the surface A_i by the radiation coming from all other surface of the system. In turn, we can write for the irradiance E_i:

$$E_i = \sum_{j=1}^{n} B_j F_{ij}, \tag{3.150}$$

where F_{ij} is the view factor from the surface A_i to the surface A_j. Combining (3.149) and (3.150), we obtain the system of n linear equations for n unknown radiosities:

$$B_i = \varepsilon_i \sigma T_i^4 + (1 - \varepsilon_i) \sum_{j=1}^{n} B_j F_{ij}, \quad i = 1,\ 2,\ \ldots,\ n. \tag{3.151}$$

If all view factors F_{ij} are known, the unknown B_i can be easily found by the standard methods for solution of linear systems or matrix inversion.

Let $q_i = B_i - E_i$ be the heat flux density supplied to the surface A_i by means other than internal radiation from inside the enclosure (e.g. by conduction or convection). Alternatively, we can consider q_i as the net radiative loss (i.e. the difference between the outgoing and incoming radiant fluxes) from A_i by radiation exchange inside the enclosure. We can write the energy balance equations in terms of temperature:

$$q_i = \frac{\varepsilon_i}{1 - \varepsilon_i} \left(\sigma T_i^4 - B_i \right) \tag{3.152}$$

or radiosity:

$$q_i = B_i - \sum_{j=1}^{n} B_j F_{ij}. \tag{3.153}$$

If the system considered is not closed, it can be closed artificially by replacing openings with cold (at 0 K) perfectly black surfaces. Such virtual, non-emitting, and non-reflecting surfaces allow to model external radiation entering the enclosure. Typical examples of such a situation is irradiation of receiving cavity of a thermal radiation detector or effect of background radiation on the radiative characteristics of blackbody cavities. For similar situations, Eq. 3.151 should be rewritten in the form:

$$B_i = \varepsilon_i \sigma T_i^4 + (1 - \varepsilon_i)\left[E_{0i} + \sum_{j=1}^{n} B_j F_{ij}\right], \quad i = 1, 2, \ldots, n, \tag{3.154}$$

where E_{0i} is the irradiance of the surface A_i by radiation coming from outside the enclosure. The summands corresponding to the virtual black surfaces disappear from the sum in Eq. 3.154, since for these surfaces $\varepsilon_i = 1$ and $\rho_i = 0$.

The approach described above is often called the radiosity method or the net radiation method for diffuse enclosures.

The assumption of uniform distribution of emitted radiation over each surface of an enclosure is easily fulfilled for isothermal surface of uniform emissivity. The use of view factors between pairs of finite surfaces contains the implicit assumption that the radiant flux leaving a surface is uniformly distributed over that surface. However, the radiosityis composed not only of emitted but also of the reflected flux. As a rule, the incident radiant flux is distributed unevenly. Therefore, the assumption of uniform outgoing radiant flux is generally not satisfied. The only way to reduce the deviation of the approximation used above from the actual heat transfer conditions is to divide the surfaces that form the enclosure into smaller subsurfaces. In the limit, we obtain the system of integral equations for the radiosities B_i at the point indicated by the vector $\boldsymbol{\xi}_i$ on the surface A_i:

$$B_i(\boldsymbol{\xi}_i) = \varepsilon_i \sigma T_i^4 + (1 - \varepsilon_i)\sum_{j=1}^{n} \int_{A_j} B_j(\boldsymbol{\xi}_j) d^2 F(\boldsymbol{\xi}_i, \boldsymbol{\xi}_j)$$

$$= \varepsilon_i \sigma T_i^4 + (1 - \varepsilon_i)\sum_{j=1}^{n} \int_{A_j} B_j(\boldsymbol{\xi}_j) K(\boldsymbol{\xi}_i, \boldsymbol{\xi}_j) dA_j, \quad i = 1, 2, \ldots, n, \tag{3.155}$$

where $d^2 F(\boldsymbol{\xi}_i, \boldsymbol{\xi}_j)$ is the differential view factor between the area elements $dA_i(\boldsymbol{\xi}_i)$ and $dA_j(\boldsymbol{\xi}_j)$,

$K(\boldsymbol{\xi}_i, \boldsymbol{\xi}_j) dA_\xi = d^2 F(\boldsymbol{\xi}_i, \boldsymbol{\xi}_j)$ is the kernel of the Fredholm linear integral equation of the second kind. In general, Eq. 3.155 requires numerical solution, for which a number of numerical methods has been developed [7]. In particular, these methods play an important role in calculation of radiation characteristics of diffuse blackbody cavities, they were reviewed in [10] and [61].

A spherical cavity Fig. 3.28 is the only 3D enclosure, which allows the closed-form solution for the radiosity of internal walls. It is easy to see that for any diffusely emitting element of the inner surface of a sphere, the spherical surface is the surface of uniform irradiance. Therefore, the view factor from an arbitrary area on the spherical surface to another arbitrary area on the same surface is equal to the ratio of the second area to the entire area of the sphere and do not depend on the mutual arrangement of these areas:

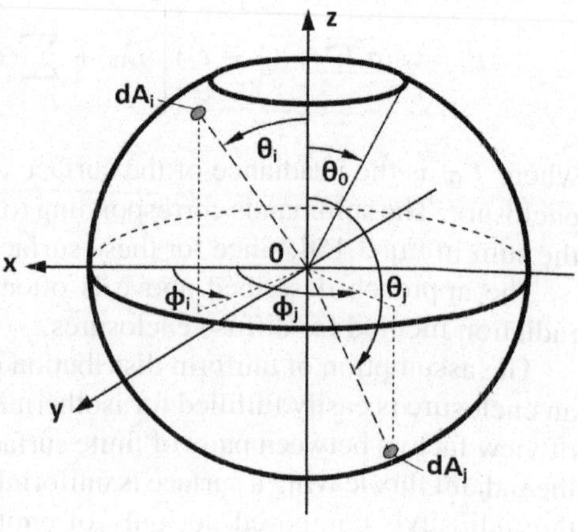

Fig. 3.28 Schematic of a spherical cavity

$$dF_{dA_1-dA_2} = \frac{dA_2}{4\pi r^2}, \tag{3.156}$$

$$F_{dA_1-A_2} = F_{A_1-A_2} = \frac{A_2}{4\pi r^2}, \tag{3.157}$$

where r is the radius of the sphere. Sparrow and Cess [77] considered the arbitrary irradiance $E_0(\theta, \varphi)$ by the incident radiation that enters the cavity through its aperture defining by the opening half-angle θ_0. The area of the spherical wall (except for the opening) is

$$A_w = 2\pi r^2(1 + \cos\theta_0). \tag{3.158}$$

The radiant flux balance at a particular location determining by the polar angle θ and the azimuth angle φ is

$$B(\theta, \varphi) = \varepsilon\sigma T^4(\theta, \varphi) + \rho\left[E_0(\theta, \varphi) + \frac{1}{4\pi r^2}\int_{A_w} B(\theta', \varphi')dA\right], \tag{3.159}$$

where A is the area of the sphere wall. A general solution for the radiosity B that is valid for any prescribed distributions of $T(\theta, \varphi)$ and $E_0(\theta, \varphi)$ is

$$B(\theta, \varphi) = G(\theta, \varphi) + \frac{\frac{1-\varepsilon}{4\pi r^2}\int_{A_w} G(\theta', \varphi')dA}{1 - \frac{(1-\varepsilon)A_w}{4\pi r^2}}, \tag{3.160}$$

where

$$G(\theta, \varphi) = \sigma T^4(\theta, \varphi) + (1 - \varepsilon)E_0(\theta, \varphi). \tag{3.161}$$

Earlier, Sparrow and Jonsson [75] defining the "apparent absorptivity" (now we would call it "effective absorptance," α_{eff}) as the ratio of the total absorbed energy to the total incoming energy, obtained a simple equation:

$$\alpha_{eff} = \alpha \big/ [1 - 0.5(1 - \alpha)(1 + \cos\theta_0)], \quad (3.162)$$

where α is the absorptance of the sphere wall. The thermal emission of the isothermal diffuse spherical cavity can be characterized by the ratio $B \big/ \left(\sigma T^4\right)$, which is now called the effective emissivity, ε_{eff}:

$$\varepsilon_{eff} = \varepsilon \big/ [1 - 0.5(1 - \varepsilon)(1 + \cos\theta_0)]. \quad (3.163)$$

It should be noted the complete similarity of Eqs. 3.162 and 3.163, which become identical, if $\alpha = \varepsilon$. The effective emissivities of a diffuse spherical cavities are plotted in Fig. 3.29 as functions of the opening half-angle θ_0 for several value of the wall emissivity ε. Sparrow and Jonsson [75] made remarkable conclusions that: (i) the effective absorptance of a diffuse sphere does not depend on angular and spatial

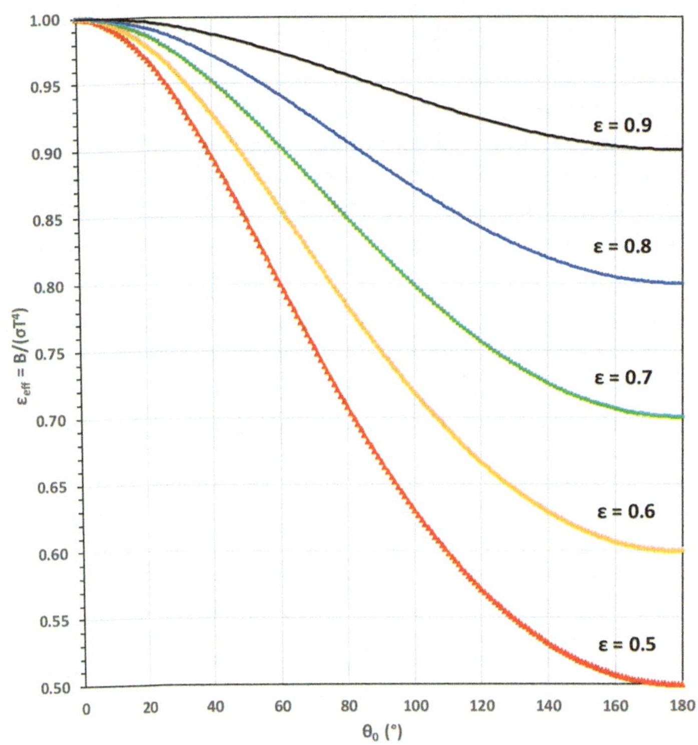

Fig. 3.29 Effective emissivity of a diffuse spherical cavity as a function of the aperture half-angle

distribution of the incoming radiant flux, (ii) the effective emissivityof the isothermal diffuse sphere (or radiosity) does not depend on the position on the sphere wall. A diffuse sphere owing to its unique properties is widely used in optical radiometry, photometry, and illuminating engineering as devices for effective integration of the radiant flux from anisotropic radiation sources, improving the cosine response of radiation detectors, as radiation diffusers, uniform radiance sources, etc. The spherical blackbody cavities are used not often due to the relative complexity of manufacturing and maintaining temperature uniformity, however, exact analytical solutions obtained for a diffuse sphere are often used to test various numerical methods for calculation of radiation characteristics of blackbody cavities.

3.4.4 *Beyond the Diffuse Gray Approximation*

All above analysis can be extended to non-gray (spectrally selective) radiative properties. By analogy with (3.154), we can introduce the spectral radiosity of i-th surface:

$$B_{\lambda,i}(\lambda) = \frac{c_1 \varepsilon(\lambda)}{\lambda^5\left[\exp\left(c_2/(\lambda T_i)\right) - 1\right]} + \rho_i(\lambda) E_{\lambda,i}(\lambda), \tag{3.164}$$

where $\varepsilon(\lambda)$ and $\rho(\lambda)$ are the spectral emissivity and spectral reflectance, respectively; $E_{\lambda,i}(\lambda)$ is the spectral irradiance of the surface A_i by radiation incident from all surfaces constituting the enclosure. For non-gray (spectrally selective) surfaces, Eq. 3.155 can be rewritten for the spectral radiosity and thus can be extended to non-gray (spectrally selective) surfaces to be solved wavelength-by-wavelength:

$$\begin{aligned} B_{\lambda,i}(\lambda, \boldsymbol{\xi}_i) &= \varepsilon_i(\lambda) L_{bb}(\lambda, T_i) + [1 - \varepsilon_i(\lambda)] \sum_{j=1}^{n} \int_{A_j} B_{\lambda,j}(\lambda, \boldsymbol{\xi}_j) d^2 F(\boldsymbol{\xi}_i, \boldsymbol{\xi}_j) \\ &= \varepsilon_i(\lambda) L_{bb}(\lambda, T_i) + [1 - \varepsilon_i(\lambda)] \sum_{j=1}^{n} \int_{A_j} B_{\lambda,j}(\lambda, \boldsymbol{\xi}_j) K(\boldsymbol{\xi}_i, \boldsymbol{\xi}_j) dA_j, \\ i &= 1, 2, \ldots, n \end{aligned} \tag{3.165}$$

The relative simplicity of the net radiation method is due to the assumption of angular independence of the surface radiative properties. Significant efforts were made in the 1960s and 1970s to extend the radiosity method to non-diffuse surfaces. Eckert and Sparrow [22] and Sparrow et al. [78] used the mirror-formed images to incorporate specularly reflecting surfaces into the radiosity method.

Introducing the exchange factor constructed as a series, each term of which is a diffuse view factor multiplied by a power of the surface specular reflectance, allowed to analyze, in particular, the radiation heat exchange in conical and cylindrical cavities with diffusely emitting and specularly reflecting walls [44].

Many works were dedicated to analysis of systems of surfaces, which have both diffuse and specular components of reflectance [13, 69, 76] or even have arbitrary bidirectional reflectance (e.g., [14]). A closer look at these methods shows that all of them are either too general or difficult to implement, or developed ad hoc, for only one particular case (e.g. [47]). These methods could not compete in universality and flexibility with the Monte-Carlo methods, which began to be developed in the second half of the 1960s [21, 29, 32, 46, 51, 84].

Computer graphics challenges in the 1980s and 1990s were very similar to that earlier solved in the radiation heat transfer. Since computer equipment made a giant leap over the years, researchers in computer graphics, image synthesis, and adjacent areas rapidly went all the way from the net radiation method for diffuse surfaces, through incorporation of specular reflection into the radiosity equations, to the Monte Carlo ray tracing (see, e.g. [33, 45, or 56]). Not all ray-tracing algorithms developed for the computer graphics applications are suitable for solution of the heat transfer problems, mainly due to performance-oriented simplifications such as limiting the number of interreflections, oversimplified BRDF models, selection of preferred processing area of an image, etc. However, many ray tracing techniques developed for computer graphics where successfully applied to solution of sophisticated problems in radiation heat transfer.

Now, we can state with confidence that the Monte Carlo ray tracing became the dominating technique for solution of problems of the radiation heat transfer and, in particular, of optical radiometry. The main applications of the Monte Carlo method in optical radiometry are absorptance characteristics of cavity detectors of optical radiation, characteristics of thermal emission of blackbody radiators, performance of integrating spheres, stray light analysis, etc. As we can see, all these applications have one defining feature – multiple reflections of optical radiation between real surfaces. We consider in detail the Monte Carlo ray tracing technique applied to calculation of radiation characteristics of blackbody cavities in Chap. 4.

References

1. J. Agassi, *Radiation Theory and the Quantum Revolution* (Birkhduser Verlag, Basel, Switzerland, 1993)
2. R.E. Alley, M. Jentoft-Nilsen, *Advanced Spaceborne Thermal Emission & Reflection Radiometer. Algorithm Theoretical Basis Document for Brightness Temperature.* (Jet Propulsion Laboratory, Pasadena, CA, 1999), https://trs.jpl.nasa.gov/bitstream/handle/2014/18629/99-2009.pdf?sequence=1&isAllowed=y. Accessed 02 Jan 2019
3. Ž Andreić, Distribution temperature calculations by fitting the Planck radiation curve to a measured spectrum. Appl. Opt. **31**, 126–130 (1992)
4. ANSI/IES RP-16–10. *Nomenclature and Definitions for Illuminating Engineering.* (The IES Nomenclature Committee, Illuminating Engineering Society of North America, New York, 2010)

5. T.M. Apostol, Zeta and related functions, in *NIST Handbook of Mathematical Functions*. ed. by F.W.J. Olver (NIST and Cambridge University Press, New York, 2010), pp. 601–616
6. G.B. Arfken, H.J. Weber, F.E. Harris, *Mathematical Methods for Physicists. A Comprehensive Guide*, 7th ed. (Acad. Press, Amsterdam, 2013)
7. K.E. Atkinson, *The Numerical Solution of Integral Equations of the Second Kind* (Cambridge University Press, Cambridge, UK, 1997)
8. M. Badino, *The Bumpy Road. Max Planck from Radiation Theory to the Quantum (1896–1906)* (Springer, Cham, Switzerland, 2015)
9. M. Bass, (ed.). *Handbook of Optics, Vol. IV. Optical Properties of Materials, Nonlinear Optics, Quantum Optics*, 3rd. edn. (McGraw-Hill, New York, 2010)
10. R.E. Bedford, Calculation of effective emissivities of cavity sources of thermal radiation, in *Theory and Practice of Radiation Thermometry*. ed. by D.P. DeWitt, G.D. Nutter (Wiley, New York, 1988), pp. 653–672
11. A. Berk, Analytically derived conversion of spectral band radiance to brightness temperature. J. Quant. Spectrosc. Rad. Transfer **109**, 1266–1276 (2008)
12. D.C. Bertilone, Stokes parameters and partial polarization of far-field radiation emitted by hot bodies. J. Opt. Soc. Am. **11**, 2298–2304 (1994)
13. J.T. Bevans, D.K. Edwards, Radiation exchange in an enclosure with directional wall properties. J. Heat Transfer **87**, 388–396 (1965)
14. R.P. Bobco, A script-F Matrix formulation for enclosures with arbitrary surface emission and reflection characteristics. J. Heat Transfer **93**, 33–40 (1971)
15. L. Boltzmann, Über eine von Hrn. Bartoli entdeckte Beziehung der Wärmestrahlung zum zweiten Hauptsatz. Wiedernannsche Annalen der Physik. **22**, 31–39 (1884)
16. S.B. Bopche, A. Sridharan, Determination of view factors by contour integral technique. Ann. Nucl. Energy **36**, 1681–1688 (2009)
17. M.A. Bramson, *Infrared Radiation. A Handbook for Applications* (Springer, New York, 1968)
18. S.L. Chang, K.T. Rhee, Blackbody radiation functions. Int. Commun. Heat Mass Transf. **11**, 451–455 (1984)
19. P.E. Ciddor, Refractive index of air: new equations for the visible and near infrared. Appl. Optics **35**, 1566–1573 (1996)
20. M.F. Cohen, J.R. Wallace, The form factor, in *Radiosity and Realistic Image Synthesis*. ed. by M.F. Cohen, J.R. Wallace (Acad. Press, Boston, MA, 1995), pp. 65–107
21. R.C. Corlett, Direct numerical simulation of thermal radiation in vacuum. J. Heat Transfer **88**, 376–382 (1966)
22. E.R.G. Eckert, E.M. Sparrow, Radiative heat exchange between surfaces with specular reflection. Int. J. Heat Mass Transfer **3**, 42–54 (1961)
23. B. Edlén, The refractive index of air. Metrologia **2**, 71–80 (1966)
24. J.L. Gardner, Uncertainties in source distribution temperature and correlated colour temperature. Metrologia **43**, 403–408 (2006)
25. T.A. Germer, J.C. Zwinkels, B.K. Tsai, et al. (eds.) Theoretical concepts in spectrophotometric measurements. in *Spectrophotometry: Accurate Measurement of Optical Properties of Materials*, ed. by T.A. Germer, J.C. Zwinkels, B.K. Tsai (Acad. Press, Amsterdam, 2014), pp. 11–66
26. M.A. Heald, Where is the "Wien peak"? Amer. J. Phys. **71**, 1322–1323 (2003)
27. D. Hoffmann, On the experimental context of Planck's foundation of quantum theory. Centaurus **43**, 240–259 (2001)
28. J. Hollandt, J. Hartmann, O. Struß, et al. Industrial applications of radiation thermometry. in eds. by Z.M. Zhang, B K. Tsai, G. Machin, *Radiometric Temperature Measurements. II. Applications*, (Academic Press, Amsterdam, 2010), pp. 1–56
29. J.R. Howell, Application of Monte Carlo to heat transfer problems, in *Advances in Heat Transfer*, vol. 5, ed. by T.F. Irvine, J.P. Hartnett (Academic Press, New York, 1968), pp. 1–54
30. J.R. Howell, M.P. Mengüç, R. Siegel, *Thermal Radiation Heat Transfer*, 6th edn. (CRC Press, Boca Raton, FL, 2016)

31. J.R. Howell, *A Catalog of Radiation Heat Transfer Configuration Factors*, 3rd edn. (University of Texas at Austin, 2018), https://www.thermalradiation.net/indexCat.html. Accessed 17 Oct 2019
32. J.R. Howell, The Monte Carlo method in radiative heat transfer. J. Heat Transfer **120**, 547–560 (1998)
33. J.F. Hughes, A. Van Dam, M. Mcguire et al. *Computer Graphics. Principles and Practice*, 3rd edn. (Addison-Wesley, Upper Saddle River, NJ, 2014)
34. R.W.G. Hunt, M.R. Pointer, *Measuring Colour*, 4th edn. (Wiley, Chichester, UK, 2011)
35. *ILV: International Lighting Vocabulary/Vocabulaire International de l'Eclairage/ Internationales Wörterbuch der Lichttechnik. CIE S 017/E:2011.* (CIE Central Bureau, Vienna, 2011)
36. J. Jablonski, C. Durell, J. LaVeigne, et al. A turn-key calibration roadmap for temperature and radiance from 0.3–14 μm. Proc. SPIE. **10178**, 1017804 (2017)
37. J.H. Jeans, Temperature-radiation and the partition of energy in continuous media. Phil. Mag. Series **6**(17), 229–254 (1909)
38. H. Kangro, *Early History of Planck's Radiation Law* (Taylor and Francis, London, 1976)
39. G.R. Kirchhoff, Über den Zusammenhang zwischen Emmission und Absorption von Licht und Wärme. Monatsberichte der Königlichen Preuß. Akademie der Wissenschaften zu Berlin aus dem Jahre. **1859**, 783–787 (1860)
40. G.R. Kirchhoff, Ueber das Verhältniss zwischen dem Emissionsvermögen und dem Absorptionsvermögen der Körper für Wärme und Licht. Annalen der Physik. **185**, 275–301 (1860)
41. G. Kirchhoff, On the relation between the radiating and absorbing powers of different bodies for light and heat. The London, Edinburgh, and Dublin Phil. Mag. J. Sci. **20**(130), 1–21 (1860)
42. T.S. Kuhn, *Black-Body Theory and the Quantum Discontinuity. 1894–1912* (The University of Chicago Press, Chicago, 1978)
43. J.H. Lambert, *Photometry, or, on the Measure and Gradations of Light, Colors and Shade* (English translation of 1760 Latin edn.) (Illuminating Engineering, New York, 2001)
44. S.H. Lin, E.M. Sparrow, Radiant interchange among curved specularly reflecting surfaces—application to cylindrical and conical cavities. J. Heat Transfer **87**, 299–307 (1965)
45. S. Marschner, P. Shirley, *Fundamentals of Computer Graphics*, 4th edn. (CRC Press, Boca Raton, FL, 2016)
46. S. Maruyama, Radiation heat transfer between arbitrary three-dimensional bodies with specular and diffuse surfaces. Num. Heat Transfer, Part a **24**, 181–196 (1993)
47. H. Masuda, Radiation heat transfer between specularly and diffusely reflecting surfaces. J. Quant. Spectrosc. Radiat. Transfer **16**, 997–1009 (1976)
48. R.J. Mathar, Refractive index of humid air in the infrared: model fits. J. Opt. A: Pure Appl. Opt. **9**, 470–476 (2007)
49. M.F. Modest, *Radiative Heat Transfer*, 3rd edn. (Acad. Press, New York, 2013)
50. J. Muelaner, *Ciddor Equation for Refractive Index of Air* (2015), https://www.mathworks.com/matlabcentral/fileexchange/52204-ciddor-equation-for-refractive-index-of-air. Accessed 21 Dec 2018
51. M.H.N. Naraghi, B.T.F. Chung, A stochastic approach for radiative exchange in enclosures with nonparticipating medium. J. Heat Transfer. **106**, 690–698 (1984)
52. F.E. Nicodemus, J.C. Richmond, J.J. Hsia, et al. *Geometrical Considerations and Nomenclature for Reflectance. NBS Monograph 160.* (Natl. Bureau of Standards, U.S. Dept. of Commerce, Washington, DC, 1977)
53. N. Ohta, A.R. Robertson, *Colorimetry Fundamentals and Applications* (Wiley, Chichester, UK, 2005)
54. C. Oleari, *Standard Colorimetry. Definitions Algorithms and Software* (Wiley, Chichester, UK, 2016)
55. J.C. Owens, Optical refractive index of air: dependence on pressure, temperature and composition. Appl. Opt. **6**, 51–59 (1967)
56. M. Pharr, W. Jakob, G. Humphreys, *Physically Based Rendering. From Theory to Implementation*, 3rd edn. (Morgan Kaufmann, Cambridge, MA, 2017)

57. M. Planck, *The Theory of Heat Radiation* (P. Blakiston's Son & Co., Philadelphia, PA, 1914)
58. M. Planck, Über eine Verbesserung der Wienschen Spektralgleichung. Verh. Dtsch. Phys. Ges. **2**, 202–204 (1900)
59. M. Planck, Zur Theorie des Gesetzes der Energieverteilung im Normalspectrum. Verh. Dtsch. Phys. Ges. **2**, 237–245 (1900)
60. M. Polyanskiy, Refractive Index Database (2018), https://refractiveindex.info/?shelf=other&book=air&page=Ciddor. Accessed 17 Oct 2019
61. A.V. Prokhorov, L.M. Hanssen, S.N. Mekhontsev, Calculation of the radiation characteristics of blackbody radiation sources. in eds. by Z.M. Zhang, B. Tsai, G. Machin, (eds.) *Radiometric Temperature Measurements. I. Fundamental*, (Acad. Press, Amsterdam, 2010), pp. 181–240
62. Rayleigh, Lord: Remarks upon the law of complete radiation. Phil. Mag. **49**, 539–540 (1900)
63. A. Resnick, C. Persons, G. Lindquist, Polarized emissivity and Kirchhoff's law. Appl. Opt. **38**, 1384–1387 (1999)
64. J.C. Richmond, F.E. Nicodemus, Blackbodies, blackbody radiation, and temperature scales, in ed. by F.E. Nicodemus, *Self-study Manual on Optical Radiation Measurements: Part 1—Concepts, Chapter 12* (U.S. Dept. of Commerce, NBS, 1985)
65. J. Roberts, *Black Body Radiation Functions* (The University of Newcastle 1971), https://ia902303.us.archive.org/25/items/Roberts1971/Roberts-1971.pdf. Accessed 17 Oct 2019
66. P. Rosenkranz, M. Matus, M.L. Rastello. On estimation of distribution temperature. Metrologia **43**, S130–S134 (2006)
67. W.D. Ross, Methods of representing radiation formulas. J. Opt. Soc. Amer. **44**, 770–771 (1954)
68. O. Sandus, A review of emission polarization. Appl. Opt. **4**, 1634–1642 (1965)
69. A.F. Sarofim, H.C. Hottel, Radiative exchange among non-Lambert surfaces. J. Heat Transfer. **88**, 37–43 (1966)
70. P. Saunders, E. Woolliams, H. Yoon, et al. *Uncertainty Estimation in Primary Radiometric Temperature Measurement* (CCT, BIPM, 2018), https://www.bipm.org/utils/en/pdf/si-mep/MeP-K-2018_Absolute_Primary_Radiometry_Uncertainty.pdf. Accessed 17 Oct 2019
71. G. Schaepman-Strub, M.E. Schaepman, T.H. Painter et al., Reflectance quantities in optical remote sensing—definitions and case studies. Remote Sens. Environ. **103**, 27–42 (2006)
72. J. Smith, *Air Refractive Index* (2011), https://www.mathworks.com/matlabcentral/fileexchange/31240-air-refractive-index. Accessed 17 Oct 2019
73. E.M. Sparrow, Heat radiation between simply-arranged surfaces having different temperatures and emissivities. AIChE J. **8**, 12–18 (1962)
74. E.M. Sparrow, A new and simpler formulation for radiative angle factors. J. Heat Transfer. **85**, 81–87 (1963)
75. E.M. Sparrow, V.K. Jonsson, Absorption and emission characteristics of diffuse spherical enclosures. J. Heat Transfer. **84**, 188–189 (1962)
76. E.M. Sparrow, S.L. Lin, Radiation heat transfer at a surface having both specular and diffuse reflectance components. Int. J. Heat Mass Transfer. **8**, 769–779 (1965)
77. E.M. Sparrow, R.D. Cess, *Radiation Heat Transfer*, Augmented edn. (Hemisphere Publ., Washington, DC, 1978)
78. E.M. Sparrow, E.R.G. Eckert, V.K. Jonsson, An enclosure theory for radiative exchange between specularly and diffusely reflecting surfaces. J. Heat Transfer. **84**, 294–299 (1962)
79. J. Stefan, Über die Beziehung zwischen der Wärmestrahlung und der Temperatur. Sitzungsberichte der kaiserlichen Akademie der Wissenschaften: Mathematisch-Naturwissenschaftliche Classe **79**, 391–428 (1879)
80. S.M. Stewart, R.B. Johnson, *Blackbody Radiation. A History of Thermal Radiation Computational Aids and Numerical Methods* (CRC Press, Boca Raton, FL, 2017)
81. J.A. Stone, J.H. Zimmerman, Index of refraction of air, in Engineering Metrology Toolbox. Refractive Index of Air Calculator (2011), https://emtoolbox.nist.gov/Wavelength/Documentation.asp. Accessed 17 Oct. 2019
82. A.S. Tenney, Radiation ratio thermometry, in *Theory and Practice of Radiation Thermometry*. ed. by D.P. DeWitt, G.D. Nutter (Wiley, New York, 1988), pp. 459–494
83. D. Ter Haar, *The Old Quantum Theory* (Pergamon Press, Oxford, 1967)

84. J.S. Toor, R.A. Viskanta, Numerical experiment of radiant heat interchange by the Monte Carlo method. Int. J. Heat Mass Transfer **11**, 883–897 (1968)
85. W. Wien, O. Lummer, Methode zur Prüfung des Strahlungsgesetzes absolut schwarzer Körper. Ann. Phys. **292**, 451–456 (1895)
86. W. Wien, Eine neue Beziehung der Strahlung schwarzer Körper zum zweiten Hauptsatz der Wärmetheorie. Sitzungsbericht Der Koniglich Preußischen Akademie Der Wissenschaften Zu Berlin **1**, 55–62 (1893)
87. W. Wien, Ueber die Energievertheilung im Emissionsspectrum eines schwarzen Körpers. Ann. Phys. **58**, 662–669 (1896)
88. E.R. Woolliams, N.P. Fox, M.G. Cox et al., Final report on CCPR K1-a: Spectral irradiance from 250 nm to 2500 nm. Metrologia. **43**, 02003 (2006). https://doi.org/10.1088/0026-1394/43/1A/02003
89. W. Zhigang, An accurate equation to determine the temperature for optimum efficiency of blackbody radiation. Infrared Phys. **33**, 313–316 (1992)

Chapter 4
Effective Emissivity

Abstract The Chapter is dedicated to issues relevant to the effective emissivity, one of the most important characteristics of artificial blackbodies. Its definitions for the isothermal and nonisothermal cavities are given; the analysis of affecting factors, the methods for calculating and measuring the effective emissivity are described. The approximate formulae for the effective emissivity of isothermal diffuse cavities, the method of integral equations for isothermal and nonisothermal diffuse cavities, and the Monte Carlo method for arbitrary cavities are described and compared. Special attention is paid to the Monte Carlo algorithms for modeling the viewing conditions and optical characteristics of the cavity walls, and the technique of tracing rays inside the cavity. Modern methods for the experimental determination of effective emissivities as a tool for verifying the results of calculations are overviewed.

Keywords Effective emissivity · Blackbody cavity · Monte Carlo · Ray tracing · Reflectance measurement

4.1 Definitions of Effective Emissivities

Theoretically, a perfect blackbody, whose radiation characteristics can be calculated on the base of fundamental physical laws, is the most suitable thermal radiation source for the calibration of optical radiation sensors, radiometers, spectroradiometers, radiation thermometers, etc. in absolute units. However, a perfect blackbody is an "ideal" object, that is no more than a physical abstraction not existing in the real world. We can use for these purposes only an artificial device that reproduces the radiation characteristics of a perfect blackbody with a certain degree of accuracy. Thermal radiation inside an isothermal enclosure of an arbitrary shape whichever opaque material is used for its walls is identical with the radiation of a perfect blackbody at the same temperature. However, the blackbody radiation inside an enclosure is accessible and observable for an internal observer only. To be registered by an external observer, the radiation should go out from the enclosure through the opening in its wall. Radiation leaving the enclosure through this opening should reproduce the radiation of a perfect blackbody, the closer, the smaller the ratio of the area of the opening to the area of the inner surface of the enclosure. The presence of an opening leads to the

V. Sapritsky and A. Prokhorov, *Blackbody Radiometry*, Springer Series in Measurement Science and Technology, https://doi.org/10.1007/978-3-030-57789-6_4

occurrence of the resulting thermal radiation fluxes between the enclosure walls and the surroundings having a different temperature (if there is no temperature difference, the resulting thermal radiation fluxes are absent and there is nothing to register or measure). The presence of the opening disturbs thermodynamic equilibrium inside the enclosure hence the radiation from it becomes depending upon radiation characteristics of the wall material, direction of observation, and the observed part of the incomplete enclosure (e.g., cavity).

Another reason for the deviation of the radiation of a real enclosure from the radiation of a perfect blackbody is the inevitable transformation of its isothermal surface into nonisothermal, unless the radiation losses through the opening are compensated by the supply of heat to the wall of the enclosure. As a result, instead of a perfect blackbody, we get a real blackbody—an artificial thermal radiator, commonly used as a standard reference source in optical radiometry and radiation thermometry.

In order to use the blackbody as a calculable material measure of radiometric quantities, it is necessary to know how big the difference is between the radiation characteristics of the blackbody and the characteristics of a perfect blackbody. Obviously, this difference depends on the cavity geometry, wall material, and deviations from a uniform temperature.

Although we have already introduced the emissivity as a quantity that describes the difference between the radiation of real bodies and the radiation of a perfect blackbody, the special term "effective emissivity" (less commonly used is the term "apparent emissivity") is introduced for artificial blackbodies. First, we will define the effective emissivity for the case of an isothermal blackbody and then generalize this definition to a blackbody with an arbitrary temperature distribution.

While the emissivity of a flat sample is, as a rule, a measurable quantity, the effective emissivity is a calculated quantity *par excellence.* The methods for computing the effective emissivity were reviewed by Bedford [8, 9] and Prokhorov et al. [125]. Many of them are of historical interest only. In this Chapter, we consider only computational methods that continue to be actively used to date. For simplicity and brevity, we will talk about blackbody cavities, although, as will be shown in Sect. 5.5 of the next Chapter, the cavity is the most familiar, but not the only embodiment of the idea of an artificial blackbody.

4.1.1 Effective Emissivity of Isothermal Cavity

Let us consider a cavity that has the uniform temperature T (here and below, the thermodynamic temperature is meant). Let the cavity is surrounded with the perfectly black environment at zero absolute temperature, so the interaction of the cavity with its environment is absent. The primary definition relates to an elementary area dA of the cavity internal surface specified by the positional vector $\boldsymbol{\xi}$ (see Fig. 4.1), in the direction specified by the vector $\boldsymbol{\omega}$, at a particular wavelength λ. The local (at the point $\boldsymbol{\xi}$ of the cavity internal surface) directional (in the direction $\boldsymbol{\omega}$) spectral effective emissivity at the wavelength λ is defined as

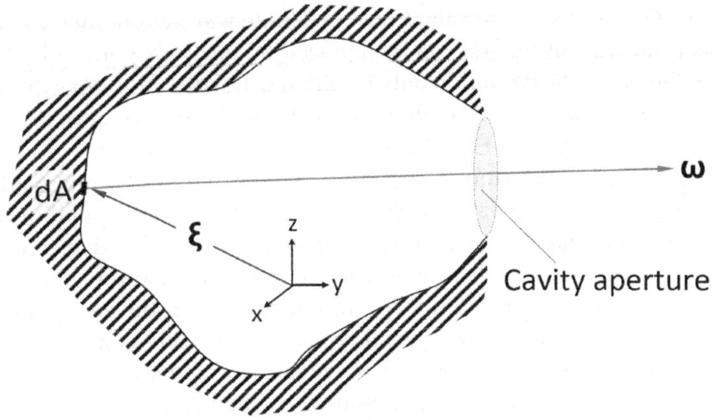

Fig. 4.1 To the definition of the directional effective emissivity of a cavity

$$\varepsilon_e(\boldsymbol{\xi}, \boldsymbol{\omega}, \lambda) = \frac{L_\lambda(\boldsymbol{\xi}, \boldsymbol{\omega}, \lambda)}{L_{\lambda,bb}(\lambda, T)}, \quad (4.1)$$

where L_λ is the spectral radiance at the point $\boldsymbol{\xi}$ of the cavity internal surface in the direction $\boldsymbol{\omega}$ at the wavelength λ and $L_{\lambda,bb}$ is the spectral radiance of a perfect blackbody that does not depend on the direction (due to Lambert's law) and position but depends on the wavelength λ and the temperature T according to the Planck law.

Along with the *spectral* effective emissivities, the corresponding *total* effective emissivities are introduced. The local directional total effective emissivity is defined via the Stefan-Boltzmann law as

$$\varepsilon_e(\boldsymbol{\xi}, \boldsymbol{\omega}) = \frac{L(\boldsymbol{\xi}, \boldsymbol{\omega})}{L_{bb}(T)} = \frac{\pi L(\boldsymbol{\xi}, \boldsymbol{\omega})}{\sigma T^4}, \quad (4.2)$$

where $L(\boldsymbol{\xi}, \boldsymbol{\omega})$ is the radiance of the element dA at the position $\boldsymbol{\xi}$ of dA in the cavity internal surface in the direction $\boldsymbol{\omega}$ of observation, $L_{bb}(T) = \sigma T^4 / \pi$ is the radiance of a perfect blackbody at the temperature T, and σ is the Stefan-Boltzmann constant.

The total effective emissivities are rarely used in the practice of modern radiometry since they correspond to the physically unrealizable case of registering the cavity radiation by a sensor responsive to all wavelengths, from zero to infinity. However, they retain their importance for less precise measurements, analysis of radiation heat transfer, etc.

A more realistic case is described by the bandwidth effective emissivities, which can be obtained from Eq. 4.1 using the relative spectral responsivity $s(\lambda)$ of the sensor:

$$\varepsilon_{e,bw}(\boldsymbol{\xi}, \boldsymbol{\omega}, s(\lambda)) = \frac{\int_0^\infty L_\lambda(\boldsymbol{\xi}, \boldsymbol{\omega}, \lambda) s(\lambda)\, d\lambda}{\int_0^\infty L_{\lambda,bb}(\lambda, T) s(\lambda)\, d\lambda}. \quad (4.3)$$

As a rule, $s(\lambda)$ takes nonzero values only in a finite wavelength interval; therefore, the improper integrals in Eq. 4.3 can be replaced by integrals with finite limits. For brevity, we shall write the definition only for the spectral effective emissivity because the corresponding equations for total and band-limited effective emissivities can be written in the same way as in Eqs. 4.2 and 4.3.

In the special case of axisymmetric cavities, the spectral local directional effective emissivity in the direction parallel to the cavity axis (and, correspondingly, normal to the cavity aperture) has the name of spectral *local normal* effective emissivity and is denoted as $\varepsilon_{e,n}(\boldsymbol{\xi}, \lambda)$. Their distribution across the cavity aperture can serve as the useful measure of the non-uniformity spectral radiance over the cavity aperture.

It is convenient to define some special types of the effective emissivity by averaging the local directional effective emissivities over certain angular and spatial domains. A fraction of the radiant flux emitted by the blackbody reaches a sensor, which yields a signal that is a function (linear, in the ideal case) of the incident radiant flux. The sensor may or may not include an optical system with diaphragms and other components, which form its field of view (FOV). Typically, if the sensor is used in conjunction with an optical system, the diameter of the target spot is much smaller than the distance between the cavity and the sensor. Therefore, the FOV envelope can be regarded as a straight circular cone with a vertex at the focal point of the optical system. In fact, the simplest pinhole model of an optical system is used here. Depending on the position of the focal point (in front of or behind the radiating surface), the sensor registers the radiation within either the diverging or converging conical beam as it is shown in Fig. 4.2. Both cases depicted in Fig. 4.2 correspond to so-called *conical* effective emissivity $\varepsilon_{e,c}$. This is a typical case of measurements in the radiance mode, including measurements carried out by radiation thermometers. The conical effective emissivities can be obtained from the local directional effective emissivities by averaging over the conical solid angle Ω and the cavity viewable surface A:

$$\varepsilon_{e,c}(\lambda) = \frac{1}{A\Omega} \int\limits_{A} \int\limits_{\Omega} \varepsilon_e(\boldsymbol{\xi}, \boldsymbol{\omega}, \lambda) d\Omega dA. \tag{4.4}$$

The integrated effective emissivity corresponds to the most common viewing conditions when radiation is transferred from disk A_a to disk A_d. It is assumed that

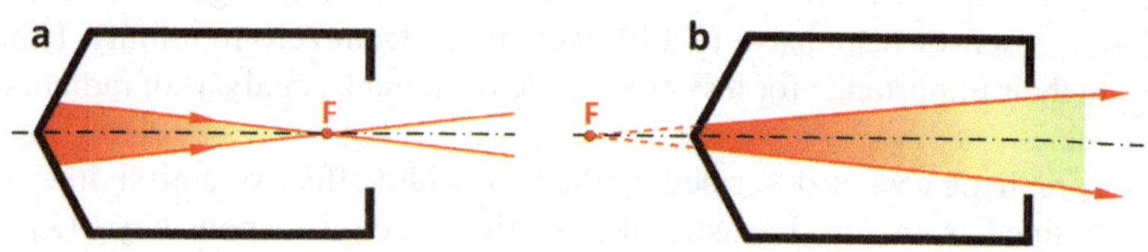

Fig. 4.2 Viewing conditions for **a** converging and **b** diverging conical beams; F denotes the focal point

these disks lie in parallel planes perpendicular to the axis of the cavity and the centers of both disks lie on that axis.

The spectral integrated effective emissivity is defined as the ratio of the spectral radiant flux Φ_λ incident on the sensor from the cavity to the spectral radiant flux $\Phi_{\lambda,bb}$ that could be collected by this sensor if the blackbody aperture would be replaced by a perfectly black surface at the temperature of the cavity walls:

$$\varepsilon_{e,int}(\lambda) = \frac{\Phi_\lambda(\lambda)}{\Phi_{\lambda,bb}(\lambda,T)} = \frac{\Phi_\lambda(\lambda)}{\pi L_{\lambda,bb}(\lambda,T) F_{A_a-A_d} A_a}, \tag{4.5}$$

where $F_{A_a-A_d}$ is the view factor between the cavity aperture and the sensor; A_a and A_d are the areas of the cavity aperture and the detector, respectively.

In the simplest cases, shown in Fig. 4.3a, b, A_a is the cavity aperture and A_d is the detector. This configuration is typical for the irradiance measurements and non-imaging radiometers.

The cavity shown in Fig. 4.3b, in contrast to the one depicted in Fig. 4.3a, has a flat lid, which prevent the unobstructed observation of the entire internal surface of the cavity from the surface of the detector. As can be seen in Fig. 4.3b, there are three regions on the radiating surface of the cavity: radiation from the first zone can reach any point of the detector surface (the rays within the shaded area),radiation from the second zone can reach only a limited subset of points on the detector surface,finally, radiation emitted by the third zone cannot reach the detector. The dependence of the visibility of the cavity internal surface upon the position of the observing point on the surface of the detector is often referred to as the penumbra effect. Due to the penumbra effect, the calculation of the integrated effective emissivity of a baffled cavity is not a trivial task even for the simplest case of purely diffuse (Lambertian) internal surface [18, 20].

The integrated effective emissivity has two important special cases. When $R_a = R_d$ and $H = 0$, it is called the *hemispherical* effective emissivity, $\varepsilon_{e,h}$. The total hemispherical effective emissivity is used for calculation of the radiant flux emitted by a cavity into the entire hemisphere, for example, to calculate the radiation losses of a cavity as a whole through its aperture. When $H \to \infty$, we obtain the *average normal* effective emissivity, $\varepsilon_{e,n}$, which, for an axisymmetric cavity, corresponds to the collection of the cavity radiation in directions parallel to the cavity axis.

From a computational point of view, the normal effective emissivity is a convenient approximation for registering the cavity radiation with a sensor, whose optical system has a very large f-number.

In practice, a pair of coaxial apertures is often located between the cavity and the detector (see Fig. 4.3c), which makes the rigorous calculation of the effective emissivity corresponding to such a configuration even more complicated.

Additionally, the integrated effective emissivity can be used to model the viewing conditions for an imaging radiometer more accurately then using the conical effective emissivity. This configuration is presented in Fig. 4.3d, where A_a corresponds to a

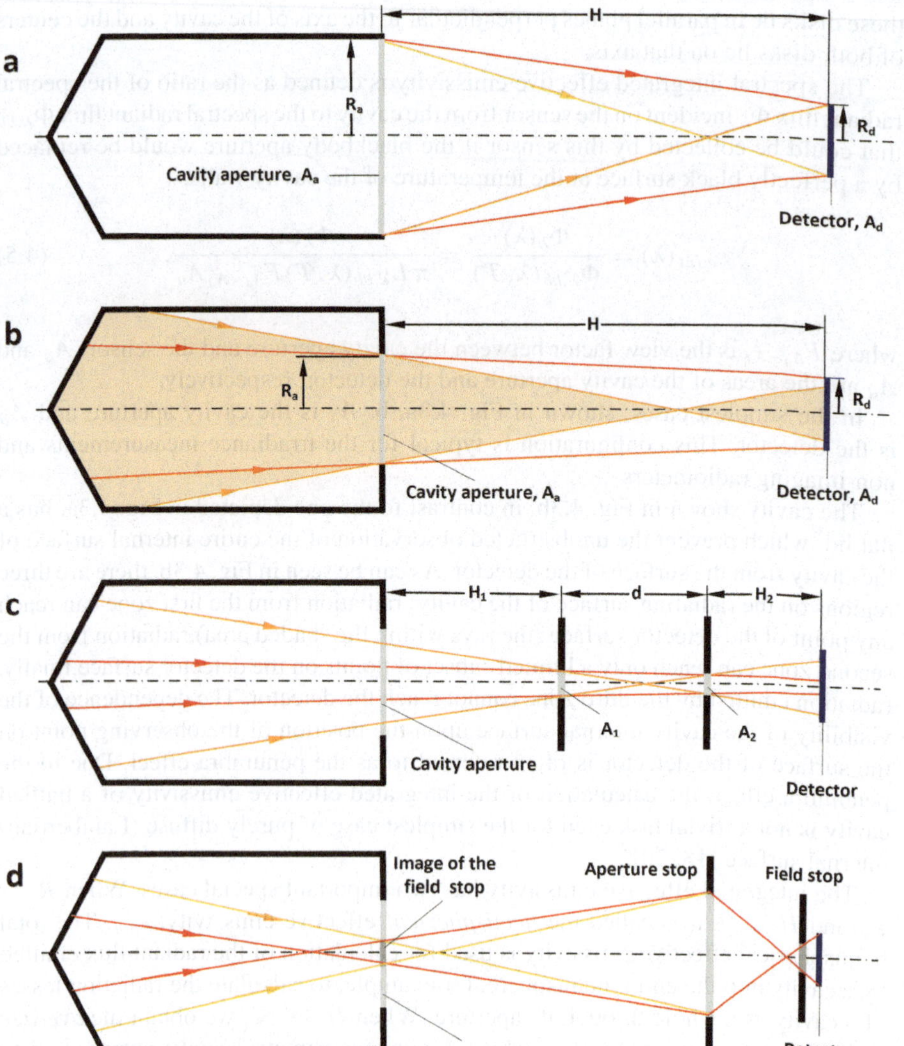

Fig. 4.3 Viewing conditions for the integrated effective emissivity

measurement spot (an image of the field stop of the imaging radiometer) and A_d—to the aperture stop of the radiometer.

In the literature dedicated to analysis of the cavities with the perfectly diffuse (Lambertian) walls, one can find so-called *local hemispherical* effective emissivities defined as

$$\varepsilon_{e,h}(\boldsymbol{\xi}, \lambda) = \frac{M_\lambda(\boldsymbol{\xi}, \lambda)}{M_{\lambda,bb}(\lambda, T)} = \frac{M_\lambda(\boldsymbol{\xi}, \lambda)}{\pi L_{\lambda,bb}(\lambda, T)}, \tag{4.6}$$

where M_λ and $M_{\lambda,bb}$ are the spectral exitance of the point specified by the vector $\boldsymbol{\xi}$ on the cavity wall and the spectral exitance of a perfect blackbody, respectively.

The local hemispherical effective emissivities of diffuse cavities do not depend on the direction of observation. Their distribution over the internal surface of diffuse cavities are the output values of some computational methods; further integration of this distribution over the viewable part of a cavity allows calculation of normal, conical, hemispherical, and integrated effective emissivities. Therefore, the local hemispherical effective emissivities have a theoretical rather than a practical application. For cavities with walls different from diffuse, the local hemispherical effective emissivities become dependent on the direction of observation, so their use loses all meaning.

One of the most important goals for designers of a blackbody radiation source is achieving temperature distribution over the internal surface of a radiating cavity as uniform as possible. The concept of isothermal cavity plays the important role in the blackbody radiometry. The isothermal cavity is not only the best approximation of a perfect blackbody; its effective emissivity can be calculated more reliably, its radiation characteristics can be determined with lower uncertainties, and the calculation results are easier to verify experimentally. Although all real cavities are to some extent non-isothermal, the isothermal approximation is undoubtedly useful. Firstly, it allows separating the influence of cavity geometry and radiation characteristics of its walls from the effects caused by temperature non-uniformity of the internal surface of the cavity. Secondly, the effective emissivities of an isothermal cavity (or cavity, the temperature non-uniformity of which can be neglected) is a quantity that can be determined experimentally by measuring their effective reflectance. Finally, some properties of the isothermal cavities can be investigated in a very general form, without calculating the effective emissivities for particular cavities.

A convenient tool for studying isothermal cavities is the Kirchhoff law applied to the effective radiation characteristics, validity of which is not at all obvious. Kelly [72] presented the rigorous proof of the "generalized Kirchhoff's law" for the effective emissivity and the effective absorptance of isothermal diffuse cavities. Takata [150] extended Kelly's conclusions to cavities with arbitrary angular characteristics of thermal emission and reflection. Unfortunately, his work published in Japanese has remained unknown to English-speaking readers for many years. Finally, Ohwada [102] presented a strict proof of the extended Kirchhoff law for direction-dependent radiation characteristics by mathematical induction, after which the question can be considered exhausted. The mathematical expression for the extended (generalized) Kirchhoff law is as follows:

$$\varepsilon_e(\boldsymbol{\xi}, \boldsymbol{\omega}, \lambda) = \alpha_e(\boldsymbol{\xi}, -\boldsymbol{\omega}, \lambda), \tag{4.7}$$

where α_e is the effective absorptance of the cavity. The minus sign in front of the direction vector $\boldsymbol{\omega}$ means adherence to the principle of reciprocity. If the cavity walls are opaque, in accordance with the energy conservation law, we can write out:

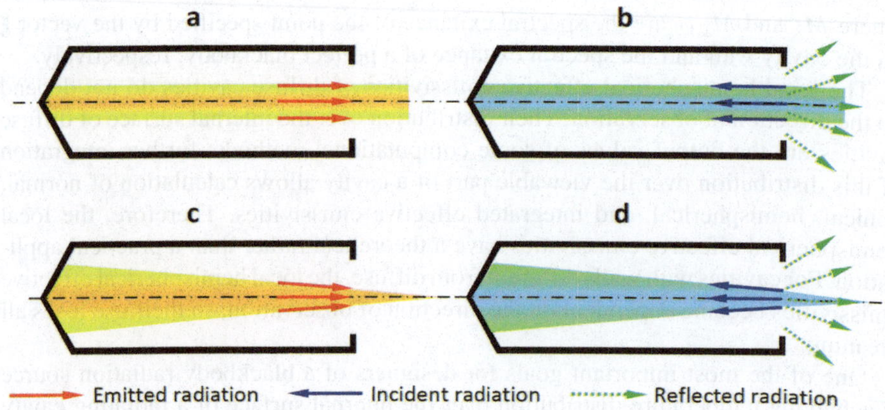

Fig. 4.4 Examples of the reciprocity principle application: schematics for calculations of **a** normal effective emissivity, **b** normal-hemispherical effective reflectance, **c** conical effective emissivity, and **d** conical-hemispherical effective reflectance

$$\varepsilon_e(\boldsymbol{\xi}, \boldsymbol{\omega}, \lambda) = 1 - \rho_e(\boldsymbol{\xi}, -\boldsymbol{\omega}, \lambda), \quad (4.8)$$

where ρ_e is the effective reflectance of the cavity.

The reciprocity principle allows extending Eq. 4.8 on the arbitrary viewing conditions by replacing the directions of observation with the directions of irradiation. Two examples of the application of the reciprocity principle are shown schematically in Fig. 4.4.

The normal or conical effective emissivity of the isothermal cavity can be easily expressed through the effective hemispherical reflectance at the irradiation of the cavity with the coaxial collimated or conical beam, respectively. The use of the generalized Kirchhoff law in conjoining with the reciprocity principle open the way to the experimental determination of the effective emissivities of isothermal cavities by measuring the effective reflectance.

4.1.2 Cavity Effect

The so-called "cavity effect" is manifested in the fact that the radiant energy emerging from an isothermal cavity always exceeds the radiant emission from a surface of the same emissivity stretched across the cavity opening. The lower the ratio of the area of the aperture to the surface area of the cavity, the more pronounced the cavity effect. This effect was first described, apparently, in the early 1960s (see, e.g. [145, 146]), there is still no clear mathematical interpretation of it.

The outgoing radiant flux from the cavity wall toward the opening consists of two parts: the flux of the surface element's own thermal radiation and the flux reflected by this element. The first part depends only on the emissivityand temperature of the cavity wall and does not depend upon the presence of the rest of the cavity. The reflected flux depends on the surface element reflectance and on the radiant flux falling onto this element from the rest of the cavity. If the cavity has no opening, each element of the cavity surface emits thermal radiation as a perfect blackbody. The net radiation flux exiting from the isothermal cavity with an opening is greater than that of a flat surface having the same temperature and emissivity as the cavity walls, but less than the radiation flux of a perfect blackbody at the same temperature.

Mathematically, the effect of the emissivity enhancement can be described most easily for isothermal cavities with homogeneous diffuse walls. In such a case, radiation characteristics of the cavity internal surface can be characterized by the single value of the reflectance ρ or emissivity $\varepsilon = 1 - \rho$ (we omit the possible dependence on wavelength λ for brief).

We can represent the effective emissivity of an isothermal cavity with arbitrarily reflecting walls in the following form:

$$\varepsilon_e = 1 - \sum_{k=1}^{\infty} f_k (1 - \varepsilon)^k, \tag{4.9}$$

where $0 \leq f_k \leq 1$ is the fraction of the radiation flux reflected by a cavity that escapes it after k successive reflections. The values of f_k are determined by the cavity geometry, angular dependences of the cavity walls radiation characteristics, and the viewing conditions.

It is clear that $\varepsilon_e = 1$, if $\varepsilon = 1$ (the case of perfectly black surface), $\varepsilon_e = 0$, if $\varepsilon = 0$ (the case of non-absorbing surface), and $\varepsilon_e > \varepsilon$ for any $0 < \varepsilon < 1$. The inequality $\varepsilon_e > \varepsilon$ is the mathematical formulation of the cavity effect, which is, in fact, the effect of multiple reflections inside a cavity. The following inequality for the first derivative of $\varepsilon_e(\varepsilon)$ with respect to ε takes place:

$$\frac{d\varepsilon_e(\varepsilon)}{d\varepsilon} = \sum_{k=1}^{\infty} k \cdot f_k (1 - \varepsilon)^{k-1} > 0 \quad \forall(\varepsilon < 1), \tag{4.10}$$

which means that $\varepsilon_e(\varepsilon)$ is the monotonically increasing function. The second derivative is also positive:

$$\frac{d^2\varepsilon_e(\varepsilon)}{d\varepsilon^2} = \sum_{k=1}^{\infty} k(k-1) f_k (1 - \varepsilon)^{k-2} > 0 \quad \forall(\varepsilon < 1), \tag{4.11}$$

which means that the curve representing $\varepsilon_e(\varepsilon)$ is convex upwards. The dependences $\varepsilon_e(\varepsilon)$ and $\rho_e(\rho)$ for different sets of f_k, $k = 1, 2, ...$ are plotted (without exact figures) in Fig. 4.5 for four cases:

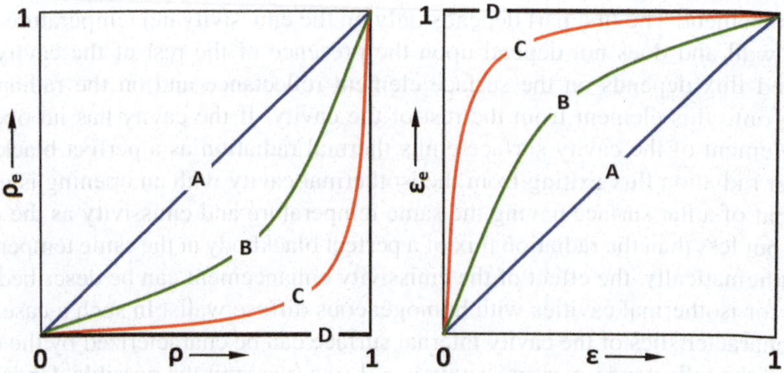

Fig. 4.5 Dependences of $\rho_e(\rho)$ and $\varepsilon_e(\varepsilon)$

Fig. 4.6 The critical points ε_B and ε_C for the dependences A through D

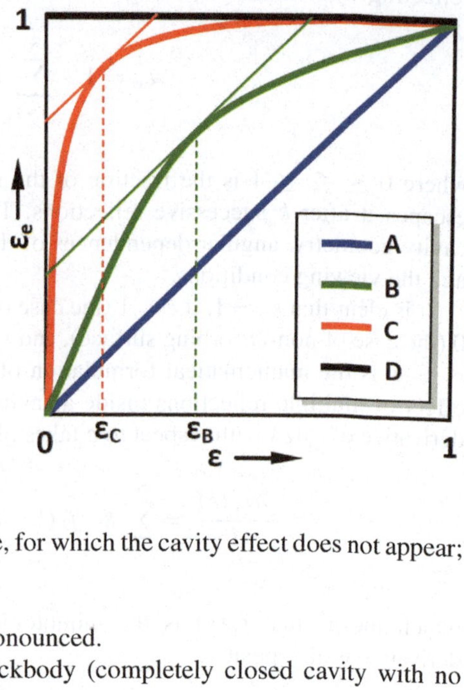

A. We deal with a flat or convex surface, for which the cavity effect does not appear; $\rho_e = \rho$ and $\varepsilon_e = \varepsilon$.
B. The cavity effect appears.
C. The cavity effect becomes more pronounced.
D. The limiting case of a perfect blackbody (completely closed cavity with no openings), $\varepsilon_e \equiv 1$ for any ε.

The further analysis of dependences of the effective emissivity on the emissivity of the cavity wall using the mean value theorem (in Lagrange's form) shows that for every curve $\varepsilon_e(\varepsilon)$, there is a critical point, at which the tangent line to the curve is parallel to its secant line, i.e. the line $\varepsilon_e = \varepsilon$. On the left from this point, $d\varepsilon_e / d\varepsilon > 1$; on the right, $d\varepsilon_e / d\varepsilon < 1$. In Fig. 4.6, the critical points have the abscissae ε_B and ε_C.

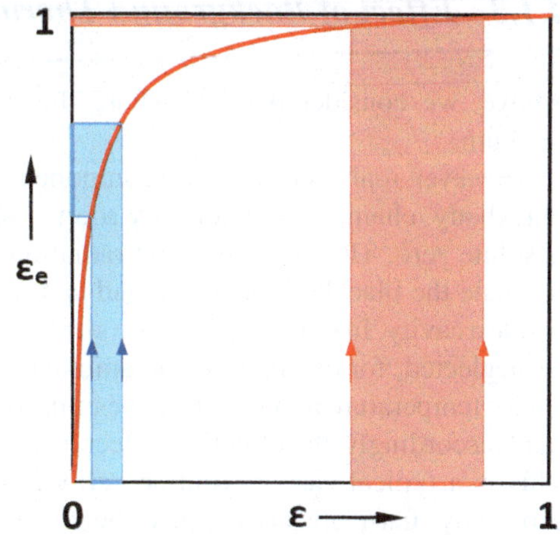

Fig. 4.7 Illustration to the effect of reducing the variation range of the spectral effective emissivity of a cavity in comparison with the cavity surface emissivity

The important consequence can be derived from the existence of these critical points: although the cavity effect always makes the effective emissivity higher than the emissivity of the cavity walls, it can decrease or increase the variation range of the effective emissivity as compared with the variation range of the emissivity of the cavity walls, depending on their values. Figure 4.7 illustrates both cases of $d\varepsilon_e / d\varepsilon < 1$ and $d\varepsilon_e / d\varepsilon > 1$.

If $d\varepsilon_e / d\varepsilon < 1$, all variations of the cavity emissivity affect the effective emissivity weakened and therefore, less affected by ageing, contamination, or local non-uniformity of the cavity wall emissivity. Besides, this means that the uncertainty in determination of ε_e will be reduced in comparison with the uncertainty in determination of the emissivity of the cavity walls.

If $d\varepsilon_e / d\varepsilon > 1$, the effective emissivity values are becoming greater than the emissivity of the cavity walls, but, at the same time, the variation range of the effective emissivity is wider than that for the emissivity of the cavity wall.

This is why materials with the low emissivity (such as some kinds of ceramics) have to be used with caution for fabrication of blackbody cavities: although the required effective emissivity, probably, can be achieved, the absolute uncertainty of its determination may increase as compared with that for the emissivity of a flat sample.

All the arguments above can be repeated almost unchanged for cavities consisting of several parts with different radiation characteristics (the only condition remains in force—these characteristics must be direction-independent; the simplest example is a diffuse cavity with a specular lid). There is still no a rigorous proof of above-mentioned conclusions for cavities with the arbitrary directional radiation characteristics of their walls. However, multiple results of numerical modeling are in favor of correctness of the above provisions.

4.1.3 Effect of Background Thermal Radiation

Above, we considered cases where the cavity is surrounded by a non-emitting medium.

However, real environments (components of a measuring system lying outside the blackbody, chamber or laboratory room walls, etc.) have temperatures greater than absolute zero. Therefore, thermal radiation from the surrounding environment will irradiate the blackbody aperture and can reach the sensor after multiple reflections inside a cavity. In practice, there are situations when the background radiation cannot be neglected, for example, if the ambient temperature is not too different from the cavity temperature and the effective emissivity of the cavity is not too close to unity (and, accordingly, the effective reflectance of the cavity differs markedly from zero). This is a typical case of routine calibrations using low-temperature blackbodies, especially, flat-plate ones. Since the spectral, spatial, and angular distributions of the background radiation is generally not known and can hardly be predicted, the simplest case of isotropic radiation of a perfect blackbody at the temperature T_{bg} of the background can be adopted as a work hypothesis. A perfectly black surface irradiating the cavity opening can have any geometrical configuration, e.g., it may be a surface tightly stretched across the cavity opening. The schematic for the calculation of the normal effective emissivity accounted for the background effect is shown in Fig. 4.8.

Let us consider the cavity having the uniform temperature T surrounded by the background at temperature T_{bg}. Let us assume that the sensor has temperature lower than T, does not distort the isotropy of the background radiation field, and is not irradiate by the background directly. The effect of the background radiation on the local directional spectral effective emissivity of an isothermal blackbody can be taken into account by the second term in the equation:

$$\varepsilon_e(\boldsymbol{\xi}, \boldsymbol{\omega}, \lambda, T, T_{bg}) = \varepsilon_e(\boldsymbol{\xi}, \boldsymbol{\omega}, \lambda) + [1 - \varepsilon_e(\boldsymbol{\xi}, \boldsymbol{\omega}, \lambda)] \cdot X(\lambda, T, T_{bg}), \quad (4.12)$$

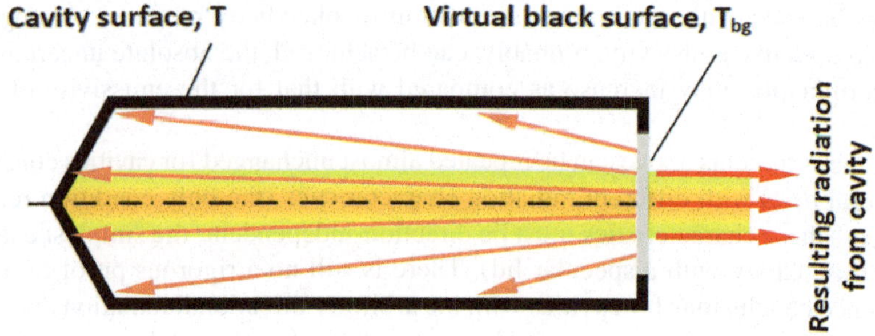

Fig. 4.8 For calculation of the effect of background radiation on the average normal effective emissivity

where $\varepsilon_e(\boldsymbol{\xi}, \boldsymbol{\omega}, \lambda)$ is the spectral effective emissivity of an isothermal blackbody for non-emitting environment, i.e. determined assuming $T_{bg} = 0$ K, $\varepsilon_e(\boldsymbol{\xi}, \boldsymbol{\omega}, \lambda)$ is the same but for the background temperature $T_{bg} \neq 0$,

$$X(\lambda, T, T_{bg}) = \frac{\exp\left(\frac{c_2}{\lambda T}\right) - 1}{\exp\left(\frac{c_2}{\lambda T_{bg}}\right) - 1} \quad (4.13)$$

is the ratio of Planckian exponents, λ is the wavelength, and $c_2 \approx 1.4388 \times 10^{-2}$ m·K is the second radiation constant.

The Planckian exponent ratio X expressed by Eq. 4.13 depends on the wavelength λ, the temperatures of the cavity T and the background T_{bg}. When the cavity temperature is much higher than T_{bg}, and the spectral range, in which measurements are carried out, are around the peak wavelength of the Planckian curve for the cavity temperature, the correction term is negligibly small. For instance, if the cavity and the background temperatures are 1000 K and 300 K, respectively, and effective emissivity of an isothermal cavity is 0.995, the second term in the right-hand part of Eq. 4.12 expressing the contribution of the background radiation is about 2×10^{-24} at 0.65 μm, 3×10^{-18} at 0.9 μm, 10^{-11} at 1.5 μm, etc. However, the effect of background radiation cannot be neglected for cavities, the effective emissivities of which are not so high and if the difference between temperatures of the cavity and the background is relatively small. The dependences of X on λ for the cavity temperature $T = 300$ K and four temperatures T_{bg} of the background are plotted in Fig. 4.9.

We must remember that the temperature of background is never known precisely. It can change over the hemispherical sold angle, so the isotropy of the background radiation may be violated. Therefore, Fig. 4.9, as well as Eqs. 4.13 and 4.14 are intended to evaluation of order rather than to precise numerical correction for the effect of background radiation. That is why we do not consider the case when $T = T_{bg} = 300$ K.

From the formal point of view, Eq. 4.14 gives $X(\lambda, T, T) \equiv 1$ and then Eq. 4.13 gives $\varepsilon_e(\boldsymbol{\xi}, \boldsymbol{\omega}, \lambda, T, T) \equiv 1$, i.e. the cavity's thermal radiation should reproduce radiation of a perfect blackbody at temperature T. However, it would be too naive to believe that true blackbody conditions can be realized in this manner.

Elementary calculations show that even a relatively small deviation of the background temperature (unknown) from the temperature of the cavity can lead to a noticeable (and uncontrollable) deviation of the effective emissivity from unity.

The analysis of these dependences allows identifying the following patterns:

(i) the absolute value of the deviation of the ratio X from unity grows when λ decreases;
(ii) if $T_{bg} < T$, then $X < 0$ and $X < 0$, if $T_{bg} > T$;
(iii) since the background temperature has the absolute zero as the lower limit, in the shortwave spectral range, the effect of the background temperature grows more rapidly for background temperatures above the cavity temperature than

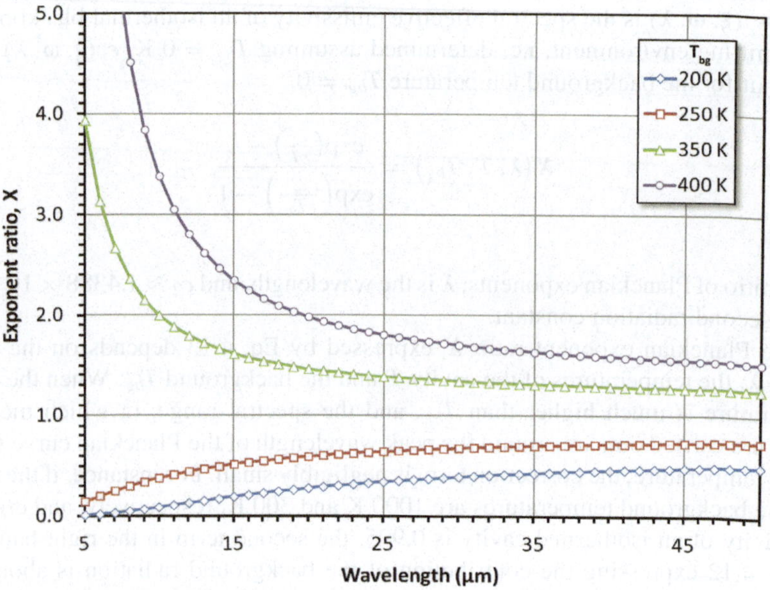

Fig. 4.9 Spectral dependences of the exponent ratios X for the temperature $T = 300$ K of the blackbody and four temperatures T_{bg} of the background

for those below it (this is expressed in the asymmetry of the curves in the figure relative to the horizontal line $X \equiv 1$);

(iv) when λ increases, the ratio of Planckian exponents $X \to T/T_{bg}$ following the Rayleigh-Jeans approximation;

(v) when $T_{bg} \to 0$, $X \to 0$.

The effect of background radiation becomes noticeable when the Planckian spectral radiance of the background radiation reflected by a cavity becomes comparable with the spectral radiance of the cavity's own thermal radiation. Correction for the background radiation is necessary for large-aperture low-temperature blackbodies unless they are used within special cryogenic chambers. However, the most effective technique is to reduce the background temperature and, at the same time, to increase the effective emissivity of the isothermal cavity.

4.1.4 Effective Emissivity of Nonisothermal Cavity

The effective emissivity is the quantity introduced to simplify calculations of the characteristics of the radiation emitted by artificial blackbodies. The spectral radiance L_λ of an isothermal blackbody can be written out as

$$L_\lambda(\lambda) = \varepsilon_e(\lambda) L_{\lambda,bb}(\lambda, T), \tag{4.14}$$

where $\varepsilon_e(\lambda)$ is the spectral effective emissivity and $L_{\lambda,bb}$ is the spectral radiance of a perfect blackbody calculated according the Planck radiation law.

The situation with nonisothermal cavities is more complicated. If the temperature is not uniform over the cavity internal surface, the Planckian spectral radiances corresponding to different temperatures are subjected to repeated reflections inside the cavity, so that the spectral radiance of the radiation emerging from the cavity is determined by a certain weighted average. One can choose a temperature T_0, at which the Eq. 4.14 is fulfilled for a single predefined wavelength λ_0:

$$L_\lambda(\lambda_0) = \varepsilon_e(\lambda_0, T_0) L_{\lambda,bb}(\lambda_0, T_0) = \varepsilon_e(\lambda_0) L_{\lambda,bb}(\lambda_0, T_0), \tag{4.15}$$

where ε_e is computed for the chosen temperature T_0 but is not a function of temperature T.

However, it is impossible to choose a temperature, for which Eq. 4.14 holds for any wavelength λ. The commonly accepted approach for nonisothermal cavities is introducing the so-called *reference* temperature T_{ref} and defining the effective emissivities, which depend on T_{ref}. Again, we consider a cavity in the perfectly black surroundings at zero absolute temperature. Under these conditions, the spectral local directional effective emissivity for a non-isothermal cavity is defined as

$$\varepsilon_e(\boldsymbol{\xi}, \boldsymbol{\omega}, \lambda, T_{ref}) = \frac{L_\lambda(\boldsymbol{\xi}, \boldsymbol{\omega}, \lambda)}{L_{\lambda,bb}(\lambda, T_{ref})}, \tag{4.16}$$

where L_λ is the spectral radiance of the element dA of the nonisothermal blackbody and $L_{\lambda,bb}$ is the spectral radiance of a perfect blackbody at a reference temperature T_{ref}. Strictly speaking, ε_e as well as the numerator in the right part of Eq. 4.16 depends also on the temperature distribution $T(\boldsymbol{\xi})$ but we omit this dependence for brief.

The reference temperature T_{ref} is introduced only to avoid ambiguity in determination of the effective emissivity of a nonisothermal cavity. The reference temperature can be chosen arbitrarily because it is a fictitious quantity to the same extent as the effective emissivity. The actually measurable quantity is the spectral radiance; its value does not depend upon the choice of the reference temperature:

$$\begin{aligned} L_\lambda(\boldsymbol{\xi}, \boldsymbol{\omega}, \lambda) &= \varepsilon_e(\boldsymbol{\xi}, \boldsymbol{\omega}, \lambda, T_{ref}) L_{\lambda,bb}(\lambda, T_{ref}) \\ &= \varepsilon_e(\boldsymbol{\xi}, \boldsymbol{\omega}, \lambda, T'_{ref}) L_{\lambda,bb}(\lambda, T'_{ref}) \end{aligned} \quad \forall(\boldsymbol{\xi}, \boldsymbol{\omega}, \lambda, T_{ref}, T'_{ref}). \tag{4.17}$$

For the reference temperature value, one can choose either some average temperature of a cavity or the temperature at a certain point of a cavity (e.g. at the point where the temperature is measured by a contact sensor). In any case, it is more practical to choose a reference temperature so that the effective emissivity of the non-isothermal cavity is of the same order as the corresponding isothermal.

4.1.5 Effect of Temperature Non-uniformity

The effective emissivity of an isothermal cavity does not depend on the wavelength if the emissivity of the cavity walls is wavelength-independent (the gray surface approximation). The effective emissivity of a nonisothermal cavity depends on wavelength even if its surface is gray. Temperature heterogeneity of the cavity can manifest itself directly (when the area of the cavity within the detector's FOV is nonisothermal) and indirectly (when a nonisothermal area of the cavity is outside the detector's FOV). In practice, we most often deal with a combination of both cases. The temperature non-uniformity of the directly viewable area of the cavity surface affects stronger on spectral dependence of the effective emissivity than that of surfaces, which are outside the detector's FOV.

Below, we give two examples of the effect of the cavity temperature non-uniformity on the spectral effective emissivity of the cavity shown in Fig. 4.10a. The cavity consists of the right circular cylinder with the radius $R = 1$ and the conical bottom with the apex angle of 120°. The total length of the cavity $L = 10$. The cavity internal surface is supposed to be perfectly diffuse, i.e. its thermal emission and reflectionobey the Lambert law. It is also supposed (for simplicity of the analysis of results) that the cavity walls are gray and have the uniform emissivity of 0.9. The cavity radiation passing through the external aperture of the radius $R_a = 0.5$ is registered by the infinitely distant detector of the same radius (the average normal effective emissivitycorresponds to these viewing conditions). Figure 4.10b, c present two families of temperature distributions. In the first family, the temperature of the conical bottom is constant and equal $T_0 = 2000$ K. The temperature changes linearly along the cylindrical part reaching temperatures of 1950, 1925, 2025, and 2050 K at the cavity's open end. In the second family, the temperature of the cylindrical part is constant and equal $T_0 = 2000$ K while the temperature changes linearly toward the center of the conical bottom reaching temperatures of 1995 K, 1990 K, 2005 K, and 2010 K. The reference temperature $T_{ref} = 2000$ K was adopted for all temperature distributions. We have performed calculations using the Monte Carlo ray-tracing program STEEP321 from Virial International, LLC (www.virial.com). The calculation results for the first family of temperature distributions are plotted against wavelength in Fig. 4.11.

Since the temperature of the conical bottom is the same for all temperature distributions of the first family, the dependence of the effective emissivity on the wavelength can occur only due to the thermal radiation emitted by the non-isothermal cylindrical part and reached the detector after reflections from the conical bottom surface. When the temperature along the cylindrical part decreases toward the cavity opening, the corresponding spectral emissivity curves lie below the horizontal line that corresponds to the effective emissivity of the isothermal cavity. Conversely,

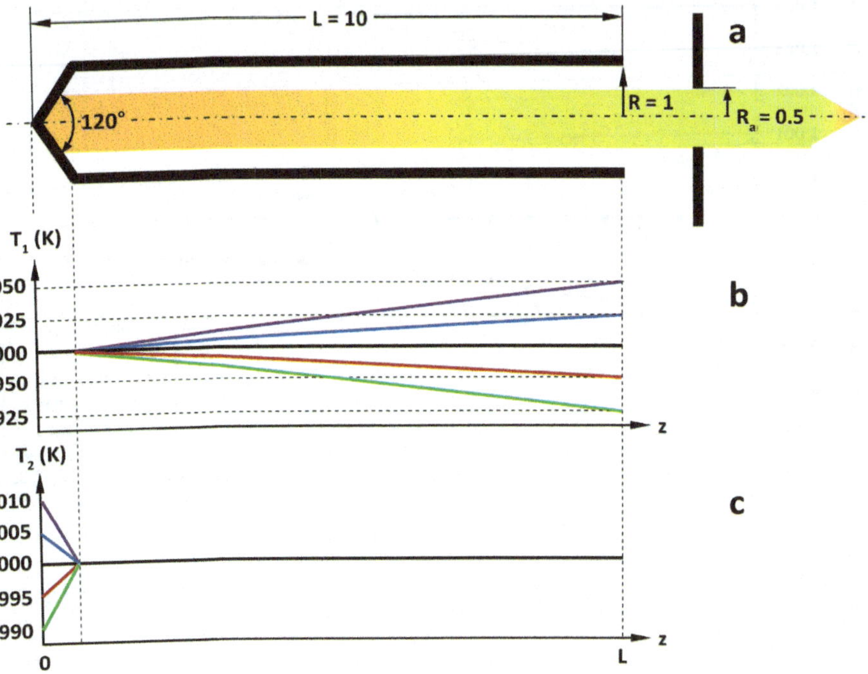

Fig. 4.10 The model data for the example on effective emissivities of nonisothermal cavity: **a** geometry of the cavity and viewing conditions, **b** the first family of temperature distributions $T_1(z)$, and **c** the second family of temperature distributions $T_2(z)$

when the temperature along the cylindrical part increases toward the cavity opening, the corresponding spectral effective emissivities lie above the effective emissivity of the isothermal cavity. The calculation results for the second family of temperature distributions are plotted against wavelength in Fig. 4.12. The effective emissivities for the temperature distributions of the second family look quite similar to the graphics of Fig. 4.11 but the range of their changes is much larger.

While the influence of temperature inhomogeneity for the distributions of the first family is weakened by reflection from the surface of the conical bottom, for the temperature distributions of the second family there is no such weakening. The own thermal radiation of the non-isothermal bottom of the cavity brings the main contribution to the spectral dependence of the effective emissivity.

A much smaller contribution (its order is the same as for the temperature distributions of the first family) belongs to the radiation of the cavity bottom, which undergoes at least one reflection before it reaches the detector.

By analogy with the effective emissivities of isothermal cavities, the corresponding total and band-limited effective emissivities, as well as the effective emissivities for certain viewing conditions can be introduced for nonisothermal cavities.

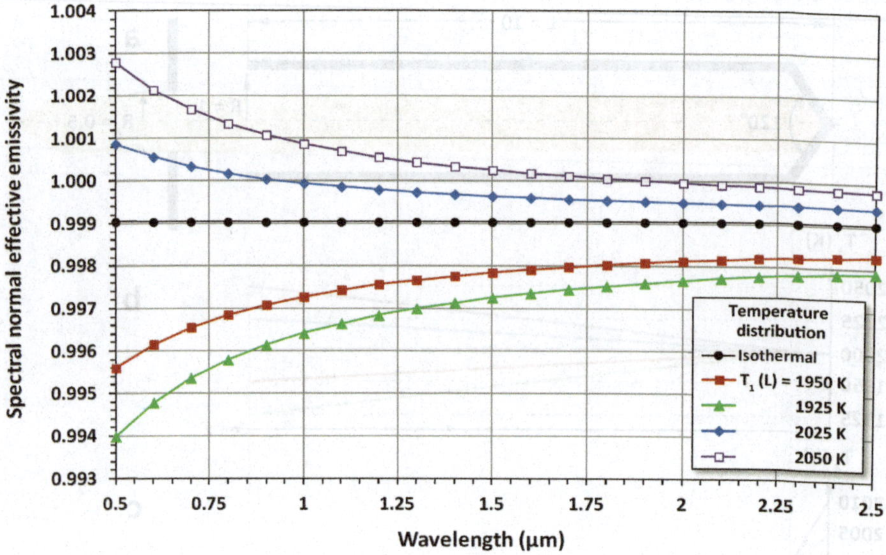

Fig. 4.11 Spectral normal effective emissivities of the nonisothermal cavity shown in Fig. 4.10a for the first family of temperature distributions (see Fig. 4.10b)

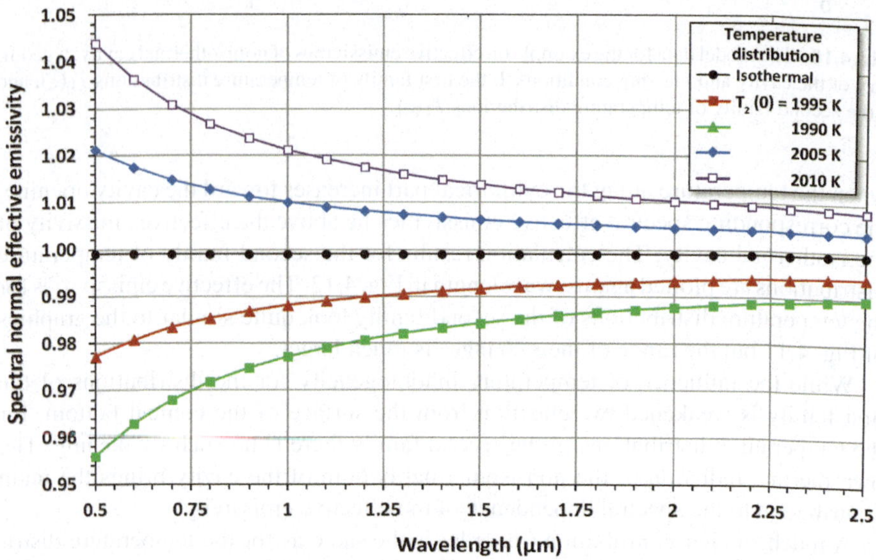

Fig. 4.12 Spectral normal effective emissivities of the nonisothermal cavity shown in Fig. 4.10a for the second family of temperature distributions (see Fig. 4.10c)

It should be noted especially, that the effective emissivities of nonisothermal cavities can be greater than unity at some choices of the reference temperature.

The effect of background radiation does not depend on the temperature distribution over the cavity wall; in this sense, it is additive effect:

$$\varepsilon_e\left(\lambda, \boldsymbol{\xi}, \boldsymbol{\omega}, T_{ref}, T_{bg}\right) = \varepsilon_e\left(\lambda, \boldsymbol{\xi}, \boldsymbol{\omega}, T_{ref}\right) + \left[1 - \varepsilon_e(\lambda, \boldsymbol{\xi}, \boldsymbol{\omega})\right] \frac{\exp\left(\frac{C_2}{\lambda T_{ref}}\right) - 1}{\exp\left(\frac{C_2}{\lambda T_{bg}}\right) - 1}, \tag{4.18}$$

where the augend in the right-hand part is the effective emissivity of the nonisothermal cavity determined for a non-radiating background and the first multiplier in the addend is the effective reflectance of the isothermal cavity.

From a physical point of view, this means that we can, at least in principle (for example, using a hypothetical device that modulates background radiation), separate the radiation emitted by the cavity itself from that reflected from the cavity to the sensor.

4.2 Approximate Formulae for Effective Emissivities of Isothermal Diffuse Cavities

Due to the necessity of predicting the effective emissivities of blackbody radiators at the initial stage of their design and the difficulties of accurate measurement of the effective emissivities of ready-to-use cavities (see Sect. 4.6), computational methods remain the major tools for their determination. For some time after the introduction of artificial blackbodies into radiometric practice, a low level of accuracy in measuring radiant energy made it possible to consider the radiation of a sufficiently small opening in any isothermal (to one degree or another) cavity as the radiation of a perfect blackbody.

Perhaps, the first mathematically rigorous analysis of the cavity thermal radiation was carried out by Buckley [16] for semi-infinite isothermal diffuse cylinders, which were used as sources of blackbody radiation in optical pyrometry. However, the mathematically rigorous integral equation could only be solved at that time by an approximate analytical method. Therefore, until the appearance in the 1960s of digital computers available to researchers of the cavity radiation, their attention was focused on approximate analytical methods.

The simplest approximate analytical formulae were derived for an isothermal diffuse cavity by Ribaud [132] assumed that all radiation is absorbed by the cavity after a single reflection:

$$\varepsilon_e = 1 - \rho \frac{\Omega}{\pi}, \tag{4.19}$$

where $\rho = 1 - \varepsilon$ is the reflectance of a cavity wall, ε is its emissivity, and Ω is the solid angle subtended by the cavity aperture from the center of the area irradiated by an infinitely thin incident ray.

Since only the first reflection is taken into account, Eq. 4.19 gives results of more or less acceptable accuracy only for cavities with very "black" surfaces. Some improvement in results can be achieved by replacing the ratio Ω/π with the view factor F_a from the initially irradiated (viewable) area to the cavity aperture.

Gouffé [46] considered a cavity of arbitrary shape, with the aperture area s and the sum S of areas of the cavity surface and the aperture. He assumed that radiation reflected by the cavity wall and remaining inside the cavity is uniformly distributed over the cavity surface after each reflection. Summing up the radiation fluxes emerging from the cavity after each reflection and using the expression for an infinite geometric progression, Gouffé obtained the following formula for effective emissivity:

$$\varepsilon_e = \frac{\varepsilon\left[1 + (1-\varepsilon)\left(\frac{s}{S} - \frac{\Omega}{\pi}\right)\right]}{\varepsilon\left(1 - \frac{s}{S}\right) + \frac{s}{S}}. \tag{4.20}$$

Equation 4.20 gives the exact value of the effective emissivity only for one 3D shape—for an isothermal diffuse sphere, for which the assumption about uniform distribution of the reflected radiation over the cavity walls, holds exactly. The exact expression for the effective emissivityof isothermal diffuse spherical cavity can be used for verification of more advanced computational methods. One can be easily shown that Eq. 4.20 coincides with the expression provided for the isothermal diffuse spherical cavity by Sparrow and Cess [148]:

$$\varepsilon_e = \frac{\varepsilon}{1 - 0.5 \cdot (1-\varepsilon)(1+\cos\psi)}, \tag{4.21}$$

where ψ is the flat half-angle having the vertex at the sphere center and subtended by a circular opening.

An interesting method was proposed by Treuenfels [156]. He represented the effective emissivity of a diffuse cavity in the form:

$$\varepsilon_e = 1 - \sum_{k=1}^{\infty} f_k \rho^k, \tag{4.22}$$

where f_k is the fraction of the radiation flux, reflected k times before emerging from the cavity and

$$\sum_{k=1}^{\infty} f_k = 1. \tag{4.23}$$

Treuenfels made a somewhat speculative assumption that the ratio of successive fractions $\beta = f_{k+1}/f_k < 1$ is constant for all $k \geq 1$. This allows deriving the following equation for the effective emissivity by summing up the geometrical progression:

$$\varepsilon_e = \frac{\varepsilon}{\varepsilon + f_1(1-\varepsilon)}, \tag{4.24}$$

where

$$f_1 = \int_A F_a^2(\xi)dA \Big/ \int_A F_a(\xi)dA, \tag{4.25}$$

where A is the cavity internal surface, $F_a(\xi)$ is the diffuse view factor from a surface element dA at a position $\boldsymbol{\xi}$ to the cavity aperture. Using analytical integration, Treuenfels obtained expressions for f_1 for several simple geometries such as spherical cavity and infinite grooves of cylindrical, triangular, and rectangular profile.

Although the method proposed by Treuenfels [156] requires more complex calculations than the Gouffé method, it does not provide a guaranteed increase in accuracy for calculating the effective emissivity of arbitrary-shaped cavities.

Quinn [127] derived the second-order expressions for the normal effective emissivities of the bottom center of an isothermal diffuse cylindrical cavity. He relied on the method of successive reflections developed by De Vos [29], a very powerful method, equally suitable for isothermal and nonisothermal cavities with an arbitrary angular reflection of the walls, but requiring extremely complex multiple analytical integration. For the center of thecylindrical cavity bottom, Quinn obtained

$$\varepsilon_e = 1 - \rho/D^2 - \rho^2 I_2, \tag{4.26}$$

where $D = L/R$, L and R are the cavity depth and radius, respectively,

$$I_2 = \frac{1}{\pi}\int_0^D \int_{-\frac{\pi}{2}}^{\frac{\pi}{2}} \left[\tan^{-1}\left(\frac{2\cos\varphi}{x}\right) - \frac{2x\cos\varphi}{x^2+4\cos^2\varphi}\right]\frac{(D-x)d\varphi\, dx}{\left[1+(D-x)^2\right]^2}. \tag{4.27}$$

For a cylinder with a diaphragm and a small radius R_0 of the aperture ($R_0 << R$), Quinn derived a simple approximate expression

$$\varepsilon_e = 1 - \rho\cdot(a/D)^2 - \rho^2(R_0/R)^2 I_2', \tag{4.28}$$

where

Table 4.1 The values of the integrals I_2 and I_2' as functions of dimensionless depth $D$of cylindrical cavities according to Quinn [127]

$D = L/R$	I_2	I_2'	$D = L/R$	I_2	I_2'	$D = L/R$	I_2	I_2'
4	0.0508	0.0488	10	0.0035	0.0031	16	0.0008	0.0007
5	0.0282	0.0267	11	0.0025	0.0022	17	0.0006	0.0005
6	0.0168	0.0155	12	0.0019	0.0017	18	0.0005	0.0005
7	0.0105	0.0096	13	0.0015	0.0013	19	0.0004	0.0004
8	0.0070	0.0063	14	0.0012	0.0010	20	0.0004	0.0003
9	0.0049	0.0043	15	0.0009	0.0008			

$$I_2' = 2\int_0^D \frac{x(D-x)dx}{\left(1+x^2\right)^2\left[1+(D-x)^2\right]^2}. \tag{4.29}$$

The integrals I_2 and I_2' computed numerically are tabulated for $D = L/R$ ranging from 4 to 20 in Table 4.1. Figure 4.13 presents the dependences of effective emissivities of cylindrical cavities on D calculated according to formulae of Gouffé (Eq. 4.21) and Quinn (Eqs. 4.26 and 4.27), as well as computed by the Monte Carlo method[1] verified by comparison with the results obtained by the method of integral equations for diffuse cavities [10]. The effective emissivities were compared for two cavities shown in the inset in Fig. 4.13.

The cavity A is the cylindrical cavity without diaphragm; the cavity B is the cylindrical cavity with the diaphragm of radius $R_0 = 0.25 \cdot R$. The emissivity of diffuse cavity walls $\varepsilon = 0.8$ for both cases. As can be seen in Fig. 4.13, Gouffé's formula gives acceptable results for not too deep ($L/R < 5$) cylinders without diaphragm but fails for deeper cavities due to assumption of uniform distribution of reflected radiant fluxes over the cavity surface. For the cylinder with the diaphragm, the error in calculated effective emissivity is much lower because the radiant flux distributions become more uniform after multiple reflections in the cavities of higher effective emissivity.

Equation 4.26 of Quinn's method gives accurate results for cylinder without diaphragm but fails for the cavity with the diaphragm. For the used aperture radius $R_0 = 0.25 \cdot R$, the condition $R_0 \ll R$ of applicability of Eq. 4.28 is not fulfilled, and the restriction to terms of only 2nd order with respect to reflectance ρ becomes insufficient.

The specular component is present in reflection of virtually any material, so the radiating cavities of practical blackbodies rarely have a simple cylindrical shape. To prevent a decrease in the normal effective emissivity, the cavity bottom is usually made conical, concave (the case of cylindro-conical cavity) or convex (so-called cylinder-inner-cone cavity). For a cylindro-conical cavity, some modifications of

[1]The Monte Carlo ray tracing code STEEP321 from Virial International, LLC (www.virial.com) was used for this purpose.

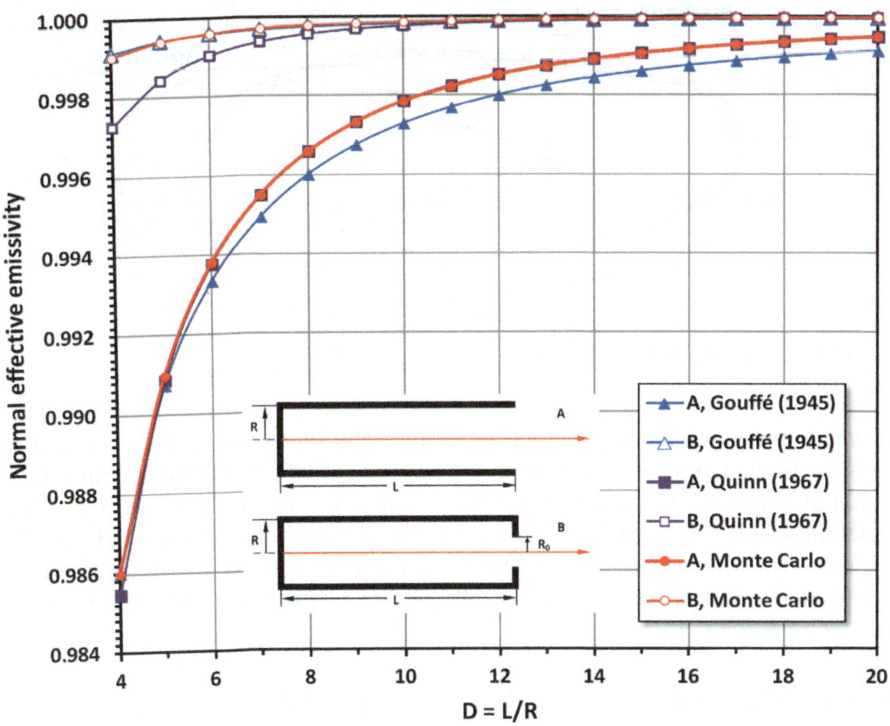

Fig. 4.13 Dependences of the normal effective emissivities of cylindrical cavities on the dimensionless depth $D = L/R$ computed according to formulae of Gouffé, Quinn, and by the Monte Carlo method for the cavity without diaphragm (A) and with the diaphragm of radius $R_0 = 0.25 \cdot R$ (B); the emissivity of cavity walls $\varepsilon = 0.8$

Eq. 4.28 have been proposed [80, 128]. However, these equations were heuristic, rather than physically based, and were not as widely used as Quinn's original formulas.

None of the approximate analytical formulae takes into account the actual temperature distribution over the radiating surface of the cavity. Nevertheless, approximate analytical formulae of Gouffé, Quinn, and the like are employed for estimations of effective emissivity up to date, mainly for applications that do not require the highest accuracy (see, e.g. [31, 169], or [32]).

4.3 Method of Integral Equations for Effective Emissivities of Diffuse Cavities

The method of integral equations is a standard method of the theory of radiation heat transfer in the system of opaque diffuse surfaces, i.e. a cavity, internal surface

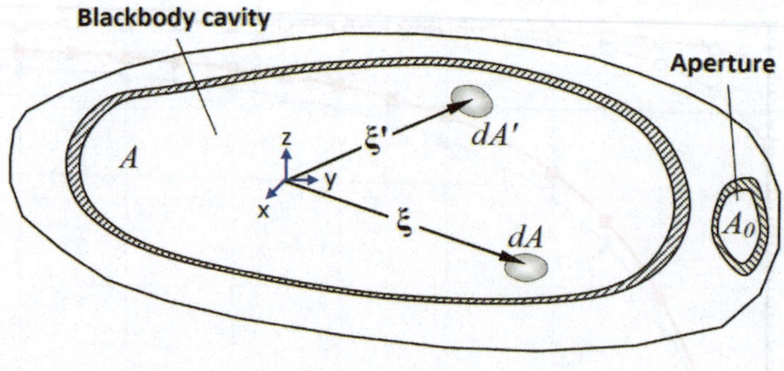

Fig. 4.14 A schematic of the diffuse cavity

of which emits and reflects thermal radiation according to Lambert's cosine law. Figure 4.14 depicts such a cavity with the internal surface A and the aperture A_0.

Let us consider two surface elements dA and dA' with positional vectors $\boldsymbol{\xi}$ and $\boldsymbol{\xi}'$, temperatures $T(\boldsymbol{\xi})$ and $T(\boldsymbol{\xi}')$, and the spectral emissivities $\varepsilon(\boldsymbol{\xi}, \lambda)$ and $\varepsilon(\boldsymbol{\xi}', \lambda)$, respectively. We assume the emissivity of the cavity walls to be temperature-independent. The spectral radiosity[2] B_λ of the element dA can be expressed as:

$$B_\lambda(\boldsymbol{\xi}, \lambda, T(\boldsymbol{\xi})) = \varepsilon(\boldsymbol{\xi},\lambda)L_{\lambda,bb}(\lambda, T(\boldsymbol{\xi})) + [1 - \varepsilon(\boldsymbol{\xi}, \lambda)] \int_A B_\lambda(\xi', \lambda, T(\xi'))\, dF_{dA'-dA}, \tag{4.30}$$

where $dF_{dA'-dA}$ is the view factor between the surface elements dA' and dA. If to divide both parts of Eq. 4.30 by the Planckian spectral radiance $L_{\lambda,bb}(\lambda, T_{ref})$ at a reference temperature T_{ref}, one can obtain the following integral equation for the distribution of the local hemispherical spectral effective emissivity $\varepsilon_{e,h}(\boldsymbol{\xi}, \lambda, T_{ref})$:

$$\varepsilon_{e,h}(\boldsymbol{\xi}, \lambda, T_{ref}) = \varepsilon(\lambda)\frac{\exp\left(\frac{c_2}{\lambda T_{ref}}\right) - 1}{\exp\left(\frac{c_2}{\lambda T(\boldsymbol{\xi})}\right) - 1} + [1 - \varepsilon(\lambda)] \int_A \varepsilon_{e,h}(\xi', \lambda, T_{ref})dF_{dA'-dA}. \tag{4.31}$$

In the same way, the integral equation can be written for the distribution of the local hemispherical total effective emissivity $\varepsilon_{e,h}(\boldsymbol{\xi}, T_{ref})$:

---

[2]The spectral radiosity of the element dA is the spectral radiant exitance for the sum of its own thermal radiation and the radiation falling onto the element dA from the rest of the surface A and then reflected by dA [148].

$$\varepsilon_{e,h}(\boldsymbol{\xi}, T_{ref}) = \varepsilon\left[\left(\frac{T(\boldsymbol{\xi})}{T_{ref}}\right)\right]^4 + (1-\varepsilon)\int\limits_A \varepsilon_{e,h}(\xi', T_{ref})dF_{dA'-dA}, \quad (4.32)$$

where ε is the total emissivity of the cavity walls.

Equations 4.31 and 4.32 become somewhat simpler for an isothermal cavity:

$$\varepsilon_{e,h}(\boldsymbol{\xi}, \lambda) = \varepsilon(\lambda) + [1-\varepsilon(\lambda)]\int\limits_A \varepsilon_{e,h}(\xi', \lambda)dF_{dA'-dA}, \quad (4.33)$$

$$\varepsilon_{e,h}(\boldsymbol{\xi}) = \varepsilon + (1-\varepsilon)\int\limits_A \varepsilon_{e,h}(\xi')dF_{dA'-dA}. \quad (4.34)$$

Equations 4.31 and 4.33 have to be resolved with respect to $\varepsilon_{e,h}(\boldsymbol{\xi}, \lambda, T_{ref})$ and $\varepsilon_{e,h}(\boldsymbol{\xi}, \lambda)$ for each λ. Mathematically, Eqs. 4.31 through 4.34 are Fredholm's integral equations of the 2nd kind that can be written in general form as

$$y(x) = g(x) + k\int\limits_a^b y(x')K(x, x')dx', \quad (4.35)$$

where $y(x)$ is the unknown function, $g(x)$ is the given (known) function, $K(x, x')$ is the given function of two variables (so-called kernel of the integral equation), $a \leq x, x' \leq b$.

A variety of methods, analytical and numerical, has been developed for solution of such integral equations (see, e.g. [76, 166]). However, Eqs. 4.31 through 4.34 have some features that make solving these equations a non-trivial task:

(i) Blackbody cavities commonly served as the standard/calibration radiation sources are usually composed of cylindrical and conical surfaces of revolution, flat, and (less often) spherical surfaces. As a result, the integral equation for the entire cavity consisting of N surfaces splits into the system of N integral equations written for each surface constituting the cavity. For example, for the widespread cavity shape consisting of a conical bottom, a cylindrical middle part, and a flat annular lid, the system of three integral equations must be written and solved.

(ii) Integral equations become singular on the lines of junction of separate surfaces, where the normal vector changes its direction abruptly; this leads to discontinuities in distribution of the local hemispherical effective emissivities.

(iii) The analytical expressions for the view factors in some cases (as for the apex of the conical surface—see Kelly [71]) have an indeterminate form and cannot be computed directly, without preliminary application of L'Hôpital's rule.

(iv) Significant efforts may be required to derive analytical expressions for view factors for the surface elements in the presence of intermediate surfaces that obstruct the radiative heat exchange between the elements considered.
(v) The standard quadrature rules applied to the integral equations for some cavities lead to the ill-conditioned systems of linear equations, solution of which may require excessive computational resources.
(vi) Solution of the integral equations (or their systems) for Lambertian cavities gives the distributions of the local hemispherical effective emissivities. They must be additionally integrated over the viewable surface of a cavity in order to obtain the effective emissivities for particular viewing conditions. For the integrated effective emissivities, the integration boundaries vary over the sensor's surface. Additional difficulties occur for baffled cavities due to the "penumbra effect."

Analytical solution can be found in the close form only for diffuse spherical cavities. Attempts to solve numerically the integral equations for cavities of at least the simplest form (conical, cylindrical, etc.) have been made since the advent of digital computers (see, e.g. [17, 108, 145]). Bedford and Ma [10] noted that most of the early calculations are not accurate enough due to not entirely correct handling of singularities and proposed a technique to overcome this problem.

The essence of the method of *zonal approximation* proposed by Bedford and Ma [10] is addressed to elimination of singularities in the kernels of the integral equations. The essence of the method is the replacement of the integral over the interval [0, L] by the sum of integrals over n equal sub-intervals:

$$\int_0^L \varepsilon_{e,h}(z')dF_{z,z'} = \sum_{i=1}^{n} \int_{z_i'}^{z_i'+\Delta z} \varepsilon_{e,h}(z')dF_{z,z'}, \tag{4.36}$$

where $z_1 = 0$, $\Delta z = L/(n-1)$, $dF_{z,z'}$ is the view factor between elemental rings having axial coordinates z and z'. By assuming $\varepsilon_{e,h}(z)$ to be a slowly varying function of z, one can write:

$$\int_0^L \varepsilon_e(z')dF_{z,z'} = \sum_{i=1}^{n} \frac{\varepsilon_e(z_i') + \varepsilon_e(z_i' + \Delta z)}{2} \int_{z_i'}^{z_i+\Delta z} dF_{z,z'}. \tag{4.37}$$

Each surface is subdivided into a large number of small areas, over which the local effective emissivity is assumed constant. After analytical integration in the right-hand part of Eq. 4.37, the view factor between two infinitesimal elements is replaced by the sum of the view factors between a finite area and an infinitesimal element. This allows eliminating singularity and replacing the integrals by the sums. The resulting system of linear algebraic equations can be easily solved by the standard methods.

This approach was applied to calculation of the distributions of the local hemispherical effective emissivities of conical and cylindrical [10], cylindro-conical [11], and double-conical [12] isothermal and non-isothermal diffuse cavities. The results obtained by Bedford and Ma are used up to date for verification of calculations carried out using other methods. Later Bedford et al. [13], applied this method to a cylinder-inner-conical cavity, in which the reentrant conical bottom acts as an obstacle for the radiative heat exchange between ring elements on the cylindrical surface.

The typical distributions of the local hemispherical effective emissivities $\varepsilon_e(\xi)$ of the isothermal diffuse cylindro-conical cavity along the dimensionless coordinate $\xi \in [0,\ 1]$ for the wall emissivities $\varepsilon = 0.7$, 0.8, and 0.9 are plotted in Fig. 4.15, in the inset of which the cavity is presented. The coordinate ξ is started at the center of the cavity bottom, where it equals 1, passes by the shortest path along the cavity surface toward the aperture, where $\xi = 1$. The functions $\varepsilon_e(\xi)$ are discontinuous at the joints of (i) the conical bottom and cylindrical part and (ii) the cylindrical part and flat diaphragm.

By themselves, these dependences cannot be directly used but the effective emissivities for specific viewing conditions are calculated with their help. Since the local effective emissivities a cavity (isothermal or not) with diffuse walls also obey the

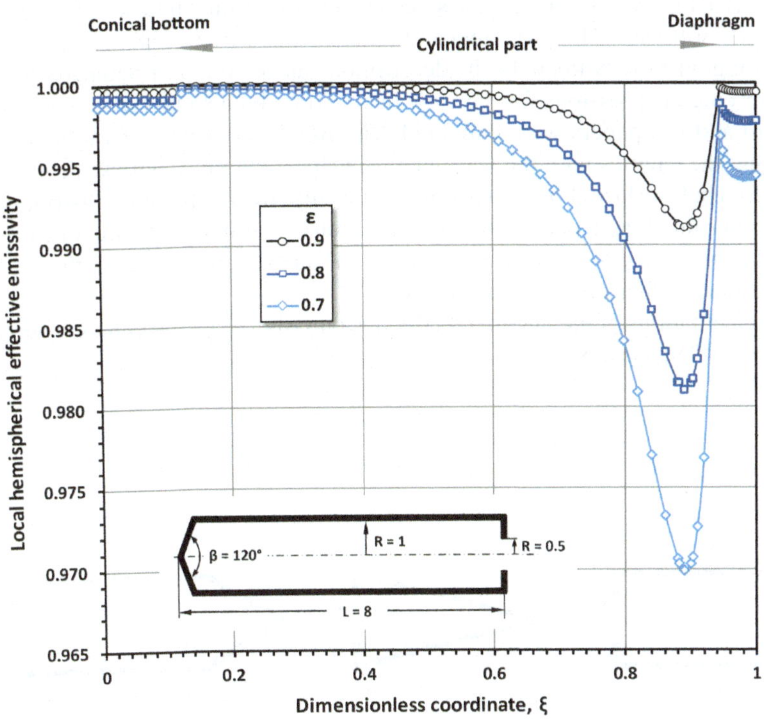

Fig. 4.15 Distributions of the local hemispherical effective emissivities for three values of wall emissivity ε along the generatrix of the isothermal diffuse cylindro-conical cavity shown in the inset

Lambert law, the calculation of the effective emissivity for specific viewing conditions is reduced to an integral averaging over the directly viewable area. No problems arise if the radiation detector can be considered as infinitely distant or infinitely small. For a detector of a finite size located at a finite distance from the cavity, the averaging becomes more complicated because different points on the detector's surface "view" different areas of the cavity (see Fig. 4.3b) for the viewing conditions that correspond to the integrated effective emissivity of the cavity with diaphragm).

There were several works dedicated to rigorous mathematical formulation of the integrated effective emissivity problem [18, 20, 22, 101]. Recently, Saunders [138, 139] formulated this problem for more realistic viewing conditions shown in Fig. 4.3d. The virtual aperture in this figure plays the same role as the detector plays in Fig. 4.3b,the role of the aperture of the cavity with the diaphragm in Fig. 4.3d plays the measurement spot (an image of the virtual aperture). The measurement spot of a radiation thermometer or radiometer, lying in the focal plane of the optical system, usually coincides with the cavity aperture (the optical system is said to focus on the cavity aperture), but may be located in front of it or even be placed inside the cavity.

Dr. Peter Saunders of Measurement Standards Laboratory of New Zealand developed the *Blackbody Emissivity* software that reproduces the Bedford and Ma's technique for isothermal and nonisothermal diffuse cavities of three simple shapes shown in Fig. 4.16. Figure 4.17 present the screenshot of the main window of the *Blackbody Emissivity* software. The program can be downloaded free of charge at https://measurement.govt.nz/download/44. Its description can be found in Saunders [139]. He also presented expressions for the view factors, in which some typos in the original Bedford and Ma papers were corrected. The *Blackbody Emissivity* software allow modeling the cavities with the wavelength-independent emissivities specified separately for each surface. For a nonisothermal cavity, the temperature distribution can be set constant along a part of the cavitygeneratrix and linear or parabolic along the rest. The effective emissivities are calculated as an average value over the waveband of the radiation thermometer (radiometer).

For calculations of the integrated effective emissivity ε_{bb}, the equation derived by Chen et al. [18] was applied:

$$\varepsilon_{bb} = \frac{\int_C \int_S \frac{\varepsilon_e(\xi)\cos\psi_\xi \cos\psi_s}{\pi d^2} dS dA_\xi}{\int_C \int_S \frac{\cos\psi_\xi \cos\psi_s}{\pi d^2} dS dA_\xi}, \tag{4.38}$$

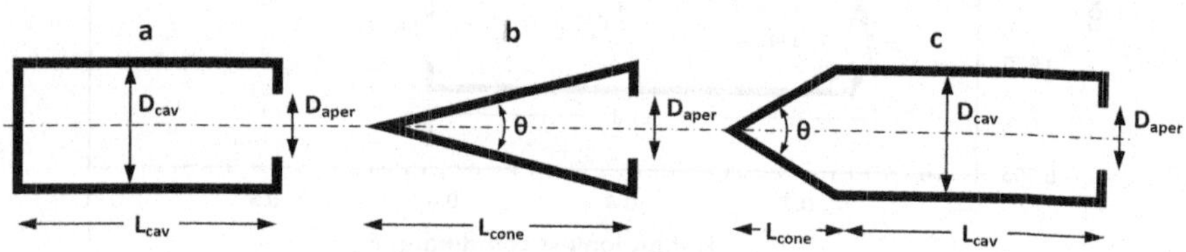

Fig. 4.16 Geometrical parameters of cavities supported by the *Blackbody Emissivity* software

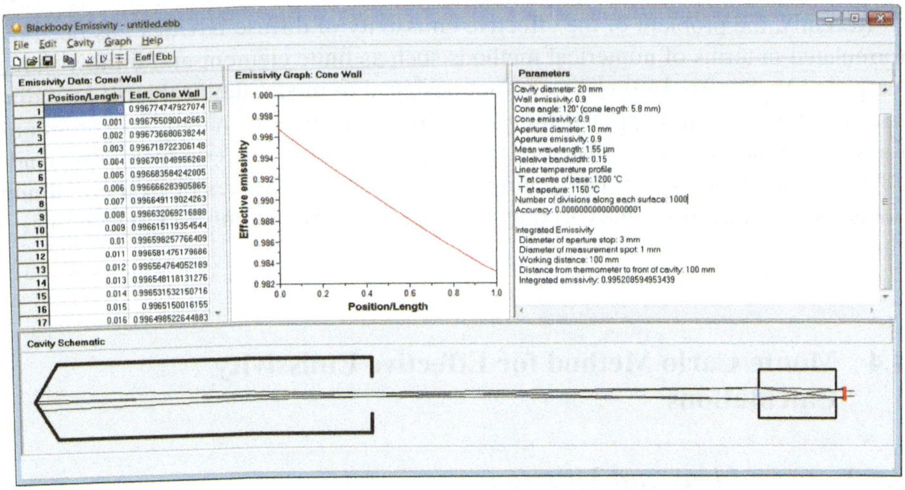

Fig. 4.17 Screenshot of the main window of the *Blackbody Emissivity* software

where ξ is the coordinate of the element dA_ξ of area on the cavity surface, dS is an element of area on the virtual aperture, d is the distance between these two elements, ψ_ξ and ψ_s are the angles between the normals to these elements and the line connecting them.

Each surface integral in Eq. 4.38 can be represented by the triple nested integral over umbral and penumbral regions; to calculate triple integrals, Saunders [139] employed the Gauss–Legendre quadrature rule for the node number $N = 20$, which gives exact values of integrals for polynomial of order up to $2N - 1$ (see, e.g. [79]).

It is now becoming clear that the technique proposed by Bedford and Ma to solve the problem of singularity is not the only possible. One of the alternative ways is to reject integral equations in favor of a system of algebraic equations for surfaces of finite area. De Paulo Nicolau [28] applied the Gebhart coefficients method [37] to the calculation of the effective emissivities of acylindrical cavity. The Gebhart method is one of the possible formulations for calculating the radiant interchange within the enclosure. Sparrow [144] proved the equivalence of Gebhart's formulation to that based on the view factors. Unfortunately, de Paulo Nicolau [28] did not provide the details of computations. It was reported only that the number of discretization rings of the cavity cylindrical part $N = 200$ is enough to calculate the effective emissivity for the cavity length-to-radius ratio of up to 10 and that "the results show a good agreement with Bedford and Ma."

Another possible approach is to use the quadrature rules of the highest accuracy, for example, the Gauss or Chebyshev formulae [15, 79] applied to the cavity surface, previously divided into the required number of finite-area regions. The nodes of these quadrature formulas, unlike the Simpson quadrature rule, do not coincide with the edges of the integration interval, so the singularity problem is completely eliminated [158].

Recently, the problem of the effective emissivity of diffuse cavities has been reformulated in terms of numerical methods such as finite element and finite volume methods [36, 82, 93, 159] originally not designed to model the problems of radiative heat transfer. This approach allows avoiding derivation of analytical expressions for view factors and opens the way to employ the solvers of commercially available software packages for calculating the effective emissivities of isothermal and nonisothermal diffuse cavities of complicated geometry, including non-axisymmetric shapes.

4.4 Monte Carlo Method for Effective Emissivity Calculations

4.4.1 Core Ideas and Milestones

The Monte Carlo method (MCM) is the generic name for various computational algorithms based on the repetitive random sampling and subsequent use of the appropriate estimator in order to obtain the expectation value as an approximation of the exact result. A short brochure written by Sobol' [143] is very suitable for the first acquaintance with MCM and at the same time inspires the continuation of this acquaintance. Many excellent monographs could be recommended for in-depth study of the MCM, among which we would highlight the books by Kroese et al. [77] and Dunn and Shultis [33]. The MCM is widely used for modeling of various physical problems including the particle transport [34, 48] and the optical radiation transfer [35, 86, 87, 164].

The application of the MCM to solution of computational problems of radiative heat transfer is based on a probabilistic treatment of the interaction of the optical radiation and matter. This approach allows building a stochastic model of the system under investigation and estimating its parameters on the base of a large number of realizations of the stochastic process, which simulates the real process of the radiation-matter interactions. The radiation transfer is treated as the tracing of separate rays (or particles—the difference is only terminological within frameworks of geometrical optics).

Many computational problems of optical radiometry [114] can be also successfully resolved using the MCM. The effective emissivitycalculation is among these problems. A large computation time required to achieve a low stochastic error is often considered as the main drawback of the MCM. However, advantages of the MCM outweigh its drawback: such difficulties as an excessive complexity of the cavity shape, the accounting for angular distributions of emitted and reflected radiation, which are unlikely to be overcome by using the deterministic methods, can be resolved by proper building of the computational model and application of the suitable MCM modeling algorithms.

Unlike deterministic ray tracing routinely used in the optical systems design, every Monte Carlo ray-tracing algorithm necessarily includes a pseudo-random number generator (PRNG), a procedure that delivers the next pseudo-random number from the sequence uniformly distributed from 0 to 1. It can be said that the use of a PRNG is a necessary feature of MCM. Below, we mention the most influential work on the application of the MCM for modeling the radiation characteristics of cavities.

Apparently, the first application of the MCM to modeling the radiative properties of cavities belongs to Polgar and Howell [112, 113]. They computed the effective reflectance, absorptance and emissivity of a diffuse conical cavity irradiated by a collimated beam. There are two possible ways to model perfectly diffuse (Lambertian) reflection. The first technique consists of generating the random direction uniformly distributed over the hemispherical solid angle and then ascribe a statistical weight proportional to the cosine of the angle between that direction and the normal to the surface. At such a uniform distribution of reflected rays, many of them will bring too small contribution to the result while the statistic for rays carrying more radiant energy will be insufficient. Polgar and Howell used another algorithm, later called importance sampling [83], which consists in generation of rays distributed not uniformly but according to the surface BRDF. Importance sampling substantially accelerates convergence of Monte Carlo computations. The analytical expressions for simulating the random direction of a diffuse reflection derived by Howell and Perlmutter [64] were adopted by Polgar and Howell [113].

The limited performance of computers in the 1960s motivated the researchers to develop time-efficient algorithms for Monte Carlo applications, including the problems of radiation heat exchange (see, e.g. [23, 65, 154]). On the other hand, advances in computer hardware have permanently expanded the scope of application and increased the complexity of the problems to be solved by the MCM.

Sparrow with co-workers [60, 141, 147] proposed a method of variance reduction in the Monte Carlo modeling of the radiative properties of a diffuse cavity, which they called *energy partitioning*. They applied it to isothermal baffled conical andcylindrical cavities. The essence of the method is the following. A random point on a radiating surface of a cavity emits a ray. Let F is the fraction of the radiation emitted at that point, which passes directly out of the aperture, numerically equal to the diffuse view factor of the aperture from that point. The energy content E^* of the i-th ray is partitioned into two portions. One portion, $F_i E^*$, passes directly out of the aperture and is accumulated in a special counter. The other portion, $(1 - F_i)E^*$, remains within the cavity and is carried by the ray along a straight path. Before the trajectory of the ray is traced, a random number η_α, having a value between 0 and 1, is selected to determine whether or not the ray will be absorbed at its intersection with the cavity wall or the baffle. Absorption takes place if $\eta_\alpha \leq \alpha = \varepsilon$. If the ray is absorbed, its trajectory ends and the tracing of the next ray is started. If the ray is not absorbed, its point of intersection is determined by constructing a straight-line trajectory based on the random angles θ and φ in the local spherical coordinate system. At the point of intersection, a second partitioning may take place. Let F_{i1} denote the view factor of the aperture as seen from the point of intersection ($F_{i1} = 0$ if the point of intersection is on the baffle). Then, the two partitioned fractions are $F_{i1}(1 - F_i)E^*$

and $(1 - F_{i1})(1 - F_i)E^*$. The first of these passes out of the aperture and is added to the corresponding counter, while the second remains in the cavity and continues its life cycle, experiencing successive partitionings until the absorption occurs.

Any ray emitted in a location on the cavity wall contributes to the counter for determination of the cavity radiant exitanceE_{out}. The total hemispherical effective emissivity is computed as

$$\varepsilon_e = \frac{E_{out}}{\sigma T^4 A_a} = \frac{\varepsilon(A_w/A_a)}{N}\left[\sum_{i=1}^{N} F_i + \sum_{i=1}^{N} G_i\right], \tag{4.39}$$

where $G_i = 0$, if no reflections occur; $G_i = (1 - F_i)F_{i1}$ for one reflection; $G_i = (1 - F_i)F_{i1} + (1 - F_i)(1 - F_{i1})F_{i2}$ for two reflections, and so forth. To avoid a direct hit of the second portions into the aperture, random directions of diffuse emission and reflection are generated not within the entire hemisphere, but in the hemisphere minus the conical solid angle subtending the aperture.

Hemispherical effective emissivities of conical andcylindrical cavities with emitting and non-emitting baffles were computed for various values of wall emissivity and geometrical parameters. The convergence of the computational process was studied. It was shown that although the energy partitioning method requires time-consuming calculations, the stochastic process with the energy partitioning converges faster than that of a conventional Monte Carlo technique.

The main drawback of the algorithm described is its practical inapplicability to the calculation of the directional effective emissivity: if a finite-sized detector is located in front of the cavity aperture, the geometric probability of a ray hitting the detector tends to zero when the distance between the aperture and the detector tends to infinity. This problem was solved for isothermal specular-diffuse cavities by Ono [103].

Instead of emission, Ono considered the reflection of radiation by a cavity and replaced the conditions of cavity observation with those of cavity irradiation. In this case, rays start at the point of observation, proceed in the direction of observation, pass through the cavity aperture, and repeatedly reflect off the cavity walls until either they escape the cavity through the aperture or the energy carried by the ray becomes negligible. Such a method is often referred to as the *backward* (or inverse) ray tracing. Ono used the ray-tracing paradigm to confirm the conclusions of Takata [150] on the applicability of Kirchhoff's law to the effective characteristics of the cavity. The reversibility of ray trajectories allows evaluating the directional effective emissivity along an infinitely thin ray.

Selection of specular or diffuse reflection was done using the random numbers. A distinctive feature of the algorithm is an estimate of the probabilities p_i that the ray would leave the cavity after $i$reflections. The estimates for p_i, $i = 1, 2, \ldots$ were obtained for cylindrical and cylindro-conical cavities with the non-absorbing walls. The directional effective emissivity $\varepsilon_e(\boldsymbol{\xi})$ of the cavity in the direction $\boldsymbol{\xi}$ can be expressed as

$$\varepsilon_e(\boldsymbol{\xi}) = 1 - \sum_{i=1}^{\infty} p_i (1-\varepsilon)^i, \tag{4.40}$$

where $\varepsilon = 1 - \rho$ is the emissivity of the cavity walls, ρ is their reflectance.

The relative frequencies (as the estimates of corresponding probabilities) were computed up to tenfold reflections using 10^4 or 10^5 rays traced. Such an approach allowed to use once computed set of p_i for calculation of the dependence $\varepsilon_e(\varepsilon)$. Later, the algorithm described was applied to a cylindro-conical cavity with the specular-diffuse conical bottom and diffuse cylindrical part [105], blackbodies formed by lateral holes in an infinitely long cylinder [104], and an opaque sample covered by a hemispherical mirror [106].

Chu et al. [21] calculated by the MCM the directional effective emissivity of an isothermal specular-diffuse cylindro-inner-cone. A collimated beam consisting of 10^5 rays was modeled to be incident normally on the cavity aperture and the fraction of the incident flux returned to the aperture was calculated. Each ray is traced until either it is absorbed or it leaves the cavity through the aperture. The calculation results were compared with those obtained by deterministic methods [13, 19], satisfactory agreement has been demonstrated.

Sapritsky and Prokhorov [136, 137] described an algorithm for the numerical modeling of isothermal and non-isothermal specular-diffuse cavities. They also employed the uniform specular-diffuse model as was used by Ono [103] and Chu et al. [21] but the algorithm had fundamental distinguishing features.

(1) To accelerate the convergence of the computational process, the method of statistical weights was used. Before the first interaction with the wall, a statistical weight $w = 1$ is assigned to the ray. After each reflection, the statistical weight is multiplied by the wall reflectance ρ to obtain a new weight. The effective emissivity of an isothermal cavity is evaluated as the ratio of the sum of the statistical weights of rays escaping the cavity to those launched into the cavity. There is a mathematically strict proof [97] of the fact that the variance of the result obtained by the method of statistical weights is always less than that obtained by the conventional method, which uses the termination of ray trajectories after the absorption event. In the method of statistical weights, each ray, having a statistical weight greater than a predefined threshold value, contributes to the result.

(2) To reduce the variance of the result for specular-diffuse cavities, an analog of the energy partitioning method was used by analytical calculation of the view factor from every point of a diffuse reflection to the cavity aperture. Unlike the original paper by Heinisch et al. [60], where the energy partitioning method was devised, a faster technique was proposed to generate the direction of the diffuse reflection: instead of time-expensive generating of the random direction within a subset of hemispherical solid angle, the rejection sampling technique was used. If the ray hits the aperture after a diffuse reflection, the last direction of reflection is ignored, and a new direction of diffuse reflection is generated.

Ray tracing of a single ray terminates when its statistical weight becomes less than a prescribed small value that determines the admissible systematic error, or when the ray leaves the cavity through the aperture after the specular reflection.

(3) The MCM was, for the first time, used to calculate *directional* effective emissivities (spectral and total) of *non-isothermal* cavities. It was shown that the effective emissivities of non-isothermal cavities can be modeled on the base of the reciprocity principle and the assuming of only the local thermodynamic equilibrium. The algorithm used combines the method of statistical weights with the backward ray tracing. We describe its latest modification in Sects. 4.4.2 and the following in more detail.

(4) To save computing time (for example, when we have a wavelength-dependent emissivity of cavity walls and need to compute the spectral effective emissivity for a large number of wavelengths), the method of dependent trials is used. According to it, a set of emissivity values is assigned to the cavity walls and a corresponding set of statistical weights is assigned to each ray. By transforming the statistical weights at each point of reflection, we use a single trajectory to obtain a variety of spectral results, although this leads to a strong correlation of their random errors.

This algorithm was applied to axisymmetric convex cavities formed by three surfaces of right circular cones. Later, the ray-tracing algorithm was improved to apply to an arbitrary cavity formed by revolution of a non-self-intersecting polygonal line around an axis. To avoid loss of generality, the energy partitioning was excluded. The increased performance of personal computers allowed tracing a greater number of rays and thus compensating for some reduction in convergence speed. Prokhorov and Martin [116] described the application of the MCM to the simulation of the radiative heat transfer from a blackbody to a cryogenic radiometer through the radiation trap, which are consist of 31, 53, and 194 surfaces of revolution, respectively (see Fig. 4.18).

Since then, the specular-diffuse model of reflection has been applied to a variety of blackbody radiators, including concentric V-grooves [122, 123], acylindrical cavity

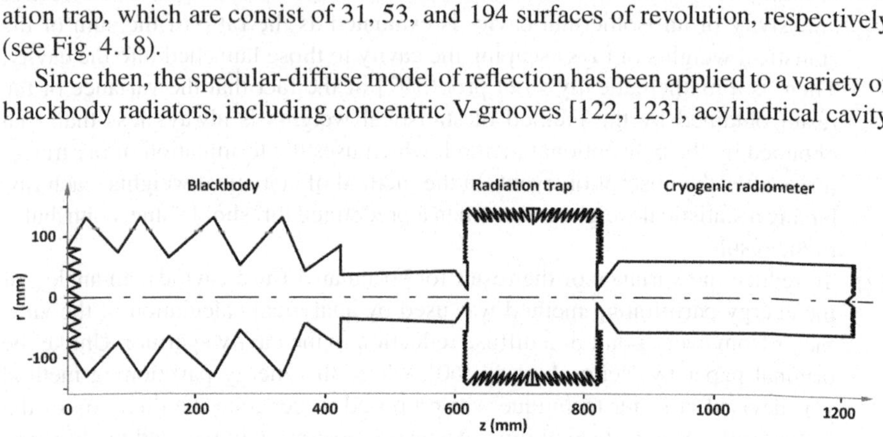

Fig. 4.18 Geometrical configuration of the radiometric system consisting of the blackbody, radiation trap, and cryogenic cavity radiometer (after [116])

with an inner surface covered with triangular grooves [151, 152], and acylindrical cavity with a flat diaphragm and a flat inclined bottom [117, 118].

Researchers of NMIJ (Japan) adapted the commercial ray tracing software (OptiCAD from OptiCAD Corp., Santa Fe, NM, https://www.opticad.com/) to calculation of the effective emissivities of isothermal specular-diffuse cavities. Ishii et al. [68] studied cylindro-conical cavities with triangular or rectangular grooves on the inner cylindrical surface. Kaneko et al. [70], using OptiCAD, assessed the effect of thermal radiation of the aperture plate on the effective emissivity of the NMIJ standard blackbody.

The diffuse (Lambertian) model continues to be used in analysis of blackbodies of medium- and high-temperature ranges. Interesting results have recently been obtained for diffuse cavities using the MCM in the series of papers by De Lucas [24–26] and De Lucas and Segovia [27]. Mei et al. [91, 92] applied the MCM to model the effective emissivities of a cavity formed by two coaxial tubes, which is used as a part of a sensor measuring the temperature of molten steel.

4.4.2 Models of Reflection

The adequacy of the model adopted for the radiation characteristics of the cavity walls is the crucial problem, which determines the reliability of results obtained by the MCM in modeling of effective emissivities. All the radiation characteristics of an opaque surface can be comprehensively described by its bidirectional reflectance distribution function (BRDF). The MCM implies stochastic simulation of angular distributions of the radiation reflected by the cavity internal surface within the hemispherical solid angle at each incidence angle and a given wavelength. The simplest way to model reflection is to generate directions of reflected rays uniformly within a hemispherical solid angle and to assign statistical weights to each ray according to the value of BRDF in this direction.

However, the real-world surfaces have non-uniform BRDFs; most often, they are highly elongated in the direction of specular reflection. The BRDF values may vary within several orders of magnitude and be equal to zero for the most directions. Therefore, such a straightforward approach would be extremely ineffective: contribution of the absolute majority of generated rays would be negligibly small. The better approach is to generate reflected rays non-uniformly, in such a way that the probability density for ray direction will be coordinated with the BRDF values. This method of accelerating the convergence of the stochastic modeling and reducing the variance of the result is known in the MCM under the name of importance sampling (see [77, 83], or [48]). The probability that the ray incident from the direction $\boldsymbol{\omega}_i$ will be reflected within a differential solid angle $d\Omega_v$ around the direction $\boldsymbol{\omega}_v$ is equal to $p(\boldsymbol{\omega}_i, \boldsymbol{\omega}_v)d\Omega_v$.

An analysis of the BRDF of some materials used for the radiating surfaces of blackbodies (some of this data is discussed in Chap. 6) shows that the uniform specular-diffuse (USD) model of reflection commonly used in Monte Carlo ray-tracing algorithms might be too crude an approximation for these data. As a result, the effective emissivities computed with the USD model may not reproduce adequately an actual behavior of radiation characteristics of a blackbody under modeling. The challenge is to develop a BRDF model that satisfies the following requirements.

(1) The BRDF model must obey the reciprocity principle:

$$f_r(\theta_i, \phi_i, \theta_v, \phi_v) = f_r(\theta_v, \phi_v, \theta_i, \phi_i). \tag{4.41}$$

(2) The BRDF model must obey the energy conservation law, i.e., DHR

$$\rho(\theta_i) = \int\limits_{\phi_v=0}^{2\pi} \int\limits_{\theta_v=0}^{\frac{\pi}{2}} f_r(\theta_i, \phi_i, \theta_v, \phi_v) \sin\theta_v \cos\theta_v d\theta_v d\phi_v \tag{4.42}$$

must be less than 1 for any incidence angle θ_i; DHR dependence on θ_i computed using Eq. 4.32 must reproduce DHR behavior of a real material.

(3) The BRDF model must have enough tuning parameters to be fitted to the measured in-plane BRDF for all incidence angles for which it was measured.

(4) The BRDF model must be physically based to predict correctly BRDF values outside the range of measured incidence and viewing angles.

(5) It is desirable that the BRDF model will be suitable for importance sampling [83], i.e., for generating randomly reflected rays with the probability density function $p(\boldsymbol{\omega}_i, \boldsymbol{\omega}_v)$ [sr^{-1}] linked with the BRDF $f_r(\boldsymbol{\omega}_i, \boldsymbol{\omega}_v)$ by the equation:

$$p(\boldsymbol{\omega}_i, \boldsymbol{\omega}_v) = f_r(\boldsymbol{\omega}_i, \boldsymbol{\omega}_v) \cos\theta_v, \tag{4.43}$$

where θ_v is the angle between $\boldsymbol{\omega}_v$ and the normal to the surface at the point of reflection.

Since the different physical mechanisms may be responsible for reflection from various materials or even be superimposed, it is hardly possible to employ one certain existing BRDF model to describe the reflection characteristics of all materials considered.

An attempt to modify the specular-diffuse model used by Ono [103] has been made by Ballico [3]. He has assumed that the specular reflectance obeys the Fresnel reflection law while the diffuse reflectance obeys Lambert's law. The data on the refractive index of graphite that is necessary for calculation of the Fresnelian reflectance at an arbitrary incidence angle was adopted from reference literature.

The selection of a BRDF that satisfies both the energy conservation law and the reciprocity principle is not a trivial problem. Since 1980s, many BRDF models has been developed for computer graphics, image synthesis, computer vision, and similar applications; the most recent reviews can be found in Marschner and Shirley [89] or Pegoraro [109]. The BRDF models developed for computer graphics, as well as remote sensing, stray light suppression, lighting technique, etc. have various degree of complexity and physical plausibility. Among these models, there are physically based (i.e., describing real physical phenomena such as a reflection of optical radiation from randomly rough surfaces or volumetric scattering in translucent media) as well as empirical or semi-empirical models, which is only mathematical formulas with a set of tunable parameters. Many of these models cannot be used in precise radiometric calculations due to their limitations. Some are only oriented to plausible visual perception while others poorly reproduce reflective behavior of real materials or reproduce it satisfactorily for not all combinations of incidence and viewing angles.

Zhou et al. [170] applied to the Monte Carlo modeling the modified version of the microfacet-based reflection model developed by Ward [160] for computer graphics applications. The original form of the Ward BRDF for isotropic surfaces is

$$f_{r,Ward}(\theta_i, \theta_v) = \frac{\rho_d}{\pi} + \frac{\rho_s \exp\left(-\frac{\tan^2\theta_h}{\sigma^2}\right)}{4\pi\sigma^2\sqrt{\cos\theta_i\cos\theta_v}}, \quad (4.44)$$

where ρ_d and ρ_s are the angle-independent diffuse and specular reflectances, respectively, σ is the roughness parameter, θ_i and θ_v are the incidence and viewing angles, respectively, θ_h is the angle between the normal to the surface and so-called halfway vector $\mathbf{h} = \frac{\omega_v - \omega_i}{\|\omega_v - \omega_i\|}$ (see Fig. 4.19).

Zhou et al. [170] modified Eq. 4.44 rewriting it in the form:

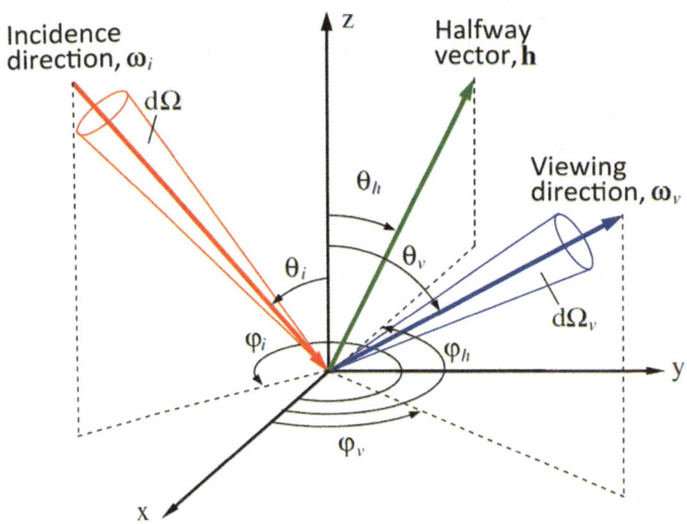

Fig. 4.19 Schematic for the definition of a mircofacet BRDF

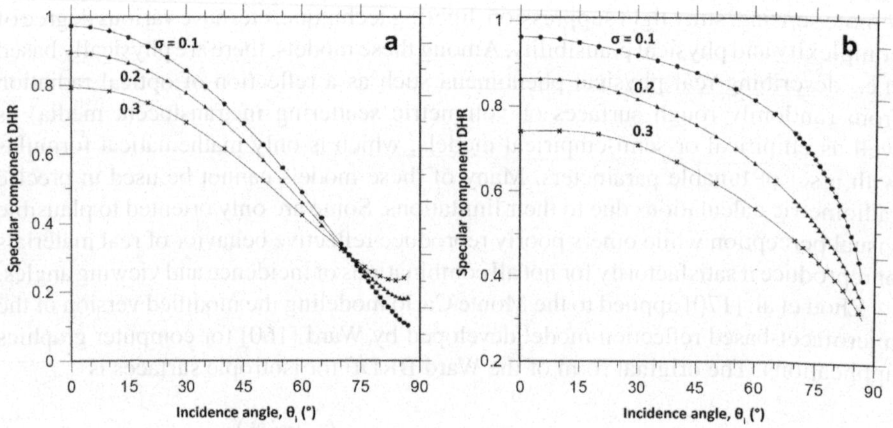

Fig. 4.20 Dependences of the specular components of DHR on the incidence angle θ_i at $\rho_s = 1$ and three values of σ for: **a** Ward [160] model and **b** Zhou et al. [170] at $g(\sigma) = 1$

$$f_{r,Zhou}(\theta_i, \theta_v) = \frac{\rho_d}{\pi} + \frac{\rho_s g(\sigma)\exp\left(-\frac{\tan^2\theta_h}{\sigma^2}\right)}{4\pi\sigma^2 \max\left(\cos^2\theta_i \cos^2\theta_v\right)}, \quad (4.45)$$

where $g(\sigma)$ is the polynomial chosen to force the energy conservation law for $\theta_i = 0°$. According to the energy conservation law, the spectral directional-hemispherical reflectance (spectral DHR) obtained by numerical integration over the hemispherical sold angle must be equal to unity at $\rho_s = 1$ and any incidence angle. Such dependences are shown in Fig. 4.20 for the specular components of the two above-mentioned BRDF models. Although the DHR as a function of the incidence angle for the model proposed by Zhou et al. [170] looks flatter even at $g(\sigma) = 1$, the deviations from the unity still remained. Nevertheless, the work by Zhou et al. [170] was a significant step toward more adequate modeling of reflection from real-world surfaces.

Prokhorov et al. [124] and Hanssen and Prokhorov [51] described a model of reflection from rough surfaces, referred to as the TETRA BRDF model. It is also based on microfacet theory and developed within the framework of geometrical optics. The TETRA BRDF model employs the generation of random tetrahedral pits, the walls of which reflect according to Fresnel's law. The model developed can be easily incorporated into the Monte Carlo ray-tracing algorithms. The tetrahedral pits are randomly generated "on the fly," for each incidence ray. The TETRA model is very flexible and physically plausible, i.e. obeys both reciprocity principle and the energy conservation law. Its biggest drawback is the difficulty of fitting model parameters to experimental BRDFs: every change of fitting parameters requires time-consuming statistical modeling.

Motivated by the need of reducing the uncertainty in determination of the effective emissivity for the blackbody-based radiometric systems of the next generation,

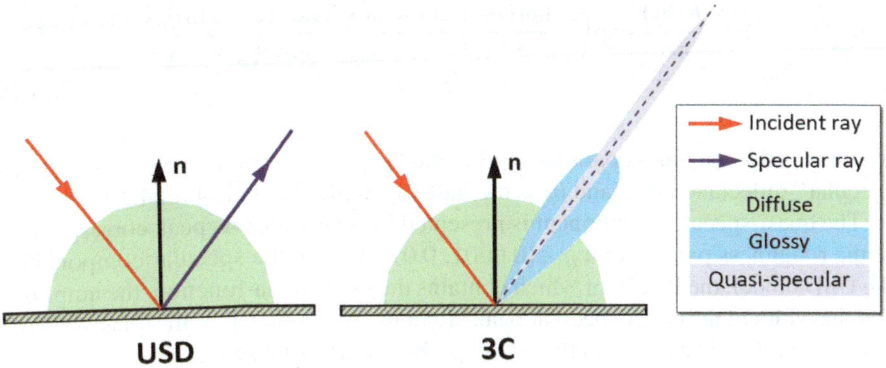

Fig. 4.21 Schematics of the uniform specular-diffuse (USD) and the three-component (3C) models of reflection

Prokhorov [115] proposed the so-called Three-Component (3C) model of reflection that represents the BRDFs f_r by the linear combination of three components: diffuse (Lambertian) $f_{r,d}$, almost specular (quasi-specular) $f_{r,qs}$, and glossy (having wide forward-scattering lobe) $f_{r,g}$:

$$f_r = k_d f_{r,d} + k_{qs} f_{r,qs} + k_g f_{r,g}, \tag{4.46}$$

where k_d, k_{qs}, and k_g are nonnegative and

$$k_d + k_{qs} + k_g = 1. \tag{4.47}$$

Here the names of the components are not related to the physical phenomena by which they are defined, but relate only to their geometric forms. For example, diffuse component might be caused by scattering on the surface roughness, or by volumetric scattering inside a translucent coating and so forth. Therefore, a phenomenological rather than the first principle-based approach was used. Figure 4.21 presents schematically the USD[3] and 3C models. Both models include the diffuse (Lambertian) component, the BRDF of which is constant:

$$f_{r,d} = \frac{R_d}{\pi}, \tag{4.48}$$

where R_d is the diffuse reflectance (DHR) of the diffuse component.

For quasi-specular and glossy components, the BRDF model proposed by Geisler-Moroder and Dür [38] for rough surfaces has been chosen in its isotropic version:

---

[3]The name for this model was coined by Ono [103] although the model itself has been used earlier.

$$f_r(\theta_i, \theta_v, \varphi) = \frac{R(\theta_h)}{\pi\sigma} \exp\left(-\frac{\tan^2\theta_h}{\sigma^2}\right) \frac{2(1 + \cos\theta_i \cos\theta_v - \sin\theta_i \sin\theta_v \cos\varphi)}{(\cos\theta_i \cos\theta_v)^4}, \tag{4.49}$$

where σ is the roughness parameter, R is the "glossy" reflectance R_g or the "quasi-specular" reflectance R_{qs}, and θ_h is the halfway angle shown in Fig. 4.19.

The quasi-specular component is presented by a very narrow peak corresponding to the roughness parameter $\sigma_{qs} \in [0.0001, 0.01]$. Unlike the specular component of the USD model, the BRDF of which contains the Dirac delta-function, the amplitude and the width of the quasi-specular peak are finite, so the BRDF of the quasi-specular component $f_{r,qs}$ is a function that is integrable in the usual sense.

The glossy component differs from the quasi-mirror component only in the values of the roughness parameter $\sigma_g \in (0.01, 0.5]$ and, therefore, in the transformation of a narrow peak into a wider lobe. The BRDFs for quasi-specular and glossy components [38] are based on the microfacet model of reflection.

For the microfacet reflectance, the approximation of the Fresnel equation for unpolarized radiation proposed by Schlick [140] is used.

It allows expressing the directional reflectance of a specularly reflecting surface of a microfacet via the normal reflectance R_0 (R_{qs} and R_g for the quasi-specular and glossy components of the 3C model, respectively):

$$R(\theta_i) = R_0 + (1 - R_0) \cdot (1 - \cos\theta_i)^5. \tag{4.50}$$

The Schlick approximation provides the relative error of less than 0.2% for $k \approx 0$ and $1.7 \leq n \leq 2.7$. Figure 4.22 presents the angular dependences of reflectance of optically smooth graphite and silicon carbide at 2.0 μm calculated according the Fresnel and Schlick equations.

It was assumed that$n = 3.96$, $k = 2.99$ for graphite and $n = 2.57$, $k = 3.98 \cdot 10^{-4}$ for α-SiC.[4] The Schlick approximation fairly reproduces the Fresnel law for the silicon carbide but for graphite, a satisfactory accuracy is achieved only when $\theta_i <$ 45°. However, in the microfacet reflection model, $\theta_h \ll 45°$ for any reasonable value of the surface roughness parameter σ.

Besides, the effect of multiple reflections among microfacets overrides Schlick's approximation errors. The only drawback with Schlick's approximation, which hampers the use of the Geisler-Moroder and Dür model in the fitting problem, is nonzero reflectance that Eq. 4.49 gives for $R_0 = 0$ and $\theta_i > 0$. To avoid this problem, the condition $R(\theta_i) \equiv 0$ for $R_0 = 0$ and $\theta_i > 0$ must be added. From the formal point of view, this condition leads to mathematical discontinuity of $R(\theta_i)$, but computational experiments did not detect any erroneous or artifactual fitting results.

It is assumed that microfacet normals have Gaussian distribution with zero mean value and standard deviation σ. The greater the σ the greater the surface roughness and, correspondingly, the wider the BRDF lobe. Elementary physical reasoning

[4]The most common polymorph of silicon carbide with a hexagonal crystal structure.

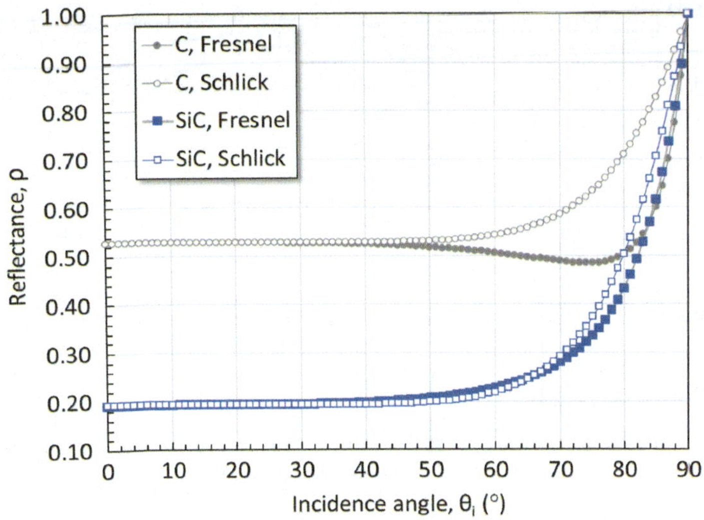

Fig. 4.22 Angular dependences of reflectance of optically smooth graphite and silicon carbide at 2.0 μm calculated according to the Fresnel and Schlick equations

suggests that DHR of a microfacet model must equal 1 for all incidence angles if $R(\theta_i) \equiv 1$. For all microfacet models that neglect multiple reflections among microfacets, DHRs deviate from unity the greater, the greater surface roughness.

The last multiplier in Eq. 4.49 is constructed in such a way that the deviation of DHR from 1 for $R(\theta_i) \equiv 1$ is minimized (see Fig. 4.23) and allows us to avoid infinite BRDF values at $\theta_i = \theta_v = \frac{\pi}{2}$. Deviations become significant only for large σ and large θ_i.

Since the reflectance of materials of blackbody cavities is essentially low, deviations of DHR computed by integration of BRDF in Eq. 4.48 from the actual DHR values are expected to be small. The DHR of the 3C BRDF model can be represented in the form

$$\rho(\theta_i) = \rho_d + \rho_{qs}(\theta_i) + \rho_g(\theta_i), \tag{4.51}$$

where every summand is the DHR of the appropriate component:

$$\rho_d = k_d \int_{\varphi_v=0}^{2\pi} \int_{\theta_v=0}^{\pi/2} f_{r,d}(R_d, \theta_i, \varphi_i, \theta_v, \varphi_v) \sin\theta_v \cos\theta_v \, d\theta_v \, d\varphi_v = k_d R_d, \tag{4.52}$$

$$\rho_{qs}(\theta_i) = k_{qs} \int_{\varphi_v=0}^{2\pi} \int_{\theta_v=0}^{\pi/2} f_{r,qs}\left(R_{qs}, \theta_i, \varphi_i, \theta_v, \varphi_v\right) \sin\theta_v \cos\theta_v \, d\theta_v \, d\varphi_v, \tag{4.53}$$

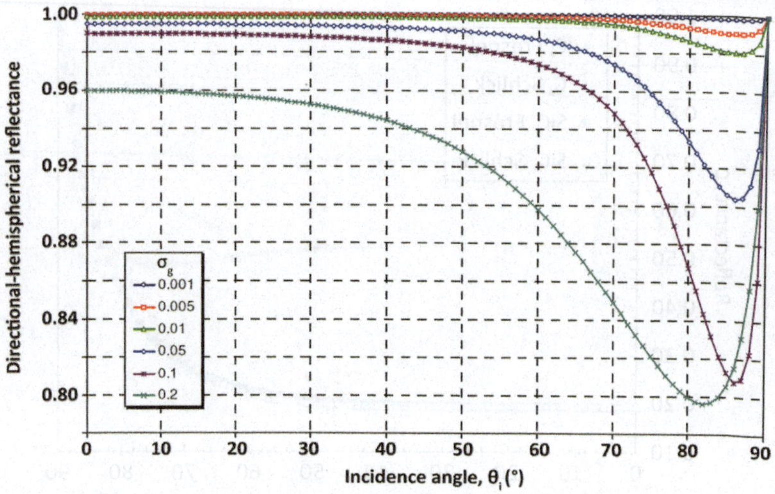

Fig. 4.23 Dependences of the DHR on the incidence angle computed of the glossy component BRDF for $R_g = 1$ and several values of σ_g

$$\rho_g(\theta_i) = k_g \int_{\varphi_v=0}^{2\pi} \int_{\theta_v=0}^{\pi/2} f_{r,g}\left(R_g, \theta_i, \varphi_i, \theta_v, \varphi_v\right) \sin\theta_v \cos\theta_v \, d\theta_v \, d\varphi_v. \tag{4.54}$$

In-plane BRDFs of the glossy component plotted in the Cartesian coordinate system for $R_g = 0.1$, $\sigma_g = 0.1$, and six values of θ_i (0°, 15°, 30°, 45°, 60°, and 75°) are shown in Fig. 4.24. The negative values of viewing angles θ_v correspond to the backscattering. The three-dimensional plots of glossy BRDFs expressed in relative units for $R_g = 0.1$ and $\sigma_g = 0.05, 0.1$ and 0.2 are presented in the spherical coordinate system in Fig. 4.25 for 45° of incidence.

The double integrals in Eqs. 4.53 and 4.54 were computed numerically using an adaptive quadratures algorithm [39, 157]. For very small σ_{qs}, the dependence $\rho_{qs}(\theta_i)$ practically coincides with appropriate Schlick's curves.

Although the 3C BRDF model is created using the model originally developed for reproducing reflection from rough surfaces, it can be applied to surfaces with another nature of scattering, e.g. paints and coatings on opaque substrates. The 3C model has eight fitted parameters: k_d, R_d, k_{qs}, R_{qs}, σ_{qs}, k_g, R_g, σ_g.

Three of them are interrelated as shown in Eq. 4.47. Besides, the following constrains must be fulfilled according to the definitions of corresponding fitting parameters:

$$0 \le k_{d,qs,g},\ R_{d,qs,g} \le 1, \tag{4.55}$$

$$0.0001 \le \sigma_{qs} \le 1, \tag{4.56}$$

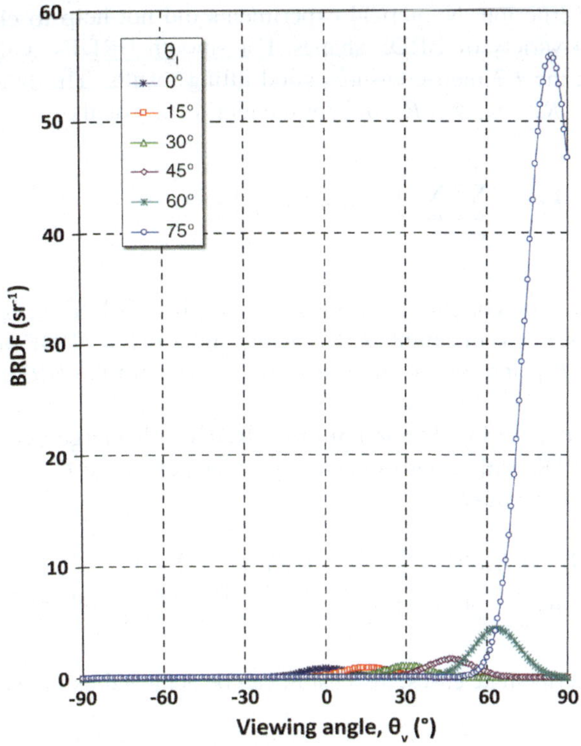

Fig. 4.24 In-plane Cartesian plots for the BRDF of the glossy component with $R_g = 0.1$, $\sigma_g = 0.1$, and six incidence angles

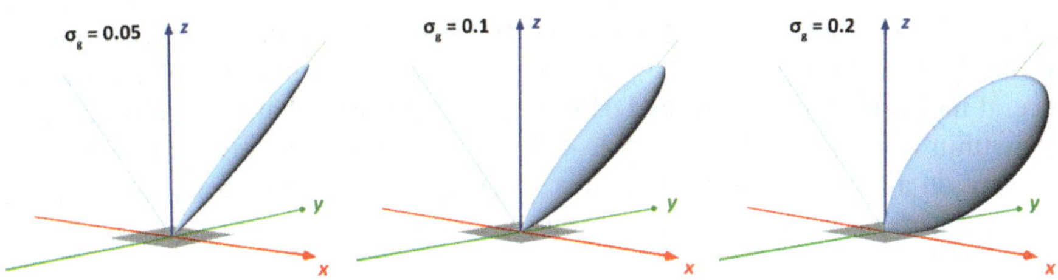

Fig. 4.25 3D plots (not to scale) in spherical coordinates for the BRDF of the glossy component at the incidence angle of 45°; $R_g = 0.1$, and three values of σ_g

$$0.001 < \sigma_g \leq 0.5. \tag{4.57}$$

Mathematically, this is the problem of constrained multidimensional optimization. It is equivalent to searching eight values of the 3C model parameters, which minimize distances between computed and measured BRDFs according to some goodness-of-fit criterion. Numerical experiments did not help to choose a unified criterion for all variety of BRDF shapes. For smooth BRDFs with the moderate dynamic range, the $L2$ metric ensures good fitting results. The objective function $F_{L2}(k_d, R_d, k_{qs}, R_{qs}, \sigma_{qs}, k_g, R_g, \sigma_g)$ for the metric $L2$ equals

$$F_{L2} = \sum_{k=1}^{n_i} \sum_{j=1}^{n_{v,k}} \left[f_{r,m}(\theta_{i,k}, \theta_{v,jk}) - f_r(\theta_{i,k}, \theta_{v,jk}) \right]^2, \tag{4.58}$$

where f_r and $f_{r,m}$ are computed and measured in-plane BRDF, n_i is the number of incidence angles, $n_{v,k}$ is the number of viewing angles in the BRDF measured at k-th incidence angle $\theta_{i,k}$ and $\theta_{v,jk}$ is the j-th viewing angle for the BRDF measured for $\theta_{i,k}$.

It was found that for highly non-uniform BRDFs with a large dynamic range (for instance, for BRDF with a narrow quasi-specular peak), the C metric for relative deviations is more suitable:

$$F_C = \begin{cases} 0, & \text{if } f_{r,m}(\theta_{i,k}, \theta_{v,jk}) = f_r(\theta_{i,k}, \theta_{v,jk}) = 0 \\ \max\limits_{k=1,\ldots,n_i} \max\limits_{j=1,\ldots,n_{v,k}} \left\{ \frac{\left| f_{r,m}(\theta_{i,k}, \theta_{v,jk}) - f_r(\theta_{i,k}, \theta_{v,jk}) \right|}{f_{r,m}(\theta_{i,k}, \theta_{v,jk}) - f_r(\theta_{i,k}, \theta_{v,jk})} \right\} & \text{otherwise} \end{cases}. \tag{4.59}$$

For some intermediate cases, the best fit can be achieved using the $L1$ metric:

$$F_{L1} = \sum_{k=1}^{n_i} \sum_{j=1}^{n_{v,k}} \left| f_{r,m}(\theta_{i,k}, \theta_{v,jk}) - f_r(\theta_{i,k}, \theta_{v,jk}) \right|, \tag{4.60}$$

Prokhorov and Prokhorova [119] proposed a method for extension of a monochromatic 3C BRDF model to a continuous spectral range. The initial data for this method are the BRDFs measured in the plane of incidence at a single wavelength λ_0 and several incidence angles and the DHR $\rho^*(\theta_{i,0}, \lambda)$ measured at one incidence angle $\theta_{i,0}$ within a finite spectral range $\lambda \in [\lambda_{\min}, \lambda_{\max}]$. The method is based on the assumption about the constancy on this spectral interval of that parameters k_d, k_{qs}, k_g, σ_{qs}, and σ_g, as well as the relative contributions γ_d, γ_d, and γ_d of three components of the DHR:

$$\begin{cases} \gamma_d = \rho_d(\lambda_0) / \rho^*(\theta_{i,0}, \lambda_0) \\ \gamma_{qs} = \rho_{qs}(\lambda_0) / \rho^*(\theta_{i,0}, \lambda_0) \\ \gamma_g = \rho_g(\lambda_0) / \rho^*(\theta_{i,0}, \lambda_0) \end{cases}. \tag{4.61}$$

Dependences $R_d(\lambda)$, $R_{qs}(\lambda)$, and $R_g(\lambda)$ can be found for a discrete set of $\lambda_k \in [\lambda_{\min}, \lambda_{\max}]$ from solution of three independent equations:

$$\gamma_d \rho^*(\theta_{i,0}, \lambda) = k_d R_d(\lambda), \tag{4.62}$$

$$\gamma_{qs}\rho^*(\theta_{i,0},\lambda) = k_{qs}\int\limits_{\varphi=0}^{2\pi}\int\limits_{\theta_v=0}^{\pi/2} f_{r,qs}(R_{qs}(\lambda_0),\theta_{i,0},\theta_v,\varphi)\sin\theta_v\cos\theta_v\,d\theta_v\,d\varphi, \quad (4.63)$$

$$\gamma_g\rho^*(\theta_{i,0},\lambda) = k_g\int\limits_{\varphi=0}^{2\pi}\int\limits_{\theta_v=0}^{\pi/2} f_{r,g}(R_g(\lambda_0),\theta_{i,0},\theta_v,\varphi)\sin\theta_v\cos\theta_v d\theta_v d\varphi. \quad (4.64)$$

After numerical integration in the right-hand parts of Eqs. 4.63 and 4.64, the relative contributions γ_d, γ_d, and γ_d of each component can be found and then the measured DHR $\rho^*(\theta_{i,0},\lambda)$ can be divided into three components, whose BRDFs retain their shapes for each wavelength $\lambda_k \in [\lambda_{\min},\lambda_{\max}]$.

The main drawback of the 3C model is its complexity that requires special software for non-linear global optimization for determination of the fitting parameters. Recently, several attempts have been made to model non-diffuse and non-specular reflection. Quang and Van [126] applied the BRDF model proposed by Phong [111] for the nascent computer graphics. In modern notation it can be written as

$$f_r(\theta_i,\theta_v) = \frac{R_d}{\pi} + R_s\frac{\nu+2}{2\pi}\cos^{\nu}\alpha, \quad (4.65)$$

where $R_d + R_s \leq 1$, α is the angle between the viewing direction and the direction of perfect specular reflection; in order to prevent any negative values of the cosine factor, α is set to $\pi/2$ when $\alpha > \pi/2$.

The Phong reflection model is the earliest and simplest model of reflection; it is still used in some applications, despite the fact that its inconsistency has been known since Lewis [81] showed that it satisfies neither the energy conservation law, nor the reciprocity principle. Prokhorov et al. [121] demonstrated that the choice of a physically correct reflectance model for the cavity walls is extremely important at the Monte Carlo calculations of effective emissivities, especially when the reciprocity principle is used to inverse ray trajectories. Therefore, although the work by Quang and Van [126] is of particular methodological interest, the results obtained with the Phong model can hardly be used in precise blackbody radiometry.

Mahan et al. [85] proposed the four-component BRDF model that includes one constant summand for the Lambertian reflection and four Gaussian components,in our notation, it takes the form

$$f_r = C + \sum_{k=1}^{4}\frac{B_k}{b_k\theta_i}\exp\left[-\pi b_k^2\theta_i^2(\theta_v-\theta_i)^2\right], \quad (4.66)$$

where C, B_k, b_k $(k = 1,\ 2,\ 3,\ 4)$ are nine fitting constants.

It can be easily seen that employing for Gaussian components of amplitudes and standard deviations, which depend on the incidence angle θ_i, makes the BRDF (4.66) non-reciprocal though improves the fitting quality.

Summarizing the content of this section, we should note that the 3C reflection model remains the only BRDF model obeying the energy conservation law, the reciprocity principle, and well-suited for the importance sampling. We can suppose that the prospective approach is the use of 2 or 3 simpler models with the option of an automatic selection of the model that is better suited for particular material.

4.4.3 *Accuracy of the Monte Carlo Results*

Since the MCM solution is the mean value of individual trials (tracing of individual rays), the results of which fluctuate about the mean value, the accuracy of the result can be increased by increasing the number of trials. If to interpret the Monte Carlo trials as numerical experiments, a rough analogy between stochastic modeling and individual independent measurements can be established and the accuracy of the MCM calculations can be assessed by performing repeated calculations and statistical processing of results. The uncertainty due to stochastic nature of the MCM should be considered as the type A measurement uncertainty. This analogy is especially clear for simplest Monte Carlo algorithms, e.g. when the effective emissivity ε_e of an isothermal cavity is computed via the MCM calculation of the effective reflectance ρ_e without applying any variance-reducing techniques. If the statistical fluctuations have a normal distribution, the standard uncertainty u_A of the stochastic error can be evaluated on the base of the central limit theorem:

$$u_A = \sqrt{\frac{1}{n-1}\sum_{k=1}^{n}\left(\rho_{e,k} - \rho_e\right)^2}, \tag{4.67}$$

where ρ_e is the true value of the effective reflectance (if it is known) or its value evaluated at $n \to \infty$ (if the true value is unknown); $\rho_{e,k}$ is the effective reflectance evaluated using the k-th ray trajectory.

For a large n, we can predict that $u_A(\rho_e) \propto 1/\sqrt{n}$ that is a precision of the Monte Carlo calculations is in inverse proportion to the square root of a number of rays traced. A following numerical experiment illustrates this statement. The effective emissivity of an isothermal diffuse spherical cavity was modeled by the MCM for the number of rays traced n up to 10^6. After every 1000 rays the result $\varepsilon_{e,k}$ was computed and compared with the effective emissivity ε_e calculated using Gouffé formula (4.20) that gives the exact result for the isothermal diffuse sphere. The data collected in the six independent runs (the PRNG was randomized before starting each computational process) are presented in Fig. 4.26 as $\Delta\varepsilon_e = \left|\varepsilon_{e,k} - \varepsilon_e\right|$ plotted against the number of rays traced with a logarithmic scale on both axes. These plots

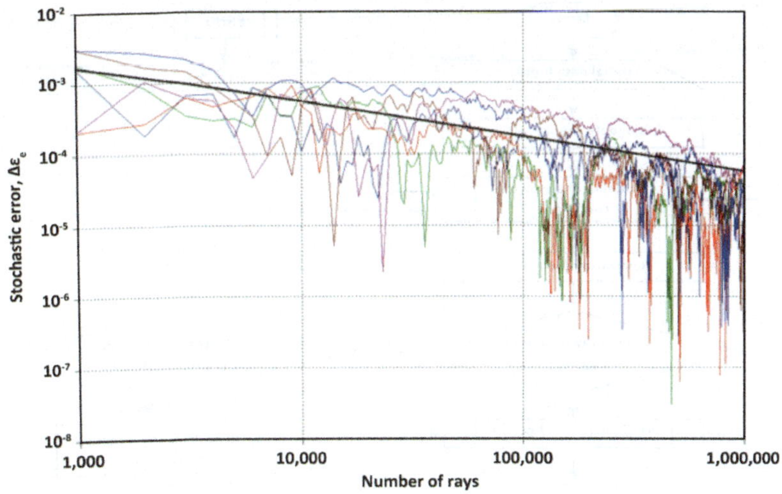

Fig. 4.26 Convergence of six values of the effective emissivity computed by the MCM to the exact value for an isothermal diffuse spherical cavity

show clearly the convergence of the results computed by the MCM to the exact value predicted by the Gouffé formula for each of six runs. For comparison, the graph of the function $0.05/\sqrt{n}$ is presented by the thick straight line. It corresponds approximately to the upper envelop of all graphs for the MCM calculations.

If the method of statistical weights or other variance-reducing techniques are employed, the envelop moves down but the $n^{-\frac{1}{2}}$ dependence remains in force. If truncation of ray trajectories is used, the convergence of the computational process finishes as soon the stochastic uncertainty approaches the prescribed radiance threshold value. Further, it will fluctuate around this value. The dependence $n^{-\frac{1}{2}}$ holds also for non-isothermal cavities if the difference between the maximal and minimal temperature of the cavity surfaces is within reasonable limits.

Truncation of ray trajectories introduces the systematic error that may lead to underestimating of the effective emissivity. This error can be characterized be the uniform distribution on the interval $[0, w_{\min}]$; its standard deviation is

$$u_B = \frac{w_{\min}}{\sqrt{12}} \approx 0.289 \cdot w_{\min}. \tag{4.68}$$

The total uncertainty can be evaluated as

$$u(\varepsilon_e) = \sqrt{u_A^2 + u_B^2} = \sqrt{u_A^2 + w_{\min}^2/12}. \tag{4.69}$$

Usually, it is enough to trace 10^6 to 10^7 rays to achieve the accuracy of $10^{-4} \ldots 10^{-6}$ in effective emissivity, if $w_{\min}$ is less than $10^{-4} \ldots 10^{-6}$, respectively. As a rule,

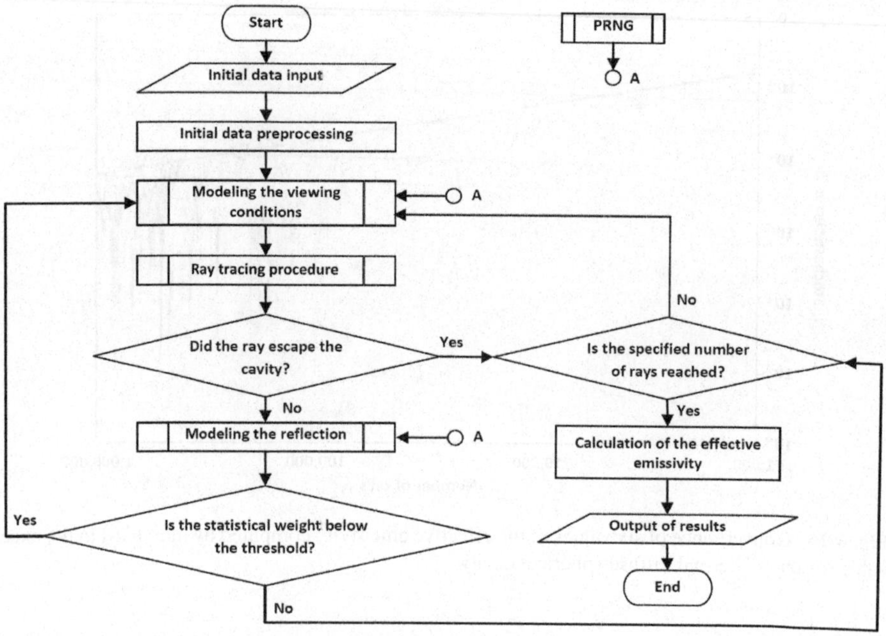

Fig. 4.27 General flowchart of the algorithm for the Monte Carlo modeling the effective emissivity of a cavity

this uncertainty satisfies requirements of major tasks of the modern blackbody radiometry.

4.5 General Monte Carlo Ray Tracing Algorithm

The flowchart of the algorithm for the Monte Carlo modeling of thespectral effective emissivity at a given wavelength is presented in Fig. 4.27. The flowchart follows the structure of the Monte Carlo ray tracing software of the 3rd generation of the STEEP3 family (STEEP321 and STEEP323).[5] Below, we comment briefly the blocks of the flowchart.

The **PRNG** (pseudo-random number generator) is the indispensable component of any MCM algorithm. This subroutine generates the next pseudo-random (or simply random, for brief) number u from the ensemble uniformly distributed on the interval (0, 1).

Without going into the subtleties of algorithms and software implementation of PRNGs, we only note that not all built-in generators of standard programming tools are capable to produce a sufficiently long pseudo-random numbers sequence, which

[5] See www.virial.com

can pass the rigorous statistical tests [74]. The Mersenne Twister [90], the *de facto* standard for modern PRNGs is used in the 3rd generation of the STEEP3 software family.

4.5.1 Data Input and Pre-processing

Input of initial data involves the formation of the initial data set: geometry of the cavity and the viewing conditions, wavelength λ of interest, radiation characteristics (spectral reflectance, BRDF, etc., depending on the reflection model adopted) for all surfaces forming the cavity, the reference temperature T_{ref}, the temperature distribution $T(\boldsymbol{\xi})$ over the cavity walls (point-by-point or via the coefficients of the interpolating function), the number n of rays being traced, and the threshold value $w_{\min}$ of the statistical weight w.

To specify geometry of an axisymmetric cavity formed by revolution of a non-self-intersecting polygonal line around an axis, the two coordinates (y and z) of generatrix nodes have to be entered. Cavity shape is defined point-by-point as shown in Fig. 4.28, where the numbered points represent the generatrix nodes, while the lines with the numbers in brackets represent the generatrix segments between neighbor points.

It is convenient to set the coordinates of the 0th point $y_0 = z_0 = 0$. In this way, the cavity can be formed by an arbitrary number of flat, right cylindrical or right conical surfaces. If necessary, the curvilinear parts of the generatrix can be represented by polygonal lines with a large number of segments. Within the same reflection model, different radiation properties can be assigned to each segment.

The temperature distribution for non-isothermal cavities can be set in various ways. For instance, the corresponding temperatures can be assigned to each node. Additional points can be introduced on each segment. The temperature between given

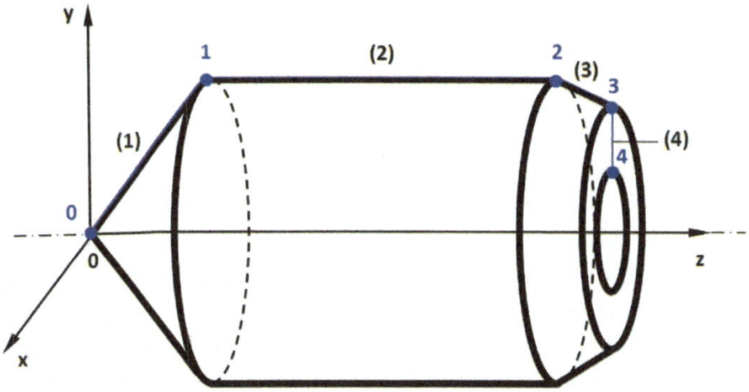

Fig. 4.28 Defining the cavity shape by the generatrix nodes. Generatrix segments are numbered with the figures in brackets

points can be easily calculated using interpolation. Alternatively, the constant, linear, parabolic, or another functional dependence on the axial coordinate can be set.

The reference temperature T_{ref}, the number n of rays traced, and the threshold value $w_{\min}$ of the statistical weight are entered as the scalar constants.

Initial data pre-processing involves the preliminary calculation of some quantities that do not change throughout the Monte Carlo modeling cycle. In particular, for each segment of the generatrix, the type of the corresponding surface (plane, cylinder, or cone), the parameters of its equation in the Cartesian coordinate system, and the components of normal vectors, which are constant throughout each surface are determined. If the 3C model is used for radiation characteristics of the cavity walls, the dependences of quasi-specular and glossy components of reflectance on the incidence angle should be pre-computed and approximation polynomial coefficients should be found.

The initial data pre-processing allows avoiding multiple repetitions of the same calculations and, thus, significantly reducing the modeling time.

Before the modeling starts, the initial assignment is performed: the counters of the spectral radiance L_λ and initial statistical weights Σw_0 reset to zero and the counter of the rays traced is set to 1.

4.5.2 *Modeling the Viewing Conditions*

Modeling the Viewing Conditions is the first stage of the MCM modeling each ray trajectory. In accordance with the Helmholtz reciprocity principle, we will discuss the modeling of viewing conditions in terms of the backward ray tracing, i.e., we will say that the ray is "launched" into the cavity instead of the ray directed from the cavity to the sensor or observer. Further, it is a question of tracing one (j-th) ray. This block is responsible for simulation of geometrical conditions of collecting the cavity radiation by a sensor with or without optical system or, in terms of the backward ray tracing, geometrical conditions for launching rays into the cavity. As examples, we consider the modeling of the three main types of observation conditions.

Normal effective emissivity To model the normal effective emissivity, the rays have to be launched into the cavity along directions that parallel to the cavity axis. We can choose a plane perpendicular to the cavity axis and spaced from the aperture A_0 at an arbitrary distance (let us say, the plane $z_b > z_{np}$, where z_{np} is the z-coordinate of the last point in the cavity generatrix) and consider the circular viewing beam of radius R_b, the axis of which is shifted by y_b from the cavity axis Fig. 4.29. To pick a random point from an ensemble uniformly distributed over the circle B, two independent random numbers u_x and u_y uniformly distributed on the interval (0, 1) are generated by the PRNG. The coordinates of the starting point for the next ray are calculated by the acceptance-rejecting method: if $u_x^2 + u_y^2 > 1$, the pair of random numbers are rejecting and the next pair is generated.

The accepted pair is transformed as

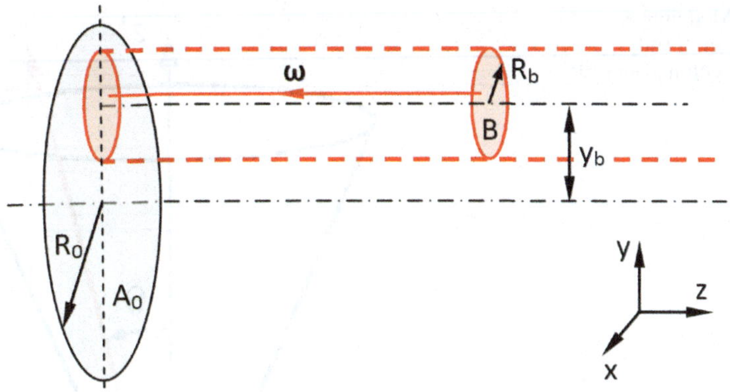

Fig. 4.29 Modeling of the normal effective emissivity

$$\begin{cases} x = R_b \cdot (2u_x - 1) \\ y = R_b \cdot \left(2u_y - 1\right) + y_b \, . \\ z = z_b \end{cases} \tag{4.70}$$

The directional vector $\boldsymbol{\omega}$ $\left(\omega_x, \omega_y \omega_z\right)$ has the components $(0, 0, -1)$. The initial statistical weight $w_0 = 1$.

Conical effective emissivity To model the conical effective emissivity, two cases have to be considered: when the focal point is in front and behind the cavity aperture, as shown in Fig. 4.30, where A_0 is the cavity aperture. Let coordinates of the focal point be $(0, y_b, z_b)$. Let us also consider the local coordinate system $\left(x', y', z'\right)$ having the origin coinciding with the focal point and the axis z' that coinciding with the axis of the cone with the span angle β (see Fig. 4.31).

To pick the direction (θ, φ) from an ensemble uniformly distributed over the conical solid angle, the next pair of the random numbers u_θ and $u_\varphi$generated by the PRNG are transformed as

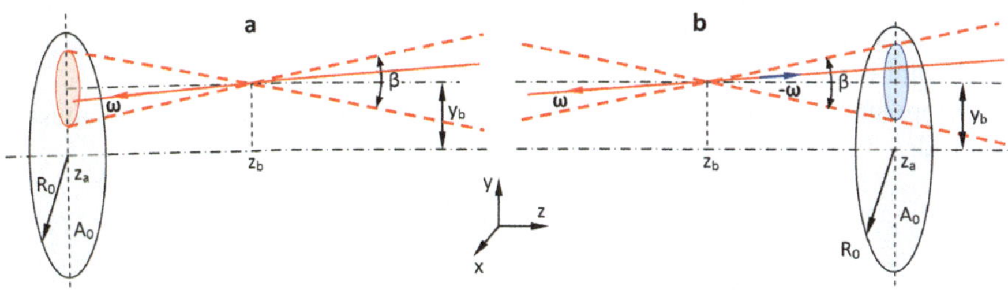

Fig. 4.30 Modeling of the conical effective emissivity: **a** the focal point is in front of the aperture and **b** the focal point is behind the aperture

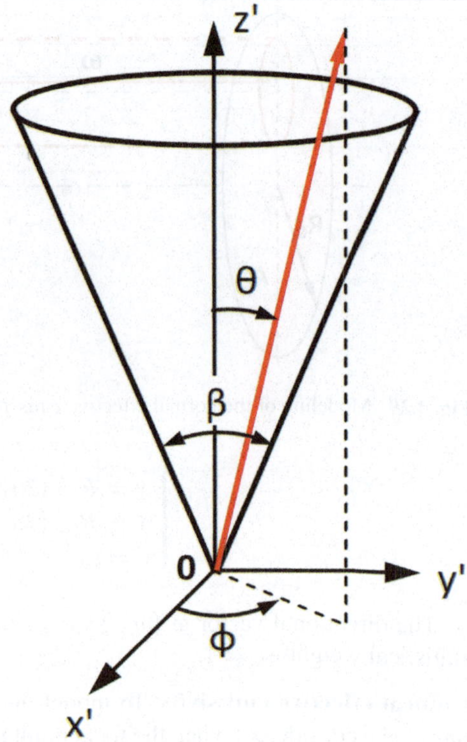

Fig. 4.31 Modeling a random ray uniformly distributed within a conical solid angle

$$\begin{cases} \cos\theta = \cos(\beta/2) + [1 - \cos(\beta/2)] \cdot u_\theta \\ \varphi = 2\pi \cdot u_\varphi \end{cases}. \quad (4.71)$$

The Cartesian coordinates $(\omega'_x, \omega'_y \omega'_z)$ of the directional vector ω' in the primed coordinate system are $(\sin\theta\cos\varphi, \sin\theta\sin\varphi, \cos\theta)$.

The transformation to the global coordinate is done by rotation $(\omega_x = \omega'_x, \omega_y = \omega'_y, \omega_z = -\omega'_z)$ for the case shown in Fig. 4.31a (the focal point is in front of the cavity aperture) and $(\omega_x = \omega'_x, \omega_y = -\omega'_y, \omega_z = \omega'_z)$, if the focal point is behind the cavity aperture Fig. 4.31b.

Before launching the next ray toward the cavity, the following inequality must be checked:

$$\omega_x^2 t^2 + (y_b + \omega_y t)^2 < R_0^2, \quad (4.72)$$

where $R_0 = y_{np}$ is the aperture radius, $t = (z_b - z_{np})/\omega_z$. If inequality (4.72) is not fulfilled, the viewing conditions are set incorrectly. For each ray, the initial statistical weight $w_0 = 1$ is assigned.

Integrated effective emissivity The simulation of integrated effective emissivity is different from the simulation of other viewing conditions, just as the definition of integrated effective emissivity is different from the definitions of all other types of effective emissivity.

Let us recall that the integrated effective emissivity is defined as the ratio

$$\varepsilon_{e,\text{int}}(\lambda) = \Phi_{\lambda}(\lambda) / \Phi_{\lambda,bb}(\lambda, T_{ref}), \tag{4.73}$$

where Φ_{λ} is the spectral radiant flux emitted by the cavity aperture A_a and falling on the detector surface A_d; $\Phi_{\lambda,bb}$ is the spectral radiant flux that will fall on A_d if A_a is the perfectly black disk at temperature T_{ref}. The denominator in Eq. 4.73 can be computed analytically as

$$\Phi_{\lambda,bb}(\lambda, T_{ref}) = \pi A_0 F_{A_a - A_d} L_{\lambda,bb}(\lambda, T_{ref}), \tag{4.74}$$

where

$$F_{A_a - A_d} = \frac{1}{2}\left(Z - \sqrt{Z^2 - 4\left(\frac{X}{Y}\right)^2}\right) \tag{4.75}$$

is the view factor between two coaxial disks [63],

$$X = \frac{R_a}{H}, \quad Y = \frac{R_d}{H}, \quad \text{and} \quad Z = 1 + \frac{1 + Y^2}{X^2}. \tag{4.76}$$

The spectral radiant flux transferred from a cavity aperture A_a to the detector A_d is expressed by the double surface integral

$$\Phi_{\lambda}(\lambda) = \int_{A_a}\int_{A_d} \frac{L_{\lambda}(\lambda)\cos^2\psi}{d^2} dA_a dA_d, \tag{4.77}$$

where L_{λ} is the spectral radiance of point P of the cavity aperture along the direction toward the point Q on the sensor surface and d is the distance between these points.

The spectral radiant flux Φ_{λ} can be evaluated by launching rays from random points Q_i uniformly distributed over the detector surface A_d and passing through random points P_i uniformly distributed over the cavity aperture A_a (see Fig. 4.32). Then every ray is directed into the cavity, and modeling of its further trajectory inside the cavity is performed. The estimator of spectral radiant flux transferred from inside the cavity to the sensor is equal to

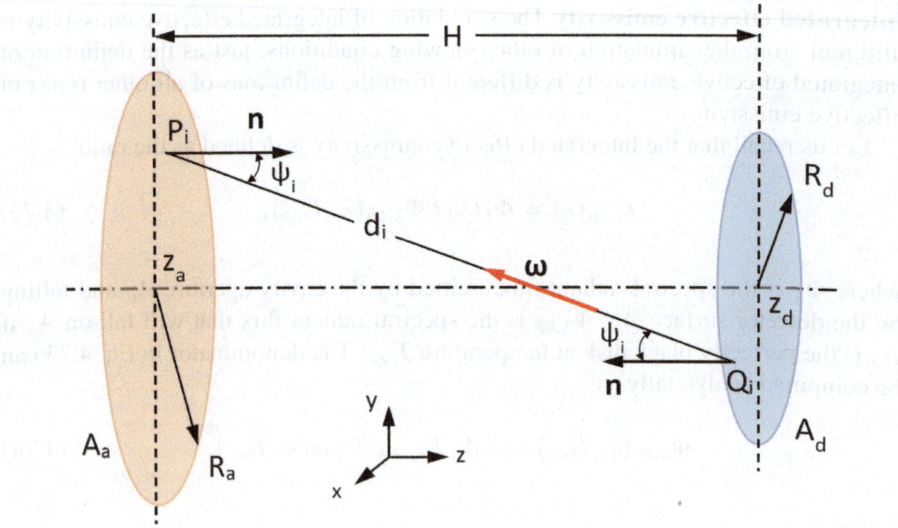

Fig. 4.32 Modeling of the integrated effective emissivity

$$\Phi_\lambda(\lambda) = \frac{A_a A_d}{n} \sum_{i=1}^{n} \frac{L_i(\lambda)\cos^2\psi_i}{d_i^2}, \tag{4.78}$$

where the index $i = 1, \ldots, n$ denotes the ith ray traced between the aperture and the detector, n is the total number of rays launched into the cavity.

According to the general procedure for modeling the integrated effective emissivity, the initial statistical weight $w_{0,i} = \cos^2\psi_i / d_i^2$ is assigned to i-th ray.

The directional vector $\boldsymbol{\omega}$ can be expressed as

$$\boldsymbol{\omega} = \left(\boldsymbol{\xi}_{P_i} - \boldsymbol{\xi}_{Q_i}\right) / d_i, \tag{4.79}$$

where

$$d_i = \sqrt{\left(x_{Q_i} - x_{P_i}\right)^2 + \left(y_{Q_i} - y_{P_i}\right)^2 + \left(z_{Q_i} - z_{P_i}\right)^2} \tag{4.80}$$

is the distance between point $Q_i \in A_d$ and the point $P_i \in A_a$, $\left(x_{Q_i}, y_{Q_i}, z_{Q_i}\right)$ and $\left(x_{P_i}, y_{P_i}, z_{P_i}\right)$ are the coordinates of these points.

However, this computational scheme exhibits slow convergence when the distance H from the cavity aperture to the detector becomes too small and fails at $H = 0$ (the case corresponding to the hemispherical effective emissivity). It was found that for $H / \max(R_a, R_d) < 2$, the usual algorithm consisting in backward ray tracing from the detector to the cavity aperture, gives more accurate result and works even for $H = 0$. In this case, the same initial statistical weight $w_0 = 1$ is assigned to each ray. Once the modeling of viewing conditions is done and the passing through the aperture is checked, the ray counter is incremented by one.

4.5.3 Ray Tracing Procedure

The goal of the ray tracing procedure is to find coordinates of the next reflection point using the coordinates of the previous reflection point (or, if the ray is launched from outside the cavity, using the coordinates of the starting point) and the direction of the previous reflection (or the direction obtained as the result of the viewing conditions modeling). Let the cavity surface consists of n_p segments of surfaces of revolution ($A = \bigcup_{i=1}^{n_p} A_i$), each described by equation $\Phi_i(x, y, z) = 0$, $i = 1, 2, \ldots, n_p$. Let the parametric equations of the straight line, along with the ray is traces are $\boldsymbol{\xi} = \xi' + \boldsymbol{\omega} \cdot t$ or

$$\begin{cases} x = x' + \omega_x t \\ y = y' + \omega_y t \, , \\ z = z' + \omega_z t \end{cases} \tag{4.81}$$

where (x', y', z') are the Cartesian coordinates of the starting point ξ', (x, y, z) are the Cartesian coordinates of the point $\boldsymbol{\xi}$, and $0 \le t < \infty$ is the time-like parameter.

Some systems may have more than one solution; the others may have no solutions at all. Then we have to select the minimal value among all positive values of t (i.e., the value that corresponds to the minimal distance between points $\boldsymbol{\xi}$ and ξ') and, at the same time, gives the point $\boldsymbol{\xi}$ (x, y, z) that lies on some surface forming the cavity within the boundaries of that surface.

The straightforward algorithm of ray tracing involves solution of the system (4.81) together with the surface equation $\Phi_i(x, y, z) = 0$ for each surface constituting the cavity to obtain t_i and (x, y, z). The value of t_i is rejected if $z \notin [z_m, z_{m+1}]$ or $y \notin [y_m, y_{m+1}]$, i.e. $\boldsymbol{\xi} \notin A_i$. Among all t_i that are not rejected, we must select the minimal positive value. That value of t as well as the values (x, y, z) computed using Eq. 4.81 for that t correspond to the actual point of the ray-cavity intersection. If all t_i, $i = 1, 2, \ldots, np$ are rejected, the ray leaves the cavity through the aperture. Let us consider as an example the rays traced from the point $\boldsymbol{\xi}$ on the bottom of a cavity shown in Fig. 4.33. This hypothetical cavity is the cylinder-inner-cone with a flat internal baffle and flat lid. Figure 4.33a depicts the cavity generatrix that is formed by 8-section polygonal line. The internal baffle is assumed to have a finite thickness and is represented by two planes and one cylindrical surface. The latter one corresponds to the end of the annular baffle. Although the cavity lid (the diaphragm with the cavity aperture) is also assumed to be of a finite thickness, two surfaces are sufficient for modeling, since the outer flat surface of the lid cannot exchange radiation with the inner surface of the cavity.

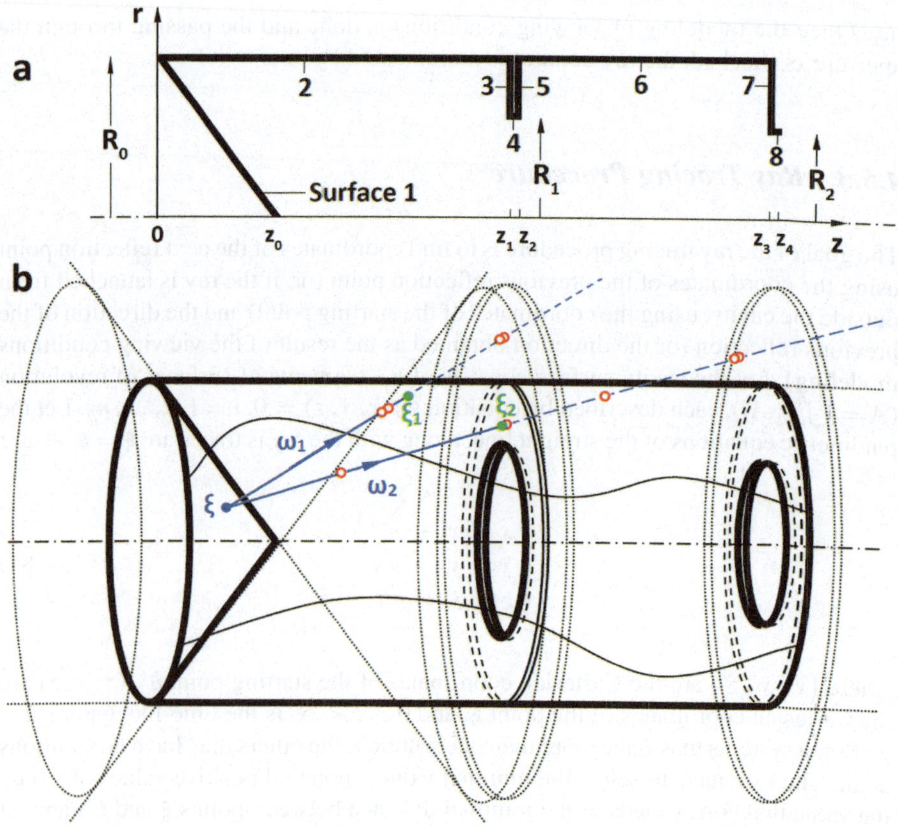

Fig. 4.33 An example of rays traced from the point $\boldsymbol{\xi}$ on the internal conical bottom of the cavity with an internal annular baffle. Filled circles are for the actual points of intersection ($\boldsymbol{\xi}_1$ and $\boldsymbol{\xi}_2$) with the cavity surfaces. Empty circles denote the points of intersection, which should be rejected

The internal conical bottom is defined by the equation $x^2 + y^2 - (z - z_0)^2 tg^2(\beta/2) = 0$ and the inequality $0 \le z < z_0$, where z_0 is the axial coordinate of the cone apex, $tg\frac{\beta}{2} = \frac{R_0}{z_0}$, and β is the apex angle of the conical bottom. The cylindrical part is defined by the equation $x^2 + y^2 = R_0^2$ and the inequality $0 \le z < z_1$, where R_0 is the cavity radius. The flat surfaces of the internal baffle are defined by equations $z = z_1$ and $z = z_2$ and the inequality $R_1^2 \le x^2 + y^2 \le R_0^2$. For the flat lid, $z = z_3$ and $R_2^2 \le x^2 + y^2 \le R_0^2$. Finally, the ends of the internal baffle and the lid are described by the equations $x^2 + y^2 = R_1^2$ and $x^2 + y^2 = R_2^2$, respectively. Their boundaries are defined by the inequalities $z_1 \le z \le z_2$ and $z_3 \le z \le z_4$, respectively.

Let us consider the point $\boldsymbol{\xi}$ lying on the conical bottom of the cavity and the rays sent from this point in directions $\boldsymbol{\omega}_1$ and $\boldsymbol{\omega}_2$. The ray defined by $\boldsymbol{\omega}_1$ (the positive values of the parameter t are implied) intersects: the imaginary part of the conical surface forming the bottom, the cylindrical section of the cavity, and two planes, which are extensions of the surfaces forming the inner baffle.

It is clear, that only intersection with the cylindrical surface should be accepted while other points must be rejected. The ray defined by $\boldsymbol{\omega}_2$ gives the actual intersection with the left-hand side of the internal baffle but four intersections with other surfaces must be rejected.

If the ray escapes the cavity and the maximal number of the rays traced is not achieved, the next ray is launched into the cavity; i.e. the unit "Modeling the viewing conditions" in Fig. 4.27 must be performed once again. If the specified number of rays is reached, control passes to the unit "Calculation of the effective emissivity" (see below). Otherwise (when the actual ray-surface intersection is found), the reflection of the ray is modeled.

Calculations performed by the ray tracing routine takes a significant part of the computational time and therefore have to be optimized. Various techniques that allow avoiding unnecessary calculations have been developed in computer graphics and adjacent disciplines during several last decades (see, e.g. [1, 45, 142]). Proper use of these techniques is a prerequisite for building an effective MCM ray tracing code.

4.5.4 Modeling the Reflection

This procedure is performed each time when the ray hits one of the surfaces, which form the cavity. If the additive model of reflection is used, the PRNG generates the next random number u_r that defines the reflection type. For the specular-diffuse model, the diffuse reflection is chosen if $u_r < D$ and specular—otherwise. For the 3C BRDF model, when a ray hits the cavity wall at an angle of incidence θ_i, if $u_r \leq \rho_d(\theta_i) = \rho_d = const$, the diffuse reflection is chosen. If $u_r \leq \rho_d + \rho_{qs}(\theta_i)$ the reflection is quasi-specular; otherwise, the reflection is glossy.

In the case of specular reflection, the unit vectors $\boldsymbol{\omega}$ describing the directions of reflection is expressed via the unit vector $\boldsymbol{\omega}'$ indicating the direction of incidence and the normal $\mathbf{n}$ to the surface at the point of reflection as

$$\boldsymbol{\omega} = \omega' - 2\mathbf{n} \cdot \left(\mathbf{n}\omega'\right), \tag{4.82}$$

where brackets mean the scalar product of two vectors.

Diffuse reflection There are two major techniques for modeling diffuse reflection. The first algorithm, usually recommended in all textbooks on radiative heat transfer (see, e.g. [63, 86, 148], etc.) is the following. The direction of the reflected ray is set by the polar angle θ_i and azimuthal angle φ_i in a local spherical system of coordinates, the polar axis of which is collinear with the normal to the surface at the point of the ray incidence. If u_θ and u_φ are two random numbers generated by the PRNG, then θ and φ can be modeled as

$$\begin{cases} \sin^2\theta = u_\theta \\ \varphi = 2\pi\, u_\varphi \end{cases}. \tag{4.83}$$

Then θ and φ are used to find the vector $\boldsymbol{\omega}$ defining the direction of reflection. In the local (primed) coordinate system linked to the spherical system (θ, φ), the components of $\boldsymbol{\omega}$ $(\sin\theta\cos\varphi, \sin\theta\sin\varphi, \cos\theta)$ have to be transformed to the global Cartesian coordinate system.

However, there is an alternative, a little bit faster algorithm. Every sphere that is tangent to the surface of a Lambertian reflector at the point of reflection is a surface of uniform irradiance. Therefore, to model the direction of a diffuse reflection, it is sufficient to generate the points uniformly distributed over such a sphere and find the unit vector connecting the point of reflection with the point generated on the unit sphere. One can use the algorithm proposed by Marsaglia [88] to obtain the points uniformly distributed on the spherical surface of unit radius. According to this algorithm, the pair of pseudo-random numbers u_x and u_y undergoes the linear transformation

$$\begin{cases} v_x = 2u_x - 1 \\ v_y = 2u_y - 1 \end{cases}. \tag{4.84}$$

The points with coordinates (v_x, v_y) are uniformly distributed within the square $(-1 < x, y < 1)$. If $s = v_x^2 + v_y^2 > 1$, then the point is outside the circle of unit radius, this pair of u_x and u_y is rejected, and a new pair is generated. Otherwise, the coordinates of the point on the unit sphere are computed as

$$\begin{cases} x_s = 2v_x\sqrt{1-s} \\ y_s = 2v_y\sqrt{1-s} \\ z_s = 1 - 2s \end{cases}. \tag{4.85}$$

Then coordinates of a random point (x_s, y_s, z_s) are transformed into the coordinates of the reflection vector $\boldsymbol{\omega}$ $(\omega_x = y_s, \omega_y = x_s, \omega_z = -z_s)$. When the USD model is employed, the statistical weight of the incident ray w' is multiplied by the reflectance ρ to obtain the statistical weight w of the reflected ray as

$$w = w' \cdot \rho. \tag{4.86}$$

For the 3C model of reflection, the transformation of the statistical weight is done differently:

$$w = w' \cdot R_d, \tag{4.87}$$

where $R_d = \rho_d / \pi$.

Glossy and quasi-specular reflection To model glossy and quasi-specular reflection, the same sampling algorithm developed by Geisler-Moroder and Dür [38] is used. First, spherical coordinates of the halfway vector **h** (see Fig. 4.19) are computed using random numbers u_θ and $u_\varphi$generated by the PRNG:

$$\begin{cases} \theta_h = \tan^{-1}(-\sigma \ln u_\theta) \\ \varphi_h = 2\pi u_\varphi \end{cases}. \tag{4.88}$$

Second, spherical coordinates are transformed to Cartesian coordinates of the vector **h** $(\sin\theta_h \cos\varphi_h, \sin\theta_h \sin\varphi_h, \cos\theta_h)$. When coordinates of the halfway vector **h** are found, we can compute coordinates of the viewing vector $\boldsymbol{\omega}_v$ specularly reflected from the microfacet for the given incidence vector $\boldsymbol{\omega}_i$:

$$\begin{cases} \omega_{vx} = \omega_{ix} - 2 \cdot (\omega_i \cdot \mathbf{h}) \cdot h_x \\ \omega_{vy} = \omega_{iy} - 2 \cdot (\omega_i \cdot \mathbf{h}) \cdot h_y \\ \omega_{vz} = \omega_{iz} - 2 \cdot (\omega_i \cdot \mathbf{h}) \cdot h_z \end{cases}. \tag{4.89}$$

According to the sampling procedure proposed by Geisler-Moroder and Dür [38], the statistical weight of the reflected ray is transformed

$$w = w' \cdot \frac{2R(\theta_h)}{1 - \omega_{iz} / \omega_{vz}}, \tag{4.90}$$

where R is R_{qs} or R_g depending on the type of reflection—quasi-specular or glossy.

Each ray is traced until it escapes the cavity through the aperture. The weight of the ray decreases after each reflection; therefore, we can stop tracing a ray if its weight becomes less than some given threshold value $w_{\min}$ that defines the allowable error in the effective emissivity. After each reflection, the counter $L_{\lambda,j}(\lambda)$ of the spectral radiance along the ray trajectory is increased according to the following scheme:

$$L_{\lambda,j}(\lambda) \leftarrow L_{\lambda,j}(\lambda) + \varepsilon_k(\lambda) \cdot w \cdot L_{\lambda,bb}(\lambda, T_k), \tag{4.91}$$

where the lower index k denotes the k-th reflection of the j-th ray, ε_k and T_k are the emissivity and the temperature of the cavity wall at the k-th point of reflection.

4.5.5 Calculation of the Effective Emissivity

The estimate of the effective emissivity is calculated as the quotient of the total spectral radiance of all rays leaving the cavity and the spectral radiance of a perfect blackbody at the given wavelength and the reference temperature. The trajectory of each ray is modeled from outside the cavity toward its aperture; the j-th ray is reflected from the cavity walls at the points $\boldsymbol{\xi}_{j,0}, \boldsymbol{\xi}_{j,1}, \ldots, \boldsymbol{\xi}_{j,m_j}$ until it escapes the cavity through the aperture or if the ray trajectory is interrupted in accordance with some predefined criterion. Then the direction of the ray propagation is reversed. The point of the last reflection becomes the point of the initial emission; the spectral radiance $L_{\lambda,j}$ along the j-th ray at the exit of a cavity consists of the spectral radiance of the thermal radiation of the cavity surface at the point $\boldsymbol{\xi}_{j,0}$ and those at the points $\boldsymbol{\xi}_{j,k}$ ($k = 1, ..., m_j$, where m_j is the number of reflections in j-th trajectory), each multiplied by reflectances at the intermediate reflection points $\boldsymbol{\xi}_{j,1}, ..., \boldsymbol{\xi}_{j,k-1}$. The spectral radiance $L_{\lambda,j}(\boldsymbol{\xi}_0, \boldsymbol{\omega}_0, \lambda)$ along the j-th ray at the exit of a cavity can be written in the compact form as

$$L_{\lambda,j}(\boldsymbol{\xi}_0, \boldsymbol{\omega}_0, \lambda) = \varepsilon(\boldsymbol{\xi}_0, \boldsymbol{\omega}_0, \lambda)L_{\lambda,bb}(\boldsymbol{\xi}_0, \boldsymbol{\omega}_0, \lambda, T(\boldsymbol{\xi}_0)) + \sum_{k=1}^{m_j} \varepsilon(\boldsymbol{\xi}_{j,k}, \boldsymbol{\omega}_{j,k}, \lambda)L_{\lambda,bb}(\boldsymbol{\xi}_{j,k}, \boldsymbol{\omega}_{j,k}, \lambda, T(\boldsymbol{\xi}_{j,k})) \prod_{l=0}^{k} w_l, \tag{4.92}$$

where w_l is the ray's statistical weight after l-th reflection in the j-th ray trajectory, $\varepsilon(\boldsymbol{\xi}_{j,k}, \boldsymbol{\omega}_{j,k}, \lambda) = 1 - \rho(\boldsymbol{\xi}_{j,k}, -\boldsymbol{\omega}_{j,k}, \lambda)$ is the directional emissivity at the point $\boldsymbol{\xi}_{j,k}$, $1 - \rho(\boldsymbol{\xi}_{j,k}, -\boldsymbol{\omega}_{j,k}, \lambda)$ is the directional-hemispherical reflectance at the same point, $\boldsymbol{\omega}_{j,k}$ is the direction from point $\boldsymbol{\xi}_{j,k}$ to $\boldsymbol{\xi}_{j,k-1}$, $\boldsymbol{\omega}_0$ is the viewing direction. After averaging of n ray trajectories and dividing the mean value by $L_{\lambda,bb}(\lambda, T_{ref})$, the local directional spectral effective emissivity is calculated as

$$\varepsilon_e(\boldsymbol{\xi}_0, \boldsymbol{\omega}_0, \lambda) = \varepsilon(\boldsymbol{\xi}_0, \boldsymbol{\omega}_0, \lambda)\frac{L_{\lambda,bb}(\boldsymbol{\xi}_0, \boldsymbol{\omega}_0, \lambda, T(\boldsymbol{\xi}_0))}{L_{\lambda,bb}(\lambda, T_{ref})} + \frac{1}{L_{\lambda,bb}(\lambda, T_{ref})}\sum_{j=1}^{n}\sum_{k=1}^{m_j} \varepsilon(\boldsymbol{\xi}_{j,k}, \boldsymbol{\omega}_{j,k}, \lambda)L_{\lambda,bb}(\boldsymbol{\xi}_{j,k}, \boldsymbol{\omega}_{j,k}, \lambda, T(\boldsymbol{\xi}_{j,k})) \prod_{l=0}^{k} w_l. \tag{4.93}$$

Although the algorithm described above was applied to a single wavelength, it can be modified to work with spectral dependences. Moreover, if the effective emissivity is considered as parameter-dependent, the estimation of the effective emissivity for different values of this parameter can be done using the same set of ray trajectories. In contrast to variance reduction methods (such as the method of statistical weights) and various methods of speeding up computations by optimizing computational processes, the time-saving techniques, which allow obtaining results for more than one dataset at once, are something like parallel computing but without actual parallelization. The idea of employing the same set of ray trajectories to estimate effective emissivities for different dataset is easily realizable for spectral dependences of radiation characteristics and for different temperature distributions.

If the spectral dependencies are calculated, and the angular distributions of the surface radiating characteristics do not depend on the wavelength λ, it is possible to associate a set of wavelengths λ_j, the corresponding sets of statistical weights w_j, and spectral radiances $L_{\lambda,j}$, $j = 1, 2, ..., n_\lambda$ to each ray. At each reflection, the elements of the array of statistical weights w_j will be multiplied by the reflectance values $\rho(\lambda_j)$ at the appropriate wavelength λ_j, and the current values of the spectral radiance are given by $w_j L_j$.

In the same way, this scheme can be extended to the n_t temperature distributions by assigning to each ray the two-dimensional arrays $w_{j,k}$ and $L_{\lambda,j,k}$, $j = 1, 2, ..., n_\lambda$, $k = 1, 2, ..., n_t$. At each reflection, the statistical weights are multiplied by $\rho(\lambda_j)$, and $L_{\lambda,j,k}$ is computed for k-th temperature distribution and j-th wavelength.

Finally, **Output of results** includes displaying the results and some post-processing actions, e.g. calculation of radiance temperatures, plotting the graphs, saving results in the files of predefined formats, generation of reports, etc.

4.6 Experimental Determination of Effective Emissivities

4.6.1 General Principles and Rationales

Since the definition of the directional effective emissivity is analogous to the definition of the directional emissivity of flat opaque samples, most techniques of measuring the emissivity (see, e.g. [40, 161]) can be used also when the effective emissivity is measured. Similar to measurements of the emissivity, there is no universal method for measuring the effective emissivity nor even a comprehensive approach to solving this problem. Experimental determination of the effective emissivity of blackbody cavities is a complex measurement task that can often be solved only for certain specific geometrical, optical, or thermal conditions. The early experimental works on this problem were reviewed by Bedford [9]. The studies of recent decades were discussed by Mekhontsev et al. [94]. Analysis of these works shows that their usefulness was and continue to be limited to verifying the results of calculations. None of the experimental determinations of the effective emissivity values was used as a definitive at the metrological characterization of precision radiometric systems without relying on the calculated values. We will not consider various relative measurements of radiometric characteristics of blackbodies, which make possible to investigate their dependences upon some parameters (e.g. the viewing angle or the detector-to-cavity distance) and check the adequacy of a model underlying the calculations but do not allow determining the actual values of the effective emissivity.

We can roughly divide methods for measuring the effective emissivity into radiometric and reflectometric. Radiometric methods follow the definition of the effective emissivity. To measure the spectral effective emissivity, we must compare the spectral radiance of the blackbody under study (BBUS) with the spectral radiance of the perfect blackbody at the same temperature. Experimental determination of the effective emissivity can be performed by means of relative radiometry, by comparing the spectral radiance of the BBUS with that of another blackbody, which has characteristics that are certainly closer to a perfect blackbody than BBUS.

This approach does not require absolute measurement of radiometric quantities but do require strict equality of temperatures of compared blackbodies. This equality can be most easily realized if the reference cavity is formed on the base of the BBUS, e.g. by modifying geometry of the cavity. The simplest solution is to drill a tiny cavity in the wall of the BBUS, just as an auxiliary cavity is created at measurement of the emissivity of flat samples (see, e.g. [61, 75, 153]). The effective emissivity is measured as the ratio of the signals of a linear spectroradiometer or radiation thermometer when it sees a surface close to the drilled micro-cavity and the micro-cavity itself. Practical implementation of this method is possible only if the walls of the BBUS are thick enough to the micro-cavity can be drilled and the thermal conductivity of these walls is sufficiently high to minimize the temperature difference between the bottom of the drilled cavity and the radiating surface of the BBUS. Another way is to increase the effective emissivity of the BBUS by reducing the diameter of its aperture (by applying replaceable diaphragm or another extra part). The effective emissivity of an isothermal cavity approaches unity as its aperture diminishes. Comparison of the spectral radiances of a cavity with a given geometry and a geometry modified in this way allows us to estimate the spectral effective emissivity of the BBUS. The main difficulty in implementing this approach is that changing the size of the aperture changes the conditions of heat exchange between the cavity and the environment, on which the temperature of the cavity depends.

The emissivity (as well as the effective emissivity) is a dimensionless quantity obtainable as the ratio of homonymous quantities (e.g., spectral radiances) measured by a linear sensor, the calibration of which is not necessary. For this reason, the use of absolute radiometric measurements to determine the effective emissivity may seem like a strange idea, since absolute measurements are always more complicated and more expensive than relative measurements. Nevertheless, the absolute radiometric measuring the spectral radiance of the BBUS may make sense if the temperature of the BBUS can be measured by a contact sensor. Then the spectral emissivity of the BBUS can be determined as the ratio of the measured spectral radiance and the spectral radiance calculated via the Planck law for the temperature of the BBUS.

In recent years, the reflectometric methods based on the Kirchhoff law has taken a leading position among all methods for measuring the effective emissivity, although Kirchhoff's law applies only to isothermal cavities. The reflectometric methods use well-established theory, measurement methodology, and easily accessible experimental equipment, which are similar to those employed for measuring the reflectance of flat diffuse samples (see [62, 107, 149]). However, the similarity is not the identity. There are at least two major differences in the reflectometry of diffuse flat samples and cavities.

The first difference concerns the measured reflectance values. The typical reflectance values for the flat diffuse samples are ranged from about 0.02 (for "black" paints) to about 0.98 (for "white" diffuse surfaces), while for blackbodies this range is less than 0.05 (for the flat-plate blackbodies) and lower than 0.01 for blackbody cavities.

The second difference concerns the areas of reflecting surfaces. For a flat sample, the reflected radiation comes from the same region that is exposed to irradiation; due to subsurface scattering, the reflecting area can be slightly greater than that initially irradiated. The reflecting region of a cavity is the cavity opening. Its diameter can be much greater than the diameter of the incident beam. For small cavities with the opening diameters of several millimeters, no special problems arise. They appear for large-area blackbodies, when their dimensions make it difficult or even impossible to place them in a measurement setup. As a rule, the reflectometric measurements are carried out for cavities that are at room temperature. One can expect that the most adequate results can be obtained for blackbodies whose operating temperature differs from room temperature not too much, so the correction for the possible temperature dependence of the emissivity of the cavity walls is not necessary. Therefore, the IR spectral range is very important for reflectometric measurements of the effective emissivities. In the following sections, we focus on the most influential studies of recent years.

4.6.2 *Radiometric Technique for Measuring Effective Emissivity*

The key idea of the radiometric technique is simultaneous measurements of the spectral radiance of the BBUS by a calibrated spectroradiometer and the temperature of the BBUS by a contact thermometer. The effective emissivity is found as the ratio of measured spectral radiance and that calculated according to Planck's law. This is the only method that can be applied (at least, theoretically) to non-isothermal blackbodies.

A practical approach to the determination of $\varepsilon_e(\lambda)$ in the IR spectral range, developed at NIST, was described by Hanssen et al. [52] and Mekhontsev et al. [96]. Three blackbodies are employed in measurements: the BBUS (the blackbody under study), the "hot" reference blackbody (HRBB), and the "cold" reference blackbody (CRBB) with the temperatures above and below the temperature of the BBUS, respectively. The optical scheme of the experimental setup is presented in Fig. 4.34. The Fourier

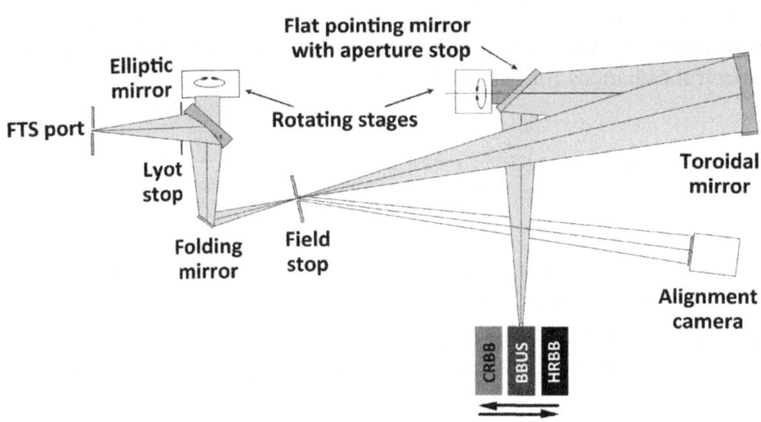

Fig. 4.34 Optical scheme of the experimental setup (after [96])

transform spectrometer (FTS) is used to measure the spectral radiances of three blackbodies.

The measurement equation for the spectral radiance $L_{\tilde{\nu}} = dL/d\tilde{\nu}$ in the scale of wavenumbers $\tilde{\nu}$ is equal to

$$L_{\tilde{\nu},BBUS}(\tilde{\nu}) = \mathrm{Re}\left[\frac{V_{BBUS}(\tilde{\nu}) - V_{CRBB}(\tilde{\nu})}{V_{HRBB}(\tilde{\nu}) - V_{CRBB}(\tilde{\nu})}\right] \cdot \left[L_{\tilde{\nu},HRBB}(\tilde{\nu}) - L_{\tilde{\nu},CRBB}(\tilde{\nu})\right] + L_{\tilde{\nu},CRBB}(\tilde{\nu}), \tag{4.94}$$

where V is the complex spectrum measured by the FTS.

Revercomb et al. [130] considered this procedure as the calibration of the FTS but we can consider the FTS as the interpolating instrument to derive the spectral radiance of the BBUS according to Eq. 4.94.

A measurement procedure used at NIST can be described as follows. A high-precision blackbody under study is the sodium or cesium heat-pipe variable-temperature blackbody (VTBB); the HRBB is the Ag, Al, or Zn fixed-point blackbody (FPBB); the CRBB is the water heat-pipe VTBB.

The radiance temperatures of the HRBB and the CRBB are measured at the wavelength λ_0 (0.9 μm or 1.5 μm) by a radiation thermometer. The temperature T, the spectral effective emissivity $\varepsilon_e(\lambda)$, and the radiance temperature T_S at the wavelength λ_0 of these blackbodies are related by equation

$$\varepsilon_e(\lambda_0) = \frac{L_{\lambda,bb}[\lambda_0, T_S(\lambda_0)]}{L_{\lambda,bb}(\lambda_0, T)}, \tag{4.95}$$

where $L_{\lambda,bb}$ is the spectral radiance of the perfect blackbody expressed by the Planck law. The temperature T for a FPBB is the corresponding fixed point of the ITS-90 (see Appendix B); for the H_2O heat-pipe VTBB, the temperature measured by a contact thermometer is used as T. The effective emissivities $\varepsilon_{e,CRBB}$ and $\varepsilon_{e,HRBB}$ calculated via Eq. 4.95 are assumed wavelength-independent and used for calculation of the spectral radiances for Eq. 4.94:

$$L_{\tilde{\nu}}\left(\tilde{\nu}, T_{CRBB,HRBB}\right) = \varepsilon_{e,CRBB,HRBB} \frac{2hc\tilde{\nu}^3}{\exp\left(\frac{hc\tilde{\nu}}{k_B T_{CRBB,HRBB}}\right) - 1}, \tag{4.96}$$

where h is the Planck constant, c is the speed of light, and k_B is the Boltzmann constant (see Appendix A).

After measuring $V_{BBUS}(\tilde{\nu})$, $V_{CRBB}(\tilde{\nu})$, and $V_{HRBB}(\tilde{\nu})$ and calculation of $L_{\tilde{\nu},BBUS}(\tilde{\nu})$ using Eq. 4.94, we can calculate $\varepsilon_{e,BBUS}(\tilde{\nu})$ using T_{BBUS} as

$$\varepsilon_{e,BBUS}(\tilde{\nu}) = \frac{L_{\tilde{\nu}}(\tilde{\nu}, T_{BBUS})}{2hc\tilde{\nu}^3}\left[\exp\left(\frac{hc\tilde{\nu}}{k_B T_{BBUS}}\right) - 1\right]. \tag{4.97}$$

Hanssen et al. [52] described determination of the effective emissivity of the sodium heat-pipe VTBB using the Al FPBB as the HRBB (660.32 °C) and the water heat-pipe VTBB at 100 °C as the CRBB. The PRT (platinum resistance thermometer) was used to measure the temperature of the Na heat-pipe VTBB. To obtain the effective emissivity, the measured spectral radiance, transformed to the representation in the wavelength scale, was divided by the Planckian spectral radiance for the temperature of 657.49 °C measured by PRT.

Similar measurements were performed by Mekhontsev et al. [96] for sodium heat pipe VTBB. The silver FPBB was used as the HRBB at 961.78 °C and the water heat-pipe VTBB at 20 °C as the CRBB. The spectral effective emissivity of Na heat-pipe VTBB is shown in Fig. 4.35 together with the error bars that correspond to the expanded uncertainty at $k = 2$. Measurements were carried out in air, so the artifacts due to atmospheric absorption lines can be seen in the spectrum in the CO_2 and H_2O absorption bands. As can be seen from Figs. 4.35, the effective emissivity of the Na heat-pipe VTBB is greater than at some wavelengths, but without additional studies, we cannot say definitely whether this is due to an inaccurate determination of temperature of the BBUS or due to non-uniform temperature distribution over the radiating surface.

The radiometric method of determining the effective emissivity is complicated, laborious, and requires expensive equipment. If we take into account the fact that usually such measurements in the IR region are typically conducted in vacuum chambers, it becomes clear that they can be performed only in leading research institutions. Application of the spectral effective emissivity of the BBUS obtained at one temperature to a continuous temperature range can be done only for blackbodies of high quality such as heat-pipe VTBBs.

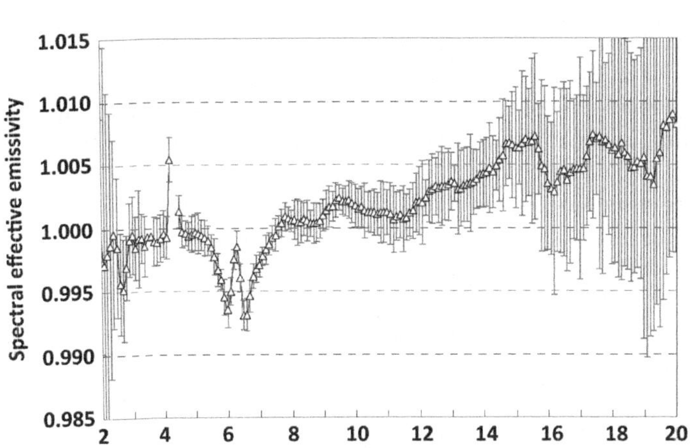

Fig. 4.35 Spectral effective emissivity of Na heat-pipe VTBB measured using comparison with Ag FPBB (after [96])

For less accurate VTBBs, the change of operating temperature leads, as a rule, to change in temperature distribution and, therefore, in the spectral effective emissivity. Thus, the assumption of the temperature independence of the spectral effective emissivity is satisfied in a very limited temperature range, which reduces the attractiveness of this method.

4.6.3 Measurement of Effective Reflectance Using Integrating Spheres

For isothermal opaque cavities, the effective reflectance ρ_e complements the effective emissivity ε_e to unity according the Kirchhoff law. A prerequisite for the applicability of Kirchhoff's law is the fulfillment of the reciprocity principle for the geometrical conditions of viewing and irradiation. In such a way, we get the opportunity to measure the directional-hemispherical effective reflectance of the blackbody instead of its directional effective emissivity.

The methods based on the measurement of effective reflectance is the most accurate method not only due to well-developed reflectometric methods and measuring equipment, but also due to the well-known relationship between relative uncertainties of the effective emissivity and the effective reflectance. Since their absolute uncertainties are equal and $\rho_e \ll \varepsilon_e$ for real-world cavities,

$$\frac{\Delta\varepsilon_e}{\varepsilon_e} = \frac{\Delta\rho_e}{1-\rho_e} \ll \frac{\Delta\rho_e}{\rho_e}, \tag{4.98}$$

that is a significant relative uncertainty $\Delta\rho_e / \rho_e$ results in a small relative uncertainty in the effective emissivity. For instance, if the effective reflectance of the cavity having the effective emissivity of 0.999 is measured with the relative uncertainty of 20%, the corresponding relative uncertainty of the effective emissivity is $0.2 \times 0.001/0.999 \approx 0.0002$.

There are two major classes of reflectometric devices for measuring thedirectional-hemispherical reflectance of flat samples: the concave mirrors (hemispherical and hemiellipsoidal) and the integrating spheres (ISs). The concave mirrors gather the radiation reflected by a sample into the hemispherical solid angle onto the radiation detector. The larger the reflection area, the more difficult it is to collect the reflected radiation at the detector. The only known setup for measuring the effective emissivity using the hemiellipsoidal mirror reflectometer [58, 59] was developed for cavities with the opening diameter of 1 cm. Perhaps, the difficulties in effective collecting the reflected flux explain the dominant position of the ISs when measuring the reflectance of cavities. An IS is the hollow sphere, whose internal surface has a high diffuse reflectance. It is used to average the radiation flux reflected by a sample placed in the port made in the IS wall. The most common material for the IS in the spectral

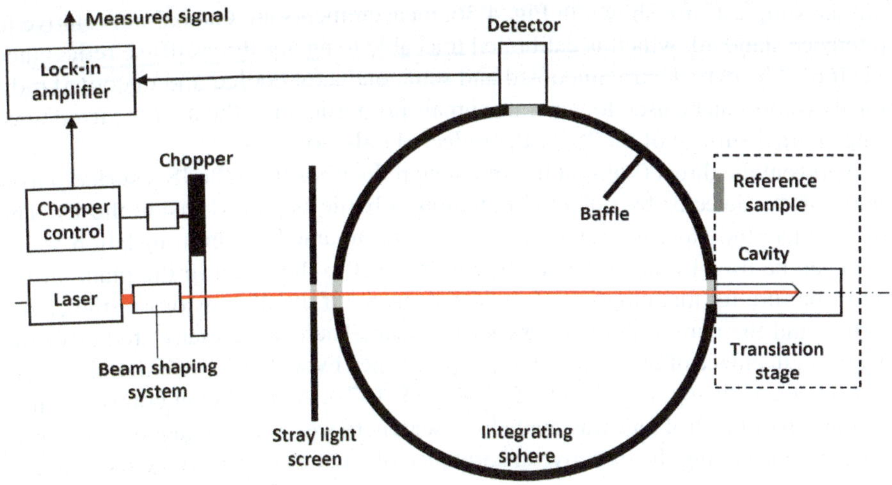

Fig. 4.36 Diagram of typical integrating sphere-based facility for measuring the directional-hemispherical effective reflectance

range from 0.25 to 2 μm is Spectralon.[6] For larger wavelength, good results were obtained using the IS coated with plasma-sprayed diffuse gold [49, 50]. The incident radiation beam is injected through another port of the IS,a radiation detector observes the IS internal surface through the third port. The radiation detector generates a signal proportional to the almost uniform irradiance of the IS wall that is established because of multiple reflections. A general scheme of the experimental setup with the IS for measuring the effective reflectance of cavities is presented in Fig. 4.36.

A most suitable radiation source for these measurements is the laser that is able to provide a monochromatic collimated beam of high intensity. The most common are the continuous wave gas lasers: He–Ne ($\lambda = 0.63$ μm, 1.13 μm, and 3.39 μm) and CO_2 (10.6 μm).

In the last decade, the quantum cascade lasers and optical parametric oscillators with the tunable wavelengths in the mid- and far-IR spectral range have been increasingly used. Through the beam shaping system, which forms the incident beam of a prescribed diameter and a uniform distribution of power density across the beam. In some cases, the use of special technique suppressing speckle pattern that typically occurs in diffuse reflection of laser radiation is needed. To separate the radiation reflected by the cavity from the thermal radiation of the cavity and its environment, the laser beam is subjected to amplitude modulation using a mechanical, electro-optical, or other type's chopper. The amplitude of the AC signal from the detector is measured via a lock-in amplifier within a small bandwidth around the applied frequency of modulation.

[6]A polymer material developed by Labsphere, Inc. (https://www.labsphere.com) as a highly reflecting Lambertian material for calibration panels, integrating spheres, and reflectance standards.

In the simplest case shown in Fig. 4.36, measurements are carried out relative to a reference standard, which is calibrated traceable to an absolute diffuse reflectance scale [66]. The pyroelectric uncooled and semiconductor cooled and uncooled radiation detectors can be used to generate signals proportional to the average irradiance of the internal surface of the IS by the reflected radiation.

To prevent the direct (without intermediate reflections from the IS's surface) irradiation of the detector by reflected radiation, a baffle is placed inside the IS. The cavity under study and the reference sample are mounted on the translation stage, which can be used to move the cavity with respect to the beam in the plane of the cavity aperture for mapping its effective reflectance if the dimensions of the cavity aperture and the port of the IS allow such a scan. There were many studies of the effective reflectance of blackbody cavities performed by with this or similar measurement scheme (see, e.g. [5, 54, 69, 95, 134, 135, 167]). A significant contribution to the solution of the problem was made by researchers who were engaged in a similar problem—measuring the effective absorptance of cavities of cryogenic radiometers [84, 98, 155, 165].

Zeng and Hanssen [167] and Hanssen et al. [57] described the Complete Hemispherical infrared Laser-based Reflectometer (CHILR) developed at NIST to measure the normal directional-hemispherical reflectance of the flat samples, radiating elements of blackbodies, and receiving cavities of radiometers. The CHILR facility employ the diffuse-gold-coated IS with the diameter of 200 mm for collecting the reflected radiation and pyroelectric and HgCdTe (mercury cadmium telluride or MCT) detectors.

A set of lasers covers the spectral range from 1 to 11 μm. The collimated laser beam of less than 4 mm diameter through the input port of 6 mm diameter irradiates the sample or cavity arranged in the sample port of 50 mm diameter. Only a small fraction of reflected radiation is lost through the input port. The cone having the input port as the base and the vertex at the center of the collecting port has the opening angle of only 1.72°. Figure 4.37 presents the CHILR setup with the cylindro-conical radiating cavity of the water-bath blackbody in the measurement position. The color map of the 2D distribution of the effective reflectance for this cavity at 10.6 μm is shown in Fig. 4.38. The central peak on this map is due to the reflection from the meniscus formed at the apex of the cone during the cavity painting.

Zeng et al. [168] and Hanssen et al. [56] described the CHILR II facility with the larger diameter IS (~500 mm) and, correspondingly, the larger sample port (can be varied up to 200 mm in diameter). Such a reflectometer allows characterization of the large-area blackbodies commonly used for calibration of remote sensing instrumentation in terms of the IR spectral radiance by complete mapping of large cavities with the effective reflectance down to 10^{-5}.

The modern blackbodies designed for IR calibrations (especially, for the remote sensing wide-FOV instrumentation) can have radiating areas much greater than the corresponding port of the IS. For instance, the V-grooved calibrator of the MODIS (Moderate-resolution Imaging Spectroradiometer, one of the most successful remote sensing instruments launched into Earth orbit in 1999 and 2002 aboard the Terra and Aqua satellites) has the radiating surface of about 37 cm × 22 cm [47]. In addition to

Fig. 4.37 Photograph of Complete Hemispherical Infrared Laser-based Reflectometer (CHILR) with the water-bath blackbody cavity in the measurement position. Courtesy of Dr. Leonard Hanssen (NIST)

the obvious advantages of a planar blackbody with V-grooves such as ease of manufacture and compactness, its grooves form separate cavities between which there is no radiation exchange. Therefore, it is enough to measure the distribution of the effective reflectance across an individual groove to evaluate the effective reflectance or effective emissivity of the blackbody calibrator as a whole. An ancestor of the MODIS is the VIIRS (Visible Infrared Imaging Radiometer Suite) instrument [163].

It was launched in October 2011 on board the Suomi National Polar Orbiting Partnership (S-NPP) satellite. On-orbit calibrations of the VIIRS sensor is performed by alternating observation in near-parallel rays of the central zone of the blackbody (see Fig. 1.13), whose design is similar to the MODIS blackbody calibrator, and the cold space.

Kloepfer et al. [73] presented the methodology of determination the the effective emissivity of the onboard blackbody calibraror of the VIIRS sensor. The diagram of

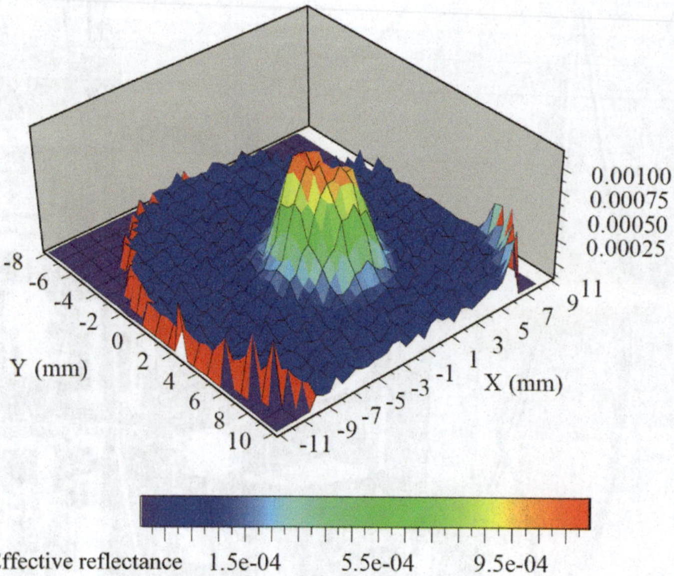

Fig. 4.38 The 3D color map of the nornal effective reflectance of the water-bath blackbody cavity at 10.6 μm. Courtesy of Dr. Leonard Hanssen (NIST)

the experimental setup is shown in Fig. 4.39. It includes the gold-coated IS with the sample port of about 5 cm diameter, the gold reference mirror, and the He-Ne CW laser emitting radiation at 3.39 μm, a part of which is directed by the beam splitter on the auxiliary detector to monitor the beam power stability. Since the sample port of the IS is smaller than the working area of the blackbody, the scanning technique was used.

Measurements of the effective reflectance were carried out at different distances from the port, so that the angle, in which the reflected radiation is collected by the IS varies from several degrees (at a maximal distance z) to almost 180° (at a minimal distance z). The results of scanning across the grooves are plotted in Fig. 4.40a for five collection angles. The averaged over the scanned area effective reflectance was plotted against the collection angle (Fig. 4.40b).

The linear trend was used to extrapolate the average effective reflectance to the collection angle of 180° that corresponds to $z = 0$. Although the relevancy of the use of linear extrapolation was not proven, it hardly can introduce a noticeable error. Recently, Kuljis et al. [78] published the results of measurements of the effective reflectance of two VIIRS onboard blackbodies at two wavelengths of laser IR radiation, 3.39 and 10.6 μm.

The effective emissivity values at 3.39 μm of 0.9982 ± 0.00013 and 0.9984 ± 0.00013 were obtained. At 10.6 μm, these blackbodies have the effective emissivities of 0.9975 ± 0.00008 and 0.9973 ± 0.00007, respectively.

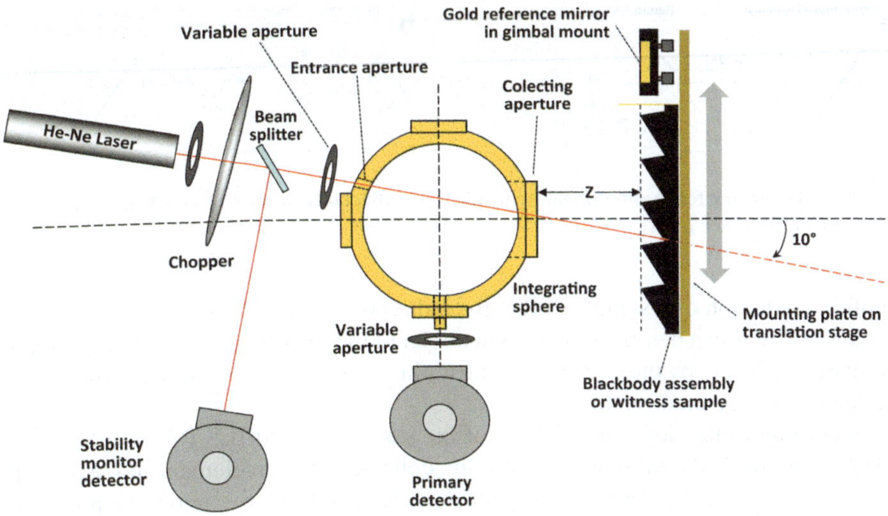

Fig. 4.39 Schematic of the experimental setup for measuring the effective emissivity of the VIIRS on-board blackbody. Reproduced from [73] with permission of Jeremy Kloepfer (Raytheon Company)

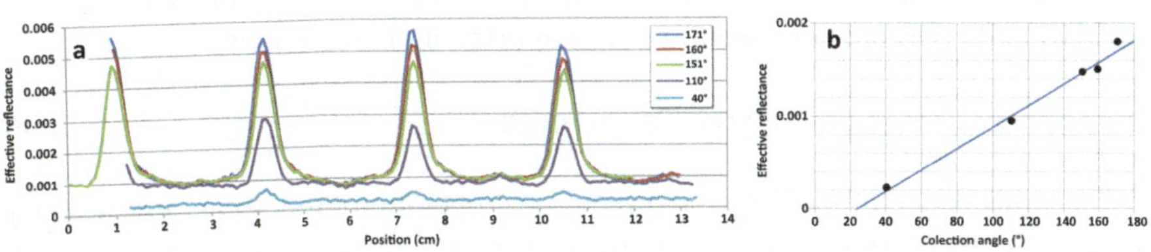

Fig. 4.40 **a** Distributions of the effective reflectance at 3.39 μm across the VIIRS calibration blackbody for five collectin angles and **b** dependence of the averaged effective reflectance on the collection angle. Reproduced from [73] with permission of Jeremy Kloepfer (Raytheon Company)

A more complex case of large cavities has not been investigated sufficiently, although separate experimental studies have been carried out. However, there is no theoretical considerations, on the base of which we could be able to derive the effective emissivity of the cavity as a whole from the measured partial reflectance collected by the IS port, which diameter is smaller than the cavity opening.

4.6.4 Hemispherical Irradiation of a Cavity

Instead of direct measurement of the directional-hemispherical reflectance (DHR), carried out in conformity with its definition, the reciprocal (or inverse) mode of operation is often used in the practice of optical reflectometry. The direct mode

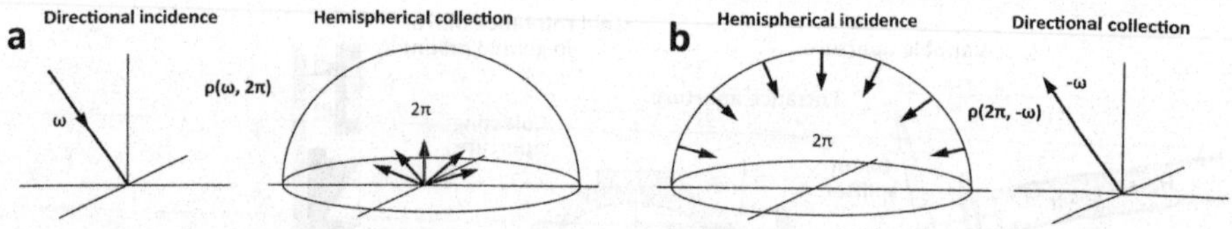

Fig. 4.41 Geometry for: **a** directional-hemispherical reflectance $\rho(\boldsymbol{\omega}, 2\pi)$ and **b** hemispherical-directional reflectance $\rho(2\pi, -\boldsymbol{\omega})$

implies irradiation of a sample in the given direction and collection of the reflected radiant flux within a hemispherical solid angle. In the reciprocal mode, the sample is irradiated diffusely by the isotropic incident flux and collected within a narrow solid angle around a given direction.

According to the definition 7.3.3.3 of the standard ANSI/IES RP-16-10 [2], the DHR $\rho(2\pi, \boldsymbol{\omega})$ is the ratio of reflected flux collected over an element of solid angle surrounding the given direction to the incident flux from the entire hemisphere. The hemispherical-directional reflectance (HDR) $\rho(\boldsymbol{\omega}, 2\pi)$ is given in the definition 7.3.3.7 of ANSI/IES RP-16-10 [2] as the ratio of reflected flux collected over an element of solid angle surrounding the given direction to the incident flux from the entire hemisphere.

With geometrical configurations of irradiation and collection shown in Fig. 4.41, the DHR and HDR are equivalent through the Helmholz reciprocity:

$$\rho(\boldsymbol{\omega}, 2\pi) = \rho(2\pi, -\boldsymbol{\omega}). \tag{4.99}$$

The equivalence of the direct and reciprocal operation modes can also be interpreted in terms of reflectance factors defined by the International Lighting Vocabulary [67]. This formalism, for instance, was used by DeWitt and Richmond [30, pp. 98–101].

Implementation of a uniform hemispherical irradiation often involves significant technical difficulties. However, the reciprocal technique was implemented in such well-known types of instruments as ISs [6, 7, 53, 120], hemiellipsoidal mirror reflectometers [99, 100, 110, 129, 162], and heated-cavity reflectometer [44, 133].

Before the advent of sources of intense IR radiation, the reciprocal mode dominated in the reflectometry of the IR spectral range, since it can be implemented more easily than the direct mode. A polychromatic (as a rule, thermal) radiation can be used for hemispherical irradiation of a sample.

The spectral selection can be performed only for a beam of reflected radiation.

Ballico [4] proposed a method based on the reciprocal technique for assessment of the effective emissivity of blackbodies in the IR spectral range that involve absolute measurements. A schematic for this method referred to as "hotplate" (also known as "scene plate" or "controlled background") technique is presented in Fig. 4.42. The hardware implementation of the method is quite simple, but requires the use of a calibrated spectroradiometer (or radiation thermometer), i.e. measurements have to

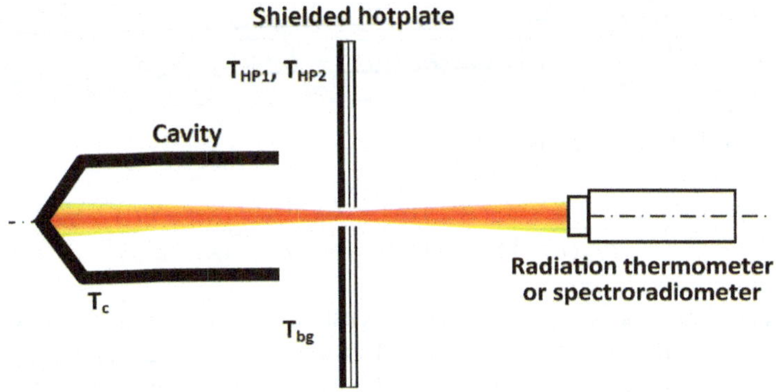

Fig. 4.42 Schematic for the "hotplate" technique

be traceable to the standard of the spectral radiance unit for the radiation reflected by the cavity. The cavity under study is irradiated with a flat plate with adjustable temperature, measured by a contact sensor (for example, a thermocouple).

A spectroradiometer measures the spectral (or bandlimited) radiance of the cavity through a small hole in the hotplate. It is supposed that the emissivity of the hotplate does not depend on the viewing angle and the hotplate is sufficiently large to simulate a uniform hemispherical irradiation of the blackbody aperture.

Ballico [4] reported an uncertainty of 0.0002 (at $k = 2$) in the effective emissivity achieved with the hotplate technique. Hanssen et al. [55] described a modification of the hotplate method that consists in measuring the radiance of the cavity radiation at two temperatures of the hotplate and referred to as the two (background) temperature method (2TM). If $\rho_e(\lambda)$ and T_{bb} are the spectral effective reflectance and temperature of the blackbody, respectively, $T_{hp,1}$ and $T_{hp,2}$ are two temperatures of the hotplate, $\varepsilon_{hp}(\lambda)$ is its spectral emissivity, and T_{bg} is the background (ambient) temperature, then the spectral radiances $L_{\lambda,1}$ and $L_{\lambda,2}$ of the blackbody under study can be expressed as:

$$\begin{aligned} L_{\lambda,1}(\lambda) =& [1 - \rho_e(\lambda)] \cdot L_{\lambda,bb}(\lambda, T_{bb}) \\ &+ \rho_e(\lambda) \left\{ \varepsilon_{hp}(\lambda) \cdot L_{\lambda,bb}\left(\lambda, T_{hp,1}\right) + \left[1 - \varepsilon_{hp}(\lambda)\right] \cdot L_{\lambda,bb}\left(\lambda, T_{bg}\right) \right\} \end{aligned} \tag{4.100}$$

and

$$\begin{aligned} L_{\lambda,2}(\lambda) =& [1 - \rho_e(\lambda)] \cdot L_{\lambda,bb}(\lambda, T_{bb}) \\ &+ \rho_e(\lambda) \left\{ \varepsilon_{hp}(\lambda) \cdot L_{\lambda,bb}\left(\lambda, T_{hp,2}\right) + \left[1 - \varepsilon_{hp}(\lambda)\right] \cdot L_{\lambda,bb}\left(\lambda, T_{bg}\right) \right\}. \end{aligned} \tag{4.101}$$

Subtracting Eq. 4.101 from Eq. 4.100, we obtain the spectral effective reflectance

$$\rho_e(\lambda) = \frac{L_{\lambda,1}(\lambda) - L_{\lambda,2}(\lambda)}{\varepsilon_{hp}\left[L_{\lambda,bb}\left(\lambda, T_{hp,1}\right) - L_{\lambda,bb}\left(\lambda, T_{hp,2}\right)\right]} \tag{4.102}$$

and the spectral effective emissivity

$$\varepsilon_e(\lambda) = 1 - \frac{L_{\lambda,1}(\lambda) - L_{\lambda,2}(\lambda)}{\varepsilon_{hp}(\lambda) \cdot \left[L_{\lambda,bb}\left(\lambda, T_{hp,1}\right) - L_{\lambda,bb}\left(\lambda, T_{hp,2}\right)\right]}. \tag{4.103}$$

The spectral radiances $L_{\lambda,1}(\lambda)$ and $L_{\lambda,2}(\lambda)$ should be measured at the wavelength λ, while $L_{\lambda,bb}\left(\lambda, T_{hp,1}\right)$ and $L_{\lambda,bb}\left(\lambda, T_{hp,2}\right)$ should be computed via Planck's law at the hotplate temperatures $T_{hp,1}$ and $T_{hp,2}$. The 2TM was implemented at NIST with a 200 mm × 280 mm temperature-controlled plate as a part of the Advanced Infrared Radiometry and Imaging (AIRI) facility. Since the expanded (at $k = 2$) uncertainty for the reflectometric measurements of the black paint's emissivity reported by Hanssen et al. [55] is 0.001, we can expect similar uncertainties for the effective emissivities of blackbodies.

The simplicity of the hardware implementation makes the hotplate technique attractive for monitoring the long-term stability of the effective emissivity of blackbodies operating aboard Earth satellites. Researchers of the Space Science and Engineering Center (University of Wisconsin – Madison) developed so-called heated halo device [14, 41–43] that differs from the hotplate mainly in geometry. The last modification of the heated halo [131] has the shape of a truncated cone with a large base facing the cavity under study, which is observed through the heated halo using the Fourier transform spectrometer calibrated in terms of spectral radiance Fig. 4.43.

Gero et al. [42] claimed that the expanded (at $k = 3$) uncertainty of the effective emissivity detemination using the heated halo is less than 0.0006 within the spectral range from 500 to 2000 cm^{-1} (5 to 20 μm). Unfortunately, the key question of applicability of the reciprocal mode—adequacy of substituting the ideal isotropic irradiation of a cavity with the irradiation created by a real heated object (hotplate

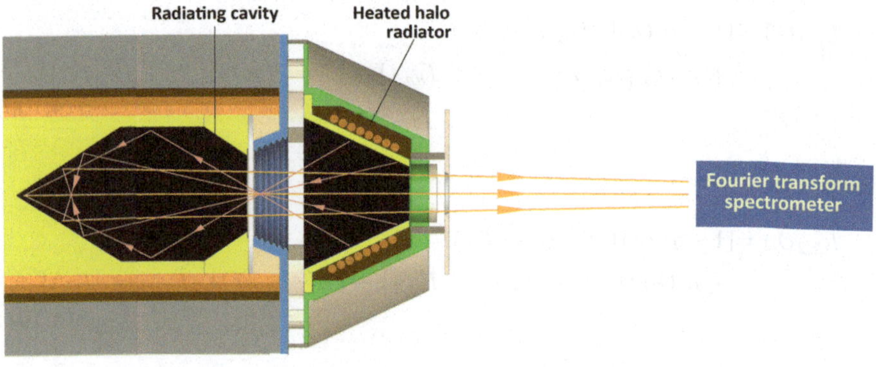

Fig. 4.43 The conically shaped heated halo mounted in front of the onboard blackbody

or halo)—remains unexplored and the uncertainty introduced by such a substitution still remains unknown.

References

1. T. Akenine-Möller, E. Haines, N. Hoffman et al., *Real-Time Rendering*, 4th edn. (CRC Press, Boca Raton, FL, 2018)
2. ANSI/IES RP-16-10, *Nomenclature and Definitions for Illuminating Engineering* (The IES Nomenclature Committee (Illuminating Engineering Society of North America, New York, 2010)
3. M.J. Ballico, Modelling of the effective emissivity of a graphite tube black body. Metrologia **32**, 259–265 (1995/96)
4. M. Ballico, A simple technique for measuring the infrared emissivity of black-body radiators. Metrologia **37**, 295–300 (2000)
5. M. Ballico, High precision measurement of the emissivity of small fixed point blackbodies, in *Proceedings of the TEMPMEKO 2001, 8th International Symposium on Temperature and Thermal Measurements in Industry and Science*, vol. 1, ed. by B. Fellmuth, J. Seidel, G. Scholz (VDE Verlag GmbH, Berlin, 2002), pp. 233–237
6. G. Bauer, K. Bischoff, Evaluation of the emissivity of a cavity source by reflection measurements. Appl. Opt. **10**, 2639–2843 (1971)
7. G. Bauer, Reflexionsmessungen an offenen Hohlräumen. Optik **40**, 603–622 (1961)
8. R.E. Bedford, Effective emissivities of blackbody cavities—a review, in *Temperature: Its Measurement and Control in Science and Industry*, vol. 4, part 1, ed. by H.H. Plumb (Instrument Society of America, Pittsburgh, PA, 1972), pp. 425–434
9. R.E. Bedford, Calculation of effective emissivities of cavity sources of thermal radiation, in *Theory and Practice of Radiation Thermometry*, ed. by D.P. DeWitt, G.D. Nutter (Wiley, New York, 1988), pp. 653–772
10. R.E. Bedford, C.K. Ma, Emissivities of diffuse cavities: isothermal and nonisothermal cones and cylinders. J. Opt. Soc. Am. **64**, 339–349 (1974)
11. R.E. Bedford, C.K. Ma, Emissivities of diffuse cavities, II: isothermal and nonisothermal cylindro-cones. J. Opt. Soc. Am. **65**, 565–572 (1975)
12. R.E. Bedford, C.K. Ma, Emissivities of diffuse cavities, III: isothermal and nonisothermal double cones. Opt. Soc. Am **66**, 724–730 (1976)
13. R.E. Bedford, C.K. Ma, Z. Chu et al., Emissivities of diffuse cavities. 4: isothermal and nonisothermal cylindro-inner-cones. Appl. Opt. **24**, 2971–2980 (1985)
14. F.A. Best, D.P. Adler, C. Pettersen et al., Results from recent vacuum testing of an On-orbit Absolute Radiance Standard (OARS) intended for the next generation of infrared remote sensing instruments. Proc. SPIE **9263**, 926314 (2014)
15. H. Brass, K. Petras, *Quadrature Theory. The Theory of Numerical Integration on a Compact Interval*. (American Mathematical Society, Providence, RI, 2011)
16. H. Buckley, LXVII. On the radiation from the inside of a circular cylinder. Lond. Edinb. Dublin Philos. Mag. J. Sci.: Ser. 7, **4**, 753–762 (1927). https://doi.org/10.1080/14786441008564378
17. R.J. Chandos, R.E. Chandos, Radiometric properties of isothermal, diffuse wall cavity sources. Appl. Opt. **13**, 2142–2152 (1974)
18. S. Chen, Z. Chu, H. Chen, Precise calculation of the integrated emissivity of baffled blackbody cavities. Metrologia **16**, 69–72 (1980)
19. Z. Chu, Y. Sun, R.E. Bedford et al., Precise calculation of effective emissivity of a diffuse cylindro-inner-cone with a specular lid. Appl. Opt. **25**, 4343–4348 (1986)
20. Z. Chu, R.E. Bedford, W. Xu et al., General formulation for the integrated effective emissivity of any axisymmetric diffuse blackbody cavity. Appl. Opt. **28**, 1826–1829 (1989)

21. Z. Chu, J. Dai, R.E. Bedford, Monte Carlo solution for the directional effective emissivity of a cylindro-inner-cone, in *Temperature: Its Measurement and Control in Science and Industry*, vol. 6, part 2, ed. by J.F. Schooley. (American Institute of Physics, New York, 1992), pp. 907–912
22. Z. Chu, J. He, J. Dai et al., Formulation for the integrated effective emissivity of any diffuse, circularly-symmetric blackbody cavity. High Temp.–High Press. **25**, 607–612 (1993)
23. R.C. Corlett, Direct numerical simulation of thermal radiation in vacuum. J. Heat Transf. **88**, 376–382 (1966)
24. J. De Lucas, Numerical optimization of the radial dependence of effective emissivity in blackbody cylindrical cavities. Metrologia **51**, 402–409 (2014)
25. J. De Lucas, Validation of a geometrical model for the calculation of the integrated effective emissivity of conical cavities with a lid. Metrologia **52**, 600–612 (2015)
26. J. De Lucas, A simple geometrical model for calculation of the effective emissivity in blackbody cylindrical cavities. Int. J. Thermophys. **36**, 267–282 (2015)
27. J. De Lucas, J.J. Segovia, Uncertainty calculation of the effective emissivity of cylinder-conical blackbody cavities. Metrologia **53**, 61–75 (2016)
28. V. de Paulo Nicolau, Numerical evaluation of effective emissivities of black-body cylindrical cavities. High Temp.–High Press. **39**, 133–149 (2010)
29. J.C. De Vos, Evaluation of the quality of a blackbody. Physica **XX**, 669–689 (1954)
30. D.P. DeWitt, J.C. Richmond, Thermal radiative properties of materials, in *Theory and Practice of Radiation Thermometry*, ed. by D.P. DeWitt, G.D. Nutter (Wiley, New York, 1988), pp. 91–187
31. A. Diril, H. Nasibov, S. Uğur, UME radiation thermometer calibration facilities below the freezing point of silver (961.78 °C). AIP Conf. Proc. **684**, 663–668 (2003)
32. R.B. Dinwiddie, M.M. Kirka, P.D. Lloyd et al., Calibrating IR cameras for in-situ temperature measurement during the electron beam melt processing of Inconel 718 and Ti-Al6-V4. Proc. SPIE **9861**, 986107 (2016)
33. W.L. Dunn, J.K. Shultis, *Exploring Monte Carlo Methods* (Academic Press, Amsterdam, Netherlands, 2012)
34. S.A. Dupree, S.K. Fraley, *A Monte Carlo Primer. A Practical Approach to Radiation Transport* (Springer, New York, 2002)
35. H. Ertürk, J.R. Howell, Monte Carlo methods for radiative transfer, in *Handbook of Thermal Science and Engineering*. ed. by F.A. Kulacki (Springer, Cham, Switzerland, 2018), pp. 1201–1242
36. W. Fei, Z. Jing, L. Bin et al., Finite-element analysis of emissivity in cylindrical blackbody cavity sensor, in *International Conference on Internet Computing and Information Services*, Hong Kong, China (2011), pp. 453–456. https://doi.org/10.1109/ICICIS.2011.118
37. B. Gebhart, *Heat Transfer* (McGraw-Hill, New York, 1961)
38. D. Geisler-Moroder, A. Dür, A new Ward BRDF model with bounded albedo. Comput. Graph. Forum **29**, 1391–1398 (2010)
39. A.C. Genz, A.A. Malik, Remarks on algorithm 006: an adaptive algorithm for numerical integration over an N-dimensional rectangular region. J. Comput. Appl. Math. **6**, 295–302 (1980)
40. T.A. Germer, J.C. Zwinkels, B.K. Tsai (eds.), *Spectrophotometry: Accurate Measurement of Optical Properties of Materials*. (Academic Press, Amsterdam 2014)
41. P.J. Gero, J.K. Taylor, F.A. Best et al., On-orbit absolute blackbody emissivity determination using the heated halo method. Proc. SPIE **7857**, 78570L (2010)
42. P.J. Gero, J.K. Taylor, F.A. Best et al., On-orbit absolute blackbody emissivity determination using the heated halo method. Metrologia **49**, S1–S8 (2012)
43. P.J. Gero, J.K. Taylor, F.A. Best et al., The heated halo for space-based blackbody emissivity measurement. Proc. SPIE **8527**, 85270O (2012)
44. J.T. Gier, R.V. Dunkle, J.T. Bevans, Measurement of absolute spectral reflectivity from 1.0 to 15 microns. J. Opt. Soc. Am. **44**, 558–561 (1954)
45. A.S. Glassner (ed.), *An Introduction to Ray Tracing* (Academic Press, London, 1989)

46. A. Gouffé, Corrections d'ouverture des corps-noirs artificiels compte tenu des diffusions multiples internes. Revue d'Optique Théorique et Instrumentale **24**, 1–10 (1945)
47. B. Guenther, W. Barnes, E. Knight et al., MODIS calibration: a brief review of the strategy for the at-launch calibration approach. J. Atmos. Ocean. Technol. **13**, 274–285 (1996)
48. A. Haghighat, *Monte Carlo Methods for Particle Transport* (CRC Press, Boca Raton, FL, 2015)
49. L. Hanssen, CLARREO traceability concept and demonstration study results. Presentation on CLARREO Meeting, 12 June 2008, https://slideplayer.com/slide/6131953/. Accessed 8 Feb 2019
50. L.M. Hanssen, K.A. Snail, Integrating spheres for mid- and near-infrared reflection spectroscopy, in *Handbook of Vibrational Spectroscopy*, vol. 2. Sampling Techniques for Vibrational Spectroscopy, ed.by J.M. Chalmers, P.R. Griffiths (Wiley, Hoboken, NJ, 2006), pp. 1175–1192
51. L.M. Hanssen, A.V. Prokhorov, A procedural model of reflection from random rough surfaces. Proc. SPIE **7065**, 70650W (2008)
52. L. Hanssen, S. Mekhontsev, V. Khromchenko, Infrared spectral emissivity characterisation facility at NIST. Proc. SPIE **5405**, 1–12 (2004)
53. L.M. Hanssen, C.P. Cagran, A.V. Prokhorov et al., Use of a high-temperature integrating sphere reflectometer for surface-temperature measurements. Int. J. Thermophys. **28**, 566–580 (2007)
54. L.M. Hanssen, S.N. Mekhontsev, J. Zeng et al., Evaluation of blackbody cavity emissivity in the infrared using total integrated scatter measurements. Int. J. Thermophys. **29**, 352–369 (2008)
55. L.M. Hanssen, S.N. Mekhontsev, V.B. Khromchenko, Validation of the infrared emittance characterization of materials through intercomparison of direct and indirect methods. Int. J. Thermophys. **31**, 1972–1978 (2010)
56. L. Hanssen, S. Kaplan, S. Mekhontsev et al., Status update of the NIST infrared optical properties of materials program. Presentation at the Conference on CALCON 2015, Logan, UT (2015), https://digitalcommons.usu.edu/cgi/viewcontent.cgi?article=1185&context=calcon. Accessed 8 Feb 2020
57. L. Hanssen, J. Zeng, J. Morrill et al., Blackbody and radiometer cavity reflectometry for satellite instrumentation, in *Proceedings of the NEWRAD 2011*, Maui, HI, ed. by S. Park, P. Kärhä, E. Ikonen, (2011), pp. 34–35, https://physics.nist.gov/newrad2011/NewRAD2011_AbstractCollection_20110704.pdf. Accessed 8 Feb 2020
58. R.P. Heinisch, R.N. Schmidt, Development and application of an instrument for the measurement of directional emittance of blackbody cavities. Appl. Opt. **9**, 1920–1926 (1970)
59. R.P. Heinisch, J.K. Andersson, R.N. Schmidt, *Emittance Measurement Study*. NASA Contractor Report CR-1583 (NASA, Washington, D. C., 1970), https://ntrs.nasa.gov/archive/nasa/casi.ntrs.nasa.gov/19700016377.pdf. Accessed 8 Feb 2020
60. R.P. Heinisch, E.M. Sparrow, N. Shamsundar, Radiant emission from baffled conical cavities. J. Opt. Soc. Am. **63**, 152–158 (1973)
61. M. Honner, P. Honnerová, Survey of emissivity measurement by radiometric methods. Appl. Opt. **54**, 669–683 (2015)
62. A. Höpe, Diffuse reflectance and transmittance, in *Spectrophotometry: Accurate Measurement of Optical Properties of Materials*. ed. by T.A. Germer, J.C. Zwinkels, B.K. Tsai (Academic Press, Amsterdam, 2014), pp. 179–219
63. J.R. Howell, M.P. Mengüç, R. Siegel, *Thermal Radiation Heat Transfer*, 6th edn. (CRC Press, Boca Raton, FL, 2016)
64. J.R. Howell, M. Perlmutter, Monte Carlo solution of thermal transfer through radiant media between gray walls. J. Heat Transf. **86**, 116–122 (1964)
65. J.R. Howell, Application of Monte Carlo to heat transfer problems, in *Advances in Heat Transfer*, vol. 5, ed. by T.F. Irvine Jr., J.P. Hartnett (Academic Press, New York, 1968), pp. 1–54

66. J. Hwang, D.-J. Shin, K.R. Jeong, Integrating sphere-based relative methods for reflection measurements. Metrologia **53**, 1231–1242 (2016)
67. ILV, *CIE S 017/E:2011. International Lighting Vocabulary* (CIE, Vienna, 2011)
68. J. Ishii, M. Kobayashi, F. Sakuma, Effective emissivities of black-body cavities with grooved cylinders. Metrologia **35**, 175–180 (1998)
69. O.C. Jones, C. Forno, Reflectance measurements on cavity radiators. Appl. Opt. **10**, 2644–2650 (1971)
70. Y. Kaneko, Y. Shimizu, J. Ishii, Numerical estimation of effective emissivities of low-temperature blackbody cavities with apertures, in *SICE Annual Conference* (Sapporo, Japan, 2004), pp. 1764–1767
71. F.J. Kelly, An equation for the local thermal emissivity at the vertex of a diffuse conical or V-Groove cavity. Appl. Opt. **5**, 925–927 (1966)
72. F.J. Kelly, On Kirchhoff's law and its generalized application to absorption and emission by cavities. J. Res. NBS **69B**, 165–171 (1965)
73. J. Kloepfer, C. Taylor, V. Murgai, V., Characterization of the VIIRS blackbody emittance. Presentation at the Conference on CALCON 2013, Logan, UT (2013), https://digitalcommons.usu.edu/cgi/viewcontent.cgi?article=1011&context=calcon. Accessed 8 Feb 2020
74. R.T. Kneusel, *Random Numbers and Computers* (Springer, Cham, Switzerland, 2018)
75. M. Kobayashi, M. Otsuki, H. Sakate et al., System for measuring the spectral distribution of normal emissivity of metals with direct current heating. Int. J. Thermophys. **20**, 289–298 (1999)
76. R. Kress, *Linear Integral Equations*, 3rd edn. (Springer, New York, 2014)
77. D.P. Kroese, T. Taimre, Z.I. Botev, *Handbook of Monte Carlo Methods* (Wiley, Hoboken, NJ, 2011)
78. D. Kuljis, V. Murgai, J. Kloepfer, Characterization of JPSS J3 and J4 blackbody emissivity. Proc. SPIE **11127**, 111270F (2019)
79. P.K. Kythe, M.R. Schäferkotter, *Handbook of Computational Methods for Integration* (CRC Press, Boca Raton, FL, 2005)
80. K.C. Lapworth, T.J. Quinn, L.A. Allnutt, A black-body source of radiation covering a wavelength range from the ultraviolet to the infrared. J. Phys. E: Sci. Instrum. **3**, 116–120 (1970)
81. R.R. Lewis, Making shaders more physically plausible. Comput. Graph. Forum **13**, 109–120 (1994)
82. D. Liu, Y.-Y. Duan, Z. Yang, Calculations of the average normal effective emissivity for nonaxisymmetric cavities using the modified finite volume method. Opt. Eng. **52**, 039702 (2013)
83. J.S. Liu, *Monte Carlo Strategies in Scientific Computing* (Springer, New York, 2004)
84. M. López, H. Hofer, S. Kück, Measurement of the absorptance of a cryogenic radiometer cavity in the visible and near infrared. Metrologia **42**, 400–405 (2005)
85. J.R. Mahan, N.Q. Vinh, K.J. Priestley, An application of the Monte Carlo ray-trace method with bidirectional reflection, in *The 3rd Thermal and Fluids Engineering Conference* (Fort Lauderdale, FL, 2018), https://www1.phys.vt.edu/~vinhnq/papers/TFEC-2018-22038.pdf. Accessed 8 Feb 2020
86. J.R. Mahan, *Radiation Heat Transfer. A Statistical Approach* (Wiley, New York, 2002)
87. G.I. Marchuk, G.A. Mikhailov, M.A. Nazaraliev et al., *The Monte Carlo Methods in Atmospheric Optics* (Springer, Berlin, 1980)
88. G. Marsaglia, Choosing a point from the surface of a sphere. Ann. Math. Stat. **43**, 645–646 (1972)
89. S. Marschner, P. Shirley, *Fundamentals of Computer Graphics* (CRC Press, Boca Raton, FL, 2016)
90. M. Matsumoto, T. Nishimura, Mersenne Twister: a 623-dimensionally equidistributed uniform pseudo-random number generator. ACM Trans. Model. Comput. Simul. **8**, 3–30 (1998)
91. G. Mei, J. Zhang, S. Zhao et al., Effective emissivity of a blackbody cavity formed by two coaxial tubes. Appl. Opt. **53**, 2507–2514 (2014)

92. G. Mei, J. Zhang, X. Wang et al., Spectral and total effective emissivity of a nonisothermal blackbody cavity formed by two coaxial tubes. Appl. Opt. **54**, 3948–3955 (2015)
93. G. Mei, J. Zhang, S. Zhao et al., Simple method for calculating the local effective emissivity of the blackbody cavity as a temperature sensor. Infrared Phys. Technol. **85**, 372–377 (2017)
94. S.N. Mekhontsev, A.V. Prokhorov, L.M. Hanssen, Experimental characterization of blackbody radiation sources, in *Radiometric Temperature Measurements. II. Applications*, ed. by Z.M. Zhang, B.K. Tsai, G. Machin (Academic Press, Amsterdam, Netherlands, 2010), pp. 57–136
95. S. Mekhontsev, V. Khromchenko, A. Prokhorov, et al., Emissivity evaluation of fixed point blackbodies, in *Proceedings of the TEMPMEKO 2004, 9th International Symposium on Temperature and Thermal Measurements in Industry and Science*, vol. 1, ed. by D. Zvizdić (LPM/FSB, Zagreb, Croatia, 2005), pp. 581–586
96. S. Mekhontsev, M. Noorma, A. Prokhorov et al., IR spectral characterization of customer blackbody sources: first calibration results. Proc. SPIE **6205**, 620503 (2006)
97. G.A. Mikhailov, *Optimization of Weighted Monte Carlo Methods* (Springer, New York, 1992)
98. J.C. Molina, J.J. Soto Bernal, H.A. Castillo M. et al., Electrical substitution radiometer cavity absorptance measurement. Measurement **64**, 89–93 (2015)
99. J.T. Neu, Measurement of hemispherical directional reflectance in the infrared. Proc. SPIE **1995**, 101–120 (1993)
100. J.T. Neu, M.T. Beecroft, R. Schramm, Extended performance infrared directional reflectometer for the measurement of total, diffuse and specular reflectance. Proc. SPIE **2260**, 62–73 (1994)
101. Y. Ohwada, Evaluation of the integrated emissivity of a black body. Jpn. J. Appl. Phys. **23**, L167–L168 (1984)
102. Y. Ohwada, Mathematical proof of an extended Kirchhoff law for a cavity having direction-dependent characteristics. J. Opt. Soc. Am. A **5**, 141–145 (1988)
103. A. Ono, Calculation of the directional emissivities of the cavities by the Monte Carlo method. J. Opt. Soc. Am. **70**, 547–554 (1980)
104. A. Ono, R.M. Trusty, D.P. DeWitt, Experimental and theoretical study on the quality of reference blackbodies formed by lateral holes on a metallic tube, in *Temperature: Its Measurement and Control in Science and Industry*, vol. 5, part 1, ed. by J.F. Schooley (American Institute of Physics, New York, 1982), pp. 541–550
105. A. Ono, Apparent emissivities of cylindrical cavities with partially specular conical bottoms, in *Temperature: Its Measurement and Control in Science and Industry*, vol. 5, part 1, ed. by J.F. Schooley (American Institute of Physics, New York, 1982), pp. 513–516
106. A. Ono, Evaluation of the effective emissivity of reference sources for the radiometric emissivity measurements. Int. J. Thermophys. **7**, 443–453 (1986)
107. J.M. Palmer, Measurement of transmission, absorption, emission, and reflection, in *Handbook of Optics. Volume II. Design, Fabrication, and Testing; Sources and Detectors; Radiometry and Photometry*, ed. by M. Bass, 3rd edn. (McGraw-Hill, New York, 2010), pp. 35.1–35.23
108. B.A. Peavy, A note on the numerical evaluation of thermal radiation characteristics of diffuse cylindrical and conical cavities. J. Res. NBS **70C**, 139–147 (1966)
109. V. Pegoraro, *Handbook of Digital Image Synthesis. Scientific Foundations of Rendering* (CRC Press, Boca Raton, FL, 2017)
110. M.J. Persky, M. Szczesniak, Infrared, spectral, directional-hemispherical reflectance of fused silica, Teflon polytetrafluoroethylene polymer, chrome oxide ceramic particle surface, Pyromark 2500 paint, Krylon 1602 paint, and Duraflect coating. Appl. Opt. **47**, 1389–1396 (2008)
111. B.T. Phong, Illumination for computer generated pictures. Commun. ACM **18**, 311–317 (1975)
112. L.G. Polgar, J.R. Howell, Directional thermal-radiative properties of conical cavities. NASA Technnical Note TN D-2904 (NASA, Washington, DC, 1965), https://ntrs.nasa.gov/archive/nasa/casi.ntrs.nasa.gov/19650017672.pdf. Accessed 8 Feb 2020
113. L.G. Polgar, J.R. Howell, Directional radiative characteristics of conical cavities and their relation to lunar phenomena, in *Thermophysics and Temperature Control of Spacecraft and Entry Vehicles*. ed. by G.B. Heller (Academic Press, New York, 1966), pp. 311–323

114. A.V. Prokhorov, Monte Carlo method in optical radiometry. Metrologia **35**, 465–471 (1998)
115. A. Prokhorov, Effective emissivities of isothermal blackbody cavities calculated by the Monte Carlo method using the three-component bidirectional reflectance distribution function model. Appl. Opt. **51**, 2322–2332 (2012)
116. A.V. Prokhorov, J.E. Martin, Monte Carlo simulation of the radiative heat transfer from a blackbody to a cryogenic radiometer. Proc. SPIE **2815**, 160–168 (1996)
117. A.V. Prokhorov, L.M. Hanssen, Effective emissivity of a cylindrical cavity with an inclined bottom: I. Isothermal cavity. Metrologia **41**, 421–431 (2004); Erratum Metrol **46**, 719 (2009)
118. A.V. Prokhorov, L.M. Hanssen, Effective emissivity of a cylindrical cavity with an inclined bottom: II. Non-isothermal cavity. Metrologia **47**, 33–46 (2010)
119. A. Prokhorov, N.I. Prokhorova, Application of the three-component bidirectional reflectance distribution function model to Monte Carlo calculation of spectral effective emissivities of nonisothermal blackbody cavities. Appl. Opt. **51**, 8003–8012 (2012)
120. A. Prokhorov, S. Mekhontsev, L. Hanssen, Evaluation of performance of integrating spheres for indirect emittance measurement, in *Proceedings of the TEMPMEKO 2001, 8th International Symposium on Temperature and Thermal Measurements in Industry and Science*, vol. 1, ed. by B. Fellmuth, J. Seidel, G. Scholz (VDE Verlag GmbH, Berlin, 2002), pp. 277–282
121. A. Prokhorov, S. Mekhontsev, L. Hanssen, Reciprocity principle and choice of the reflectance model for physically correct modeling of effective emissivity. AIP Conf. Proc. **684**, 729–734 (2003)
122. A.V. Prokhorov, S.N. Mekhontsev, L.M. Hanssen, Emissivity modeling of thermal radiation sources with concentric grooves. High Temp.–High Press. **35/36**, 199–207 (2003/2004)
123. A.V. Prokhorov, L.M. Hanssen, S.N. Mekhontsev, Radiation properties of IR calibrators with V-grooved surfaces. Proc. SPIE **6205**, 620505 (2006)
124. A.V. Prokhorov, S.N. Mekhontsev, L.M. Hanssen, Radiative properties of blackbody calibration sources: recent advances in computer modeling. Int. J. Thermophys. **28**, 2128–2144 (2007)
125. A.V. Prokhorov, L.M. Hanssen, S.N. Mekhontsev, Calculation of the radiation characteristics of blackbody radiation sources, in *Radiometric Temperature Measurements: I. Fundamentals*, ed. by Z.M. Zhang, B.K. Tsai, G. Machin (Academic Press, Amsterdam, 2010), pp. 181–240
126. M.N. Quang, B.N. Van, Evaluation of average directional effective emissivities of isothermal cylindrical-inner-cone cavities using Monte Carlo method. Commun. Phys. **27**, 357–367 (2017)
127. T.J. Quinn, The calculation of the emissivity of cylindrical cavities giving near black-body radiation. Br. J. Appl. Phys. **18**, 1105–1113 (1967); Corrigendum: Br. J. Appl. Phys. **18**, 1659 (1967)
128. T.J. Quinn, M.C. Ford, On the use of the NPL photoelectric pyrometer to establish the temperature scale above the gold point (1063 °C). Proc. R. Soc. A **312**, 31–50 (1969)
129. N.M. Ravindra, B. Sopori, O.H. Gokce et al., Emissivity measurements and modeling of silicon-related materials: an overview. Int. J. Thermophys. **22**, 1593–1611 (2001)
130. H.E. Revercomb, H. Buijs, H.B. Howell et al., Radiometric calibration of IR Fourier transform spectrometers: solution to a problem with the High-Resolution Interferometer Sounder. Appl. Opt. **27**, 3210–3218 (1988)
131. H. Revercomb, F. Best, D. Tobin et al., Monitoring climate from space: a metrology perspective. Proc. SPIE **9881**, 98810F (2016)
132. G. Ribaud, *Traite De Pyrometrie Optique* (Éd. de la Revue d'optique théorique et instrumentale, Paris, 1931)
133. T. Ricolfi, M. Battuello, F. Lanza, Heat-pipe black-body reflectometer for room temperature emissivity measurements. J. Phys. E: Sci. Instrum. **19**, 697–700 (1986)
134. F. Sakuma, L. Ma, Cavity emissivity of fixed-point blackbody, in *SICE Annual Conference in Fukui*, Fukui University, Japan, 4–6 Aug 2003, pp. 2187–2190
135. F. Sakuma, L. Ma, Evaluation of the fixed-point cavity emissivity at NMIJ, in *TEMPMEKO 2004, 9th International Symposium on Temperature and Thermal Measurements in Industry and Science*, vol. 1, ed. by D. Zvizdić (LPM/FSB, Zagreb, Croatia, 2005), pp. 563–568

136. V.I. Sapritsky, A.V. Prokhorov, Calculation of the effective emissivities of specular-diffuse cavities by the Monte Carlo method. Metrologia **29**, 9–14 (1992)
137. V.I. Sapritsky, A.V. Prokhorov, Spectral effective emissivities of nonisothermal cavities calculated by the Monte Carlo method. Appl. Opt. **34**, 5645–5652 (1995)
138. P. Saunders, Optimising blackbody cavity shape for spatially uniform integrated emissivity. Int. J. Thermophys. **38**, 9 (2017)
139. P. Saunders, *Emissivity of blackbody cavities*. MSL Technical Guide 35. Version 3 (Measurement Standards Laboratory of New Zealand, 2018), https://measurement.govt.nz/assets/Uploads/TG35-Version-3-November-2018.pdf. Accessed 8 Feb 2020
140. C. Schlick, An inexpensive BRDF model for physically based rendering. Comput. Graph. Forum **13**, 233–246 (1994)
141. N. Shamsundar, E.M. Sparrow, R.P. Heinish, Monte Carlo radiation solutions—effect of energy partitioning and number of rays. Int. J. Heat Mass Transf. **16**, 690–694 (1973)
142. P. Shirley, R.K. Morley, *Realistic Ray Tracing*, 2nd edn. (A K Peters, Natick, MA, 2003)
143. I.M. Sobol', *A Primer for the Monte Carlo Method* (CRC Press, Boca Raton, FL, 1994)
144. E.M. Sparrow, On the calculation of radiant interchange between surfaces, in *Modern Developments in Heat Transfer*. ed. by W. Ibele (Academic Press, New York, 1963), pp. 186–211
145. E.M. Sparrow, L.U. Albers, E.R.G. Eckert, Thermal radiation characteristics of cylindrical enclosures. J. Heat Transf. **84**, 73–79 (1962)
146. E.M. Sparrow, V.K. Jonsson, Radiant emission characteristics of diffuse conical cavities. J. Opt. Soc. Am. **53**, 816–821 (1963)
147. E.M. Sparrow, R.P. Heinisch, N. Shamsundar, Apparent hemispherical emittance of baffled cylindrical cavities. J. Heat Transf. **96C**, 112–114 (1974)
148. E.M. Sparrow, R.D. Cess, *Radiation Heat Transfer*. Augm. edition (Hemisphere, Washington, DC, 1978)
149. A. Springsteen, Reflectance spectroscopy: an overview of classification and techniques, in *Applied Spectroscopy. A Compact Reference for Practitioners,* ed. by J. Workman Jr., A.W. Springsteen (Academic Press, San Diego, CA, 1998), pp. 193–224
150. S. Takata, A formulation of the theory of interreflection. J. Illum. Eng. Inst. Jpn. **51**, 702–713 (1967) (in Japanese). https://doi.org/10.2150/jieij1917.51.12_702
151. Y. Té, P. Jeseck, C. Camy-Peyret et al., Balloonborne calibrated spectroradiometer for atmospheric nadir sounding. Appl. Opt. **41**, 6431–6441 (2002)
152. Y. Té, P. Jeseck, C. Camy-Peyret et al., High emissivity blackbody for radiometric calibration near ambient temperature. Metrologia **40**, 24–30 (2003)
153. L.K. Thomas, An apparatus for the measurement of the normal and off-normal spectral emissivity of incandescent materials. J. Sci. Instrum. (J. Phys. E). Ser. 2, **1**, 311–316 (1968)
154. J.S. Toor, R.A. Viskanta, Numerical experiment of radiant heat interchange by the Monte Carlo method. Int. J. Heat Mass Transf. **11**, 883–897 (1968)
155. O. Touayar, B. Rougié, J.M. Coutin et al., Measurement of the reflectance of the INM cryogenic radiometer cavity at several wavelengths. Metrologia **35**, 387–391 (1998)
156. E.W. Treuenfels, Emissivity of isothermal cavities. J. Opt. Soc. Am. **53**, 1162–1171 (1963)
157. P. van Dooren, L. de Ridder, An adaptive algorithm for numerical integration over an n-dimensional cube. J. Comput. Appl. Math. **2**, 207–217 (1976)
158. E. Venturino, Simple quadrature for singular integrals. J. Integral Equ. Appl. **5**, 257–276 (1993)
159. J. Wang, J. Chen, J. Wang et al., Calculation and study of the integrated emissivity of cylindrical cavity. Appl. Mech. Mater. **148–149**, 1442–1448 (2012)
160. G.J. Ward, Measuring and modeling anisotropic reflection. Comput. Graph. **26**, 265–272 (1992)
161. H. Watanabe, J. Ishii, H. Wakabayashi et al., Spectral emissivity measurements, in *Spectrophotometry: Accurate Measurement of Optical Properties of Materials*. ed. by T.A. Germer, J.C. Zwinkels, B.K. Tsai (Academic Press, Amsterdam, 2014), pp. 333–366

162. B.E. Wood, J.G. Pipes, A.M. Smith et al., Hemi-ellipsoidal mirror infrared reflectometer: development and operation. Appl. Opt. **15**, 940–950 (1975)
163. X. (J.) Xiong, J. Butler, A. Wu et al., Comparison of MODIS and VIIRS on-board blackbody performance. Proc. SPIE **8533**, 853318 (2012)
164. W.-J. Yang, H. Taniguchi, K. Kudo, *Radiative Heat Transfer by the Monte Carlo Method*. Advances in Heat Transfer, vol. 27 (Academic Press, San Diego, CA, 1995)
165. X. Yi, W. Fang, Y. Luo et al., Correction of cavity absorptance measure method for cryogenic radiometer. IET Sci. Meas. Technol. **10**, 564–569 (2016)
166. S.M. Zemyan, *The Classical Theory of Integral Equations. A Concise Treatment* (Springer, New York, 2012)
167. J. Zeng, L. Hanssen, An infrared laser-based reflectometer for low reflectance measurements of samples and cavity structures. Proc. SPIE **7065**, 70650F (2008)
168. J. Zeng, S. Mekhontsev, L. Hanssen, New capability for evaluating the emissivity of large aperture infrared blackbodies. Presentation at CALCON 2013. SDL, Utah State University, Logan, UT (2013), https://digitalcommons.usu.edu/cgi/viewcontent.cgi?article=1016&context=calcon. Accessed 8 Feb 2020
169. Y.F. Zhang, Z.F. Shao, Y.Q. Wu et al., Calibration method for spectral responsivity of infrared detector based on blackbody at multiple temperature. Proc. SPIE **9623**, 96230U (2015)
170. Y.H. Zhou, Z.M. Zhang, D.P. DeWitt et al., Effects of radiative properties of surfaces on radiometric temperature measurement, in *Proceedings of the 9th International Conference on Advanced Thermal Processing of Semiconductors (RTP '2001)* (2001), pp. 179–188. https://doi.org/10.1109/RTP.2001.1013764

Chapter 5
Elements of Blackbodies Design

Abstract Some thermophysical and optical aspects of designing blackbodies are considered in this chapter. A general set of characteristics and parameters of blackbodies are derived from the typical requirements contained in the corresponding technical documentation. The most frequently used classification systems for blackbodies are examined. The issues discussed in this chapter include the techniques of blackbody heating, cooling, and isothermalization for fixed-point and various types of variable-temperature blackbodies and methods of improving the radiation characteristics of blackbody radiators.

Keyword Blackbody design · Heating · Cooling · Isothermalization · Effective emissivity

5.1 Introductory Notes

The concept of a blackbody radiator as an isothermal cavity with high-emissivity walls and a small opening is firmly entrenched in the textbooks and entry-level special literature (see, e.g. [5, 97, 158]). The blackbody depicted in Fig. 5.1 contains most elements of the actual design of a modern blackbody. The highly absorbing surface (as a consequence, it must have a high emissivity) forms the radiating cavity; the small opening in it provides the cavity's effective emissivity close to unity; the heating element warms up the metal core having high thermal conductivity to reduce temperature non-uniformity of cavity walls. Thermal insulation minimizes the heat loss from the cavity external surface and, at the same time, improves temperature uniformity of the cavity. The polished reflecting plate in front of the cavity opening plays the role of the external aperture defining the FOV of a radiometer or radiation thermometer and increases the contrast facilitating the proper sighting.

A blackbody must reproduce as accurate as possible the Planckian spectrum at a given temperature and within a given wavelength range. There is a common opinion that has become almost axiomatic: all that is necessary to create a good approximation of a perfect blackbody is (i) an isothermal cavity having small ratio of the aperture area to the area of internal surface and (ii) high emissivity of the cavity wall material.

V. Sapritsky and A. Prokhorov, *Blackbody Radiometry*, Springer Series in Measurement Science and Technology, https://doi.org/10.1007/978-3-030-57789-6_5

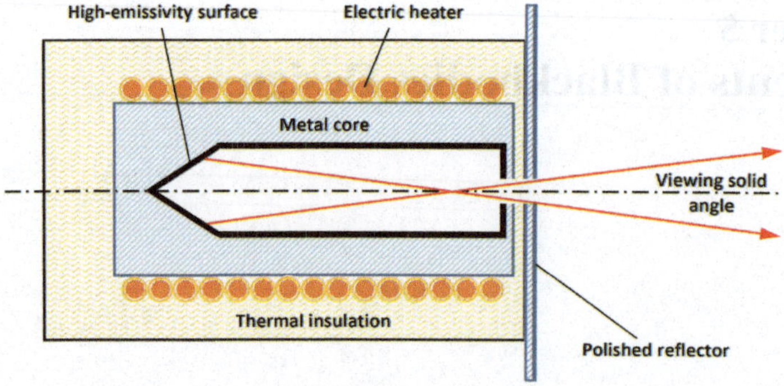

Fig. 5.1 Schematic diagram of an experimental realization of a blackbody radiator

However, these rules of thumb are sufficient but not always necessary. Unlike simplified scheme depicted in Fig. 5.1, the modern blackbody radiators: (i) are not always cavities, (ii) may include surfaces that have low emissivity, and (iii) not always have small aperture. In addition, the use of a metal equalizing block is not the only, universal, or the best method for obtaining the temperature uniformity.

The designer of a blackbody for a modern radiometric system has to solve two interrelated tasks: optical (achieving the near-unity effective emissivity within the desired spectral range) and thermophysical (providing the best possible temperature uniformity of the radiating surface). These problems can be considered as independent only in a rough approximation. The actual designing process includes several stages.

At the first stage, the radiating surface is supposed to be isothermal; the effective emissivity of the blackbody is maximized by varying the shape of the radiator with the account of the radiation characteristics of its material and for given viewing conditions. The second stage consists of selection and practical implementation of technical means ensuring temperature uniformity of the radiator. It should be remembered that residual non-uniformities of temperature could nullify all efforts for maximizing the effective emissivity undertaken in the first stage. Therefore, after the choice of the technical solution for the heating (cooling) and isothermalization of a blackbody radiator, the effective emissivity must be re-computed taking into account temperature non-uniformities evaluated experimentally and/or by numerical simulation. This also allows estimating the uncertainty of the effective emissivity determination due to residual temperature non-uniformity.

5.2 Parameters and Characteristics of Blackbodies

The choice of the blackbody construction is based on the technical and metrological requirements to that blackbody, optical radiation it emits, and conditions, in which it operates. The document OIML R 147 [118] recommending calibration and verification procedures for blackbody radiators operating at temperatures from −50 to

2500 °C provides the list of characteristics that should be given in the corresponding technical documentation. Since OIML R 147 [118] concerns only thermometric blackbodies and does not consider radiometric ones, we can use it only as a basis for drawing up an indicative list of parameters and characteristics that must be specified before designing a modern blackbody. We divide them into several groups and give brief comments to each item.

The first group includes the most general characteristics concerning:

- The purpose of the blackbody. This item specifies the normal conditions, in which the blackbody has to be used (laboratory, field, cryo-vacuum chamber, etc.) There is a fuzzy classification of the operating conditions by the radiation background surrounding the blackbody. The low-background conditions correspond to the background that radiates as a perfect blackbody at temperatures of a few kelvins (so-called liquid helium temperatures). The medium-background (or reduced background) conditions correspond to the background radiating as a perfect blackbody at temperatures about 70 K (so-called liquid nitrogen temperatures). The ambient-background conditions correspond to the room temperature and atmospheric pressure. As well, the blackbody can be portable or stationary installed.
- The radiometric quantity that the blackbody reproduces (e.g. the spectral radiance for the wavelength range from 0.25 to 2.5 μm, or the band-limited irradiance over the wavelength range from 8 to 12 μm, etc.)

The second group contains parameters and characteristics related to the temperature of the radiator:

- The fixed operating temperatures or an extended temperature range. This item corresponds to the traditional division of blackbodies into fixed-point blackbodies (FPBBs) and variable-temperature blackbodies (VTBBs). We will consider the most important features of the FPBB and VTBB constructions below and in more detail in the relevant chapters of Volume 2.
- The temperature instability. This parameter indicates a permissible temporal instability of temperature control at specified temperature levels.
- The temperature non-uniformity. Although this parameter is often laid down in technical requirements, it can be inferred from the requirements for the effective emissivity. It makes sense to specify it only for the simplest blackbodies with the planar radiators (so-called flat-plate blackbodies).
- The temperature drift. This is the allowable long-term (during operation in the specified stable modes) slow change of temperature.
- The temperature uncertainty. This parameter indicates the admissible expanded uncertainty of the blackbody temperature at a specified confidence level and is must be specified for the primary thermometric blackbodies. For radiometric blackbodies, the temperature uncertainty can be derived from the given uncertainty of the reproduced radiometric quantity.
- The warm-up and/or cool-down time, i.e. the time required to reach a specified steady-state temperature for a VTBB at the lower and upper levels of the operating temperature range. For an FPBB, the duration of the phase change and the time interval between consecutive phase transitions are more important.

- The transition time. For the VTBB, this parameter indicates the time required for the blackbody to pass from one stable mode to another. For the FPBB, see the remark to the previous item.

The third group contains characteristics related to the effective emissivity:

- The mean, lower, and upper values of the effective emissivity over the spectral range of interest. The effective emissivity is evaluated for the prescribed arrangement of the blackbody and the radiometer or radiation thermometer. For precision spectroradiometric measurements, specifying of the spectral effective emissivities for at least several critical wavelengths is more suitable. It is important to indicate the reference temperature, for which the spectral emissivity of the nonisothermal radiator is determined.
- The spatial and directional non-uniformity of the effective emissivity. These characteristics are required if the blackbody is intended for comparison of radiometers or radiation thermometers with substantially different viewing conditions.

The fourth group contains geometrical characteristics:

- Specification of the geometric configuration of the radiating element. OIML R 147 [118] consider the source of thermal radiation with an effective emissivity $\varepsilon_e \geq 0.95$ as necessarily cavity radiators and with the emissivity ranged from 0.9 as the radiators with an extended flat surface. This classification should obviously be revised in connection with the latest developments of carbon nanotube coatings (see Sect. 6.2.4), which allow flat-plate radiators to have effective emissivity significantly exceeding 0.95. Besides, the subdivision into the cavity and flat-plate blackbodies is incomplete; it does not include radiators with more complex geometry, which will be discussed in Sect. 5.5.3 through 5.5.6.
- The radiating area. It is determined by the FOV of the radiometer or radiation thermometer that is intended to use with the blackbody. The basic radiometric quantities (radiance and irradiance, as well corresponding spectral and band-limited quantities) reproduced by blackbodies or in terms of which blackbody-based calibrations are carried out are determined substantially by geometric conditions for collecting the blackbody radiation. Since the radiance mode implies measurements within a narrow solid angle (ideally, along an infinitely thin ray), a blackbody with a small aperture is suitable for calibration in terms of radiance, spectral radiance, or band-limited radiance. However, the radiant power emitted by such a blackbody may not be sufficient to obtain the required level of irradiance (spectral irradiance or band-limited irradiance) at a given distance from the blackbody aperture. As a rule, the large-aperture blackbodies (LABBs) are used to operate in the irradiance mode. Sometimes, the same acronym is used for the large-area blackbodies, which are designed to calibrate radiometric instrumentation with wide FOVs, not necessarily in irradiance mode.
- Overall dimensions. The overall dimensions of the blackbody, with and without the power supply and, perhaps, other accessories are important for use in specific conditions (e.g. inside a thermovacuum chamber, on board an aircraft or satellite, etc.)

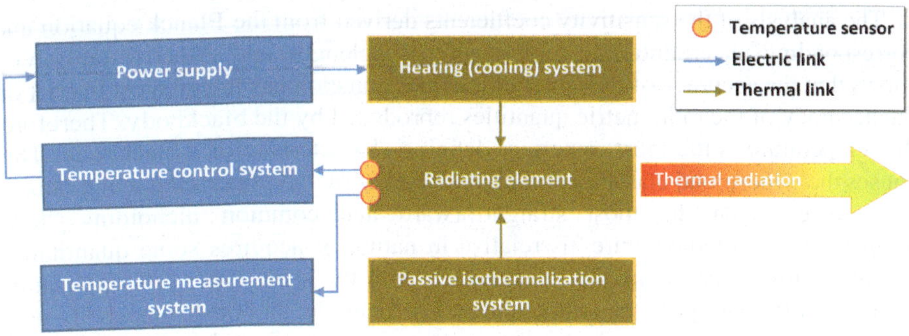

Fig. 5.2 A generalized diagram of the blackbody structure

The last, fifth group (for radiometric applications only), contains overall metrological parameters for the radiometric quantities reproduced by the blackbody. The allowable expanded uncertainties of spectral, band-limited, or total radiometric quantities and the uncertainty of the radiance temperature are connected via the Planck or Stefan-Boltzmann equations. For the spectral radiometric quantities, it is customary to indicate their uncertainties for each wavelength (or at least, for a set of critical wavelengths). Finally, it should be indicated in which rank (primary standard, secondary standard, etc.) the blackbody can be used. If the blackbody is not the primary standard, the calibration method and appropriate equipment should be specified.

The above-mentioned characteristics and parameters help us draw up a generalized blackbody diagram, which includes the radiating element, heating (cooling) system fed by the power supply via the temperature control system, and the temperature measurement system with a possible analogue and/or digital output connected to a personal computer or microprocessor-based system. Such a diagram is presented in Fig. 5.2. In addition to the above mentioned, it includes also the system of the radiator passive isothermalization, which presents in many modern blackbodies, although the temperature equalization capability differs greatly for their various types.

5.3 Classification Systems for Blackbodies

We already mentioned the subdivision of blackbodies into FPBBs and VPBBs, as well as introduced categories of large-area and large-aperture blackbodies. There is a wide variety of feature and characteristics, according to which the blackbodies can be classified. A number of blackbody taxonomies can be proposed: by the operating temperature, heating or cooling method, technique of the radiator isothermalization, spectral range, metrological characteristics, operating conditions, reproduced radiometric quantities, etc. Different taxonomies can be partially overlapped; therefore, a combined approach is often needed to categorize comprehensively a particular blackbody.

The analysis of the sensitivity coefficients derived from the Planck equation and corresponding uncertainties of temperature, wavelength, and effective emissivity shows that the accuracy of the temperature determination has the greatest impact on the accuracy of the radiometric quantities reproduced by the blackbody. Therefore, the temperature is the most important defining characteristic of a blackbody. This causes the widespread classification of blackbodies by their operating temperature.

However, even the most straightforward and common dichotomy "low-temperature/high-temperature" is relative in nature. It acquires some quantitative meaning only when comparing blackbodies with the same types and application areas but different operating temperatures. For example, an 800-K blackbody used for calibrating the sensors typically designed for remote sensing of the Earth should be undoubtedly classified as high-temperature since the typical temperatures of blackbodies used for this purpose are around 300 K. The same blackbody can hardly be called high-temperature, if it is used to calibrate a silicon photodiode.

The boundaries of commonly used temperature subrange cannot be strictly defined and change with time (this especially concerns the blackbodies that are classified as high-temperature). With sufficient degree of certainty, we only can attribute blackbodies operating at temperatures below approximately 100 K to cryogenic ones. They are usually cooled with liquid nitrogen (LN_2, the boiling temperature is 77 K) and are used for calibration of wide-band IR radiometric instrumentation or as a radiometric zero reference when measurements with hotter blackbodies are carried out. The boundaries between low-, medium, and high-temperature ranges depend on the context. We do not intend to change the established terminology, but we need to supplement the temperature-based classification with additional clarifying attributes.

The lower bound of operating temperature for the modern blackbodies is around 100 K; some applications such as calibration of spaceborne sensors may require even lower temperatures. The upper bound (about 3500 K at present and tends to increase) is conditioned by properties of existing high-temperature materials.

The operating temperature substantially determines the spectral range of radiometric quantities reproduced by the blackbody. Modern blackbodies are capable to emit, depending on their temperature, the optical radiation of near-Planckian spectra with maxima taking positions from the near UV to the far IR. The natural desire of experimenters to work near the maximum of the Planck curve is more feasible for longwave than for shortwave part of the Planckian distribution due its asymmetry.

5.4 Overview of Thermal Designs of Blackbodies

As radiometric standards, blackbodies are subdivided into primary and secondary ones. Primary blackbodies reproduce radiometric quantities without reference to other standards of the same quantities. A secondary standard blackbody has to be calibrated against a primary one, as a rule, in terms of radiance or radiance temperature.

The FPBBs operating at temperature of the ITS-90 fixed points are always considered as primary radiation sources. Their readiness for measurements corresponds to the duration of the phase transition (freezing or melting) of the fixed-point material.

In contrast to the FPBBs, the VTBBs operate within continuous temperature ranges; their operational times are not limited by the duration of some physical process but are determined by the balance of the heat supplied to and dissipated by the radiator. All VTBBs had to be either calibrated against the FPBBs or linked with the temperature scale using contact of non-contact temperature sensors. However, the concept of the primary blackbody began to blur over time. Development of radiometric methods of blackbody thermodynamic temperature determination independent of the current defined temperature scale (see Chap. 9 for detail) allows employing some VTBBs as the primary radiation sources.

Realization of the spectral radiometric quantities on the base of Planck's law requires measurement of blackbody temperature traceable to the thermodynamic temperature scale, measurement of wavelength traceable to the primary standard of length, and determination of the effective emissivity by computational and/or experimental means. The owners of primary standards are mainly National Metrology Institutes (NMIs). Certification of a blackbody as a primary standard source is the subject of not only scientific but also legal metrology.

As we have already mentioned, the most common blackbody taxonomies based on the dichotomy "low-temperature—high temperature," trichotomy "low-temperature—medium-temperature—high-temperature," etc. are hardly productive. Considering the need to incorporate into the classification such well-established names of VTBB types as the heat-pipe and fluid-bath (liquid-bath) blackbodies [61, 65], we present the taxonomy based mainly on the design features of blackbody together with their typical operating temperature ranges in Table 5.1. Below, we provide brief comments to each row of this table.

Table 5.1 Typical temperature ranges for blackbodies of various types

Type of blackbody	Typical temperature range
Blackbodies with indirect resistance heating	Above ambient
Heat-pipe blackbodies	−60 °C …200 °C 450 °C …1200 °C
Fluid-bath blackbodies	−50 °C …150 °C ~ 77 K
Fluid-circulation blackbodies	−50 °C …150 °C ~ 77 K
Blackbodies with thermoelectric cooling and heating	−50 °C …50 °C
Blackbodies with direct resistance heating	1000 K …3500 K
Inductively heated blackbodies	1000 K …3500 K

5.4.1 Indirect Resistance Heating

The indirect resistance (or Joule) heating proceeds when a current is passed through an electrical resistor (heater) and electrical energy is converted to thermal energy, which is transferred to the heat load (the radiating element in our case) by heat conduction, thermal radiation, and, in some cases, convection. The indirect resistance heating is used in many different constructions of VTBBs since all VTBBs operating above the ambient temperature require heat to be supplied in order to reach the specified radiator temperature.

The easiest way to heat up a radiator is to place it in an electric resistance furnace. There is a wide selection of commercially available electric furnaces of tube and split tube (clamshell) types with operating temperatures up to 3000 °C. Some of them are suitable for placing the assembly with a radiator (usually of cavity type); others cannot be used in blackbody design, mainly due to lack of temperature uniformity of the heated zone and geometric constraints making impossible unobstructed observation of the radiator. The most advanced researcher teams prefer to develop their own blackbodies with resistance heaters.

Electrical energy in the electrical heating element is converted into heat according to Joule's law. The power P of heating can be expressed via the current I flowing through the heater, the voltage drop V across the heater, and its resistance R:

$$P = IV = I^2 R = V^2/R. \tag{5.1}$$

For the AC circuit, we can formally replace P, I, and V in Eq. 5.1 by the mean power $\overline{P}$ and the root mean square values I_{rms} and V_{rms} calculated for one or more AC periods.

The Joule heat released in the electric heater depends on electrical resistivity ρ [Ω·m], the intrinsic property of the material of the heater (electric conductor) determining its ability to resist the flow of electric current:

$$\rho = RA/l, \tag{5.2}$$

where A and l are the cross-sectional area and the length of the conductor, respectively.

The material of an electrical heating element must have a high electrical resistivity throughout operating temperature range to avoid the use of high current (and low voltage) power supplies. Other requirements for the heater materials include good resistance to oxidation in air, sufficient strength, and ductility that allows forming the heaters of desired shapes. Resistivities of most widespread heater materials are shown in Fig. 5.3 as functions of temperature.

Some of suitable materials are Fe–Cr-Al and Ni–Cr-Fe alloys (1000 to 1400 °C), refractory metals such as W, Mo and Ta (from 1500 to 2000 °C), Pt and Pt–Rh alloys (1200 to 1800 °C), nonmetals as silicon carbide (SiC) and molybdenum disilicide ($MoSi_2$) for temperatures from 1200 to 1750 °C [36].

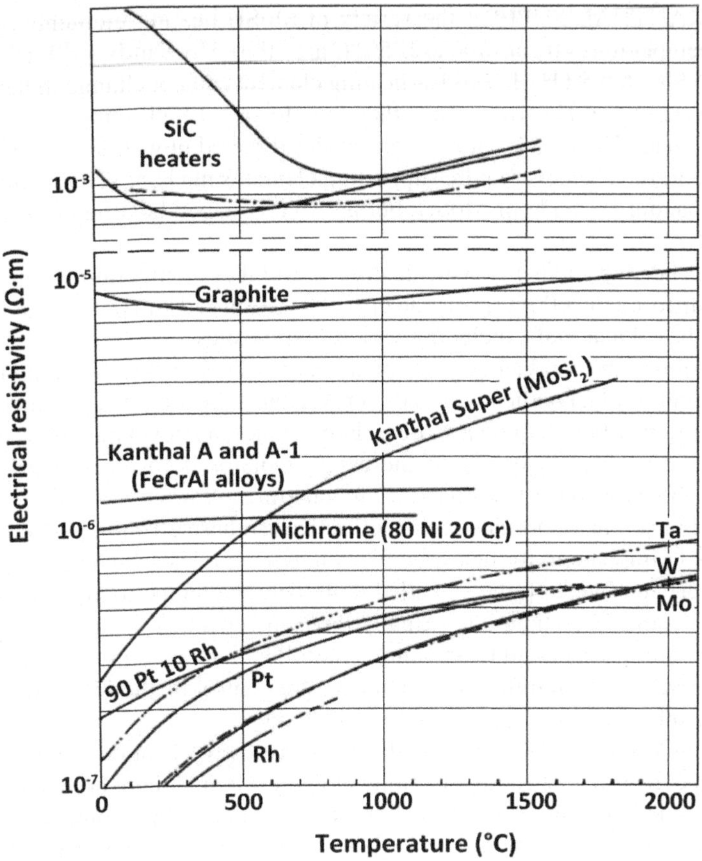

Fig. 5.3 Resistivity as a function of temperature for some materials of heating elements. Reproduced from [110] with permission of John Wiley & Sons

Silicon carbide for heating elements is made usually in the form of rods, although it can be shaped arbitrarily for special purposes. In air, silicon carbide is covered by a layer of glassy silica (SiO_2), which protects the heater against further oxidation. In dry air, silicon carbide heaters can be used up to about 1650 °C. In the presence of water vapor, the temperature must be lower due to accelerated oxidation. The use of SiC heaters in vacuo is not recommended.

Unlike most other resistor materials, silicon carbide is subject to ageing: the resistance of the SiC heater under load permanently increases with time during its service life.

Molybdenum disilicide ($MoSi_2$) is a refractory ceramics with the melting temperature of 2030 °C. It is electrically conductive and withstands high temperatures due to the formation of a protective layer of glassy SiO_2 on the surface. Under the name of KANTHAL SUPER,[1] the variety of $MoSi_2$ heating elements operated at maximal temperatures from 1590 to 1850 °C are offered by Sandvik AB (Sweden). In contrast to SiC, KANTHAL SUPER heating elements do not change their resistance during serving time. However, since they tend to sag at high temperature, they are preferably made U-shaped, with two electrodes directed upwards. At the horizontal arrangement, such heaters require supporting elements made of suitable refractories. The wire cannot be made of $MoSi_2$, but the spiral-shaped heating elements can be fabricated.

Electric heating elements withstanding the highest temperatures are made of graphite (up to about 2000 °C in air and up to 2500 °C in vacuum or argon atmosphere) or carbon-based materials, including various composites.

In addition to the conductive heat spreader, passive isothermalization of the radiating element in blackbodies with indirect resistance heating can be carried out by non-uniform distribution of heat sources. For instance, a known method for reducing the temperature inhomogeneity of the cavity radiator is to increase the winding density of the wire heater toward the cavity aperture.

There is also an important option of active isothermalization, which is especially easy to implement for blackbodies with indirect resistance heating. Approved by Hishikari and Ide [64], the so-called multiple-zone temperature control method consists in using more than one (usually three) independent heaters, each with its own temperature sensor and temperature controller, combined in a common control system. Presently, the multiple-zone temperature control method is widely used for radiators isothermalization of precision blackbodies.

The simplest (i.e. with no special passive isothermalization system) VTBBs with indirect resistance heating operating at highest temperature are widely used as secondary standard sources for the calibration of radiation thermometers. However, they have a very limited application in radiometry, much narrower than blackbodies with direct resistance heating (see Sect. 5.4.6), which allow achieving the higher temperatures, and, therefore, operating with shorter wavelengths.

As a rule, blackbodies with indirect resistance heating operating in the low-temperature range are miniature radiators such as those for satellite applications. The heating element for these blackbodies can be integrated with the radiating element to minimize the weight, overall dimensions, and dissipated heat. Custom-made wire-wound heaters or commercially available flexible heaters are suitable for this purpose. At the highest temperatures (above approximately 1500 °C), blackbodies with direct resistance heating (see Sect. 5.4.6) compete with blackbodies with indirect resistive heating and above 2000 °C become dominant.

---

[1]Not to be confused with Kanthal that is the trademark (also owned by Sandvik AB) for a family of iron-chromium-aluminum (FeCrAl) alloys.

5.4.2 *Fixed-Point Blackbodies*

A fixed-point blackbody (FPBB) is a blackbody that operates at a single temperature. This name is used exclusively for blackbodies operating at temperatures of phase transitions (freezing or melting) of pure substances (metals, alloys, or compounds).

A schematic of the FPBB and the typical melting/freezing curve (temperature plotted against time) are depicted in Fig. 5.4. Since each phase transition proceeds at a constant temperature, the melting/freezing curve has flat portions (melting and freezing plateaux), whose durations allows performing the necessary measurements.

For a long time, only FPBBs operating at the fixed points of the current defined temperature scale were considered as primary radiation sources. Presently, the blackbodies at the defining fixed points of the ITS-90 continue to be used in conventional radiation thermometry.

The ITS-90 is defined above the freezing point of silver by means of Planckian extrapolation from any of the freezing points of silver (1234.93 K), gold (1337.33 K), or copper (1357.77 K). The modern technique for realization of these fixed points is outlined in the official Guide issued by the CCT [183]. The full list of the fixed points is provided in Table B.1 in Appendix B. The ITS-90 [129] does not provide for the use of the radiation thermometers (RTs) below the freezing point of silver. In the practice, however, other fixed points (mainly, at temperatures of or above the melting point of gallium) are often used for measurements with RTs, when contact measurements of temperatures below the silver point is not desired. In such cases, the RT is used for temperature interpolation.

The FPBBs at the fixed points of the ITS-90 are often used as standards of spectral radiance. However, in order to be used in the Planck equation, the temperature must be thermodynamic, that is, the correction must be introduced in temperatures T_{90}. We used the data on $T - T_{90}$ and the corresponding standard uncertainties $u(T - T_{90})$ provided by Fischer et al. [45] to compose the upper part of Table 5.2. Recent redefinition of the kelvin [93] introduces only imperceptibly small changes in the thermodynamic temperatures given in Table 5.2.

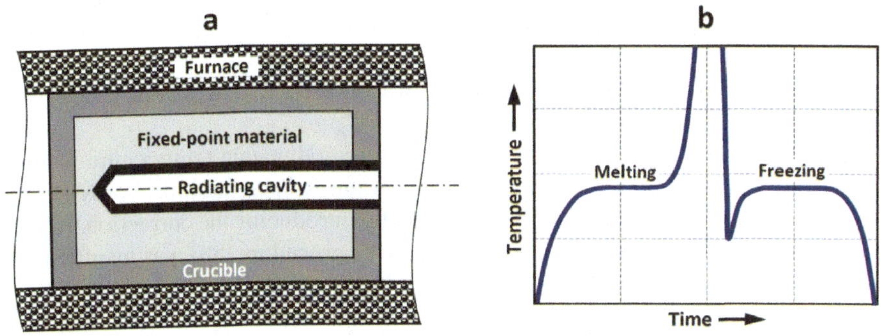

Fig. 5.4 **a** Schematic of the fixed-point blackbody and**b** the melting/freezing curve of pure substance

Table 5.2 Fixed points of the ITS-90 and high-temperature fixed-points with assigned melting temperatures used for the FPBBs

Material	State	Thermodynamic temperature T (K)	Uncertainty $u(T)$ (mK)	λ_{max} (μm)
Ga (gallium)	Melting	302.9584	0.4	9.6
In (indium)	Freezing	429.7586	0.8	6.7
Sn (tin)	Freezing	505.090	1.3	5.7
Zn (zinc)	Freezing	692.691	6.9	4.2
Al (aluminum)	Freezing	933.502	6.6	3.1
Ag (silver)	Freezing	1235.39	14	2.4
Au (gold)	Freezing	1337.73	20	2.2
Cu (copper)	Freezing	1357.82	20	2.1
Co-C (eutectic alloy)	Melting	1597.39	65	1.8
Pd–C (eutectic alloy)	Melting	1764.85	350	1.6
Pt-C (eutectic alloy)	Melting	2011.43	90	1.4
Re-C (eutectic alloy)	Melting	2747.84	175	1.0

The FPBBs at the freezing temperatures of Ag, Au, and Cu allow, in principle, realization of the unit of the spectral radiance, at best, in the visible part of the optical spectrum using the ascending branches of Planckian curves. The scales of the spectral radiance and spectral irradiance at shortest wavelength are realized, as a rule, using VTBBs, temperatures of which are higher than the freezing temperature of copper, the hottest fixed point of the ITS-90, and can be measured, according to the ITS-90, using extrapolation from either silver, gold, or copper FPBB.

The mise en pratique for the definition of the kelvin (*MeP*-K, CCT [24]) legalized the use of high-temperature fixed points (HTFPs) with assigned thermodynamic temperatures (the four lower rows in Table 5.2) for the relative primary radiometric thermometry [96]. This allows using interpolation between the hottest fixed-points of the ITS-90 and HTFPs instead of temperature extrapolation usual for conventional radiation thermometry. The temperatures of the point of inflection T_{PoI} on the melting curve in the middle part of the melting plateau (see [178]) are given as the temperature T.

The last column in Table 5.2 contains the wavelengths λ_{max} that correspond to the maxima of the Planckian curves at temperatures T.

The complete list of HTFPs with assigned thermodynamic temperature is given in Table B.2 in Appendix B. Along with temperatures T_{PoI}, Table B.2 provides the equilibrium liquidus temperature T_{ELq} derived by introducing the correction for non-equilibrium melting. Perhaps, since the correction procedure [88] was insufficiently tested to the date of the *MeP*-K publication [96], two slightly different values of temperature (T_{PoI} and T_{ELq}) were published. It should be noted that the difference between them is small enough to be neglected in most practical applications. The use of HTFPs can improve significantly the accuracy of the radiometric scale realization due to the decrease in the uncertainty of determining the HTBB temperature. Besides, the displacement of the Planck distribution maximum from 2.1 μm for Co to 1.0 μm for Re-C together with an increase of spectral radiance values of an FPBB at the Re-C melting temperature as compared to that at the Cu freezing temperature opens an opportunity of the spectral radiance scale realization directly using FPBBs with HTFP radiating cells.

The FPBBs are considered as the most accurate blackbodies due to following their features:

(1) The FPBBs cells have radiating cavities with the highly isothermal internal surface. This is ensured by an extremely homogeneous heating of the cavity surrounded by a metallic melt and a low radiation loss through a small aperture (typically 3 to 4 mm in diameter).
(2) The graphite cavities of the FPBBs have geometrical configurations providing the effective emissivities of 0.999 or higher.
(3) During the melting or solidification of the fixed-point material in the crucible, the radiation of the FPBB is extremely stable in time.
(4) The values of the fixed points are virtually independent of external conditions. This ensures the highest repeatability and reproducibility of radiation of the FPBBs.

At the same time, the FPBBs are not free from some inherent drawbacks:

(1) The limited operating time. The duration of the freezing plateau is determined by the amount of metal, the furnace temperature uniformity, and the heating/cooling schedule. The typical duration of the freezing plateau of gold realized in the FPBB at NIST is about 20 min [182].
(2) The FPBBs as the most accurate standard radiation sources are also the most expensive. Reliable realization of fixed points requires the highest purity of fixed-point materials. The presence of impurities adversely affects the reproducibility of the fixed points. Above-mentioned gold-point FPBB [182] contains 1.3 kg of expensive high-purity (99.999% mass fraction) gold. Even the copper-point FPBB, due to price of precision equipment and the maintenance costs, can be too expensive for most laboratories except the largest NMIs.
(3) Small apertures of radiating cells of FPBBs allow to measure conveniently the spectral radiance and the radiance temperature. However, the use of FPBBs as standards of spectral irradiance is problematic because of impossibility to create a sufficient level of spectral irradiance from a small aperture.

The FPBBs at the melting point of gallium and freezing point of indium find application in the IR radiometry for the calibration of remote sensing instrumentation and thermal imagers. Other fixed points of the ITS-90 (as well as various VTBBs operating at temperatures between 500 and 1000 K) are demanded mainly in the radiation thermometry.

A certain interest in FPBBs operating around the human body temperature is manifested by the medical non-contact thermometry and IR thermography. For these purposes, the most important are the closeness of fixed points to the measured temperatures, duration of the phase transition plateau, reliability, compactness, and ease of use. Machin and Simpson [95] described a prototype of a body-temperature FPBB at the freezing temperature of the ethylene carbonate ($(CH_2O)_2CO$), an organic compound whose triple point was earlier proposed as the reference point for calibration of contact thermometers. This FPBB traceable to ITS-90 was designed to validate the IR tympanic and ear canal thermometers that measure the IR emitted by the surface of the tympanic membrane and by the ear canal as a whole, respectively. Simpson et al. [160] described design, testing, validation, and results of the field tests in several clinical centers of a multi-fixed-point source that serves as an in-image calibration and validation system for medical thermal imagers. These fixed-point sources employ the phase transition of gallium–zinc eutectic (~24.3 °C), gallium (~29.8 °C), and ethylene carbonate (~34.3 °C), plateau durations are of about 3 h at ambient conditions, a stability over that period is about or better than 0.1 °C, a repeatability is also not exceed 0.1 °C.

The low-temperature (this is a natural way to call them, by analogy with high-temperature) FPBBs could significantly improve the accuracy of IR radiance measurements in the remote sensing of the Earth, by both pre-flight (ground) and in-flight (on-orbit) calibrations. The use of the binary eutectic alloys of gallium with some metals for calibration and verification of contact temperature measurements of blackbodies in remote sensing applications was discussed in many works [17, 18, 50, 81, 149, 150]. Development of gallium FPBB for on-orbit calibration was reported by Burdakin et al. [19]. However, development of low-temperature FPBBs on the base low-temperature eutectic alloys of gallium is still under progress.

5.4.3 *Heat-Pipe Blackbodies*

The heat pipe is a capillary-driven two-phase system (see, e.g. [39, 85, 139]) that transports heat from a heat source to a heat sink via the latent heat of working fluid vaporization. A conventional heat pipe has three sections: an evaporator, an adiabatic (or transport) section, and a condenser, all inside a sealed container. Working liquid turns into vapor at the evaporator, reaches the condenser through the transport section, and comes back to the evaporator along the wick structure using capillary forces. A conventional heat pipe transfers the heat and working fluid from evaporator to condenser in axial direction. However, the coaxial heat pipes with radial heat and mass transfer also exist. Figure 5.5 presents the schematics of heat pipes with axial and with radial transfer.

Various substances can be used as working fluids in proper combination with the heat pipe housing materials depending on operating temperature range (see Table 5.3. Some organic liquids and water are working fluids in low-temperature heat pipes. For operation below ambient temperature, the working fluid should be pre-cooled, then its temperature is controlled using the resistance heater. Various porous and fibrous

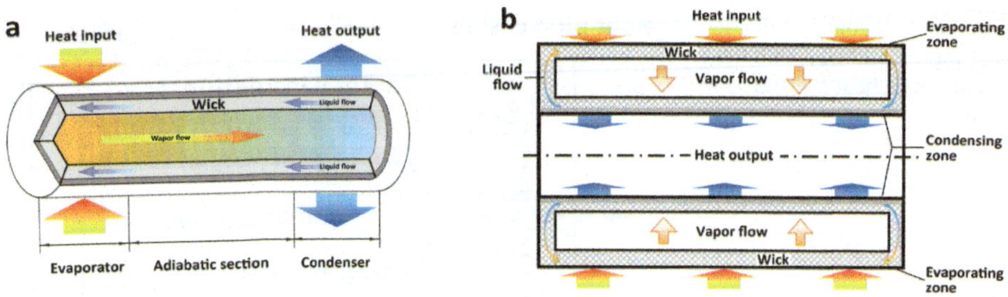

Fig. 5.5 Schematics of heat pipes with: **a** axial and **b** radial heat and mass transfer

Table 5.3 Combinations of working fluids and housing materials of heat pipes used in blackbody constructions

Working fluid	Housing material	Operating temperature (°C)	
		Min	Max
Ammonia (NH_3)	Aluminum, nickel, carbon steel, stainless steel	−60	100
Heptane (C_7H_{16})	Stainless steel	0	150
Water (H_2O)	Copper, Monel, 347 stainless steel	30	200
Cesium (Cs)	Titanium, Niobium + 1% Zirconium	450	900
Potassium (K)	Stainless steel, Inconel	500	1000
Sodium (Na)	Stainless steel, Inconel, Haynes 230	600	1200

materials can be used as a wick structure. High-temperature heat pipes use the melts of some alkali metals as working fluids; a metal mesh can be used as a wick.

A condensing zone of a heat pipe has a very small temperature gradient. The effective thermal conductivity of the heat pipe can exceed thousand times that of copper (about 400 $W{\cdot}m^{-1}\ K^{-1}$ at room temperature) and hence can equalize temperature much better than blackbody with metal heat spreader.

There are two possibilities to employ the coaxial heat pipes as the passive isothermalization system in the blackbody constrictions. First, it can be used as an isothermal liner inside a resistance furnace. This design is often used to achieve uniform heating of crucibles with melts in the FPBBs (see, e.g. Hollandt et al. [66]). Second, the condensation zone of the coaxial heat pipe can form a radiating cavity.

Figure 5.6 depicts the schematic of the heat-pipe blackbody with the liquid–metal heat transfer agent similar to the sodium heat-pipe blackbody described by Bliss et al. [12].

The heat-pipe blackbodies are employed mainly in NMIs as the reference standards of units of radiometric quantities and radiance temperatures.

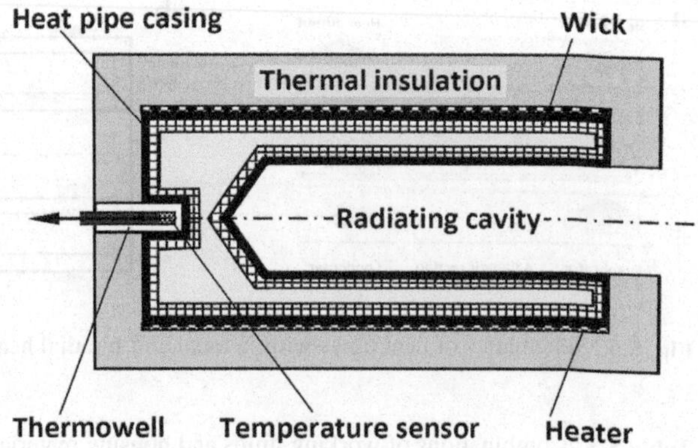

Fig. 5.6 Schematic of the heat-pipe blackbody with the liquid–metal heat transfer agent

5.4.4 Fluid-Bath and Fluid-Circulation Blackbodies

An easily realizable possibility of heating and cooling the radiator is immersing it into a large volume of almost still or slowly stirred fluid at a given temperature. If the radiating element (usually, the cavity) has thin walls and is made of a material with a high thermal conductivity (e.g., copper), the homogeneous temperature is rapidly set throughout the radiator and maintained by convective and conductive heat transfer. A schematic of the fluid-bath blackbody with the radiating cavity is shown in Fig. 5.7. The sensors located near the radiating cavity serve to control temperature uniformity of fluid, their readings are used in the feedback system to manage the power of the heating/cooling element and thus to set the desired temperature of liquid. The simpler in fabrication and cheaper than heat-pipe blackbodies, the fluid-bath blackbodies are used in many NMIs for precise calibrations.

Temperature ranges of the fluid-bath blackbodies are determined by the temperature of thermostated fluid. The most common heat transfer medium is water that allows stable operating of blackbodies within temperature range from 5 °C to about

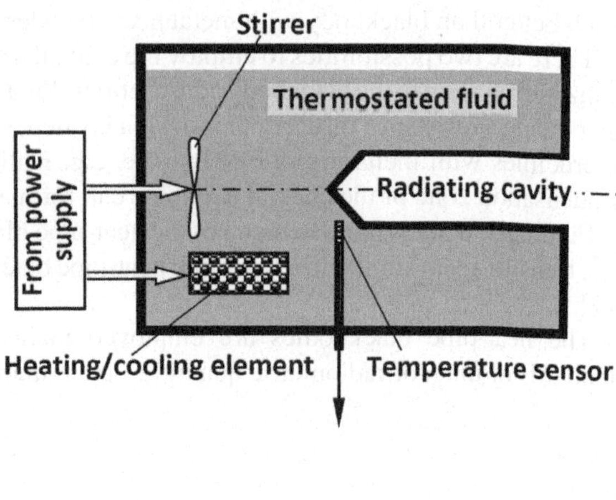

Fig. 5.7 A simplified diagram of the fluid-bath blackbody

70 °C. The temperature range from 60 to 250 °C is covered using various grades of silicon oil.

The lower bound can be decreased down to about −10 °C with addition of some glycol to water. The pure ethanol or organosilicon compounds are served as thermofluids down to about −100 °C. To avoid falling dew or frost during low-temperature operation of the blackbody in the open air, one can use a "curtain" from the nitrogen or dry air flowing in front of the aperture and preventing the penetration of atmospheric air into the cavity [180].

Temperature uniformity and stability of radiating elements of the fluid-bath VTBBs depends not only on the quality of the thermostat used but also on the scrupulousness of the radiator thermal design. Its foundations were laid by Geist and Fowler of NIST [47–49].

One more application of fluid-bath blackbodies should be mentioned. When a very cold blackbody source is needed to provide a zero reference at two-point radiometric calibrations, a cavity immersed in full or in part in the liquid nitrogen (the boiling temperature is about 77 K at atmospheric pressure) can be used for this purpose [153, 154]. In this case, the evacuated space must be on the side of the radiating surface.

If the requirements for the blackbody overall dimensions do not allow the use of a large volume of fluid surrounding the radiator, it can be mounted in a compact fluid–solid heat exchanger of tube (coil) or shell type, through which thermostated fluid flows being pumped from an external thermostat. After passing the heat exchanger, the fluid is returned to thermostat forming a closed loop (see Fig. 5.8). For the fluid-circulation VTBBs, special attention should be paid to the heat exchanger design. A well-designed heat exchanger should ensure a constant temperature and fluid flow rate and have a small contact thermal resistance with the radiator. The temperature

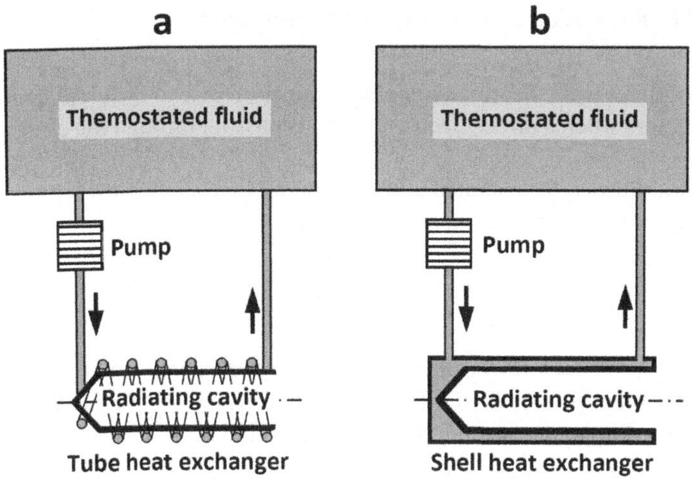

Fig. 5.8 Schematics of the fluid-circulation blackbodies: **a** with a tube heat exchanger and **b** fluid-circulation blackbody with a shell heat exchanger

ranges and the heat transfer media of the fluid-circulation blackbodies are generally the same as for the fluid-bath blackbodies. The cryogenic blackbodies operating in the FIR spectral range usually employ liquid nitrogen as coolant. Auxiliary heaters allow varying the radiator temperature and use cold nitrogen vapor as thermofluid.

The fluid circulation principle allows creating blackbodies with aperture diameters much larger than the typical aperture diameters of fluid-bath blackbodies (20 to 100 mm). The large-aperture blackbodies are needed, in particular, for calibration of thermal imagers. However, in order to increase the diameter of the aperture while preserving the effective emissivity, it is necessary to increase the overall cavity size, which inevitably entails an increase in the volume of the fluid in the bath.

Miklavec et al. [104, 105] described the blackbody with the aperture diameter of 261 mm operating from 5 to 90 °C (water is the working fluid) and from −40 to 20 °C (with ethanol). In fact, this blackbody has the hybrid design occupying an intermediate position between the fluid-bath and the fluid-circulation blackbodies. The radiation cavity is immersed in the fluid bath, into which the fluid is supplied from the auxiliary (preparation) tank, whose volume is 20 L. In total, two baths contain 150 L of heat transfer fluid circulating with the flowrate up to 50 L/min. In the preparation tank, fluid is regulated to the set temperature and then is pumped through six electrically controlled valves in the working volume. This design is similar to the multiple-zone resistance furnace, but the role of separately controlled electric heaters plays separately controlled valves. Park et al. [126] described a water-circulation blackbody operating at 10 to 90 °C with the radiance temperature uniformity of 0.11 °C over a target area of 1 m in diameter and temperature stability better than 0.05 °C.

5.4.5 Thermoelectric Cooling and Heating

The Peltier effect is the heating or cooling a junction of two different conductors or semiconductors when a direct electric current is passing through the circuit with this junction.

The Peltier power P generated at the junction of two materials is proportional to the flowing electric current I(see, e.g. [52, 86]:

$$P = \Pi \cdot I, \tag{5.3}$$

where Π [$W{\cdot}A^{-1} = V$] is the differential Peltier coefficient for two given materials.

It is convenient to express Π through the corresponding Seebeck coefficient S [$V{\cdot}K^{-1}$] according to the second Thomson relation [35]:

$$\Pi = S \cdot T = \left(S_p - S_n\right) \cdot T, \tag{5.4}$$

where T is the temperature of the junction, S_p and S_n are the Seebeck coefficients of materials with p- and n-type of conductivity, respectively.

A thermoelectric cooler (TEC) is a solid-state heat pump operating with direct current. Since the Peltier effect is reversible, the TEC can be used for heating or cooling by reversing the direction of electric current. A typical single-stage TEC consists of two p- and n-type semiconductor material as shown in Fig. 5.9. In the real-world thermoelectric couple, the Peltier effect is always accompanied by Joule heating (due to nonzero electric resistance) and conductive heat transfer (due to nonzero thermal conductivity).

We can write the heat balance equation by supposing that Joule's heat is divided in half between the hot and the cold junctions and by assuming the linear temperature change along the length of the thermoelectric legs.

Incoming thermal power P_c and outgoing thermal power P_h are equal.

$$P_c = ST_cI - \tfrac{1}{2}RI^2 - \sigma_t(T_h - T_c), \tag{5.5}$$

$$P_h = ST_hI + \tfrac{1}{2}RI^2 - \sigma_t(T_h - T_c), \tag{5.6}$$

where $R = R_p + R_n$, R_p and R_n are the electric resistance of p- and n-leg, respectively; $\sigma_t = \frac{A_p k_p}{L_p} + \frac{A_n k_n}{L_n}$ [W·K^{-1}] is the total thermal conductance of the Peltier couple, A_p and A_n are cross-sectional areas of p- and n-leg, respectively; L_p and L_n are their lengths; k_p and k_n are their thermal conductivities.

Figure 5.10 shows that the incoming (cooling) power P_c as a function of current I is positive within a limited range of electric current. Consumed electric power of the Peltier element is

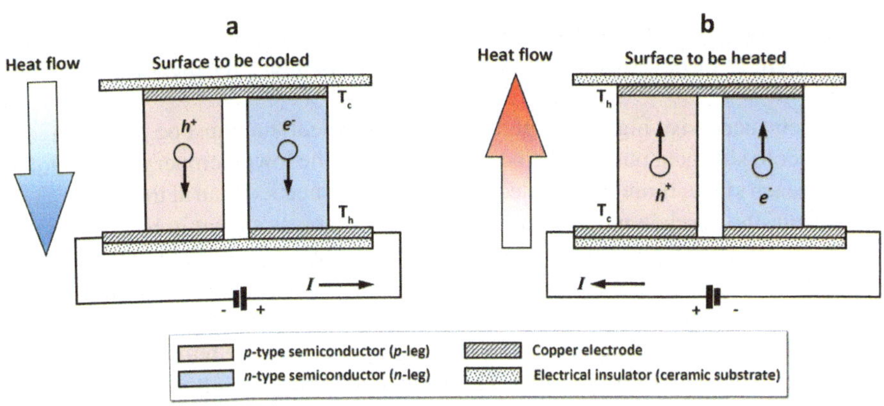

Fig. 5.9 Semiconductor thermoelectric couple used for: **a** cooling and **b** heating

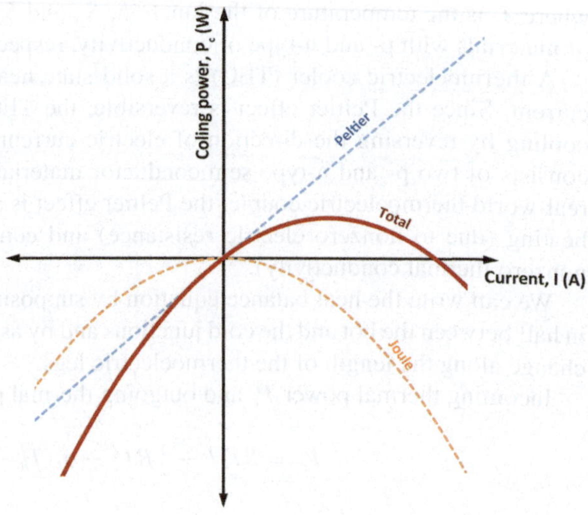

Fig. 5.10 The cooling power of the Peltier element plotted against the electric current

$$P = P_h - P_c = S(T_h - T_c) \cdot I + RI^2. \tag{5.7}$$

Most commercially available TECs work around room temperature and have thermoelectric elements made of tellurium compounds (n-type Bi_2Te_3 and p-type Sb_2Te_3). The semiconductor legs are connected in series electrically and thermally in parallel. Due to low efficiency of the TECs, they are more suitable for cooling and temperature control of small objects.

Larger cooling capacity can be achieved by using several TECs connected thermally in parallel. When a single module cannot provide required performance, one can use a multi-stage cascade that consists of two or more single thermoelectric modules stacked in series.

The first stage of the cascade provides a low-temperature heat sink for the second stage which, in turn, provides a sink at a lower temperature for the third stage, and so forth.

At each successive higher temperature stage, the heat that must be pumped is not only the cold side heat but also the heat dissipated by the lower temperature modules. Thus, at each stage, more thermoelectric columns are needed than at the upper stage. As a result, the cascade module has ziggurat-like shape as shown in Fig. 5.11.

The heat pumping power for various types of commercially available TECs may lie between a few milliwatts and hundreds of watts. The TEC advantages are the absence of moving parts and liquid or gaseous heat transfer media, the possibility of cooling and heating using the same device, and the ease of performing temperature control. Therefore, it is not surprising that TECs early attracted the attention of creators of blackbodies for remote sensing applications. Karoli [70] proved applicability of TECs to thermal management of a compact large-area (65 cm^2) blackbody with an operating temperature range from −40 to 60 °C for laboratory and satellite IR radiometry.

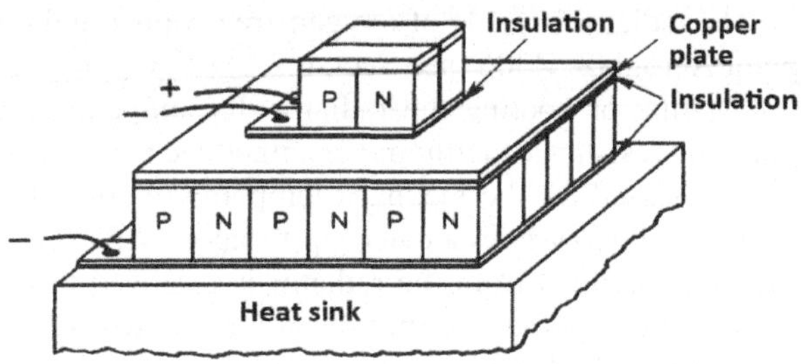

Fig. 5.11 Schematic of the two-stage thermoelectric cascade. Reproduced from [51] with permission of Springer Nature

There were attempts ([187, 189]) to employ semiconductor TECs in the design of the ammonia/stainless steel heat-pipe blackbody with the operating temperature range from −50 to 50 °C. The two-stage TECs served as heaters or coolers when the heat pipe needs fast transitions between its lower and upper temperature limits. The electric power for the fine temperature control was supplied by a thin-film Constantan heater attached to the lower part of the heat pipe. At present, the thermoelectric modules find the greatest application in the thermal design of miniature (including embedded) IR calibrators ([42–44]). A natural niche for thermoelectrically cooled blackbodies is temperatures around 0 °C and relatively low thermal mass of a radiating element, as a rule, of flat-plate type. Figure 5.12 presents the cross-sectional and 3-D views of the VTBB operating in the temperature range from 268 to 333 K and designed to calibrate the IR radiometric systems operating under vacuum conditions (see [87]). Diameter of its planar radiator is 80 mm,its thickness is of 10 mm.

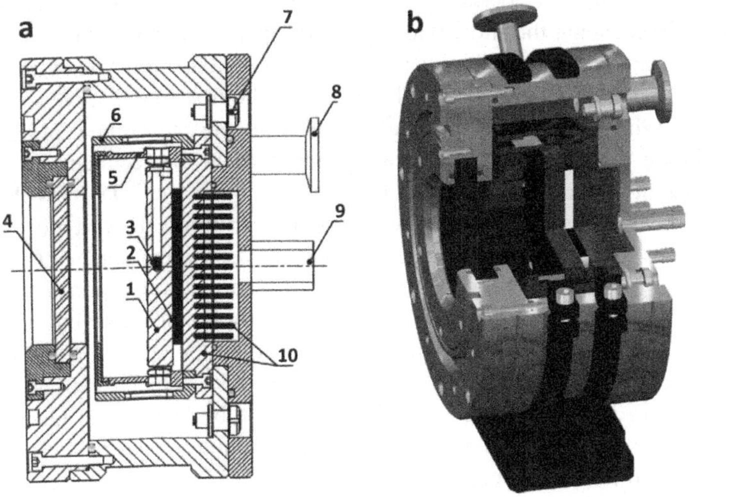

Fig. 5.12 **a** A cross-sectional view and **b** 3D layout of vacuum thermoelectrically cooled VTBB: 1 Radiator made of oxygen-free copper; 2 Thermoelectric element; 3 Platinum resistance thermometer; 4 Aperture window; 5 and 6 Thermal shields; 7 Thermoelectric control connector; 8 Vacuum port; 9 Refrigerant port; 10 Heat sink. Reproduced from [87] with permission of Elsevier

The radiator is made of oxygen-free copper and coated with the high-emissivity paint Aeroglaze Z306 (see Sect. 6.2.2) of 30 μm thick. The Peltier element is used for heating or cooling the radiator. Its temperature is measured by the thin-film platinum resistance thermometer embedded in the radiator. The signal of temperature sensor is used for the feedback temperature control. The radiator under the sealed aperture window is evacuated up to about $2.67 \cdot 10^{-2}$ Pa through the vacuum port. The heat pumped from the radiator is removed through the heat sink cooled with the liquid refrigerant. To improve temperature uniformity over the radiator surface, two thermal shields prevent the radiative heat exchange between the radiator and the VTBB housing.

The greater the thermal mass of the blackbody, the higher heat power needs to be pumped out to achieve the same temperature of the radiator. This requires effective cooling of the heat sink with fluid coolants. Moreover, the liquid-circulation cooling (without TECs) can be easier in realization and more effective for radiators that have greater thermal masses.

5.4.6 Direct Resistance Heating

In the blackbodies with direct resistance heating, the functions of the heater and the radiating element are combined. This heating technique allows achieving temperatures, which are unattainable with the indirect resistance heating. The temperature range of blackbodies with the direct resistance heating extends to about 3500 K, i.e. to the upper limit of operating temperatures achievable by modern blackbodies. Figure 5.13a presents the most widespread design of tubular VTBBs of this type. A flat partition (or septum) divides the tube into two axisymmetric cylindrical cavities, apertures of which are oppositely oriented. The first cavity serves as a working radiator, and the second is used to measure the temperature of the septum by means of a radiation thermometer, which can be included in the feedback loop to maintain a constant temperature of the working cavity. Since the VTBBs with the direct resistance heating operate at extremely high temperatures, the end electrodes must be cooled (usually, by running water) to prevent destroying solder or weld joints. The location of the septum can be chosen so that the viewable part of the working cavity is arranged in the uniform zone of the temperature distribution along the tube (see Fig. 5.13b). In the alternative configuration depicted in Fig. 5.13c, a non-axisymmetric cavity is formed by the walls of the tube, usually vertical. A circular or slit aperture is made in the lateral wall of the tube, which can have additional partitions preventing the radiative heat exchange among zones of the tube internal surfaces with different temperatures.

The direct resistance heating is a part of many industrial processes; this is a well-developed technology, to which extensive literature is devoted see, e.g., [32, 36, 92]. Unlike all blackbody types discussed above, which use surface heating/cooling, the blackbodies with direct resistance heating employ the volumetric heating of radiating

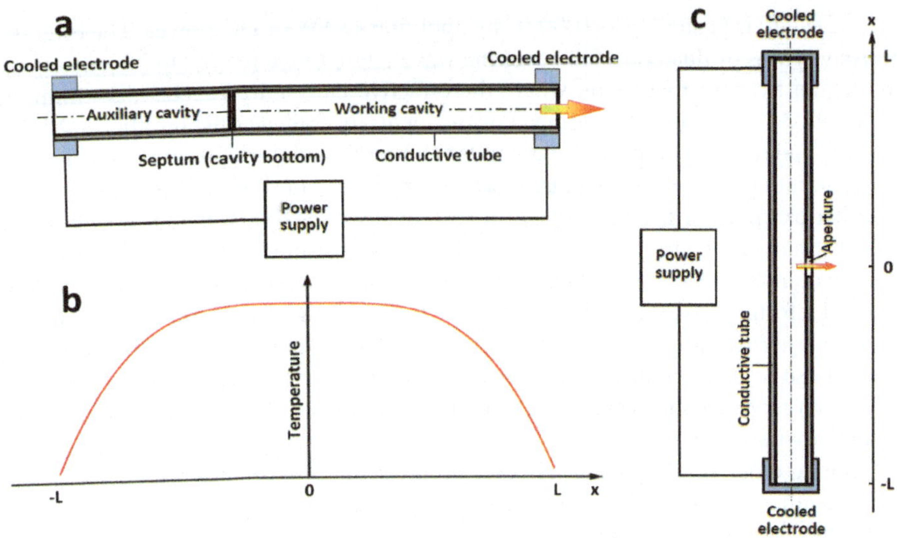

Fig. 5.13 Blackbodies with the direct resistance heating: **a** Double-cavity design, **b** Temperature distribution along the tube length, and **c** Vertical tube design

elements. Materials for such blackbodies should not only have high emissivity, but also be electric conductors.

An overview of electrical conductors capable of withstanding high temperatures allows dividing them into three groups: refractory metals, some ceramics, and carbon-based materials (graphite, first and foremost).

Despite a low emissivity of refractory metals, there were attempts to make the blackbody radiators of tungsten [3, 133] and tantalum [78]. However, these lamp-type radiators operating at temperature from 2500 to 3000 K inside evacuated or argon-filled glass bulbs were not widely disseminated for a number of reasons and are not used today.

It is also difficult to find a suitable material among refractive ceramics: most of them also have low emissivities; only a few of them have acceptable electrical and thermal conductivities; production of many of them is expensive. An exception is silicon carbide (SiC), which has long been used for the manufacturing the infrared emitter known as Globar. Only in the last years, many studies have been carried out for silicon carbide on dependences of emissivity on temperature, viewing angle, etc. (see, e.g. [62, 107, 178]). The structural and physical properties of SiC can be found in [128] or [127].

The Globar shortcomings are spectral selectivity of radiation and thermal degradation of material. Nevertheless, long-term experience of using Globars motivated researchers to create cavity radiators made of SiC. Nagasaka and Suzuki [112] described the SiC blackbody in the form of the hollow cylinder with the slit aperture in the lateral wall.

The cavity is heated up to 1000 °C by applying 450 W of DC power. The effective emissivity was evaluated as 0.99 over the wavelength range from 2 to 20 μm, except the region from 11 to 14 μm, where the effective emissivity reaches its minimum value of 0.97. The spectral radiance of the cavity becomes constant after a 20-min burning and remains constant for about 500 h for the radiator operating at a constant wattage but decreases (or increases) when it run at a constant voltage (or current). Another example of the SiC-based directly heated blackbody is that developed at SIPAI (Shanghai Institute of Process Automation Instrumentation, China) for calibration of industrial radiation thermometers at temperature between 400 and 1500 °C [187]. Unfortunately, the main disadvantages of SiC radiators (low emissivity and poor long-term stability) seems to be irremovable. The SiC cavity radiators have limited use to date, mainly in blackbodies designed for routine measurements and in custom-made facilities (see. e.g. [34]) not so critical to the long-term stability of blackbody radiation.

Among other ceramic materials ever used for the blackbodies with direct resistance heating, zirconia (ZrO_2), lanthanum chromite ($LaCrO_3$), and niobium carbide (NbC) deserve to be mentioned.

Cabannes and Vutien [20] described the blackbody with the cavity made of yttria-stabilized zirconia that can be used for operation in the air up to 2500 K but require preheating to about 1400 K, when it begins to conduct electricity.

Lanthanum chromite ($LaCrO_3$) is the well-known ceramic material for electric heaters used in air up to temperatures of about 1600 K. Zuev et al. [191] described the blackbody with cylindrical cavity made of lanthanum chromite doped with CaO (non-stoichiometric compound $La_{0.975}Ca_{0.025}CrO_3$) for operation at temperatures from 800 to 1500 °C in the air. It was reported that the emissivity of the cavity material is about 0.95 in the spectral range from 2.5 to 15 μm.

In studies aimed at achieving the highest operating temperature of blackbodies, researchers of VNIIOFI (Moscow) tried a number of high-temperature materials including niobium carbide (NbC) and proved its suitability for operations in temperature range from 2500 to 3000 K in vacuum. Bacherikov et al. [2] described the blackbody made of thin-walled (1.5 mm thick) tubes fabricated by sintering of NbC powder. To achieve the effective emissivity of the order of 0.995, the deep cylindrical cavity (350 mm depth, 19 mm diameter, and 12 mm diameter of circular aperture) were formed.

The most widespread and almost universal material for high-temperature blackbodies with direct resistance heating is artificial graphite that has no competitors in the field due to its outstanding properties such as resistance to high temperatures, good electrical and thermal conductivity, chemical stability, machinability, and comparatively low cost. In general, the use of other materials for this purpose is a rare, if not exceptional case. Therefore, we do not discuss in this book high-temperature blackbodies with direct resistance heating made of other materials than graphite.[2] We briefly overview the properties of graphite in Sect. 7.4.

---

[2]We consider pyrolytic graphite (see below) as a form of graphite.

Lummer and Pringsheim [91], who built the first graphite tube blackbody operating in the air up to 2300 K, identified technical problems common to the design of high-temperature blackbodies and proposed some specific solutions to overcome the difficulties encountered.

Graphite begins to oxidize in atmospheric air at about 400 °C forming primarily CO and CO_2. Since all the products of graphite oxidation are volatile gases, graphite loses mass. The oxidation rate rapidly growths when temperature increases. This erodes the surface of the radiator and eventually may lead to its destruction. The service life of the heated graphite tube in the air is limited: burnout of graphite leads to the need for frequent replacement of the radiator. However, due to the low cost of graphite, this is not a serious problem.

It is possible to place the blackbody into the sealed vacuum vessel with a sighting window that should have high transmittance within the spectral range of interest, but the uncertainty in measurement of the window spectral transmittance increases the combined uncertainty of any measurement performed with that blackbody. Besides, a high sublimation rate of graphite above 2000 °C leads to continuously growing contamination of the window, alters its transmittance, and, accordingly, worsens measurements reproducibility.

Presently, this problem is solved using a flow of inert gas (usually, argon) that minimizes graphite sublimation at higher temperatures and oxidation at lower temperatures. The gas flow rate is maintained at a level, which prevents the ingress of air into the cavity but excludes the development of turbulence that may sharply increase the convective heat transfer and result in temperature instabilities on the radiating surface. At present, the radiator lifetime of the best directly heated graphite blackbodies with argon purging is several tens of hours at 3000 K.

Providing reliable electric contacts between the current leads and graphite tube become a complicated technical problem at high temperatures. As a rule, contacts are achieved via copper or brass rings (caps) tightly put on the tube both ends. To prevent destroying solder or weld joints, the caps must be water-cooled. At 3000 K, thermal expansion of graphite may result in 1 to 3% increase of radiator's dimensions. If the length of the graphite tube is 300 mm at a room temperature, the length may increase by 3 to 9 mm at 3000 K. Therefore, the water-cooled electrodes cannot be rigidly fixed on the radiator's tube; some measures should be taken to compensate the radiator thermal expansion. There are various solutions of this problem.

The simplest solution is to arrange the graphite tube vertically, when it is standing on the lower electrode and can be freely expanded up together with the upper one (see, e.g., [77, 159, 181]). The VTBB of such a design resistively heated in an argon atmosphere up to a temperature between 800 and 2400 °C and shown in Fig. 5.14a was used for establishing the NBS scales of the spectral radiance and spectral irradiance [170, 171].

Electric current is supplied to the graphite tube through water-cooled electrical connections at each end. The graphite tube length is about 200 mm; its inner diameter is about 11 mm. The outer surface has the profile providing the tube variable thickness to improve temperature uniformity along the tube length by the gradually changing tube resistance. The tube wall is about 4 mm thick at the central part.

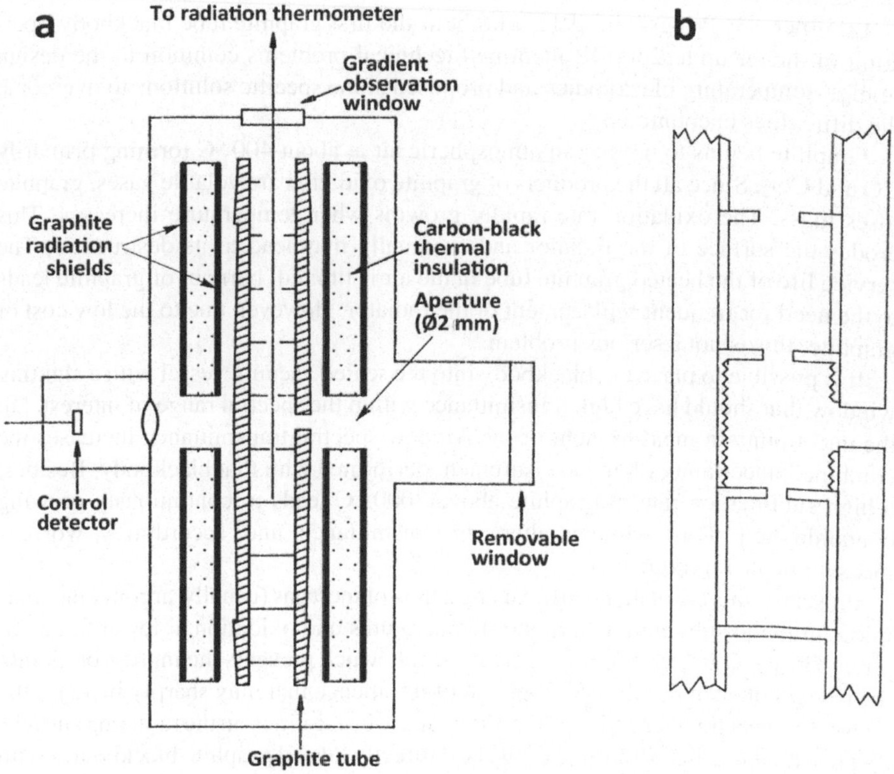

Fig. 5.14 The graphite VTBBs developed at the NBS: **a** General scheme and **b** The radiating cavity in the central section [171]. Reprinted courtesy of the National Institute of Standards and Technology, U.S. Department of Commerce. Not copyrightable in the United States

The graphite tube is surrounded by two cylindrical radiation shields made of graphite, with the carbon black thermal insulation between them. This assembly is placed inside a water-cooled metal housing, with an observation port, which can be sealed during evacuation of the atmospheric air within the housing before flushing with argon. The window at the top of the housing allows for evaluating the temperature distribution along the tube interior using the radiation thermometer. Another window at the rear of the housing allows radiation from the rear wall of the graphite tube to be registered by a silicon or germanium photodiode, whose signal provides automatic control of the power supply when operating at high or low temperatures, respectively. The tube is subdivided into cylindrical sections by a series of thin graphite disks. Figure 5.14b shows a cross-sectional view of the radiating cavity in the central part of graphite tube. The central cylindrical section forms the radiating cavity of 9 mm high, 10 mm in diameter, and the aperture diameter of 2 mm. The cavity walls are threaded to suppress specular reflections and increase effective emissivity, which has been assessed as 0.999 by approximate computational method.

The holes in the graphite disks of diameters varying from 6 mm for the upper disk to 0.75 mm for the disk above the central section allow measurement temperatures of the disks using a radiation thermometer. The major drawback of this blackbody as well as other blackbodies of design similar to that depicted in Fig. 5.13c, is the difficulty in achieving sufficiently high effective emissivity at a reasonable diameter of graphite tube. Therefore, such a scheme is rarely used now. A higher effective emissivity can be achieved with the graphite tube situated horizontally as shown in Fig. 5.13a.

For this scheme, the problem of ensuring reliable electric contacts becomes dominant. A number of technical solutions were proposed after the pioneering work of Lummer and Pringsheim [91].

Schumacher [155] described the graphite tube blackbody designed for the field calibration of missile-borne and air-borne radiometers. The blackbody works regularly up to 2300 °C, but has been tested with enhanced cooling system and electrical power supply at temperatures up to 2800 °C. The possibility of thermal expansion of the graphite tube is provided by a fixed front electrode and the movable rear electrode screwed into a copper piston soldered to a bellows and then to the current lead.

Lapworth et al. [84] used two tapered graphite rods to support the small (the length and the outer diameter of the graphite tube are 90 mm and 5 mm, respectively) horizontal radiator that is a push-fit into the holes at the top of each rod. The rods have sufficient flexibility to allow for expansion of the graphite tube. The reduced cross-section of the rods ensures gradual reduction of their temperature toward the water-cooled stainless steel electrodes screwed to the base plate. It is obvious that this technique is suitable only for lightweight radiators.

Groll and Neuer [54] described a graphite blackbody developed at the IKE (Institut für Kernenergetik, Universität Stuttgart, Germany) for operation between about 900 and 2000 K in vacuo and up to 2900 K in a protective atmosphere of pressured argon to reduce the sublimation of graphite. The clamping spring provides the reliable contact of rear electrode to the graphite dual-cavity radiator (see Fig. 5.15). The temperature uniformity of tubular radiating element of the HTBB heated by the flowing electric current may be insufficient due to the outflow of heat to the cold water-cooled end electrodes.

The profile of the cross-section of the graphite rod in the IKE blackbody was empirically optimized to ensure the best possible temperature uniformity of the working cavity over the entire range of operating temperatures. With a decrease in cross-sectional area, the longitudinal electrical resistance increases, which leads to an increase in the Joule heat release. Despite the main drawback of the IKE blackbody—the need for the output window—it was used in some laboratories until the beginning of the 21th century.

Profiling the cross section of the radiating element to improve its temperature uniformity is one of the most common techniques in the design of HTBBs with direct resistance heating. Chahine et al. [25] described a simple one-dimensional

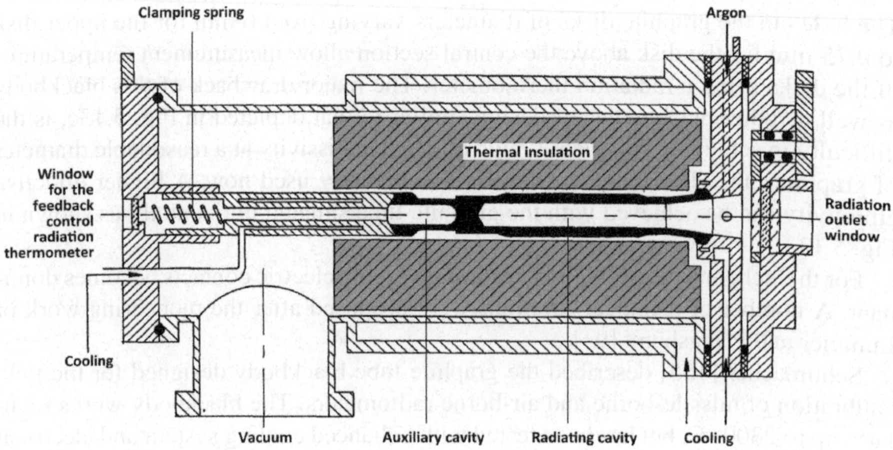

Fig. 5.15 A schematic cross-sectional view of the IKE graphite blackbody. Reproduced from [13] with permission of Elsevier

finite-element model of the double-cavity radiating element of the 48-kW Thermogage[3] graphite tube blackbody and showed a marked improvement in longitudinal temperature uniformity after a creating a thinner region with higher electrical resistance near the ends of the tube. Recently, a similar technique was employed by Lowe et al. [89] to improve temperature uniformity of the graphite tube of the Thermo Gauge HTBB, which was used as a furnace to realize the Co–C HTFPs (~1324 °C). The photograph of the graphite heaters before and after optimization is presented in Fig. 5.16.

Guided by the desire to place cold electrodes as far as possible from the radiating cavity, scientists of VNIIOFI (Moscow) in the early 1990s developed so-called coaxial design of dual-cavity graphite HTBBs [143, 144]. As can be seen in Fig. 5.17, a blackbody includes a pair of coaxial graphite tubes connected in series. The use of a coaxial design significantly increases the distance between the visible part of the working cavity and the water-cooled electrode. In addition, the outer tube plays the role of radiation shield, improving further the temperature uniformity of the inner tube and reducing the consumption of electric energy.

To compensate the radiative heat losses near the open end of the radiating cavity, the walls of both tubes are made thinner toward the aperture of the radiating cavity to increase their electric resistances and, correspondingly, the heat release.

Although manufacturers of graphite blackbodies often claim the upper limit of the operating temperature as high as 3000 °C, actually a graphite blackbody can work at 3000 °C during a very limited time. Due to intensive sublimation of graphite, operation above 2750 °C in vacuo reduces radiator's lifetime so much that they are needed in unacceptably frequent replacements. However, some modern applications

[3]Now, Thermo Gauge Instruments, Inc. (https://thermogauge.com/).

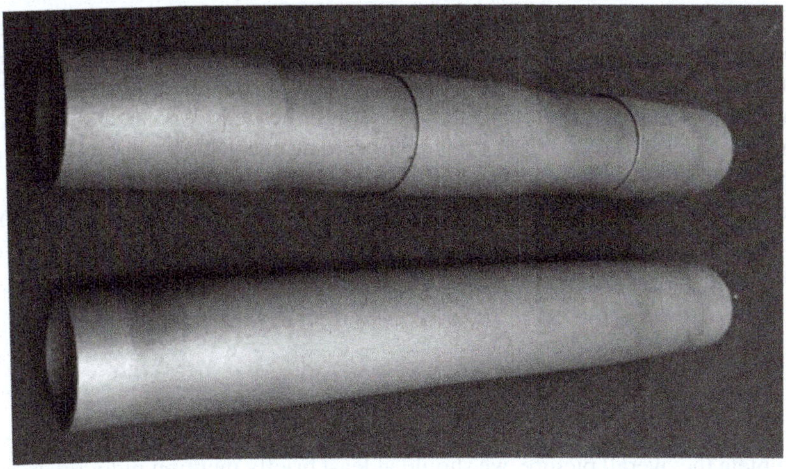

Fig. 5.16 The graphite radiating element optimized for realization of the Co–C HTFP at 1324 °C (top) in comparison with the original radiating element (bottom). Reproduced from [89] under the Creative Commons Attribution License 4.0

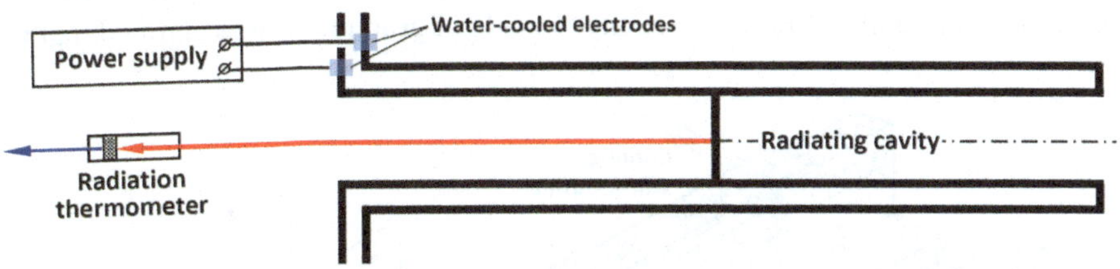

Fig. 5.17 A schematic of the coaxial design of a directly heated graphite HTBB

require long-term operation of the blackbody at temperatures up to 3000 °C or even higher.

At present, such operating temperatures can be achieved only by using pyrolytic graphite blackbodies, which were developed at VNIIOFI (Moscow) in the 1990s.

Pyrolytic graphite (PG) is a polycrystalline variation of artificial graphite obtained by slow deposition of the carbon on the substrate during thermal decomposition of hydrocarbons at temperatures not less than 2500 K. Due to the formation conditions of PG, it has a layered, highly anisotropic structure. PG has a high purity, a virtually zero porosity, a density higher than that of ordinary graphite, and, consequently, a lower sublimation rate. Physical and mechanical properties of PG vary greatly with respect to orientation. Thermal and electrical conductivities along layers are much greater than across layers. We discuss physical properties of PG in Sect. 7.5. The unique properties of PG as an anisotropic refractory material make it possible to increase the operating temperatures of blackbodies with direct resistance heating above 3000 K and significantly improve their metrological characteristics. The principle of designing the PG blackbodies follows the conventional tubular double-cavity graphite blackbodies but the monolithic graphite tube is substituted by the tube assembled from PG rings cut from plates 5–6 mm thick. Anisotropy of transport properties of PG makes electric resistance of the tube composed of PG rings much higher than that made of ordinary graphite; a high thermal resistance along the tube length decrease conductive heat transfer toward water-cooled electrodes. The PG blackbodies with aperture diameter up to 35 mm developed at VNIIOFI [146–148] are well suited for operation in the irradiance mode. In addition, these blackbodies with very large cavities (inner diameter up to 57 mm) can be used as furnaces for the realization of high-temperature fixed points up to temperature of 3500 K [71].

5.4.7 Induction Heating

To complete the overall picture, we should at least briefly mention inductively heated blackbodies, although the time of their popularity remains in the past, and now the use of induction heating of blackbodies is rare exceptions despite its widespread use in industry.

An example of the inductively heated VTBB [103] is shown in Fig. 5.18. The blackbody core with the radiating cavity is made of graphite. It was claimed that the

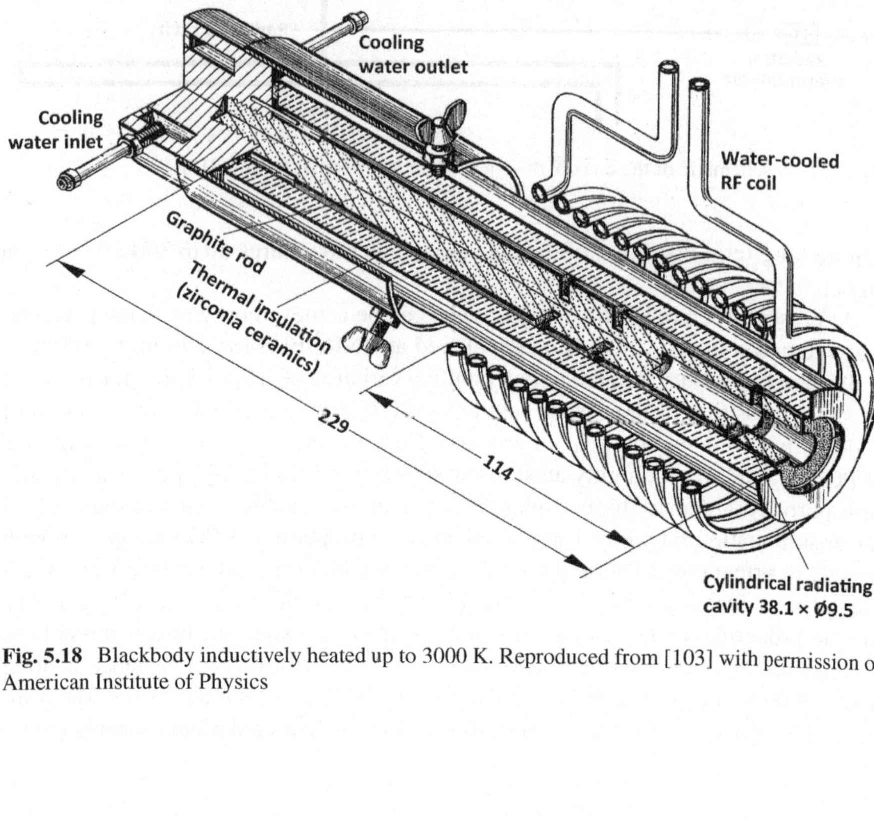

Fig. 5.18 Blackbody inductively heated up to 3000 K. Reproduced from [103] with permission of American Institute of Physics

VTBB is capable to operate in air at 1800 K for up to 6 h and up to 3000 K for shorter periods. The time constant of the blackbody is approximately 2.5 min.

Induction heating employs the Joule heat created by eddy currents induced inside the conductive body (load) exposed to an alternating magnetic field produced by the AC flowing in the coil (inductor). The typical frequency of the AC used in induction heating is 10^3–10^5 Hz. The advantages of induction heating are the contactless transfer of electromagnetic energy between the inductor and the load and a possibility of fast attaining high temperatures of the load. The main disadvantage is inhomogeneity of the induced current density and the associated volumetric density of the heat release.

As a result, the temperature distribution over the load cross-section is determined by the magnitude and frequency of the inducing magnetic field and the physical properties of the load.

Due to the skin effect, heat tends to release in the load subsurface layer, the depth of which is inversely proportional to the square root of the frequency. Technique of induction heating is well elaborated since it has a variety of applications in ferrous and non-ferrous metallurgy, heat treatment of metals, and similar fields. There is an extensive literature on this area (see, e.g., [32, 92, 140]). Graphite VTBBs with induction heating were in use until the end of 1980th (see, e.g. [55, 56, 67, 83, 103, 163]).

Induction heating was used in early designs of FPBBs, including the blackbody at the freezing temperature of platinum (~2041.4 K) employed in the primary standard of the SI unit of luminous intensity, candela, according to its old definition, which was changed in 1979 because of difficulties in practical realization [31, 141, 142].

Due to such drawbacks as insufficient temperature uniformity, coil vibration, and electromagnetic noises that interfere with the work of precise measurement instrumentation, inductively heated blackbodies have been superseded by blackbodies with the resistance heating. In recent years, interest in the inductively heated blackbodies has been resumed in connection with the problem of in situ calibration of radiation thermometers at extremely high temperatures. For such industrial applications, in which the highest level of accuracy is not necessary but other criteria such as a high heating rate play the key role, it was proposed to use small-size rugged and reliable HTFP cells together with simple induction furnaces [125, 157].

5.5 Methods for Improving Effective Emissivity

5.5.1 *Flat-Plate Blackbodies*

The simplest calibration sources are the flat-plate blackbodies (a.k.a. surface calibrators). Market is saturated with calibrators of the IR sensors, thermometers, and cameras that are simply temperature-controlled cooled or heated surfaces (temperatures of individual sources may vary from −50 to 600 °C) coated with “black” paint to increase their emissivity and designed to perform routine calibrations.

The usual application fields of flat-plate blackbodies are routine calibrations of radiation thermometers including those with wide FOVs, thermal imagers, and radiometric instrumentation of moderate accuracy. Simplified design of flat-plate blackbodies allows development of compact and low-cost calibrators with a large target size. The flat-plate blackbodies require calibration in terms of radiance temperature or other radiometric quantities against more accurate blackbodies.

The reverse side of a simple design of flat-plate blackbodies is a limited accuracy [99], for which several reasons exist:

(1) Relatively high uncertainty of the spectral emissivity determination leads to low accuracy of determining the Planckian spectral radiance and the radiance temperature.
(2) Moderate values of the spectral emissivity (as a rule, not exceeding 0.95–0.98) make noticeable a contribution of the ambient radiation reflected by the flat radiating surface into the radiant flux registered by a sensor. Although there are measurement methods designed to exclude the effect of ambient radiation such as that described by Clausen [29], their application complicates calibrations and increases measurement duration.
(3) The flat-plate blackbodies usually operate in laboratory conditions, so their radiating surfaces are note protected against dust contamination and water vapor absorption. These factors, as well as the ageing, lead to changing the spectral emissivity of radiating surfaces, directly affect their accuracy, and reduce time intervals between calibrations of flat-plate blackbodies.
(4) In order to keep a low cost of flat-plate blackbodies, they often have a simplified temperature control that cannot ensure high uniformity of temperature over the large radiating area.

The development of carbon nanotube coatings (see Sect. 6.2.4) can lead to a breakthrough in the creation of flat-plate blackbodies. Santa Barbara Infrared (USA) in cooperation with Surrey NanoSystems (UK) offer extended-area flat-plate blackbody sources coated with VANTABlack®-S CNT-based coating [153]. The claimed emissivity is greater than 0.998 in MWIR and greater than 0.995 in the LWIR spectral range. It is also expected that the CNT-coated blackbodies will be used for in-flight calibrations of satellite radiometric instrumentation (see, e.g. [123]).

5.5.2 *Blackbody Cavities*

The oldest known method for increasing the emissivity of a thermal radiator goes back to Kirchhoff [74], who not only introduced the concept of a perfect blackbody, but also pointed out a practical method of obtaining blackbody radiation inside an enclosure formed by opaque bodies of the same temperature. Wien and Lummer [175] proposed to employ the near-isothermal *Hohlraum* (cavity – Germ.) for the experimental verification of the Stefan-Boltzmann law and first mentioned the need to take into account the effect of the cavity opening. Lummer and Pringsheim [90] described a series of blackbody cavities for operation at temperatures from 100 and 1300 °C. The accuracy of existing measuring instruments did not yet require an obligatory determination of the effective emissivity of cavity radiators. However, already Mendenhall [101] in 1911 was able to assess the effective emissivity of the wedge-shaped cavity assuming specular reflections from its internal surface. Since works of Buckley [15, 16] the efforts of numerous researchers have been directed toward a computational evaluation of effective emissivities of diffuse cavities. Before the advent of digital computers, only approximate analytical methods can be used for the calculations, and they have been used for the cavities of the simplest forms - spherical, conical, and cylindrical. The exact, closed-form expression for the effective emissivity can be derived only for a diffuse spherical cavity [21, 161]. However, the spherical cavity is far from the best choice for the practical embodiment of a blackbody with respect to overall dimensions and providing uniform heating. In addition, the presence of a specular reflection component of the internal surface of the cavity leads to a marked decrease in its normal effective emissivity. The spherical cavities are used much rarely than cavities composed of coaxial cylindrical and conical segments. The problem for diffuse cavities of such shapes was solved, at least in principle, by applying the rigorous numerical technique to solution of singular integral equations of the theory of radiation heat transfer [7–9]. These cavities are the surfaces of revolution, their axial symmetry makes easier the fabrication of a cavity or its parts from a single piece of a homogeneous material by precision turning (except for thin-walled cavities, which can be made by stamping a metal sheet or folding a foil). For simple axisymmetric cavities, the total number of surfaces is usually at most three (see Fig. 5.19. Some practical cavities may have many more surfaces including internal diaphragms or an external sighting tube with a set of diaphragms.

If the components of the sighting tube or a part of the furnace can exchange radiation with the viewable area of the cavity (even after multiple reflections), these parts must be taken into account at calculation of the effective emissivity.

The Monte Carlo ray tracing removes all restrictions imposed on the complexity of radiating surfaces geometry. It allows accounting for radiation heat exchange between a radiating cavity and the furnace construction elements, which may differ in temperature from the cavity and affect significantly the effective emissivity. This allows us to take into account the radiation heat transfer between the radiating cavity and the furnace structural elements, temperatures of which can differ from the temperature of the cavity and significantly affect the effective emissivity. Khlevnoy et al. [72] used the Monte Carlo-based software to compute the effective emissivity of the complicated system depicted in Fig. 5.20. The system consists of the cylindro-conical cavity embedded in the Re-C HTFP cell and the front part of the high-temperature pyrolytic graphite furnace tube with internal diaphragms. The temperature varies from 2747.9 K in the Re-C cell to 300 K at the opposite end of the system. Calculations of the effective emissivities performed separately with and without account of the furnace (see Table 5.4 show that its influence is non-negligible and, as expected, is more pronounced for cavities with lower own effective emissivities.

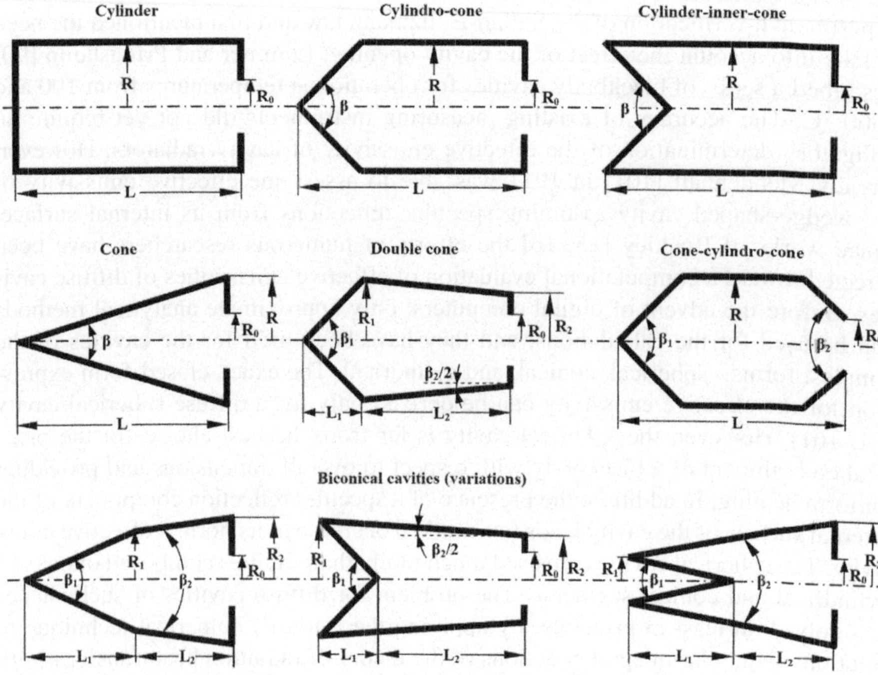

Fig. 5.19 The most common axisymmetric cavity shapes

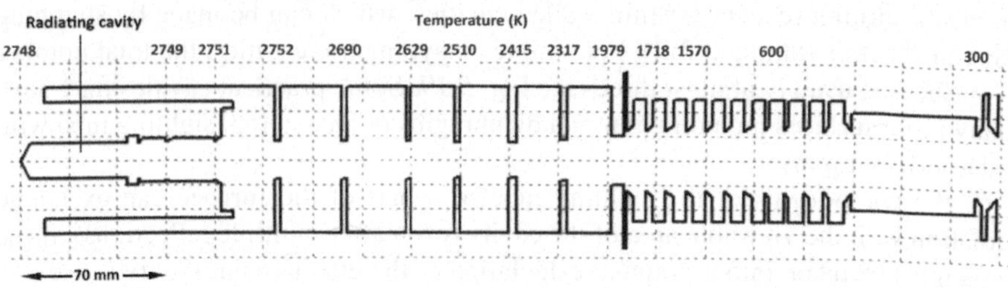

Fig. 5.20 The computational model of the HTFP blackbody cavity with the front end of the furnace (after [72]). Data courtesy of Dr. Boris Khlevnoy (VNIIOFI)

Table 5.4 Effective emissivities of cavities of different size with and without account of radiative heat exchange with high-temperature furnace [72]

Cavity dimensions Depth/Diameter (mm)	Eff. emissivity of cavity only	Eff. emissivity of cavity inside furnace
44/14	0.9962	0.99967
68/14	0.9984	0.99971
88/14	0.9991	0.99974
34.5/3	0.99973	0.99982

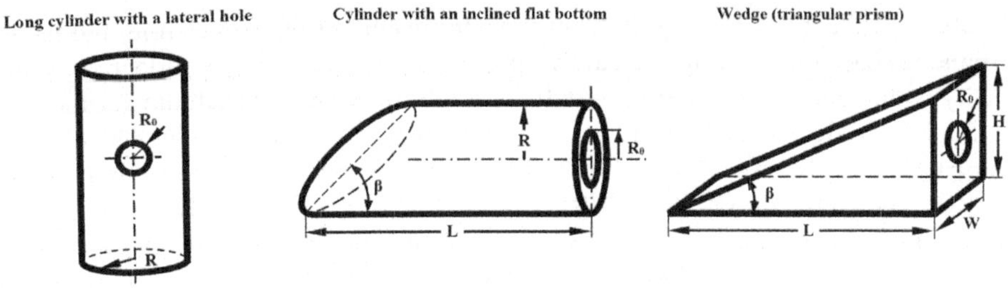

Fig. 5.21 Some common non-axisymmetric cavity shapes

Along with axisymmetric cavities, the cavities of non-axisymmetric shapes also have long been used as artificial blackbodies. Some of these shapes are presented in Fig. 5.21.

A cylindrical cavity with the lateral hole was first employed by Worthing in 1917 [179] at measuring the emissivity of tungsten. The tube had to be carefully polished to reflect specularly. The observation of the hole in the inclined direction ensures its near-blackbody behavior.

However, the effective emissivity in the slanted direction of the diffuse cavity of such a shape is always lower than that of the specular cavity, all other things being equal. Aforementioned cavity shapes are not the only applied in blackbody radiometry and radiation thermometry. From time to time, the researchers propose new forms of cavities (see, e.g. [40, 41]).

Diffuse cylindrical cavities with lateral holes or slits were widely used in directly heated high-temperature blackbodies (see, e.g. [20, 27, 77, 121, 170, 181]) because this configuration allows a simple solution of the thermal expansion problem (see Sect. 5.4.6). Now, the diffuse cavities of these shapes are used rarely because of impossibility of obtaining sufficiently high effective emissivity.

On parity with the axisymmetric conical and cylindro-conical cavities, the cylindrical cavity with the inclined flat bottom is often used in low-temperature blackbodies, the cavities of which have to be coated with the high-emissivity paint [14, 37, 53, 66, 102]. In contrast to conical and cylindro-conical cavities, which require special measures to obtain uniform effective emissivity over the cavity bottom, the cavities with flat bottom can be manufactured and finished easier. Besides, fewer efforts are required to achieve uniformity of temperature over a flat bottom by an appropriate arrangement of heating or cooling (e.g. thermoelectric) elements.

The internal surface of a cylindrical cavity with an inclined flat bottom may have near-diffuse as well as near-specular reflection.

The wedge-shaped cavities find application as the pre- and in-flight low-temperature calibration sources for the satellite instrumentation intended for remote sensing of the Earth in the thermal infrared (TIR) range. Most materials and coatings have near-specular reflection at long wavelengths; therefore, one can expect very high effective emissivities if these cavities work as radiation traps providing large number of reflections for a collimated incident beam. At the same time, the background radiation irradiating the cavity aperture hemispherically is reflected by the cavity within entire hemisphere, so only its small fraction can reach the sensor.

Such a wedge-shaped cavity was used in a narrow field-of-view blackbody (NFBB) in the Radiometric Calibration Facility for pre-flight calibration of the CERES scanning sensor. The CERES (Clouds and the Earth's Radiant Energy System) has been launched by NASA into a Sun-synchronous orbit in December 1999 as a part of a payload of the Terra (AM-1) multi-national scientific research satellite, a flagship of the Earth Observing System (EOS). The NFBB is used for calibration in the 3 to 100 μm wavelength range. The wedge-shaped cavity has the form of a right triangle with the larger cathetus of 20 cm length. To be compatible with the FOV of the CERES scanning sensors, the NFBB aperture is elliptical, with transverse and conjugate diameters of 3.8 cm and 4.7 cm, respectively. A calibrated sensor views the flat surface, which corresponds to the hypotenuse of the triangle. The wedge angle of about 27.5° provides more than 7 reflections before a ray within the sensor field-of-view could exit the cavity. The copper cavity walls were coated with the Chemglaze Z-302 near-specular black paint. According to [46], the directional specular reflectance of Z-302 varies from 0.01 to 0.2 at the incidence angles between 10° and 60° over the wavelength range from 0.3 μm to 50 μm. The BRDF of the paint was measured only at 0.633 μm. The NFBB has approximately 21.45 cm depth and a 3.8 cm by 4.7 cm elliptical aperture coinciding with the FOV of the sensor under calibration. On the base of these data, it was deduced [69] that the effective emissivity of the NFBB is not less than 0.999952. We believe that this value is somewhat overestimated because the computational model has not been accounted for the diffuse component at higher-order reflections and finite width of the BRDF specular peak. In general, it is hard to achieve a uniform effective emissivity across a large aperture of the prismatic cavity. Therefore, it is rarely used in precision calibration.

5.5.3 Regular Grooving of Radiating Surfaces

The general way to increase the emissivity of a flat surface is to change the surface profile in such a manner that the incident radiation undergoes multiple reflections from the modified surface. It is known that surface roughness (random irregularity at the microscopic level) increases the emissivity in comparison with the smooth surface of the same material. The application of regularly alternating recesses and protrusions is the well-known method of increasing the emissivity of a surface at the macro level. Most often, straight or concentric grooves of a simple triangular profile (so-called V-grooves) are applied to the surface by turning or milling (see Fig. 5.22). To increase the emissivity of the grooves surface, it can be further chemically treated or painted with a high emissivity paint.

Rectilinear grooves are identical and, therefore, simpler in calculating their effective emissivity. Since there is no radiant exchange between different grooves, it is

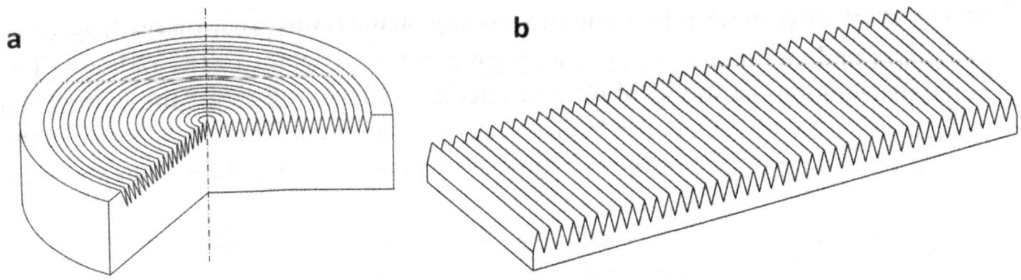

Fig. 5.22 Radiating surfaces with triangular grooves (V-grooves): **a**, Concentric and **b**, Rectilinear

enough to evaluate the effective emissivity of a single groove. Infinitely long V-shaped diffuse groove was studies by [33, 132, 162, 166, 188], specular grooves were considered by Zipin [190] and recently by Mulford et al. [111].

The parallel V-grooves were used in the design of large-area on-board blackbody calibrators of several generations of satellite sensors. The MODIS (Moderate-resolution Imaging Spectroradiometer, see [58]) is a key instrument aboard the Terra and Aqua satellites launched into Earth orbit in 1999 and 2002, respectively. The MODIS on-orbit calibration blackbody thermal emissive bands (wavelengths from 3.7 to 14.4 μm) is an aluminum plate with the twelve V-grooves cut at a 40.5° included angle. The V-grooves are polished, anodized, painted with the specular paint and then re-polished. To be observed in the direction parallel to the bisectors of the grooves, a ray must undergo at least four specular reflections. According to [6], the effective emissivity of the MODIS blackbody determined from pre-launch calibration against a precision NIST-traceable blackbody calibration source is equal to 0.997. Blackbody operates on orbit within a temperature range from 270 to 315 K. Its temperature is monitored and controlled by 12 thermistors also referenced to the NIST temperature standard.

The MODIS blackbody temperature uniformity is about 0.03 K and less than 0.08 K for ambient temperature and 315 K, respectively.

Since the MODIS on-board blackbody showed excellent on-orbit performance, long-term stability, and reliability, its design was used in the next-generation instrument, the Visible/Infrared Imager Radiometer Suite (VIIRS) for the Joint Polar Satellite System (JPSS). Two first satellites of the JPSS were launched in 2011 and 2017. Measurement of the blackbody reflectance performed by [75] using the integrating sphere coated with diffuse gold gave the effective emissivity exceeding 0.9982 at the laser radiation wavelength of 3.39 μm,the ray tracing model predicts the effective emissivity greater than 0.997 across all bands of interest.

Concentric V-grooves are used more widely, in particular, since their axisymmetric radiation field is less sensitive to polarization effects. The effective emissivities of concentric V-grooves began to be studied only in the 2000s, when the Monte Carlo ray tracing was widely used for radiometric calculations. Diffuse grooves were modeled in [184–186]. Prokhorov et al. [130, 131] studied isothermal and non-isothermal concentric grooves within the framework of the specular-diffuse model. The main conclusions that can be drawn from these studies are as follows.

- The local effective emissivity varies depending on the point position on the groove generatrix and on the distance of the groove from the center of the radiator. The radial distributions of the local normal effective emissivities over the isothermal diffuse radiator with concentric V-grooves shown in Fig. 5.23a varies between the maximum values at the bottoms of the grooves and the minimum values at their tops.
- If V-grooves encompassed by the FOV of the radiometer or radiation thermometer are few in number, the effective emissivity of the central part of the radiator encompassed by the FOV depends noticeably on the FOV radius R(see Fig. 5.23b as well as on the placement in the center of the peak (re-entrant cone) or valley (conical cavity).
- The directional effective emissivity of concentric V-grooves reaches a maximum under normal observation and decreases with oblique observation.
- The normal effective emissivity of concentric specular V-grooves is higher than that of diffuse grooves everything else being equal.

The inherent drawbacks of V-grooved radiators are difficulty of manufacturing ideal triangular grooves and inevitable inhomogeneity of temperature of their surfaces. Machining of soft or tough metals results in forming of rather rounded tops and bottoms of V-grooves.

Oxidizing or coating with the high-emissivity paint also may lead to further deviations of the groove profiles from ideal triangles.

Figure 5.24 presents the computational model of the radiator with concentric grooves of the trapezoidal profile as an imitation of the worst case of the manufacturing imperfectness.

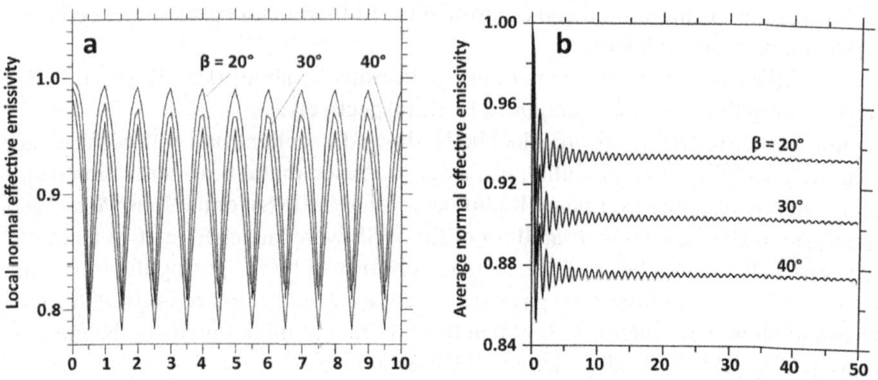

Fig. 5.23 Effective emissivities of isothermal concentric diffuse V-grooves: **a** dependences of the local normal effective emissivities on the ratio r/p and **b** dependences of the average normal effective emissivities on the ratio R/p, where r is the radial coordinate, R is the FOV radius, β is the apex angle of isosceles triangular grooves, and p is their period. The surface emissivity is 0.7. It was assumed that the center of the radiator is concave in all cases (after [130])

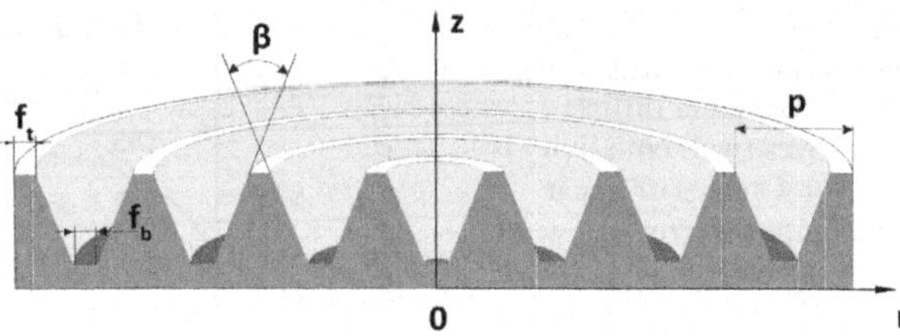

Fig. 5.24 Computational model of a radiator with trapezoidal grooves: p is the pitch, β is the opening angle, f_t and f_b are the widths of the flat areas at the top and bottom, respectively (after [131])

Dependences of average normal effective emissivities of isothermal specular-diffuse grooves on the angle β are shown in Fig. 5.25 for triangular grooves ($f_t = f_b = 0$) and for trapezoidal grooves ($f_t = f_b = 0.1$). As can be seen, the effective emissivities of trapezoidal grooves are noticeably lower than of triangular grooves. This difference is more pronounces for the cases of larger specular components. At nonzero specular component, the dependence on the angle β becomes stepwise due to the abrupt changes in the number of specular reflections.

The difficulty in achieving uniform temperature of radiating surface is another drawback of a V-grooved radiator. Even if its base is isothermal, the heat exchange of the radiator with the environment inevitably leads to temperature non-uniformity of the radiator surface. It is not possible to eliminate this effect even if the radiator operates in vacuo, when the convective heat loss from the grooved surface is absent. In any case, the grooved surface remains non-isothermal due to the change of the radiative heat loss along the groove depth. The effect of temperature inhomogeneity of the triangular diffuse V-grooves on the spectral average normal effective emissivity is presented in Fig. 5.26. The spectral dependences shown are for four values of the

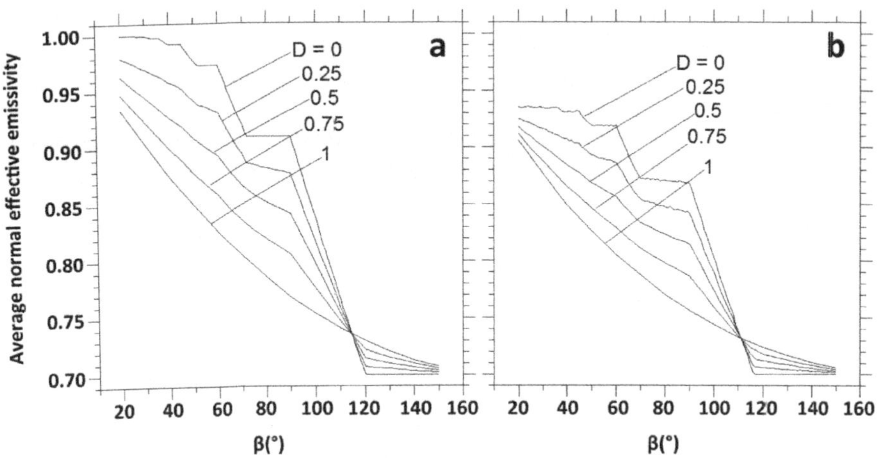

Fig. 5.25 Dependences of the average normal effective emissivities of isothermal specular-diffuse grooves on the angle β: **a** for triangular grooves ($f_t = f_b = 0$) and **b** for trapezoidal grooves ($f_t = f_b = 0.1 \cdot p$). For both cases, the surface emissivity $\varepsilon = 0.7$, $R = 50 \cdot p$, the diffusity $D = 0,\ 0.25,\ 0.5,\ 0.75,\ 1$ (after [130])

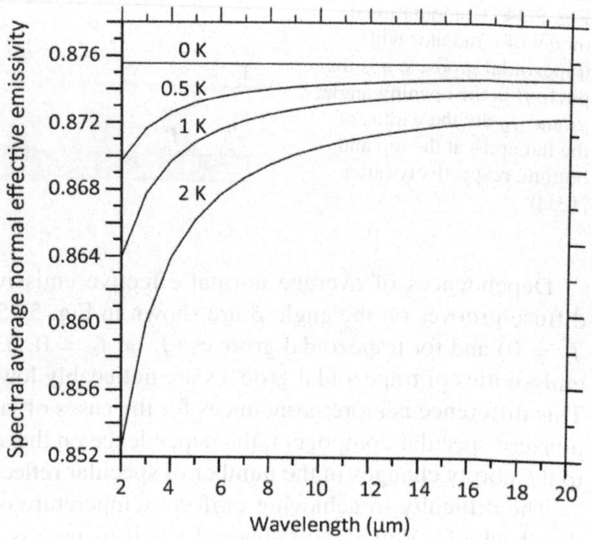

Fig. 5.26 Spectral average normal effective emissivity of 30° triangular diffuse V-grooves (wall emissivity is 0.7) for 4 values of linear temperature decrease toward the groove vertices (0, 0.5, 1, and 2 K). Temperature of the groove bases is 1000 K (after [131])

linear temperature decreases toward the groove vertices (0, 0.5, 1, and 2 K), the apex angle β =30°, the surface emissivity is 0.7, the temperature of the groove bases is 1000 K [131].

The radiative heat loss from V-grooves can be reduced if they are applied to an internal surface of a cavity. In such a case, the V-grooved surface is surrounded by other surfaces having approximately the same temperature, and only a small fraction of the emitted radiation flux escapes the cavity through its aperture.

The grooving of the cavity flat bottom is the long-known technique of suppressing specular reflections and increasing the effective emissivity [63, 100, 113, 115, 116]. Sometimes, V-grooves are applied to the curvilinear, most often, cylindrical surfaces [68, 76, 165].

Wang et al. [172] described the liquid-bath VTBB with the cavity immersed into the water-ethylene glycol mixture at temperature from −30 to 80 °C. The copper cylindrical cavity (80 mm diameter and 520 mm depth) has 31-degree V-grooves on the entire internal surface (see Fig. 5.27) and is painted with the black coating to achieve the effective emissivity above 0.9999.

5.5.4 *Use of Pyramid Arrays*

An alternative to the V-grooves is another type of regular structure – thee array of convex pyramids. The bases of pyramids must cover the plane without gaps. There are only three types of regular convex polygons (equilateral triangles, squares, and regular hexagons), which yield regular tessellations. Only pyramids with square bases are used for the surface regular structuring due to simplicity of their fabrication:

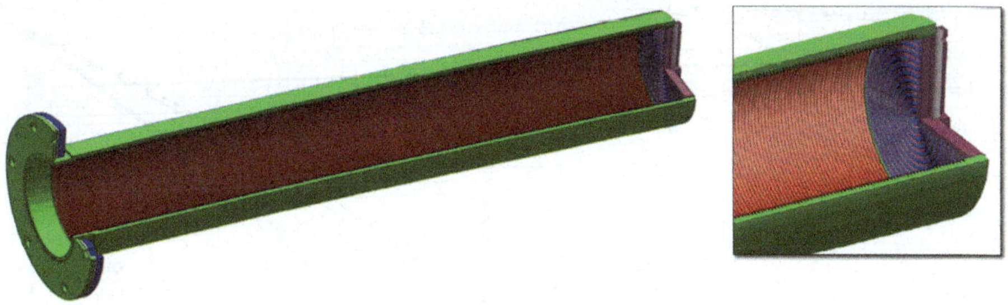

Fig. 5.27 The 3D representation of entirely V-grooved cavity; the right-hand picture is the zoomed image of the cavity bottom and adjacent area. Reproduced from [172] with permission of Springer Nature

an array of square pyramids can be fabricated by superimposing of two mutually perpendicular sets of rectilinear V-grooves.

Despite the fact that the arrays of pyramids have long been used in the construction of blackbodies [4, 22, 28, 59, 60, 94, 152], radiation properties of pyramid arrays are almost not investigated. Exceptions are some early works (e.g. [1]) considering radiative properties of pyramid arrays in connection with the approximate modeling of emissivity of a rough surface.

Serious study of radiation characteristics of pyramid arrays becomes possible only with development of the Monte Carlo ray tracing technique. Figure 5.28a presents the geometric model of a right pyramid with a square base. This model is idealized: a real pyramid cannot be with a perfectly acute apex.

An array of pyramids also cannot be made with perfectly sharp peaks and infinitely narrow valleys between the pyramids. The simplest way to model imperfectness of the pyramid array is to specify flat areas at the top of pyramids and between them

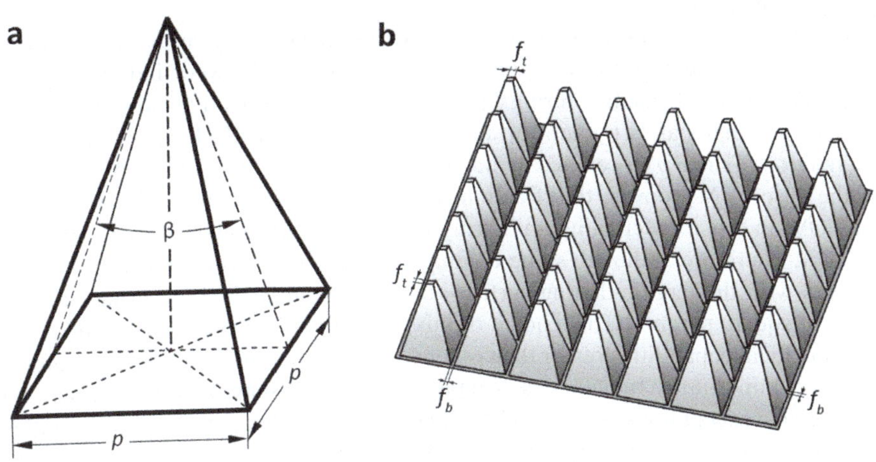

Fig. 5.28 Geometrical model of: **a** Pyramid without flats and **b** An array of pyramids with flats

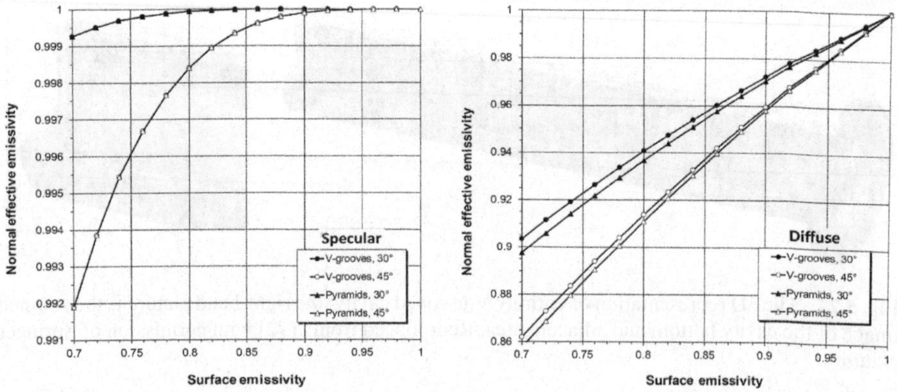

Fig. 5.29 A comparison of dependences of normal effective emissivities on the surface emissivity for purely specular and purely diffuse a V-grooved radiator and a pyramid array for apex angles of 30° and 45°

as shown in Fig. 5.28b. Pyramid array does not form true cavities. Some emitted or reflected rays can pass between pyramids. Therefore, emissivity of the pyramid array should be lower than that of the V-grooves having the same apex angle and surface emissivity. The normal effective emissivity of purely specular V-grooves coincides with that of a purely specular regular pyramid array at the same apex angle.

A comparison of computed dependences for normal effective emissivities on the surface emissivity for purely specular and purely diffuse V-grooved radiator and a pyramid array is made in Fig. 5.29 for the apex angles of 30° and 45°. Calculations were performed using the Monte Carlo modeling code PyramidA [169]. Numerical experiments with the intermediate values of the diffusity D (the diffuse fraction of the reflectance) and various apex angles always predict certain advantage of V-grooves over pyramid arrays.

For the same reasons as for V-grooves, the production of an array of ideal pyramids is impossible: this concern pyramid peaks and space between individual pyramids. The effect of manufacturing imperfections on the effective emissivities of a V-grooved radiator and a pyramid array is manifested differently. For a pyramid array, the fraction of the FOV occupied by flat areas remains constant if the FOV increases. In the case of V-grooves, this fraction decreases tending to zero with increasing the FOV. In this respect, a pyramid array has also no advantages over a V-grooved radiator. Therefore, although some companies manufacture commercial blackbody sources with the large radiating surface (up to 500 mm × 500 mm) shaped as an array of square pyramids, the choice of this type of radiators can be determined by technological capabilities rather than other reasons. The same can be said about the use of pyramid array radiators as compact calibration sources for remote sensing instruments except few cases when developers tried to employ pyramids of complex shapes (see, e.g., [177]), irregular arrays, and so on.

An interesting solution was found at the University of Wuppertal (Germany) for the large-area (126 mm × 126 mm) blackbody for calibration of the airborne imaging

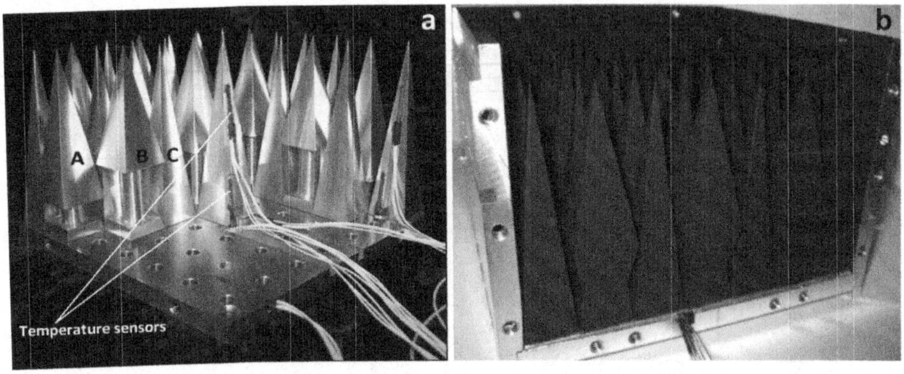

Fig. 5.30 The GLORIA in-flight calibration blackbody: **a**, partly assembled array of uncoated pyramids with the position of platinum resistance thermometers embedded in the pyramids and the base plate marked in blue and **b**, assembled and blackened radiator. Reproduced from [119] under the Creative Commons Attribution 3.0 License

Fourier Transform Spectrometer GLORIA (Gimballed Limb Observer for Radiance Imaging of the Atmosphere). The GLORIA in-flight calibration system consists of two identical blackbodies equipped with the Peltier coolers providing the two-point calibration at temperatures 10 K below and 30 K above the ambient temperature. The GLORIA blackbody, presented in Fig. 5.30 is an array of 49 individual pyramids made of aluminum.

To avoid surfaces perpendicular to the line-of-sight, the pyramids of three types (marked as A, B, and C in Fig. 5.30a) having different square bases and placed at different levels are used [118, 119].

According to calculations performed by Olschewski et al. [119] within frameworks of an approximate diffuse model, the average normal effective emissivity of the GLORIA blackbody is about 0.9996 for the spectral range of interest (about 7 to 13 μm). The GLORIA blackbody is used as a secondary standard,before placing the GLORIA instrument onboard a high-altitude aircraft, the blackbody was calibrated against the Vacuum Low-Temperature Blackbody (VLTBB, see [109]) using the PTB Reduced Background Calibration Facility (RBCF, see [108]). The VLTBB, in turn, was compared to the PTB ammonia heat-pipe blackbody. GLORIA allowed obtaining novel information on small-scale atmospheric dynamics at the border of the stratosphere and troposphere. In order to explore long-term phenomena in the stratosphere for up to two weeks, GLORIA has to be installed into a gondola of a stratospheric balloon. Because of strong restriction on the payload mass, the more compact and lightweight blackbody is needed to calibrate the GLORIA balloon instrument. For this purpose, the pyramid array blackbody was developed and described in [120]. The radiating surface of the 125 mm × 125 mm blackbody is formed by 225 small pyramids machined by wire eroding of aluminum plate (see Fig. 5.31).

Dimension of the base of each pyramid is 5 mm × 5 mm; the pyramid height is 9 mm. The pyramid array is coated with NEXTEL Velvet Black (see Sect. 6.2.1),

Fig. 5.31 Surface of the pyramid array blackbody before blackening. Reproduced from [120] under the term of Creative Commons Attribution 4.0 License

which provides the effective emissivity of higher than 0.997. The blackbody is operated at a temperature of about 240 K (~ −33 °C). The calibration of the pyramid array blackbody is carried out also using the PTB RBCF.

5.5.5 *Multiple-Cavity Blackbodies*

The large-area blackbodies (calibration targets) are needed to perform radiometric calibration of satellite remote sensing instrumentation, including wide-FOV and scanning sensors. These blackbodies must operate either in ground cryo-vacuum facilities (during pre-launch calibration) or in space environment (at in-flight calibration onboard a satellite). In both cases, severe restrictions are imposed on the overall dimensions of blackbodies, so the use of the cavity (e.g., with the ratio of the depth to the aperture diameter of at least 2) to achieve the effective emissivity of 0.99 or higher sometimes does not seem possible. One of simplest solutions of this problem is to form an extended-area radiator by grouping together a number of cavities of smaller apertures. The effective emissivity $\overline{\varepsilon}_e$ of such a radiator is equal to the weighted average

$$\overline{\varepsilon}_e = (1 - r)\varepsilon_e + r\varepsilon, \tag{5.8}$$

where ε_e is the effective emissivity of an individual cavity, ε is the emissivity of flat surfaces between cavities, and r is ratio of the flat surface area to the total area of a multiple-cavity blackbody.

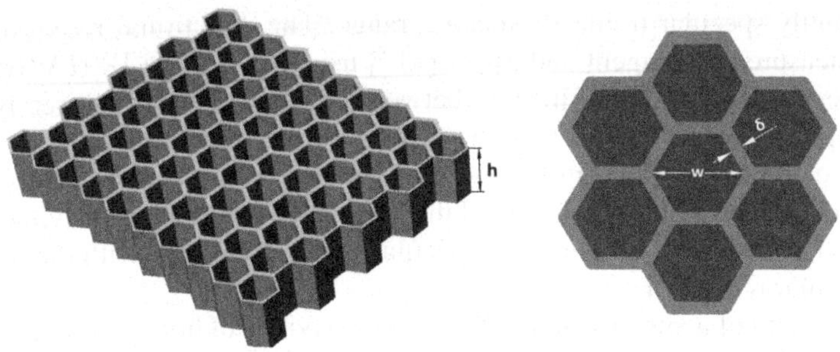

Fig. 5.32 Honeycomb cavity array

The choice of geometric parameters should be made as a compromise between mutually contradictory requirements. First, an individual cavity should be deep enough to provide the large value of ε_e but at the same time, it must be sufficiently shallow to be considered isothermal. Second, the ratio r should be as low as possible. However, a denser arrangement of cavities leads to thinner cavity walls and increasing the cavity temperature non-uniformity.

A contiguous array of hexagonal (honeycomb) cavities provides the best value of r. Karoli et al. [70] described the multiple-cavity honeycomb blackbody with the square emitting area of 65 cm^2, $w = 6$ mm, $h = 25$ mm, and $\delta = 0.05$ mm (see designations in Fig. 5.32), and $r = 0.024$. The radiating surfaces were made of aluminum and coated with the black paint. Instead of the effective emissivity of a single hexagonal cavity, the effective emissivity of an inscribed cylinder was calculated.

The effective emissivity of the multiple-cavity blackbody made of aluminum and coated with black paint was assessed as not less than 0.996 in the wavelength range from 4 to 10 μm. This blackbody was designed for calibrating satellite radiometers and spectrometers. It operated at temperature from −40 to + 60 °C using the thermo-electric heat pump. Temperature uniformity across the target area of multiple-cavity blackbody is ±0.25 °C; the temperature drop to the open end of each cavity is about 2 K.

Pang and Wen [124] presented the copper honeycomb array with the geometrical parameters chosen as in the work above but with the greater working area (625 cm^2). The blackbody mounted on the ammonia heat pipe operates between −50 and + 50 °C. The temperature uniformity of the base plate was assessed as ±0.2 °C. The temperature drop toward the open ends of cavities is about 3 °C, from which the conclusion about its negligible effect on the effective emissivity ($\varepsilon_e > 0.995$) was made.

The results of calculation of the effective emissivities of diffuse, isothermal cavities with regular polygonal cross sections (including the hexagonal cavity) were published for the first time by Bedford and Ma [10, 11]. However, the diffuse approximation is hardly applicable to the spectral range, for which multiple-cavity blackbodies are designed. Reflection of most of the known black coatings becomes predominantly specular in the IR spectral range. The directional radiation characteristics measured by Nugent and Shaw [114] using long-wave IR (LWIR) camera with the response band approximately between 7.5 and 13.5 μm indirectly confirm the assumption about near-specular character of the coating reflection.

The directional effective emissivity of the honeycomb blackbody falls off more slowly than the directional emissivity of the flat-plate blackbody as the viewing angle increases. This behavior is typical for specular-diffuse coatings, with the dominance of the specular reflection.

The difficulty of achieving high effective emissivities of honeycomb blackbodies in the normal direction forced the researchers of the Shenyang Institute of Technology and Northeast University of Technology (R. P. China) to test another cavity shape for the multiple cavity blackbody [57, 82, 173, 174]. The individual cylindro-conical cavities were formed by drilling the bores of 4 mm in diameter in each 5×5 mm^2 of the working area of 150 mm in diameter. The rear side of the radiator is the condensing zone of the water/copper gravity-assisted heat pipe (thermosiphon) operating in the temperature range from 40 to 150 °C.

Although satisfactory temperature homogeneity was obtained, significant improvement of the emissivity was not achieved.

The researchers at the Space Dynamics Laboratory (SDL) of the Utah State University (USU, Logan, UT) again turned to multiple-cavity honeycomb design when developing the HAES15, the High Accuracy Extended Source blackbody [158] intended for ground-based calibration of satellite IR sensors. HAES15 radiator is a cylindrical cavity, the bottom of which is covered by honeycomb cells. The surfaces of cylindrical walls and honeycombs are blackened providing the normal spectral effective emissivity of HAES15 in the wavelength range from 1 to 20 μm is not less than 0.996 and higher than 0.994 from 20 to 25 μm. The temperature range of the HAES15 is 100 to 350 K; the exit aperture diameter is up 381 mm, depending on application and calibration geometry.

5.5.6 Use of Specular Enclosures

A practical blackbody radiator must reproduce radiation of a perfect blackbody only within a narrow solid angle surrounding the FOV of a sensor (radiometer, radiation thermometer, etc.) that collects thermal radiation of the blackbody. This requirement is not obligatory for directions of *all* rays leaving the cavity. We can achieve a higher effective emissivity for the most important directions at the expense of others.

The key idea of this technique is straightforward: the diffuse surface is observed through the hole in the mirror enclosure above that surface. Its radiation is repeatedly reflected between the mirror and the surface. This technique has been used for a long time in radiation thermometry and in radiometric emissivity measurements [26, 30, 38, 79, 80, 121, 122]. The most widely known configurations of mirror enclosures (hemispherical, cylindrical, and conical) are shown in Fig. 5.33.

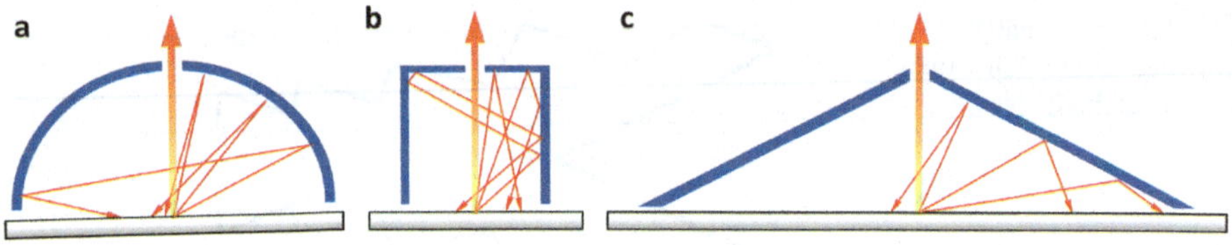

Fig. 5.33 Some geometric configurations of mirror enclosures: **a** Hemispherical, **b** Cylindrical and **c** Conical (double-wedge)

An approximate analytical expression for the effective emissivity of all these configurations can easily be obtained by summing multiple reflections:

$$\varepsilon_e = \frac{\varepsilon}{1 - (1 - \varepsilon) \cdot (1 - \varepsilon_m) \cdot (1 - F_0)}, \tag{5.9}$$

where F_0 is the view factor between the flat surface and the aperture in the mirror enclosure, ε and ε_m are the emissivities of the diffuse surface and the mirror enclosure, respectively.

When deriving Eq. 5.9, the mirror enclosure was assumed non-radiating. This is possible only if its temperature is 0 K. At the same time, the temperature of the diffuse surface under specular enclosure is non-zero.

As the result, the cavity formed by combining both surfaces cannot be considered as the perfect blackbody even if the hole in the mirror is infinitesimal and there is no gap between two surfaces ($F \approx 0$). According to approximate Eq. 5.9,

$$\lim_{F \to 0} \varepsilon_e = \frac{\varepsilon}{1 - (1 - \varepsilon) \cdot (1 - \varepsilon_m)} \neq 1. \tag{5.10}$$

The Monte Carlo modeling of real-world configurations with gradually decreasing radii of the hole in the specular enclosure also shows that the effective emissivities converge to the values less than unity though higher than those predicted by Eq. 5.10.

An ingenious design for specular enclosures was developed at the NPL (UK) by Quinn and Martin [134–138]. The mirror enclosure is composed from segments of truncated conical surfaces,their temperatures may differ slightly from temperature of the diffuse target area with high emissivity.

The mirror surfaces are shaped such that almost all the thermal radiation emitted by the target (with the exception of radiation passing directly through the aperture) is reflected back to the target surface after one or two consecutive reflections from the specular walls. The external detector (radiometer or radiation thermometer) can view only the target area; therefore, only for this region it is necessary to fulfill the requirement of precise temperature uniformity. As an example, the schematics of the blackbody design with the specular enclosure (sometimes referred to as "Christmas tree" design) are presented in Fig. 5.34.

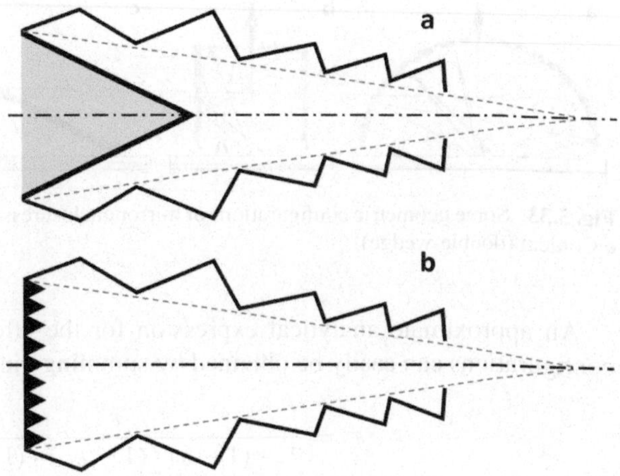

Fig. 5.34 Schematic drawings of the "Christmas tree" blackbody design

It was assumed that the re-entrant conical (or V-grooved) bottom is isothermal at temperature T; ΔT is the difference in temperature between the conical bottom and the specular enclosure. Quinn and Martin [136] have tried to link the effective emissivity ε_e of the blackbody with its geometrical parameters, temperature difference ΔT, and surface emissivities ε and ε_m of the conical bottom and the mirror, respectively.

The approximate expression for the spectral effective emissivity with the account of two first reflections was derived:

$$\varepsilon_e(\lambda) = 1 - [1 - \varepsilon(\lambda)] \cdot \frac{\Delta\Omega}{\pi} \cos\theta - 2 \cdot \varepsilon_m(\lambda) \cdot [1 - \varepsilon(\lambda)] \cdot c_2 \frac{\Delta T}{\lambda T^2}, \quad (5.11)$$

where λ is the wavelength, Ω is the average solid angle subtended by the aperture at points on the target zone, θ is the angle between the normal to the target surface and the direction toward the aperture, and c_2 is the second radiation constant in the Planck law.

One of possible algorithms of forming the generatrix of the specular enclosure can be described as follows.

The semicircle (cross-section of the hemispherical mirror) is divided into an odd number $2n + 1$ of sectors by $2n$ rays originating in the center of the diffuse flat surface. The angle formed by two central rays is subtended by the aperture. The other pairs of neighboring rays divide the semicircle into non-equal zones.

There was no indication in the works of Quinn and Martin of how the generatrix shape of the specular enclosure was obtained. The Monte Carlo modeling shows that Eq. 5.11 relies on oversimplified base: the effective emissivity ε_e calculated according Eq. 5.11 for the given value of ΔT is overestimated the effective emissivity for some blackbodies while it is underestimated for others.

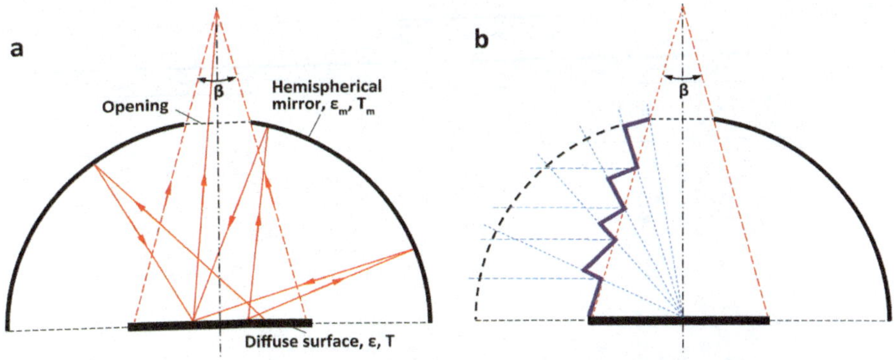

Fig. 5.35 **a** A hemispherical mirror over a diffuse flat surface and **b** A principle of building the polygonal generatrix

Researchers of VNIIOFI adopted the "Christmas Tree" design for large-aperture blackbodies [98, 145], however, the algorithms of constructing the mirror generatrices seemingly differ from the "Christmas Tree" design. As can be seen from Fig. 5.35, the initial point of the design is the hemispherical mirror.

The technical specifications include 100 mm aperture diameter, ±12° FOV, the effective spectral emissivity of 0.999 over the wavelength range from 3 to 15 μm, the temperature range from 100 to 450 K, and the radiance temperature non-uniformity across the aperture of less than 50 mK.

These requirements were fulfilled by a combination of V-grooved graphite bottom heated by a resistive heater and specially profiled gold-plated stainless steel reflector.

The computer modeling of the blackbody was performed as follows.

First, the effective emissivities of the isothermal blackbody was computed by the Monte Carlo ray tracing technique for various conditions of observation. Optimal choice of geometry and coatings was made.

Their sizes can be chosen, for instance, on the base of equality of radiation fluxes diffusely reflected by the target center within the solid angles corresponding to the ring segments of the hemisphere. Then the arcs cut off by rays are replaced by chords that are moved toward marginal rays limiting the FOV. As in the works of Quinn and Martin, the target (blackbody bottom) can be not only flat, but also, for example, V-grooved or even to be a cavity (see [73]). Figure 5.36 present the design of the extended-area medium-temperature blackbody BB900 developed at the VNIIOFI [145] for the calibrations of IR radiometers and imaging systems in terms of radiation temperature, pre-launch characterization of spaceborne optical sensors in the low- and medium-background test facilities.

At the second stage, the steady-state temperature distributions were evaluated considering conductive and radiative heat transfer. The diffuse view factors for the points on the surface of trapezoidal grooves were computed by the Monte Carlo method with the account of multiple reflections. Calculations of view factors show their significant decrease with the use of reflector that acts as a radiation screen.

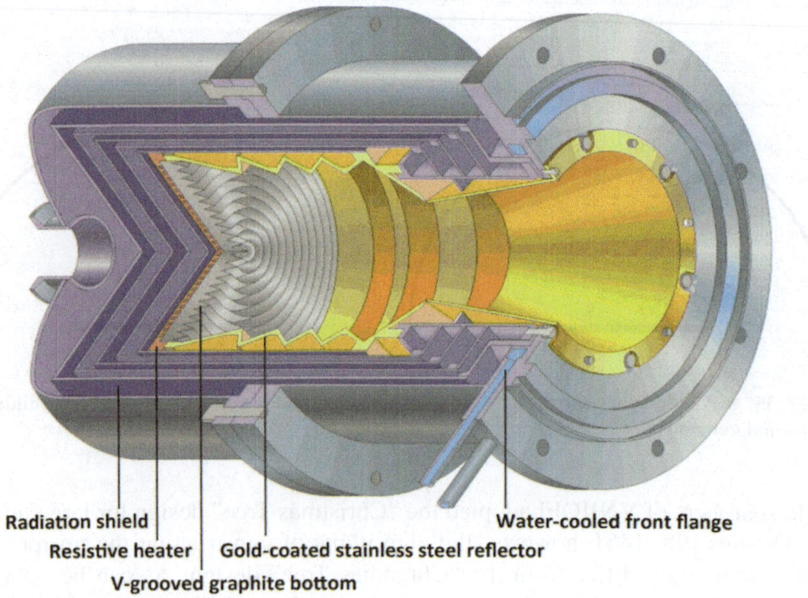

Fig. 5.36 Medium-temperature blackbody BB900

Then, the radial temperature distributions along the grooved bottom were evaluated using finite-element analysis software. Finally, the Monte Carlo method was applied to compute the effective emissivities of the blackbody with obtained earlier temperature distributions. The normal, conical, and directional effective emissivities for wavelength from 3 to 15 μm were computed.

Distributions of the local normal effective emissivities across the aperture of the BB900, with and without reflector, computed using Monte Carlo ray tracing software for the worst case (the surface emissivity of the blackbody graphite bottom with isothermal diffuse trapezoidal grooves is $\varepsilon = 0.7$ and the surface emissivity of the reflector is $\varepsilon_m = 0.05$) are shown in Fig. 5.37.

The graphs in this figure show that the specular enclosure not only increase the effective emissivity but also makes it more uniform across the aperture.

Further development of the design of blackbodies with specular enclosures was made at NPL [23, 167], and summarized by Usadi [168]. In this design called "Sydney Opera House" (SOH), unlike the "Christmas Tree" design, half of the surfaces are spherical (see Fig. 5.38). The principle of choosing the angles θ_i was not disclosed. It remains unclear whether the SOH design gives any advantages over the "Christmas Tree" design in the effective emissivity or in uniformity of its distribution over the aperture.

The SOH design was employed in construction of miniature blackbody for on-board satellites calibrations in the framework of the TIFRI (Technology Innovations for Radiometer Instruments) project supported by European Space Agency

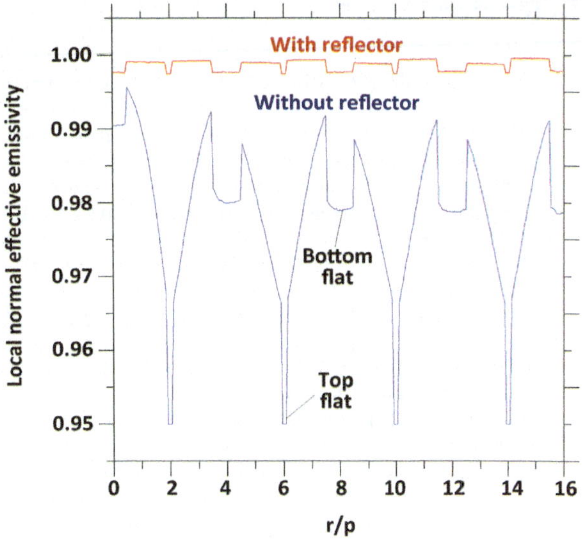

Fig. 5.37 Distributions of local normal effective emissivities along dimensionless radial coordinate for a radiator with isothermal diffuse trapezoidal grooves with and without reflector

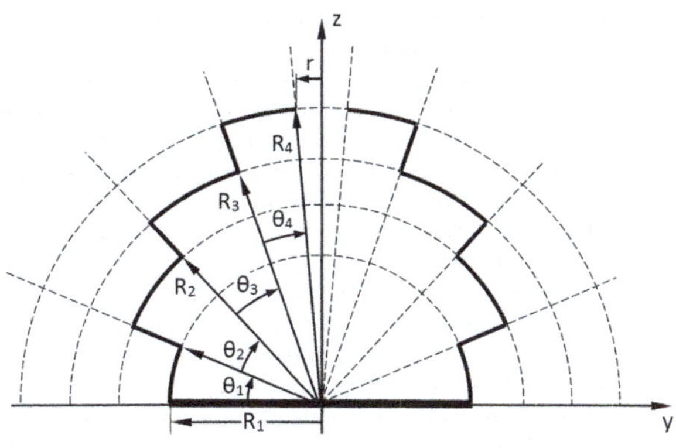

Fig. 5.38 A principle of building the curvilinear generatrix of the "Sidney Opera House" design for specular enclosure

(ESA). Unfortunately, information about the TIFRI, with the exception of temperature sensors and control system, is extremely scarce. The mass of the TIFRI blackbody is less than 300 g, the overall diameter is about 90 mm, the effective emissivity is greater than 0.999. The range of operating temperatures is ±30 K around the ambient temperature; the temperature stability is better than 50 mK; the maximal power consumption is 1 W and up to 10 W for short periods of heating/cooling.

Calibrating of the TIFRI blackbody was planned using the AMBER (Absolute Measurement of Black body Emitted Radiance) facility (see [164]) at NPL since it was expected that the traceability to a radiometric standard allows better accuracy to be obtained than in the case of traceability to the ITS-90 and computational determination of the effective emissivity.

References

1. A. Abdulkadir, R.C. Birkebak, Effects of large pyramidal surface roughness on spectral directional emittance. Warme-Und Stoffubertragung **10**, 23–32 (1977)
2. V.V. Bacherikov, L.V. Vlasov, N.A. Morozov et al., High-temperature radiators for standards in energy photometry. Meas. Tech. **26**, 988–992 (1983)
3. C.R. Barber, T.J. Quinn, E.B. Power et al., A full-radiator lamp designed to replace the tungsten strip lamp as a pyrometric standard. Nature **202**, 686 (1964)
4. M. Battuello, V. Chimenti, G. Machin, et al. A comparison of the primary standard zinc point blackbody cavities of NPL and CEM with IMGC, in *Proc. TEMPMEKO 2001, 8th Int. Symp. on Temperature and Thermal Measurements in Industry and Science,* vol. 2, eds. by Fellmuth, B., Seidel, J., Scholz, G., pp. 857–862 (VDE Verlag GmbH, Berlin, 2002)
5. H.D. Baehr, K. Stephan, *Heat and Mass Transfer*, 3rd edn. (Springer, Berlin, 2011)
6. J. Bauer, J. Mehrten, J. Young et al., MODIS on board calibrator blackbody performance. Proc. SPIE **3439**, 269–276 (1998)
7. R.E. Bedford, C.K. Ma, Emissivities of diffuse cavities: isothermal and nonisothermal cones and cylinders. J. Opt. Soc. Am. **64**, 339–349 (1974)
8. R.E. Bedford, C.K. Ma, Emissivities of diffuse cavities, II: isothermal and nonisothermal cylindro-cones. J. Opt. Soc. Am. **65**, 565–572 (1975)
9. R.E. Bedford, C.K. Ma, Emissivities of diffuse cavities, III: isothermal and nonisothermal double cones. Opt. Soc. Am. **66**, 724–730 (1976)
10. R.E. Bedford, C.K. Ma, Effective emissivities of diffuse, isothermal cavities with regular polygonal cross section: mathematical formulation, in *Proc. TEMPMEKO 2001, 8th Int. Symp. on Temperature and Thermal Measurements in Industry and Science,* vol. 1, ed. by B. Fellmuth, J. Seidel, G. Scholz (VDE Verlag GmbH, Berlin, 2002), pp. 239–244
11. R.E. Bedford, C.K. Ma, Effective emissivities of diffuse, isothermal cavities with regular polygonal cross section: computational details and results, in *Proc. TEMPMEKO 2001, 8th Int. Symp. on Temperature and Thermal Measurements in Industry and Science*, vol. 1, ed. by B. Fellmuth, J. Seidel, G. Scholz (VDE Verlag GmbH, Berlin, 2002), pp. 245–251
12. F.E. Bliss, S. Davis, B. Stein, Isothermal cavity, blackbody radiation source. Appl. Opt. **9**, 2023–2027 (1970)
13. P. Bloembergen, R.E. Bedford, C.K. Ma, A study of the effects of non-isothermal source conditions on the transfer of radiance temperatures in the temperature-wavelength domain from 180 to 30 000 μm K. Measurement **10**, 157–164 (1992)
14. S. Boles, I. Pušnik, D.M. Lochlainn et al., Development and characterisation of a bath-based vertical blackbody cavity calibration source for the range −30 °C to 150 °C. Measurement **106**, 121–127 (2017)
15. H. Buckley, On the radiation from the inside of a circular cylinder. Phil. Mag. Series 7. **4**, 753–762 (1927)
16. H. Buckley, On the radiation from the inside of a circular cylinder—Part II. Phil. Mag. Series 7. **6**, 447–457 (1928)
17. A. Burdakin, B. Khlevnoy, M. Samoylov et al., Melting points of gallium and of binary eutectics with gallium realized in small cells. Metrologia **45**, 75–82 (2008)
18. A. Burdakin, B. Khlevnoy, M. Samoylov et al., Development of gallium and gallium-based small-size eutectic melting fixed points for calibration procedures on autonomous platforms. Int. J. Thermophys. **30**, 20–35 (2009)
19. A. Burdakin, B. Khlevnoy, V. Krutikov, et al. Upcoming space experiments for developing space-borne low-temperature fixed-point blackbody, in *Proc. NEWRAD 2014*, ed. by S. Park, P. Kärhä, E. Ikonen, (2014), pp. 57–58, https://newrad2014.aalto.fi/Newrad2014_Proceedings.pdf. Accessed 10 Feb 2020
20. F. Cabannes, L. VuTien, A zirconia near-blackbody radiation source, 2500 K in air. Int. J. Thermophys. **16**, 277–287 (1995)
21. M.J. Caola, Radiation from a nonisothermal spherical cavity: an exact solution. Appl. Opt. **40**, 3232–3234 (2001)

22. D. Cárdenas-García, E. Méndez-Lango, Use of radiometrically calibrated flat-plate calibrators in calibration of radiation thermometers. Int. J. Thermophys. **36**, 1775–1783 (2015)
23. R. Carvell, E. Usadi, N. Fox et al., Innovative black body for on-board calibration. Proc. SPIE **5234**, 411–422 (2004)
24. CCT, Mise en pratique for the definition of the kelvin in the SI. SI Brochure, 9th ed. Appendix 2 (2019), https://www.bipm.org/utils/en/pdf/si-mep/SI-App2-kelvin.pdf. Accessed 10 Feb 2020
25. K. Chahine, M. Ballico, J. Reizes et al., Optimization of a graphite tube blackbody heater for a Thermogage furnace. Int. J. Thermophys. **29**, 386–394 (2008)
26. X. Che, Z. Xie, Development of ReFaST pyrometer for measuring surface temperature with unknown emissivity: methodology, implementation, and validation. IEEE Trans. Instrum. Meas. **66**, 1845–1855 (2017)
27. S.M. Chernin, High-temperature miniature blackbody radiation sources. Appl. Opt. **36**, 1580–1591 (1997)
28. B. Chu, H.C. McEvoy, J.W. Andrews, The NPL reference sources of blackbody radiation. Meas. Sci. Technol. **5**, 12–19 (1994)
29. S. Clausen, Spectral emissivity of surface blackbody calibrators. Int. J. Thermophys. **28**, 2145–2154 (2007)
30. P. Coates, D. Lowe, *The Fundamentals of Radiation Thermometers* (CRC Press, Boca Raton, FL, 2017)
31. I.E. Cottington, Platinum and the standard of light. A selective review of proposals which led to an international unit of luminous intensity. Platinum Metal Rev. **30**, 84–95 (1986)
32. E.J. Davies, *Conduction and Induction Heating* (Peter Peregrinus Ltd., London, 1990)
33. L.F. Daw, The emissivity of a groove. Brit. J. Appl. Phys. **5**, 182–187 (1954)
34. J. De Lucas, J.J. Segovia, Measurement and analysis of the temperature gradient of blackbody cavities, for use in radiation thermometry. Int. J. Thermophys. **39**, 57 (2018)
35. S.R. De Groot, P. Mazur, *Non-Equilibrium Thermodynamics* (Dover, New York, 1984)
36. Y.V. Deshmukh, *Industrial Heating. Principles, Techniques, Materials, Applications, and Design* (CRC Press, Boca Raton, FL, 2005)
37. A. Diril, H. Nasibov, S. Uğur, UME radiation thermometer calibration facilities below the freezing point of silver (961.78 °C), in *Temperature: Its Measurement and Control in Science and Industry*, vol. 7, ed. by D.C. Ripple (Am. Inst. Phys., New York, 2003), pp 663–668
38. M.D. Drury, K.P. Perry, T. Land, Pyrometers for surface temperature measurement. J. Iron St. Inst. **169**, 245–250 (1951)
39. A. Faghri, Heat pipes and thermosyphons, in *Handbook of Thermal Science and Engineering*, ed. by F.A. Kulacki (Springer, Cham, Switzerland, 2018), pp. 2163–2211
40. Q. Fang, W. Fang, Y. Wang et al., New shape of blackbody cavity: conical generatrix with an inclined bottom. Opt. Eng. **51**, 086401 (2012)
41. W. Fei, Z. Jing, L. Bin, et al. Finite-element analysis of emissivity in cylindrical blackbody cavity sensor. in *Int. Conf. on Internet Computing and Information Services*, (Hong Kong, 2011), pp. 453–454. https://doi.org/10.1109/ICICIS.2011.118
42. D.K. Finfrock, Thermoelectric Thermal Reference Sources (TTRS) for calibration of infrared detectors and systems. Proc. SPIE **4093**, 435–444 (2000)
43. D.K. Finfrock, Calibration of starring infrared systems using Thermoelectric Thermal Reference Sources. Proc. SPIE **4772**, 86–94 (2002)
44. D.K. Finfrock, W.L. Kolander, Third generation infrared system calibration using dual band thermoelectric thermal reference sources and test systems to calibrate uncooled IRFPAs. Proc. SPIE **6940**, 69400V (2008)
45. J. Fischer, M. de Podesta, K.D. Hill et al., Present estimates of the differences between thermodynamic temperatures and the ITS-90. Int. J. Thermophys. **32**, 12–25 (2011)
46. M.A. Folkman, P.J. Jarecke, L.A. Darnton, Enhancements to the radiometric calibration facility for the Clouds and the Earth's Radiant Energy System instruments. Proc. SPIE **1493**, 255–266 (1991)

47. J.B. Fowler, A third generation water bath based blackbody source. J. Res. Natl. Inst. Stand. Technol. **100**, 591–599 (1995)
48. J.B. Fowler, An oil-bath-based 293 K to 473 K blackbody source. J. Res. Natl. Inst. Stand. Technol. **101**, 629–637 (1996)
49. J. Geist, J.B. Fowler, *A Water Bath Blackbody for the 5 to 60 °C Temperature Range: Performance Goal, Design Concept, and Test Results.* Technical Note 1228. (Natl. Bureau of Standards, U.S. Department of Commerce, Washington, DC, 1986)
50. P.J. Gero, J.A. Dykema, J.G. Anderson et al., A blackbody design for SI-traceable radiometry for Earth observation. J. Atm. Oceanic Technol. **25**, 2046–2054 (2008)
51. H.J. Goldsmid, *Thermoelectric Refrigeration* (Springer, New York, 1964)
52. H.J. Goldsmid, *Introduction to Thermoelectricity*, 2nd edn. (Springer, Berlin, 2016)
53. J. Gröbner, Operation and investigation of a tilted bottom cavity for pyrgeometer characterizations. Appl. Opt. **47**, 4441–4447 (2008)
54. M. Groll, G. Neuer, A new graphite cavity radiator as blackbody for high temperatures, in *Temperature: Its Measurement and Control in Science and Industry*, vol. 4, part I, ed. by H.H. Plumb (Instrument Society of America, Washington, DC, 1972), pp. 449–454
55. S. Gu, G. Fu, Q. Zhang, 3500 K high frequency induction heating blackbody sources. in *AIAA/ASME 4th Joint Thermophysics and Heat Transfer Conference. Paper AIAA-86–1282*, (Boston, MA, 1986). https://doi.org/10.2514/4.1986-1282
56. S. Gu, G. Fu, Q. Zhang, 3500-K high-frequency induction-heated blackbody source. J. Thermophys. **3**, 83–85 (1989)
57. S. Guanghui, G. Kuiming, Development of the radiation characteristics of a compound-cavity blackbody. Proc. SPIE **1311**, 420–427 (1990)
58. B. Guenther, W. Barnes, E. Knight et al., MODIS calibration: a brief review of the strategy for the at-launch calibration approach. J. Atm. Oceanic Technol. **13**, 274–285 (1996)
59. B. Gutschwager, H. Driescher, J. Herrmann et al., Characterization of the 300K and 700K calibration sources for space application with the Bepicolombo mission to Mercury. Int. J. Thermophys. **32**, 1429–1439 (2011)
60. X. Hao, H. McEvoy, G. Machin et al., A comparison of the In, Sn, Zn and Al fixed points by radiation thermometry between NIM and NPL and verification of the NPL blackbody reference sources from 156 °C to 1000 °C. Meas. Sci. Technol. **24**, 075004 (2013)
61. J. Hartmann, J. Hollandt, B. Khlevnoy, et al. Blackbody and other calibration sources. in *Radiometric Temperature Measurements. I. Fundamentals,* eds. by Z.M. Zhang, B.K. Tsai, G. Machin (Acad. Press, Amsterdam, 2010), pp. 241–294
62. S. Hatzl, M. Kirschner, V. Lippig et al., Direct measurements of infrared normal spectral emissivity of solid materials for high-temperature applications. Int. J. Thermophys. **34**, 2089–2101 (2013)
63. C.L. Hepplewhite, R.E.J. Watkins, F. Row et al., HIRDLS instrument radiometric calibration black body targets. Proc. SPIE **5152**, 223–230 (2003)
64. I. Hishikari, T. Ide, Precision practical blackbody furnaces by a 3-zone temperature control method. in *Temperature. Its Measurement and Control in Science and Industry*, vol. V, part I, ed. by J.F. Schooley, (Amer. Inst. Physics, New York, 1982), pp. 551–558
65. J. Hollandt, J. Seidel, R. Klein et al., Primary sources for use in radiometry, in *Optical Radiometry,* ed. by A.C. Parr, R.U. Datla, J.L. Gardner (Acad. Press, Amsterdam, 2005), pp. 213–290
66. J. Hollandt, R. Friedrich, B. Gutschwager, et al. High-accuracy radiation thermometry at the National Metrology Institute of Germany, the PTB. High Temp.-High Pres. **35/36**, 379–415 (2003/2004)
67. C.T. Hua, R. Peyturaux, Réalisation d'un corps noir a haute témperature par chauffage haute fréquence. Nouv. Rev. D'Optique Appliquée **3**, 31–36 (1972)
68. J. Ishii, M. Kobayashi, F. Sakuma, Effective emissivities of black-body cavities with grooved cylinders. Metrologia **35**, 175–180 (1998)
69. P.J. Jarecke, M.A. Folkman, T.R. Hedman et al., Clouds and the Earth's radiant energy system (CERES): long-wave calibration plan and radiometric test model (RTM) calibration results. Metrologia **30**, 223–230 (1993)

70. A.R. Karoli, J.R. Hickey, R.E. Nelson, An absolute calibration source for laboratory and satellite infrared spectrometers. Appl. Opt. **6**, 1183–1188 (1967)
71. B.B. Khlevnoy, M.L. Samoylov, I.A. Grigoryeva et al., Development of high-temperature blackbodies and furnaces for radiation thermometry. Int. J. Thermophys. **32**, 1686–1696 (2011)
72. B. Khlevnoy, I. Grigoryeva, K. Anhalt et al., Development of large-area high-temperature fixed-point blackbodies for photometry and radiometry. Metrologia **55**, S43–S51 (2018)
73. V.B. Khromchenko, S.N. Mekhontsev, L.M. Hanssen, Design and evaluation of large-aperture gallium fixed-point blackbody. Int. J. Thermophys. **30**, 9–19 (2009)
74. G. Kirchhoff, Ueber das Verhältnis zwischen dem Emissionsvermögen und dem Absorptionsvermogen der Körper für Wärme und Licht. Ann. Phys. **109**, 275–301 (1860)
75. J. Kloepfer, C. Taylor, V. Murgai, Characterization of the VIIRS blackbody emittance. Conference on Characterization and Radiometric Calibration for Remote Sensing, Utah State University, Space Dynamics Laboratory (2013), https://digitalcommons.usu.edu/cgi/viewcontent.cgi?article=1011&context=calcon. Accessed 10 Feb 2020
76. H.Y. Ko, B.J. Wen, S.F. Tsa, et al. A high-emissivity blackbody with large aperture for radiometric calibration at low-temperature. Int. J. Thermophys. **30**, 98–104 (2009)
77. H.J. Kostkowski, D.E. Erminy, A.T. Hattenburg, High-accuracy spectral radiance calibration of tungsten-strip lamps. in *Precision Radiometry*. Advances in Geophysics, vol. 14, ed. by A.J. Drummond (Acad. Press, New York, 1970), pp. 111–127
78. B.P. Kozyrev, Yu.K. Gornostaev, Standard multichamber high-temperature radiation sources. J. Appl. Spectrosc. **7**, 541–545 (1967)
79. J.C. Krapez, C. Belanger, P. Cielo, A double-wedge reflector for emissivity enhanced pyrometry. Meas. Sci. Technol. **1**, 857–864 (1990)
80. J.C. Krapez, P. Cielo, M. Lamontagne, Reflective-cavity temperature sensing for process control, in *Temperature. Its Measurement and Control in Science and Industry, vol. 6, part 2*, ed. by J.F. Schooley (Am. Inst. Physics, New York, 1992), pp. 877–882
81. V.N. Krutikov, V.I. Sapritsky, B.B. Khlevnoy et al., The Global Earth Observation System of Systems (GEOSS) and metrological support for measuring radiometric properties of objects of observations. Metrologia **43**, 94–97 (2006)
82. G. Kuiming, Z. Wei, X. Zhi et al., Development of an extended blackbody source for the calibration of infrared systems (II). Proc. SPIE **940**, 131–136 (1988)
83. G.T. Lalos, R.J. Corruccini, H.P. Broida, Design and construction of a blackbody and its use in the calibration of a grating spectroradiometer. Rev. Sci. Instrum. **29**, 505–509 (1958)
84. K.C. Lapworth, T.J. Quinn, L.A. Allnutt, A black-body source of radiation covering a wavelength range from the ultraviolet to the infrared. J. Phys. E: Sci. Instrum. **3**, 116–120 (1970)
85. H. Lee, *Thermal Design. Heat Sinks, Thermoelectrics, Heat Pipes, Compact Heat Exchangers, and Solar Cells* (Wiley, Hoboken, NJ, 2010)
86. H. Lee, *Thermoelectrics Design and Materials* (Wiley, Chichester, UK, 2017)
87. S.-Y. Lee, G.-H. Kim, Y.-S. Lee et al., Thermal performance analysis of vacuum variable-temperature blackbody system. Infrared Phys. Technol. **64**, 97–102 (2014)
88. D.H. Lowe, A.D.W. Todd, R. Van den Bossche et al., The equilibrium liquidus temperatures of rhenium-carbon, platinum-carbon and cobalt-carbon eutectic alloys. Metrologia **54**, 390–398 (2017)
89. D. Lowe, L. Wright, C. Liller, Improving fixed-point cell realization by modifying furnace heater shape. in *Proceedings of the 19th Int. Congress of Metrology, 24002* (2019), https://cfmetrologie.edpsciences.org/articles/metrology/pdf/2019/01/metrology_cim2019_24002.pdf. Accessed 8 Feb 2020
90. O. Lummer, E. Pringsheim, Die Strahlung eines “schwarzen” Körpers zwischen 100 und 1300° C. Ann. D. Phys. **299**, 395–410 (1897)
91. O. Lummer, E. Pringsheim, Die strahlungstheoretische Temperaturskala und ihre Verwirklichung bis 2300° abs. Verh. D. Physik. Ges. **5**, 3–13 (1903)

92. S. Lupi, *Fundamentals of Electroheat. Electrical Technologies for Process Heating* (Springer, Cham, Switzerland, 2017)
93. G. Machin, The kelvin redefined. Meas. Sci. Technol. **29**, 022001 (2018)
94. G. Machin, B. Chu, A transportable gallium melting point blackbody for radiation thermometry calibration. Meas. Sci. Technol. **9**, 1653–1656 (1998)
95. G. Machin, R. Simpson, Tympanic thermometer performance validation by use of a body-temperature fixed point blackbody. Proc. SPIE **5073**, 51–56 (2003)
96. G. Machin, K. Anhalt, P. Bloembergen, et al. *MeP-K Relative Primary Radiometric Thermometry* (2017), https://www.bipm.org/utils/en/pdf/si-mep/MeP-K-2018_Relative_Primary_Radiometry.pdf. Accessed 10 Feb 2020
97. W.R. McCluney, *Introduction to Radiometry and Photometry*, 2nd edn. (Artech House, Boston, MA, 2014)
98. S.N. Mekhontsev, V.I. Sapritsky, A.V. Prokhorov et al., Modeling, design, and characterization of medium background blackbodies for full aperture calibration of spaceborne infrared systems and imagers. SPIE **3553**, 247–258 (1998)
99. S.N. Mekhontsev, V.B. Khromchenko, L.M. Hanssen, NIST radiance temperature and infrared spectral radiance scales at near-ambient temperatures. Int. J. Thermophys. **29**, 1026–1040 (2008)
100. N. Melzack, E. Jones, D.M. Peters et al., Variable temperature blackbodies via variable conductance: thermal design, modelling and testing. Int. J. Thermophys. **38**, 30 (2017)
101. C.E. Mendenhall, On the emissive power of wedge-shaped cavities and their use in temperature measurements. Astrophys. J. **33**, 91–97 (1911)
102. U. Mester, P. Winter, New blackbody calibration source for low temperatures from −20 °C to +350 °C. Proc SPIE **4360**, 372–380 (2001)
103. A.J. Metzler, J.R. Branstetter, Fast response, blackbody furnace for temperatures up to 3000 °K. Rev. Sci. Instrum **34**, 1216–1218 (1963)
104. A. Miklavec, I. Pušnik, V. Batagelj et al., Calibration of thermal imagers by evaluation of the entire field-of-view. Int. J. Thermophys. **32**, 2600–2609 (2011)
105. A. Miklavec, I. Pušnik, V. Batagelj et al., A large aperture blackbody bath for calibration of thermal imagers. Meas. Sci. Technol. **24**, 025001 (2013)
106. S. Moaveni, *Finite Element Analysis: Theory and Applications with ANSYS*, 2nd edn. (Prentice Hall, Upper Saddle River, NJ, 2003)
107. C. Monte, J. Hollandt, The measurement of directional spectral emissivity in the temperature range from 80 °C to 400 °C at the Physikalisch-Technische Bundesanstalt. High Temp. High Press. **39**, 151–164 (2010)
108. C. Monte, B. Gutschwager, S. Morozova et al., Radiation thermometry and emissivity measurements under vacuum at the PTB. Int. J. Thermophys. **30**, 203–219 (2009)
109. S. Morozova, N. Parfentiev, B.E. Lisiansky et al., Vacuum Variable Temperature Blackbody VLTBB100. Int. J. Thermophys. **29**, 341–351 (2008)
110. K. Motzfeldt, *High Temperature Experiments in Chemistry and Materials Science* (Wiley, Chichester, UK, 2013)
111. R.B. Mulford, N.S. Collins, M.S. Farnsworth et al., Total hemispherical apparent radiative properties of the infinite V-groove with specular reflection. Int. J. Heat and Mass Transfer **124**, 168–176 (2018)
112. T. Nagasaka, M. Suzuki, Silicon carbide infrared cavity radiator. Appl. Opt. **16**, 2358–2359 (1977)
113. T.J. Nightingale, J. Crawford, A radiometric calibration system for the ISAMS remote sounding instrument. Metrologia **28**, 233–237 (1991)
114. P.W. Nugent, J.A. Shaw, Large-area blackbody emissivity variation with observation angle. Proc. SPIE **7300**, 73000Y (2009)
115. S.A. Ogarev, M.L. Samoylov, N.A. Parfentyev et al., Low-temperature blackbodies for IR calibrations in a medium-background environment. Int. J. Thermophys. **30**, 77–97 (2009)
116. S.A. Ogarev, S.P. Morozova, A.A. Katysheva et al., Blackbody radiation sources for the IR spectral range. AIP Conf. Proc. **1552**, 654–659 (2013)

117. OIML R 147, *Standard blackbody radiators for the temperature range from −50 °C to 2500 °C. Calibration and verification procedures. International Recommendation* (OIML 2016), https://www.oiml.org/en/files/pdf_r/r147-e16.pdf. Accessed 10 Feb 2020
118. F. Olschewski, C. Rolf, P. Steffens et al., In-flight blackbody calibration sources for the GLORIA interferometer. Proc. SPIE **8511**, 85110I (2012)
119. F. Olschewski, A. Ebersoldt, F. Friedl-Vallon et al., The in-flight blackbody calibration system for the GLORIA interferometer on board an airborne research platform. Atmos. Meas. Tech. **6**, 3067–3082 (2013)
120. F. Olschewski, C. Monte, A. Adibekyan et al., A large-area blackbody for in-flight calibration of an infrared interferometer deployed on board a long-duration balloon for stratospheric research. Atmos. Meas. Tech. **11**, 4757–4762 (2018)
121. A. Ono, Evaluation of the effective emissivity of reference sources for the radiometric emissivity measurements. Int. J. Thermophys. **7**, 443–453 (1986)
122. A. Ono, Methods for reducing emissivity effects, in *Theory and Practice of Radiation Thermometry*, ed. by D.P. DeWitt, G.D. Nutter (Wiley, New York, 1988), pp. 565–623
123. D.P. Osterman, S. Collins, J. Ferguson et al., CIRiS: compact infrared radiometer in space. Proc. SPIE **9978**, 99780E (2016)
124. S.-J. Pang, Y.-P. Wen, A low-temperature large-area blackbody radiant source with heat pipe. Infrared Phys. **28**, 203–214 (1988)
125. C.J. Parga, C. Journeau, A. Tokuhiro, Development of metal-carbon eutectic cells for application as high temperature reference points in nuclear reactor severe accident tests: results on the Fe-C, Co-C, Ti-C and Ru-C alloys' melting/freezing transformation temperature under electromagnetic induction heating. High Temp.–High Press. **41**, 423–448 (2012)
126. C.-W. Park, Y.S. Yoo, B.-H. Kim et al., Construction and characterization of a large aperture blackbody for infrared radiometer calibration. Int. J. Thermophys. **32**, 1622–1632 (2011)
127. G. Pensl, F. Ciobanu, T. Frank, et al. SiC material properties. Int. J. High Speed Electron. Syst. **15**, 705–745 (2005)
128. H.O. Pierson, *Handbook of Refractory Carbides and Nitrides. Properties, Characteristics, Processing and Applications.* (Noyes Publ., Westwood, NJ, 1996)
129. H. Preston-Thomas, The International Temperature Scale of 1990 (ITS-90). Metrologia **27**, 3–10 (1990)
130. A.V. Prokhorov, S.N. Mekhontsev, L.M. Hanssen, Emissivity modeling of thermal radiation sources with concentric grooves. High Temp.–High Press. **35/36**, 199–207 (2003/2004)
131. A.V. Prokhorov, L.M. Hanssen, S.N. Mekhontsev, Radiation properties of IR calibrators with V-grooved surfaces. Proc. SPIE **6205**, 620505 (2006)
132. J. Psarouthakis, Apparent thermal emissivity from surfaces with multiple V-shaped grooves. AIAA J **1**, 1879–1882 (1963)
133. T.J. Quinn, C.R. Barber, A lamp as a reproducible source of near black-body radiation for precise pyrometry up to 2700 °C. Metrologia **3**, 19–23 (1967)
134. T.J. Quinn, J.E. Martin, A radiometric determination of the Stefan-Boltzmann constant and thermodynamic temperatures between −40 °C and +100 °C. Phil. Trans. Roy. Soc. London **A316**, 85–189 (1985)
135. T.J. Quinn, J.E. Martin, A black-body cavity for total radiation thermometry. Metrologia **23**, 111–114 (1986/87)
136. T.J. Quinn, J.E. Martin, Blackbody source in the −50 to +200°C range for the calibration of radiometers and radiation thermometers. Appl. Opt. **30**, 4486–4488 (1991)
137. T.J. Quinn, J.E. Martin, Cryogenic radiometry, prospects for further improvements in accuracy. Metrologia **28**, 155–161 (1991)
138. T.J. Quinn, J.E. Martin, Total radiation measurements of thermodynamic temperature. Metrologia **33**, 375–381 (1996)
139. D.A. Reay, P.A. Kew, R.J. McGlen, *Heat Pipes. Theory, Design and Applications*, 6th ed. (Elsevier, Amsterdam, 2014)
140. V. Rudnev, D. Loveless, R.L. Cook, *Handbook of Induction Heating*, 2nd edn. (CRC Press, Boca Raton, FL, 2017)

141. C.L. Sanders, Thermodynamic considerations in realizing the primary standard of light. Metrologia **3**, 119–129 (1967)
142. C.L. Sanders, O.C. Jones, Problem of realizing the primary standard of light. J. Opt. Soc. Am. **52**, 731–746 (1962)
143. V.I. Sapritsky, B. Khlevnoi, B. Khromchenko, et al. High-temperature blackbody sources for precision radiometry. Proc. SPIE. **2815**, 2–10 (1996)
144. V.I. Sapritsky, B.B. Khlevnoy, V.B. Khromchenko et al., Precision blackbody sources for radiometric standards. Appl. Opt. **36**, 5403–5408 (1997)
145. V.I. Sapritsky, S.N. Mekhontsev, A.V. Prokhorov et al., Precision large area low and medium temperature blackbody sources. Proc. SPIE **3437**, 434–445 (1998)
146. V.I. Sapritsky, B.B. Khlevnoy, V.B. Khromchenko et al., Blackbody sources for the range 100 K to 3500 K for precision measurements in radiometry and radiation thermometry. AIP Conf. Proc. **684**, 619–624 (2003)
147. V.I. Sapritsky, B.B. Khlevnoy, S.A. Ogarev et al. New high-temperature variable-temperature and fixed-point blackbodies for precision measurements in radiation thermometry and radiometry. MĀPAN—J. Metrol. Soc. India. **20**, 193–204 (2005)
148. V.I. Sapritsky, B.B. Khlevnoy, S.A. Ogarev et al., Variable and fixed-point blackbody sources developed at VNIIOFI for precision measurements in radiometry and thermometry within 100 K…3500 K temperature range. Proc. SPIE **6297**, 629710 (2006)
149. V.I. Sapritsky, A.A. Burdakin, B.B. Khlevnoy et al., Metrological support for climatic time series of satellite radiometric data. J. Appl. Remote Sens. **3**, 033506 (2009)
150. V.I. Sapritsky, V.N. Krutikov, V.S. Ivanov et al., Current activity of Russia in measurement assurance of Earth optical observations. Metrologia **49**, S9–S16 (2012)
151. SBIR, *Vantablack®-S Infinity Series Infrared Blackbody Sources* (Santa Barbara Infrared, Inc. Santa Barbara, CA, 2017), https://www.sbir.com/datasheets/319-001-988_NC.pdf. Accessed 11 Feb 2020
152. S. Scheiding, H. Driescher, I. Walter et al., Compact blackbody calibration sources for in-flight calibration of spaceborne infrared instruments. Proc. SPIE **10563**, 105635P (2017)
153. J. Schreiber, T. Blumenstock, H. Fischer, Effects of the self-emission of an IR Fourier-transform spectrometer on measured absorption spectra. Appl. Opt. **35**, 6203–6209 (1996)
154. J. Schreiber, T. Blumenstock, F. Hase, Application of a radiometric calibration method to lunar Fourier transform IR spectra by using a liquid-nitrogen-cooled high-emissivity blackbody. Appl. Opt. **36**, 8168–8172 (1997)
155. P.E. Schumacher, A high-temperature circular-aperture blackbody radiation source. in ed. by S. Katzoff, *Symposium on Thermal Radiation of Solids. NASA SP-55 (Air Force ML-TDR-64–159),* (NASA, Washington, DC, 1965), pp. 233–240
156. SDL, *HAES15—High Accuracy Extended Source* (Space Dynamics Laboratory, Utah State University Research Foundation, North Logan, UT, 2014), https://www.sdl.usu.edu/downloads/haes15.pdf. Accessed 11 Feb 2020
157. M. Seifert, K. Anhalt, C. Baltruschat, S. Bonss, B. Brenner, Precise temperature calibration for laser heat treatment. J. Sens. Sens. Syst. **3**, 47–54 (2014)
158. R. Siegel, J.R. Howell, *Thermal Radiation Heat Transfer*, 4th edn. (Taylor & Francis, New York, 2002)
159. F.S. Simmons, A.G. DeBell, Q.S. Anderson, A 2000 °C slit-aperture blackbody source. Rev. Sci. Instrum. **32**, 1265–1266 (1961)
160. R.C. Simpson, H.C. McEvoy, G. Machin et al., In-field-of-view thermal image calibration system for medical thermography applications. Int. J. Thermophys. **29**, 1123–1130 (2008)
161. E.M. Sparrow, V.K. Jonsson, Absorption and emission characteristics of diffuse spherical enclosures. J. Heat Transfer **84**, 188–189 (1962)
162. E.M. Sparrow, S.H. Lin, Absorption of thermal radiation in a V-groove cavity. Int. J. Heat Mass Transfer **5**, 1111–1115 (1962)
163. R. Stair, R.G. Johnston, E.W. Halbach, Standard of spectral radiance for the region of 0.25 to 2.6 microns. J. Res. NBS. **64A**, 291–296 (1960)

164. E. Theocharous, N.P. Fox, V.I. Sapritsky et al., Absolute measurements of black-body emitted radiance. Metrologia **35**, 549–554 (1998)
165. Y. Té, P. Jeseck, C. Camy-Peyret et al., High emissivity blackbody for radiometric calibration near ambient temperature. Metrologia **40**, 24–30 (2003)
166. E.W. Treuenfels, Emissivity of isothermal cavities. J. Opt. Soc. Amer. **53**, 1162–1173 (1963)
167. E. Usadi, S. Bruce, N. Fox, et al. A novel black body for on-board calibration. Calcon 2004 CD-ROM (2004)
168. E. Usadi, Reflecting cavity blackbodies for radiometry. Metrologia **43**, S1–S5 (2006)
169. Virial, *PyramidA. Monte Carlo Modeling of Pyramid Array Radiation Characteristics. Manual* (Virial International, LLC, Gaithersburg, MD, 2018), https://www.virial.com/zip/PyramidA_Manual.zip. Accessed 10 Feb 2020
170. J.H. Walker, R.D. Saunders, A.T. Hattenburg, The NBS scale of spectral radiance. Metrologia **24**, 79–88 (1987)
171. J.H. Walker, R.D. Saunders, J.K. Jackson et al., The NBS scale of spectral irradiance. J. Res. Natl. Inst. Stand. Technol. **93**, 7–20 (1988)
172. J. Wang, Z. Yuan, X. Hao et al., A −30 °C to 80 °C stirred-liquid-bath-based blackbody source. Int. J. Thermophys. **36**, 1766–1774 (2015)
173. Z. Wei, G. Kuiming, X. Zhi, Development of an extended blackbody source for the calibration of infrared systems (I). Proc. SPIE **940**, 125–130 (1988)
174. Z. Wei, X. Zhi, G. Kuiming, Precise calculation of the integrated emissivity of a multiple-celled large-area blackbody source. Proc. SPIE **1311**, 428–436 (1990)
175. W. Wien, O. Lummer, Methode zur Prüfung des Strahlungsgesetzes absolut schwarzer Körper. Ann. Phys. **292**, 451–456 (1895)
176. B. Wilthan, L.M. Hanssen, S. Mekhontsev, Measurements of infrared spectral directional emittance at NIST—a status update, AIP Conf. Proc. **1552**, 746–751 (2013)
177. E.J. Wollack, R.E. Kinzer Jr., S.A. Rinehart, A cryogenic infrared calibration target. Rev. Sci. Instrum. **85**, 044707 (2014)
178. E.R. Woolliams, K. Anhalt, M. Ballico et al., Thermodynamic temperature assignment to the point of inflection of the melting curve of high-temperature fixed points. Phil. Trans. R. Soc. **374**, 20150044 (2016)
179. A.G. Worthing, The true temperature scale of tungsten and its emissive powers at incandescent temperatures. Phys. Rev. **X**, 377–394 (1917)
180. S.G. Wu, C. Zhong, M. Zheng, et al. An anti-frosting blackbody cavity below 0 °C. in *Proceedings of the 2015 Int. Conf. on Mechanics and Mechanical Engineering (MME2015)* (Chengdu, China, 2015). https://doi.org/10.1142/9789813145603_0079
181. H.Y. Yamada, A high-temperature blackbody radiation source. Appl. Opt. **6**, 357–358 (1967)
182. Yoon, H. W., Allen, D. W., Gibson, et al. Thermodynamic-temperature determinations of the Ag and Au freezing temperatures using a detector-based radiation thermometer. Appl. Opt. **46**, 2870–2880 (2007)
183. H. Yoon, P. Saunders, G. Machin et al.*Guide to the realization of the ITS-90. Radiation thermometry* (Consultative committee for thermometry under the auspices of the International Committee for Weights and Measures, 2018), https://www.bipm.org/utils/common/pdf/ITS-90/Guide_ITS-90_6_RadiationThermometry_2018.pdf. Accessed 10 Feb 2020
184. H. Zhang, J. Dai, Evaluation on effective emissivity of a surface with homocentric V-grooves by Monte-Carlo method. in *Proceedings Asia Simulation Conf./6th Int. Conf. on System Simulation and Sci. Computing*, vol. 1, (Int. Acad. Publishers/World Publ. Corp., Beijing, 2005), pp. 124–128
185. H. Zhang, J.M. Dai, X.G. Sun, Research on a radiant source for infrared image calibration. J. Phys. Conf. Series. **48**, 1053–1057 (2006)
186. H. Zhang, J. Dai, A novel radiant source for infrared calibration by using a grooved surface. Chin. Opt. Let. **4**, 306–308 (2006)
187. J. Zhang, Calibration facilities for industrial radiation thermometers at SIPAI, in *Temperature. Its Measurement and Control in Science and Industry*, vol. 6, part 2, ed. by J.F. Schooley (Am. Inst. Phys., New York, 1992), pp. 781–784

188. Z. Zhimin, A method for calculating the effective emissivity of a groove structure. Proc. SPIE **810**, 270–277 (1987)
189. Y. Zhu, H. Ma, R. Wang et al., −50 to + 150 °C heat pipe blackbody sources for radiation thermometer calibration. in *Temperature. Its Measurement and Control in Science and Industry*, vol. 5, part 1, ed. by J.F. Schooley (Am. Inst. Phys., New York, 1982), pp. 559–564
190. R.B. Zipin, The apparent thermal radiation properties of an isothermal V-groove with specularly reflecting walls. J. Res. NBS **70C**, 275–280 (1966)
191. A.V. Zuev, V.A. Chistyakov, V.A. Rozhkov, A cylindrical model of an absolutely black body of lanthanum chromite. Meas. Tech. **53**, 70–73 (2010)

Chapter 6
Materials for Blackbody Radiators

Abstract The most common types of materials used in the design of radiating elements of blackbodies are considered in this chapter. Along with traditional high-emissivity paints, the carbon nanotube arrays are examined as promising coatings for low-temperature blackbodies. For variable-temperature blackbodies of medium-temperature range, the radiation properties of some oxidized metals are reviewed. Thermophysical and radiation characteristics of ordinary synthetic graphite and pyrolytic graphite are discussed in connection with their use in high-temperature blackbodies with the direct resistance heating.

Keywords High emissivity · Black paint · Carbon nanotubes · Oxidized metal · Graphite · Pyrolytic graphite

6.1 Preliminary Remarks

6.1.1 Principles of Materials Selection

The choice of material for the radiating surface is an integral step in the design of any blackbody. All materials used in the design of blackbody radiators can be roughly subdivided into several groups. The radiating surfaces of blackbodies operating from cryogenic temperatures to about 200 °C are typically layers of paint or coating with high emissivity, applied to a metal substrate that forms a radiating element (e.g., a cavity). There are paints suitable for operating up to 1000 °C but the usual materials for medium-temperature blackbodies are metals and alloys, the surface of which is oxidized to enhance its emissivity. The blackbodies with the direct resistance heating, which are the most widespread type of blackbodies for the highest temperatures, the dominant position belongs to graphite (ordinary graphite for 1500 °C to about 2500 °C and pyrolytic graphite up to 3500 °C). A separate, new, and little explored group consists of so-called carbon nanotube coatings. The choice of material in each group is carried out in accordance with three matter-of-course criteria common to all materials:

V. Sapritsky and A. Prokhorov, *Blackbody Radiometry*, Springer Series in Measurement Science and Technology, https://doi.org/10.1007/978-3-030-57789-6_6

- the material must withstand a temperature within the operating range of the blackbody and in the working medium (air, vacuum, inert gas) for a time sufficient to perform all necessary measurements;
- the radiation properties of the material should provide an effective emissivity close to unity for an isothermal radiating element in the spectral range of interest for any operating temperature;
- the radiation properties of the material must not change during measurements and should have an admissible long-term stability.

Usually, these criteria are enough to select material at the initial stage of designing. Reference books and other resources provide some indicative data on the individual material radiative properties, which can be used for a crude estimate of the effective emissivity. At the next stage, the deeper development of the blackbody structure, including the choice of the radiator configuration and approximate thermophysical modeling, should be made. To choose the shape of the radiating element and determine its geometrical parameters, the effective emissivity calculation has to be performed within the framework of the adopted model of radiation characteristics. For this, the radiation characteristics should be known to the needed extent and with an accuracy that allow calculating the effective emissivity in and with a given uncertainty. For modeling the steady-state temperature distributions and dynamic characteristics in this stage of designing blackbodies, the thermophysical characteristics of materials in the temperature range of interest are needed. Since the high accuracy is not required at this stage, the thermophysical characteristics, as a rule, can be adopted from reference literature.

The final stage is the metrological investigation of the finished blackbody. If the blackbody will be used as a primary standard, precise calculation of the blackbody effective emissivity and determination of its uncertainty are necessary. The calculations must be performed with the highest available accuracy; therefore, the most accurate and reliable data on radiation characteristics for the material of the radiating element will be needed. At this stage, literature data are usually not sufficient by the following reasons:

- Information on the material, its measured characteristics, and measurement conditions may be incomplete.
- The spectral and temperature ranges, for which the radiation characteristics are given in the special literature, can and, as a rule, do not coincide with the required. Interpolation, or even more extrapolation, can be a source of hardly detectable errors.
- There are many uncontrollable technological parameters often not provided in publications but affecting the radiation characteristics of materials. For this reason, published data on seemingly the same material often have a huge scattering or inconsistency. The probability that nondisclosed parameters will coincide with those corresponding to a specific material under investigation is extremely small. For instance, a method for applying a paint coating to a substrate (by brushing or spraying), a substrate material, a substrate treatment method, the presence or

absence of a primer, etc. is often omitted when published but can have a significant effect on the radiation characteristics of the coating.

The radiation characteristics of materials depend on too many parameters and may vary within wide ranges for different samples of the same material. Chemical composition, manufacturing technology, roughness of the surface—this is far from a complete list of factors that can affect the radiative characteristics of materials and coatings. Therefore, the information taken from the literature can be used with caution. Typically, an individual study of the radiation characteristics of the same material as in the design of a specific blackbody is necessary to obtain a more reliable input for accurate calculation of the effective emissivity. For this reason, all the graphs in this chapter serve for illustrative purposes only. They can be used as a guide for the choice of material for radiating surface but not for precise calculations.

For the most responsible application, an individual characterization of materials is required. That is why the results given in this chapter serve mainly for illustrative purposes, allowing the readers to navigate in the choice of materials, and can be used only at the initial stage of designing blackbodies. At the final stage of the blackbody development, when the individual characterizing the radiator's materials is required, the best solution is to prepare the witness sample and to perform a thorough study of their radiation characteristics necessary for calculations. The witness sample, as its name implies, is the representative sample of material or coating (for coatings, it is often called "coupon") that is fabricated in exactly the same conditions as the radiating element of the blackbody. Witness samples allow their investigation without handling the actual blackbody radiator. They can easily be transported to a laboratory that specializes in the type of measurements required.

6.1.2 Availability of Radiation Characteristics Data

Before we discuss the availability of data on the radiation characteristics of the materials traditionally used or promising for radiating elements of blackbodies, we need to make some important remarks.

The spectroradiometry is the central part of the modern optical radiometry. It rarely deals with the total and band-limited radiometric quantities, mainly when working with low-level IR radiation from low-temperature radiation sources. For realization and dissemination of spectral radiometric quantities by means of blackbodies, it is necessary to know *spectral* effective emissivities of those blackbodies, as a rule, within a continuous spectral range. In reality, it can be a dense set of discrete wavelengths. It is understood that this set is *sufficiently* dense to allow linear interpolation between adjacent points without loss of accuracy. An example of an insufficiently dense regular wavelength set is presented in Fig. 6.1a. Adding a number of points allows the use of linear interpolation to reproduce all the details of the original curve as it is shown in Fig. 6.1b. Instead of regular set of wavelengths with a smaller

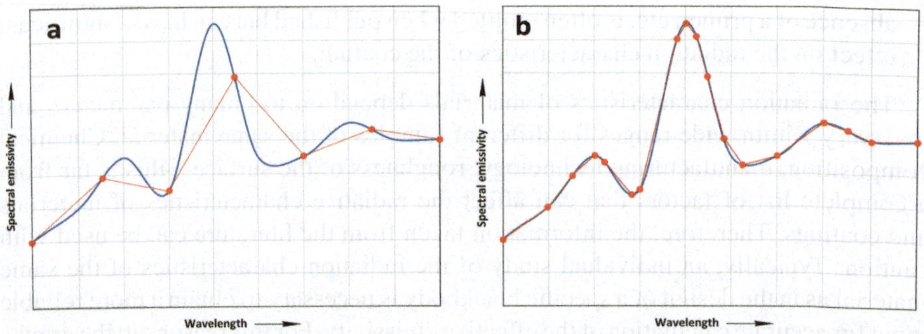

Fig. 6.1 Linear interpolation of the spectral emissivity for: **a**, insufficiently dense regular set of wavelengths and **b**, sufficiently dense irregular wavelength set. Interpolation nodes are indicated by red points

increment, an irregular wavelength set is used to decrease the total number of interpolation nodes. Comparing the results of measurements carried out with instruments having different spectral resolutions makes it possible to control the correctness of the choice of interpolation nodes. The noise component of high-resolution measurements (usually performed using Fourier transform spectrometers) can be eliminated by numerical filtering or smoothing.

A substantial part of published emissivity-related data refers to integral (total) rather than spectral values, while spectral measurements are more typical for the reflection characteristics. The vast majority of reflectance measurements are performed near room temperature, while the emissivity measurements are most often carried out at elevated or high temperatures.

The unknown temperature dependences of the radiation characteristics can be a source of uncertainty if they are measured at a temperature different from the temperature at which the blackbody is operated. The question of the correctness of extrapolation of the results obtained, for example, at room temperatures, to a higher (or alternatively, lower) temperature, has not yet been solved in general form. However, a number of observations suggest that if the material does not change the chemical composition during heating or cooling (due to, for example, oxidation in air or gas release in a vacuum) and does not change the surface state (e.g., due to temperature erosion or sublimation), then the temperature changes of radiation characteristics are small and reversible.

Many published data are concerned to the total emissivity, which cannot be converted to the spectral emissivity. However, we can obtain certain qualitative information from the temperature dependence of the total emissivity (see [101]).

For our purposes, both the emissivity and reflectance data of opaque materials can be of interest. The Kirchhoff law and the reciprocity principle establish relationships between these data. The directional emissivity can be derived from the directional-hemispherical reflectance (i.e. the ratio of reflected flux collected over the entire hemisphere to essentially collimated incident flux). However, the transformation of the directional emissivity to the hemispherical-directional reflectance (i.e. the ratio of reflected flux collected over an element of solid angle surrounding the given direction to the incident flux from the entire hemisphere) is useless for us.

The basic methods and techniques that have been developed for measurement of different radiation characteristics of opaque surfaces at large ranges of temperatures and other affecting factors were reviewed by DeWitt and Richmond [48]. The overview of recent developments can be found in Höpe [87] and Watanabe et al. [228]. Since the 1950s, a huge volume of measurement data on radiation characteristics has been accumulated. These data are extremely heterogeneous. The most successful attempt to consolidate them was made by Touloukian and DeWitt [214, 215, 216]. Up to date, this work continues to be an important source of reference information on radiation characteristics of metals, alloys, some nonmetals and coatings.

The theoretical approach to predicting the emissive and reflective properties of materials (except for Fresnel reflection from polished metals) used in designing the blackbody radiators is unproductive since it implies a high degree of idealization, in which most of the factors affecting the characteristics of real surfaces, remains outside the scope of consideration. However, understanding the features of the interaction of optical radiation with opaque, transparent and translucent bodies, at least at the phenomenological level, is highly desirable for reference data interpretation, interpolation, and extrapolation, as well as for selection of the reflection model suitable for the effective emissivity calculation. We will not specifically address theoretical issues, but we strongly encourage readers at least briefly familiarize themselves with the relevant chapters in Modest [142] and Howell et al. [88], as well as the monograph by Hapke [75]. The reflection model chosen for the effective emissivity calculation determines the type of data that we need, while the Kirchhoff law and the reciprocity principle determine how to obtain (if possible) them from the data we usually have.

The simplest model of reflection is the uniform specular model. It implies that all radiation reflected from an opaque surface is concentrated in the infinitely small solid angle around the direction of specular reflection. The uniform specular model assumes independence of reflectance upon the angle of incidence and, as the Kirchhoff law and the reciprocity principle dictate, the angle-independent emissivity. This means that thermal emission of such a surface obeys the Lambert law. A real surface has reflectance that depends on the incidence angle. If to consider such a surface, we must also consider non-Lambertian emission.

The Fresnel law makes it possible to predict the angular dependence of the reflectivity of optically smooth surfaces, for which it is sufficient to know the spectral refractive index $n(\lambda)$ and the spectral absorption index $k(\lambda)$ (the imaginary part of the complex refractive index often called the is called the extinction coefficient) at a wavelength λ of interest. For unpolarized incident radiation, the reflectance ρ at the interface of a vacuum and a medium with a refractive index n and an absorption index k can be written in the compact form [193] as

$$\rho(u) = \frac{1}{2}\frac{(a-u)^2+b^2}{(a+u)^2+b^2}\left[\frac{(a+u-1/u)^2}{(a-u+1/u)^2}+1\right], \tag{6.1}$$

where $u = \cos\theta_i$, θ_i is the angle of incidence,

$$a^2 = \frac{1}{2}\left(\sqrt{\left(n^2 - k^2 + u^2 - 1\right)^2 + 4n^2k^2} + n^2 - k^2 + u^2 - 1\right), \quad (6.2)$$

$$b^2 = \frac{1}{2}\left(\sqrt{\left(n^2 - k^2 + u^2 - 1\right)^2 + 4n^2k^2} - n^2 + k^2 - u^2 + 1\right). \quad (6.3)$$

The wavelength dependences are omitted in Eqs. (6.1) to (6.3) for brief. The spectral directional emissivity ε of an opaque surface can be expressed via the spectral directional reflectance ρ using the Kirchhoff law:

$$\varepsilon(\theta_v) = 1 - \rho(\theta_i), \quad (6.4)$$

where θ_v is the viewing (observation) angle and $\theta_i = \theta_v$ in accordance with the reciprocity principle.

There is voluminous literature containing optical constants n and k data for various materials from far UV to far IR (see, e.g. [1, 2, 158]). As an example, Fig. 6.2 presents the dependences of reflectance on the incidence angle at the wavelength of 1.0 μm calculated according to Eqs. (6.1) to (6.3) for silver, aluminum, gold, and tungsten.

The Fresnel law gives satisfactory results for optically smooth pure substances but it fails for rough or oxidized metals and for dielectrics with noticeable subsurface scattering. Besides, for the spectral range that we are interested in, the optical constants are mainly derived from the angular dependences of the reflectance at room temperature, so the data on the spectral reflectance or the spectral emissivity

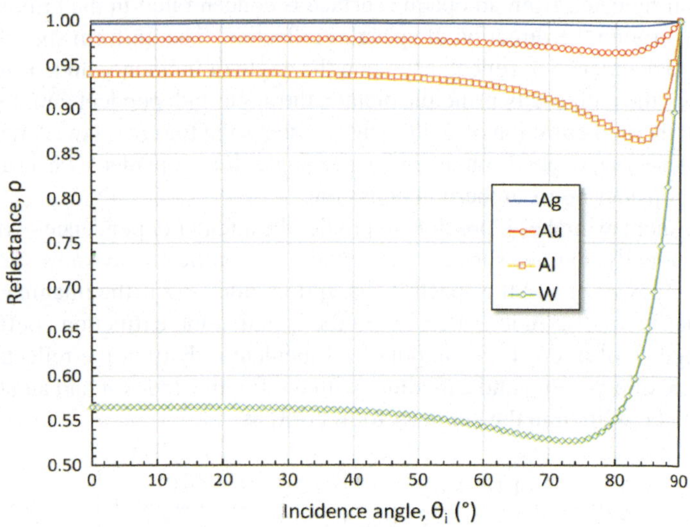

Fig. 6.2 Fresnelian reflectance of four metals at 1.0 μm

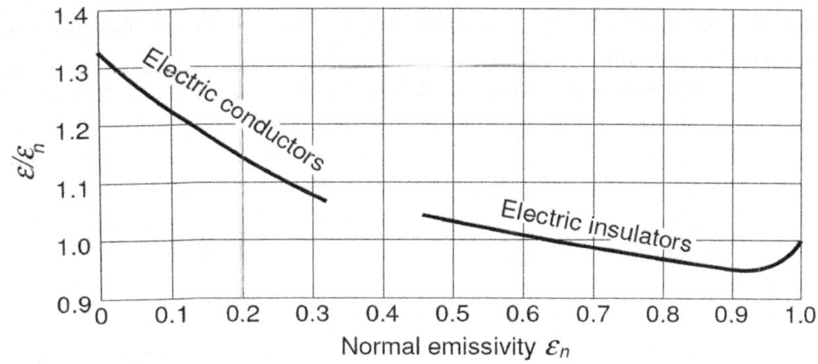

Fig. 6.3 Ratio of hemispherical to normal emissivity of radiating surfaces. Reproduced from [222] with permission of Springer Nature

for the desired temperature may be easier to access than those for $n(\lambda)$ and $k(\lambda)$. However, if only the value of the hemispherical emissivity ε of an optically smooth surface is known, the Fresnel law allows us to evaluate its normal emissivity by successive application of the Kirchhoff law and the integration over the hemispherical solid angle. The theoretical ratios of hemispherical to normal emissivity for a typical conductive material and a typical dielectric are plotted in Fig. 6.3.

In some cases, especially when the model of radiation characteristics should be built on the base of a limited amount of experimental data, the use of the Schlick approximation [193] of the Fresnel equation for unpolarized radiation might be preferable because it allows expressing the directional reflectance of a specularly reflecting surface via the normal reflectance $\rho(0)$:

$$\rho(\theta_i) = \rho(0) + [1 - \rho(0)] \cdot (1 - \cos\theta_i)^5. \tag{6.5}$$

Generally, the specular model of reflection should apply to calculations of the effective emissivities with caution. In its pure form, the specular model can be used mainly for polished metals, e.g., in tungsten cavities or in specular enclosures. When there is radiation scattering (subsurface and/or due to surface roughness), more complex reflection models should be used.

The use of specular model of reflection may lead to overestimating of the effective emissivity.

Geometrical ray tracing shows that the conical cavity with the apex angle of 30° provides 6 successive specular reflections for each ray escaping the cavity in a direction parallel to its axis. Within framework of the specular model, the normal effective emissivity of such an isothermal cavity is expressed as $\varepsilon_e = 1 - (1 - \varepsilon)^6$, where ε is the emissivity of the cavity wall.

If $\varepsilon = 0.9$ (the typical value for some "specular black" paints), calculation according to Eq. (6.6) gives $\varepsilon_e = 0.999999$, while the experimental values derived from measurements of the effective reflectance give much less values. This might be explained by both presence of small amount of diffusely scattered radiation in each reflection and by the finite width of the specular lobe.

A perfectly diffuse or Lambertian model of reflection, commonly referred to as a diffuse model, is the most popular model used to describe non-specular reflection over many years. Due to its simplicity, it is applied to surfaces that scatter incident radiation irrespective of the nature of this scattering. If there is no pronounced direction of the preferred scattering (usually around the direction of specular reflection), or if the scattering is represented by a very wide lobe, the diffuse model usually gives adequate results.

The minimalist data requirements that we need to use a diffuse model are the directional-hemispherical reflectance $\rho(\theta_i)$ at any incidence angle θ_i or of the directional emissivity $\varepsilon(\theta_v)$ at any viewing angle θ_v. A rationale for applying the diffuse model may be a weak angular dependence of the directional-hemispherical reflectance or of the directional emissivity. It is necessary to take into account the fact that a surface that behaves as diffuse at shorter wavelengths will most likely not be diffuse for longer wavelengths.

The specular-diffuse model of reflection is a combination of two above-mentioned models. It is commonly used for parametric investigation of the dependence of the effective emissivity on the diffuse and specular reflection components of surfaces forming the radiating elements of blackbody radiators and optimization of their geometry. In order to assign the effective emissivity to a blackbody radiator, the parameters of the specular-diffuse model must be established. In addition to the spectral directional-hemispherical reflectance $\rho(\lambda)$ (supposed to be independent of the incidence angle) or the spectral directional or hemispherical emissivity $\varepsilon(\lambda) = 1 - \rho(\lambda)$ (supposed to be independent of the viewing angle), application of the specular-diffuse model requires knowledge of the diffusity D, which expresses the fraction of diffusely reflected radiation. The problem of subdividing the reflection indicatrix (BRDF in modern terminology) into diffuse and specular components was posed by Sarofim and Hottel [191] but does not have a satisfactory solution until the present time. Figure 6.4 illustrates the principle of a subdivision without indication of its quantitative criterion. Instrumental (experimental) separation of components also does not have a unique solution. Its idea is in measuring the radiation reflected into hemispherical solid angle and repeating measurements, but with the excluded specular component.

Both measurements are carried out at a fixed angle of incidence resulting in the hemispherical reflectance ρ and the diffuse reflectance ρ_d values. In the visible and NIR spectral regions, the integrating sphere is the most widely used instrument for measuring the reflectance. The scheme for excluding the specular component at measurements using the integrating sphere is presented in Fig. 6.5. For simplicity, measurements performed are shown for the relative method based on comparison of the radiation detector signals when the specimen under investigation and the diffuse reference are subsequently irradiated with incident radiation. The baffle prevents direct irradiation of a detector by reflected radiation. The specularly reflected radiation can be brought out of the sphere through a special port made in its wall to be completely absorbed in a radiation trap. The specular component of the reflected

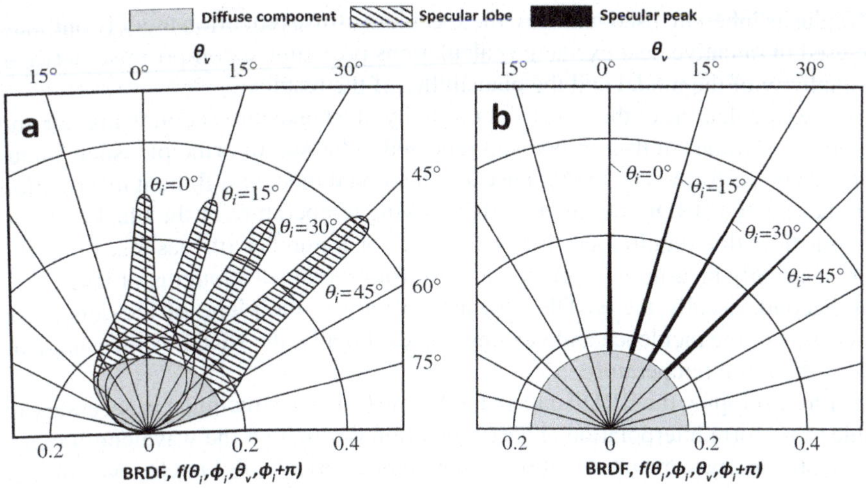

Fig. 6.4 Conversion of the 3D BRDF into the equivalent specular-diffuse model: **a** the in-plane BRDF and **b** its specular-diffuse approximation. Reproduced from [142] with permission of Elsevier

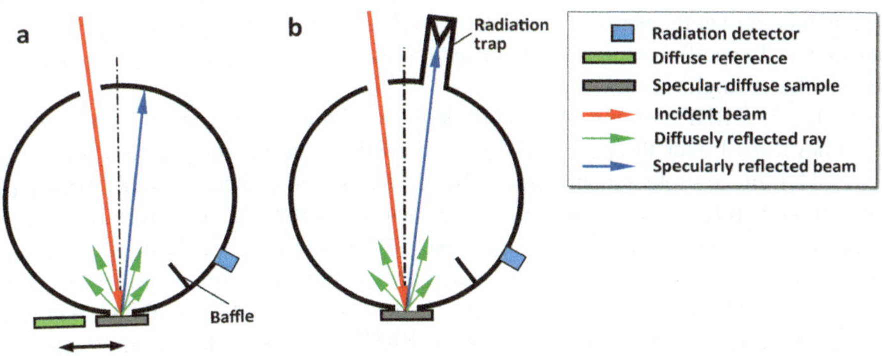

Fig. 6.5 Diagram explaining measurements of the directional-hemispherical reflectance (**a**) and its diffuse component (**b**) using the integrating sphere

flux is included to determine the hemispherical reflectance (Fig. 6.5a) or excluded to measure only the diffuse component (Fig. 6.5b). Obviously, the fraction of reflected radiation flux passing through the specular exclusion port depends on the ratio of the diameters of the port and the sphere, as well as on the angular width of the specular lobe or peak. Therefore, unique separation of the specular components is possible only for a narrow reflection lobe (or, better to say, the peak). For wider lobes, the result of such a separation depends on geometry of the measuring instrument. The foregoing applies not only to integrating spheres, but also to other reflectometric devices (see, e.g. [39–41, 52, 150]).

Despite its inherent disadvantages, the specular-diffuse reflection model continues to be used in radiative heat exchange calculations providing a compromise between the simplicity of the model and the plausibility of the results.

Precise calculations of the effective emissivity must take into account angular and spectral distributions of the emitted and reflected radiation. In principle, such calculations can be performed by the Monte Carlo method if there is sufficient information about the spectral BRDF measured at the working temperature of the blackbody.

In practice, this requirement is not feasible for a number of reasons. First, the BRDF is usually measured for one or, at best, for several wavelengths of laser radiation. Second, usually the BRDF is measured only in the plane of incidence, for discrete sets of the incidence and viewing angles. Finally, the BRDF is measured, as a rule at the room temperature.

In order to employ the BRDF data in the Monte Carlo ray tracing, it is necessary to be able to perform interpolation and extrapolation throughout the wavelength range of interest, for any viewing direction in the hemispherical solid angle, and for any incidence angle from 0 to 90°. The interpolant (the BRDF model) must be physically based to avoid obtaining solutions that may violate the energy conservation law and the reciprocity rule. The BRDF model may depend on a number of adjustable parameters, which allow the model to be fitted to measured data. Integration of the fitted BRDF model over the hemispherical solid angle should result in physically plausible and material-specific angular dependence of directional-hemispherical reflectance on the incidence angle.

Different BRDF models of varying degrees of complexity and adequacy have been developed over the past few decades mostly in computer graphics (see, e.g. [68, 144]) and remote sensing (see [107, 199], and references therein). Many of these models refer to scattering of a certain nature: scattering by surface roughness or subsurface scattering. Other models (like the Lambertian model) are not relevant to any particular physical phenomenon, and are merely mathematical functions. The additivity of BRDF allows us to combine models extending their application areas.

The common strategy of application of BRDF models to the Monte Carlo ray tracing is the analysis of all the data available, building the physically plausible BRDF model depending upon several numerical parameters, performing the fitting of the model to the experimental data and deriving the optimal values of those parameters. One of the possible models, the three-component (3C) model of reflection, was proposed by Prokhorov [170]. This is a moderately complicated model developed especially for fast generation of rays in the Monte Carlo ray tracing. The 3C BRDF model is the sum of diffuse (Lambertian), glossy, and quasi-specular components. The name of the first component is easily understandable. The glossy component represents scattering on surface roughness but can be applied to forward scattering of physically different nature such as subsurface scattering in partially transparent layers of metal oxides or paints. The model contains 8 fitting parameters: the weights k_d, k_g, and k_{qs} of the diffuse, glossy, and quasi-specular component, respectively; the partial reflectances R_d, R_g, and R_{qs} of these components, and the width parameters σ_g and σ_{qs} of the glossy and quasi-specular component, respectively. The 3C model was discussed in more detail in Sect. 3.4.2. Two examples of 3C models fitted to the in-plane BRDFs of the rough graphite sample measured at the wavelength of 0.658 μm and Z306 black paint at 10.6 μm are shown in Fig. 6.6. The metrics $L2$ and C were used for the fitting for graphite and Z306 fitting, respectively. The fitted parameters are presented in Table 6.1. Figure 6.7 shows the dependences of the directional-hemispherical reflectance (DHR) on the incidence angle for the 3C models with fitted parameters. Graphite sample exhibits reflection behavior typical for randomly rough surfaces.

The glossy component is responsible for single reflection from rough surface (the normal reflectance $\rho_g(0°) \approx 0.0236$); the diffuse component approximately

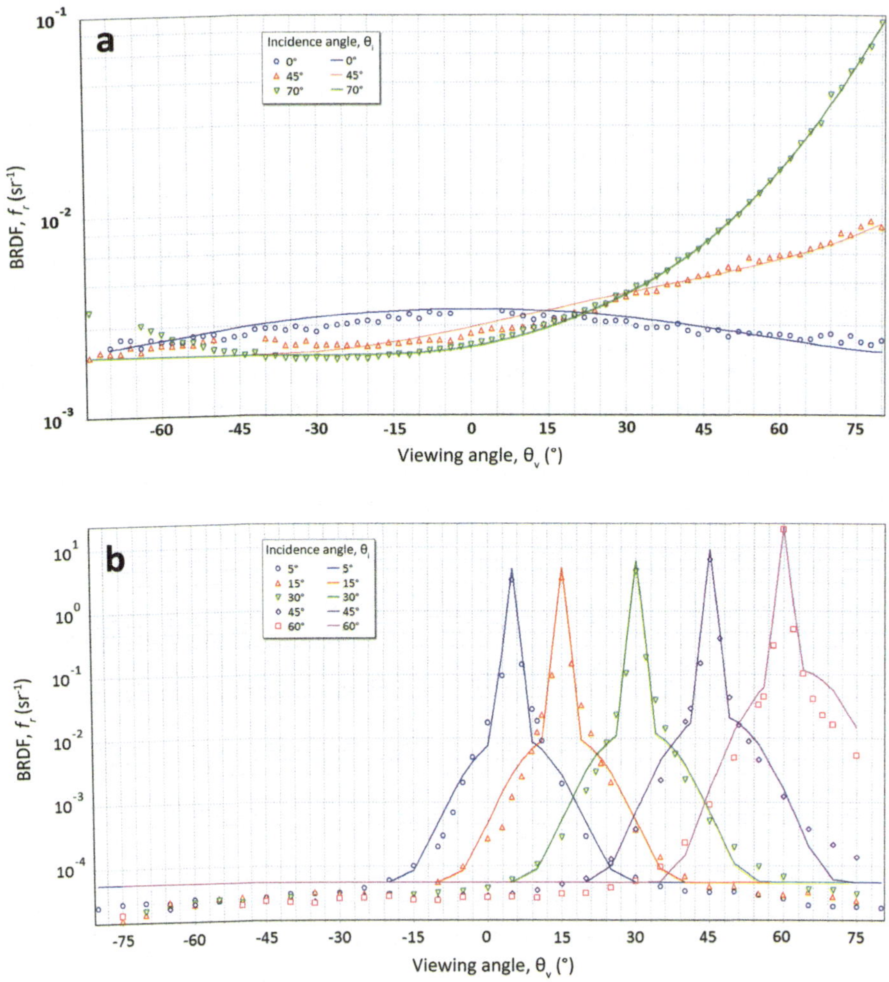

Fig. 6.6 Results of fitting the 3C BRDF model to the BRDF data measured for: **a**, rough graphite at the wavelength of 0.658 μm and **b**, Z306 paint at the wavelength of 10.6 μm

Table 6.1 Parameters of the 3C models of BRDF fitted to graphite (Fig. 6.6a) and Z306 paint (Fig. 6.6b) data

Component	Model parameter	Graphite at 0.658 μm	Z306 at 10.6 μm
Diffuse	k_d	0.5626	0.3453
	R_d	0.1116	0.0050
Glossy	k_g	0.4374	0.3453
	R_g	0.06387992537	0.0206
	σ_g	0.39640799245	0.0723
Quasi-specular	k_{qs}	0	0.3093
	R_{qs}	-	0.1858
	σ_{qs}	-	0.0101
Metric		*L*2	*C*

models multiple reflections (the corresponding reflectance $\rho_d = k_d \cdot R_d \approx 0.0636$); the quasi-specular component is absent. The normal emissivity of graphite sample $\varepsilon(0°) = 1 - \rho(0°) \approx 0.9135$.

The reflection from the sample of the Z306 paint has more complicated nature. The largest component is the quasi-specular that models the reflection from relatively smooth air-paint interface. Since there is no off-specular shift [212] of the maximum of the glossy reflection lobe, the glossy component has nothing to do with scattering on a rough surface but is due to subsurface forward scattering on the shallow scatterers. A small diffuse component may express volumetric scattering on the deeper scatterers.

The 3C model, parameters of which are defined in this way, can be used to calculate the effective emissivities at the same wavelengths for which the fitting was carried out. To extend the model to a continuous wavelength range, Prokhorov and Prokhorova [171] proposed a procedure that uses the spectral hemispherical reflectance to obtain the wavelength-dependent parameters of the 3C model.

The basic assumption underlying this procedure is the preservation of the shape of the BRDF throughout the spectral range of interest. The experimental angular dependences of the spectral directional-hemispherical reflectance can serve for validation of the model built. It should be noted that the 3C BRDF model is not the only possible model that uses measured BRDFs. However, the minimum set of experimental data remains unchanged: the in-plane BRDF measured at one or more wavelengths and the spectral DHR measured for one or more angles of incidence.

6.1.3 Required Accuracy of Emissivity Data

We will try to answer a simple question: how big can the uncertainty in determining the radiation characteristics of a material be so that the uncertainty in calculating the effective emissivity of the blackbody radiator made of this material is less than

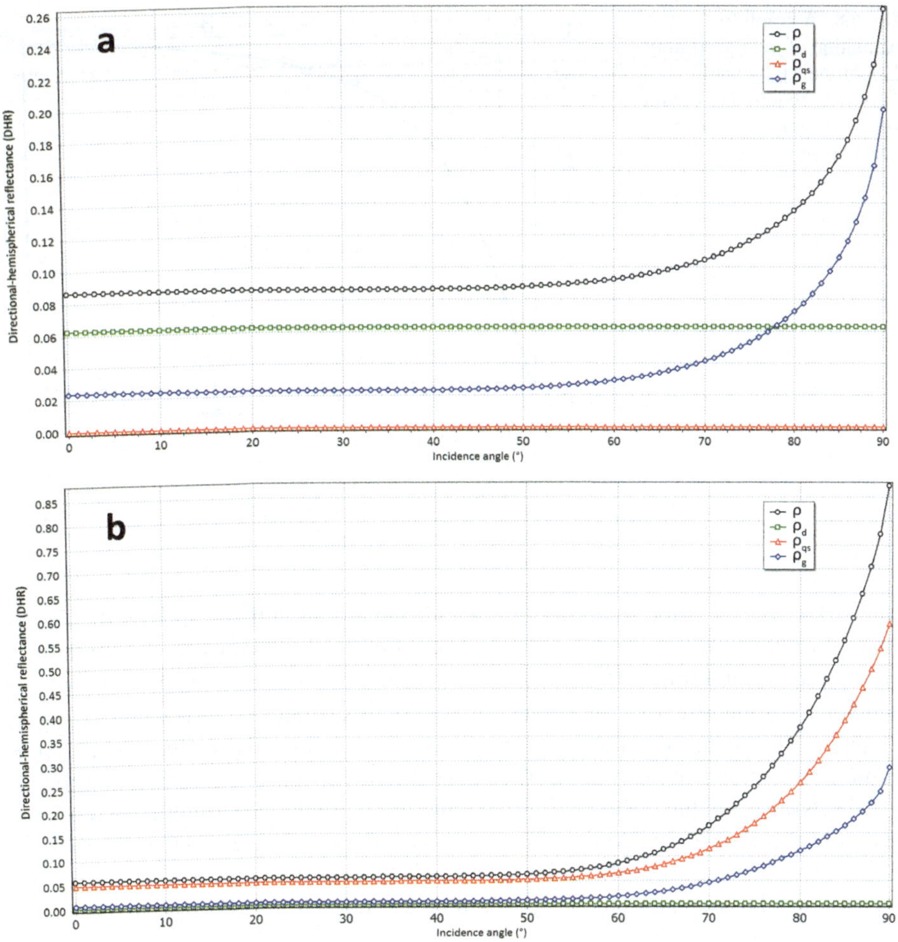

Fig. 6.7 Results of calculation of the DHR for the 3C BRDF model of: **a**, rough graphite at the wavelength of 0.658 μm and **b**, Z306 paint at the wavelength of 10.6 μm

a given value? Strictly speaking, it is impossible to separate the thermal, optical, and geometric effects in the consideration of this question. Since the adequacy of the mathematical model used for calculations has a decisive impact on dependences of the effective emissivity upon various affecting factors, we must accept the most general model. Let us consider an isothermal cavity whose inner walls are made of the same material characterized by the emissivity ε. By applying the generalized Kirchhoff law [111], we can express the effective emissivity ε_e through the effective reflectance ρ_e, which, in turn, can be expressed as a series, regardless of the BRDF of cavity walls:

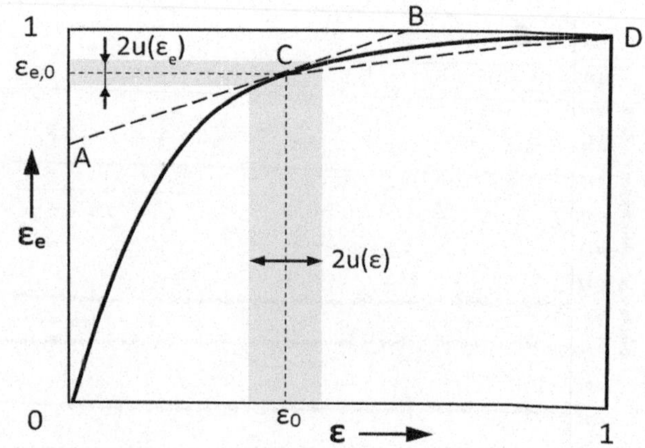

Fig. 6.8 A graph for evaluation of the uncertainty $\Delta\varepsilon_0$ of the cavity wall emissivity required to obtain the uncertainty $\Delta\varepsilon_{e,0}$ of the effective emissivity

$$\varepsilon_e = 1 - \rho_e = 1 - \sum_{i=1}^{\infty} f_i \rho^i = 1 - \sum_{i=1}^{\infty} f_i (1 - \varepsilon)^i, \tag{6.6}$$

where $0 < f_i \leq 1$ is the fraction of the incidence radiation flux escaping the cavity after i-th reflection ($i = 1, 2, \ldots$). According to the reciprocity principle, the viewing direction for the effective emissivity must coincide with the direction of the incidence for the effective reflectance. It is easy to see that $\frac{d\varepsilon_e}{d\varepsilon} \geq 0$ while $\frac{d^2\varepsilon_e}{d\varepsilon^2} \leq 0$ for any ε. This means that the function $\varepsilon_e(\varepsilon)$ is convex upwards and has no inflection points as it is shown in Fig. 6.8. Let C be the point on the curve $\varepsilon_e(\varepsilon)$ corresponding to the abscissa ε_0 and the ordinate $\varepsilon_{e,0}$; $u(\varepsilon)$ and $u(\varepsilon_e)$ are the uncertainties in determination of the emissivity and the effective emissivity, respectively. AC is the tangent line to the curve $\varepsilon_e(\varepsilon)$ at the point C. By substituting the tangent line AB for the secant line CD, we can write the approximate equation for the derivative:

$$\frac{d\varepsilon_e}{d\varepsilon} \approx \frac{1 - u(\varepsilon_e)}{1 - u(\varepsilon)}. \tag{6.7}$$

After replacing the differentials $d\varepsilon_e$ and $d\varepsilon$ by finite differences $u(\varepsilon_e)$ and $u(\varepsilon)$, respectively, we obtain:

$$u(\varepsilon_e) \approx \frac{1 - \varepsilon_e}{1 - \varepsilon} u(\varepsilon) \tag{6.8}$$

or

$$u(\varepsilon) \approx \frac{1 - \varepsilon}{1 - \varepsilon_e} u(\varepsilon_e). \tag{6.9}$$

Equation 6.8 (in other notation) appeared in the CCT documents [60, 61] without indication of the way of its derivation. Although Eq. 6.9 indicates only the lower limit of uncertainty, this is usually sufficient for most practical cases. For instance, if $\varepsilon = 0.85$ and $\varepsilon_e = 0.995$, in order to achieve $u(\varepsilon_e) = 0.001$, the uncertainty $u(\varepsilon)$ has to be about 0.03 (relative uncertainty $u(\varepsilon)/\varepsilon \approx 3.5\%$).

The state-of-the-art level of measurement uncertainties for the spectral directional emissivity and spectral directional-hemispherical reflectance of diffuse samples is a few tenths of a percent. The spectral reflectance of near-specular samples can be measured with an uncertainty of less than 0.1%. Thus, the uncertainty in the measurement of the emissivity of the cavity walls does not make a decisive contribution to the uncertainty of calculating the effective emissivity of the blackbody radiator. Nothing definite can be said about influence of uncertainties in measured BRDFs of the cavity walls material on the uncertainties of the effective emissivity calculation without numerical modeling performed for each specific case. Moreover, to obtain a reliable evaluation of the uncertainty in the effective emissivity determination, it is necessary to perform statistical modeling of the propagation of probability distributions of all the affecting variables according to [100]. An example of this complex strategy is given by De Lucas and Segovia [44] for diffuse cylindro-conical blackbody cavities.

6.2 Black Paints and Coatings

So-called "black" paints are conventional materials used for radiating surfaces of blackbodies operating at temperatures from cryogenic to approximately 200 °C; some paints are capable to withstand higher temperatures. The adjective "black," originally referred to materials with low reflectance in the visible part of the optical spectrum, has later been spread to invisible radiation. Thus, a "black" material means a material that has a low reflectance (and, correspondingly, a high absorptance and high emissivity) within a spectral range of interest.

"Black" paints and coatings have long been used as absorbing layers in thermal detectors of optical radiation and solar power collectors, for the stray light suppression in optical systems, and as radiating surfaces of space heat exchangers. Further, we will use the term "black" without quotation marks.

A paint is a dispersion of tiny insoluble pigment particles suspended in the macromolecular organic film-forming agent dissolved or dispersed in a solvent medium called "binder." The most common binders are acrylic, alkyd, and epoxy polymers. Paints can be applied by simple techniques such as brush, roll, dip, and spray coating. The variations of the surface roughness and paint layer thickness depend on the application technique. The size of pigment particle, their clustering, concentration of pigment, thickness of the paint layer, and the paint application technique affect the radiation property of painted surfaces.

A variety of black paints used in low-temperature radiating cavities is similar to those used as absorbing coatings of the thermal sensors of optical radiation. These paints may have BRDFs ranged from near-specular to near-diffuse depending on chemical composition, layer microstructure and thickness, substrate material, and so forth. Most of the early optical radiation detectors were designed to measure visible and NIR radiation; therefore, it should be borne in mind that materials with high emissivity in the visible and NIR spectral ranges may have low emissivity for longer wavelengths and vice versa. In a like manner, when we talk about a "diffuse" or "specular" material, the spectral range implied has to be specified because most of the surfaces reflecting visible light diffusely are converted into almost specular reflectors with an increase of the wavelength of the incident radiation. Specular character of the reflection is not an obstacle to the use of material in the construction of blackbody radiators. On the contrary, specular black paints look especially attractive due to seeming simplicity of obtaining very high effective emissivities. Even if a possible imperfectness of the specular reflection and a dependence of the paint reflectance on the incidence angle are taken into account, the effective emissivity remains very high. Fortunately, a high spectral emissivity of black paints (typically, greater than 0.95 for the near and middle IR spectral ranges) makes calculation of the effective emissivities for low-temperature blackbody cavities not too sensitive to a reflection model adopted.

An alternative for black paints is black coatings that can be applied to a solid surface using a large number of techniques such as electrodeposition, physical vacuum deposition, and chemical vacuum deposition. The thicknesses of the deposited layers can vary from monoatomic to submillimeter. It is known that ultrafine metal particles produced by evaporation in an inert gas form a spongy structure that has high emissivities within large spectral ranges. For instance, gold black obtained by deposition of evaporated gold in helium or nitrogen atmosphere at low pressures was widely used as an absorptive coating for thermal detectors of optical radiation.

Unfortunately, the application area of gold black is limited because it is very fragile and can easily be damaged by heat and mechanical vibration.

The morphology of metal blacks characterized by particle size and their clustering can be controlled by the evaporation rate of metals, pressure of the gas, and its composition. These coatings are intensively used as the tools of thermal control in thermal engineering and in optical instrumentation for suppressing the stray light. However, they were never used for radiating surfaces of blackbodies (at least, such cases are not known to us). Perhaps this is due to the lack of reproducibility of optical characteristics, the difficulty of achieving uniform coating of the interior surfaces of cavities and/or the excessive cost of the required process equipment. However, technologies that have appeared recently allow application of high-emissivity coatings to extended surfaces of any curvature (see https://www.acktar.com), which makes them quite promising for blackbody design.

Overviews of black paints and coatings can be found in Persky [164], Dury et al. [52], Persky and Szczesniak [165], Zeng and Hanssen [238], and Marshall et al. [135].

Some of these materials have found application in the designs of low-temperature blackbodies, operating, as a rule, in vacuum. In addition to the common requirement of high emissivity, materials for blackbodies operating at low temperatures in vacuum must withstand low temperatures without changes of radiation characteristics and structure (in particular, the coefficient of thermal expansion of a coating must match to that of the substrate) and have low outgassing. It is naturally to require good adhesion of the coating layer to the substrate and matching their coefficients of thermal expansion (CTE) in order to prevent detachment of the coating when the temperature changes.

One more criterion has to be added for the case of black paints and coatings applied to the surfaces of radiating elements heated or cooled from the rear side: the thermal resistance R_T [W· K^{-1}] of the paint or coating layer should be small to minimize the temperature drop effect. In brief (the effect of temperature drop is considered in more detail in Sect. 8.4.2), the essence of this effect is as follows. The temperature of low- and medium-temperature blackbodies is measured, as a rule, with a contact thermometer placed at the backside of the radiator wall. The difference in the heat exchange conditions of the temperature sensor and the blackbody radiating surface leads to the systematic error in measuring the blackbody temperature that is proportional to the thermal resistance R_T of the blackbody wall with the black paint or coating applied. The thermal resistance, in turn, is approximately proportional to the thickness that the heat flux must pass and inversely proportional to the thermal conductivity (in W·m^{-1}·K^{-1}) of the wall material. From this point of view, the vacuum-deposited black coatings have definite advantage over black paints both due to the higher thermal conductivity and due to the smaller thickness.

Below, we consider some black paints suitable for radiating surfaces of blackbodies.

6.2.1 NEXTEL Velvet-Coating 811–21

This black paint originally developed by the Minnesota Mining and Manufacturing Company as 3M 101-C10 [24] for temperature range from about –50 to 150 °C contains microscopic silica sphere coated with carbon to ensure near-diffuse low IR reflection. Later, silica microspheres were substituted by polyurethane microspheres to increase the spectral emissivity around 9 μm. Until recently, the paint of such a composition was available from Mankiewicz GmbH & Co. KG (Germany) under the trading name of "Nextel-Velvet-Coating 811–21". Now, it is offered as "NEXTEL suede coatings" in various shades. For brevity, in this section, we will call it Nextel Black.

The Nextel black is used since 1990s, so its radiation characteristics are well-studied [4, 52, 116, 128]. The paint is usually applied by compressed air spraying a layer of up to 100 μm thick. A recommended nozzle size is 1.5 to 1.8 mm; the spraying pressure is about 3 bars; the recommended thickness of the dry film is 45 to 55 μm. The microstructure of the surface coated with Nextel Black is shown in Fig. 6.9 obtained using a scanning electron microscope by Song et al. [205]. This image reveals microspheres with a diameter of 10 to 20 μm in a viscous binder that form a rough surface containing some pores. A high emissivity of Nextel Black can

Fig. 6.9 The of microstructure image of the Nextel Black coating obtained using scanning electron microscopy. Reproduced from [205] with permission of Springer Nature

be attributed, at least partially, to multiple reflections occurred in the open pores on the coating surface.

Early measurements of BRDF for the Nextel black performed at a wavelength of 0.6328 μm for only a few angles of incidence and with a low angular resolution showed near-Lambertian behavior. In this regard, the Nextel black gained a reputation as the diffuse pain. The only known study of IR BRDF of the Nextel black was undertaken by Balling [12], who measured the in-plane BRDFs at 3.39 μm and revealed the behavior typical for scattering from randomly rough surfaces combined with subsurface forward scattering. Unfortunately, the thickness of the paint layer on the machined brass substrate was not indicated, and the BRDFs were not measured at different thicknesses. This makes it impossible to attribute the BRDF characteristics to reflections on two rough interfaces (air-paint and paint-substrate) or to volumetric scattering in the paint layer.

There are extensive studies on the spectral emissivity (or spectral DHR) of the Nextel black in the visible and near IR wavelength ranges; however, the need for a high-emissivity coating for low-temperature blackbodies has led researchers to extend measurements to the longer wavelengths. Such studies of various materials in air and in vacuo were conducted at the PTB [3, 4, 85, 143].

The spectral directional (at the viewing angle of 10°) emissivity of Nextel black measured at the PTB at a temperature of 120 °C is presented in Fig. 6.10.

Figure 6.11 presents the spectral normal emissivities of the Nextel black sample measured at the PTB[1] and CENAM[2] during bilateral comparison on spectral directional emissivity. Measurements were carried out in the wavenumber range between about 500 and 3000 cm^{-1}, at the viewing angle of 5° (PTB) and at 0° (CENAM); both

[1] PhysikalischTechnische Bundesanstalt (Germany).

[2] Centro Nacional de Metrología (Mexico).

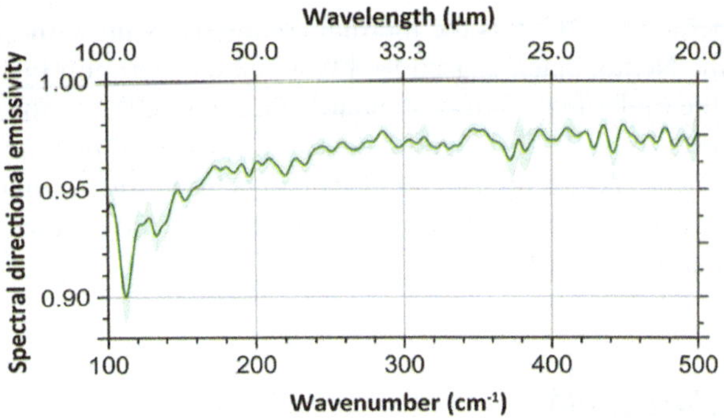

Fig. 6.10 Spectral directional (at the viewing angle of 10°) emissivity of Nextel 811–21 measured at a temperature of 120 °C. The shaded zone indicates one standard uncertainty. Reproduced from [4] under the Creative Commons CC BY license

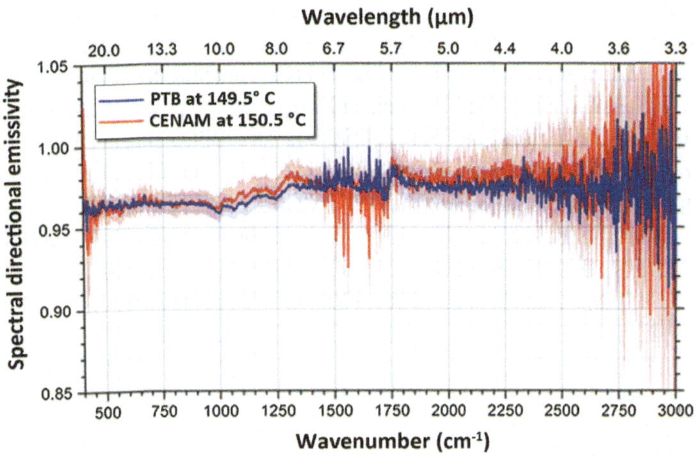

Fig. 6.11 Directional spectral emissivity observed at 5° (PTB), and at 0° (CENAM), of Nextel 811–21 at approximately 150 °C. Shaded area around each curve indicates the range of combined standard uncertainty. Reproduced from [33] with permission of Springer Nature

participants used the samples of Nextel black at approximately 150 °C. Shaded area around each curve in Fig. 6.11 indicates the range of combined standard uncertainty.

The Nextel black is one of the few black paints, for which thermal conductivity is reliably known. Lohrengel and Todtenhaupt [128] reported an approximate equation for the thermal conductivity of the Nextel-Velvet-Coating 811–21 at a temperature of 25 to 125 °C:

$$k = 0.2042 - 0.0001 \cdot t, \tag{6.10}$$

where t is temperature in °C, k is the thermal conductivity in $W \cdot m^{-1} \cdot K^{-1}$.

Up to date, the Nextel black is used for blackbodies operating in air, from simple blackbodies in the rank of secondary standards (see, e.g. [50]) to the blackbodies of highest precision [114, 160, 224]. After outgassing performed in a furnace for 60 min at 80 °C or 30 min at 120 °C, the Nextel black can be used for coating the blackbody cavities designed for pre-flight calibration of space-borne radiometric instruments in cryo-vacuum chambers [79, 153, 154, 147].

6.2.2 Aeroglaze Z306

The Aeroglaze (formerly, Chemglaze) Z306 is a polyurethane-based paint with silica microspheres and carbon black filler. Aeroglaze Z306 and its similar Z302 have been produced for a long time by Lord Corp. The paint Z306 was considered as diffuse, while Z302 as specular, or more precisely, glossy paint. The main area of Z306 application was the stray light control (see [57]), in particular, in space-based optical system. Currently, production of Z302 is discontinued,the technology of Z306 fabrication is transferred to the Socomore Group (https://www.socomore.com/shop/coatings/anti-erosion-coating/aeroglaze-z306/). The Z306 paint should be applied over special primers onto metal substrate following the manufacturer's specifications for painting and curing. Usual thickness of the dry paint film is 25 to 38 μm.

Ames [5] studied the spectral directional-hemispherical reflectance (DHR) at near-normal incidence of Z306 samples with and without microspheres and found that adding microspheres reduces the DHR in the spectral range from 3 to 25 μm by an average of 3.5%.

The measurements made at wavelength of 0.6328 μm showed significant flattening (i.e. reduced glossiness) of BRDF for the sample with added microspheres; the flattening is less pronounced for wavelength of 10.6 μm.

Barilli and Mazzoni [13] measured the BRDF of the Z306-coated aluminum alloy 7075 at 0.65 μm and 10° of incidence. A small specular peak was detected even at these conditions of irradiation. Unfortunately, the lack of information on samples does not allow discussing the agreement of results obtained by Barilli and Mazzoni [13] with those of Ames [5].

The paint Z306 has a high emissivity in the broad spectral range [229, 231], excellent mechanical properties, and low outgassing [53, 137]. Owing to this, Z306 is one of the most popular coatings for radiating surfaces of low-temperature blackbodies designed for pre-flight calibration and characterization of spaceborne sensors in the cryo-vacuum environment.

In this concern and within the framework of the project "Infrared Optical Properties of Materials Measurement", the extensive studies of IR radiative properties of high-emissivity materials and coatings were carried out at the NIST [72, 73, 238, 239].

The DHRs of the Z306 sample at the incidence angles θ_i of 16, 30, 45, and 60° are presented in Fig. 6.12. The BRDFs of Z306 for various angle of incidence at wavelength of 1.55 and 10.6 μm are presented in Fig. 6.13. Measurement were performed

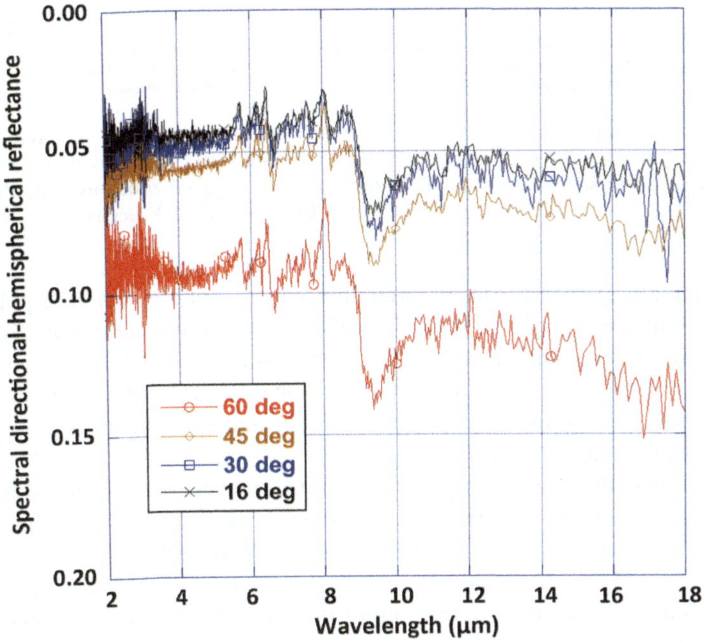

Fig. 6.12 Spectral DHR at four incidence angles of the Z306 sample at room temperature. Courtesy of Dr. Leonard Hanssen (NIST)

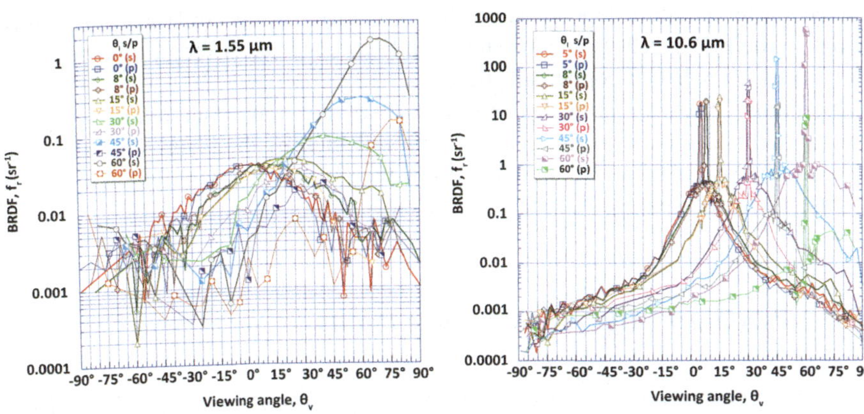

Fig. 6.13 The in-plane BRDFs $f_r(\theta_i, \theta_v)$ of Z306 measured at wavelengths λ of 1.55 and 10.6 μm for several incidence angles θ_i and for s and p polarization states. Courtesy of Dr. Leonard Hanssen (NIST)

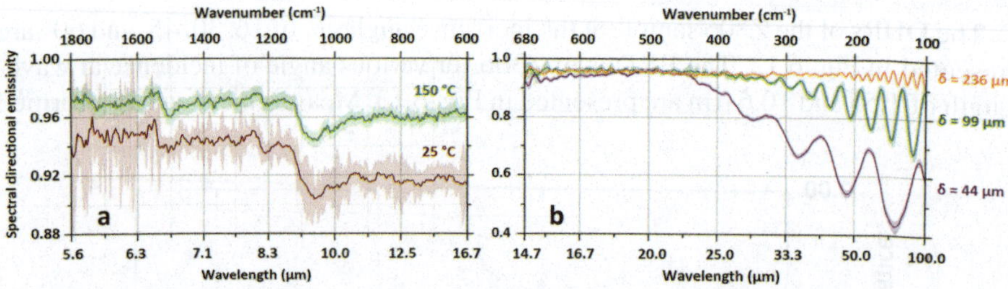

Fig. 6.14 Spectral directional emissivities at the viewing angle of 10° of the Aeroglaze Z306 coating: **a**, with thickness of 99 μm at temperatures of 25 and 150 °C and **b**, with three thicknesses δ on polished copper substrates at a temperature of 25 °C. Shaded zones indicate the standard uncertainty. Reproduced from [4] under the Creative Commons CC BY license

for two mutually perpendicular polarization planes, *s* and *p*, where *s* polarization corresponds to the electric field of the incident electromagnetic wave oscillating perpendicular to the plane of incidence. Comparison of the BRDF data at different wavelengths shows that the glossy reflection lobes at 1.55 μm are transformed at 10.6 μm into a composition of the glossy lobes and specular peaks containing a larger part of reflected energy.

It is obviously that the microspheres in the Z306 layers provide near-Lambertian reflection in the visible and, perhaps, in the near IR spectral ranges but they are ineffective at larger wavelengths, which become comparable with the microsphere diameters. This allows characterizing reflection from the Z306 at 10.6 μm as predominantly specular; therefore, we must admit incorrect the frequent references to the Z306 as diffuse paint.

Recently, the IR spectral DHR of the Z306 was studied by Adibekyan et al. [4]. Figure 6.14a presenting the spectral (5.6 to 15.7 μm) directional (at the viewing angle of 10°) emissivities of the Z306 paint for temperature of 25 and 150 °C shows a noticeable temperature dependence of the emissivity. The spectral directional emissivities in the FIR spectral range is shown in Fig. 6.14b for different thicknesses of Z306 layers on polished copper substrates.

Oscillatory behavior indicates interference of the wavefront reflected by the external surface of the coating layer and that reflected by the layer-substrate interface. An increase in the oscillation amplitude with increasing wavelength means an increase in the transparency of the Z306 layer.

6.2.3 *Pyromark High-Temperature Paints*

Although in general, the use of paint and varnish coatings is not typical for blackbodies with operating temperatures that exceed 150 °C, two silicon-based polymer

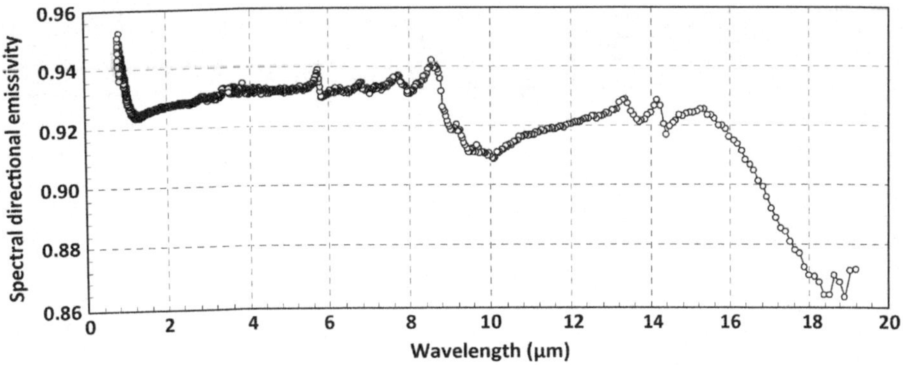

Fig. 6.15 Room-temperature spectral directional (at the viewing angle of 8°) emissivity of the Pyromark 1200 paint on the copper alloy substrate. Reproduced from [152] with permission of SPIE Press

high-temperature paints from the Tempil® company[3] (https://markal.com/pages/tempil), Pyromark 1200 and Pyromark 2500 are an exception. These paints were developed for the aerospace industry needs but were also applied to the IR heaters and solar energy receiver systems. The Pyromark paints consists of finely divided particles of transition metal oxides in a silicate binder, which is cured at 600 °C after application. Flammable in the liquid state, Pyromark 1200 and Pyromark 2500, after application to the surface by brush, conventional spray, or high volume low-pressure spray and thorough drying, withstand temperatures up to ~538 and ~1093 °C, respectively.

Song et al. [204] measured the spectral near-normal emissivity of the Pyromark 1200 on the copper alloy substrate at room temperatures and found that it varies from 0.85 to 0.95 in the wavelength range from 3 to 14 μm. Dupont and Avenas [51] found that the spectral emissivity of the Pyromark 1200 depends weakly on temperature and the paint layer thickness in the temperature range from room temperature to about 200 °C.

The Pyromark 1200 paint has been applied to the design of the water heat-pipe blackbody operating in air between 50 and 250 °C [152] and the water-bath blackbody for calibration of IR radiation thermometers in the temperature range from −30 to 150 °C [27]. The room-temperature spectral directional (at the viewing angle of 8°) emissivity of the Pyromark 1200 paint on the copper alloy substrate measured at the NIST [152] is presented in Fig. 6.15.

Pyromark 2500 is commonly used for of concentrating solar power receivers. The spectral near-normal emissivities of the Pyromark 2500 on the Nickel 201 alloy substrate for several surface temperatures up to 800 °C presented in Fig. 6.16 show their obvious temperature dependence.

[3] Tempil®, a division of Illinois Tool Works Inc. was acquired by LA-CO Industries, Inc. (USA) in 2013.

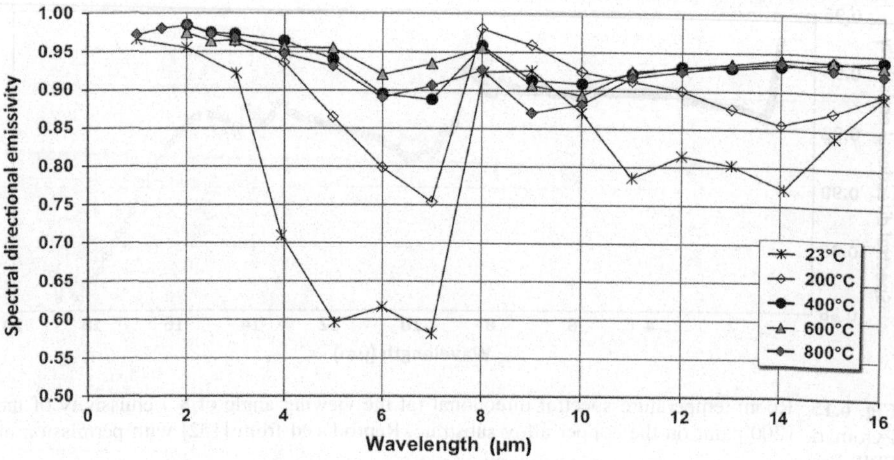

Fig. 6.16 Spectral directional (at the viewing angle of 12°) emissivity of the Pyromark 2500 paint on the Nickel 201 substrate. Reproduced from [71] with permission of Springer Nature

Battuello et al. [14], comparing measurements performed at different NMIs for materials commonly used in blackbody radiators, also noted a pronounced temperature dependence of the spectral emissivity of the Pyromark 2500 layers on aluminum substrates in the wavelength range between 4 and 7 μm. Persky and Szczesniak [165] also stated a noticeable temperature dependence of the Pyromark 2500. According to Ho et al. [83], temperature dependences of the Pyromark 2500 spectral emissivity are different for different substrate materials. Mekhontsev et al. [138] described application of the Pyromark 2500 to the radiating surface of the blackbody operating in vacuum at temperatures from 400 to 1000 K. Clausen [42] studied the suitability of the Pyromark 2500 for the surface blackbody calibrators (flat-plate and with structured surface) operating in air at 75 to 550 °C.

Unfortunately, the available information on radiation and thermophysical characteristics of Pyromark 1200 and Pyromark 2500 paints is far from complete. Judging by the published data, additional investigation of temperature dependence of the Pyromark 2500 emissivity is required to wider use of this paint in the design of blackbodies operating at temperature above 200 °C.

6.2.4 Carbon Nanotube Coatings

Nanoscience is the branch of science that studies and operates material systems on atomic, molecular and supramolecular scales (the sub-micrometer or nanometer scale). On such a scale of length, quantum and surface boundary effects become relevant, imparting to materials the properties, which are not observable on larger,

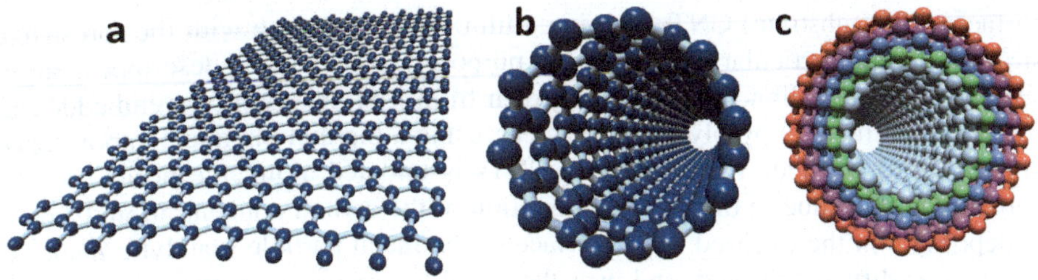

Fig. 6.17 The "ball-and-stick" models of carbon allotropes: **a**, graphene sheet, **b**, single-walled carbon nanotube (SWCNT), and **c**, multi-walled carbon nanotube (MWCNT). Spheres designate carbon atoms; thin cylinders denote covalent bonds

macroscopic length scales. In the first decade of the new millennium, nanoscience began to affect various aspects of applied science, technology, and even mass production. Measurement technique also did not escape this influence. Perhaps, the most important for optical radiometry was the creation of black coatings based on carbon nanotubes (CNTs). This is one of carbon allotropes with a cylindrical nanostructure, which can be considered as graphene sheet (a single layer of carbon atoms arranged in a hexagonal lattice) rolled up into a seamless cylinder with nanometer-sized diameter and micrometer-sized length (the length-to-diameter ratio may exceed 10^3). The atoms in graphene are arranged in hexagons similar to that of graphite. Depending on the way the graphene is wrapped into a seamless cylinder, the CNT may have three different designs (armchair, chiral, and zigzag). The CNTs are divided into single-walled and multi-walled carbon nanotubes (SWCNTs and MWCNTs, respectively). The "ball-and-stick" models of graphene, SWCNT, and MWCNT are shown in Fig. 6.17, where spheres depict carbon atoms and thin cylinders denote covalent bonds.

One can consider the MWCNT as a series of SWCNTs of gradually decreasing diameters nested one inside the other. The typical diameter of the MWCNT is greater than 100 nm. Both SWCNTs and MWCNTs are widely used in modern technologies. They can be synthesized by a variety of methods such as arc discharge, chemical vapor deposition, laser ablation, liquid pyrolysis, and some others reviewed, e.g. by Prasek et al. [169] or Tehrani and Khanbolouki [209]. The unusual electrical, magnetic, mechanical, thermal, and optical properties of CNTs (see, e.g. Monthioux et al. [145] and references therein) have been extensively studied theoretically, numerically, and experimentally. These properties make CNTs highly attractive for a variety of applications [55, 45], including (but not limited to) electron field emission, electrochemical energy storage, high-performance electronics, membranes and filters, etc. Many of these applications require the preparation of well-aligned CNTs. Thermal growth usually leads to curved, chaotically oriented, and randomly entangled CNTs.

Only under certain conditions, closely spaced CNTs can maintain a defined direction of growth. In parallel with the first reports on the mass synthesis of CNTs, were undertaken the early attempts to grow CNTs aligned perpendicularly or parallel to the surface of the substrate. The vertically aligned (i.e. grown perpendicular to the surface of the substrate) CNTs forming a film on the substrate with the forest-like structures are of particular interest for our purposes. Li et al. [127] described a large-scale synthesis of aligned MWCNTs grown from iron nanoparticles embedded in mesoporous silica by pyrolysis of acetylene using a thermal chemical vapor deposition (CVD) method. The growth of CNTs is initiated along the normal to the substrate surface due to the gradient direction of the carbon concentration. Carbon is deposited on the exposed upper surface of the metal particle (catalytic nucleus) and then is diffused through and over the metal to precipitate at the opposite face in the form of CNT. Once the initial growth of CNT on or under catalyst particle is established, it tends to continue in this same direction. The growth is reinforced by the presence of surrounding CNTs, which limit the possibilities for the CNT growth in other directions. The thermal CVD method allows growing very long vertically aligned CNTs with a tube spacing of about 100 nm.

A straight, well-aligned arrays of CNTs over areas up to several square centimeters on nickel-coated glass were first successfully grown by Ren et al. [177] using a plasma-enhanced CVD (PECVD) method at temperature around 700 °C. The DC PECVD method provides alignment of CNTs using the electric field in the glow discharge, enabling CNTs with lengths of less than 10 μm to be relatively well aligned.

The diameter of the aligned CNTs could be controlled from 20 to 400 nm with the length varied from 0.1 to 50 μm. The VACNTs were grown from catalytic nickel nanoparticles randomly deposited by radio frequency magnetron sputtering on the glass substrate. Since then, various methods have been developed for growing VACTN on various substrates and ensuring controllability and reproducibility of their parameters [178]. Among the many applications of aligned CNTs [120, 109], there are especially intriguing for optical engineering and optical radiometry as black coatings: for thermal (first of all, pyroelectric) detectors of optical radiation, systems suppressing stray light, and blackbody radiators.

Yang et al. [234] reported investigation of reflection characteristics of the VACNT arrays and announced the creation of the "darkest" substance ever made. The VACNT films they investigated were prepared by a water-assisted CVD process [77] yielding MWCNTs with a typical tube diameter of 8 to 11 nm and the tube-to-tube spacing in the range of 10 to 50 nm. The arrays of well-aligned CNTs has a density of 0.01–0.02 g/cm^3 forming a low-density porous nanostructure. Once grown on a substrate, the VACNT arrays were peeled off to get freestanding films with thicknesses varied in the range of 10 to 800 μm depending on growth time. Reflectance (directional-hemispherical and its diffuse component) of the 300 μm thick VACNT films was measured in the visible spectral range using the laser-based facility with the integrating sphere and the sample placed in its center. It was found that the spectral DHR of samples is of order of 0.045% with the diffuse reflectance of about 10^{-7} and weakly depends on wavelength from 457 to 633 nm and on the incidence angle for up to ± 70°. It was also shown that the reflection characteristics of a VCNT could be controlled by varying the CNT diameter and the tube-to-tube spacing. The conclusions made by Yang et al. [234] were confirmed by subsequent research.

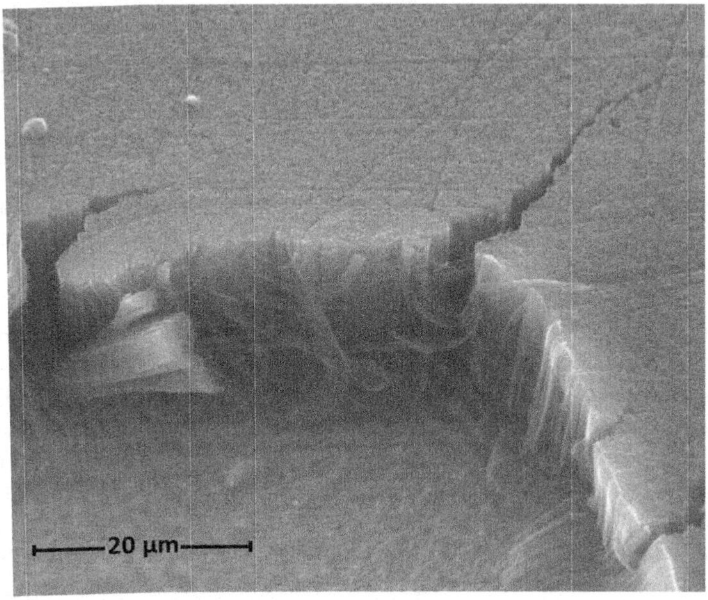

Fig. 6.18 Side-view SEM images of vertically aligned MWCNT arrays grown on aluminum substrate. The nanotubes are partially scraped off the substrates for inspection purposes. Reproduced from [210] under Creative Commons Attribution 4.0 International License

Figure 6.18 is the scanning electron microscope (SEM) image of vertically aligned MWCNT array grown on an aluminum substrate. It would be wrong to think that the low reflection of the forest is due to multiple reflections of the incident radiation from individual nanotubes. Their diameters and the distances between them are too small in comparison with the wavelengths of the incident radiation to use geometric optics. A simple qualitative explanation of unusual radiative properties of VACNT provides the effective medium theory [64, 65, 232, 240, 241] treating the VACNT film as a homogenous but anisotropic medium. The optical constants of the VACNT are modeled on the base of the anisotropic permittivity and dispersion relations for carbon crystal (graphite). By its properties, the VACNT forest resembles an aerogel, in which the nanotubes occupy only a few percent of the total volume. Owing to their inherent high porosity, the average effective refractive index n_{eff} of VACNTs is very close to unity.

The normal Fresnelian reflectance ρ_n can be evaluated as

$$\rho_n = \left(\frac{n_{eff} - n_0}{n_{eff} + n_0}\right)^2, \tag{6.11}$$

where n_0 is the refractive index of vacuum ($n_0 = 1$) or air ($n_0 \approx 1.0003$).

If $n_{eff} = 1.0005$ (an easily attainable value, see, e.g. Yang et al. [234]), then the normal Fresnelian reflectance at the air-VACNTs interface as low as 10^{-8}. The

Table 6.2 Reflectance measurements for the SWNT forest (after [141])

Spectral range	UV to NIR	NIR to MIR	MIR to FIR
Wavelengths (μm)	0.2–2	2–20	25–200
Reflection geometry	Directional-hemispherical	Directional-hemispherical	Specular
Angle of incidence (°)	8	5	10
Average reflectance	0.0160	0.0097	0.0017
Standard deviation	0.0048	0.0041	0.0027

homogeneous sparseness and alignment are the key properties for achieving the low reflectance and high emissivity of a VACNT array.

Mizuno et al. [141] studied a vertically aligned array of SWCNTs (SWCNT forest) synthesized by water-assisted CVD on silicon substrates at 750 °C. Ethylene was used as a carbon source. The height of the SWNT forest was 460 μm, with the mean density of 0.07 g/cm^3. It was found that the SWCNT forest absorbs light almost perfectly across a very wide spectral range from 0.2 to 200 μm (see Table 6.2).

Only at wavelengths larger than 25 μm, the transmittance of the SWCNT forest becomes noticeable; at shorter wavelengths, it is below the detection limit. Along with the spectral reflectance, the direct measurements of the normal spectral emissivity at room temperature in vacuum were carried out for the wavelengths from 5 to 12 μm. This allowed the internal consistency of results to be checked (with the accuracy limited mainly by the 1%-relative standard uncertainty of the spectral emissivity measurements) according to the energy conservation law:

$$\varepsilon(\lambda) = 1 - \rho(\lambda) - \tau(\lambda), \tag{6.11}$$

where λ is the wavelength; ε, ρ, and τ are the emissivity, reflectance and transmittance, respectively.

The emissivity measurements showed a nearly constant emissivity (absorptance) of 0.98–0.99 of a SWNT forest over a wide spectral range from UV (200 nm) to FIR (200 μm); the forest with heights of only 2 μm has the emissivity higher than 0.97; the emissivities of forests taller than 50 μm are greater than 0.98. It was revealed that the emissivity of an SWNT forest depends more on the area mass density (height × density) than on the density itself: an increase of emissivity to 0.987 was found at the area mass density of to 2.33 mg·cm^{-2}.

Wang et al. [225] conducted a complex study of vertically aligned MWCNTs grown on silicon substrates using the CVD method [118]. Measurements of the spectral reflectance in the visible and NIR region, as well as the polarization-dependent BRDF at the wavelength of 635 nm, were performed for the vertically aligned MWCNT arrays obtained by two different growth mechanisms: tip grown and base growth (catalyst particles are at the top and in the base of the grown nanotubes, respectively). The base and tip growth are characterized by a strong and weak catalyst-substrate adhesion, respectively. Parameters of six MWCNT samples under investigation are listed in Table 6.3.

The spectral directional-hemispherical reflectances of these samples measured using an integrating sphere with a grating monochromator in the wavelength range from 400 to 1800 nm are shown in Fig. 6.19. To reduce the fraction of specularly reflected radiation that escapes the integrating sphere through the entrance port, the samples were tilted to form an incidence angle of 8°. The transmittance of the samples is found to be negligible. The tip-growth samples (#4, 5 and 6) exhibited the reflectance peaks near $\lambda = 600$ nm attributed to the iron impurities on top of the MWCNTs. Samples also showed the difference in the visual appearance: samples 4 to 6 looked brownish, while samples 1–3 clearly looked black. The spectral directional-hemispherical reflectance slightly decreases toward longer wavelength. The sample

Table 6.3 Parameters of the MWCNT samples used by Wang et al. [225]

Sample #	Growth mechanism	MWCNT film height (μm)	Density (g/cm^{-3})	Volume (filling) fraction (%)
1	Base growth	86 ± 10	0.079	3.6
2	Base growth	107 ± 18	0.072	3.3
3	Base growth	141 ± 9	0.046	2.2
4	Tip growth	58 ± 13	0.180	8.2
5	Tip growth	54 ± 8	0.330	15
6	Tip growth	88 ± 14	0.025	1.1

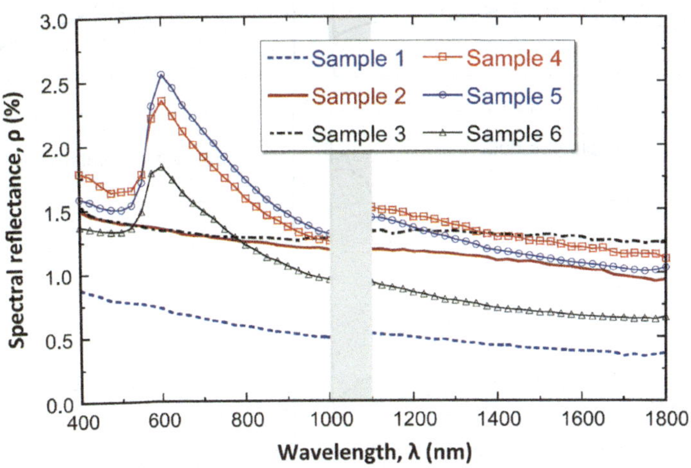

Fig. 6.19 The spectral DHRs of the six MWCNT samples at 8° of incidence measured in the spectral ranges from 400 to 1000 nm and from 1100 to 1800 nm, using Si and Ge photodetectors and different diffraction gratings. Reproduced from [225] with permission of IOP Publishing

#1 exhibits the lowest reflectance, which is less than 0.01 in the spectral range from 400 to 1000 nm and less than 0.005 for $\lambda > 1200$ nm.

The sample-to-sample variations were explained by the entanglement of MWCNTs, nonuniformity of the sample surface, and impurities in Fe catalyst. The BRDFs of two MWCNT samples #1 (base growth) and #4 (tip growth) were measured at the wavelength of 635 nm of the power-stabilized diode laser. The change in the plane of polarization of the collimated laser beam was accomplished by rotating a linear polarizer. The BRDFs of two these samples are shown in Fig. 6.20 for *s* and *p* polarization states. The BRDFs were measured in the plane of incidence for three incidence angles (0, 30 and 60°). The reflection from the MWCNT samples is not purely Lambertian or purely Fresnelian but exhibits more complicated behavior. A small specular peak that grows with the growing incidence angle and more pronounced for *s* than for *p* polarization can be recognized as Fresnelian reflection from the smooth interface of air and the MWNT film. It was hypothesized that bifurcation of specular peaks observed for the sample #1 for both polarization states has diffraction nature.

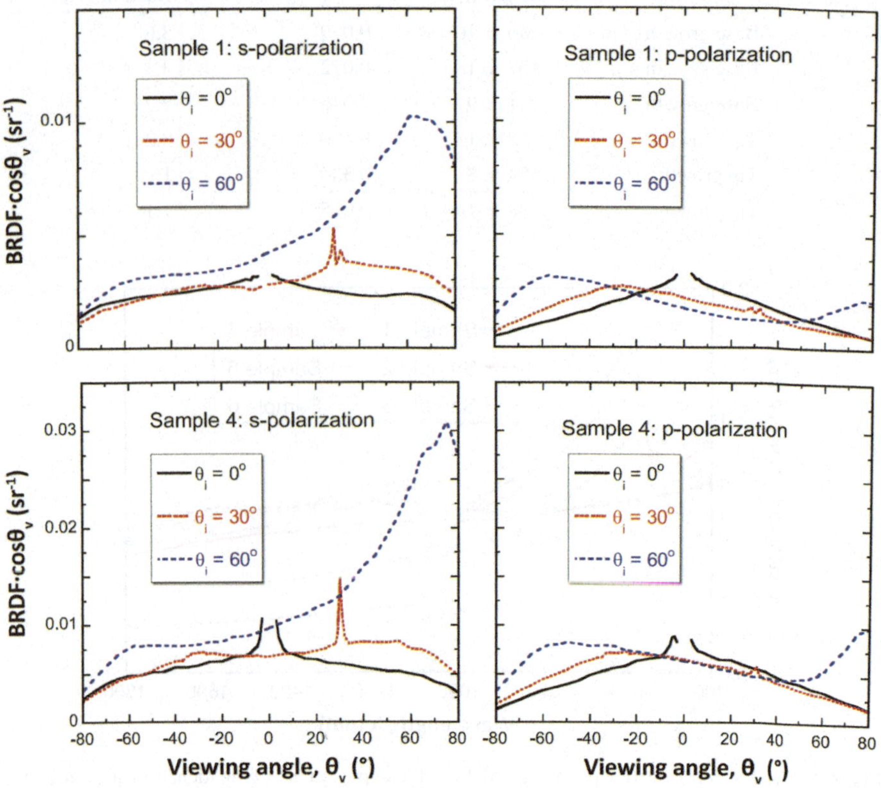

Fig. 6.20 Cosine-weighted BRDFs of six MWCNT samples measured at $\lambda = 635$ nm for samples #1 and #4 at three incidence angles θ_i for both *s*- and *p*-polarization. Reproduced from [225] with permission of IOP Publishing

The lobes around directions of specular reflection belong to reflection from "roughness" formed by the MWCNT tips. The wide lobes around the directions of retro-reflection can be caused by multiple reflections from very large roughness, or are associated with Rayleigh scattering by catalyst nanoparticles and/or the enclosed cups on the CNT tips.

It should be noted that a comprehensive theory of reflection properties for CNT coatings is not developed up to date. One can suppose that statistical distributions of the CNT heights and distances between neighboring nanotubes play the important role in explanation of low reflection throughout a wide spectral range [124].

The 2010s were a period of active accumulation of experimental data. Wang et al. [226, 227] found that BRDFs of vertically aligned MWCNT coatings can vary from near-diffuse to near-specular depending on film thickness and density. Although these observations are of a particular nature (the BRDF measurement was carried out at the wavelength of 635 nm only), one can conclude that more specular reflection corresponds to VACNT arrays with high thickness and low density and, vice versa, BRDFs of arrays with the lower thickness and higher density have only small specular peaks.

The VACNT coatings have been applied for stray light suppression in the satellite optical instrumentation, which required the development of technologies for growing films on various materials, including titanium, Inconel, alumina, etc. [70, 30, 172].

The reflectometric measurements were extended to the IR spectral range [235, 38, 232], which is especially important for the use of CNT coatings in the blackbody radiometry. Virtual benefits of the use of VACNT-based coatings in the blackbody radiometry are obvious.

If the blackbody radiator is employed as the precise calibration source for thermal imagers, radiometer or radiation thermometers with a wide FOV, it must have a large opening. In order to obtain a sufficiently high and uniform emissivity over a large aperture, a conventional cavity radiator should have unacceptably large size even if its internal surface has emissivity of order of 0.9. This leads to a large, cumbersome, and massive source. Besides, it is difficult to achieve the temperature uniformity over the walls of a large cavity. Researchers of the National Metrology Institute of Japan (NMIJ) developed the VTBB with the cylindrical cavity and CNT-coated disk as in its base [201]. The blackbody is made of graphite, with the VACNT grown on the graphite substrate, and heated by air circulated with fans to maintain the operation temperature range between 100 and 500 °C. The cavity has the aperture diameter of 40 mm and the length of 412 mm. The SWCNT forest is grown on a circular graphite plate of 0.5 mm thick and 38 mm in diameter coated with a Fe-catalyst layer. The diameter of each SWCNT is 10 nm; its height is 100 μm, the area density of the forest is 3 mg/cm^2. The effective emissivity was modeled by the Monte Carlo ray tracing and confirmed experimentally. The spectral effective emissivities calculated for three wavelengths are listed in Table 6.4.

The effective emissivity increases for the blackbody with the CNT-coated bottom as compared to the all-graphite cavity up to 0.993–0.994 in the spectral range from 1.6 to 10 μm.

Table 6.4 Spectral effective emissivities of the air-bath VTBB with the VACNT-coated bottom (after [201])

	Wavelength (μm)		
	1.6	5	10
All-graphite cavity	0.9912	0.9923	0.9935
CNT bottom and graphite side wall	0.9984	0.9990	0.9994
Effective emissivity increase	0.0072	0.0067	0.0059

Temperature of the blackbody was monitored and controlled with contact sensors; however, for precise determination of the bottom temperature, it was measured using three radiation thermometers with the operating wavelengths of 1.6, 5, and 10 μm.

The VTBB was used calibrate thermal imagers and radiation thermometers operating at wavelengths of around 10 μm. On the base of this VTBB, the NMIJ in cooperation with Chino Corp. (https://www.chino.co.jp) developed a commercial version of the blackbody with the aperture diameter of 40 mm [155]. The blackbody can be operated in the temperature range from 50 to 300 °C. The declared effective emissivity is 1.000 ± 0.001 in the wavelength range from 1.6 to 10 μm.

Commercialization of VACNTs has begun when the British company Surrey NanoSystems, Ltd. (https://www.surreynanosystems.com) made them available in the market under the trade name "Vantablack". Exceptional light absorption properties in conjunction with robustness against shock and low outgassing makes Vantablack highly suitable for space applications.

Zeidler et al. [237] measured the in-plane BRDFs of a Vantablack-coated aluminum plate at the wavelength of 1.064 μm of a solid-state laser. The cosine-weighted BRDFs are presented in Fig. 6.21 for 4 different angles of incidence. At oblique incidence, no pronounced specular component was found. However, the wide angle shading the laser beam by the scanning detector does not exclude the presence of significant backscattering, especially at large incidence angles.

Theocharous et al. [210] reported results of the partial space qualification of vertically aligned MWCNTs coating grown on aluminum substrates. The spectral normal reflectances of a sample of vertically aligned MWCNT array grown on the aluminum substrate measured before and after the space qualification test are shown in Fig. 6.22 together with the spectral reflectance change that is shown to be less than 0.005 for wavelengths shorter than 18 μm and less than 0.01 for wavelengths up to 25 μm.

In order to improve in-flight calibration capability of satellite sensors, Ball Aerospace (https://www.ball.com/aerospace) developed a flat-plate blackbody [157, 62] coated with vertically aligned MWCNTs. Each nanotube of this coating is of 10–30 nm in diameter and from 100 to 300 μm in length. The surface density of the nanoforest is greater than 10^{10} CNTs/cm^2. The Ball's satellite flat-plate blackbody has the area of 152.4×152.4 mm^2 formed by four 75.2×75.2 mm^2 VACNT-coated coupons. The blackbody can operate in the temperature range from 270 to 330 K and exhibit emissivity greater than 0.996 in the 5.5 to 13.5 μm wavelength range.

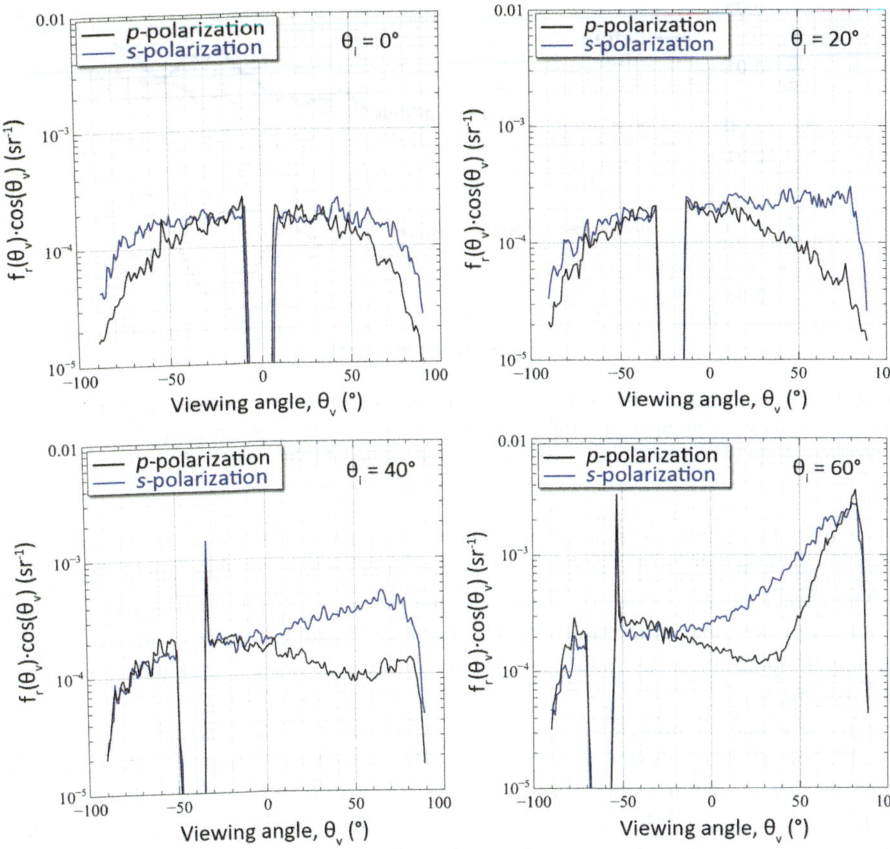

Fig. 6.21 Cosine-weighted in-plane BRDFs of the VantaBlack sample measured at the wavelength of 1.064 μm and the angles of incidence $\theta_i = 0$, 20, 40, and 60° Reproduced from [237] under Creative Commons Attribution 4.0 International License

Therefore, this planar radiator has emissivity comparable with the effective emissivity of cavity-type blackbodies but has much lower size and weight.

Unfortunately, the existing fabrication methods make it difficult to grow VACNT coatings on non-planar, especially concave surfaces. It would be highly desirable to develop sprayed VACNT coatings for simple application on any surface. All that is required is to stir VACNTs in a liquid, which should evaporate, leaving the sprayed VACNT coating behind. Of course, the use of randomly aligned CNT coatings instead of VACNTs should result in some increase in reflectance (decrease in emissivity); however, this can be a reasonable trade-off.

Surrey NanoSystems Ltd. developed black coatings Vantablack S-VIS (Surrey NanoSystems 2017a) and Vantablack S-IR (Surrey NanoSystems 2017b) based on CNT technology and a proprietary spray process allowing their application to a wide range of substrate materials and to complex shapes. Since the coating process

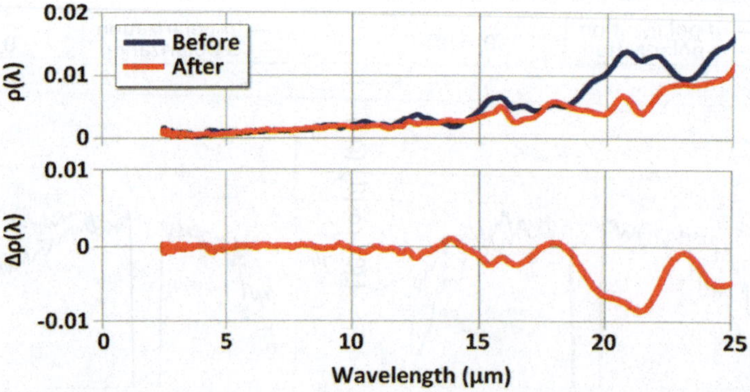

Fig. 6.22 Spectral normal reflectance $\rho(\lambda)$ and its change $\Delta\rho(\lambda)$ for vertically aligned MWCNT array grown on the aluminum substrate measured before and after the space qualification tests. Reproduced from [210] under Creative Commons Attribution 4.0 International License

proceeds at low temperature of about 280 °C, the only requirements to the substrate material are non-volatility and the melting point above 280 °C. The typical thickness of Vantablack sprayed coating is about 40 μm (may reach up to 100 μm locally). The lower operating temperature declared by the manufacturer is -196 °C (77 K); the upper operating temperature is 300 °C in air and 700 °C in vacuum or inert atmosphere.

An additional advantage of these coatings is low outgassing, which makes them suitable for vacuum and space applications. The main disadvantage of CNT sprayed coatings is relatively low durability: they are soft, brittle, and difficult to clean.

The Santa Barbara Infrared, Inc. (SBIR, https://sbir.com) in collaboration with the Surrey NanoSystems Ltd. have developed the series of flat-plate blackbodies with the radiating plates coated with spray-on Vantablack S-Vis. According to very scanty documentation, the average emissivity of blackbodies is greater than 0.998 in MWIR and greater than 0.995 in LWIR spectral range. Operating temperature range is from −40 to 175 °C.

Until the possibility of creating the VACNT coatings grown on refractory substrates and withstanding high temperatures without significant changes in morphology and radiation characteristics is proved, the maximum operating temperatures of blackbodies with VACNT coatings will be below 500 °C [201].

The contact temperature sensors (resistance thermometers and thermocouples) are the most common for such temperature range and provide the lowest measurement uncertainties when used properly. Obviously, a contact measurement of the temperature of a virtual surface, which is simply a locus of the tops of CNTs, is impossible.

Therefore, the temperature of the blackbody must be measured from the substrate side. Achieving the highest accuracy in the contact temperature measurement of blackbodies is associated with the possibility of introducing an accurate correction for the temperature difference between the temperature sensor and the radiating surface (so-called "temperature drop" effect—see Sect. 8.4.2). The temperature difference is proportional to the thermal resistance $R_T = S/(\kappa \cdot d)$ [K·W^{-1}], where κ [W·m^{-1}·K^{-1}] is the thermal conductivity, S is the cross-sectional area of one-dimensional heat flux, and d is the thickness of the coating. The main difficulty lies in the measurement of the thermal conductivity of the VACNT layer.

The early theoretical studies [21] predicted unusually high thermal conductivity of individual carbon CNTs—up to 6600 W·m^{-1}·K^{-1} at room temperature. This gave ground for numerous speculations and overestimated expectations. However, later theoretical and experimental studies yielded values for individual SWCNTs and MWCNT pipes in the range from few dozen to several thousand of W·m^{-1}·K^{-1} [11, 133]. Prediction of thermal conductivity for VACNT arrays are more difficult problem than for an individual CNT: apart from multiple parameters (tube length, number of walls, their diameters, chirality, etc.) characterizing a single CNT, many parameters characterizing the array should be taking into account. In addition, it is necessary to take into account collective effects and the effect of thermal contact resistance of mutually intersecting tubes. It is not surprising that the experimentally determined values of thermal conductivity for VACNT arrays also scatter over a wide range, from 0.145 to 217 W·m^{-1}·K^{-1} [129]. In the presence of such a scatter of values, blackbodies with VACNT coatings can be used only as secondary standards with temperatures traceable to the temperature scale via radiation thermometer.

It is still too early to make any categorical conclusions about the degree of influence of nanotechnology on blackbody radiometry, since studies of nano-coatings have begun quite recently. There is still a set of unresolved problems, solution of which is necessary to clarify the range of possible applications of blackbodies with VACNT coatings.

The interested reader can obtain more information on CNTs growth and alignment, methods of calculation and measurement of reflectance for CNT-based coatings, problems of their space qualification, etc. from the recent review by Lehman et al. [125] and reference therein.

6.3 Oxidized Metals and Alloys

Temperature uniformity of blackbody radiating elements (including cavities) can be achieved most easily if they are fabricated from materials with high thermal conductivity such as metals and alloys. The common materials for this purpose are copper, stainless steels, various nickel-, and aluminum-base alloys. The maximum temperatures of their use are limited only by the loss of mechanical strength when approaching their melting points. However, the emissivities of metals and alloys are low; therefore, radiating surfaces of radiators made of them need in high emissivity coatings. At temperatures below approximately 150 °C, application of black paints is the usual method of increasing the surface emissivity. A few high-temperature black paints appeared not so long ago. Besides, some surfaces (for example, inside deep cavities) are often difficult to paint. For a long time, the problem of increasing the emissivity of the surfaces of metals and alloys was solved by their oxidation, often after the radiator assembly.

At elevated temperatures, metals react easily with atmospheric oxygen and form metal oxides. During oxidation, the metal or conductor is converted into a dielectric. The emissivity of the oxide layer strongly depends on the thickness of the oxide. As the oxidation progresses, the emissivity increases, and the angular distributions of the emitted radiation become more uniform [6].

The relative simplicity of obtaining natural oxide layers on metal surfaces at high temperature in air makes thermal oxidation an attractive method of the emissivity increasing.

Copper has one of the highest thermal conductivity among metals and therefore is of special interest as a material of blackbody radiators. The first documented use of a cavity made of copper oxidized by heating in air refers to the end of the nineteenth century [161]. High-temperature oxidation of copper results in formation of a thick (up to hundreds of micrometers) layer of Cu_2O and a thin (about several micrometers) layer of CuO over it [20]. Fully oxidized copper has the spectral emissivity varying from 0.6 to 0.85 for wavelength from 2 to 10 μm and slightly increases at temperature from 400 to 700 °C. At the same time, the angular dependences of the spectral directional emissivity become almost angle-independent [102, 104]. However, the further use of oxidized copper for these purposes was very limited (mainly in simple and low-cost laboratory blackbodies) because of insufficient durability: a cyclic temperature change can lead to detachment of the oxide layer.

A very small number of metals and alloys is used in the constructions of blackbody radiators to date. Battuello et al. [14] reported the results of the spectral emissivity measurements conducted in the wavelength and temperature ranges up to 20 μm and up to 800 °C, respectively, by several European NMIs for some typical materials of reference blackbody radiators. This paper can serve as a rough guide to the selection of materials that we need to consider in this Section. These are anodized aluminum, oxidized stainless steel SS-310, and oxidized Inconel 600.

Thermal oxidation of alloys is more complex process than of pure metals. The oxides formed may either solve in each other or form separate phases. Contents of metals in the oxide scale differ from the alloy composition due to different affinities of metals for oxygen, different rates of oxidation for different metals, different diffusion coefficients in the oxide and alloy, etc.

For many metallic materials, conditions for the formation of a stable oxide layer can be achieved, and, in most instances, the spectral emissivity remains constant during subsequent heating in air. The structure and oxide layer thickness depend in a complex manner upon atmospheric and surface conditions, temperature and duration of oxidizing, and surface roughness. The reflection from oxidized metals and alloys is determined by not only the surface scattering but also volumetric scattering in the semitransparent oxide layer and, probably, by reflection from metallic substrate, if the oxide layer is insufficiently thick. There are no simple relationships for estimating the emissivity of oxidized metals and alloys [96], therefore, the researcher should rely on literature data or on the witness sample measurements. The literature data on oxidized metals and alloys are usually limited to emissivity (sometimes only total) and/or reflectance at near-normal incidence. There is virtually no data on either the thermal conductivity of the oxide layers or the BRDF of their surfaces. However, in practice, this does not pose any serious problems. Either the blackbodies with such radiating surfaces serve as secondary standards or their temperatures are measured by means of a radiation thermometer.

6.3.1 *Anodized Aluminum*

Aluminum is one of the most common materials in industry and engineering, including various aerospace applications. The emissivity of pure aluminum is low and cannot be increased by thermal oxidation in air. Upon contact of pure aluminum with air, the aluminum atoms in the surface layer instantly interact with air oxygen and form a thin and durable oxide film of Al_2O_3, which protects aluminum from further oxidation. The thickness of the natural oxide layer can be increased by an electrolytic process called anodization [185], electrochemical oxidation of the aluminum surface to obtain a stable thicker alumina film. Anodizing is not an application of a coating but an electrochemical process, through which a metal surface is converted into an oxide ceramic layer. The metal in a suitable electrolyte, usually sulfuric or phosphoric acid, is applied to the anode potential (hence the name of the process) and a thick oxide film is formed on the metal. The anodized film consists of a thin barrier layer directly adjacent to the substrate and a micro-porous layer above it. A porous anodic aluminum oxide coating consists of closely packed hexagonal-shaped cells of oxide; each cell contains a single pore. The specific characteristics of the oxide is a strong function of the forming voltage and only moderately affected by the electrolyte parameters [110].

After the pores absorb dyes, they can be sealed using post-anodizing treatment. As a result, an impermeable durable film is formed. The anodizing is very flexible process: by varying process parameters (electrolyte composition and temperature, voltage, current type, shape, and density, substrate roughness, post-anodizing treatment, etc.) one can obtain anodized layers of different thickness, composition, porosity, roughness, and therefore, with different radiation characteristics, for example, with the emissivity from 0.1 to 0.9 and the reflection behavior from diffuse to specular.

Aluminum has a low density and sufficiently high thermal conductivity (~200 $W{\cdot}m^{-1}{\cdot}K^{-1}$ for aluminum vs. ~390 $W{\cdot}m^{-1}{\cdot}K^{-1}$ for copper at room temperature). Anodized coatings withstand up to 300 °C, has a high thermal conductivity, at least several times higher than that of black paints [31, 122]. A high emissivity is easily attainable with anodizing. All this makes aluminum very attractive material for compact and lightweight portable and onboard blackbody calibrators [15, 42, 80, 163, 211]. Since anodizing is widely used for finishing articles and goods made of aluminum and its alloys, measurement results for the emissivity of anodized aluminum are widely published. However, only a few of these publications are related to the IR spectral range (see, e.g. [16, 22, 34, 93, 115]).

Although there are industry, national, and international standards for aluminum anodizing process (MIL-A-8625F, ASTM B580–79, ISO 7599:2018), the authors rarely indicate exhaustively the determining parameters of the anodizing process and the state of the substrate. This fact, as well as differences in resolution of the used spectral equipment and different, often unevaluated, measurement uncertainties complicate the analysis and comparison of published data on spectral emissivity (or spectral reflectance).

Stierwalt [206] measured the spectral emissivity of black anodized aluminum at the viewing angles of 0 and 60° at the sample temperatures of 4.2 K and at 77 K. It was found that the minimum value of the spectral emissivity of about 0.75 is reached at about 5 μm. The difference in emissivity values measured at two angles turned out to be small; no measurable temperature dependence was found. Jaworske [99] obtained the values of spectral normal-hemispherical reflectance of black anodized aluminum using the hemiellipsoidal mirror reflectometer described by Neu et al. [150]. Figure 6.23 shows the room-temperature spectral normal reflectance of anodized aluminum sample in the wavelength range from 2 to 25 μm measured by Jaworske and Skowronski [98]. This graph shows a high (above 0.9) spectral emissivity in the important wavelength range between 8 and 12 μm.

The systematic investigations of angular distributions of optical radiation reflected from the surface of anodized aluminum were never carried out. Design of blackbody radiators made of this material and calculation of their effective emissivities are often based on measured spectral emissivity of flat samples and their visual appearance instead of actual BRDFs. To our knowledge, only a few studies of the anodized aluminum BRDF were published and only one of them was conducted in the IR spectral range.

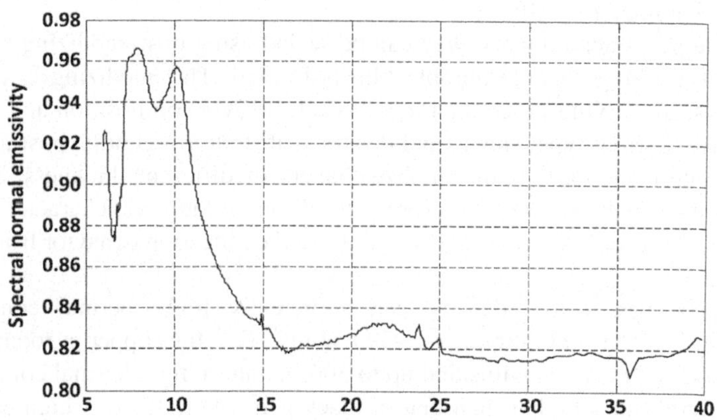

Fig. 6.23 Spectral normal reflectance of anodized aluminum at room temperature. Reproduced from [98] with permission of American Institute of Physics

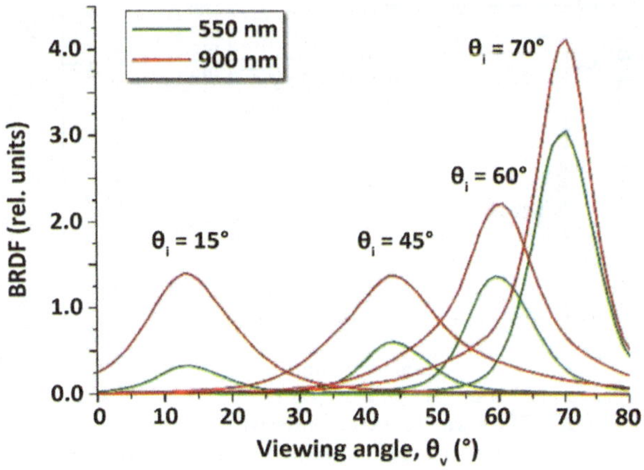

Fig. 6.24 Relative in-plane BRDFs of a black anodized aluminum sample at wavelengths of 550 and 900 nm and incidence angles of 15, 45, 60, and 70°. Reproduced from [23] according to Creative Commons Attribution 3.0 license

Berni et al. [23] presented the relative BRDFs depicted in Fig. 6.24 for two wavelengths in the visible spectral range and four incidence angles. Similar BRDFs were obtained by Hoover and Gamiz [86] for two incidence angles in the plane of incidence. Measurements were performed at the wavelengths of 1.064 and 10.6 μm and normalized by dividing by their maximum values.

Unfortunately, this form of the BRDF presentation does not allow direct comparison of the BRDF maxima heights, allowing only evaluating the dynamic range of BRDF at a fixed incidence angle. The measured BRDFs demonstrate wide specular lobes typical for reflectance from a semi-transparent medium involving reflection from rough surface of the alumina film, absorption according to Beer's law, and volumetric scattering in the oxide layer.

The experimental data are clearly insufficient. To calculate more accurately the effective emissivity of blackbody radiators made of anodized aluminum, complex investigation of the BRDFs is necessary, primarily in the important IR spectral range.

6.3.2 *Oxidized Stainless Steel*

By its definition, stainless steel is a steel with a minimum chromium content of 10.5 or more and a maximum carbon content of less than 1.20 by mass percent (ASTM A941–18). At normal or elevated temperatures in atmospheric air, the stainless characteristics are developed through the formation of an invisible and adherent chromium-rich oxide film. The Cr_2O_3 protective oxide layer grows on the stainless steel surface, the faster the higher the temperature, to a typical thickness of 1–3 nm [156].

When stainless steel is heated to a temperature below a certain limiting temperature T^*(about 1100 °C for high-temperature types and about 900 °C for others), the

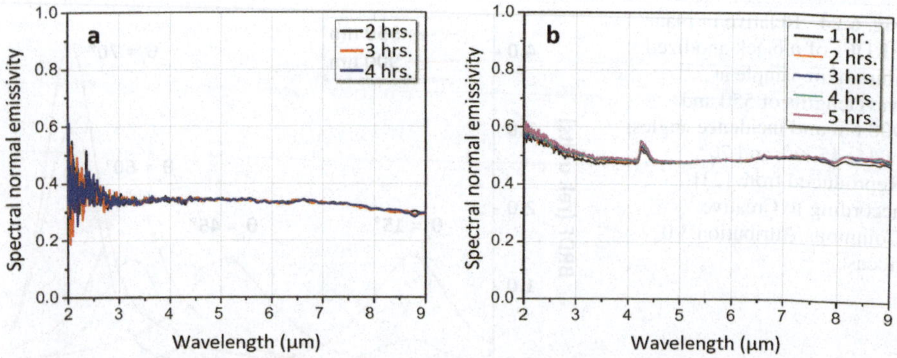

Fig. 6.25 The normal spectral emissivity measured for 316 stainless steel samples oxidized in air for time indicated in legends. Oxidation and measurements were made at temperature of **a** 500 °C and **b** 700 °C. Reproduced from [32] with permission of Elsevier

oxide layer prevents further oxidation. Above T^*, the chromia layer starts decomposing into volatile CrO_3 and the protective properties of the steels are no longer retained. If a stainless steel sample is heated to a temperature above T^* for a predetermined time in air, a relatively thick (of the order of 1 μm) and strong layer consisting of a mixture of oxides and having a high emissivity is formed. After 15–20 min of oxidation, the emissivity almost does not change. This state of the surface is often called "fully oxidized." However, the emissivity values of the "fully oxidized" surface depend on the temperature of oxidation. This fact is illustrated by Fig. 6.25 adopted from Cao et al. [32].

Figure 6.25a shows the spectral normal emissivity of the 316 stainless steel sample measured at 500 °C after 2–4 h oxidation in air at the same temperature. The emissivity remains low due to significant oxidation resistance of this steel. Owing to a high Cr content, the surface forms a passive oxide, which limits further growth of the oxide layer. Figure 6.25b shows the spectral normal emissivity of the same 316 stainless steel measured at 700 °C for 1–5 h exposure in air, other things being equal. The emissivity values at 700 °C is to about 0.5 versus 0.35 at 500 °C. As expected, this is due to a thicker oxide layer formed at 700 °C than at 500 °C.

After cooling, a heavily oxidized specimen can be employed up to the limiting temperature T^* without risk of further growth of the oxide layer and changing its radiation characteristics. Austenitic stainless steels (i.e. stainless steels having a face-centered cubic primary crystalline structure) are the most popular grades due to their excellent formability and corrosion resistance. The series 300 (according to the AISI[4] steel grades system) stainless steels are used in various high-temperature applications. Some widespread types of the 300 series stainless steels are listed in Table 6.5.

High-chromium austenitic stainless steels Type 310 exhibiting superior corrosion resistance is highly demanded in high-temperature applications. Spectral directional emissivities of oxidized 310S stainless steel (the low-carbon version of Type 310)

[4]AISI is for American Iron and Steel Institute.

Table 6.5 Chemical composition and suggested maximum service temperatures in air for some austenitic stainless steels (after [194])

AISI Type	Nominal Composition (%)					Service temperature (°C)	
	C (max)	Mn (max)	Si (max)	Cr	Ni	Intermittent	Continuous
304	0.08	2.00	1.00	18.00–20.00	8.00–12.00	870	925
310	0.25	2.00	1.50	24.00–25.00	19.00–22.00	1035	1150
316	0.08	2.00	1.00	15.00–18.00	10.00–14.00	870	925

at 1440 K (1167 °C) for several viewing angles were reported by Sans and Guillot [187]. It was found that the directional emissivities averaged over the spectral range from 1.5 to 14 μm drops from 0.8 at $\theta_v = 0°$ down to 0.5 at $\theta_v = 80°$.

Zhang et al. [240, 241] published the measurement results for two 304 austenitic stainless steel samples, polished and exposed in air for 1 h at 773 and 973 K, respectively. As shown in Fig. 6.26, a peak is observed on the spectral emissivity curve in wavelength of 0.8–1.0 μm. This phenomenon was explained based on interference principle (see [46]).

The position and amplitude of this peak are due to interference of the radiant fluxes reflected at the air-oxide and oxide-metal interfaces. The results provided by Zhang et al. [240, 241] are in accordance with those reported by Cao et al. [32].

The spectral hemispherical emissivity of Type 316 molybdenum-alloyed stainless steel with low roughness, unoxidized and thermally oxidized, is shown in Fig. 6.27a. Figure 6.27b presents spectral directional emissivities plotted against the viewing angle for fully oxidized sample at 773 K.

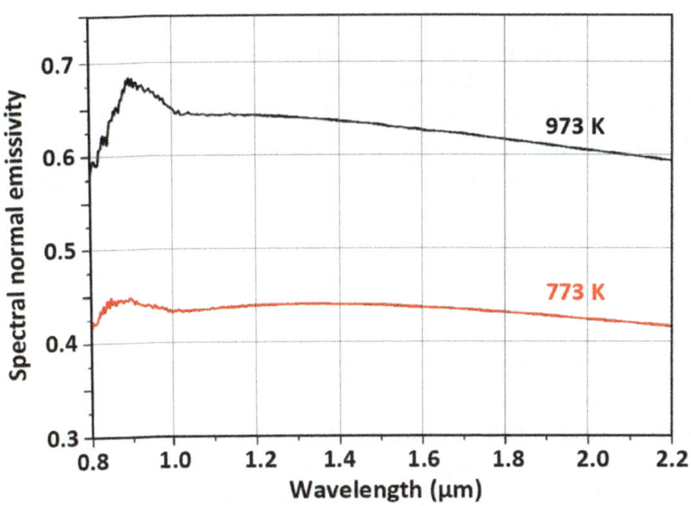

Fig. 6.26 Spectral normal emissivity of 304 austenitic stainless steel. Reproduced from [240, 241] with permission of Elsevier

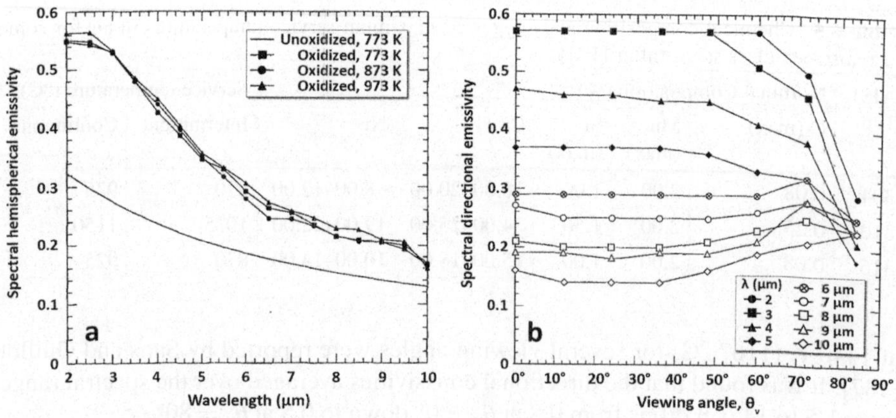

Fig. 6.27 Emissivities of 316 stainless steel: **a** spectral hemispherical, for unoxidized sample and thermally oxidized samples measured at three temperatures and **b** spectral directional emissivity plotted for 9 wavelengths against the viewing angle θ_i for the sample temperature of 773 K. Reproduced from [103] with permission of Springer Nature

Seemingly, measurements of BRDF for oxidized stainless steel in the IR spectral range were never published. We can only make some inferences based on the structure of the oxide layers, mechanisms of the incident radiation scattering, and known variations of the emissivity with the viewing angle reported by Demont et al. [47], Huetz-Aubert and Sacadura [90], and Valkonen and Karlsson [219]. One can expect the shapes of BRDF lobes similar to those for anodized aluminum (Fig. 6.23) but the actual BRDF should be greatly affected by roughness of metal and oxide surfaces, oxide layer composition, thickness, and density. The diffuse approximation can be quite acceptable for calculating the effective emissivity for most practical blackbodies. There is no reliable data on the thermal conductivity of oxide layers on stainless steel surfaces; it depends on the percentage of different types of oxides present and the compactness of the oxide layers. Different authors provide values from 0.5 W to 3 W·m^{-1}·K^{-1} for the thermal conductivity at temperatures of 800–1000 °C; however, this scattering of values is not so important, taking into account the small thickness of the oxide layer.

Oxidized stainless steel was used as material of radiating cavities of the early blackbodies with indirect resistance heating operating at temperature up to 1100 °C [56, 173, 174] and some later commercial and custom-made blackbodies (see, e.g. [126]). The radiating cavities of the early sodium heat-pipe blackbodies operating up to 800 °C [28, 181, 220, 221] were also made of stainless steel mechanically treated to increase roughness and subsequently oxidized. Presently, Inconel and other superalloys completely supersede stainless steel in the design of the high-temperature heat-pipe blackbodies but various types of stainless steel continue to be used in commercial blackbodies intended for routine calibrations of radiation thermometers and for fabrication of radiating cavities of simple custom-made blackbodies.

6.3.3 *Oxidized Inconel*

Inconel belongs to the class of so-called superalloys that are nickel-, iron-nickel-, and cobalt-base alloys generally used at temperatures above 500 °C and exhibiting high mechanical strength, excellent resistance to corrosion and oxidation, good surface stability, and resistance to thermal creep deformation [49, 66, 176]. The main components of some superalloys used in the design of the liquid metal heat pipes are listed in Table 6.6.

The Inconel, one of the most widespread alloys for high-temperature applications was developed in 1940s. Although there is a family of Inconel alloys, first, we mean the Inconel 600. It came to replace the stainless steel in the construction of heat-pipes with the alkali metals as the heat transfer fluid [37, 59, 76, 84, 130, 182, 236].

A large volume of studies of oxidized Inconel has been carried out since the end of 1950s [25, 54, 132, 179, 180, 202]. Until the early 1980, sandblasted and then oxidized Inconel was used as the NBS Standard Reference Material (SRM) of normal spectral emittance [149]. SRMs of Inconel have been calibrated for spectral normal emissivity at 800 K, 1100 and 1300 K for 156 wavelengths in the range from 1 to 15 μm (see Fig. 6.28). Richmond et al. [180] found that standard uncertainty varies between about 0.5 and 3% depending on the wavelength and temperature. In the earliest works (before approximately 1980), the Inconel is indicated simply as the nickel-chromium-iron alloy; in the best case, the percentage of alloy is specified. Richmond and Steward [179] specifies the nominal Inconel composition as 80% nickel, 14% chromium, and 6% iron. The investigations of 1980s are more specific: Jorgensen et al. [105, 106] studied the spectral hemispherical reflectance at unspecified incidence angle (perhaps, near-normal) of sandblasted and air-oxidized at 1000 °C for 10 h Inconel 600 as a function of the sample temperature within the solar

Table 6.6 The main components of some superalloys (after [49])

Alloy	Composition (%)									
	Cr	Ni	Co	Mo	W	Nb	Ti	Al	Fe	C
Incoloy 800H	21	33							45.8	0.08
Haynes 230	22.0	55.0	5.0 max	2.0	14.0			0.35	3.0 max	0.10
Inconel 600	15.5	75.0	5.0 max						8.0	0.08
Inconel 617	22.0	55.0	12.5	9.0				1.0		0.07
Inconel 718	19.0	52.5		3.0		5.1	0.9	0.5	18.5	0.08 max
Hastelloy N	7.0	72.0		16.0			0.5 max		5.0 max	0.06

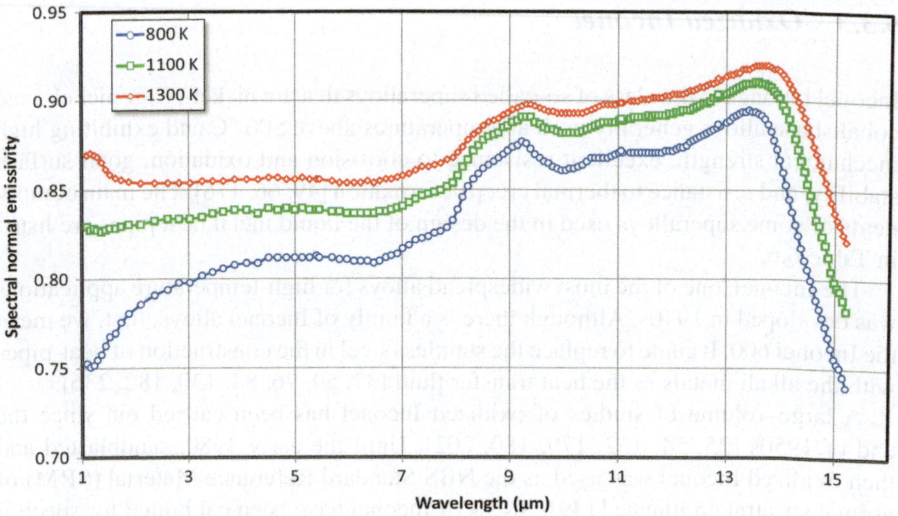

Fig. 6.28 Spectral normal emissivity of the oxidized Inconel NBS Standard Reference Material [180]. Reprinted courtesy of the National Institute of Standards and Technology, U.S. Department of Commerce. Not copyrightable in the United States

radiation spectral range (0.3 to 2.2 μm). Cunnington et al. [43] measured the spectral normal hemispherical reflectance of oxidized Inconel 617 at room temperature.

Hill and Woods [81] presented the results of the spectral directional-hemispherical reflectance measurements of the oxidized Inconel 600 sample at ambient temperature. The sample was cut from a baffle of a high-temperature furnace and, in fact, was the witness sample for the cesium- and sodium-filled heat-pipe blackbodies operating in the temperature ranges 300–660 and 600–1000 °C, respectively. Figure 6.29 presents the combined curve: for the wavelengths from 0.5 to 2.5 μm, measurements were performed at the incidence angle $\theta_i = 8°$; for greater wavelengths, the incidence angle $\theta_i = 15°$ was used. The measurement uncertainty for the shortwave spectral range was assessed as ± 0.005 and for remaining spectral range as ± 0.045.

Unfortunately, as in the case of stainless steel, we do not have data on BRDF of oxidized Inconel or other similar alloys. In the absence of such data, a certain interest is represented by Mastryukov et al. [136], who investigated the dependences of angular distributions of directional emissivity and the degree of polarization of emitted radiation on oxidation conditions for chromium-nickel alloys including X15H60 (Cyrillic notation), which is close in composition to Inconel 600. An increase in the oxidation temperature from 800–1000 °C at the duration of 20 h makes it possible to trace the transition of the spectral radiance indicatrices from the typical for conductors to that typical for dielectrics.

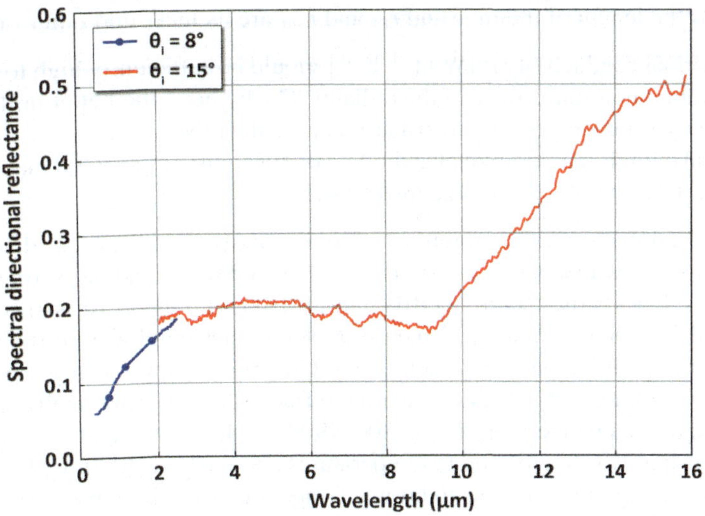

Fig. 6.29 Spectral directional reflectance of the oxidized Inconel 600 sample. The graph is combination of measurement results obtained with different instruments for two different incidence angles θ_i. Reproduced from [81] with permission of Springer Nature

6.4 Graphite

For high-temperature blackbodies with the direct resistance heating, the radiating element is, at the same time, a heating element. Therefore, the additional requirements are imposed on the physical properties of the material for a directly heated radiator:

- Material of the radiating element must withstand high temperature without loss of functionality and provide sufficient lifetime. This means, in particular, the melting temperature much higher than the blackbody operation temperature, low oxidation rate when the blackbody operates in atmospheric air and low sublimation rate when it operates in vacuum or in inert gas atmosphere.
- The electrical resistivity ρ [Ω·m] should be sufficiently high to avoid operation with the high current/low voltage power supply. The maximum temperature achievable at the graphite tube heating is determined by the balance of the Joule power released in the tube (the Joule power is proportional to the tube electrical resistance R) and the heat losses via various mechanism of heat transfer. The tube resistance equals

$$R = \frac{\rho L}{\pi\left(r_{\text{out}}^2 - r_{\text{in}}^2\right)}, \tag{6.12}$$

where L is the length of the tube and r_{in} and r_{out} are its inner and outer radii.

- The thermal conductivity κ [W·m^{-1}·K^{-1}] should be sufficiently high to minimize temperature non-uniformity of the radiator. The higher k the better uniformity of the steady-state temperature distribution along the tube.
- Material of high-temperature blackbody radiators must have a good machinability in order to give it the desired geometric shape.

If we recall the general requirement of high emissivity, it turns out that graphite is the most accessible and almost universal material for radiators of the high-temperature blackbodies (both VTBBs and FPBBs) with direct resistance heating, although the graphite radiating elements are sometimes employed in the design of other types of high-temperature blackbodies. Graphite occupies a very special place in high-temperature technology due to high refractoriness. Carbon melts only under high pressures at a temperature from 4800–4900 °C [192]. In vacuum or inert atmosphere, it is highly refractory up to temperatures approaching 3000 °C. However, in atmospheric air, graphite begins to burn at temperatures of about 600–700 °C.

The designing of blackbody radiators made of graphite usually includes the numerical modeling of its temperature distributions, both stationary and transient. There are a lot of commercial software packages allowing "multiphysics" approach to numerical modeling of complex systems, which in practice means simultaneous solution of problems of electrophysics and heat transfer (see, e.g. [119, 186, 184]). Calculation of transient temperature fields that is necessary for assessment of dynamic characteristics of a blackbody requires the knowledge of the thermal diffusivity a [m^2·s^{-1}]:

$$a = \kappa/(c_p d), \tag{6.13}$$

where c_p [J·K^{-1}·kg^{-1}] is the isobaric specific heat capacity (in the lumped-capacity approximation, the product of c_p and the thermal resistance is equal to the constant of the thermal inertia) and d [kg·m^{-3}] is the density (it, together with the porosity, affects the rates of oxidation in air and of sublimation in vacuo or in non-oxidizing atmosphere).

The thermal diffusivity is the parameter characterizing the rate of temperature change in transient thermal processes. This is a natural measure of the thermal inertia of a solid, which is part of the heat equation:

$$\partial T/\partial \tau = a\nabla^2 T, \tag{6.14}$$

where τ is the time. The thermal expansion becomes playing the important role at the design of high-temperature blackbodies. The linear coefficient of thermal expansion (CTE) α_L [K^{-1}] is defined via the length L and temperature T as

$$\alpha_L = (1/L) \cdot (dL/dT). \tag{6.15}$$

If α_L does not change much over the change in temperature ΔT and the fractional change in length is small ($\Delta L << L$), the change in the linear dimension can be evaluated as

$$\Delta L = \alpha_L L \Delta T. \tag{6.16}$$

Otherwise, Eq. 6.16 must be integrated:

$$L_2 = L_1 \exp\left(\int_{T_1}^{T_2} \alpha_L(T) dT\right), \tag{6.17}$$

where L_1 and L_2 are the initial and final lengths at corresponding temperatures T_1 and T_2.

Knowledge of κ, d, c_p, α_L, and ρ for at least several temperatures from the room temperature to the upper limit of operation temperatures as well as the emissivity data for graphite allows performing complete numerical modeling of the high-temperature graphite blackbody.

6.4.1 Physical Properties of Synthetic Graphites

Every graphite consists entirely of pure carbon, but not every pure carbon is graphite. Graphite is only one, the most widespread and the most stable under standard conditions allotrope of carbon, among others (diamond, fullerenes, graphene, etc.) Graphite has a layered structure. The individual layers are planar sheets of carbon atoms (graphene sheets), where each atom bounds to three neighbors in a honeycomb lattice. Single-crystal graphite demonstrates a high degree of anisotropy of physical properties. Graphite exists in nature. Although it can be mined in large quantities, it is of little use due to mechanical weakness and a large dispersion of properties. Synthetic graphite is an artificial material manufactured by high-temperature processing in an inert atmosphere of an amorphous carbonaceous precursor that can be derived from petroleum, coal, or natural and synthetic organic materials. A high temperature is needed for the solid-state phase transition, which results in the formation of graphite crystallites. Synthetic graphite dominates in science and technology since the middle of the twentieth century [92]. Therefore, when we talk about application of graphite in the blackbody radiometry, we are discussing synthetic graphite, the man-made crystalline form of carbon. For a reader unfamiliar with manufacturing issues, the features of measuring physical characteristics, and areas of application of synthetic graphite, we recommend the reviews written by Burchell [29, 242], or a brief but informative introduction to a topic from one of the world leading manufacturers of synthetic graphite, Poco Graphite, Inc. (https://poco.com) [168]. From earlier publications is worth mentioning the handbook by Pierson [166].

The physical characteristics of graphite were intensively studied in 1950 and 1960s in connection with the increasing use of graphite in nuclear power engineering and rocket technology. The reader, however, should be aware that the characteristics of graphite varieties in older publications might not correspond to modern grades of graphite. Modern technologies allow better grinding and mixing of raw materials and, consequently, to obtain graphite with a smaller grain size. Most of the graphites produced until the mid-1970s were obtained by extrusion, which resulted in the preservation of the preferred orientation of the crystallites (grains). The ATJ grade of fine-grain (~1 μm grain size) graphite from the Union Carbide Corp. (it will later become GrafTech, https://www.graftech.com) has been an industry standard for many years. According to Kachur [108], the CTE of the ATJ graphite in the across-grain direction is about 40% greater than that in the with-grain direction. Although the anisotropy of physical properties does not directly affect the applicability of graphite as a material for high-temperature blackbody radiators, there is a strict correlation between the degree of isotropy of graphite and its resistance to high temperatures. The properties of polycrystalline graphites strongly depend on the graphite manufacturing method. For most properties (e.g., electrical and thermal conductivities, thermal expansion, strength, etc.), this dependence is based on the extremely anisotropic properties of the crystallites that form graphite. Properties not related to the crystallite anisotropy (e.g., bulk density) also depend upon the manufacturing process. Orientation of crystallites during manufacture may vary from almost unidirectional to random and results in graphites with distinctly different properties. As a rule, isotropic graphites are spatially more uniform, denser, less porous, and have a smaller grain size. Therefore, they have lower rates of burnout in air and sublimation in vacuum. Presently, almost isotropic graphite with a very fine grain size, with excellent machinability, and enhanced strength and density can be manufactured by isostatic or vibrational pressing technique [175, 123]. The difference between microstructures of “old-fashioned” anisotropic graphite and modern isotropic fine-grained graphite can be seen in polarized-light microphotographs (Fig. 6.30), where the pores look as dark areas.

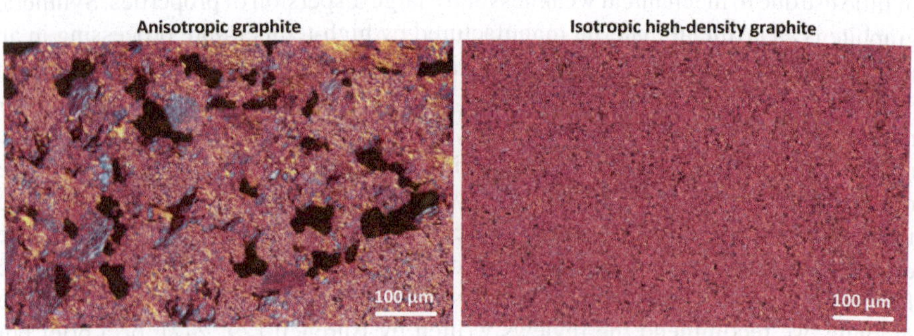

Fig. 6.30 Microphotographs in polarized light of anisotropic ordinary and isotropic high-density graphites. Reproduced from [217] with permission of Toyo Tanso Co, Ltd

There is no commonly accepted taxonomy of graphites or even an industry-wide system of graphite types, sorts, and grades. While the general characteristics of "better" graphite are understood, many "historical" grades no longer exist. So, production of the commercial graphite POCO AXM-5Q, which was the NBS Standard Reference Material of thermophysical properties from 5 to 2500 K [91] and de-facto industrial standard in 1980s, is discontinued now.

A large amount of "historical" data was recently presented by Shabalin [198] and Savvatimskiy [192]. Since the physical properties of isotropic polycrystalline graphite may vary in very wide ranges with the raw material, brand, grade, method of fabrication, etc., it makes sense to specify only the boundaries of these ranges as it was done in Table 6.7 for room temperature.

To assess the value of the physical quantity at a desired temperature, one can use the data presented in Fig. 6.31. For more precise and detailed information, user should refer to manufacturer's documentation.

Table 6.7 Typical characteristics of modern isotropic graphites at room temperature

Characteristics	From	To
Bulk density ($g \cdot cm^{-3}$)	1.5	1.95
Electrical resistivity ($\mu\Omega \cdot m$)	9	17
Thermal expansion coefficient (K^{-1})	$3 \cdot 10^{-6}$	$15 \cdot 10^{-6}$
Thermal conductivity ($W \cdot m^{-1} \cdot K^{-1}$)	70	140
Heat capacity at constant pressure ($kJ \cdot kg^{-1} \cdot K^{-1}$)	0.65	0.75

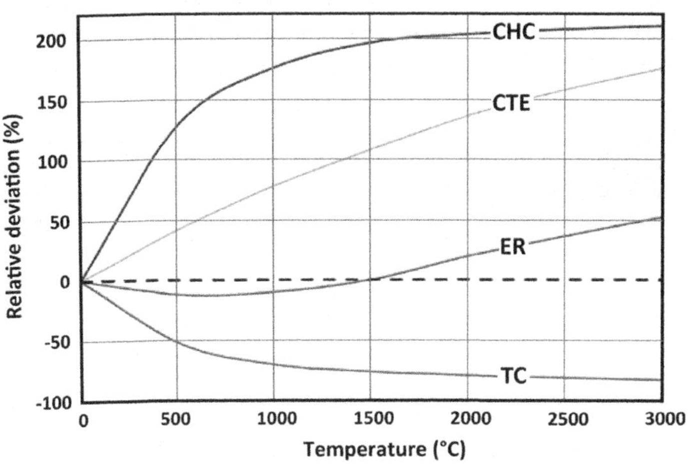

Fig. 6.31 Relative deviations of the specific heat capacity at constant pressure (CHC), coefficient of linear thermal expansion (CTE), electrical resistivity (ER), and thermal conductivity (TC) of a typical isotropic polygranular graphite from their nominal values at 0 °C plotted against temperature. Reproduced from [97] with permission of John Wiley and Sons

6.4.2 Radiation Properties

For many years, the main source of data on thermal radiative properties of graphite was the reference book by Touloukian and DeWitt [215]. Now this book is a bibliographic rarity, but contemporary researchers continue to use its data for some materials. However, graphite definitely does not belong to such materials because:

(i) most of graphites exactly specified in [215] are no longer produced;
(ii) a change in the technology of graphite manufacturing that occurred in the 70s–80s of the last century might affect their emissivities;
(iii) the introduction of new types of radiation detectors has made possible significant improvement of the measurement accuracy, especially in the IR spectral range;
(iv) most of "old" spectral measurements were carried out with dispersive instruments and without automatic systems of data acquisition, which limited both the spectral resolution and the number of collected spectral data.

Moreover, it can be assumed, that the change in graphite manufacturing technology has led to a noticeable change in its spectral emissivity. Lapworth et al. [121] described the high-temperature VTBB operating in argon at temperatures up to 3000 K covering the spectral range from 0.2 to 5 μm. The radiating cavity, directly heated by the AC power, was made from an electrode for spectrographic analysis, i.e., from one of the most qualitative grades of graphite at that time. According to the data provided by Lapworth et al. [121], $\varepsilon_n(\lambda) = 0.35$ at 0.2 μm and grows almost linearly to $\varepsilon_n(\lambda) = 0.85$ at 5 μm.

If to compare these data with the spectral normal emissivity data obtained by Neuer et al. (1995) for the fine-grained isotropic graphites manufactured by isostatic pressing (molding) and shown in Fig. 6.32, the first thing that attracts our attention is

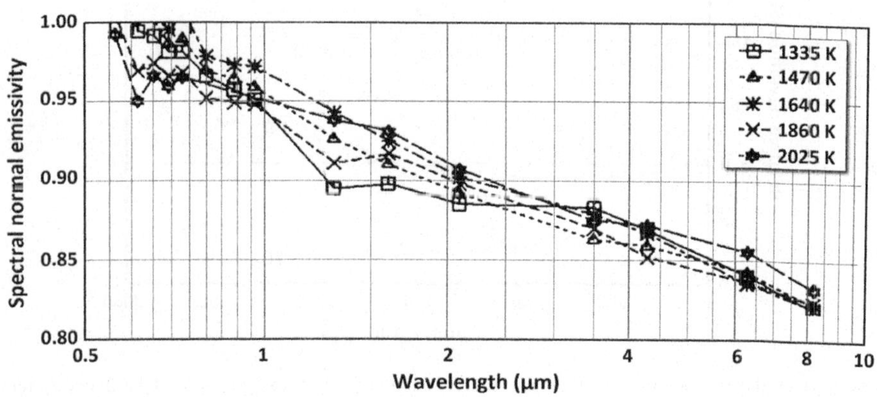

Fig. 6.32 Spectral normal emissivity of isostatically molded graphite at five temperatures. Reproduced from [151] with permission of Springer Nature

noticeably higher values of spectral emissivity of the modern graphite. The second feature is the different direction of the spectral trend.

Recently, Urban et al. (2017) measured the spectral normal emissivity of the graphite sample of similar density and surface roughness and from the same supplier (SGL Carbon; https://www.sglcarbon.com/). The modified laser-flash method [117] was used; the spectral normal emissivity at a temperature of 1529 K was measured twice: for the first time, when the sample is pre-heated in the high-temperature furnace and repeatedly, when the sample cools after heating by a solid-state laser short pulse. The results of measurement in the spectral range from 550–1100 nm are presented in Fig. 6.33 along with data of Neuer [151], interpolated for 1529 K. The measurement method used is direct, i.e. emissivity is measured according its definition but not via Eq. 6.4. Therefore, the uncertainty in the measurement of the spectral emissivity depends critically on the uncertainty in the measurement of the sample temperature. It is measured by the radiation thermometer with an absolute uncertainty of about 3.8 K. This explains the values of the spectral emissivity greater than unity.

The data obtained by Urban et al. (2017) and presented in Fig. 6.33 are in satisfactory agreement with the data obtained by Neuer [151], taking into account the measurement uncertainties.

It should be noted that the published data on the spectral emissivity have large dispersion. Since exhaustive data on the brand and grade of graphite, the method of sample preparation, the state of its surface, the measurement conditions etc. are specified infrequently, it is hard to compare the spectral emissivity data.

Graphite is not a material with strictly defined characteristics, but rather a common name for a large family of materials. Presently, the nomenclature of graphites and the number of their manufacturers throughout the world have increased significantly in comparison with the 1970s.

Differences in the state of the surface (its roughness, open porosity, degree of oxidation, etc.) additionally enhance diversity of radiation characteristics for

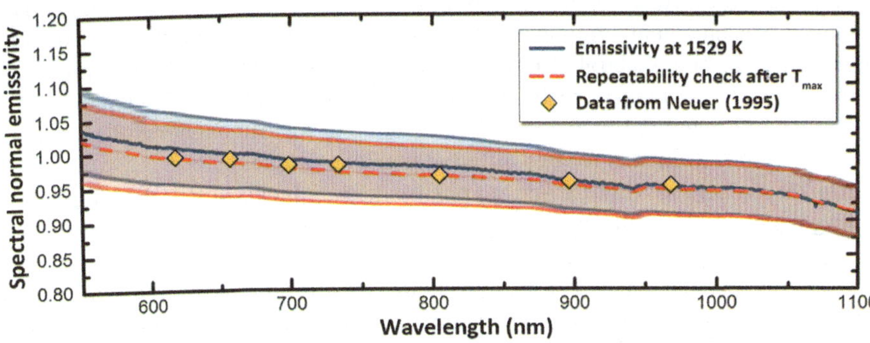

Fig. 6.33 Spectral normal emissivity measured at a temperature of 1529 K and repeatedly measured at the same temperature after reaching a maximum temperature during the heating process. The shaded strips denote combined standard uncertainties. Reproduced from [218] with permission of Springer Nature

graphite. There are no a single type or grade of graphite used exclusively for blackbody radiators: the developers of blackbodies employ the most suitable and accessible workpieces of graphite.

In connection with the emergence of reliable technique for calculating the effective emissivities of diffuse cavities [17, 18, 19] and the Monte Carlo method [188, 189] for specular-diffuse cavities, the use of so-called "reference data" on the emissivity of graphite became clearly insufficient. Therefore, the most interesting are the works performed in the context of a new approach to studying the radiation characteristics of primary blackbody standards, first described by Galal Yousef et al. [63].

The essence of this approach is the sequential implementation of the four steps:

(i) measuring the radiation characteristics of the graphite witness sample,
(ii) incorporation of measured data into the computational model for the effective emissivity,
(iii) calculating the effective emissivities for isothermal and non-isothermal conditions, and
(iv) measuring the effective reflectance of the isothermal cavity at room-temperature or (if possible) evaluation of the blackbody effective emissivity by comparison with a knowingly more accurate blackbody.

This approach guarantees the use for calculations of data for exactly the same material as in the blackbody design and allows performing an experimental validation of the adequacy of the computational model for the radiation characteristics. The analogous approach was declared in the article by Bloembergen et al. [26] in respect with the program for determining the effective spectral emissivities of graphite radiating cavities of the high-temperature fixed-point cells within the context of the "Mise en pratique of the definition of the kelvin (*MeP*-K)" at high temperature [131]. In the framework of this program, comprehensive studies of graphite (at room temperature yet) have been performed at NIST [162].

Figure 6.34 shows the spectral DHRs at the incidence angle of 8° for four samples of graphite with different degrees of roughness, from polished to very rough. The specimen was an actual component of the fixed-point cell assembly. The results of measurements of the spectral DHR obtained by Hill and Woods [81] for graphite and shown in Fig. 6.35 can serve as the confirmation for the results presented in Fig. 6.33. The graph presented in Fig. 6.35 is a combination of measurement results obtained with different instruments for two different incidence angles θ_i.

The researchers of NIST carried out precise measurements of BRDFs for graphite samples prepared from the crucibles of the radiating fixed-point cells. The measurements performed at the wavelength of 1.32 μm of laser radiation in the plane of incidence are presented in Fig. 6.36. Each point is the mean value of results measured for s and p polarization states of the incident laser radiation [73]. At the wavelength of 1.32 μm, BRDFs of the graphite sample exhibits wide specular lobes typical for moderately rough surface. The maximum of the lobe for the incidence angle $\theta_i =$ 71.5° corresponds to the viewing angle $\theta_v \approx 80°$. This shift toward the larger viewing angles known as the off-specular reflection is in accordance with the microfacet theory [213, 212].

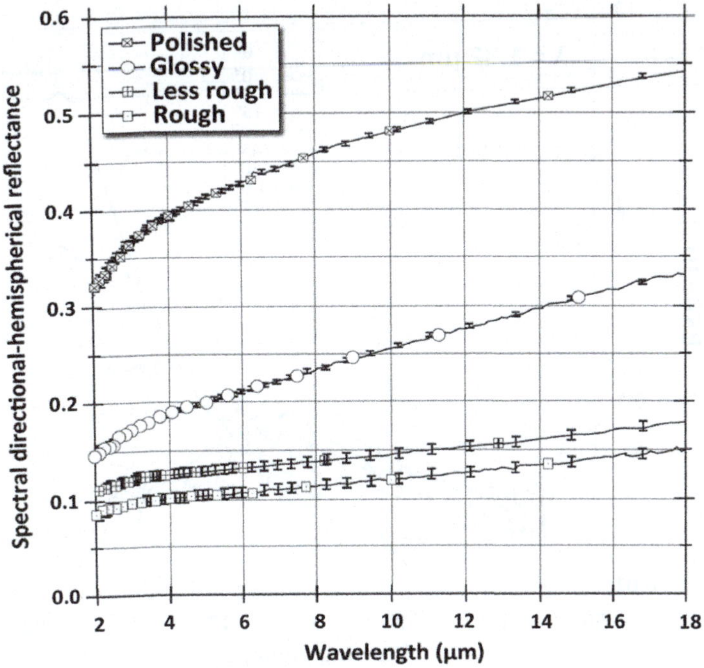

Fig. 6.34 Room-temperature spectral directional-hemispherical reflectances at near-normal incidence of graphite used in the fixed-point radiating cells for varying levels of surface roughness. Reproduced from [73] with permission of Springer Nature

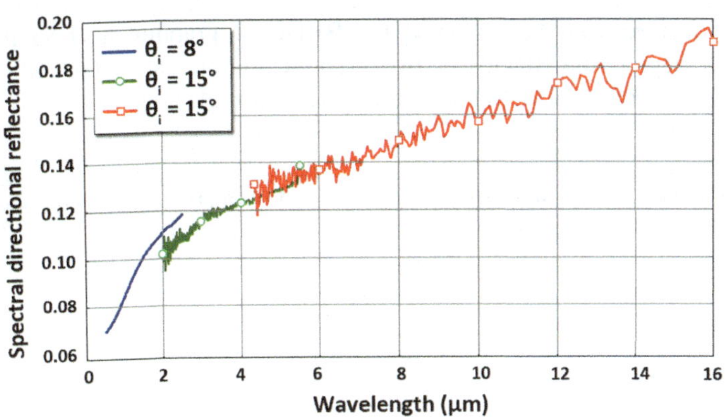

Fig. 6.35 A combined graph of the spectral directional reflectance of the graphite sample at room temperature. Reproduced from [81] with permission of Springer Nature

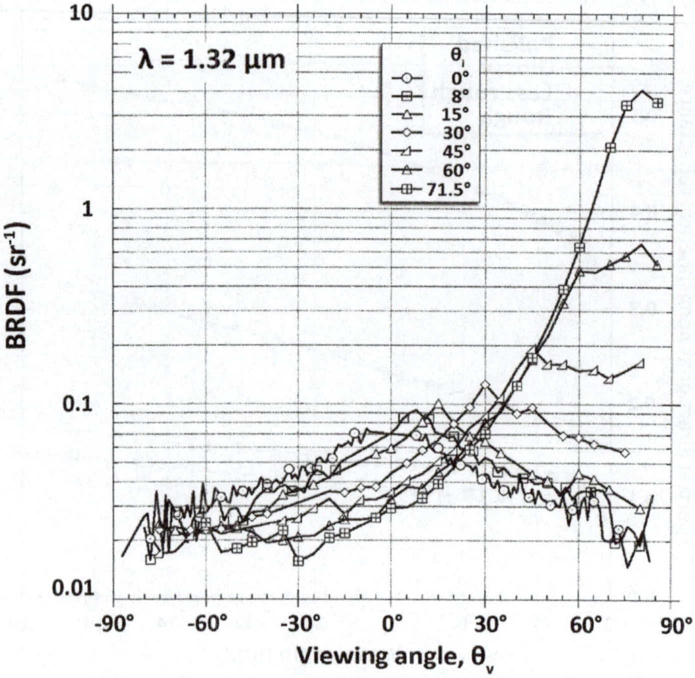

Fig. 6.36 In-plane BRDFs of a graphite sample at the wavelengths $\lambda = 1.32$ and different incidence angles θ_i. Reproduced from [73] with permission of Springer Nature

Similar features can be observed for BRDF of graphite at 1.55 μm [238]. Figure 6.37 depicts the BRDFs of two graphite samples at 0.405 and 0.658 μm measured by Patrick et al. [162] for three incidence angles.

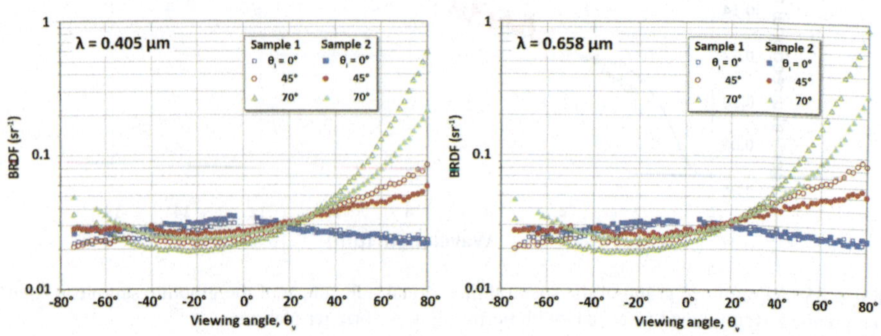

Fig. 6.37 In-plane BRDF of two graphite samples for unpolarized radiation at two wavelengths (0.495 and 0.658 μm) and three incident angles θ_i. Reproduced from [162] with permission of IOP Publishing

Analysis of the in-plane BRDFs depicted in Figs. 6.36 and 6.37 allows us identifying the mechanism of scattering that is typical for reflection from randomly rough opaque surface. Even at the normal incidence, there are significant deviations from the constant (angle-independent) Lambertian BRDF (note the logarithmic scale of the ordinates in Fig. 6.37).

Patrick et al. [162] and Germer et al. [67] provided not only in-plane BRDFs of graphite but also the color maps of the hemispherically scanned BRDFs. For representation of hemispherical BRDFs, two auxiliary variables (coordinates) were introduced:

$$x = \sin\theta_v \cos\varphi_v, \tag{6.18}$$

$$y = \sin\theta_v \sin\varphi_v, \tag{6.19}$$

where θ_v and ϕ_v are the polar angle and the azimuth of the direction of incidence, respectively.

The BRDFs at 0.658 μm for the incidence of unpolarized radiation at the angle $\theta_i = 70°$ obtained with different resolution on the axes x and y (305 and 1245 points in total) were used to plot the maps in Fig. 6.38a and Fig. 6.38b, respectively. The BRDF color maps are symmetrical with respect to the line $y = 0$. This indicates the isotropy of reflection from graphite samples. The most widely known theory of reflection from rough surfaces is the microfacet theory, according to which the rough surface is represented as an ensemble of microscopic specular flat areas, whose normals form random angles with the global (macroscopic) normal to the surface. The microfacet theory is based on geometric optics and valid when the root-mean-square surface

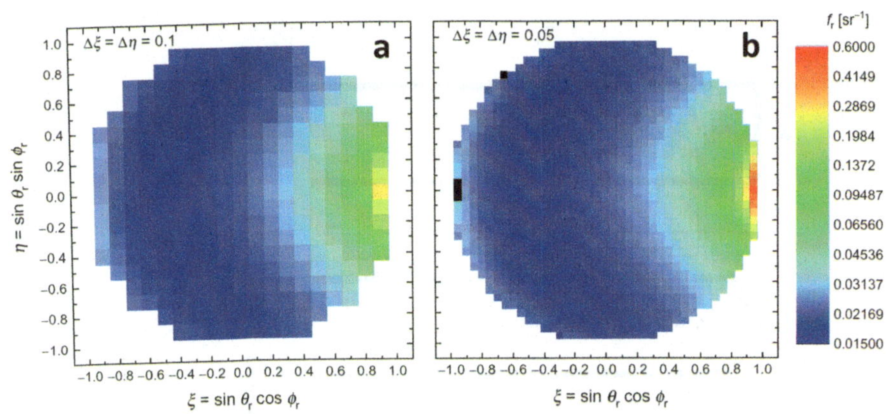

Fig. 6.38 2D color maps of BRDF measured for a sample of graphite at 658 nm of unpolarized radiation and the incidence angle of 70°. The maps are plotted for: **a** $\Delta\xi = \Delta\eta = 0.1$ (305 points) and b $\Delta\xi = \Delta\eta = 0.05$ (1245 points). The color scale is logarithmic. Reproduced from [67] with permission of Elsevier

roughness σ_m is greater than the wavelength λ of the incident radiation ($\sigma_m \gg \lambda$). Unfortunately, all known measurements of BRDFs for graphite are carried out at room temperature; we have no information on angular characteristics of reflected radiation for graphite at higher temperatures.

Of course, neglecting the real angular distributions of the emitted and reflected radiation or replacing it with a simplified diffuse or specular model can lead to considerably larger errors than in the example considered above.

We have already presented in Table 6.1 the results of fitting the parameters of the 3C BRDF model to the BRDFs of the graphite sample measured at 0.658 μm. The fitting results are presented graphically in Fig. 6.6a; dependences of the DHR and its diffuse and glossy components at 0.658 μm on the incidence angle are shown in Fig. 6.7a. According to this graph, the normal emissivity of graphite $\varepsilon(0°) = 1 - \rho(0°) = 0.91$. The natural question arises: how big the difference in the calculated effective emissivity could be, if instead of the 3C BRDF model a simpler diffuse model with $\varepsilon = 0.91$ would be applied? We have performed the computer modeling of the normal effective emissivities for two isothermal cavities of different geometry shown in Fig. 6.39. Calculations were carried out using the Monte Carlo modeling packages STEEP321 and STEEP323 from Virial International, LLC (www.virial.com). The 3C BRDF model was used with the BRDF model parameters indicated in Table 6.1 and the program STEEP323; for modeling of the diffuse cavity, the emissivity value ε was accepted to be $1 - \rho(0°) = 0.91$ for the 3C model (see Fig. 6.7a). The uncertainty of calculation for both cases was found to be less than ± 0.00001. The results of the modeling are presented in Table 6.8.

The results allow concluding that while for a cylindro-cone the error (~0.0002) can still be tolerated, it is clearly unacceptable for a cylindrical cavity (~0.0008). In general, there is no simple answer to the question posed above; the answer to it depends largely upon the cavity geometry and conditions of its observation.

a

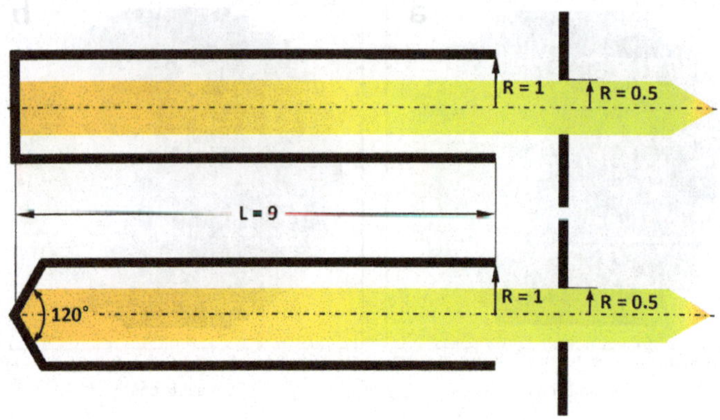

b

Fig. 6.39 Geometry data for calculation of the normal effective emissivities of: **a** cylindrical and **b** cylindro-conical isothermal cavities

Table 6.8 Results of the Monte Carlo modeling of the normal effective emissivities using two reflection models

Reflection model	Isothermal cylinder (Fig. 6.36a)	Isothermal cylindro-cone (Fig. 6.36b)
3C BRDF	0.99810	0.99874
Diffuse, $\varepsilon = 0.91$	0.99887	0.99893
$\varepsilon_{e,Dif} - \varepsilon_{e,3C}$	0.00077	0.00019

6.4.3 *Oxidation and Sublimation of Graphite*

Graphite oxidation in atmospheric air is the reaction of carbon atoms with oxygen to form carbon monoxide (CO) and carbon dioxide (CO_2). Besides, many other atmospheric components (CO_2, H_2O, N_2O, etc.) are involved in oxidation reaction with carbon. The chemical compounds formed during oxidation of graphite are volatile substances thence the oxidation reactions result in a loss of carbon atoms from the graphite surface. This affects graphite emissivity. The oxidation of graphite is highly temperature-dependent. No detectable reaction occurs at temperatures up to about 350 °C. As the temperature is increased, the rate of reaction increases rapidly [233]. The oxidation rate of any graphite at the same temperature and atmospheric conditions depends on the amount of impurities, the density of the material, and the surface area involved in oxidation reaction.

The oxidation rate increases as the surface area increases, e.g. due to large number of open pores or after coarse grinding. In this respect, the series of papers by Seo et al. [195, 196, 197] dedicated to influence of oxidation on the emissivity of nuclear graphite, including effects of density, porosity, and microstructure, is of particular interest. Purification of graphite significantly reduces the oxidation rate by removing catalytic metallic impurities.

The high-temperature characterization of modern graphites becomes the topic of the hour in connection with development of generation IV nuclear reactors to meet the goals of sustainable, safe, and competitive energy production [78]. Recently, extensive studies of nuclear graphites oxidation have been undertaken with an emphasis on their microstructure [10, 35, 36, 89]. Visual inspection makes it easy to detect the oxidized layer on graphite in the form of a black velvet coating.

An additional factor affecting the performance of a blackbody is the erosion resulting from oxidation. It increases the resistance of the graphite tube element and, therefore, decreases the temperature of a blackbody under constant voltage operation. To minimize oxidation of the graphite, argon is used to purge the interior and exterior of the graphite tube. Usually, argon is introduced through special ports at the rear part of blackbody and escapes through its opening. The radiating element is flushed thoroughly before heating, and a laminar flow is maintained during blackbody operation flowing outward. This provides a nearly inert atmosphere for the hot graphite parts even when the aperture is open to the atmosphere and eliminates the need for a window excluding problems related to measuring of its transmittance.

When heated in vacuum or inert gas atmosphere, graphite evaporates (sublimes) transforming from the solid directly to gaseous phase. Sublimation increases sharply at temperatures ranging from 2800 K in vacuum to 3600 K. The sublimation rate in argon at the pressure about atmospheric is slightly lower that in vacuum [134]. At highest temperatures, sublimation of graphite proceeds not only from the surface but also inside the internal pores, which may lead to loss of mechanical strength. Wang et al. [223] studied the effect of surface microstructure on the graphite emissivity between 1300 and 1900 K. Unfortunately, the emissivity was measured only in the wavelength range from 5.5 to 17 μm, while the shorter wavelengths are typical for the blackbody radiometry at those temperatures. Therefore, we have only indirect evidence of a weak dependence of the emissivity of graphite on temperature in vacuum (or argon).

6.5 Pyrolytic Graphite

The typical service life of graphite radiators in vacuum or argon is 500 and 50 h at temperatures of 2600 and 2900 K, respectively. When the radiator's resource is exhausted, it should be replaced [190]. However, a relatively short lifetime is not the only limitation of graphite radiators. By the end of the twentieth century, there was an urgent need for precise measurements of spectral irradiance in the shortwave part of the optical spectrum, namely, in the UV-A spectral range starting from 200 nm. The most common secondary standard sources of the spectral irradiance in the UV and visible ranges are powerful (500 W and 1 kW) tungsten filament quartz halogen lamps emitting at the wavelengths larger than 250 nm and deuterium (D_2) gas-discharge lamps used in the spectral range from 200–350 nm (the curves 1 and 2 in Fig. 6.40).

The spectral irradiance scale for this wavelength range could be conveniently established on the base of the primary blackbody source. However, in order to minimize the calibration uncertainties, the blackbody must provide the spectral irradiance levels close to those produced by the secondary standard lamps. To perform such calibrations, the temperature of the blackbody should be of 3000–3500 K, the external precision diaphragm should have a diameter of 15–20 mm, and, accordingly, the cavity opening should have a slightly larger diameter. The maximum attainable temperature of the graphite radiator is clearly insufficient to this purpose. The curves 3 and 4 in Fig. 6.38 are the spectral irradiances produced at a distance of 1 m by a perfect blackbody at 2800 K with the aperture diameter of 8 mm and at 3200 K with the aperture diameter of 16 mm, respectively. Therefore, the problem of development of the primary standard blackbody for the shortwave spectral irradiance calibration consists in increase of the blackbody operating temperature and the diameter of the radiator aperture. After many years of research, by the mid-1990s, several models of high-temperature blackbodies with radiating cavities made of pyrolytic graphite were developed at VNIIOFI (Moscow). Compared to conventional graphite, pyrolytic graphite is capable to withstand higher temperatures, less chemically active, mechanically stronger, and highly anisotropic.

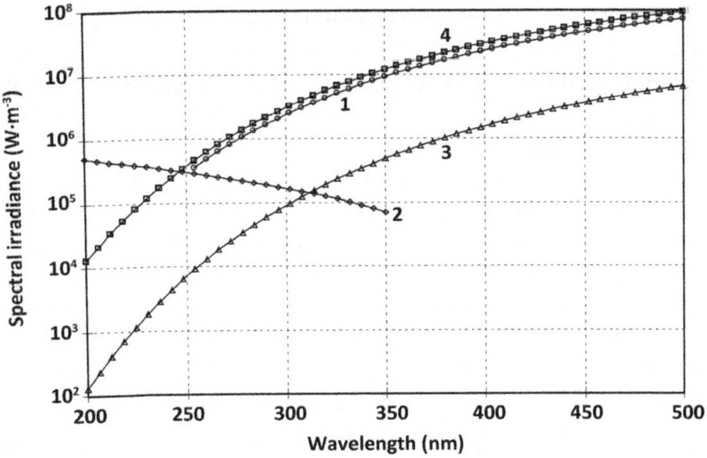

Fig. 6.40 Spectral irradiance produced by: **1** tungsten filament quartz halogen lamp FEL-1000 at a distance of 0.5 m, **2** deuterium lamp at a distance of 0.5 m, **3** blackbody at 2800 K with the aperture diameter of 8 mm at a distance of 1 m, and **4** blackbody at 3200 K with the aperture diameter of 16 mm at a distance of 1 m

Pyrolytic graphite is not a new material. Since the late 1950s, pyrolytic graphite in the form of thin-film deposit was of great interest for rocket and nuclear technology. In the 1960s, it was possible for the first time to produce pieces of almost perfect pyrolytic graphite that could be used to study more closely its bulk physical and chemical properties.

Perhaps, because of the main areas of application of pyrolytic graphite, we have a relatively small number of recent publications on this issue. Most publications on bulk pyrolytic graphite date back to the 1960 and 1970s [58, 82, 112, 113, 159, 203].

6.5.1 *Manufacturing and Properties*

Like ordinary graphite, pyrolytic graphite is pure carbon. However, they differ in structure, precursor, manufacturing methods, and most physical properties. Sometimes, the name of pyrolytic graphite is reduced to "pyrographite," which is incorrect because this is a trademark. The process involved in the production of pyrolytic graphite (PG) is known as chemical vapor deposition (CVD), a vapor-phase process, which relies on the chemical reaction of a vapor near the surface of a substrate to form solid deposits and gaseous by-products. PG is deposited by pyrolysis (thermal decomposition) of hydrocarbons.

The most common precursors are chemical compounds C_nH_{2n+2} such as methane (CH_4) or ethylene (C_2H_6). They are generally pyrolyzed over a wide range of pressure at temperatures from 750 to 2500 °C. The chemical reaction is as follows:

$$C_nH_{2n+2} \rightarrow nC + (n+1)H_2. \tag{6.20}$$

Another common precursor is acetylene (C_2H_2). The deposition of the PG film over the substrate depends on the gas temperature, pressure, composition, purity, the substrate temperature, and the angle between the gas molecules and the substrate. This process results in crystallites deposited with the basal planes (*ab* direction) essentially parallel to the deposition surface and having their *c* axes predominantly normal to the substrate. Deposited PG has the laminar structure consisting of parallel layers. The density of the final product is more than 2.2 $g{\cdot}cm^{-3}$ (cf. 1.5 to 1.95 $g{\cdot}cm^{-3}$ for ordinary graphites). Modern anisotropic PGs have high thermal and electrical conductivities in the direction parallel to the deposition surface (*ab* direction), while they are thermal and electrical insulators in the direction *c* perpendicular to the deposition plane. Some physical properties of PG are listed in Table 6.9 for temperatures up to 2500 °C.

The sublimation rate of pyrolytic graphite PGs than that of conventional high-quality graphite. Consequently, the lifetime of PG radiator at temperatures above 2900 K should be several times longer than that of similar graphite blackbody radiators. However, manufacturing a radiating cavity from PG is a non-trivial task. There is no possibility of fabrication a sufficiently large cavity directly from PG. Most manufacturers offer PG only in form of thin layers deposited or the PG plates a few millimeters thick. The PG blackbody, heated by a flowing electric current, is fabricated as follows. The rings of approximately the same thickness (5–6 mm) are cut from PG plates and then are assembled in a pile (stack) as shown in Fig. 6.41 Since the emissivity of PG is low, the bottom of the radiating cavity is made of ordinary graphite and can be flat or have a conical shape.

The high electric resistivity in the transversal direction (direction *c* in Fig. 6.40a) of PG (~100 times greater than that for ordinary graphite) makes the total electric resistance of the tube composed of PG rings higher than the typical resistance of directly heated graphite blackbodies of equivalent dimensions. Low thermal conductivity in the same direction weakens the thermal influence of the water-cooled electrodes on the viewable part of the tube playing the role of the radiating cavity. Moreover, the negative coefficient of specific electric resistivity helps to equalize the temperature distribution along the radiating cavity.

6.5.2 Spectral Emissivity

Data on the radiative characteristics of PG are extremely scarce and mainly refer to samples that are difficult to identify today, since measurements were made in the 1960 and 1970s. Moreover, because of the features of PG manufacturing, most of studies of emissivity were carried out for PG surfaces parallel to the deposition plane (*ab* plane) while the *c* surfaces form the radiating cavities of PG blackbodies (see Fig. 6.38). The main volume of data on radiation characteristics of pyrolytic graphite were published in 1960–1970s. The earliest study of spectral emissivity (see, e.g. [167]) were performed for the PG layer deposited on the graphite substrate.

Table 6.9 Physical properties of PG at temperatures from 20 to 2500 °C [198]

Property and unit	Temperature (°C)	Direction to deposition plane	
		Perpendicular	Parallel
Density ($g \cdot cm^{-3}$)	20	1.4–2.24	
Specific heat capacity at constant pressure ($kJ \cdot kg^{-1} \cdot K^{-1}$)	20	0.962–1.0	
	250	1.25	
	500	1.55	
	750	1.8	
	1000	2.0	
	1500	2.05	
	2000	2.08	
	2500	2.1	
Thermal conductivity ($W \cdot m^{-1} \cdot K^{-1}$)	20	1–5.5	190–1100
	200	1.5–5.4	290–500
	700	1.2–4.6	150–310
	1200	0.96–3.9	120–210
	1700	0.7–3.8	97–170
	2200	0.5–2.0	70–140
Coefficient of linear thermal expansion ($10^{-6} \cdot K^{-1}$)	20–100	15–25	– 1.3–1.0
	20–500	20–24	– 0.4–2.1
	20–800	–	0.5–2.6
	20–1000	22–25.5	0.7–1.1
	20–1500	22–25.5	1.2–1.6
	20–2000	24–25.5	1.2–1.9
	20–2500	24–27	2
Electrical resistivity ($\mu\Omega \cdot m$)	20	$(0.25–10) \cdot 10^3$	0.35–10
	500	$3.4 \cdot 10^3$	2.6
	700	$(2.8–3.5) \cdot 10^3$	2.5–2.7
	1700	$1.25 \cdot 10^3$	3.5
	2200	$0.7 \cdot 10^3$	4.0

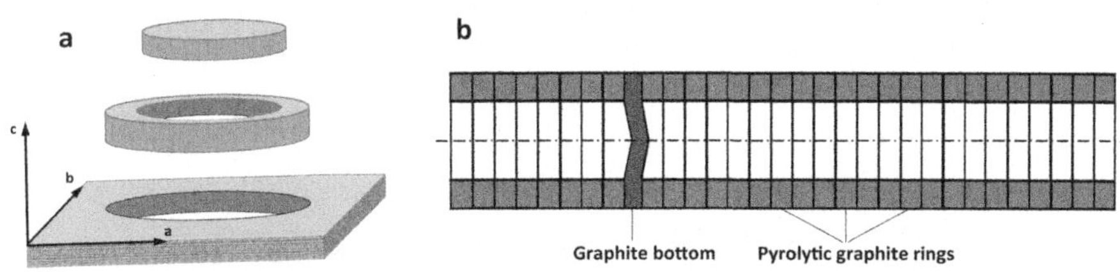

Fig. 6.41 Schematic for: **a** cutting a ring from PG plate and **b** assembling the blackbody from PG rings

Of course, measurements were carried out in the deposition plane. It was found that the spectral (at 0.665 μm) normal emissivity in the temperature range from 850 to 1800 °C is from 0.78 to 0.92, weakly varying with temperature. The band-limited (in the wavelength range from 0.2 to 3.5 μm) normal emissivity is between 0.64 and 0.88 for the same temperature range.

The results of measurements of the spectral emissivity of PG were presented by Autio and Scala (1968). Measurements were carried out over a wavelength range from 0.60 to 11 μm in atmospheres of argon or hydrogen at 830 °C. The size of the sample made it possible to measure the spectral emissivities not only in the deposition plane, but also across the layers. It was found that the spectral emissivity of PG in the plane perpendicular to the deposition plane decreases from ~0.75 around 1 μm down to ~0.6 around 10 μm (see Fig. 6.42). Although these results refer to the sample temperature near 800 °C, they do not contradict the results obtained by Wilson and Spitzer [230] for higher temperatures.

To our knowledge, there are no BRDF measurements for PG, even at room temperature. However, apparently, there is no need in such measurements because of the following reasons. First, the bottom a PG blackbody is usually made of conventional graphite; therefore, its radiation characteristics determine primarily the effective emissivity of the cavity.

The contribution of radiation characteristics of the cylindrical part of the cavity is less important. Second, the lateral walls of a cavity are formed by internal surfaces of PG rings, which are extremely rough and should have angular characteristics of reflection and thermal emission that vary from ring to ring. Apparently, the best way

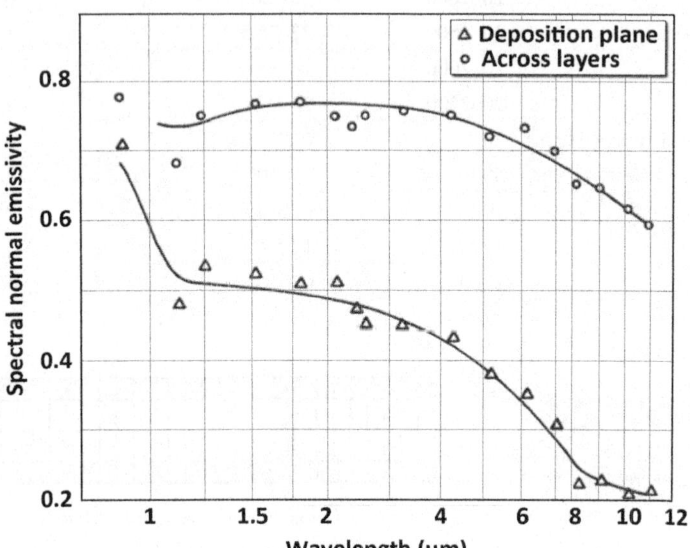

Fig. 6.42 Spectral normal emissivity of pyrolytic graphite at 830 °C. Reproduced from [9] with permission of Springer Nature

to model the effective emissivity and evaluate its uncertainty is to assume diffuse behavior of cylindrical walls and perform calculation for several admissible values of their emissivity.

References

1. S. Adachi, *Optical Constants of Crystalline and Amorphous Semiconductors. Numerical Data and Graphical Information* (Springer, New York, 1999)
2. S. Adachi, *The Handbook on Optical Constants of Metals in Tables and Figures* (World Scientific, Hackensack, NJ, 2012)
3. A. Adibekyan, C. Monte, M. Kehrt et al., Emissivity measurement under vacuum from 4 μm to 100 μm and from −40 °C to 450 °C at PTB. Int. J. Thermophys. **36**, 283–289 (2015)
4. A. Adibekyan, E. Kononogova, C. Monte et al., High-accuracy emissivity data on the coatings Nextel 811–21, Herberts 1534, Aeroglaze Z306 and Acktar Fractal Black. Int. J. Thermophys. **38**, 89 (2017)
5. A.J. Ames, Z306 Black Paint Measurements. Proc. SPIE **1331**, 299–304 (1990)
6. J.M. Ane, M. Huetz-Aubert, Stratified media theory interpretation of measurements of the spectral polarized directional emissivity of some oxidized metals. Int. J. Thermophys. **7**, 1191–1208 (1986)
7. ASTM A941–18. Standard Terminology Relating to Steel, Stainless Steel, Related Alloys, and Ferroalloys (ASTM International, West Conshohocken, PA, 2018)
8. ASTM B580–79: Standard Specification for Anodic Oxide Coatings on Aluminum. (ASTM International, West Conshohocken, PA, 2014)
9. G.W. Autio, E. Scala, Anisotropy in emissivity of single-crystal refractory materials, in *Anisotropy in Single-Crystal Refractory Compounds*. Proc. Int. Symp. Anisotropy in Single-Crystal Refractory Compounds held on June 13–15, 1967, in Dayton Ohio, vol. 1, ed. by F.W. Vahldiek, S.A. Mersol (Springer Science+Business Media, New York, 1968), pp. 357–381
10. H. Badenhorst, W. Focke, Comparative analysis of graphite oxidation behaviour based on microstructure. J. Nucl. Materials **442**, 75–82 (2013)
11. A.A. Balandin, Thermal properties of graphene and nanostructured carbon materials. Nat. Mater. **10**, 569–681 (2011)
12. B. Balling, A comparative study of the bidirectional reflectance distribution function of several surfaces as a mid-wave infrared diffuse reflectance standard. Master's thesis, AFIT/GE/ENP/09-M01 (Air Force Inst. Technol., 2009), https://apps.dtic.mil/dtic/tr/fulltext/u2/a495933.pdf. Accessed 11 Feb 2020
13. M. Barilli, A. Mazzoni, An equipment for measuring 3D bi-directional scattering distribution function of black painted and differently machined surfaces. Proc. SPIE **5962**, 59620L (2005)
14. M. Battuello, S. Clausen, J. Hameury et al., The spectral emissivity of surface layers, currently applied in blackbody radiators, covering the spectral range from 0,9 to 20 μm. An international comparison, in *Proc. TEMPMEKO '99 (The 7th International Symposium on Temperature and Thermal Measurements in Industry and Science)*, vol. 2, ed. by J.F. Dubbeldam, M.J. de Groot (IMEKO/NMi Van Swinden Laboratorium, Delft, Netherlands, 1999), pp. 601–602
15. J. Bauer, J. Mehrten, J. Young, MODIS on board calibrator blackbody performance. SPIE **3439**, 269–276 (1998)
16. W. Bauer, A. Moldenhauer, H. Oertel et al., Thermal radiation properties of different metals. Proc. SPIE **6205**, 62050E (2006)
17. R.E. Bedford, C.K. Ma, Emissivities of diffuse cavities: isothermal and nonisothermal cones and cylinders. J. Opt. Soc. Am. **64**, 339–349 (1974)
18. R.E. Bedford, C.K. Ma, Emissivities of diffuse cavities, II: isothermal and nonisothermal cylindro-cones. J. Opt. Soc. Am. **65**, 565–572 (1975)

19. R.E. Bedford, C.K. Ma, Emissivities of diffuse cavities, III: isothermal and nonisothermal double cones. Opt. Soc. Am. **66**, 724–730 (1976)
20. V.V. Belousov, A.A. Klimashin, High-temperature oxidation of copper. Russian Chem. Rev. **82**, 273–288 (2013)
21. S. Berber, Y.-K. Kwon, D. Tománek, Unusually high thermal conductivity of carbon nanotubes. Phys. Rev. Let. **84**, 4613–4616 (2000)
22. P. Berdahl, Spectral emissivity of anodized aluminum and the thermal transmittance of aluminum window frames. Nordic J. Building Phys. **3**, 1–12 (2001)
23. L.A. Berni, M.S. Ribeiro, T.F. Paes, et al., System for measuring the spatial reflectance distribution of material surfaces. J. Phys.: Conf. Series **605**, 012003 (2015)
24. D.B. Betts, F.J.J. Clarke, L.J. Cox et al., Infrared reflection properties of five types of black coating for radiometric detectors. J. Phys. E: Sci. Instrum. **18**, 689–696 (1985)
25. H.H. Blau Jr., J.R. Jasperse, Spectral emittance of refractory materials. Appl. Opt. **3**, 281–286 (1964)
26. P. Bloembergen, L.M. Hanssen, S.N. Mekhontsev et al., A Determination study of the cavity emissivity of the eutectic fixed points Co-C, Pt-C, and Re-C. Int. J. Thermophys. **32**, 2623–2632 (2011)
27. S. Boles, I. Pušnik, D. Mac Lochlainn et al., Development and characterisation of a bath-based vertical blackbody cavity calibration source for the range −30 °C to 150 °C. Measurement **106**, 121–127 (2017)
28. Brost, O., Neuer, G.: Wärmerohre zur Verbesserung der Temperaturhomogenität in Öfen. Brennsl.-Wärme-Kraft **29**, 444–449 (1977)
29. T.D. Burchell, Graphite: Properties and Characteristics, in *Comprehensive Nuclear Materials*, vol. 2, ed. by R.J.M. Konings (Elsevier, Amsterdam, Netherlands, 2012), pp. 285–302
30. J.J. Butler, G.T. Georgiev, J.L. Tveekrem et al., Initial studies of the bidirectional reflectance distribution function of carbon nanotube structures for stray light control applications. Proc. SPIE **7862**, 78620D (2010)
31. A. Cai, L. Yang, J. Chen et al., Thermal conductivity of anodic alumina film at (220 to 480) K by laser flash technique. J. Chem. Eng. Data **55**, 4840–4843 (2010)
32. G. Cao, S.J. Weber, S.O. Martin et al., Spectral emissivity measurements of candidate materials for very high temperature reactors. Nucl. Eng. Des. **251**, 78–83 (2012)
33. D. Cárdenas-García, C. Monte, Bilateral intercomparison of spectral directional emissivity measurement between CENAM and PTB. Int. J. Thermophys. **35**, 1299–1309 (2014)
34. S.L. Carman, Deep concentric grooves enhance blackbody spectral and spatial uniformity. Proc. SPIE **416**, 178–186 (1983)
35. D. Chen, Z. Li, W. Miao et al., Effects of porosity and temperature on oxidation behavior in air of selected nuclear graphites. Materials Trans. **53**, 1159–1163 (2012)
36. W.-K. Choi, B.-J. Kim, E.-S. Kim, Oxidation behavior of IG and NBG nuclear graphites. Nucl. Eng. Design **241**, 82–87 (2011)
37. B. Chu, G. Machin, The upgraded NPL blackbody calibration facility, in *Proc. TEMP-MEKO'99, 7th International Symposium on Temperature and Thermal Measurements in Industry and Science*, ed. by J.F. Dubbeldam, M.J. de Groot (IMEKO/NMi Van Swinden Laboratorium, Delft, Netherlandws, 1999), pp. 543–548
38. C.J. Chunnilall, J.H. Lehman, E. Theocharous et al., Infrared hemispherical reflectance of carbon nanotube mats and arrays in the 5–50 μm wavelength region. Carbon **50**, 5340–5350 (2012)
39. C.J. Chunnilall, E. Theocharous, Infrared hemispherical reflectance measurements in the 2.5 μm to 50 μm wavelength region using a Fourier transform spectrometer. Metrologia **49**, S73–S80 (2012)
40. F.J.J. Clarke, Measurement of the radiometric properties of materials for building and aerospace applications. Proc. SPIE **234**, 40–47 (1980)
41. F.J.J. Clarke, J.A. Larkin, Measurement of total reflectance, transmittance and emissivity over the thermal IR spectrum. Infrared Phys. **25**, 359–367 (1985)

42. S. Clausen, Spectral emissivity of surface blackbody calibrators. Int. J. Thermophys. **28**, 2145–2154 (2007)
43. G.R. Cunnington, A.I. Funai, T.K. McNab, Radiative properties of advanced spacecraft heat shield materials. NASA Contractor Report 3740 (1983), https://ntrs.nasa.gov/archive/nasa/casi.ntrs.nasa.gov/19840003190.pdf. Accessed 11 Feb 2020
44. J. De Lucas, J.J. Segovia, Uncertainty calculation of the effective emissivity of cylinder-conical blackbody cavities. Metrologia **53**, 61–75 (2016)
45. M.F.L. De Volder, S.H. Tawfick, R.H. Baughman et al., Carbon nanotubes: present and future commercial applications. Science **339**, 535–539 (2013)
46. L. del Campo, R.B. Pérez-Sáez, M.J. Tello, Iron oxidation kinetics study by using infrared spectral emissivity measurements below 570 °C. Corrosion Sci. **50**, 194–199 (2008)
47. P. Demont, H.T. Nguyen, J.F. Sacadura, Influence de l'oxydation et de la rugosité sur les caractéristiques radiatives des aciers inoxydables. Journal de Physique Colloques **42**(C1), C1–161–C1–171 (1981)
48. D.P. DeWitt, J.C. Richmond, Thermal radiative properties of materials, in *Theory and Practice of Radiation Thermometry*. ed. by D.P. DeWitt, G.D. Nutter (Wiley, New York, 1988), pp. 91–187
49. M.J. Donachie, S.J. Donachie, *Superalloys. A Technical Guide*, 2nd edn. (ASM International, Materials Park, OH, 2002)
50. C.J. Donlon, W. Wimmer, I. Robinson et al., A second-generation blackbody system for the calibration and verification of seagoing infrared radiometers. J. Atm. Oceanic Technol. **31**, 1104–1127 (2014)
51. L. Dupont, Y. Avenas, Evaluation de paramètres thermosensibles basés sur la mesure de la tension directe pour une utilisation dans des conditions d'usage. Symposium de Génie Électrique 2014 (Cachan, France, 2014), https://hal.archives-ouvertes.fr/file/index/docid/1065323/filename/SGE2014_PETS_resume.pdf. Accessed 11 Feb 2020
52. M.R. Dury, T. Theocharous, N. Harrison, et al., Common black coatings—reflectance and ageing characteristics in the 0.32–14.3 μm wavelength range. Optics Communications **270**, 262–272 (2007)
53. J.S. Dyer, R.C. Benson, T.E. Phillips et al., Outgassing analyses performed during vacuum bakeout of components painted with Chemglaze Z306/9922. Proc. SPIE **1754**, 177–194 (1992)
54. S.F. Edwards, A.G. Kantsios, J.P. Voros, et al., Apparatus description and data analysis of a radiometric technique for measurements of spectral and total normal emittance. NASA Technical Note TN D-7798 (1975), https://ntrs.nasa.gov/archive/nasa/casi.ntrs.nasa.gov/19750007880.pdf. Accessed 11 Feb 2020
55. M. Endo, M.S. Strano, P.M. Ajayan, Potential applications of carbon nanotubes, in *Carbon Nanotubes. Advanced Topics in the Synthesis, Structure, Properties and Applications*, ed. by A. Jorio, G. Dresselhaus, M.S. Dresselhaus (Springer, Berlin, 2008), pp. 13–62
56. M. Eppley, A.R. Karoli, Absolute radiometry based on a change in electrical resistance. J. Opt. Soc. Amer. **47**, 748–755 (1957)
57. E.C. Fest, *Stray Light Analysis and Control* (SPIE Press, Bellingham, WA, 2013)
58. A.S. Fialkov, A.I. Baver, N.M. Sidorov et al., Pyrographite (preparation, structure, properties). Russian Chem. Rev. **34**, 46–57 (1965)
59. J. Fischer, J. Seidel, B. Wende, The double-heatpipe black body: a radiance and irradiance standard for accurate infrared calibrations in remote sensing. Metrologia **35**, 441–445 (1998)
60. J. Fischer, M. Battuello, M. Sadli et al., *Uncertainty Budgets for Realisation of Scales by Radiation Thermometry*. CCT/03–03. (CCT-WG5 on radiation thermometry, 2003), https://www.bipm.org/cc/CCT/Allowed/22/CCT03-03.pdf. Accessed 11 Feb 2020
61. J. Fischer, P. Saunders, M. Sadli et al., Uncertainty Budgets for Calibration of Radiation Thermometers below the Silver Point. Version 1.71. (CCT-WG5 on Radiation Thermometry, 2008), https://www.bipm.org/wg/CCT/CCT-WG5/Allowed/Miscellaneous/Low_T_Uncertainty_Paper_Version_1.71.pdf. Accessed 11 Feb 2020
62. J.C. Fleming, S. Collins, B. Kelsic et al., Carbon nanotube flat plate blackbody calibrator, in *33rd Space Symp.*, Technical Track. Colorado Springs, CO. Presented on Apr.

3, 2017 (2017), https://2019.spacesymposium.org/wp-content/uploads/2017/10/Fleming_John_Carbon_Nanotube_Flat_Plate_Blackbody_Calibrator.pdf. Accessed 11 Feb 2020
63. S. Galal Yousef, P. Sperfeld, J. Metzdorf, Measurement and calculation of the emissivity of a high-temperature black body. Metrologia **37**, 365–368 (2000)
64. F.J. García-Vidal, J.M. Pitarke, J.B. Pendry, Effective medium theory of the optical properties of aligned carbon nanotubes. Phys. Rev. Let. **78**, 4289–4292 (1997)
65. F.J. García-Vidal, J.M. Pitarke, Optical absorption and energy-loss spectra of aligned carbon nanotubes. Eur. Phys. J. B **22**, 257–265 (2001)
66. B. Geddes, H. Leon, X. Huang, *Superalloys. Alloying and Performance.* (ASM International, Materials Park, OH, 2010)
67. T.A. Germer, J.C. Stover, S. Schröder, Angle-resolved diffuse reflectance and transmittance, in *Spectrophotometry: Accurate Measurement of Optical Properties of Materials*. ed. by T.A. Germer, J.C. Zwinkels, B.K. Tsai (Elsevier, Amsterdam, Netherlands, 2014), pp. 291–331
68. D. Guarnera, G.C. Guarnera, A. Ghosh et al., BRDF representation and acquisition. Computer Graphics Forum **35**, 625–650 (2016)
69. T. Guo, P. Nikolaev, A. Thess et al., Catalytic growth of single-walled nanotubes by laser vaporization. Chem. Phys. Lett. **243**, 49–54 (1995)
70. J.G. Hagopian, S.A. Getty, M. Quijada et al., Multiwalled carbon nanotubes for stray light suppression in space flight instruments. Proc. SPIE **7761**, 77610F (2010)
71. J. Hameury, A. Koenen, B. Hay et al., Identification and characterization of new materials for construction of heating plates for high-temperature guarded hot plates. Int. J. Thermophys. **39**, 16 (2018)
72. L. Hanssen, S. Mekhontsev, V. Khromchenko, Infrared spectral emissivity characterization facility at NIST. Proc. SPIE **5405**, 1–12 (2004)
73. L.M. Hanssen, S.N. Mekhontsev, J. Zeng et al., Evaluation of blackbody cavity emissivity in the infrared using total integrated scatter measurements. Int. J. Thermophys. **29**, 352–369 (2008)
74. L. Hanssen, B. Wilthan, J.-R. Filtz et al., Infrared spectral normal emittance/emissivity comparison. Metrologia **53**, 03001 (2016)
75. B. Hapke, *Theory of Reflectance and Emittance Spectroscopy*, 2nd edn. (Cambridge University Press, Cambridge, UK, 2012)
76. J. Hartmann, J. Fischer, Radiator standards for accurate JR calibrations in remote sensing based on heatpipe blackbodies. Proc. SPIE **3821**, 395–403 (1999)
77. K. Hata, D.N. Futaba, K. Mizuno et al., Water-assisted highly efficient synthesis of impurity-free single-walled carbon nanotubes. Science **306**, 1362–1364 (2004)
78. B. Hay, K. Anhalt, L. Chapman, et al., Traceability improvement of high temperature thermal property measurements of materials for new fission reactors. Proc. 3rd Int. Conf. Advancements in Nuclear Instrumentation, Measurement Methods and their Applications (ANIMMA). (Marseille, France, 2013). https://doi.org/10.1109/ANIMMA.2013.6727928
79. C.L. Hepplewhite, R.E.J. Watkins, F. Row et al., HIRDLS instrument radiometric calibration black body targets. Proc. SPIE **5152**, 223–230 (2003)
80. C.B. Herron, S.L. Steely, R.P. Young, Arnold Engineering Development Center low-background blackbody calibration. Report AEDC-TR-93–17. (Arnold Engineering Development Center, Arnold Air Force Base, Tennessee, Air Force Materiel Command, USAF, 1993)
81. K.D. Hill, D.J. Woods, Characterizing the NRC blackbody sources for radiation thermometry from 150 °C to 962 °C. Int. J. Thermophys. **30**, 105–123 (2009)
82. T. Hirai, S. Yajima, Structural features of pyrolytic graphite. J. Materials Sci. **2**, 18–27 (1967)
83. C.K. Ho, A.R. Mahoney, A. Ambrosini et al., Characterization of Pyromark 2500 paint for high-temperature solar receivers. J. Solar Energy Engineering **136**, 014502 (2014)
84. J. Hollandt, R. Friedrich, B. Gutschwager, et al., High-accuracy radiation thermometry at the National Metrology Institute of Germany, the PTB. High Temp.–High Press. **35/36**, 379–415 (2003/2004)

85. J. Hollandt, C. Monte, The determination of the uncertainties of spectral emissivity measurements in air at the PTB. Metrologia **47**, S172–S181 (2010)
86. B.G. Hoover, V.L. Gamiz, Diffractive bidirectional reflectance distributions of surfaces with large effective roughness in one dimension. Proc. SPIE **5575**, 137–142 (2004)
87. A. Höpe, Diffuse reflectance and transmittance, in *Spectrophotometry: Accurate Measurement of Optical Properties of Materials*. ed. by T.A. Germer, J.C. Zwinkels, B.K. Tsai (Acad. Press, Amsterdam, Netherlands, 2014), pp. 179–219
88. J.R. Howell, M.P. Mengüç, R. Siegel, *Thermal Radiation Heat Transfer*, 6th edn. (CRC Press, Boca Raton, FL, 2016)
89. W.-H. Huang, S.-C. Tsai, C.-W. Yang et al., The relationship between microstructure and oxidation effects of selected IG- and NBG-grade nuclear graphites. J. Nucl. Materials **454**, 149–158 (2014)
90. M. Huetz-Aubert, J.F. Sacadura, Mesure des émissivités et des réflectivités monochromatiques directionnelles des matériaux opaques. Revue Phys. Appl. **17**, 251–260 (1982)
91. J.G. Hust, Standard Reference Materials: A Fine-Grained, Isotropic Graphite for Use as NBS Thermophysical Property RM's From 5 to 2500 K. NBS Spec. Publ. 260–89. (U.S. Dept. of Commerce, Natl. Bureau of Standards, Washington, DC, 1984)
92. M. Inagaki, Applications of polycrystalline graphite, in *Graphite and Precursors*, ed. by Delhaès (Gordon and Breach Sci. Publ., Amsterdam, Netherlands, 2001), pp. 179–198
93. J. Ishii, A. Ono, A Fourier-transform spectrometer for accurate thermometric applications at low temperatures. AIP Conf. Proc. **684**, 705–710 (2003)
94. ISO 7599:2018: International Standard: Anodizing of aluminium and its alloys—Method for specifying decorative and protective anodic oxidation coatings on aluminium (ISO, Geneva, Switzerland, 2018)
95. H. Itami, Isotropic graphite for electric discharge machining, in *Handbook of Advanced Ceramics. Materials, Applications, Processing, and Properties*, 2nd ed., ed. by S. Somiya (Acad. Press, Amsterdam, 2013), pp. 135–141
96. T. Iuchi, T. Furukawa, S. Wada, Emissivity modeling of metals during the growth of oxide film and comparison of the model with experimental results. Appl. Opt. **42**, 2317–2326 (2003)
97. H. Jäger, W. Frohs, M. Banek et al. Carbon, 4. Industrial carbons, in *Ullmann's Encyclopedia of Industrial Chemistry*, vol. 6, 7th ed., ed. by V.C.H. Wiley V. C. H. (John Wiley and Sons, 2011), pp. 731–770. https://doi.org/10.1002/14356005.n05_n03
98. D.A. Jaworske, T.J. Skowronski, Portable infrared reflectometer for evaluating emittance, in *Space Technology and Applications International Forum-2000*, ed. by M.S. El-Genk. AIP Conf. Proc. **504**, 791–796 (2000)
99. D.A. Jaworske, Optical and calorimetric evaluation of Z-93-P and other thermal control coatings. Thin Solid Films **290–291**, 278–282 (1996)
100. JCGM 101:2008. Evaluation of measurement data—Supplement 1 to the "Guide to the expression of uncertainty in measurement"—Propagation of distributions using a Monte Carlo method (BIPM Joint Committee for Guides in Metrology, Paris, 2008)
101. J.M. Jones, P. Mason, A. Williams, A Compilation of data on the radiant emissivity of some materials at high temperatures. J. Energy Institute (2018). https://doi.org/10.1016/j.joei.2018.04.006
102. P.D. Jones, D.E. Dorai-Raj, D.G. McLeod, Spectral-directional emittance of oxidized copper. J. Thermophys. Heat Transfer **10**, 343–349 (1996)
103. P.D. Jones, E. Nisipeanu, Spectral-directional emittance of thermally oxidized 316 stainless steel. Int. J. Thermophys **17**, 967–978 (1996)
104. P.D. Jones, G. Teodorescu, R.A. Overfelt, Spectral-directional emittance of CuO at high temperatures. J. Heat Transfer **128**, 382–388 (2006)
105. G. Jorgensen, P. Schissel, R. Burrows, Optical properties of high-temperature materials for direct absorption receivers. Proc. SPIE **562**, 215–222 (1985)
106. G. Jorgensen, P. Schissel, R. Burrows, Optical properties of high-temperature materials for direct absorption receivers. Solar Energy Materials **14**, 385–394 (1986)

107. Jupp, D. L. B.: *A compendium of kernel and other (semi-) empirical BRDF models* (CSIRO Office of Space Science Applications—Earth Observation Centre, 2000), https://www.eoc.csiro.au/tasks/brdf/k_summ.pdf. Accessed 11 Feb 2020
108. V. Kachur, Properties of ATJ Graphite. Astronuclear Laboratory, Westinghouse Electric Corp. (1964), https://www.osti.gov/servlets/purl/4211375. Accessed 11 Feb 2020
109. L.X. Kang, D. Li, Z.Z. Yong et al., Growth of aligned carbon nanotubes and their applications, in *Industrial Applications of Carbon Nanotubes*. ed. by H. Peng, Q. Li, T. Chen (Elsevier, Amsterdam, Netherlands, 2017), pp. 381–401
110. F. Keller, M.S. Hunter, D.L. Robinson, Structural features of oxide coatings on aluminum. J. Electrochem. Soc. **100**, 411–419 (1953)
111. F.J. Kelly, On Kirchhoff's law and its generalized application to absorption and emission by cavities. J. Res. NBS **69B**, 165–171 (1965)
112. A.J. Kennedy, Graphite: a future structural material. The Aeronautical Quarterly **11**, 309–332 (1960)
113. C.A. Klein, Electrical properties of pyrolytic graphites. Rev. Modern Phys. **34**, 56–80 (1962)
114. H.-Y. Ko, B.-J. Wen, S.-F. Tsa et al., A high-emissivity blackbody with large aperture for radiometric calibration at low-temperature. Int. J. Thermophys. **30**, 98–104 (2009)
115. S. Kohara, Y. Niimi, Infrared radiation properties of anodized aluminum. Materials Sci. Forum **217–222**, 1623–1628 (1996)
116. L.R. Koirala, FTIR-Spectroscopic Measurement of Directional Spectral Emissivities of Microstructured Surfaces. Dissertation, (Helmut-Schmidt University, University of the Federal Armed Forces, Hamburg, Germany, 2004), https://www.hsu-hh.de/thermodynamik/wp-content/uploads/sites/741/2017/09/Dissertation-Koirala-1.pdf. Accessed 11 Feb 2020
117. S. Krenek, D. Gilbers, K. Anhalt et al., A dynamic method to measure emissivity at high temperatures. Int. J. Thermophys. **36**, 1713–1725 (2015)
118. M. Kumar, Y. Ando, Chemical vapor deposition of carbon nanotubes: a review on growth mechanism and mass production. J. Nanosc. Nanotechnol. **10**, 3739–3758 (2010)
119. Y.W. Kwon, *Multiphysics and Multiscale Modeling.Techniques and Applications*. (CRC Press, Boca Raton, FL, 2016)
120. Y. Lan, Y. Wang, Z.F. Ren, Physics and applications of aligned carbon nanotubes. Adv. Phys. **60**, 553–678 (2011)
121. K.C. Lapworth, T.J. Quinn, L.A. Allnutt, A black-body source of radiation covering a wavelength range from the ultraviolet to the infrared. J. Phys. E: Sci. Instrum. **3**, 116–120 (1970)
122. J. Lee, Y. Kim, U. Jung et al., Thermal conductivity of anodized aluminum oxide layer: the effect of electrolyte and temperature. Materials Chem. Phys. **141**, 680–685 (2013)
123. S.-M. Lee, D.-S. Kang, J.-S. Roh, Bulk graphite: materials and manufacturing process. Carbon Letters **16**, 135–146 (2015)
124. J. Lehman, A. Sanders, L. Hanssen et al., Very black infrared detector from vertically aligned carbon nanotubes and electric-field poling of lithium tantalite. Nano Lett. **10**, 3261–3266 (2010)
125. J. Lehman, C. Yung, N. Tomlin et al., Carbon nanotube-based black coatings. Appl. Phys. Rev. **5**, 011103 (2018)
126. A. Leupin, H. Vetsch, F.K. Kneubühl, Investigation, comparison and improvement of technical infrared radiators. Infrared Phys. **30**, 199–258 (1990)
127. W.Z. Li, S.S. Xie, L.X. Qian et al., Large-scale synthesis of aligned carbon nanotubes. Science **274**, 1701–1703 (1996)
128. J. Lohrengel, R. Todtenhaupt, Wärmeleitfähigkeit, Gesamtemissionsgrade und spektrale Emissionsgrade der Beschichtung Nextel-Velvet-Coating 811–21 (RAL 900 15 tiefschwarz matt). PTB-Mitteilungen **106**, 259–265 (1996)
129. W. Ma, T. Miao, X. Zhang et al., Thermal performance of vertically-aligned multi-walled carbon nanotube array grown on platinum film. Carbon **77**, 266–274 (2014)
130. G. Machin, B. Chu, High-quality blackbody sources for infrared thermometry and thermography between –40 and 1000 °C. The Imaging Sci. J. **48**, 15–22 (2000)

131. G. Machin, P. Bloembergen, K. Anhalt et al., Practical implementation of the Mise en Pratique for the definition of the Kelvin above the silver point. Int. J. Thermophys. **31**, 1779–1788 (2010)
132. A.G. Maki, R. Stair, R.G. Johnston, Apparatus for the measurement of the normal spectral emissivity in the infrared. J. Res. NBS **64C**, 99–102 (1960)
133. A.M. Marconnet, M.A. Panzer, K.E. Goodson, Thermal conduction phenomena in carbon nanotubes and related nanostructured materials. Rev. Modern Phys. **85**, 1296–1327 (2013)
134. É.N. Marmer, L.F. Mal'tseva, L.G. Barabanova, Investigation of the evaporation rate of graphite. Soviet Powder Metallurgy and Metal Ceramics **2**, 407–412 (1963)
135. J.L. Marshall, P. Williams, J.-P. Rheault et al., Characterization of the reflectivity of various black materials. Proc. SPIE **9147**, 91474F (2014)
136. B.S. Mastryukov, V.V. Zubov, V.A. Krivandin, Spatial distribution of radiation from oxidized heat-resistant alloys. J. Eng. Phys. **29**, 1139–1143 (1975)
137. D.M. McCroskey, G.C. Abell, M.H. Chidester, A study of Aeroglaze Z306 black paint for cryogenic telescope use. Outgassing and water vapor regain. Proc. SPIE **4096**, 119–128 (2000)
138. S.N. Mekhontsev, A.V. Prokhorov, V.B. Khromchenko, *Medium Background Blackbody BB1000. Preliminary Design Review* (Vega International, 1999), https://www.researchgate.net/publication/284173830_MEDIUM_BACKGROUND_BLACKBODY_BB1000. Accessed 11 Feb 2020
139. A. Miklavec, I. Pušnik, V. Batagelj et al., A large aperture blackbody bath for calibration of thermal imagers. Meas. Sci. Technol. **24**, 025001 (2013)
140. MIL-A-8625F: Amendment 1. Military Specification: Anodic Coatings for Aluminum and Aluminum Alloys. (US Department of Defense, 2003)
141. K. Mizuno, J. Ishii, H. Kishidac et al., A black body absorber from vertically aligned single-walled carbon nanotubes. PNAS **106**, 6044–6047 (2009)
142. M.F. Modest, *Radiative Heat Transfer*, 3rd edn. (Elsevier, New York, 2013)
143. C. Monte, B. Gutschwager, S.P. Morozova et al., Radiation thermometry and emissivity measurements under vacuum at the PTB. Int. J. Thermophys. **30**, 203–219 (2009)
144. R. Montes, C. Ureña, An overview of BRDF models. Techn. Report LSI-2012–001 (Universidad de Granada, 2012), https://digibug.ugr.es/bitstream/handle/10481/19751/rmontes_LSI-2012-001TR.pdf. Accessed 11 Feb 2020
145. M. Monthioux, P. Serp, B. Caussat et al., Carbon nanotubes, in *Springer Handbook of Nanotechnology*, 4th ed., ed. by B. Bhushan (Springer, Berlin, 2017), pp. 193–247
146. S.P. Morozova, N.A. Parfentiev, Lisiansky, et al., Vacuum variable-temperature blackbody VTBB100. Int. J. Thermophys. **29**, 341–351 (2008)
147. S.P. Morozova, B.E. Lisyanskiy, A.A. Stakharny et al., Low-temperature blackbodies for temperature range from −60 °C to 90 °C. Int. J. Thermophys. **32**, 2544–2559 (2011)
148. S.P. Morozova, A.Y. Dunaev, A.A. Katysheva et al., Facility for measuring the spatial uniformity of the radiation power of the surface of the large-area blackbody. Int. J. Thermophys. **38**, 74 (2017)
149. NBS: NBS Standard Reference Materials Catalog 1979–80 Edition. Special Publication 260 (National Bureau of Standards, Washington, DC, 1979), https://nvlpubs.nist.gov/nistpubs/Legacy/SP/nbsspecialpublication260e1979.pdf. Accessed 11 Feb 2020
150. J.T. Neu, M.T. Beecroft, R. Schramm, Extended performance infrared directional reflectometer for the measurement of total, diffuse and specular reflectance. SPIE **2260**, 62–73 (1994)
151. G. Neuer, Spectral and total emissivity measurements of highly emitting materials. Int. J. Thermophys. **16**, 257–265 (1995)
152. M. Noorma, S. Mekhontsev, Khromchenko, et al., Water heat pipe blackbody as a reference spectral radiance source between 50 °C and 250 °C. Proc. SPIE **6205**, 620502 (2006)
153. S.A. Ogarev, M.L. Samoylov, N.A. Parfentyev et al., Low-temperature blackbodies for IR calibrations in a medium-background environment. Int. J. Thermophys. **30**, 77–97 (2009)
154. S.A. Ogarev, S.P. Morozova, A.A. Katysheva, et al., Blackbody radiation sources for the IR spectral range, in *Temperature: Its Measurement and Control in Science and Industry*, vol. 8,

ed. by C.W. Meyer. Proc. 9th International Temperature Symposium. AIP Conf. Proc. **1552**, part 2, pp. 654–659 (2013)
155. H. Oikawa, Y. Shimizu, Y. Yamada, et al., A compact high-emissivity variable-temperature blackbody furnace with carbon-nanotube coated bottom, in *Proc. 56th Annual Conf. Soc. Instrument and Control Eng. of Japan (SICE)* (Kanazawa, Japan, 2017), pp. 420–423. https://doi.org/10.23919/SICE.2015.8105729
156. C.-O.A. Olsson, D. Landolt, Passive films on stainless steels—chemistry, structure and growth. Electrochim. Acta **48**, 1093–1104 (2003)
157. D.P. Osterman, S. Collins, J. Ferguson et al., CIRiS: compact infrared radiometer in space. Proc. SPIE **9978**, 99780E (2016)
158. E.D. Palik (ed.), *Handbook of Optical Constants of Solids, Volumes 1–5* (Acad. Press, San Diego, CA, 1998)
159. J. Pappis, S.L. Blum, Properties of pyrolytic graphite. J. Amer. Ceramic Soc. **44**, 592–597 (1961)
160. C.-W. Park, Y.S. Yoo, B.-H. Kim, S. Chun, S.-N. Park, Construction and characterization of a large aperture blackbody for infrared radiometer calibration. Int. J. Thermophys. **32**, 1622–1632 (2011)
161. F. Paschen, On the distribution of the energy in the spectrum of the black body at low temperatures. Astrophys. J. **10**, 40–57 (1899)
162. H.J. Patrick, L.M. Hanssen, J. Zeng et al., BRDF measurements of graphite used in high-temperature fixed point blackbody radiators: a multi-angle study at 405 nm and 658 nm. Metrologia **49**, S81–S92 (2012)
163. J. Pérez-Izquierdo, E. Sebastián, G.M. Martínez et al., The Thermal Infrared Sensor (TIRS) of the Mars Environmental Dynamics Analyzer (MEDA) instrument onboard Mars 2020, a general description and performance analysis. Measurement **122**, 432–442 (2018)
164. M.J. Persky, Review of black surfaces for space-borne infrared systems. Rev. Sci. Instrum. **70**, 2193–2217 (1999)
165. M.J. Persky, M. Szczesniak, Infrared, spectral, directional-hemispherical reflectance of fused silica, Teflon polytetrafluoroethylene polymer, chrome oxide ceramic particle surface, Pyromark 2500 paint, Krylon 1602 paint, and Duraflect coating. Appl. Opt. **47**, 1389–1396 (2008)
166. H.O. Pierson, *Handbook of Carbon, Graphite, Diamond and Fullerenes. Processing and Applications* (Noyes. Publ., Park Ridges, NJ, Properties, 1993)
167. J.D. Plunkett, W.D. Kingery, The spectral and integrated emissivity of carbon and graphite, in *Proc. of the Fourth Conference on Carbon* (Pergamon Press, Oxford, UK, 1960), pp. 457–472
168. POCO: Properties and Characteristics of Graphite for Industrial Applications. Poco Graphite, Inc. Decatur, TX (2015), https://poco.com/Portals/0/Literature/Semiconductor/IND-109441-0115.pdf. Accessed 11 Feb 2020
169. J. Prasek, J. Drbohlavova, J. Chomoucka et al., Methods for carbon nanotubes synthesis—review. J. Mater. Chem. **21**, 15872–15884 (2011)
170. A. Prokhorov, Effective emissivities of isothermal blackbody cavities calculated by the Monte Carlo method using the three-component bidirectional reflectance distribution function model. Appl. Opt. **51**, 2322–2332 (2012)
171. A. Prokhorov, N.I. Prokhorova, Application of the three-component bidirectional reflectance distribution function model to Monte Carlo calculation of spectral effective emissivities of nonisothermal blackbody cavities. Appl. Opt. **51**, 8003–8012 (2012)
172. M.A. Quijada, J.G. Hagopian, S. Getty et al., Hemispherical reflectance and emittance properties of carbon nanotubes coatings at infrared wavelengths. Proc. SPIE **8150**, 815002 (2011)
173. T.J. Quinn, A practical black-body cavity for the calibration of radiation pyrometers. J. Sci. Instrum. **44**, 221–222 (1967)
174. T.J. Quinn, M.C. Ford, On the use of the N. P. L. photoelectric pyrometer to establish the temperature scale above the gold point (1063 °C). Proc. Roy. Soc. A. **312**, 31–50 (1969)

175. S. Ragan, H. Marsh, Science and technology of graphite manufacture. J. Materials Sci. **18**, 3161–3176 (1983)
176. R.C. Reed, *The Superalloys. Fundamentals and Applications* (Cambridge University Press, Cambridge, UK, 2006)
177. Z.F. Ren, Z.P. Huang, J.W. Xu et al., Synthesis of large arrays of well-aligned carbon nanotubes on glass. Science **282**, 1105–1107 (1998)
178. Z. Ren, Y. Lan, Y. Wang, *Aligned Carbon Nanotubes. Physics, Concepts, Fabrication and Devices* (Springer, Heidelberg, Germany, 2013)
179. J.C. Richmond, J.E. Stewart, Spectral emittance of ceramic-coated and uncoated specimens of inconel and stainless steel. J. Amer. Ceramic Soc. **42**, 633–640 (1959)
180. J.C. Richmond, D.P. DeWitt, W.D. Hayes, Jr, Procedures for Precise Determination of Thermal Radiation Properties. Technical Note 252 (National Bureau of Standards, Washington, DC, 1964), https://nvlpubs.nist.gov/nistpubs/Legacy/TN/nbstechnicalnote252.pdf. Accessed 11 Feb 2020
181. T. Ricolfi, M. Battuello, F. Lanza, Heat-pipe black-body reflectometer for room temperature emissivity measurements. J. Phys. E: Sci. Instrum. **19**, 697–700 (1986)
182. T. Ricolfi, F. Lanza, Simple method for the absolute calibration of radiant heat flux meters, in *Proc. TEMPMEKO 2004, 9th Int. Symp. on Temperature and Thermal Measurements in Industry and Science*, vol. 2, ed. by D. Zvizdić (University of Zagreb, Zagreb, Croatia, 2005), pp. 903–908
183. A. Roos, C.G. Ribbing, Interpretation of integrating sphere signal output for non-Lambertian samples. Appl. Opt. **27**, 3833–3837 (1988)
184. M. Rosu, P. Zhou, D. Lin et al., *Multiphysics Simulation by Design for Electrical Machines, Power Electronics, and Drives* (Wiley, Hoboken, NJ, 2018)
185. J.M. Runge, *The Metallurgy of Anodizing Aluminum. Connecting Science to Practice* (Springer, Cham, Switzerland, 2018)
186. G. Ruocco, *Introduction to Transport Phenomena Modeling. A Multiphysics, General Equation-Based Approach* (Springer, Cham, Switzerland, 2018)
187. J.-L. Sans, E. Guillot, Assessment of emissivity measurements with a spectroradiometer at high temperature. *Solar Facilities for the European Research Area (SFERA II project)—Second Phase* (2014), https://sfera2.sollab.eu/uploads/images/jra/Files/SFERAII-D13.4.pdf. Accessed 11 Feb 2020
188. V.I. Sapritsky, A.V. Prokhorov, Calculation of the effective emissivities of specular-diffuse cavities by the Monte Carlo method. Metrologia **29**, 9–14 (1992)
189. V.I. Sapritsky, A.V. Prokhorov, Spectral effective emissivities of nonisothermal cavities calculated by the Monte Carlo method. Appl. Opt. **34**, 5645–5652 (1995)
190. V.I. Sapritsky, B.B. Khlevnoy, V.B. Khromchenko et al., Precision blackbody sources for radiometric standards. Appl. Opt. **36**, 5403–5408 (1997)
191. A.F. Sarofim, H.C. Hottel, Radiative exchange among non-Lambert surfaces. J. Heat Transfer **88**, 37–43 (1966)
192. A. Savvatimskiy, *Carbon at High Temperatures* (Springer, Cham, Switzerland, 2015)
193. C. Schlick, An inexpensive BRDF model for physically based rendering. Comput. Graph. Forum **13**, 233–246 (1994)
194. P.A. Schweitzer, *Fundamentals of Metallic Corrosion. Atmospheric and Media Corrosion of Metals.* (CRC Press, Boca Raton, FL, 2007)
195. S.-K. Seo, J.-S. Roh, E.-S. Kim et al., Thermal emissivity of a nuclear graphite as a function of its oxidation degree (1) - Effects of density, porosity, and microstructure. Carbon Letters **10**, 225–229 (2009)
196. S.-K. Seo, J.-S. Roh, E.-S. Kim et al., Thermal emissivity of a nuclear graphite as a function of its oxidation degree (2) - Effect of surface structural changes. Carbon Letters **10**, 300–304 (2009)
197. S.-K. Seo, J.-S. Roh, E.-S. Kim et al., Thermal emissivity of a nuclear graphite as a function of its oxidation degree (3) - Structural study using scanning electron microscope and X-ray diffraction. Carbon Letters **12**, 8–15 (2011)

198. I.L. Shabalin, *Ultra-High Temperature Materials. I Carbon (Graphene/Graphite) and Refractory Metals. A Comprehensive Guide and Reference Book* (Springer, Dordrecht, Netherlands, 2014)
199. J.R. Shell, II, Bidirectional reflectance: an overview with remote sensing applications & measurement recommendations. (Rochester Inst. of Technology, Rochester, NY, 2004), https://web.gps.caltech.edu/~vijay/Papers/BRDF/shell-04.pdf. Accessed 11 Feb 2020
200. T. Shibata, Nuclear graphite, in *Handbook of Advanced Ceramics. Materials, Applications, Processing, and Properties*, ed. by S. Somiya, 2nd ed. (Acad. Press, Amsterdam, 2013), pp. 113–123
201. Y. Shimizu, J. Ishii, Blackbody thermal radiator with vertically aligned carbon nanotube coating. Jap. J. App. Phys. **53**, 068004 (2014)
202. W.S. Slemp, Effects of Preoxidation Treatments on Spectral Normal and Total Normal Emittance of Inconel, Inconel-X, and Type 347 Stainless Steel. Technical Note NASA TN D-2300 (National Aeronautics and Space Administration, Washington, DC, 1964)
203. W.H. Smith, D.H. Leeds, Pyrolytic graphite, in *Modern Materials. Advances in Development and Applications*, ed. by B.W. Gonser, **7**, 139–221 (1970)
204. X. Song, K. Huan, W. Dong et al., Research on infrared radiation characteristics of Pyromark1200 high-temperature coating. Proc. SPIE **9300**, 93001S (2014)
205. J. Song, X.P. Hao, Z.D. Yuan et al., Research of ultra-black coating emissivity based on a controlling the surrounding radiation method. Int. J. Thermophys. **39**, 85 (2018)
206. D.L. Stierwalt, Infrared absorption of optical blacks. Opt. Eng. **18**, 147–151 (1979)
207. Surrey NanoSystems: Vantablack S-VIS BRDF data. Surrey NanoSystems Ltd, Newhaven, UK (2017), https://www.surreynanosystems.com/assets/media/brdf-s-vismidir-v0-2017.pdf. Accessed 11 Feb 2020
208. Surrey NanoSystems: Vantablack® S-IR Data Sheet (Surrey NanoSystems Ltd, Newhaven, UK 2017), https://www.surreynanosystems.com/assets/media/vantablack-s-ir-a4-data-brochure-2017-014.pdf. Accessed 11 Feb 2020
209. M. Tehrani, P. Khanbolouki, Carbon nanotubes: synthesis, characterization, and applications, in *Advances in Nanomaterials. Fundamentals, Properties and Applications*, ed. by G. Balasubramanian (Springer, Cham, Switzerland, 2018), pp. 3–35
210. E. Theocharous, C.J. Chunnilall, R. Mole et al., The partial space qualification of a vertically aligned carbon nanotube coating on aluminium substrates for EO applications. Opt. Express **22**, 7290–7307 (2014)
211. S. Thomas, K.J. Priestley, M. Shankar et al., Pre-launch sensor characterization of the CERES Flight Model 5 (FM5) instrument on NPP mission. Proc. SPIE **8153**, 815313 (2011)
212. K.E. Torrance, E.M. Sparrow, Theory for off-specular reflection from roughened surfaces. J. Opt. Soc. Am. **57**, 1105–1114 (1967)
213. K.E. Torrance, E.M. Sparrow, R.C. Birkebak, Polarization, directional distribution, and off-specular peak phenomena in light reflected from roughened surfaces. J. Opt. Soc. Amer. **56**, 916–925 (1966)
214. V.S. Touloukian, D.P. DeWitt, *Thermophysical Properties of Matter*, vol. 7. Thermal Radiative Properties. Metallic Elements and Alloys. (IFI/Plenum, New York, 1970)
215. V.S. Touloukian, D.P. DeWitt, *Thermophysical Properties of Matter*, vol. 8: Thermal Radiative Properties. Nonmetallic Solids. (IFI/Plenum, New York, 1972)
216. V.S. Touloukian, D.P. DeWitt, *Thermophysical Properties of Matter*, vol. 9: Thermal Radiative Properties. Coatings (IFI/Plenum, New York, 1972)
217. Toyo Tanso, *Carbon-Graphite Products* (Toyo Tanso Co, Ltd., 2018), https://www.toyotanso.com/Products/catalog.en.full.pdf. Accessed 11 Feb 2020
218. D. Urban, S. Krenek, K. Anhalt et al., Improving the dynamic emissivity measurement above 1000 K by extending the spectral range. Int. J. Thermophys. **39**, 10 (2018)
219. E. Valkonen, B. Karlsson, Spectral selectivity of a thermally oxidized stainless steel. Solar Energy Mater. **7**, 43–50 (1982)
220. L.V. Vlasov, V.I. Sapritskii, Modern cavity emitters for photometry. Meas. Techniques **34**, 156–159 (1991)

221. L.V. Vlasov, N.A. Ibragimov, A.O. Pominov, Standard infra-red sources. Proc. SPIE **2161**, 64–71 (1993)
222. P. von Böckh, T. Wetzel, *Heat Transfer* (Basics and Practice. Springer, Berlin, 2012)
223. F. Wang, L. Cheng, H. Mei et al., Effect of surface microstructures on the infrared emissivity of graphite. Int. J. Thermophys. **35**, 62–75 (2014)
224. J. Wang, Z. Yuan, X. Hao et al., A −30 °C to 80 °C stirred-liquid-bath-based blackbody source. Int. J. Thermophys. **36**, 1766–1774 (2015)
225. X.J. Wang, J.D. Flicker, B.J. Lee et al., Visible and near-infrared radiative properties of vertically aligned multi-walled carbon nanotubes. Nanotechnology **20**, 215704 (2009)
226. X.J. Wang, O.S. Adewuyi, L.P. Wang et al., Reflectance measurements for black absorbers made of vertically aligned carbon nanotubes. Proc. SPIE **7792**, 77920R (2010)
227. X.J. Wang, L.P. Wang, O.S. Adewuyi et al., Highly specular carbon nanotube absorbers. Appl. Phys. Let. **97**, 163116 (2010)
228. H. Watanabe, J. Ishii, H. Wakabayashi et al., Spectral emissivity measurements, in *Spectrophotometry: Accurate Measurement of Optical Properties of Materials*. ed. by T.A. Germer, J.C. Zwinkels, B.K. Tsai (Elsevier, Amsterdam, 2014), pp. 333–365
229. R.R. Willey, R.W. George, J.G. Ohmart, et al., Total reflectance properties of certain black coatings (from 0.2 to 20.0 micrometers). Proc. SPIE **0384**, 19–26 (1983)
230. R.G. Wilson, G.R. Spitzer, Visible and near-infrared emittance of ablation chars and carbon. AIAA J. **6**, 665–671 (1968)
231. E.J. Wollack, R.E. Kinzer Jr., S.A. Rinehart, A cryogenic infrared calibration target. Rev. Sci. Instrum. **85**, 044707 (2014)
232. B.D. Wood, J.S. Dyer, V.A. Thurgood et al., Optical reflection and absorption of carbon nanotube forest films on substrates. J. Appl. Phys. **118**, 013106 (2015)
233. L. Xiaowei, R. Jean-Charles, Y. Suyuan, Effect of temperature on graphite oxidation behavior. Nucl. Eng. Design **227**, 273–280 (2004)
234. Z.-P. Yang, L. Ci, J.A. Bur et al., Experimental observation of an extremely dark material made by a low-density nanotube array. Nano Let. **8**, 446–451 (2008)
235. Z.-P. Yang, M.-L. Hsieh, J.A. Bur et al., Experimental observation of extremely weak optical scattering from an interlocking carbon nanotube array. Appl. Opt. **50**, 1850–1855 (2011)
236. C. Yinghang, L. Yaping, L. Yongqian et al., A medium temperature radiation calibration facility using a new design of heatpipe blackbody as a standard source. Meas. Sci. Technol. **12**, 491–494 (2001)
237. S. Zeidler, T. Akutsu, Y. Torii et al., Measuring scattering light distributions on high-absorptive surfaces for stray-light reduction in gravitational-wave detectors. Opt. Express **27**, 16890–16910 (2019)
238. J. Zeng, L. Hanssen, An infrared laser-based reflectometer for low reflectance measurements of samples and cavity structures. Proc. SPIE **7065**, 70650F (2008)
239. J. Zeng, L. Hanssen, Development of an infrared optical scattering instrument from 1 μm to 5 μm. Proc. SPIE **7453**, 74530Q (2009)
240. F. Zhang, K. Yu, K. Zhang et al., An emissivity measurement apparatus for near infrared spectrum. Infrared Phys. Technol. **73**, 275–280 (2015)
241. R.Z. Zhang, X. Liu, Z.M. Zhang, Modeling the optical and radiative properties of vertically aligned carbon nanotubes in the infrared region. J. Heat Transfer **137**, 091009 (2015)
242. J.W. Zondlo, Graphite: structure, properties, and applications, in *Graphite, Graphene, and Their Polymer Nanocomposites*. ed. by P. Mukhopadhyay, R.K. Gupta (CRC Press, Boca Raton, FL, 2013), pp. 1–58

Chapter 7
Contact Measurements of Blackbody Temperatures

Abstract The chapter is devoted to measuring the temperature of blackbodies using contact thermometers with traceability to the International Temperature Scale of 1990 (ITS-90). The main types of contact temperature sensors (platinum resistance thermometers, thermistors, and thermocouples) applicable to measuring the blackbody temperature are reviewed. Some sources of systematic errors in contact temperature measurements of blackbodies are discussed. The techniques for measuring the temperature distributions over the radiating surfaces of blackbodies by means of movable and fixed contact thermometers are considered.

Keywords Contact thermometry · Blackbody radiator · Platinum resistance thermometer · Thermistor · Thermocouple · Temperature distribution

7.1 Introductory Notes

Temperature is a key parameter of a blackbody, as was argued in Sect. 1.3.1. The working temperatures of modern blackbodies serving as reference sources in the optical radiometry cover the range from the temperature of liquid nitrogen (77 K) to about 3500 K. Depending on the operating temperature of the blackbody, its temperature can be measured using a contact or non-contact (radiation) thermometer.

For practical reasons, the boundary between the areas of applicability of the contact and radiation thermometry is determined, as a rule, by the current defined temperature scale. Since 1990, the ITS-90 has been a defined temperature scale that provides a convenient set of fixed points to calibrate contact thermometers below the freezing point of silver (1234.93 K). As we already wrote in Sect. 1.3.2, the ITS-90 is an approximation of the thermodynamic temperature scale that is defined by a set of fixed points (see Table B.1 in Appendix B) and prescribed interpolation/extrapolation thermometers. In particular, from the triple point of equilibrium hydrogen (13.8033 K) to the freezing point of silver (1234.93 K), the ITS-90 is defined in terms of the temperature dependence of the electrical resistance of platinum resistance thermometers (BIPM 2018). For less accurate temperature measurements, other types of thermometers, such as thermocouples and thermistors [105], can be used to approximate the ITS-90 [12]. Temperature T_{90} is not equal to the

V. Sapritsky and A. Prokhorov, *Blackbody Radiometry*, Springer Series in Measurement Science and Technology, https://doi.org/10.1007/978-3-030-57789-6_7

thermodynamic temperature T, which is necessary for the Planck equation. The thermodynamic temperature of defining fixed points of the ITS-90 were determined by the methods of primary thermometry with the best accuracy available at the time of the scale adoption. However, about thirty years have passed since the adoption of the ITS-90. In this regard, important questions arise: what are the real deviations of the T_{90} from thermodynamic temperature T and is it possible to use the ITS-90 infrastructure for contact measurements of blackbody temperatures?

Recent progress in various techniques of primary thermometry (acoustic, dielectric-constant gas, Coulomb blockade, Johnson-noise, etc.) allowed to determine thermodynamic temperatures of the most of the ITS-90 defining fixed points more accurately than this was done to the date of the ITS-90 adoption. These techniques made possible assessing the actual differences $T - T_{90}$ as well as uncertainties $u(T - T_{90})$ of their determination. The critical analysis of data published since 1990 was carried out by Fischer et al. [36]. They presented the estimated values for the deviations of the ITS-90 from the thermodynamic temperature scale and their standard uncertainties obtained by various methods of primary thermometry. According to these data, both relative deviation $(T - T_{90})/T$ and relative uncertainty $u(T)/T$ do not exceed a few thousandths of a percent for temperatures between 70 K and the freezing point of silver (see Fig. 7.1). According to the old definition of kelvin, the difference $T - T_{90}$ was zero at the triple point of water T_{TPW}. The recent redefinition of the kelvin [20] makes only subtle changes to the thermodynamic temperatures of the ITS-90 fixed points, including T_{TPW} [70].

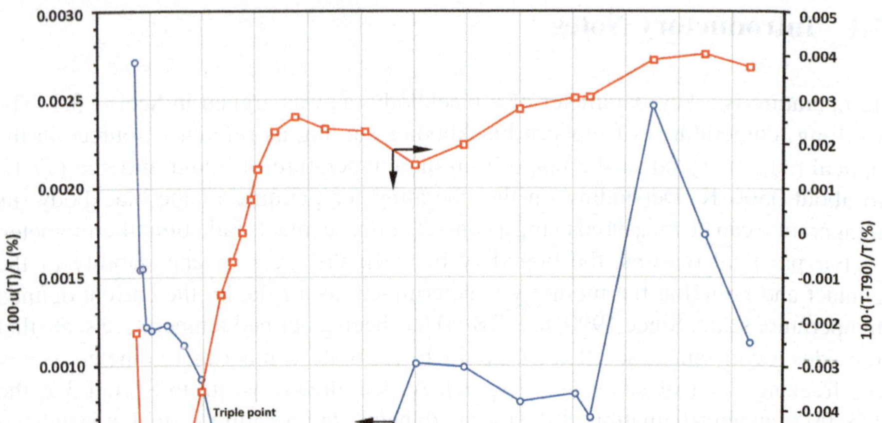

Fig. 7.1 Relative standard uncertainties of determination of thermodynamic temperature T (left axis) and percentage relative deviation of T_{90} from T (right axis)

The relative uncertainty $u(L_{\lambda,bb})/L_{\lambda,bb}$ in the spectral radiance $L_{\lambda,bb}$ of a perfect blackbody is $|C_T|$ times greater than the relative uncertainty $u(T)/T$ temperature according to Eq. 3.40. The sensitivity coefficient C_T for the Planck equation decreases from 10 to 1 when the wavelength grows from half the wavelength calculated according to the Wien displacement law (see Fig. 3.8). Therefore, if we require an uncertainty of several tenths of a percent, typical for spectroradiometric calibrations in the IR spectral range (with the possible exception of the most precision laboratory measurements), we can neglect both the correction for deviation from thermodynamic temperature and its uncertainty. In any case, uncertainties related to the deviation of temperature T_{90} from T do not make a decisive contribution to the uncertainty budget at realization of radiometric units by means of blackbodies operated at temperature below the freezing point of silver. This allows for the traceability of the ITS-90 instead of the more correct, but also more complex traceability for the SI base unit, kelvin, and the use of different contact thermometers along with well-developed measuring standards and calibration chains. Contact measurements of blackbody temperatures T_{90} are necessary only for VTBBs; for FPBBs operating at the defining fixed points of the ITS-90, measurements of temperature are not required, except for control purpose.

In the IR radiometry, it is customary to express the spectral radiance $L_\lambda(\lambda)$ in terms of radiance temperature T_S (often called "brightness temperature" in the literature on remote sensing), which can be defined for a single wavelength λ or finite wavelength range $[\lambda_1, \lambda_2]$ via the spectral radiance $L_{\lambda,bb}$ of a perfect blackbody:

$$L_\lambda(\lambda) = L_{\lambda,bb}(\lambda, T_S), \tag{7.1}$$

$$\int_{\lambda_1}^{\lambda_2} s_r(\lambda) L_\lambda(\lambda) d\lambda = \int_{\lambda_1}^{\lambda_2} L_{\lambda,bb}(\lambda, T_S) d\lambda, \tag{7.2}$$

where $s_r(\lambda)$ is the relative spectral responsivity of a sensor or its channel. Temperature of low-temperature blackbodies employed for pre-flight and in-flight calibration of remote sensing instrumentation is measured, as a rule, by contact thermometers, which, in turn, are traceable to the ITS-90. In these cases, the uncertainty $u(T_S)$ of the radiance temperature reproduced by the calibration blackbody is of the same order as the uncertainty $u(T) \approx u(T_{90})$ of the contact thermometer measuring the blackbody temperature.

One more question concerns the maximum allowable non-uniformity of blackbody's temperature, which does not result in excessive uncertainty of the spectral radiance. In other words, when a blackbody can be considered isothermal? More precisely this question can be formulated as follows: how large the temperature non-uniformity is allowed that the difference between spectral radiance of a real blackbody and spectral radiance computed at a reference temperature does not exceed a given value? The answer to this question is of great practical importance, since it determines how many temperature sensors should measure blackbody temperature, whether the hardware correction of the temperature non-uniformity should be applied, and how the temperature non-uniformity should be taken into account when calculating the effective emissivity.

Unfortunately, there is no simple answer to this question because it involves too many factors of different nature. The answer depends on geometry of a blackbody radiator, emissivity and BRDF of its radiating surface, the shape of temperature distribution and conditions of collecting the blackbody radiation (viewing conditions). As for two last factors, one can assert only that temperature non-uniformity of the directly viewable part of a radiator has a greater impact on registered spectral or bandlimited radiance than that over the areas, whose radiation reaches the detector only after one or more reflections from radiator's wall. In the general case, computer modeling of the effective emissivities corresponding to various physically admissible temperature distributions is necessary.

7.2 Overview of Contact Thermometers

7.2.1 Principles of a Contact Thermometer Selection

Let us consider the most important issues that should be considered when choosing a temperature measuring instrument. Definitions of parameters affecting the choice of a contact thermometer such as the range of measured temperatures, instrumental measurement uncertainty, resolution, sensitivity, and step response time, are given in the International Vocabulary of Metrology (JCGM 200:2012).

The **temperature range** largely determines the choice of the physical principle of converting temperature into an electrical signal, and hence the type of temperature sensor. The sensor must survive the temperatures to it is exposed. Figure 7.2 shows the temperature range capabilities of various contact sensors. Actually, two types of

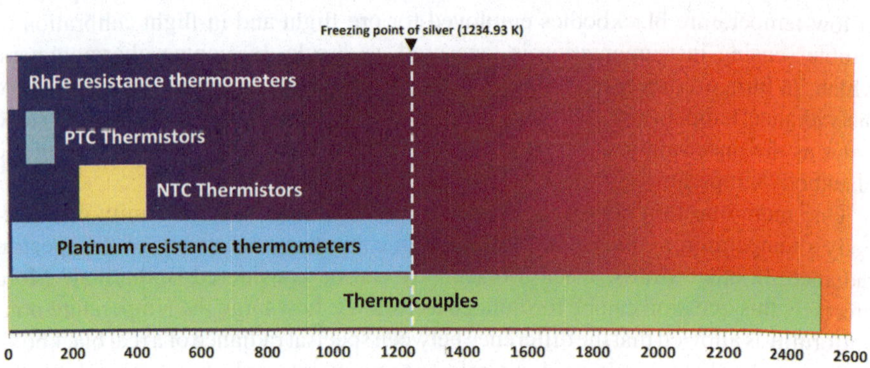

Fig. 7.2 Operating temperature ranges of various contact thermometers

contact sensors are commonly used to measure the temperature of blackbodies: thermocouples and resistance thermometers (including platinum resistance thermometers, NTC thermistors, PTC thermistors, rhodium-iron resistance thermometers, etc.). Since temperature measurements themselves are not the subject of this book, we only recall the basic principle of operation of these sensors in order to outline the area of their use in blackbody radiometry.

The resistance thermometers employ reproducible dependence of electrical resistivity of a sensing element on temperature. The resistivity of platinum like other metals depends on temperature almost linearly within a large temperature range. Platinum has intrinsic immunity to most impurities and does not oxidize up to high temperatures. The platinum resistance thermometers (PRTs) occupy a leading position among other resistance thermometers due to a wide operating temperature range, excellent reproducibility and stability. Among other type of resistance thermometers, the RhFe (0.5% Fe in Rh) resistance thermometers used in temperatures below 30 K should be mentioned.

Dependence of electrical resistance on temperature for semiconductors has more complicated character than for metal; the thermal coefficient of resistance than can be positive as for metals or negative. Within a narrow temperature ranges, semiconductor resistance thermometers (thermistors) may exceed metals in sensitivity by several orders of magnitude.

The NTC (negative temperature coefficient) thermistors measuring temperatures in the range from −50 to 150 °C are the popular contact thermometers for small low-temperature blackbodies, especially those designed for remote sensing applications. The PTC (positive temperature coefficient) thermistors are widely used in cryogenic applications.

The thermocouples employ the Seebeck effect, a phenomenon in which a temperature difference between two dissimilar electrical conductors produces the electromotive force (EMF) resulting in measurable currents or voltages. Unlike resistance thermometers that react to the temperature value of the sensor, the EMF of thermocouples depends on temperature difference. Unlike resistance thermometers, the thermocouples do not require power supply and this, in some cases, gives them a definite advantage.

The PRTs are used in most temperature measurements that are critical to accuracy. Individually calibrated PRTs operating below 600 °C have undoubted superiority in sensitivity, repeatability, stability, and accuracy over thermocouples. PRTs are passive sensors: to measure resistance of a PRT, an electric current has to be passed through its sensing element; therefore, an external power supply is needed.

One of the main disadvantages of PRTs is the self-heating effect occurred because even a vanishingly small current passing through a sensor creates a certain amount of heat and, therefore, introduces an error in the temperature measurement. This effect can be non-negligible at measurements of highest accuracy and usually requires three or four leads running to the sensor for compensation. The PRT of highest precision uses in the sensor a platinum wire wound in a strain-free suspension, which makes them fragile and vibration-sensitive. Such sensors cannot be used in air-born or satellite-born blackbody calibrators. Thick-film or thin-film PRTs are less accurate and stable but rugged, cheaper, and can be easily used for surface temperature measurements.

Selecting either a PRT or a thermistor depends completely on measurement conditions and required accuracy. If temperature range desired matches relatively narrow operating range of a thermistor, it ensures a greater sensitivity. The higher sensitivity allows two-wire resistance measurement to be performed. The nomenclature of commercially available thermistors is wide enough to choose optimal resistance and sensitivity for a desired temperature range. In addition, thermistors are typically much cheaper than other sensors.

In comparison with PRTs, thermocouples are typically more rugged, have broader temperature ranges, higher upper temperature limits, and faster response times. A thermocouple is a self-powered device and thus does not require a power supply. However, the output signals of thermocouples are only of order of millivolts, so precision amplifiers are required for further processing.

The self-heating problem does not arise with thermocouples but, unlike PRTs, special extension wires and cold junction compensation are needed for thermocouples. Thermocouples are mainly less accurate and have worst long-term stability than PRTs. Due to corrosion of the wires they require additional maintenance and more frequent calibrations. Thermocouples are more noise-sensitive and have lower signal-to-noise ratio (SNR) than PRTs. The signal drift of thermocouples is erratic, unpredictable, and may growth due to contamination and oxidation at high temperatures in contrast to PRTs that have predictable drift.

Contact temperature measurements play a significant role in our life thus a lot of books on (as well as chapters in collective monographs specifically devoted to) this subject is written for readers of various background and spheres of interest. As an introduction to contact temperature measurements as a whole, we can recommend the brochure by Rusby [94]. There are excellent monographs by Quinn [91] and Michalski et al. [75], the collective monograph edited by Lipták [67] and the three-volume set edited by Bentley [7–9] and dedicated to general questions of temperature measurements. The metrological aspects of contact thermometry and, especially, the traceability issues are highlighted by Nicholas and White [80].

The **instrumental measurement uncertainty** of a thermometer is a component of measurement uncertainty that arises from a thermometer or an entire measuring system.

Figure 7.3 shows the uncertainty ranges of the most common temperature sensors. The bars show the lowest achievable uncertainties and a more general practical assumed uncertainty. The thermometer uncertainty can be obtained from its calibration or adapted from manufacturer's specification.

The instrumental measurement uncertainty of a thermometer is evaluated as Type B uncertainty when drawing up the temperature uncertainty budget. We shall assume hereinafter that we deal with temperature sensor already calibrated, so the measured temperature is traceable to the ITS-90. We will also assume that the measuring device (digital voltmeter or multimeter, resistance bridge, etc.) is fully compatible with the temperature sensor and does not introduce noticeable errors in the measured

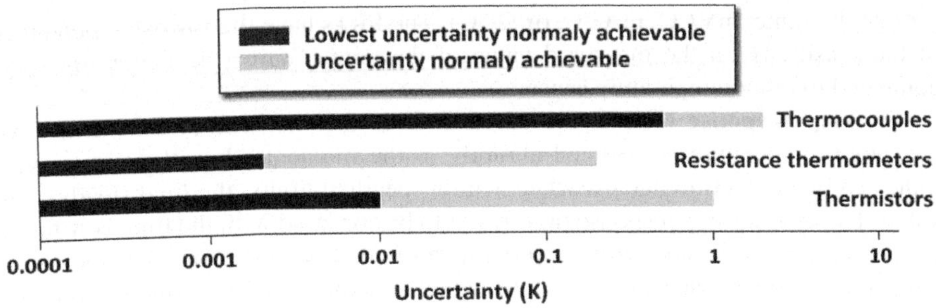

Fig. 7.3 Typical instrumental uncertainty ranges of various temperature sensors

temperature. In the best case, the sensor and the indicating instrument must be calibrated together. The digital contact thermometers, i.e. devices that measure temperature through direct contact with a sensor (PRT, thermistor, or thermocouple) and provide a digital output or display, are commonly used in laboratory temperature measurements. The standard ASTM E2877-12$^{\varepsilon 1}$ [4] introduces the nine accuracy classes based on the tolerance, which is the maximum permissible measurement error according to JCGM 200:2012 [60], for the digital contact thermometers operating in the temperature range from −200 to 500 °C (see Table 7.1). Since many digital thermometers allow changing the sensor (probe) by simple disconnecting one sensor from the measuring unit and connecting another one, it must be emphasized that a digital thermometer accuracy class refers to a particular combination of a sensor and a measuring instrument, and that changing this combination may change the accuracy class. A tolerance can be roughly considered as an uncertainty with a coverage factor of $k = 3$ (for the confidence level of 98%) or equal to an uncertainty with a coverage factor of $k = 2$ (confidence level of 95%) multiplied by 1.7. For instance, a calibrated digital thermometer with an expanded measurement uncertainty of 0.01 °C (at $k = 2$) has the tolerance value of 0.015 °C. This thermometer belongs to the Accuracy Class B.

The temperature **resolution** is the smallest change in temperature being measured that causes a perceptible change in the corresponding indication (as a rule, the value of an electric quantity—voltage, electric current or electrical resistance). Resolution of a thermometer can depend on measured temperature, its random fluctuations, or on noise in the thermometer itself. The resolution defines the best achievable precision.

The **sensitivity** of a temperature measuring system is the quotient of the change in an indication and the corresponding change in a value of a measurand. The sensitivity of a thermometric system can have dimension of, e.g. $\mathrm{mV \cdot K^{-1}}$, $\mathrm{mA \cdot K^{-1}}$, or $\Omega \cdot \mathrm{K}^{-1}$

Table 7.1 Accuracy classes and their respective tolerances for digital thermometers measured temperature from −200 to 500 °C by contact method (after [4])

Class	A	B	C	D	E	F	G	H	I
Tolerance, ≤ (°C)	0.01	0.02	0.05	0.1	0.2	0.5	1.0	2.0	7.0

(or that the same, mV/°C, mA/°C, or Ω/°C). The PRTs have the lowest dependence of the sensitivity on the measured value of the temperature (the better linearity) compared to other temperature sensors.

The **step response time** is the duration from the moment of time when the measured temperature is changed abruptly to the moment when the temperature indicated by the thermometer reaches, within specified limits, the final steady-state value. The thermal step response time τ_x not to be confused with the time constant τ describing thermometers within approximate lumped-parameter model, for which temperature dependence of the output signal is expressed by a single exponential function and τ is the time required for a thermometer to reach about 63.2% $(1 - 1/e)$ of the steady-state reading. A real-world thermometer is a system with distributed parameters; its reaction on the step change of temperature is described by an infinite series of the system-dependent parameters multiplied by exponents with different time constants. The response time τ_x for a thermometer depends on incoming and outcoming heat fluxes, so it is influenced by the thermometer's design, conditions of heat exchange between thermometer's elements with its surrounding, as well as thermophysical parameters of that surrounding (e.g. thermal diffusivity and contact thermal resistance for the case of the solid body; viscosity, density, and flow velocity for fluids, emissivity for the case of radiative heat exchange, etc.) The thermal response time for each thermometer is usually measured from thermal equilibrium at ambient temperature ≈ 23 °C to about 60 °C in the silicone oil bath. Response times for 10%, 50%, 63.2%, and 90% of the steady-state reading ($\tau_{0.1}$, $\tau_{0.5}$, $\tau_{0.632}$, and $\tau_{0.9}$, respectively) are the most common though not mandatory characteristics of thermometer's reaction rate. Making the choice of the thermometer, one should not lose sight of the fact that a contact temperature sensor performs correct measurements only when its temperature reaches or, at least, approaches the temperature of the measured object. Therefore, the thermal step response time is not so important for thermometers measured the steady-state reference temperature of blackbodies as for temperature sensors monitoring blackbody temperature's random fluctuations or sensors that define the radiator's heating rate.

The **size and shape** of a sensor or probe, together with its thermal properties, determine the step response time and the distortion, which can be introduced into the of the temperature field of a measured object. Specific measurement conditions also influence the selection of a contact thermometer. When the internal temperature of a solid body (e.g. inside a thick wall of the blackbody) is measured, the temperature sensor should be immersed into the bore or thermometric well made in the blackbody to arrange the sensor as close to its radiating surface as possible. For thin-walled blackbodies, the temperature sensor located on the external surface of the radiating cavity must ensure reliable thermal contact with it and, at the same time, not violate its thermodynamic equilibrium.

Among other things, selection of a contact thermometer for measuring blackbody temperature must be made between detachable and permanently mounted sensors. The first ones can be removed from the blackbody for separate calibration or replacement without disassembling of a blackbody; others are non-removable in principle or require full or partial destruction of a blackbody. In any case, the choice of a contact sensor should be made on the base of clear understanding of heat transfer conditions in each specific case, of all advantages and disadvantages inherent to each sensor.

Conclusion is very predictable: the right choice of the contact thermometer must be done on the base of the temperature range of interest, desired accuracy, applicability of a certain sensor to solution of tasks stated, and, finally yet importantly, the cost of a thermometer.

7.2.2 *Standard Platinum Resistance Thermometers*

According to the ITS-90 [89], in the range of temperatures from the triple point of equilibrium hydrogen to the freezing point of silver ($13.8033\ \mathrm{K} \le T_{90} \le 1234.93\ \mathrm{K}$), the temperature T_{90} is defined by means of a PRT calibrated at specified sets of defining fixed points using specified reference and deviation functions for interpolation at intervening temperatures. Temperatures T_{90} are determined in terms of the ratio $W(T_{90})$ of the resistance $R(T_{90})$ at a temperature T_{90} and the resistance $R(273.16\ \mathrm{K})$ at the TPW:

$$W(T_{90}) = R(T_{90}) / R(273.16\ \mathrm{K}). \tag{7.3}$$

A PRT made from pure, strain-free platinum is acceptable for the ITS-90 realization if it satisfies at least one of the following two inequalities:

$$W(29.7646\ ^\circ\mathrm{C}) \ge 1.11807, \tag{7.4}$$

$$W(-38.8344\ ^\circ\mathrm{C}) \le 0.844235. \tag{7.5}$$

Such a PRT is called standard PRT (or SPRT, for short) and are usually not designed for use above 660 °C. An SPRT, designed especially for use up to the freezing point of silver, is called high-temperature SPRT (HTSPRT). It must satisfy the third relation:

$$W(961.78\ ^\circ\mathrm{C}) \ge 4.2844. \tag{7.6}$$

All PRTs, which do not meet the above inequalities, are called industrial PRTs (IPRTs) Division of PRTs into standard and industrial ones was not stipulated in the original text of the ITS-90 [89] but was legitimized in accompanied materials. Presently, these terms are in common use despite the fact that the SPRT is not a measurement standard[1] as well as the use of IPRTs is not limited by industrial applications only, but extends into many other fields of science, technology, and everyday life.

[1] Let us recall that VIM [60] defines measurement standard as "realization of the definition of a given quantity, with stated quantity value and associated measurement uncertainty, used as a reference."

Currently, there is a wide range of commercially available SPRTs that differ in accuracy, long-term stability, and reliability. Only few of them can be selected for realization of the ITS-90 in the NMIs. Unfortunately, poorly chosen qualifier "standard" may mislead uninformed experimenters who may believe they deal with the most precise PRT available.

Designs of SPRTs vary widely; the interested reader can find a lot of useful information on this subject in [7, 80]. Here, we only outline the three main design types of SPRTs, whose sensing elements are wound from reference grade platinum wire.

Capsule SPRTs (CSPRTs) are used in the temperature range from 13.8 K to 157 °C. They have length of about 0.5 m, diameter of 5 mm, and resistance of 27.5 Ω at the TPW. Sensor with platinum lead wires is placed in a glass capsule usually filled with helium to increase the sensor's heat dissipation and thus decrease its thermal inertia. SPRTs of this type are fragile, complicated in use, and expensive. The accuracy provided by the capsule-type SPRTs is excessive for low-temperature and cryogenic blackbodies radiating in the IR spectral range; therefore, a cheaper substitute of the CSPRT can be found at the expense of measurement accuracy.

The long-stem SPRTs are used in the temperature range from 84 K to 450 °C (with mica supports) or to 660 °C (with quartz supports). These are the most demanded SPRTs. Their sensing elements also have a nominal resistance of 27.5 Ω. They are mounted together with support element in the glass or quartz sheath of 450 mm length and about 8 mm diameter. To control the oxidation state and prevent impurity oxides from penetration into the platinum wire of the sensing element, the sheath is filled with an especially chosen composition and concentration of inert gas and oxygen. The typical design of the long-stem SPRT is shown in Fig. 7.4. The sensing element is a platinum coil with a typical length of 35–50 mm.

The high-temperature SPRTs (HTSPRTs) are long-stem SPRTs, in which all the insulating components are made of quartz to withstand prolonged exposure of high temperatures. Operating temperature range of the HTSPRTs is from 0.01 to 962 °C; the nominal resistance at 0 °C is between 0.25 and 2.5 Ω. The three typical designs of HTSPRT are shown in Fig. 7.5.

To calibrate individual SPRTs, the ITS-90 specifies a reference function characterizing the relationship between the resistance ratio and the temperature for an

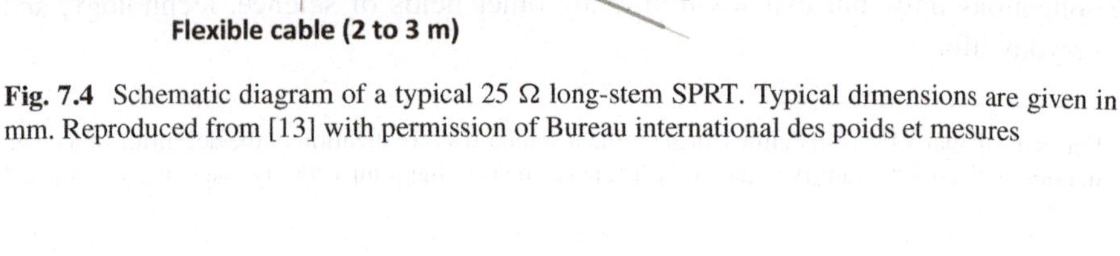

Fig. 7.4 Schematic diagram of a typical 25 Ω long-stem SPRT. Typical dimensions are given in mm. Reproduced from [13] with permission of Bureau international des poids et mesures

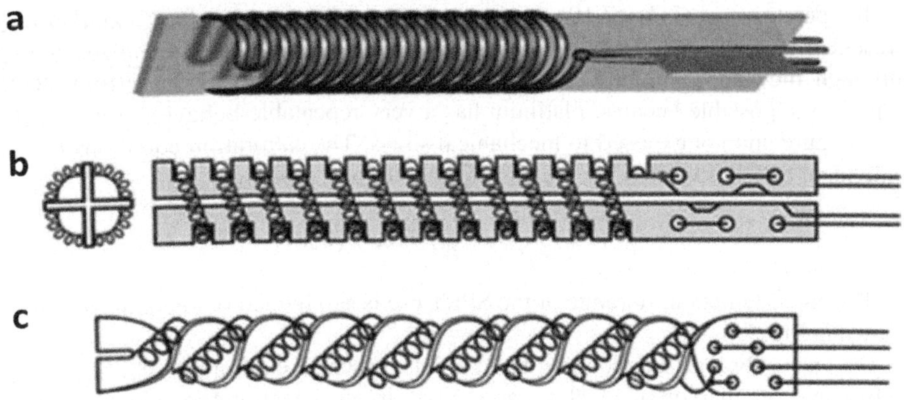

Fig. 7.5 Schematic designs of sensitive elements of SPRTs: **a** high-temperature SPRT with a simple bifilar platinum coil supported on a notched high-purity silica plate, R(0.01 °C) ~ 0.25 Ω; **b** long-stem SPRT with the platinum coil laid in notches in a silica cross, R(0.01 °C) ~ 25 Ω; **c** long-stem SPRT with the platinum coil laid in the helical turns of a twisted silica plate, R(0.01 °C) ~ 25 Ω. In all cases two platinum wires are welded to the coils at each end to create four lead connections ("Y" joints). Reproduced from [13] with permission of Bureau international des poids et mesures

idealized (or an averaged) PRT. The reference function $W_r(T_{90})$ is defined in the range 13.8033 K $\leq T_{90} \leq$ 273.16 K by the equation

$$W_r(T_{90}) = \exp\left\{A_0 + \sum_{i=1}^{12} A_i\left[\frac{\ln(T_{90}/273.16) + 1.5}{1.5}\right]^i\right\}. \tag{7.7}$$

Inversion of (7.7) with accuracy not exceeding 0.1 mK can be done using the following approximate formula:

$$T_{90} = 273.16 \cdot \left\{B_0 + \sum_{i=1}^{15} B_i\left[\frac{W_r^{1/6}(T_{90}) - 0.65}{0.35}\right]^i\right\}. \tag{7.8}$$

In the range from 273.16 K $\leq T_{90} \leq$ 1234.93 K the reference equation is

$$W_r(T_{90}) = C_0 + \sum_{i=1}^{9} C_i \cdot \left(\frac{T_{90} - 754.15}{481}\right)^i. \tag{7.9}$$

Inversion of (7.9) accurate within 0.13 mK can be done using

$$T_{90} = 273.15 + D_0 + \sum_{i=1}^{9} D_i \cdot \left[\frac{W_r(T_{90}) - 2.64}{1.64}\right]^i. \tag{7.10}$$

Temperatures in (7.7)–(7.10) are in kelvins; coefficients A_i, B_i, C_i and D_i are presented in Table 7.2. The two reference functions represent an idealized SPRT although they approximately describe the calibration curve of any SPRT. This approach is possible because platinum has a very repeatable behavior if it is sufficiently pure and not exposed to mechanical stress. The calibration equations for an individual SPRT is written via the deviation functions $\Delta W(T_{90})$:

$$W(T_{90}) = W_r(T_{90}) + \Delta W(T_{90}), \tag{7.11}$$

The entire temperature range of the SPRT use is subdivided onto eight subranges shown in Fig. 7.6.

For each subrange, the defining fixed points, in which calibration of a SPRT has to be made, are marked by black squares. Very simple deviation functions $\Delta W(T_{90})$ assigned by the ITS-90 to each subrange are presented in the last column of Table 7.3, where coefficients a_k and b have to be determined for each SPRT by least square fitting to the results of calibration at the defining fixed points.

The standard uncertainty of a long-stem SPRT calibrated over the temperature range from the TPW to the silver point does not exceed 1.2 mK [104]. Even if we make the correction for deviation of temperature T_{90} assigned to the freezing temperature of silver from the thermodynamic temperature and take into account the uncertainty in determination of this difference, the resulting relative standard uncertainty of the spectral radiance at 0.65 μm will be of about 0.04%. This is more than enough

Table 7.2 Coefficients A_i, B_i, C_i and D_i of the SPRT reference functions (7.7) through (7.10)

i	A_i	B_i	C_i	D_i
0	−2.135 347 29	0.183 324 722	2.781 572 54	439.932 854
1	3.183 247 20	0.240 975 303	1.646 509 16	472.418 020
2	−1.801 435 97	0.209 108 771	−0.137 143 90	37.684 494
3	0.717 272 04	0.190 439 972	−0.006 497 67	7.472 018
4	0.503 440 27	0.142 648 498	−0.002 344 44	2.920 828
5	−0.618 993 95	0.077 993 465	0.005 118 68	0.005 184
6	−0.053 323 22	0.012 475 611	0.001 879 82	−0.963 864
7	0.280 213 62	−0.032 267 127	−0.002 044 72	−0.188 732
8	0.107 152 24	−0.075 291 522	−0.000 461 22	0.191 203
9	−0.293 028 65	−0.056 470 670	0.000 457 24	0.049 025
10	0.044 598 72	0.076 201 285		
11	0.118 686 32	0.123 893 204		
12	−0.052 481 34	−0.029 201 193		
13		−0.091 173 542		
14		0.001 317 696		
15		0.026 025 526		

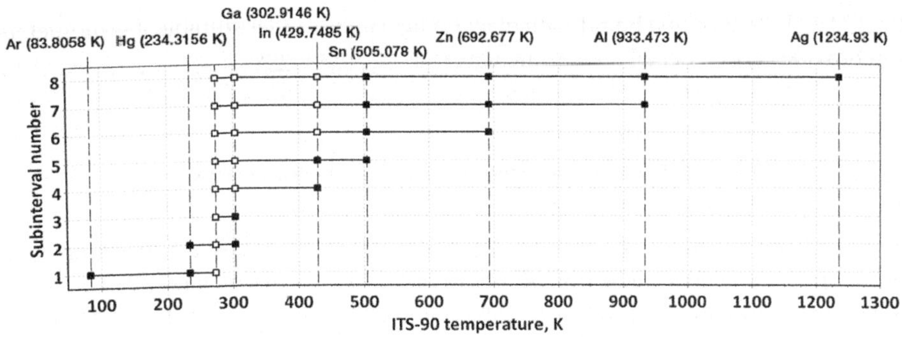

Fig. 7.6 A schematic representation of the ITS-90 subintervals and participating in calibrations of the SPRTs fixed points denotes by black squares

Table 7.3 Subranges, calibration points, and deviation function for SPRT

	Subrange	Defining fixed points	Deviation function $\Delta W(T_{90})$
1	Ar to H_2O	Ar, Hg	$a_1(W-1)+b\cdot(W-1)\cdot\ln W$
2	H_2O to Ga	Hg, Ga	$\sum_{k=1}^{2} a_k(W-1)^2$
3	H_2O to Ga	Ga	$a_1(W-1)$
4	H_2O to In	In	$a_1(W-1)$
5	H_2O to Sn	In, Sn	$\sum_{k=1}^{2} a_k(W-1)^2$
6	H_2O to Zn	Sn, Zn	$\sum_{k=1}^{2} a_k(W-1)^2$
7	H_2O to Al	Sn, Zn, Al	$\sum_{k=1}^{3} a_k(W-1)^2$
8	H_2O to Ag	Sn, Zn, Al, Ag	$\sum_{k=1}^{3} a_k(W-1)^2+b\cdot[W-W(933.473\text{ K})]^2$

even for most accurate blackbodies (such as heat-pipe and fluid-bath blackbodies) operating below the silver point. In order to realize the full potential of SPRTs, the resistance of the platinum wire must be measured very precisely. For instance, to achieve accuracy in temperature measurement of about 1 mK, the resistance must be measured with accuracy better than 0.5 mΩ. Fortunately, measurement of resistance is a well-developed area of measurement technique, for which the uncertainties are studied thoroughly and the ways to achieve highest accuracy are generally known. We will not describe methods for resistance measurement referring the interested reader to, e.g. Connolly [27].

Presently, there are commercially available high-resolution digital multimeters, precision automatic balancing DC and AC resistance bridges, as well as more complicated measurement systems allowing signals from several temperature sensors of different types to be obtained and processed in real time. The SPRTs have the highest accuracy (up to several millikelvins) and stability among all contact thermometers, so mainly SPRTs are used to measure reference temperatures for high-precision variable-temperature blackbodies operating from about −60 °C up to about 1000 °C (including heat-pipe and liquid-bath blackbodies) unless other suitable thermometers can provide certain benefits for some temperature subranges.

7.2.3 Industrial Platinum Resistance Thermometers

The niche not occupied by SPRTs is filled by other types of contact thermometers, primarily, by so-called industrial platinum resistance thermometers (IPRTs). They are designed to withstand industrial environments and have better durability at the expense of accuracy. For IPRTs, the resistance of 100 Ω is the most widespread, although IPRTs with the resistance values of 10, 500, 1000 Ω, etc. are also present in the market. As a rule, IPRTs are designated by their nominal resistance, $R_0 = R(0\,°C)$; for instance, Pt100 means the IPRT has the sensing element resistance of 100 Ω at 0 °C. Depending on the application, IPRTs can use different constructions of the platinum sensing elements (see Fig. 7.7), which are made of less pure platinum or its alloys and less susceptible to strain. In the coil- (or wire-) wound IPRTs, the sensing element can be embedded partially or entirely in a glass or refractory material. Sheath material can be chosen from stainless steel or Inconel alloy to ceramics, also depending on applications.

The IPRTs with film sensing element are more vibration- and thermal shock-resistant, have lower thermal inertia (and, correspondingly, fast response time), smaller dimensions, and lower cost than wire-wound IPRTs. Platinum thick-film IPRTs are made of a platinum paste applied to a ceramic substrate using a silkscreen process and then fused. Resistance of a film resistor is trimmed to the nominal value,

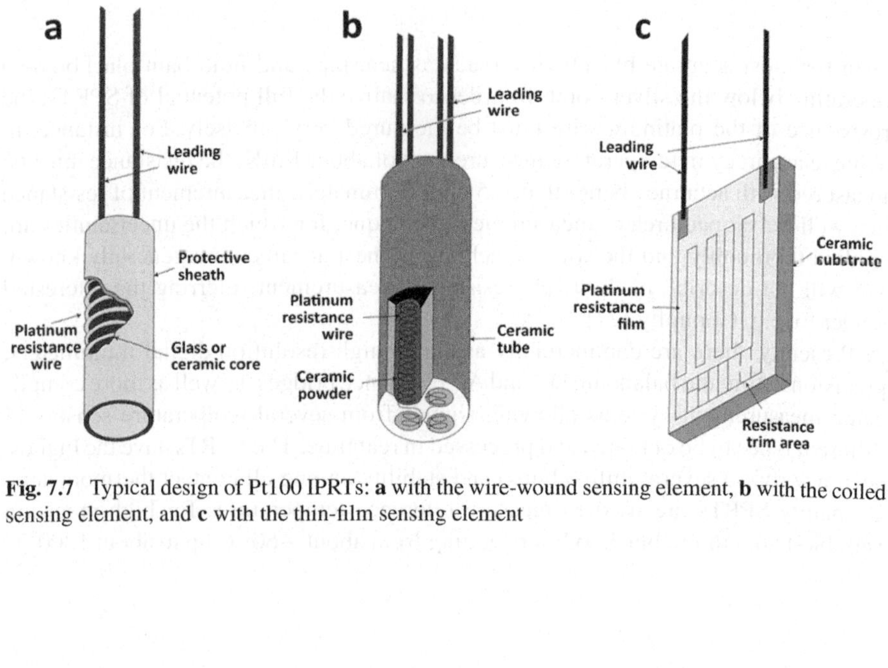

Fig. 7.7 Typical design of Pt100 IPRTs: **a** with the wire-wound sensing element, **b** with the coiled sensing element, and **c** with the thin-film sensing element

then a glass (or other) protection layer and connection leads are attached. The thickness of the platinum layer is between 10 μm and 15 μm. Thick-film IPRTs has lower response time that the wire-wound sensors but also lower long-terms stability.

Platinum thin-film IPRT has the platinum layer 1–2 μm thick made by vapor deposition or sputtering onto a ceramic substrate. The desired geometric shape is formed by cutting with a laser or structured using photolithography. The platinum resistor path may have a wide between 7 and 30 μm. A laser trimmer is used to adjust the resistance to the nominal value. For protection against mechanical damage (such as scratches), a 10 μm thick glass ceramic insulation is applied. After attaching the connection leads by welding, the connection spots are covered by a fused glass coating applied in a stress-free manner. The geometric dimensions of the flat thin film IPRTs vary from 2 to 20 mm^2; the substrate thickness varies from 0.25 to 0.65 mm. The thin-film IPRTs have very low thermal inertia and high sensitivity. There is contradictory information about their precision, accuracy, and stability: for instance, Firebaugh et al. [33] noted the significant hysteresis effect for platinum film of 0.1 μm thick; other sources informed on noticeable statistical fluctuations of resistance of the metal film resistors even for room temperatures if their thickness is less about 1 μm.

From the other side, many manufacturers guarantee the highest precision and long-term stability of temperature measurements using thin-film IPRTs. We were able to find only one work [16], where the thin-film flexible IPRT has been used for temperature measurement of the small blackbody designed for calibration of satellite onboard radiometric instruments. Obviously, this issue requires further investigations.

Up to date, the wire-wound IPRTs dominate among the sensors for measuring temperatures of the moderate-accuracy blackbodies. Unlike the SPRT, the IPRT may not be sealed hermetically. An IPRT can therefore be manufactured in a more robust manner, although it may be accurate only to within a few tenths of kelvin or even better.

The international standard IEC 60751 [56] specifies the requirements and relationship between temperature and resistance for sensors of the IPRTs, whose temperature coefficient α equals

$$\alpha = \frac{R_{100} - R_0}{R_0 \cdot 100\,°\mathrm{C}} \approx 3.851 \cdot 10^{-3}\,°\mathrm{C}^{-1}, \tag{7.12}$$

where R_{100} is the resistance at $t = 100$ °C.

The corresponding national standards have been introduced in most industrialized countries. In the United States, the ASTM E1137/E1137M-08 [2] is used. The standard IEC 60751 [56] covers IPRTs for the temperature range from −200 to +850 °C and employs the Callendar-Van Dusen equations [101] for the nominal temperature/resistance dependence $R_t(t)$:

Table 7.4 Tolerance classes for IPRTs according to IEC 60751 [56]

Tolerance class	Temperature range of validity (°C)		Tolerance values (°C)
	Wire-wound resistors	Film resistors	
AA	−50 to +250	0 to +150	± (0.1 + 0.0017· |t|)
A	−100 to +450	−30 to +300	± (0.15 + 0.002· |t|)
B	−196 to +600	−50 to +600	± (0.3 + 0.005· |t|)
C	−196 to +600	−50 to +600	± (0.6 + 0.01· |t|)

$$R_t = \begin{cases} R_0 \cdot \left[1 + At + Bt^2 + C \cdot (t - 100\,°\mathrm{C}) \cdot t^3\right], & \text{if } \; -200\,°\mathrm{C} \le t < 0\,°\mathrm{C} \\ R_0 \cdot \left[1 + At + Bt^2\right], & \text{if } \; 0\,°C \le t \le 850\,°\mathrm{C} \end{cases}, \tag{7.13}$$

where $A = 3.9083 \cdot 10^{-3}\,°\mathrm{C}^{-1}$, $B = -7.777 \cdot 10^{-7}\,°\mathrm{C}^{-2}$, and $C = -4.183 \cdot 10^{-12}\,°\mathrm{C}^{-4}$.

The standard establishes four tolerance classes (AA, A, B, and C) for IPRTs with wire-wound and film resistors of any value of R_0 (see Table 7.4 and Fig. 7.8, where the value of −196 °C is chosen as temperature close to the boiling point of liquid nitrogen). Tolerance values are defined as initial (i.e. obtained at the first calibration before any use of IPRT) maximum allowable deviations expressed as $\Delta t(t)$ in °C from the nominal temperature-to-resistance relationship expressed by (7.13). In other words, the tolerance values specify the maximum dispersion of sensitivity that is allowable within many IPRTs belonging to the same tolerance class. If necessary, the tolerance value can be used as the upper bound of uncertainty in measurement of temperatures using the IPRT of a given tolerance class. Tolerances and ranges of validity, which differ from those provided in Table 7.4, are allowed by mutual agreement between manufacturer and user but must be clearly stated.

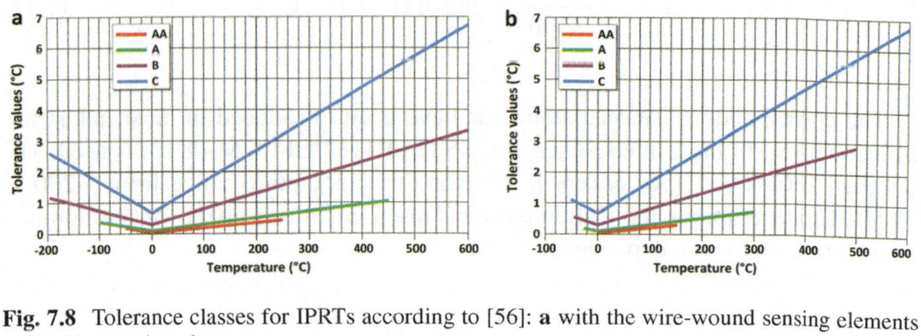

Fig. 7.8 Tolerance classes for IPRTs according to [56]: **a** with the wire-wound sensing elements and **b** film sensing elements

Recommended special tolerance classes are constructed as multiples or fractions of class B tolerance values. The manufacturers and users can establish tolerances for their IPRTs, restrict or extend temperature ranges specified in Table 7.4, and even select temperatures outside these ranges. It should be noted that assessment of measurement uncertainty via the tolerance class is too crude for precise temperature measurements that are needed in blackbody radiometry.

In each particular case of the contact thermometer use for measurement of blackbody temperature, uncertainty must be determined individually via thorough investigation of its specific components. Presently, IPRTs are the "workhorses" of the contact thermometry. An individually calibrated IPRT is often inferior to an SPRT only in the long-term stability. It should be re-calibrated more frequently, but may have much smaller size, lower thermal inertia, and better mechanical strength. In many cases, especially at lower temperatures, the IPRT can provide accuracy in measuring blackbody temperatures that is sufficient for radiometric calibrations in the IR spectral range.

From −196 °C (boiling point of the liquid nitrogen) to −50 °C, the IPRT is prevalent temperature sensor in blackbody radiometry. An SPRT having an excessive accuracy is too expensive, fragile, and require more space to be correctly positioned; signal generated by a thermocouple at low temperature is vanishingly small; an NTC thermistor can work reliably only within the narrow range of higher temperatures. Therefore, the IPRT may be the best choice as a temperature sensor for cryogenic blackbodies. From temperature of about 250 °C to the silver point, the IPRT is the most widespread temperature sensor for measuring reference temperature of inexpensive blackbodies of various types if exceptional accuracy is not necessary but compactness matters.

IPRTs are also used as auxiliary temperature sensors for monitoring the short-term stability of blackbodies, in feedback loops of blackbody temperature controllers, and at measurements of temperature distributions over blackbody radiators, heating or cooling elements (furnaces, liquid baths, and liquid-circulating systems).

7.2.4 NTC Thermistors

Thermistors, similar in function to metal resistance temperature sensors, are low-cost temperature-sensitive resistors fabricated of oxides of various metals (rhodium, manganese, nickel, cobalt, copper, etc.), pure or doped germanium, silicon, or their compositions. Thermistors can exhibit a positive or negative temperature coefficient of resistance. Positive temperature coefficient (PTC) thermistors are used to measure extremely low temperatures (around the Curie temperature of the thermistor's material; usually, below 80 K). The temperature range of negative temperature coefficient (NTC) thermistors, where they have maximal sensitivity, is about −50 to 150 °C, although there are NTC thermistors that are able to operate at lower or higher temperatures. The normal resistance $R(25\,°\mathrm{C})$ lies between 100 Ω and 100 kΩ for most NTC thermistors. The resistance-temperature behavior of a thermistor can be roughly approximated by

$$R(T) = R(T_0)\exp\left[\beta\left(\frac{1}{T} - \frac{1}{T_0}\right)\right], \qquad (7.14)$$

where T_0 is some conveniently chosen reference temperature, usually, 298.15 K (25 °C); $\beta > 0$ is the index of the thermal sensitivity, the thermistor's material characteristic, whose typical values lie between 2000 and 6000 K. The sensitivity S of a thermistor defined as the relative change of resistance at a change of temperature is given by

$$S = \frac{1}{R}\frac{dR}{dT} = \frac{d\ln R}{dT} = -\frac{\beta}{T^2}, \qquad (7.15)$$

i.e. the sensitivity of a thermistor is inversely proportional to the squared temperature; the "minus" sign means that the resistance of a thermistor decreases when temperature increases.

Figure 7.9 shows the resistance-temperature characteristic of typical NTC thermistors ($\beta \approx 3900$ K) in comparison with the PRT Pt100. Thermistors are highly nonlinear but provide a noticeable advantage in sensitivity to temperature (typically

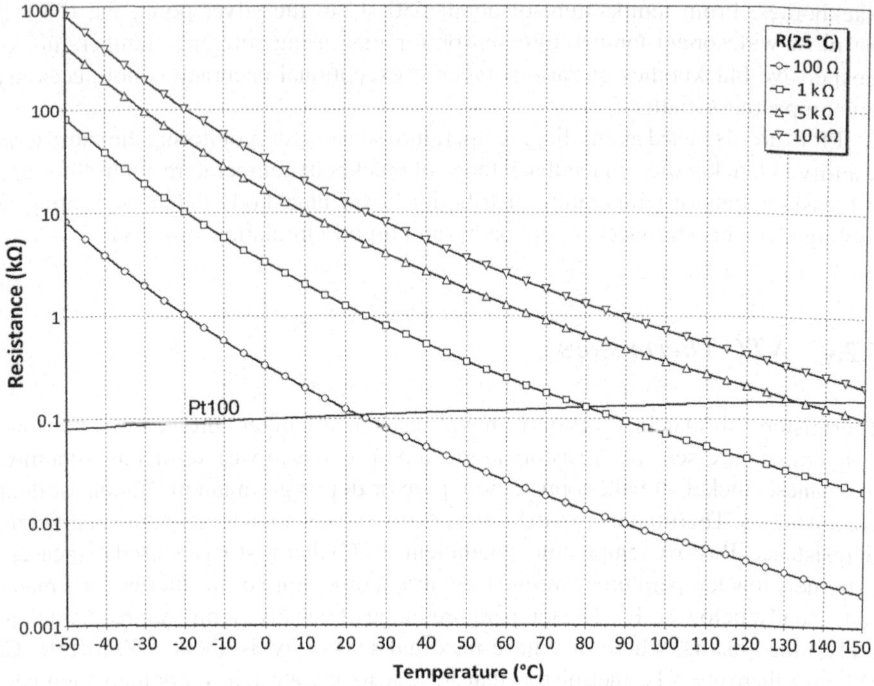

Fig. 7.9 Comparison of dependences $R(t)$ of the four typical thermistors and the Pt100

a 4% change in resistance per kelvin) as compared with PRTs. Equation 7.14 is considered as calibration equation for thermistors in the International Standards IEC 60539-1 [53]. However, there are experimental evidences that accuracy of Eq. 7.14 is insufficient to achieve temperature uncertainty better than 0.1 K for most types of thermistors throughout their operation temperature range. The β value for thermistors made of doped semiconductors varies with temperature, which means that (7.14) can be used only for very narrow temperature ranges. For high-accuracy calibration in wide temperature ranges, a more complex equation has to be employed.

The following series expansions are most commonly used:

$$T^{-1} = \sum_{k=1}^{n} A_{k-1} \ln^{k-1}\left(\frac{R}{R_0}\right) \tag{7.16}$$

and

$$\ln\left(\frac{R}{R_0}\right) = \sum_{k=1}^{n} \frac{B_{k-1}}{T^{k-1}}, \tag{7.17}$$

where R_0 is some conveniently chosen resistance value. The number n of terms in each series is chosen according to the temperature range and the accuracy required; it should be noted that (7.15) corresponds to the case of $n = 2$. Usually, four or five terms are enough to ensure any reasonable accuracy. Calibration of a thermistor should be performed at n known temperatures, for instance, in the fluid bath, against a SPRT. Coefficients A_k and B_k in Eqs. 7.16 and 7.17 can be found using standard least-squares fitting software. In the USA, the standard ASTM E879-20 [5] prescribes to use the four-term equations.

Steinhart and Hart [95] proposed to use (7.16) with $n = 4$ but with the second-order term omitted ($A_2 = 0$). This approximation has been called the Steinhart-Hart equation and is usually written in the form

$$\frac{1}{T} = A + B \ln R + C \cdot \ln^3 R, \tag{7.18}$$

where R is in ohms, T is in kelvins, and coefficients A, B, and C have dimension of K^{-1}. Unfortunately, the Steinhart-Hart equation remains widespread up to date and it is commonly used by thermistor manufacturers despite criticism by Chen [24], Matus [74], and others against the Steinhart-Hart equation accuracy. White et al. [105] in the Guide on Thermistor Thermometry published on behalf of the CCT noted that the performance of the Steinhart-Hart equation is only sometimes better than the usual 3-term version of (7.19) while its four-term version always gives more accurate fitting than the Steinhart-Hart equation. Moreover, the accuracy of the Steinhart-Hart equation depends on the normal (at 25 °C) resistance of the thermistor, which makes fitting quality hardly predictable.

Most of modern NTC thermistors used for precise temperature measurements are fabricated by sintering of metal oxide powders and may have the shapes of beads, discs, rods, etc. A glass, epoxy, or phenolic coating is often applied to provide strain relief to the interface between lead wires and ceramics and to seal hermetically the thermistor for long-term stability. Typical diameters of most widespread glass bead thermistors range from 0.25 to 1.5 mm, thermistors of other shapes may have sized from about 1 mm to several millimeters. Due to very small sizes and low thermal masses, the NTC thermistors with small enough beads are very fast: a response time of about 3 ms is possible. Generally, the NTC thermistors successfully compete in most parameters with IPRTs within temperature range from about −50 to 150 °C providing accuracy as low as 1 mK. The long-term stability of the best thermistors approaches a few millikelvins per year. Sensitivity of thermistors in the above-mentioned temperature range is substantially higher than that of PRTs in the same range.

Thermistors, due to their small sizes and weights, fast response, sensitivity, stability, durability, and operational temperature range, are highly demanded in temperature measurement of blackbodies placed on board satellites, aircrafts, and balloons to calibrate remote sensing radiometric instrumentation for Earth observation, as well as laboratory blackbodies operating at temperatures from −50 to 150 °C. Thermistors are often used as auxiliary sensors to control temperature of individual components (e.g. interference filters of filter radiometers) in the blackbody-based radiometric systems.

Non-linearity of thermistor's response can hardly be considered as a disadvantage today, when almost every measuring device contains a microprocessor or is connected to a computer. The only disadvantage of thermistors is very limited temperature range: the use of the thermistor at elevated temperatures leads to its excessive ageing and significant drift.

Although the first commercial thermistors appeared in the 1930s, for a long time they were not considered as precision thermometric devices. The "Techniques for Approximating the International Temperature Scale of 1990" [12] devoted only two of the more than two hundred pages to thermistors. The situation has been improved by the BIPM, which released a brief guide on thermistor thermometry [105]. The International standards IEC 60539-1 [53], IEC 60539-2 [54], and corresponding national standards provide a guidance in the applications of NTC thermistors.

7.2.5 Thermocouples

Thermocouples are the most widely used temperature sensors in industrial applications due to their basic simplicity and reliability. However, as soon as accuracy better than normal industrial requirements is desired, their simplicity in use is lost and their reliability becomes questionable.

The platinum-rhodium/platinum thermocouple was used to realize in the temperature range from 660 °C (later, from 630 °C) to the freezing point of gold in designated temperature scales preceding the ITS-90. In the ITS-90, the platinum resistance thermometer becomes the only interpolation instrument from the triple point of equilibrium hydrogen (13.8033 K) to the freezing point of silver (961.78 °C). Achieved superiority of platinum resistance thermometers over thermocouples in accuracy, reproducibility, and long-term stability "downshifted" the rank of thermocouples and, correspondingly, the leading role they played in contact measurement of blackbody temperatures was lost. Good reasons are needed to choose a thermocouple thermometer for measurements of blackbody reference temperatures; we will consider these cases later. However, thermocouples retain many auxiliary functions such as temperature monitoring of blackbody heaters or furnaces, or some accessorial components (e.g. aperture diaphragms or background reference targets), indicating stabilization of the radiator's temperature, evaluation of temperature non-uniformity for the blackbody radiating surface, etc. Here, we only recall some salient points in the thermocouple-based thermometry.

The operation principle of a thermocouple is based on the Seebeck effect that manifests itself in developing the electric potential difference (electromotive force or EMF), in every conductor experienced a temperature gradient. Different types of conductors exhibit this effect with varying intensity. Figure 7.10 shows an example of measuring temperature using a thermocouple. A thermocouple consists of two dissimilar thermoelement wires A and B that are joined at one end (the "hot" junction) to measure temperature T_1. The wires are insulated from each other along their lengths. The other end (the "cold" junction) is maintained at a constant reference temperature T_2 (usually the melting point of ice). Thermocouple wires can be connected directly to a voltmeter that is equipped with internal cold junction circuitry. Such a configuration is usually less accurate than when using an external cold junction maintained in the ice bath (the Dewar flask filled with the homogeneous mixture of distilled water and the crushed ice also prepared from distilled water).

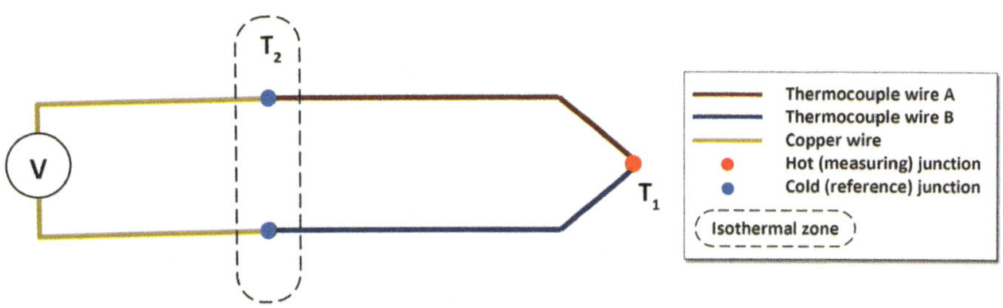

Fig. 7.10 A simplified schemes for measurement of temperature using a thermocouple

In the cases, when implementing an ice bath is impractical, the hardware (analogue or digital) cold junction compensation technique can be applied but it requires additional temperature sensor (e.g. IPRT) measuring the reference junction temperature. The difference between the measured temperature T_1 and reference temperature T_2 is corrected electronically in the instrument measuring the thermocouple to indicate the actual temperature of T_1. This adjustment is referred to as cold-junction compensation (CJC).

An EMF voltage is generated between the cold junction wires at temperature T_2 if the hot junction temperature $T_1 \neq T_2$. An instrument connected to lead wires from the cold junction is used to read the thermocouple voltage. Theoretically, this voltage measurement depends only on the temperature difference $T_1 - T_2$. As T_1 changes, the voltage output of the thermocouple changes proportionally to the change in temperature, but not linearly. The interrelation of temperature and EMF establishes a relationship that is unique to each thermocouple type. These relationships are summarized in reference tables, which provide the basis for thermocouple calibration. The thermocouple designs may vary widely depending on the operating temperature range, environment, and functions performed. Figure 7.11 presents several typical designs. The simplest thermocouple consists of two insulated wires of dissimilar metals and has welded (less frequently—brazed) junctions that must provide mechanical strength and not to introduce impurities into the thermocouple wires. Thermocouples may be encased in a protective sheath made of stainless metal, glass, mineral fibers, plastic polymers, ceramics, etc. The sheath can be filled with insulating material, which provides mechanical support and insulates electrically the thermocouple wires.

A sheathed thermocouple is a type of thermocouple where the wires are set in and insulated by a high-density ceramic compound, typically magnesium oxide, and then encased within a metal sheath. Common sheath materials are stainless steel or Alloy 600, but any metal that can be cold-worked could be used. For maximum

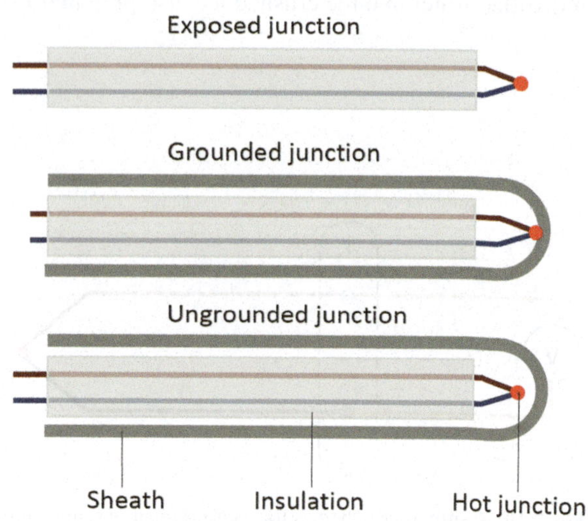

Fig. 7.11 Different designs of thermocouple junctions

sensitivity and fastest response, the measuring junction may be unsheathed. This design, however, makes the thermocouple more fragile. Sheathed tips are typical for industrial applications and available in grounded and ungrounded configurations.

Grounded junctions provide better thermal contact with protection from the environment. Ungrounded junctions provide electrical isolation from the sensor sheath. Grounded-tip thermocouples exhibit faster response times and greater sensitivity than ungrounded-tip thermocouples, but in the case of conductive sheath, they are vulnerable to ground loops: circuitous paths for electric current between the sheath and other points of the thermocouple circuit.

Apart from thermocouple wires, which cannot be too long because this is impractical and often too expensive, the special extension and compensation wires are used. We will not delve into details of thermocouple material technology and processing, constructions of thermocouples of particular types, materials for sheathes, wire insulation, etc. Interested readers may refer to Pollock [86], ASTM [1], Bentley [9], Kerlin and Johnson [63]. Metrological aspects of the thermocouple thermometry are lucidly described by Nicholas and White [80]. However, it should keep in mind that a part of materials containing in these books might be somewhat outdated.

Since thermocouples are the most popular industrial contact temperature sensors, there are pairs of metals and alloys, which are used for fabrication of thermocouples for many years. Specifications for some of them are included in national and international standards. The standard types of thermocouples established by the international standard IEC 60584-1 [55] are presented in Table 7.5. For all these types, the standard provides dependences of the output EMF on the temperature of the work junction at the reference junction temperature of 0 °C. The thermocouple types in this table can be divided into three groups: the precious metal thermocouples are designated by letters R, S, and B; the base metal thermocouples are designated by letters J, T, E, K, and N; recently standardized high-temperature tungsten-rhenium thermocouples are designated by letters C and A. The EMF-to-temperature dependences were determined by NMIs and presented in the form of reference polynomial equations and tables. Reference data included in IEC 60584-1 [55] obliges manufacturers to supply thermocouples with a guaranteed accuracy of the EMF-to-temperature characteristics within known tolerances.

The user decides whether calibration of a thermocouple is needed or not. If the accuracy required is outside the allowed tolerances of the standard reference data, or if the EMF drift outside of these tolerances is expected, the thermocouple must be individually calibrated.

The type T thermocouple is one of the oldest thermocouples for low-temperature measurements. It exhibits good thermoelectric homogeneity and is still commonly used in the temperature range from the triple point of neon (−248.5939 °C) to about 370 °C. Above this temperature, the oxidation rate of the thermocouple materials increases rapidly. This thermocouple is not suitable for precision measurements because of excessive heat losses due to high thermal conductivity of thermocouple's materials.

Table 7.5 Thermocouple types according to [55]

Letter	Elements and nominal alloy composition by weight		Temperature range (°C)	Note
	Positive conductor	Negative conductor		
T	Copper	Copper–nickel	−270 to 400	1
J	Iron	Copper–nickel	−210 to 1200	1
K	Nickel–chromium	Nickel–aluminum	−270 to 1300	
E	Nickel–chromium	Copper–nickel	−270 to 1000	1
N	Nickel–chromium – silicon	Nickel–silicon	−270 to 1300	2
C	Tungsten–5% rhenium	Tungsten–26% rhenium	0 to 2315	3
A	Tungsten–5% rhenium	Tungsten–20% rhenium	0 to 2500	3
S	Platinum–10% rhodium	Platinum	−50 to 1768.1	
R	Platinum–13% rhodium	Platinum	−50 to 1664.5	
B	Platinum–30% rhodium	Platinum–6% rhodium	0 to 1820	

Notes
1. The negative conductors of types J, T, and E are generally not interchangeable with each other
2. For the type N thermocouple, the following composition (percentages of total by weight) is recommended in order to obtain the desired properties like good stability and oxidation resistance
3. Positive conductor (known as Nicrosil): 13.7–14.7% Cr%, 1.2–1.6% Si, less than 0.15% Fe, less than 0.05% C, less than 0.01% Mg, balance Ni
4. Negative conductor (known as Nisil): less than 0.02% Cr, 4.2–4.6% Si, less than 0.15% Fe, less than 0.05% C, 0.0 5–0.2% Mg, balance Ni
5. Positive conductors of Types C and A are not necessarily interchangeable

Thermocouples of types J and K, due to their high sensitivity, relatively large defined temperature range, and low material costs, are the most commonly used in today's industry.

Thermocouple of the type E has good homogeneity, relatively small heat conductivity, and large enough defined temperature range. It is the most common thermocouple for low-temperature measurements. Above 0 °C, it has the highest sensitivity of all the thermocouples defined in the IEC 60584-1 [55]. However, the oxidation rate in air for both conductors is high for temperatures over 750 °C.

Thermocouples of the type N have distinct advantages due to their higher thermoelectric stability at temperatures over 870 °C and lower tendency to oxidize as compared with thermocouples of types J, K and E. Relatively high silicon content lead to the rapid surface oxidation that protects the thermocouple material from further corrosion.

Tungsten-rhenium (W-Re) thermocouples of types C and A are designed for use in high vacuums and inert gases up to 2315 °C and 2500 °C, respectively (see [3]).

The W-Re thermocouples are used primarily for measuring high temperatures in nuclear reactors and cannot be applied to precision measurements in blackbody radiometry due to their low accuracy and poor stability.

Precious metal thermocouples of types S, R, and B are suitable for continuous use in oxidizing or inert atmospheres at temperatures up to about 1600 °C and for short-term or intermittent use in vacuum or inert gas at higher temperatures. Precious metal thermocouples are very susceptible to contamination and can be used only being inserted into ceramic (in particular, aluminum oxide, Al_2O_3, with a purity of better than 99%) protection tube. Metallic protection tubes cannot be used at temperatures higher than 1200 °C.

Thermocouples employing platinum in combination with platinum-rhodium alloys, gold, or palladium are the most reproducible among all the types of thermocouples. Due to high melting point of platinum (~1768 °C) and its resistance to oxidation, they can be used up to very high temperatures. A great while, the type S thermocouple was considered the most accurate temperature sensor and served as an interpolating thermometer from the freezing point of antimony (630.74 °C) to the freezing point of gold of the temperature scales that were predecessors of the ITS-90, after which it was substituted by the SPRT. Correspondingly, the type S thermocouples are best studied. The type R thermocouple is very similar in its properties to the type S; it has a little higher sensitivity and reproducibility. Typically, the precision of types S and R thermocouples, especially above 500 °C, is worse than ±0.2 °C.

As the melting temperature of platinum-rhodium alloy increases with increasing content of rhodium, thermocouples comprising platinum-rhodium elements of higher rhodium content are usually more stable at high temperatures. The positive conductor of the type B thermocouple contains 30 wt% of rhodium allowing this thermocouple to operate under oxidizing conditions at temperatures higher than those for the types S and R thermocouples. The type B thermocouple has a remarkable property: its EMF varies only slightly (approximately from −2.5 to 2.5 μV) in the temperature range from 0 to 50 °C, which means that there is no necessity to keep the temperature of the reference junction constant; often its effect can be neglected or corrected for.

In addition to standardized (i.e. included in international or national standards) thermocouples, there are many non-standard thermocouple types, among which some precious metal thermocouples mainly intended for high-temperature applications are of certain interest for blackbody radiometry. These are, primarily, various platinum-rhodium/platinum thermocouples with different rhodium content in the positive conductor. Second, the gold versus platinum (Au/Pt) thermocouple has certain benefits, particularly, of simplicity and economic practicability over the HTSPRT and approaches to it in accuracy. Substitution of gold by more refractory palladium (the melting temperature of about 1555 °C) results in development of the Pt/Pd thermocouple in the early 1990s. In Fig. 7.12, the EMF-temperature reference functions for eight standard thermocouple types of IEC 60584-1 [55] are plotted together with those for the Pt/Pd thermocouple obtained by Burns et al. [15] and for the Au/Pt thermocouple obtained by Ripple and Burns [93].

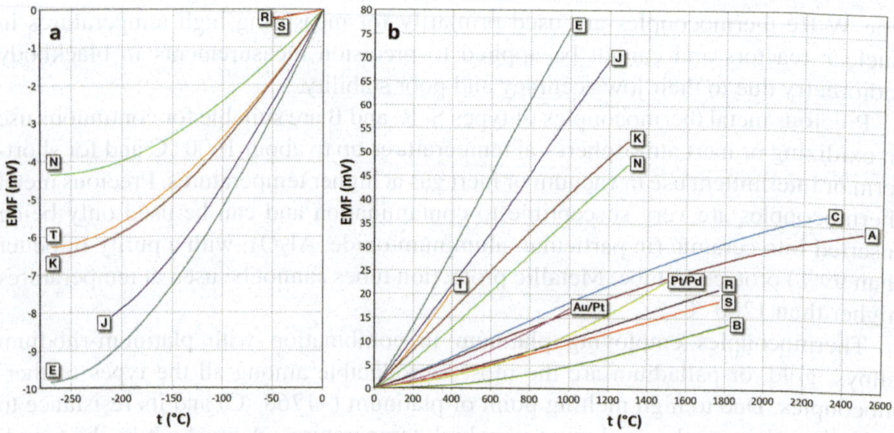

Fig. 7.12 Reference curves (EMF vs. temperature) plotted for eight standard thermocouple types, Au/Pt, and Pt/Pd thermocouples: **a** for temperatures below 0 °C; **b** for temperatures above 0 °C. Temperature of the reference junction is 0 °C in all cases

The lack of fixed points of the ITS-90 with temperatures higher than the copper freezing point did not allow unlocking metrological potential of the Pt/Pd thermocouple, so it could not compete in accuracy with radiation thermometers at temperatures above the silver point or with the platinum resistance thermometers below it. This situation should change after the adoption of phase transition temperatures for some metal-carbon eutectics as high-temperature fixed points [31, 85].

We conclude this Section by several words about thin-film thermocouples, first attempts to create which were made in early 1960s. Although they could be very useful as contact sensors for surface temperature measurements that do not require a power supply, apparently, thin-film thermocouples will never be able to compete with the best ordinary thermocouples and thin-film PRTs in accuracy and stability [65, 73, 99, 109]. Thin-film thermocouples are presently used for transient temperature measurement in nanosecond range, where high accuracy is not as important as fast time response.

7.3 Systematic Errors in Contact Thermometry of Blackbodies

7.3.1 Main Sources of Systematic Errors

The systematic errors of contact thermometry are well studied, though these studies are scattered in many books and papers dedicated, as a rule, to a specific type of temperature sensors. A kind of a general approach can be found in Sect. 4.4 “Errors in the Use of Thermometers” in [80] and in Chapters 16 and 17 of [75]. It is important for us to identify the sources of uncertainty common to all contact sensors used to measure the temperature of blackbodies.

First, the contact sensors do not allow us to measure the temperature of the radiating surface. With contact sensors, we always forced to measure some temperature, roughly speaking, below the radiating surface. The difference between the temperature set point (in this case, the temperature measured by the contact thermometer) and the temperature of the radiating surface is often referred to as the temperature drop. In the most common situations, the blackbody radiating surface gives off heat to the environment at a lower temperature and this decreases the surface temperature compared to the temperature below the radiator surface. We discuss the temperature drop effect in Sect. 7.3.2.

Second, the contact measurement of temperature implies the thermal equilibrium between the temperature sensor and the object, whose temperature has to be measured. Contrary to conventional wisdom, a contact thermometer always indicates not the body, surface, or medium temperature that has to be measured, but the equilibrium temperature of its own sensing element.

This temperature is determined, among other things, by conditions of heat exchange with the environment and generally differs from the temperature we have to measure. Additional systematic error is introduced, if these conditions differ from conditions, at which the sensor was calibrated. In many cases, the systematic errors related to the incomplete fulfillment of the thermal equilibrium conditions can be attributed to the improper location of the contact sensors.

7.3.2 *Temperature Drop Effect*

The temperature drop effect is the most common source of systematic errors in the contact measurement of blackbody temperature. It manifests itself at any time when we need to measure the temperature of the radiating surface, but, at best, we can measure the temperature from the opposite side of the radiator, where conditions of heat exchange differ significantly. Therefore, the corresponding correction should be made. Since the temperature of a radiating surface is usually lower than of the radiator's rear side, this systematic effect is known as temperature drop. Here, we outline the approach to correction for this systematic effect on an example of the one-dimensional temperature field inside an infinitely extended thermally conductive slab shown in Fig. 7.13. Let us suppose that the wall has the thickness δ and the thermal conductivity k [$\mathrm{W \cdot m^{-1} \cdot K^{-1}}$]. The left side of the wall is maintained at a constant temperature T_0; on the right side, heat is removed by convection and radiation into the environment having a temperature T_a. The ordinary differential equation for the steady-state temperature $T(x)$ can be written as

$$\frac{d^2T(x)}{dx^2} = 0. \tag{7.19}$$

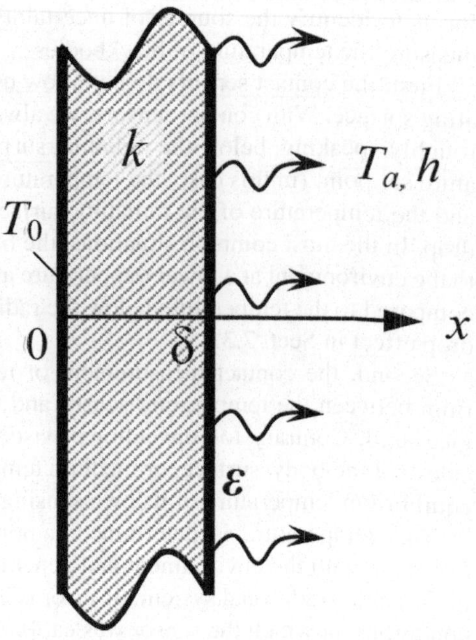

Fig. 7.13 A simple model explaining the temperature drop effect

Equation 7.19 has the general solution $T(x) = a_0 + a_1 x$, where the constants a_0 and a_1 can be found using the boundary conditions at $x = 0$ and $x = \delta$:

$$T(x)|_{x=0} = T_0, \tag{7.20}$$

$$-k\left.\frac{dT(x)}{dx}\right|_{x=\delta} = h[(T(\delta) - T_a)] + \varepsilon\sigma\left[T^4(\delta) - T_a^4\right], \tag{7.21}$$

where $h[\mathrm{W \cdot m^{-2} \cdot K^{-1}}]$ is the convective heat transfer coefficient, ε is the emissivity of the right-hand surface, and σ is the Stefan-Boltzmann constant.

Due to non-linearity in the boundary condition for the emitting right-hand side of the wall, this boundary problem requires numerical solution even in such an oversimplified formulation. An analytical solution can be found if it is assumed that the ambient temperature T_a is not much different from the temperature of the radiating surface and therefore the Stefan-Boltzmann law can be linearized, i.e. assuming

$$T^4(\delta) - T_a^4 \approx 4T_a^3[T(\delta) - T_a]. \tag{7.22}$$

In this case, the effective heat transfer coefficient $h_\Sigma = h + 4\varepsilon\sigma T_a^3$ can be introduced and the solution takes the form

$$T(x) = T_0 - \frac{h_\Sigma x}{k}[T(\delta) - T_a], \tag{7.23}$$

from which

$$\Delta T = T_0 - T(\delta) = \frac{h_\Sigma \delta}{k}(T_0 - T_a). \tag{7.24}$$

Therefore, the greater the wall thickness δ, temperature difference $T_0 - T_a$, and heat losses from the radiating side of the wall, the higher the temperature drop (i.e. the temperature difference between two sides of the wall). To minimize the temperature drop effect, the blackbody wall must be as thin as possible and made of highly conductive material. It should be noted that this conclusion should not be a guide for action but serves only to explain one particular source of systematic error. For instance, for many blackbodies except the immersed ones, too thin cavity walls may result in unacceptable temperature non-uniformity.

For high-temperature blackbodies ($T_0 >> T_a$) operating in vacuo or in a slow laminar flow of rarefied gas ($h \approx 0$), (7.21) takes the form

$$-k \left.\frac{dT(x)}{dx}\right|_{x=\delta} = \varepsilon\sigma T^4(\delta). \tag{7.25}$$

In this case, the solution of the boundary problem (7.19)–(7.21) is

$$T(x) = T_0 - \frac{\varepsilon\sigma\delta}{k} T^4(\delta). \tag{7.26}$$

At $x = \delta$, $T(\delta)$ can be found numerically; an approximate expression for the temperature drop

$$\Delta T \approx \frac{\varepsilon\sigma\delta}{k} T_0^4 \tag{7.27}$$

depends strongly on the operating temperature of a blackbody. The boundary problem (7.19)–(7.21) was stated for an oversimplified model, so (7.23) and (7.26) can be recommended only for rough evaluation of the temperature drop effect and the corresponding uncertainty due to inaccurate knowledge of the affecting parameters.

An example of numerical evaluation of the temperature drop effect is given by Ogarev et al. [82] for the precision large-area low-temperature vacuum blackbody BB100-V1 designed for use as a reference radiation source for the calibration of space-borne radiometers and radiation thermometers in the IR wavelength range from 1.5 to 15 μm. A 3D view of the blackbody is shown in Fig. 7.14. The radiating copper cavity of the BB100-V1 has the length of 200 mm and internal diameter of 120 mm. Temperature of the cavity coated inside with the black paint is controlled using the circulation-type thermostat that drives the working fluid at temperature from −45 to 200 °C through the helical-coil heat exchanger. The temperature is measured using five Pt100 temperature sensors, three of which are incorporated in the backside of the V-grooved copper bottom, and two others in the cylindrical walls. Four sensors are connected to an external multimeter via an electrical connector placed at the rear

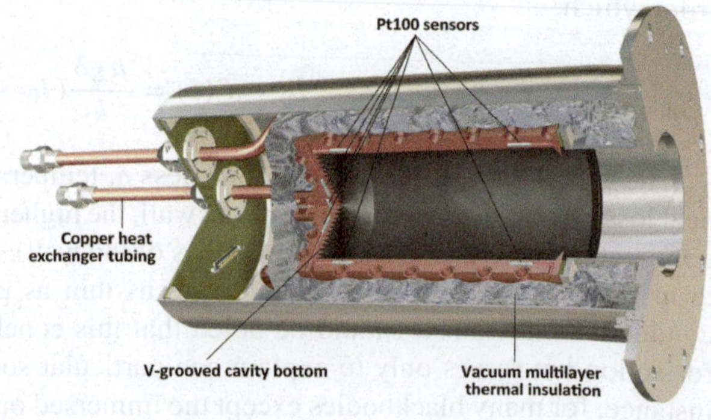

Fig. 7.14 A cutaway drawing of the BB100-V1 blackbody

flange of BB100-V1, with a four-wire scheme for measurement of the electrical resistance. The fifth sensor is in the feedback loop of the temperature stabilization system. The reference Pt100 temperature sensor placed in the center of the backside of the V-grooved bottom is made removable to allow recalibration.

To evaluate the temperature distribution over the V-grooved bottom surface, the finite element modeling was performed for the following model. It was assumed that the temperature of the external side of the cavity bottom is constant, and the heat loss from its V-grooved surface is carried out via the radiative heat transfer through the aperture that was assumed perfectly black. For the BB100-V1 operating in vacuo, no convective heat loss from the cavity interior presents.

However, radiation heat loss from the bottom of the cavity to a colder background through the cavity aperture cannot be eliminated.

To assess temperature non-uniformity over the grooved surface of the bottom, the radial distribution of the view factors from an element of the grooved surface of the bottom of the cavity to its aperture was modeled using the Monte Carlo ray tracing with the account of multiple reflection. To assess the temperature non-uniformity on the grooved bottom, the distribution of the viewing factors from the element of the bottom of the cavity to its aperture was simulated using the Monte Carlo ray tracing taking into account multiple reflections. The generatrix of the blackbody BB100-V1 is shown in Fig. 7.15. All the internal surfaces of the blackbody were considered as diffuse emitters and diffuse-specular reflectors. Figure 7.16 presents

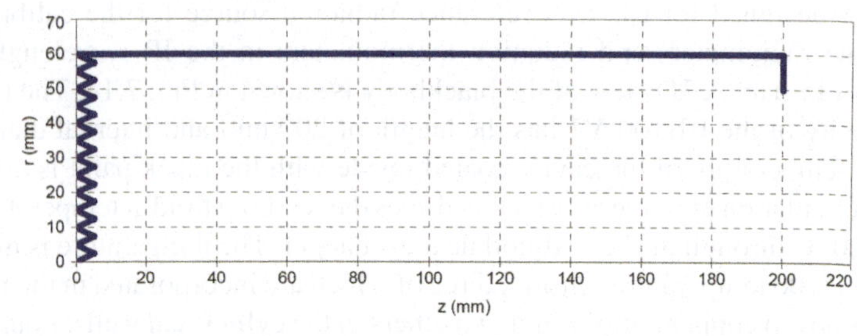

Fig. 7.15 The generatrix of the BB100-V1 radiating cavity

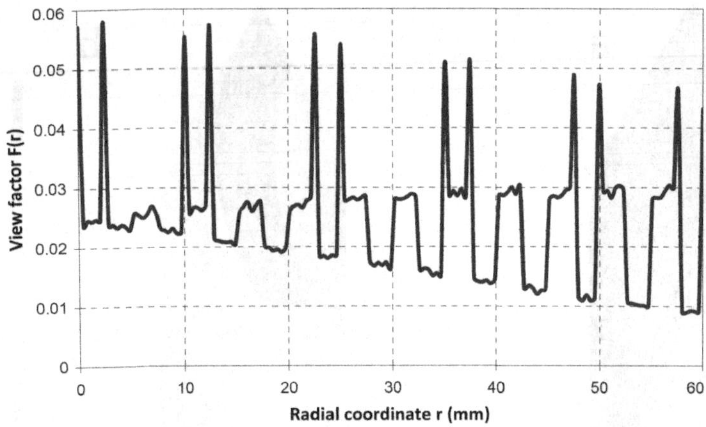

Fig. 7.16 Radial distribution of the view factors $F(r)$ over the V-grooved bottom of the BB100-V1 radiating cavity

the computed radial distribution $F(r)$ of the view factors across the cavity bottom. The minima of the distribution are at the base of V-grooves; the peaks correspond to the lateral conical surfaces of V-grooves. It was found by the finite-element modeling that the temperature gradient along the cavity bottom radius not exceeds 4 mK and can be considered negligible. Therefore, the further modeling was performed for isolated "teeth" (selected annular areas of the grooved bottom). The back side of the bottom is considered isothermal; the distribution of negative heat flux (heat loss) $q(r)$ [$\text{W}\cdot\text{m}^{-2}$] over a surface with grooves is expressed through the Stefan-Boltzmann law (the effective emissivity of a grooved surface is assumed to be equal to unity):

$$q(r) = -\sigma F(r)\left[T^4(r) - T_{bg}^4\right], \tag{7.28}$$

where σ is the Stefan-Boltzmann constant and T_{bg} is the temperature of the background.

The steady-state temperature distribution in the selected "tooth" was computed by the finite element method using the commercial software (ANSYS® from Ansys, Inc., USA; see https://www.ansys.com/products/structures/thermal-analysis).

The finite-element model of an individual "tooth" was built as shown in Fig. 7.17a. The cylindrical surfaces of a "tooth" were considered adiabatic. For the "worst" case, when the temperature nonuniformity is greatest, it was assumed that the temperature of a liquid agent is 350 K and the background temperature is 77 K. The total thickness of 0.1 mm was assumed for the black coating together with the ground. The thermal conductivity of this layer of 1 $\text{W}\cdot\text{m}^{-1}\cdot\text{K}^{-1}$ were accepted. Calculation of steady-state temperature field in an individual "tooth" by the finite element method (see Fig. 7.17b) shows that the average temperature of the grooved surface differs by about 19 mK from the temperature of the liquid agent. Computed temperature nonuniformity along the groove surface does not exceed 6 mK.

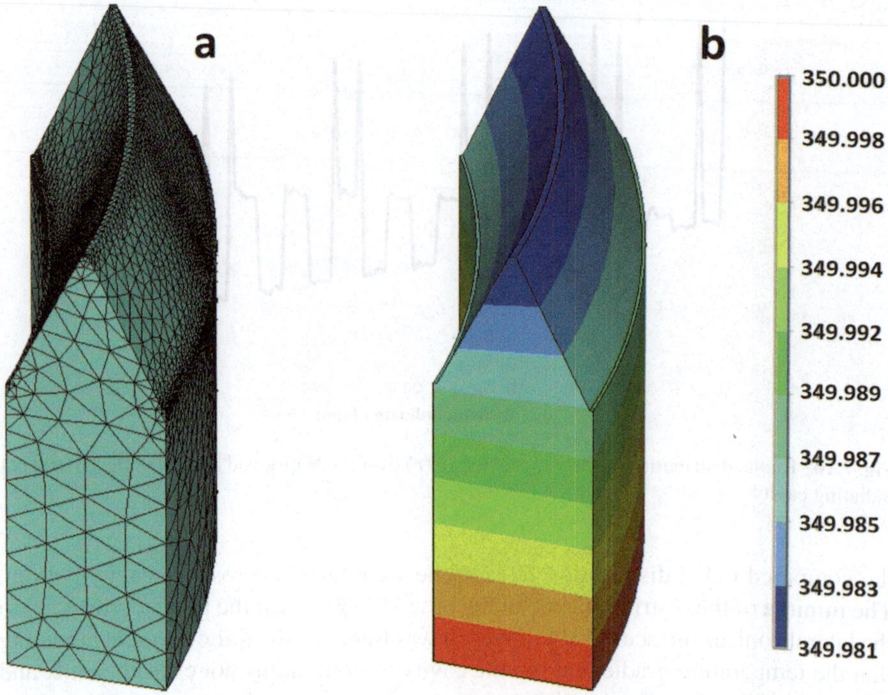

Fig. 7.17 Thermal modeling of concentric trapezoidal grooves performed with the ANSYS® (Ansys, Inc., USA): **a** finite-element model and **b** the steady-state temperature distribution

A rough estimate of the uncertainty in the spectral radiance that can be made by simple calculations using Eq. 3.38 show that the uncertainty of 19 mK at 350 K results in the uncertainty of the spectral radiance of about 0.015% at the wavelength of 15 μm and of 0.09% at 1.5 μm.

Figure 7.18 presents the spectral normal effective emissivities of the BB100-V1 computed for two values of wall emissivities and three temperatures distributions

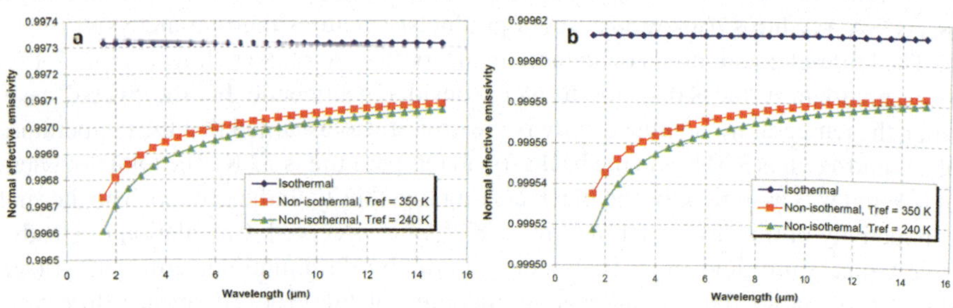

Fig. 7.18 Spectral normal effective emissivities of the BB100-V1 computed for 3 temperatures distributions for wall emissivity of: **a** 0.935 and **b** 0.990

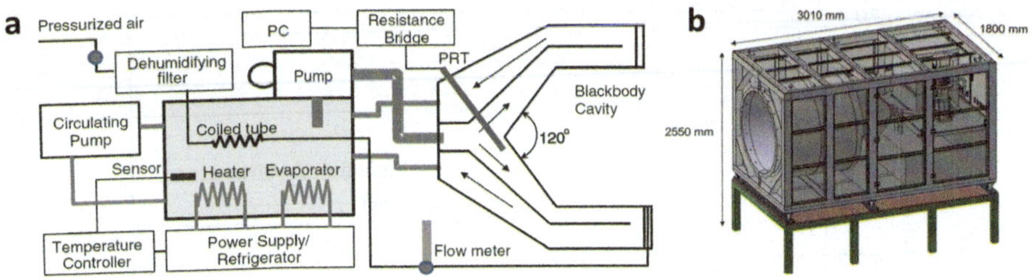

Fig. 7.19 Large-aperture water-circulating blackbody: **a** schematic thermal and electrical diagram and **b** panoramic view. Reproduced from [84] with permission of Springer Nature

(isothermal and modeled for two temperatures, 240 and 350 K, of the bottom rear side).

Sometimes, the temperature drop effect can be directly observed. An example of such a case is given by Park et al. [84], who described the large-aperture blackbody (LABB) developed at KRISS[2] for calibrating radiation thermometers and infrared radiometers with a wide FOV. The cylindro-conical radiating cavity has a diameter of 1.1 m and a length of 1 m. Its conical bottom has an apex angle of 120°. The cavity is integrated into a water bath, which is supplied with water from a tank. The LABB operates in air at temperatures from 10 to 90 °C. The convection heat loss from the radiating surface of the cavity to the ambient air is reduced by purging the cavity with dry air that passes through a coiled tube immersed in the tank and, therefore, has the same temperature. The schematic diagram and panoramic view of the LABB are shown in Fig. 7.19.

The thermostated water using a pump circulates between the external reservoir and the double-walled and partitioned cylindro-conical cavity. Figure 7.20 shows typical results of a stability test for the cavity temperature measured simultaneously using the radiation thermometer and the SPRT, which is installed near the apex of the cavity to measure the water flow temperatures that is assumed equal to the cavity temperature.

The stability test performed at the cavity temperature around 70 °C during 40 min after temperature stabilization. The temperature stability measured by the PRT ranges from 0.01 to 0.02 °C, while the stability measured by the radiation thermometer ranges from 0.02 to 0.05 °C.

This difference can be explained by four factors: (i) fluctuations of the heat transfer coefficient that accompany the forced convection of the air inside the cavity, (ii) variations of the air refraction index due to the same reason, (iii) smoothing the temperature fluctuations by the PRT that has greater thermal inertia than the radiation thermometer, and (iv) the difference between trendlines is due to the temperature drop effect.

The effect of temperature drop is very general. It manifests itself in not only contact measurements of the blackbody temperatures, but also when the temperature of a

[2] Korea Research Institute of Standards and Science, the NMI of South Korea.

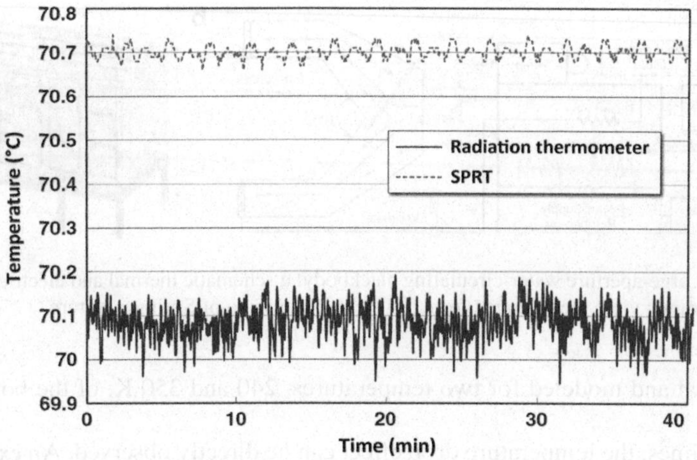

Fig. 7.20 Temperature stability of LABB simultaneously measured by SPRT and radiation thermometer at the cavity radiance temperature of 70 °C. Reproduced from [84] with permission of Springer Nature

blackbody is set by immersing the radiating cavity into the melt or liquid at a constant temperature or the liquid from a thermostat passes over the cavity. As a rule, the correction for the temperature drop effect can be made only by computational means. The examples of application of computer modeling to evaluation of the temperature drop effect in fixed-point blackbodies are given by Jimeno-Largo et al. [61], Castro et al. [17, 18].

7.3.3 *Positioning Effect*

Contact measurement of temperature implies thermal equilibrium between the sensor and the object, temperature of which should be measured. Ideal thermal equilibrium is unattainable in practice; however, a number of technical measures can be taken to minimize deviations from this state. After the transient process, the net heat flux between the sensor and the body under measurement becomes negligible and their temperatures become almost equal.

The presence of heat inflow or outflow violates thermal equilibrium and leads to a systematic error, which must be corrected if it cannot be completely excluded. Calibration of the SPRTs is performed by immersion in the fixed-point sells. Secondary contact thermometers are calibrated typically by comparison with the standard thermometers using liquid baths, dry-well blocks, and analogous calibration equipment.

Calibration conditions are strictly regulated for each type of contact thermometers in order to minimize most components of systematic error that depend on deviation of these conditions from thermodynamic equilibrium. A lot of useful information on calibration enclosures can be found in Quinn [91] and Horrigan [52].

The heat leakage through the leads is one of the most common and significant sources of systematic errors at contact measurement of temperatures inside solids and liquids. A long-stem contact thermometric probe has the sensing element at the immersed end; the part of stem protruding from the thermowell acts as a cooling fin, through which the heat flux is removed from the body under measurement and is transferred to the colder environment by convection and/or radiation. This leads to a decrease in the temperature of the sensor relative to the walls of the thermowell enclosing it. In thermometric literature, this source of systematic error often called the error due to immersion depth. Calibration of contact temperature sensors in fixed-point cells, or by comparison with the reference thermometer in the liquid bath or dry-well block is performed at a strictly regulated immersion depth.

The problem of dependence of contact thermometer readings upon immersion depth is well known, thoroughly studied by both analytical and experimental methods for most of practically important cases. These results are summarized in the compendium edited by Bernhard [10]. Nicholas and White [80] considered a simple model shown in Fig. 7.21.

The thermometer is inserted in the thermometric well inside a solid body having uniform temperature T_0. A part of the thermometer stem is outside the body, where it participates in the heat exchange with the environment having temperature $T_a < T_0$. Due to temperature difference, a continuous heat flow propagates along the stem of a thermometer between the solid body under measurement and the environment. These heat losses through the thermometer stem result in lower temperature of the thermometer's tip than the body under measurement.

The model applied to calculation of the temperature distribution is as follow. The thermal contact between the body and the thermometer is assumed ideal (zero thermal resistance), the length of thermometer stem is assumed infinite, and the

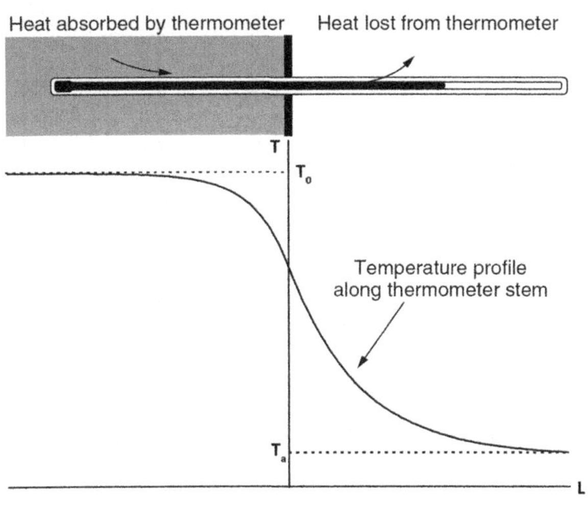

Fig. 7.21 The heat flow along the stem of a thermometer causes the thermometer to indicate temperatures slightly lower than that of the medium of interest. Reproduced from [80] with permission of John Wiley and Sons

thermal conductivity distribution in the thermometer is assumed uniform. Finally, the convective heat losses from the protruding part of the thermometer to the air is assumed obeying the Newton law (the heat flux is proportional to the difference of stem and air temperatures) gives the following dependence of the thermometer reading error on the immersion depth L:

$$\Delta T = (T_\mathrm{a} - T_0)\, K \exp\left(-L / D_\mathrm{eff}\right), \tag{7.29}$$

where D_eff is the "effective" diameter of the thermometer; both K and D_eff depend on the thermal resistance between the thermometer and the measured body and on the ratio of their heat capacities.

Equation 7.29 plotted in logarithmic scale in Fig. 7.22 for $K = 1$, can be used for determining the minimum immersion depth that ensures a negligible error. Equation 7.29 is approximately valid also for thermometer immersed into the still liquid.

When measuring the temperature inside a solid or liquid, the end of the thermometer should be immersed at least at the "minimum" (or "acceptable") depth, which ensures that the heat flux arising in the probe stem does not cause a significant error. Only with the proper immersion of the thermometer, the sensor can measure the correct temperature with an accuracy corresponding to its tolerance or calibration uncertainty.

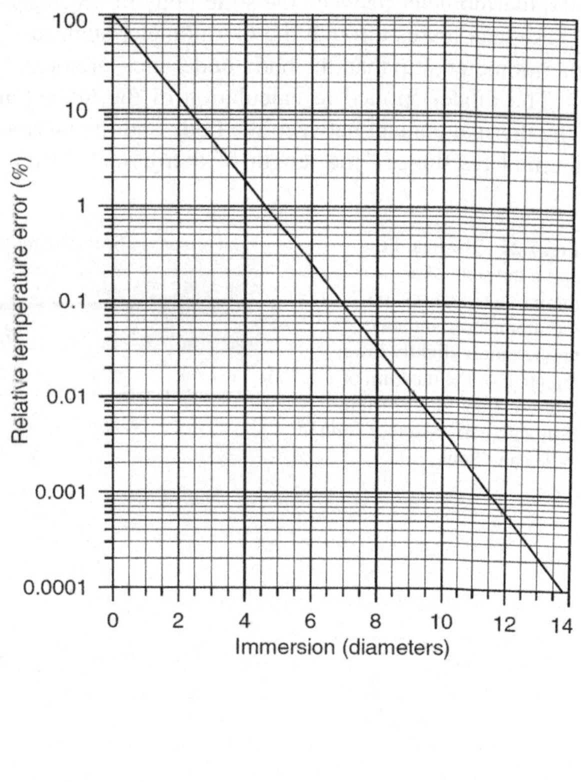

Fig. 7.22 The relative temperature error $|T_m / (T_0 - T_a)|$ plotted versus thermometer immersion length in stem diameters. Reproduced from [80] with permission of John Wiley and Sons

The value of the minimum immersion depth varies considerably and depends on the type of sensor, the design of the probe, the sheath diameter (if any), the contact thermal resistances, the diameter of the leads (for PRTs and thermistors) or the diameter of the thermoelements (for thermocouples), as well as on the body or medium, in which the probe is immersed. This depth can be determined experimentally during thermometer calibration by slow immersion of the probe into a calibration device until the indicated temperature no longer changes significantly. The minimum immersion depth is increased with temperature due to intensification of the radiant heat losses from the thermometer stem. The analytical solutions help us to understand basic physical processes occurred in the system of a body under measurement and thermometer.

However, they are not always suitable for practical calculations of correction factors or evaluation of associated uncertainties because contain many "effective" values introduced for simplification of the computational model. Nicholas and White [80] recommend the following three simple rules of thumb to determine the suitable immersion depth of a thermometer:

1. For the error not exceeding 1% (this is accuracy level of the typical industrial temperature measurements), the immersion depth must be not less than 5 probe diameters plus the length of the sensing element.
2. For the error not exceeding 0.01% (this is accuracy level of the typical calibration laboratory), the immersion depth must be not less than 10 probe diameters plus the length of the sensing element.
3. For the error of about 0.0001% (this value is typical for the highest-accuracy laboratories or the NMIs and measurements with the fixed-point cells), the immersion depth must be greater than 15 probe diameters plus the sensing element length.

If the probe is inserted into a thermowell, the gap between the sensor and well's walls should be as small as possible. This allows the immersed part of the thermometer to be at more uniform temperature, which reduces heat losses of the sensing element. At the measurement of the blackbody temperature, the length of the thermometer part, inserted into the measuring channel, thermowell, or guide pipe, may be less than the immersion depth used at calibration in the fixed-point sell, liquid bath, or dry-well calibrator. Different immersion depths at calibration and measurements lead to non-equivalent heat dissipation and may result in the systematic error, which we refer to improper positioning or mounting of the temperature sensor.

Moreover, some type of blackbody radiators such as flat-plate blackbodies or fluid-bath blackbodies with thin-walled radiating cavities may not provide a sufficient place for proper arrangement of the temperature sensor. The systematic errors arising from this should be taken into account when budgeting uncertainties.

A special case is the measurement of surface temperature. Most of resistance thermometers are stem sensing devices, which has a small length of sensing elements and, therefore, are best suited for immersion use. However, there are certain types of thin-film and miniature wire-wound resistance thermometers designed for the surface temperature measurements being in close contact with the surface but thermally insulated from the surrounding medium. Although thermocouple assemblies are mainly tip sensing devices, the use of protection sheaths makes impractical surface sensing with thermocouples. A thermocouple intended for surface sensing must has exposed welded junction with very small thermal mass or be housed in a construction, which permits reliable thermal contact with the surface under measurement.

The contact thermal resistance can be an additional source of a systematic error. If the end of the thermometer and the bottom of the thermowell have an uneven or rough surface, the contact area may be significantly less than the nominal [72]. Gaps between these two surfaces is filled with air, a very effective thermal insulator when measurements are conducted in atmospheric conditions. In vacuo, heat flux in the gaps can be transferred only via thermal radiation. The absence of reliable thermal contact between the thermometer and the measured object results in a systematic error, which is absent or accounted for at the calibration. Filling the gap with a thermally conductive material, which has a suitable thermal expansion, viscosity, and durability is the common method of reducing the contact thermal resistance. The very attractive thermal interface materials (see, e.g. [87, 108]) are metal alloys solders that may have high thermal conductivities of ~30 to ~90 W $\cdot$ m^{-1} $\cdot$ K^{-1} and applicable up to temperature of about 350 °C. In many cases, indium and indium alloys may be the best solder for thermal interface from cryogenic temperatures to about 200 °C [23]. Pure indium has a thermal conductivity of 86 W $\cdot$ m^{-1} $\cdot$ K^{-1} and can wet non-metallic materials; therefore, it can be used to seal, with excellent integrity, metallic and non-metallic parts. Thermally conductive greases (emulsions of ceramic or metal particles in organic or silicon fluids) are designed to flow into surface imperfections, minimizing the interface resistance [32, 88, 100, 106]. The bulk thermal conductivities of the greases are low (less than 4 W $\cdot$ m^{-1} $\cdot$ K^{-1}) so the contact layer must be very thin.

7.3.4 *Proper Positioning of a Contact Thermometer: Case Studies*

In this Section, we consider some typical situations with contact measurement of temperatures of VTBBs of different types. For blackbodies, whose cavity's temperatures are equalized by a massive core made of metal with high thermal conductivity (e.g. aluminum or copper), we have to measure the temperature inside a solid as close as possible to the viewable part of the radiating cavity.

As a rule, it is sufficient to insert a temperature sensor (PRT, thermocouple, or thermistor) in a single deep axial bore as shown in Fig. 7.23a. Such a positioning of an SPRT was used, for instance, in the VTBB BB7 described by Litorja and Tsai [68]. The radiating cavity of the BB7 was made of graphite and pressed into the copper (melting temperature is 1084.62 °C; thermal conductivity at room temperature is about 400 W $\cdot$ m^{-1} $\cdot$ K^{-1}) core heated by the main heater up any temperature within

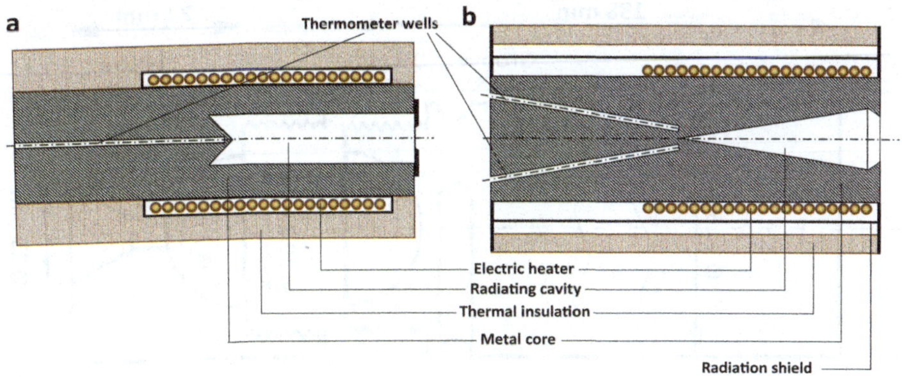

Fig. 7.23 Possible arrangement of thermometer wells in blackbodies with indirect resistance heating: **a** single well and **b** multiple wells

the range from 170 to 700 °C. Two auxiliary heaters serve for temperature non-uniformity correction. A thermowell is made so that the sensing element of the SPRT is close to the center of the bottom of the radiating cavity.

In some cases, two or more thermowells can be used to average the readings of the sensors inserted into them (see Fig. 7.23b).

The practical examples of such a design were given by Little et al. [69], Jarratt [59], and Herron et al. [48]. They described the Standard IR sources (SIRS II, SIRS IIa, and SIRS IIb) operated in the vacuum environment at 20 K of the 7 V Aerospace Chamber at the Arnold Engineering Development Center (AEDC). These VTBBs were designed to calibrate Earth orbiting satellite-based IR sensors. Radiating cavities of the SIRS-type blackbodies are made inside a massive aluminum alloy (the melting temperature is about 600 °C; the thermal conductivity at room temperature is around 160 $W \cdot m^{-1} \cdot K^{-1}$). Two long narrow channels are made in the core for insertion of thermometers to arrange their sensing elements in close proximity to the radiating cavity bottom. The SIRS blackbody can operate at temperatures from 275 to 375 K; two miniature, high-resolution thermistors in the ceramic tube sheath are used to measure the cavity temperature.

For better thermal contact, a thermally conductive epoxy is used. Blackbodies SIRS II and SIRS IIa operate from 100 to 400 K; SIRS IIb can operate up to 500 K. The IPRTs in thin-wall stainless steel sheaths are used for temperature measurements in SIRS IIb.

Another practical example of a well thought out placement of the temperature sensor has been given by Cárdenas-García and Méndez-Lango [19] for a blackbody designed for calibration of medical ear IR thermometers. As a rule, such blackbodies are the thin-walled cavities immersed into water baths. Cárdenas-García and Méndez-Lango [19] described an original construction of a blackbody (see the schematic cross-section in Fig. 7.24), which can be placed in an isothermal dry block oven. The blackbody manufactured from aluminum rod consists of a body and a cap that form the spherical cavity.

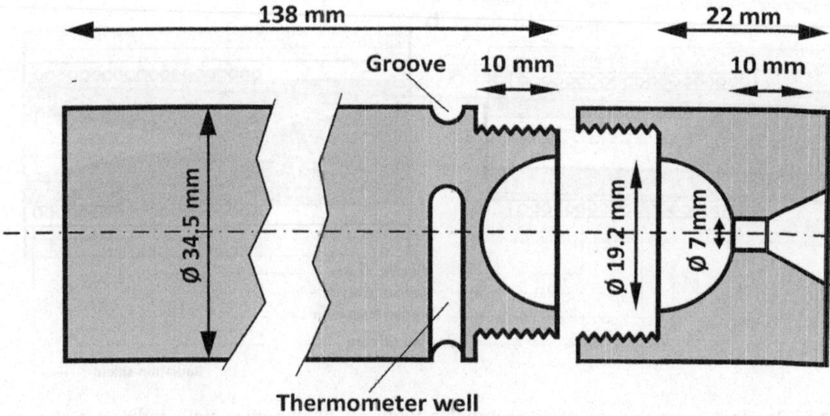

Fig. 7.24 Cross-section of the blackbody cavity for IR ear thermometers calibration. Reproduced from [19] with permission of American Institute of Physics

To improve thermal contact of the parts when joined together, the indium wire is used to form a cold weld. The inner surface of the cavity is coated with a high-emissivity diffuse black paint. A Pt100 with the resolution of 1 mK is used as a reference thermometer. It has a 15 mm length, 3 mm diameter and 4 leads. The PRT is placed approximately 2 mm below the cavity and some grease was added between the well and the PRT to enhance thermal contact. The leads of the PRT are wrapped around the cavity and laid in the annular groove. The cavity with the reference PRT is fully immersed vertically in a commercial dry block oven with calibration volume 35 mm diameter by 160 mm deep. The dry block is able to cover the temperature range from −55 to 140 °C, has the nominal uniformity of 18 mK and absolute stability of ±0.03 °C over 30 min.

The axial uniformity of the blackbody was checked by measuring the temperature placing the cavity block one centimeter up and one down from its regular position. A temperature change of less than 3 mK was found. The radial uniformity of the blackbody was checked by attaching an additional PRT to one side of the cavity and comparing its temperature readings with those from the PRT located near the bottom of the cavity. The wires of the additional PRT were also laid in the groove and brought to the front of the blackbody, which was placed inside the dry-well block. Measurements were carried out at four different positions, as shown in Fig. 7.25. The difference among these four temperatures was less than 2 mK, while the difference in temperature readings between the PRT located near the bottom of the cavity and that attached to the side was less than 6 mK.

Best et al. [11] described a blackbody designed at the University of Wisconsin (Madison) for the on-board calibration of the Geosynchronous Imaging Fourier Transform Spectrometer (GIFTS). The temperature of the radiating cavity is measured by a thermistor mounted in the wall of blackbody as shown in Fig. 7.26a. Two identical GIFTS blackbodies with aperture diameter of 27.4 mm were supposed to operate at nominal set points of 255 and 290 K. The blackbody temperature is

Fig. 7.25 Photographs of the front of the cavity block placed at four different positions to check the radial uniformity. Reproduced from [19] with permission of American Institute of Physics

measured using four glass-bead NTC thermistors; two redundant sensors are used for temperature control, and one for over-temperature protection. Thermistors are placed in threaded aluminum housings.

This allows the selection and acceptance tests (including thermal cycling) to be carried out on the thermistor assembly as a whole.

The thermistor-mounting configuration shown in Fig. 7.26a provides a minimal heat leaking through the thermistor lead wires, which is one of the major sources of temperature systematic error. An axisymmetric finite element thermal model together with the superimposed computed temperature map is shown in Fig. 7.26b. For the worst case (blackbody at 313 K and radiative environment at 140 K), computer modeling predicts a less than 8 mK temperature difference occurring between the thermistor bead and surrounding aluminum cavity walls due to heat losses through the thermistor lead wires.

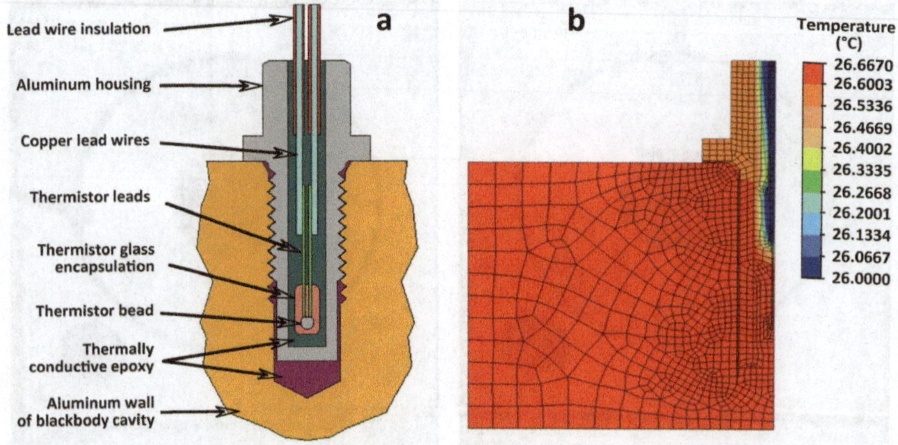

Fig. 7.26 Thermistor mounting of the GIFTS on-board calibration blackbody: **a** a section view and b axisymmetric finite element thermal model with the superimposed temperature map. Reproduced from [11] with permission of author and SPIE Press

Similar scheme of contact measurement of the temperature is applicable to heat-pipe blackbodies. A heat pipe can be roughly considered as a solid body with an effective thermal conductivity up to two orders of magnitude greater than copper. Hill and Woods [49] described the heat-pipe blackbodies developed at the NRC (Canada). Figure 7.27 shows the sections of the water-filled and sodium-filled heat-pipe blackbodies with the front arrangement of thermowells.

It is clear that this arrangement simplifies the manufacture of a heat pipe, because most welded joints are located on one side of it.

Traceability to the ITS-90 is achieved with a calibrated Au/Pt thermocouple inserted into the thermowell passing parallel to the radiating cavity and reached the middle of the viewable part. The thermowell entrance is located in front of the heat pipe. However, more common is the rear arrangement of the thermowell coaxial to the cavity and reaching almost to the center of the cavity bottom as in the ammonia-filled heat-pipe blackbody [42] shown in Fig. 7.28a. Temperature of the ammonia heat-pipe blackbody operating at −60 to 50 °C is measured by the long-stem SPRT. Another version of the design of the ammonia heat-pipe blackbody is presented in Fig. 7.28b. Unlike the previous one, it contains two thermowells for two independent PRTs.

The case of a contact thermometer located on a solid surface occurs when the wall thickness of the cavity is much smaller than all other dimensions, and therefore there is no room for the minimum acceptable immersion depth. Forgione et al. [37] described an advanced flat-plate blackbody with the carbon nanotube Vantablack [98] coating. The blackbody was designed for calibration and validation of the enhanced MODIS and ASTER airborne simulators.

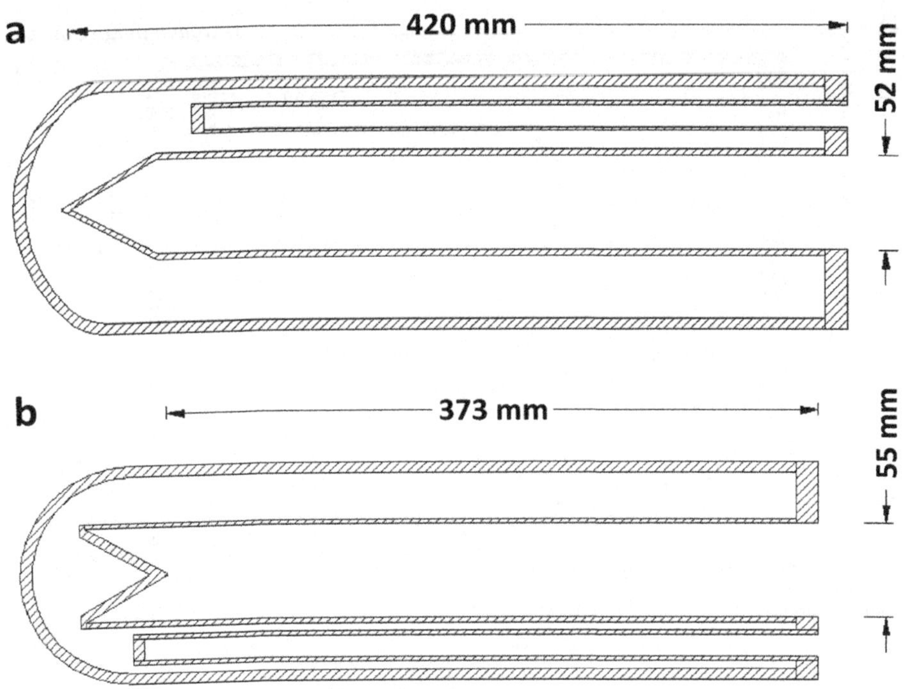

Fig. 7.27 The front arrangement of the thermowells in the heat-pipe VTBBs: **a** water-filled heat-pipe blackbody and **b** sodium-filled heat-pipe blackbody. Reproduced from [49] with permission of Springer Nature

The Moderate Resolution Imaging Spectroradiometer (MODIS) and Advanced Spaceborne Thermal Emission and Reflection Radiometer (ASTER) are imaging instruments onboard the Aqua and Terra satellites launched into Earth orbit by NASA in 1999 [92]. In the TIR spectral range, every scan of these instruments includes infrared imagery of two reference flat-plate blackbodies, held at different temperatures. Knowledge of plates' temperatures allows calculation of the radiant flux within instrumental spectral bands. To calibrate airborne simulators, the flat-plate blackbody was manufactured as a 177.8 mm diameter slab of pure, oxygen-free, high-conductivity copper. Thermally conductive Vantablack layer was used to achieve an emissivity as close to 1 as possible and reduce the temperature drop across the coating. The flag-mount NIST-traceable Pt100 temperature sensor and seven thermoelectric coolers are arranged on the rear side of the plate. The thermal gap-filler is used to reduce contact thermal resistance. The entire assembly is bolted to a heat sink, which provides structural strength and determines the path for heat flow.

Temperature measurement of thin-walled cavities is always fraught with considerable difficulties. Such cavities are typical for liquid-bath blackbodies, where the uniform temperature of the radiating cavity is achieved by its immersion into a uniform-temperature stirred bath. A foundation of the modern approach to the

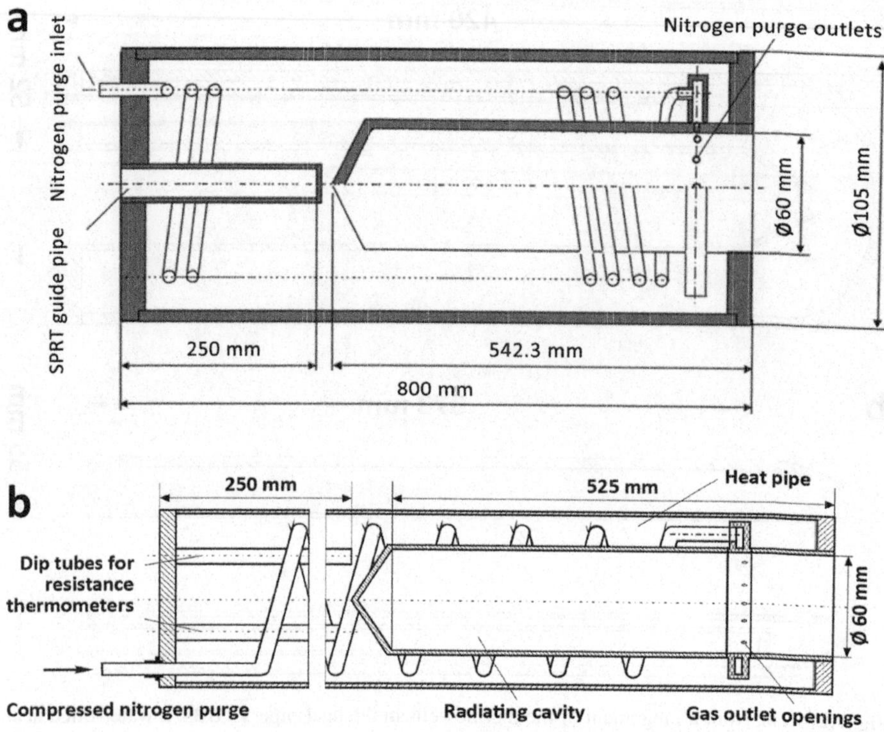

Fig. 7.28 The variant of design of the ammonia-filled heat-pipe blackbody developed at the PTB: **a** with a single thermowell (reproduced from [42] with permission of IOP Publishing) and **b** with two thermowells (reproduced from [47] with permission of Springer Nature)

precision liquid-bath blackbody design and characterization was laid by Geist and Fowler [40]. Later, the water-bath [38] and oil-bath [39] blackbodies of improved design were developed at the NIST (USA). Both blackbodies have identical large-aperture cylindro-conical cavities immersed into the working fluid of the commercial temperature-controlled baths specially modified for the horizontal arrangement of the cavity. The cavities with the wall thickness of 4 mm are made of oxygen-free copper. A small thickness and high thermal conductivity (388 W · m^{-1} · K^{-1}) of copper walls of the cavity contribute to rapid temperature equalization of the cavity.

The water-bath blackbody (WBBB) operates in the temperature range from 293 to 353 K. Its temperature stability was found to be ±2 mK within the whole operating temperature range for any ambient temperature. Without the cavity inserted, the water bath temperature uniformity is typically ±1 mK; with the cavity inserted, the uniformity degraded to ±2 mK for the range 278 to 313 K and to ±5 mK for the range 313–353 K. Temperature of the water surrounding the cavity was measured using two thermistor probes connected with the two input channels of the electronic thermometer. One thermistor probe was placed at a fixed reference point near the cavity bottom apex; another probe can move around the cavity outer surface. Special measures have been taken to minimize heat conduction along the probe and thus minimize the immersion depth effects. Measurements showed that the maximum deviation of the water temperature from the temperature measured by the fixed probe varies from 0.0 mK near the cavity bottom apex to +4 mK near the bath wall adjacent to the cavity at 278 K, has no variation at 303 K, varies from 0.0 to −5 mK at 333 K, and varies from −2 to −7 mK at 353 K. Later, a series of improvement in the design of the WBBB was made; however, the base principle of its design and characterization remained unchanged. Jung et al. [62] presented the results of the two-dimensional scans of the temperature field in the water surrounding the cavity. This is a "snapshot" of a non-stationary temperature field in a stirred liquid that shows the size of the expected temperature non-uniformity over the cavity walls conditioned by the temperature inhomogeneity of surrounding water. It was found that the temperature inhomogeneities in the water surrounding the viewable part of the cavity do not exceed several millikelvins. The initial design of the WBBB did not provide measuring the temperature of the cavity itself, but only measuring the temperature of the water near the cavity external surface. Recently, Hanssen et al. [43] reported the next improvement in the design of the WBBB cavity. The configuration of the conical bottom of the radiating cavity was changed as shown in Fig. 7.29, which allowed to not only improve the uniformity of the effective emissivity distribution across the cavity bottom but also to embed a contact thermometer (Pt100 or thermistor) into the inverted cone insert.

An original design of the WBBB described by Fowler [38] was used in a series of similar blackbodies of various degrees of accuracy. Fischer [34] presented two liquid-bath blackbodies of similar design covered the temperature ranges from −20 to 100 °C (with a mixture of 50% ethylene glycol and 50% de-ionized water) and from 100 to 200 °C (with a silicone oil as the working fluid). Both blackbody cavities are manufactured from copper and are fully immersed in the fluid baths, whose temperature is measured by SPRTs. Similar ideas were laid in the design of the water-bath blackbody described by Ko et al. [64]. A radiating cylindrical cavity

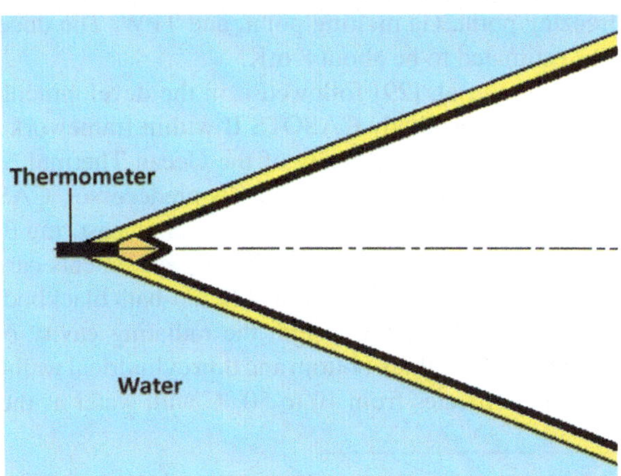

Fig. 7.29 A sectional view of the modified bottom of the WTBB cavity with the embedded thermometer

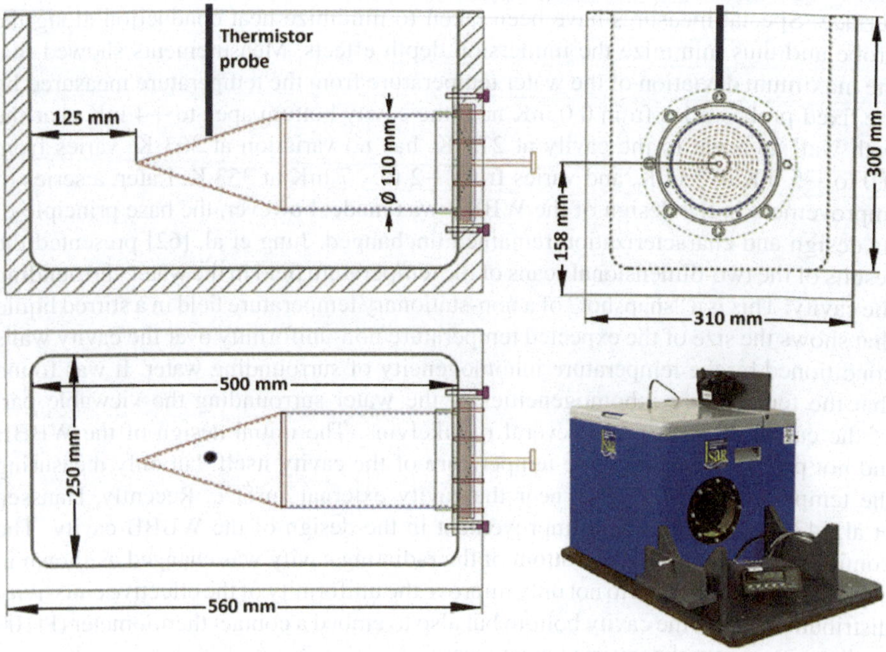

Fig. 7.30 CASOTS-II water-bath blackbody: three projections showing nominal position of the thermometer probe in the water bath and a photograph of the general arrangement of the blackbody system with the thermistor probe protruding from the water bath lid, the water bath pump (it is fitted internally when the blackbody is in operation) shown on top of the water bath lid, the radiometer-mounting jig, and the digital thermometer (readout) shown in front of the aperture. Reproduced from [30] with permission of Elsevier

with V-grooved internal surface made of oxygen-free copper of less than 3 mm thick was immersed in water, temperature of which can be controlled in the range from 30 to 70 °C. The PRTs placed at three locations close to the cavity are traceable to the In freezing point, Ga melting point, and TPW. The uncertainty of the PRT calibration was estimated to be about 8 mK.

Donlon et al. [29] followed it in the development of the low-cost transportable water-bath blackbody CASOTS II within framework of the European Union "Concerted Action for the Study of the Ocean Thermal Skin" (CASOTS) project. This blackbody shown Fig. 7.30, like its predecessor, CASOTS I [28], was intended for calibration of ship-borne IR radiometers measuring the radiance temperature of the sea surface and validate analogous measurements carried out from Earth orbit.

An alternative construction of liquid-bath blackbodies was developed at the NPL[3] [26]. Unlike Fowler's design, the radiating cavity of the NPL stirred liquid-bath blackbody has a thick bottom and thin cylindrical walls. It is entirely made of stainless steel and operates from 10 to 50 °C with water as the working fluid. Above 50 °C,

[3]National Physical Laboratory, the NMI of the United Kingdom.

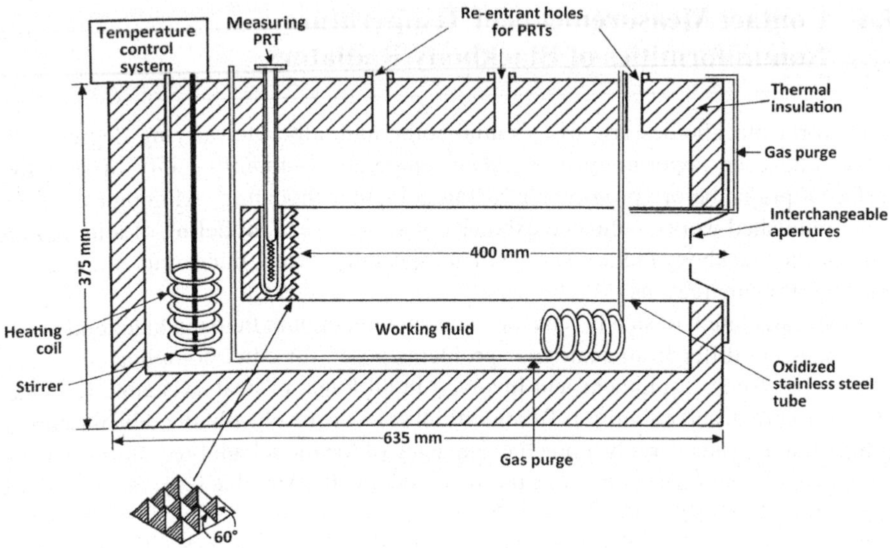

Fig. 7.31 A section view of the NPL liquid-bath blackbody. Reproduced from [71] with permission of Taylor & Francis

silicone oil is used up to the maximum operating temperature of 180 °C. The bottom is a stainless steel block grooved to form an array of small pyramids.

The cavity internal surface is oxidized by heating in air at temperature higher 900 °C for several hours to increase its surface emissivity up to approximately 0.7. The temperature of the cavity is measured using a calibrated 100 Ω PRT mounted in the bottom of the cavity. Machin et al. [71] described the similar liquid-bath blackbody (see Fig. 7.31) designed to assess the performance of thermal imagers that are in current use in medical and research clinics. The operating range of the blackbody cavity was 0–80 °C, with provision for dry gas purging below the dew point (~10 °C) to prevent condensation within the blackbody. The traceability to ITS-90 was provided through a PRT calibrated against the UK primary realization of ITS-90.

Wang et al. [102] described a VTBB with the radiating cavity (see Fig. 5.27) immersed in a stirred liquid bath containing 44 L of water-ethylene glycol mixture that provides the operating temperature range from −30 to 80 °C. A bottom disk of the cylindrical cavity made of copper and having a bore in it for mounting the reference capsule SPRT. It measures the bottom temperature using a digital thermometer. The bath temperature control system is a PID[4] closed loop with setting and displaying resolutions of 0.01 °C. A Pt100 precision sensor is used as a temperature sensor in the PID control.

[4]Proportional-Integral-Derivative (see, e.g. [103]).

7.4 Contact Measurements of Temperature Nonuniformities of Blackbody Radiators

Real-world blackbodies are always non-isothermal, albeit to varying degrees. We need to have an opportunity to evaluate temperature inhomogeneity of radiating surface a priori, before prototyping, using computer modeling. Although there are many published works dedicated exclusively to numerical modeling of temperature fields in the blackbody radiators, we will not specially review, nor discuss their results for the following reasons.

First, calculation of steady-state or transient temperature fields in blackbody radiators does not differ from analogous problems posed for other complicated objects such as electronic equipment and involved in all kinds of heat exchange (conductive, convective, and radiative). Trustworthiness of the results obtained by numerical simulation depends, largely, upon the adequacy of the model adopted. Some general recommendations on selection of a thermal model with a suitable level of complexity can be found in Ballico [6].

Second, the choice of a particular computational method depends on the task facing designers and researchers, as well as on their thermophysical and mathematical skills. We recommend readers, little or no familiar with computational problems of heat transfer, refer to Kutz [66] or such well-proven textbooks as that by Incropera et al. [57] or by Holman [51]; more experienced readers will, perhaps, find a lot of useful information in Minkowycz et al. [76].

Third, in the last decade, most calculations of temperature distributions for blackbody radiators were carried out using commercial software. The state-of-the-art thermal modeling and analysis software tools (e.g. ANSYS Fluent from ANSYS, Inc., Abaqus from Dassault Systèmes, Sinda™ from MSC Software Corp., etc.) are, in greater or lesser extent, integrated with the CAD software. Besides, such modeling packages as COMSOL Multiphysics® from COMSOL, Inc. and CAD software such as SolidWorks from SolidWorks Corp. include the modules capable to solve thermophysical problems of different levels of complexity. Although special skills in applied mathematics is not necessary to employ commercial software, the adequate model of the problem cannot be built and the right solution method cannot be chosen without a deep understanding of the heat transfer mechanisms and certain training in work with these packages. General recommendations can hardly be given here; users must rely on their own experience and consult with software documentation.

Fourth, we should always remember that some influencing factors might remain outside the computer model we built due to incompleteness of our knowledge about physical processes we simulate. For instance, emission of gaseous compounds at heating some materials or coatings is not always predictable a priori and can be accounted for in the model. However, the presence of this compound can intensify convective and conductive heat transfer, add absorption and scattering to radiative heat transfer and may result in appreciable change in conditions of heat exchange. Therefore, measurement of temperature distributions is desirable even if computer modeling is performed. Finally, we see no possibility of issuing common recommendations equally applicable to blackbodies operating in different temperature ranges and having different designs; the best thing we can do is to consider some practical examples. Approximately the same things with measurements: in the next sections, as well as throughout the book, we discuss several most commonly used techniques, their areas of applicability, and pitfalls, but avoid making universal recommendations.

The use of contact sensors is not the only method of temperature distribution measurements in blackbody radiometry. As soon as blackbody temperature becomes higher than some limit defined mainly by accuracy requirements, methods of radiation thermometry become preferable since they do not introduce into temperature field any distortions unavoidable if contact sensors are applied. Nonetheless, contact measurements dominate among cryogenic, low-, and elevated-temperature blackbodies. There are two ways of performing contact measurements of temperature distributions. The first way implies the use of a movable temperature sensor for scanning the temperature field; the second one is based on the arrangement of multiple sensors at the most critical points of a radiator and simultaneous (or almost simultaneous) registration of their output signals. The technique based on a movable sensor (usually, thermocouple) has an advantage in easier attainable self-consistency of measured temperatures. It was predominantly used until 1980s, when miniature PRTs and NTC thermistors with highly reproducible characteristics became readily available.

7.4.1 Using a Movable Temperature Sensor

Measurement of temperature distributions for the blackbody radiator in the classical experiment on determination of the Stefan-Boltzmann constant performed by Blevin and Brown [14] can be considered as a typical example of the use of movable temperature sensors. The essence of the method applied is measuring, using an electrical substitution radiometer, the radiance of the radiation emitted by a blackbody with known temperature and effective emissivity.

Figure 7.32 depicts schematically the blackbody used in this experiment. The main cylindro-conical cavity and the auxiliary cavity that serves for reducing heat loss of the main cavity are both made of graphite. Using Kanthal wire heater cavities are heated up to the freezing temperature of gold $T_{Au} \pm 0.5$ K (according to the IPTS-68, the predecessor of the ITS-90, T_{Au} was defined as 1337.58 ± 0.20 K). Four longitudinal bores of 9 mm in diameter were drilled in the graphite block and served for the insertion of Pt versus Pt/10% Rh thermocouples. Three of them centered 27 mm from the cavity axis pass through the entire block. The fourth, axial bore pass until a point 30 mm behind the apex of the cavity bottom and has a tapering through channel with diameter of 1 mm at the apex. This tapering channel is closed by the blackened end of a silica tube sheath of the central thermocouple. It does not distort radiation characteristics of the main cavity. The cold junctions of the thermocouples are immersed in the ice bath. In order to reduce the uncertainties due to variable

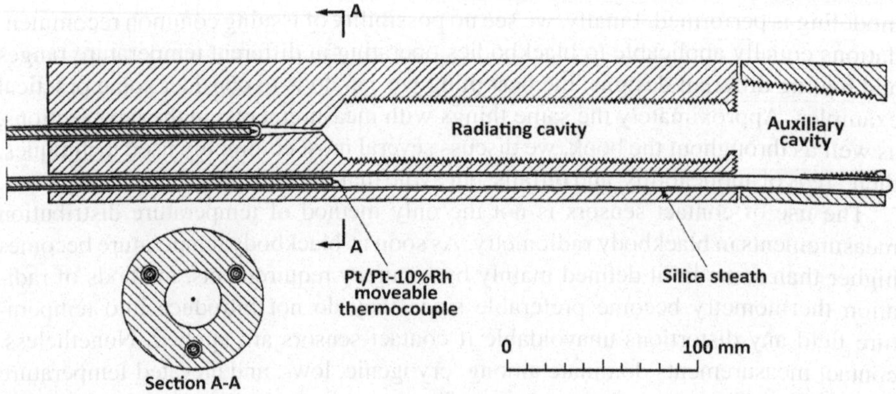

Fig. 7.32 The graphite block with the blackbody radiator and temperature measurement system. Reproduced from [14] with permission of IOP Publishing

immersion depths of the thermocouples and due to the measuring circuit, the most important calibrations were performed in the apparatus itself. For an initial check, one of thermocouples was also calibrated against national temperature standards maintaining the IPTS-68. The resulting uncertainty of the thermocouple calibration did not exceed 0.2 K.

During measurements, temperature distributions along the full length of the graphite block were periodically scanned by outer thermocouples. The mean value of temperatures they indicated was used to plot the graph presented in Fig. 7.33.

Temperatures of the main cavity bottom apex measured by the axial thermocouple and deduced from the scanned temperature distributions differ significantly; however,

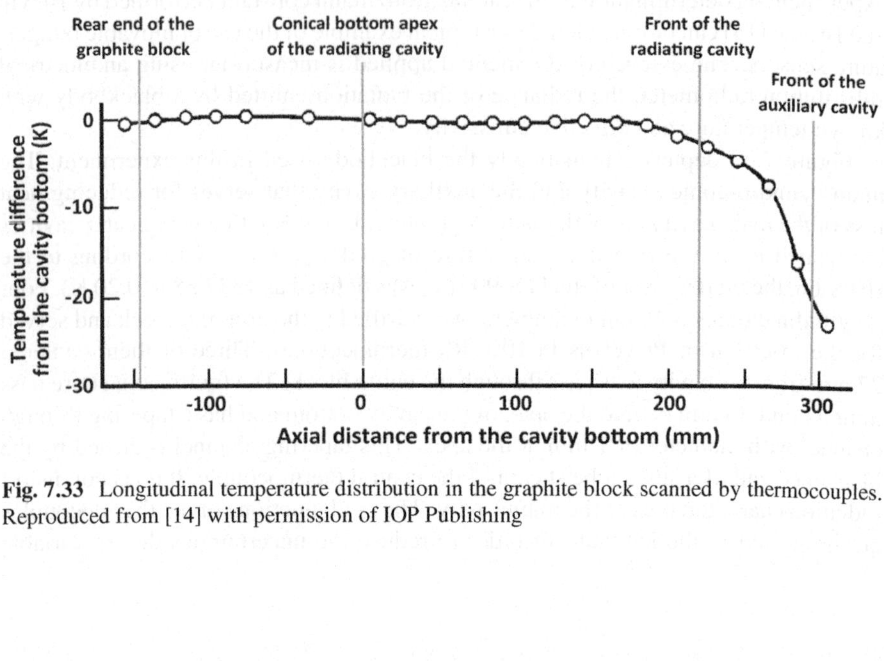

Fig. 7.33 Longitudinal temperature distribution in the graphite block scanned by thermocouples. Reproduced from [14] with permission of IOP Publishing

after correction for the temperature drop effect (this term was not used in the original paper being coined about twenty years later), the difference was not exceed 50 mK, which was comparable with measurement uncertainty of thermocouples). This result was experimentally confirmed by insertion of the thermocouple into the cavity through its opening until it almost touched the cavity bottom.

Another case of employing movable temperature sensors for temperature distribution measurements is described by Chahine et al. [21, 22]. Temperature uniformity of the 48 kW Thermogage (now, Thermo Gauge Instruments, Inc., http://thermogauge.com/) blackbody radiator heated by the flowing electric current was measured and modeled numerically at the studies focused on the minimization of temperature inhomogeneity.

The Thermo Gauge blackbodies were designed in the 1990s and was not changed much since then. This blackbody has two cylindrical cavities formed by a graphite tube with the central partition (septum). One cavity is working; another is auxiliary and is employed to measure and control the septum temperature using a radiation thermometer. A simplified cutaway drawing of the Thermo Gauge blackbody is shown in Fig. 7.34. To prevent the oxidation of graphite at high temperature, the graphite tube is purged with a laminar flow of argon introduced through water-cooled copper electrodes and exiting the tube through a small hole near the septum.

One of the main features of this design is the rapid achievement of the desired temperature. The reverse side of this design is a significant heterogeneity in the temperature of the radiating cavity.

The radiator made of graphite has the length of 289 mm and the internal diameter of 27.5 mm. Outer surface of the radiator has a stepped profile to compensate heat losses by reduction in the cross-sectional area and thereby increase of Joule heat generation. The middle septum forms two independent cylindrical cavities: the main, working cavity and the auxiliary one, monitored by a radiation thermometer included in the feedback loop of the power supply. Axial temperature distribution along the main cavity was measured by the type R thermocouple with the insulator removed

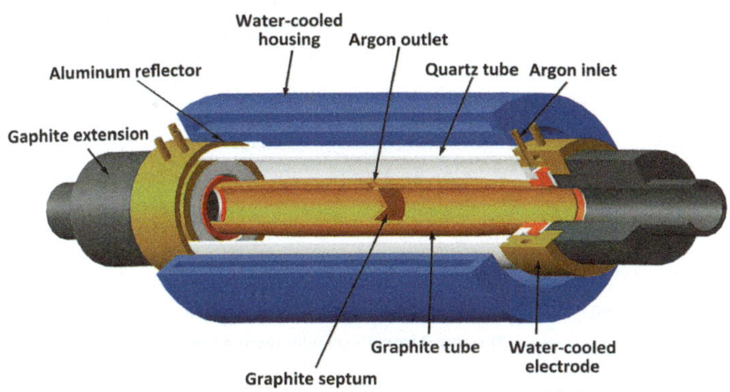

Fig. 7.34 Schematic of the Thermo Gauge graphite blackbody. Adapted from [90] with permission of Elsevier

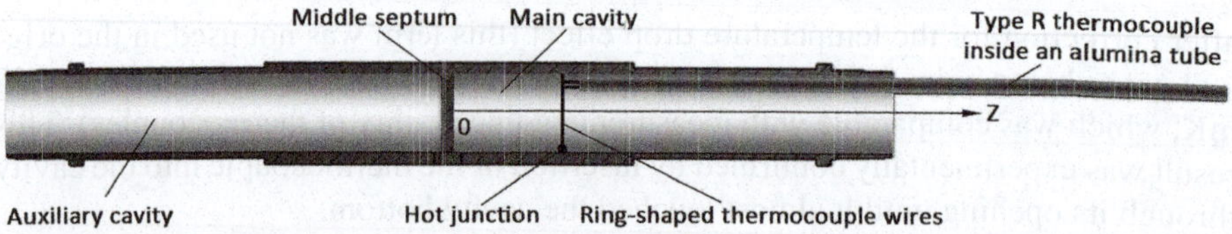

Fig. 7.35 Schematic of measuring temperature distributions of the Thermogage blackbody radiating cavity using the ring-shaped thermocouple. Reproduced from [22] with permission of Springer Nature

from its tip, so the wires connected at the hot junction formed a circular loop (ring). When the thermocouple is inserted into the working cavity, this ring together with the hot junction is tightly pressed against the circumference of the cavity's internal surface (see Fig. 7.35).

The EMF generated by a thermocouple depends not on the temperature difference between thermocouple junctions but on temperature distribution (or temperature gradient) in the thermocouple wires. The ring-shaped thermocouple allows minimizing the error due to wires temperature non-uniformity because their temperature should be uniform if the temperature distribution over the cavity internal surface is axisymmetric. Figure 7.36 shows the temperature profiles $T(z) - T(0)$ measured using the ring-shaped type R thermocouple for the cavity bottom (middle septum) temperature $T(0)$ equal to 1000 and 1500 °C.

The most important drawback of this method is impossibility to measure temperature distribution during the main measurement cycle. Besides, this technique allows scanning temperature distributions only along the part of cylindrical wall not too close to the cavity opening. When the ring approaches to the cavity's open end,

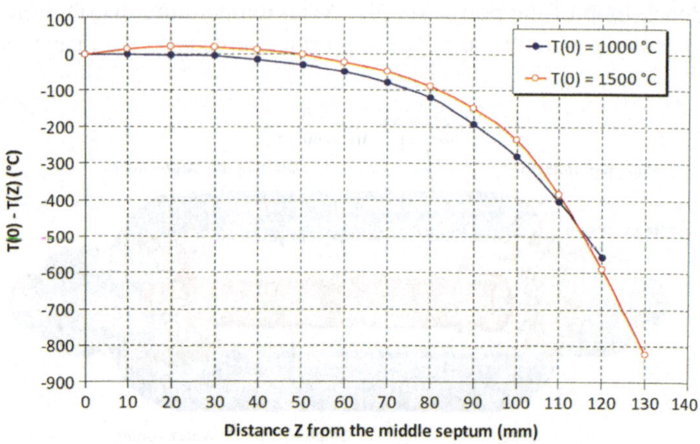

Fig. 7.36 Temperature profiles measured using the ring-shaped R-type thermocouple of the working cavity of the 48 W Thermogage blackbody at the septum temperature of 1000 and 1500 °C. Reproduced from [22] with permission of Springer Nature

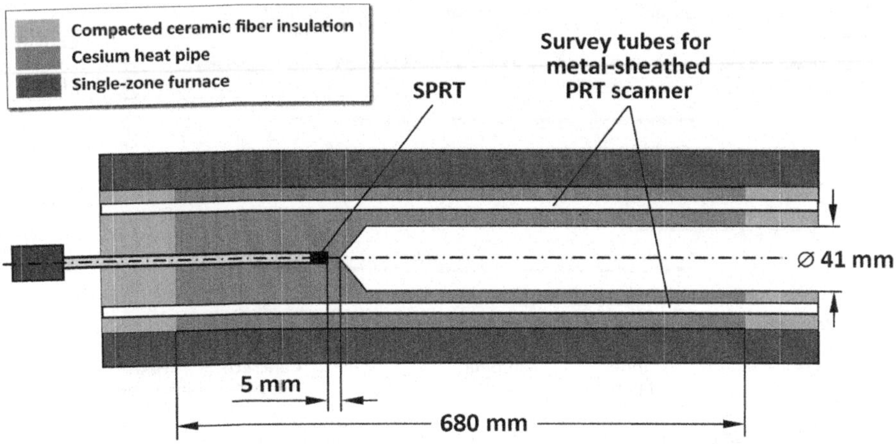

Fig. 7.37 Schematic section view of the cesium heat-pipe blackbody designed at the NPL. Reproduced from [50] with permission of Elsevier

temperature gradient in the thermocouple wires becomes too high to provide acceptable accuracy of measurements. Scanning of the axial temperature distribution by a movable contact sensor can also be applied to the heat-pipe blackbodies because the thermal behavior of a heat pipe reproduces in many respects the behavior of a solid body with extremely high thermal conductivity. Chu and Machin [25] described a cesium heat-pipe blackbody whose design is shown schematically in Fig. 7.37.

A single-zone tubular furnace heats the heat pipe with the radiating cavity up to operating temperature lying between +300 °C and +600 °C. The reference temperature of the cavity is measured by a 25 Ω SPRT calibrated according to the ITS-90. The SPRT is mounted in a thermowell approximately 5 mm behind the cavity bottom inside a silica tube preventing contamination of the sensor. The two survey tubes are embedded into the heat pipe to measure axial temperature distribution along the cavity by means of a movable metal-sheathed PRT calibrated in terms of ITS-90 with the expanded (at $k = 2$) uncertainty not exceeding 50 mK.

For accurate realization of the spectral irradiance scale in the red and near-IR spectral ranges, IR calibration of the remote sensing instrumentation, and research in radiometric measurement of thermodynamic temperatures, the large-area blackbody (LABB) has been developed at the PTB (Germany) on the base of double heat pipe [35, 44, 96, 97]. It consists of two concentric sodium heat pipes, both operating up to 1000 °C, and nested one within another. The inner heat pipe forms the radiating cavity as shown in Fig. 7.38. Four HTSPRTs are inserted into the through channels made in the bottom of the inner heat pipe; one of the thermometers is used as a sensor of the fine-control feedback loop. Each HTSPRT is shielded with a thin platinum tube and a tube of high-purity alumina.

The thickness of the bottom part of the inner heat pipe that defines the immersion depth of HTSPRTs is 229 mm. A 3.6 V potential difference is maintained between the inner heat pipe and the platinum tubes to avoid thermometers poisoning by

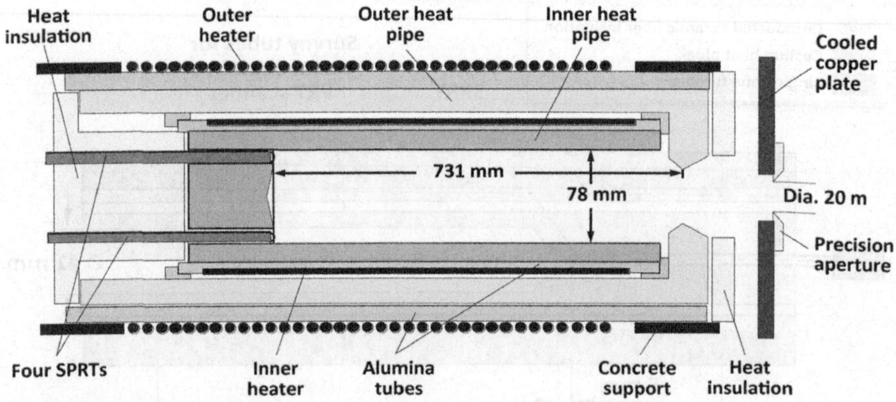

Fig. 7.38 Schematic section of the double heat-pipe large-area blackbody. Adapted from [50]with permission of Elsevier

positive ions. The thermometer used in the feedback loop is always left inside the heat pipe, while the others can be extracted for calibration and inserted back before each measurement. The short-time stability of thermometers does not exceed 2 mK, which is of the same order as the difference in the bottom temperatures measured by the three HTSPRTs.

Their rearrangement showed that these differences are due to thermometers themselves; therefore, the bottom can be considered as isothermal within the measurement accuracy. The deviation of temperature of the cavity cylindrical walls from the temperature of the bottom was measured by moving a 1.2 m long HTSPRTs inside the cavity toward its aperture. However, the measured temperature distribution is strongly influenced by radiation losses from HTSPRT into the environment through the cavity opening.

The temperature of the sensing element can be found from the condition of the balance of the absorbed radiant heat flux falling onto the sensing element from the entire cavity surface and the radiant flux leaving the sensor through the cavity opening. Such a balance must be expressed by the integral equation for the unknown temperature distribution along the cavity walls. Hartmann et al. [45] simplified the computational problem assuming that the radiation falling onto the temperature sensor consists of a uniform temperature of the cavity and a variation of the sensor's temperature with the varying immersion depth.

The latter term was assumed small enough to linearize the Stefan-Boltzmann law. This simplified model was applied to calculation of the cavity wall temperatures from temperatures measured by HTSPRT. Figure 7.39 presents the calculated temperature distributions for several temperatures of the cavity bottom. The temperature uniformity for the bottom temperatures above 500 °C is within 10 mK. The temperatures below 480 °C are insufficient for proper work of the sodium heat pipe, so significant temperature non-uniformities occur.

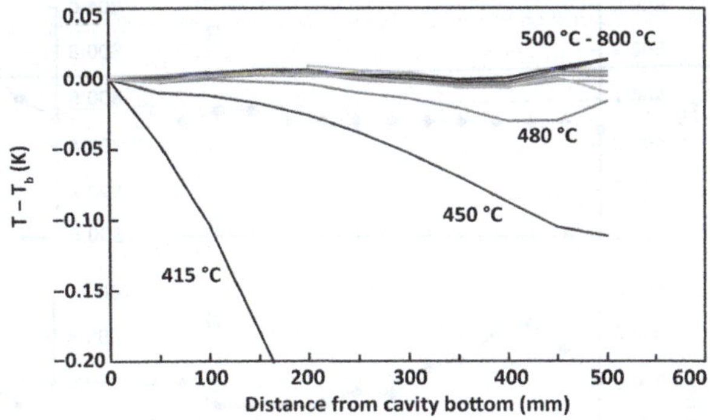

Fig. 7.39 Temperature distribution along the inner cavity wall of the large-area blackbody. Reproduced from [50] with permission of Elsevier

Yinghang et al. [107] described design and characterization of heat-pipe blackbodies capable to operate with sodium as the working fluid in the temperature range from 800 to 1200 K and with a mixture in the proportions of 73:24:3 of cesium, potassium, and sodium in the temperature range from 500 to 800 K. Both heat-pipe blackbodies have similar constructions shown in Fig. 7.40. Blackbody temperature can be measured using either the PRT or a thermocouple inserted into special tubes at the back of the heat pipe.

A long thermocouple tube allows scanning the temperature inside the heat pipe along two-thirds of the cavity length. The type S (Pt/Pt–10% Rh) and Pt/Au thermocouples were used for these measurements, the stability of which was provided by inserting the reference junctions of the thermocouples into an ice bath. From the typical results presented in Fig. 7.41, it was concluded that the temperature uniformity was better than 0.05% of the operating temperature for the CsKNa blackbody and 0.04% for the sodium heat-pipe blackbody.

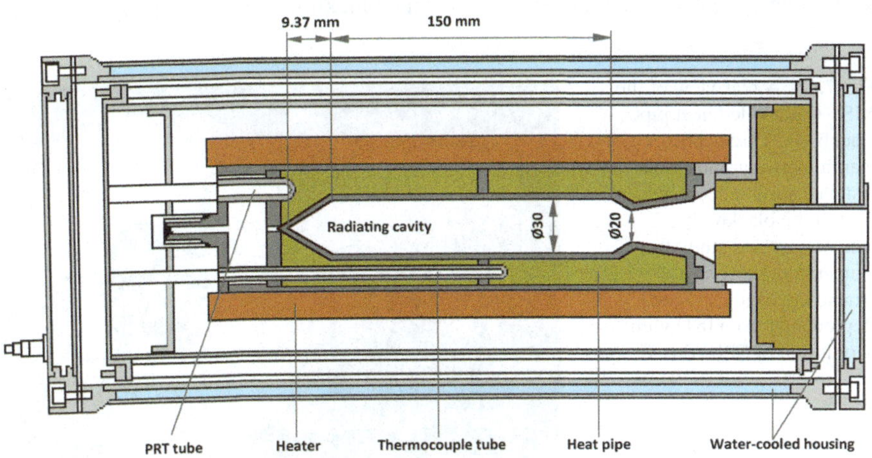

Fig. 7.40 Schematic of the CsKNa and Na heat-pipe blackbodies. Reproduced from [107] with permission of IOP Publishing

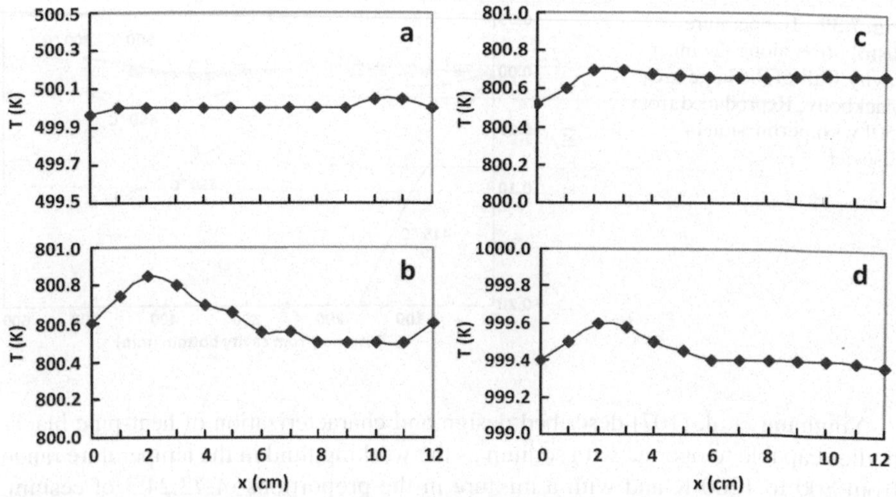

Fig. 7.41 Temperature uniformity of the heat-pipe blackbodies with heat transfer agents: **a** CsKNa at temperature of 500 K, **b** CsKNa at temperature of 800 K, **c** Na at temperature of 800 K, **d** Na at temperature of 1000 K. Reproduced from [107] with permission of IOP Publishing

Figure 7.42 presents the photograph of the rear side of the water-filled heat pipe blackbody [81] manufactured at KE Technologie GmbH according to NIST design and specifications.

Since 1990s, the NIST's scales of the spectral radiance and radiance temperature from 15 to 70 °C and from 70 to 180 °C were established on the base of the water-bath and oil-bath blackbodies (see [38, 39]). In order to improve the accuracy of the spectral radiance and radiance temperature calibrations, as well as to expand the temperature range, a water-filled heat-pipe blackbody was designed, constructed

Fig. 7.42 A rear view of the NIST water-filled heat pipe blackbody. Three contact thermometers are inserted installed simultaneously to control the blackbody temperature and study the temperature uniformity around the cavity. Reproduced from [81] with permission of SPIE Press

and characterized at NIST. The water heat-pipe blackbody has a 500-mm cylindro-conical cavity with a 62.7-mm diameter. The cavity is painted with Pyromark 1200 high-emissivity paint (see Sect. 6.2.3). The heat pipe containing 168 g of water was manufactured at KE Technologie GmbH according to NIST specifications. The rear side of the heat pipe has three thermowells, whose ends are positioned near the cavity bottom. The temperature gradients measured between the different contact sensors were used to determine the quantity of water required or the best operation of the heat pipe.

7.4.2 *The Use of Fixed Sensors*

Presently, scanning of temperature distributions by contact sensors is performed rarely. This is caused by several reasons. For blackbodies operating at temperatures above approximately 700 °C, methods of radiation thermometry are less time-consuming, require simpler correction for systematic errors, and therefore more reliable and capable to provide better accuracy.

Miniaturization of IPRTs and NTC thermistors achieved in the last decade gives a possibility to arrange enough of them to almost concurrently register their signals and obtain temperature distributions for blackbodies operating at lower temperatures. As a rule, an analog multiplexer performs sampling of temperatures sequentially measured by sensors in a repetitive manner, and sends their output signals to A/D converter or directly to a digital measuring instrument.

Figure 7.43 presents a schematic of the variable-temperature blackbody (VTBB) developed at VNIIOFI (Moscow) to operate in the cryo-vacuum calibration chambers. The operation temperature range of this VTBB is −60 to 90 °C inside the

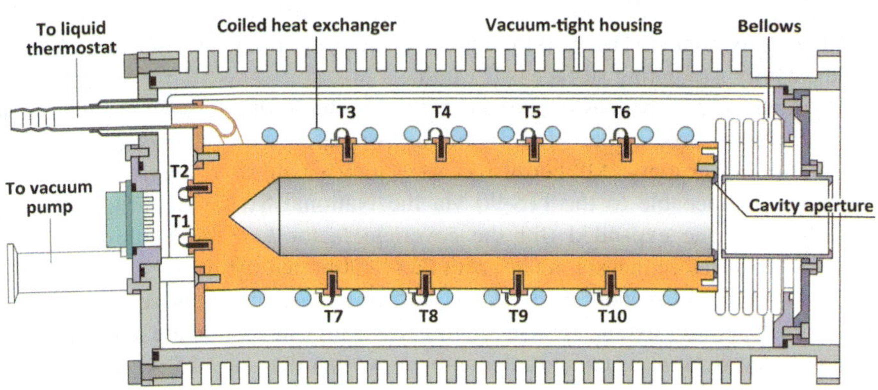

Fig. 7.43 A schematic of the VTBB radiator with the system of temperature measurement and control; ten PRTs are designated by T1 through T10. Reproduced from [46] with permission of Elsevier

LN_2-cooled shrouds and within temperature range from 20 to 90 °C in open air [58, 79, 83]. The cylindro-conical radiating cavity with a 30 mm diameter aperture is made in the copper block surrounded by the tubular coil heat exchanger. Temperature of the cavity is set by a precision external thermostat providing circulation of the working fluid in the heat exchanger. Temperature of the cavity is measured by five Pt100 sensors (the number of the temperature sensors may vary for various modifications of the VTBB). Thermometers are connected to the digital multimeter via the multiplexer. Temperature uniformity along the cavity is better than 0.1 °C and temperature stability is ±0.02 °C at −10 °C setpoint. The readouts of temperature sensors are averaged in real time to derive the VTBB reference temperature that is used in calculation according to the Planck law. Gröbner [41] reported results of development and characterization of the blackbody BB2007 developed at PMOD/WRC[5] (Switzerland) for calibration of pyrgeometers, instruments measured irradiation coming from the atmosphere within wavelength range from approximately 3 to 50 μm.

A pyrgeometer is a relatively simple device that consists of a blackened thermopile radiation detector with a silicon dome protecting the detector from the environment. Calibration of pyrgeometers in laboratory conditions was performed using blackbody radiators operated at temperatures close to the effective atmospheric temperatures. Unlike the blackbodies considered earlier, BB2007 operates in the vertical position and provides hemispherical irradiation of the pyrgeometer dome. To operate at different temperatures in the range of −30 to +30 °C, the radiating cavity is placed into the coiled tube heat exchanger with the silicon oil circulating using refrigerating/heating external thermostat.

Since the cavity must operate at temperatures below ambient, the cavity is flushing with dry air or nitrogen to reduce water vapor content and prevent condensation and icing. A humidity sensor continuously measures the relative humidity inside the cavity. Schematics of the BB2007 is shown in Fig. 7.44a. The temperature of the cavity is sampled by seven thermistors. Each thermistor (10 kΩ resistance at 25 °C) is mounted into a small aluminum holder, which, in turn, is screwed in the cavity wall from the outside. The wall thickness between the thermistor sensing element and the cavity's radiating surface is 0.1 mm.

Thermistors can be easily extracted for recalibration. It was assumed that the temperature drop effect is negligible as compared with other components of temperature uncertainty. All thermistors were jointly calibrated relative to a digital thermometer traceable to the ITS-90 via the National Swiss Metrological Institute METAS. The expanded (for the coverage factor $k = 2$) combined uncertainty of temperature measurements with these calibrated thermistors is ±0.024 K. Figure 7.44b shows deviations of temperatures measured by the seven thermistors from the temperature averaged for all thermistors. These deviations are minimal when the cavity temperature is equal to the ambient temperature of +15 °C and increase gradually with the growth of the difference between the cavity mean temperature and the ambient temperature. Improvement of the blackbody radiator's temperature homogeneity by the use of multiple heating elements implies also employing

[5] Physikalisch-Meteorologisches Observatorium Davos/World Radiation Center

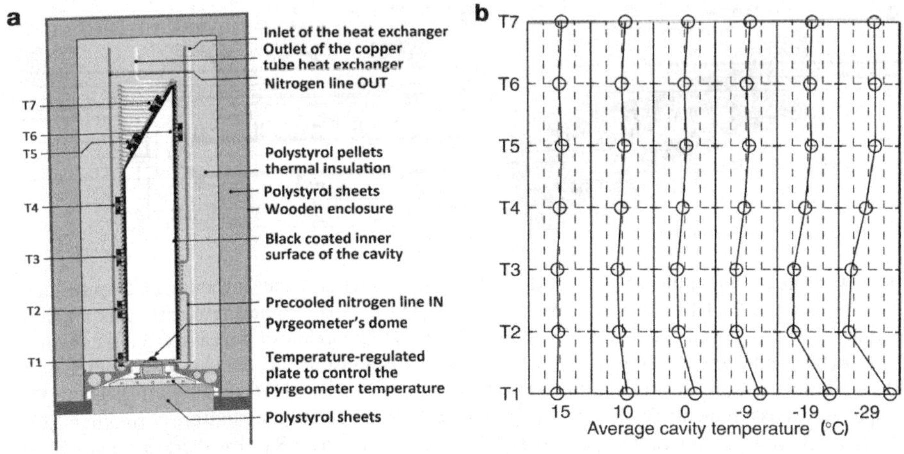

Fig. 7.44 The blackbody BB2007 for calibration of a pyrgeometer: **a** schematic of the blackbody BB2007 and **b** deviations of temperatures measured by the seven thermistors (denoted by T1 through T7) from the temperature averaged for all thermistors. The abscissa's dashed grid lines are plotted every 0.1 K. Reproduced from [41] with permission of Optical Society of America

of multiple temperature sensors monitoring radiator's temperature in each heating zone.

A typical example the vacuum variable medium-temperature blackbody is the VMTBB designed at VNIIOFI to serve as a reference source in the chamber with LN_2-cooled shroud, within temperature range from 150 to 430 °C in the moderate vacuum conditions (~10^{-3} Pa) and from −173 °C in a high vacuum [78].

The VMTBB was incorporated in the PTB's reduced-background calibration facility (RBCF) that is used for calibration of space-based IR remote-sensing experiments in terms of radiation temperature and spectral radiance and measurements of spectral emissivity in the in the wavelength range from 1 to 1000 μm [77]. A cutaway drawing of the VMTBB shown in Fig. 7.45 illustrates its design features. The internal surface of the cylindro-conical radiating cavity made of oxygen-free copper is chemically coated with nickel before applying high-emissivity paint.

The coarse temperature regulation is carried out by the LN_2-cooled and electrically heated outer thermostat; the fine regulation of the cavity temperature is carried out by three-zone electric heaters. The rear flange of the VMTBB is equipped with an electrical connector and two tubes for cooling gas that reduces the time of the cavity cooling. The VMTBB temperature is determined by 15 precision PRTs. Six of them are used for the cavity and heat exchanger temperature control, and others are used for the cavity temperature measurement and correction. To achieve desirable temperature uniformity of the cavity, the two-stage temperature control based on two precision PID controllers is used. The VMTBB tests showed that the temperature uniformity and stability of the VMTBB cavity is better than ±100 mK and ±50 mK, respectively, throughout the whole temperature range.

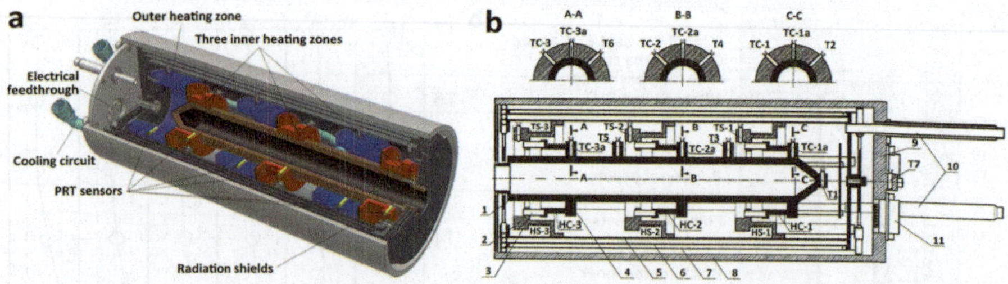

Fig. 7.45 The design of the VMTBB and its system of temperature measurement and control **a** the 3D view of the VMTBB and **b** Schematic of the VMTBB: 1 is a radiating cavity; 2 is a heating module of the thermostat; 3 is a post of heat link; 4 is a heating module of the cavity; 5 is a case of the thermostat; 6 and 7 are radiation screens; 8 is the VMTBB housing; 9 is the rear flange of the body; 10 are tubes for cooling gas; 11 is the electrical connector; T1 through T6 are PRTs for the VMTBB temperature measurement; TC-1a, TC-2a, TC-3a are PRTs for cavity-temperature correction; TC-1, TC-2, TC-3 are PRTs for cavity controller; TS-1, TS-2, TS-3 are PRTs for thermostat controller; HC-1, HC-2, HC-3 cavity heaters; HS-1, HS-2, HS-3 thermostat heaters; T7 is PRT for measuring temperature of the VMTBB housing. Reproduced from [78] with permission of Springer Nature

As we have already mentioned, contact measurement of the temperature of thin-walled cavities (which are usual for liquid-bath blackbodies) is a non-trivial problem. In order to be sure that it is the temperature of the cavity external wall that is measured, and not the temperature of the surrounding liquid, the temperature sensor must, firstly, be in close thermal contact with the wall and, secondly, thermally insulated from the liquid.

Additionally, the heat losses from the sensor (mainly, through the wires) must be minimized to avoid distortion of the temperature field in the cavity walls. This can be done using modern thin-film PTRs; however, we could not find appropriate practical examples. The information presented in the known works (e.g. [26, 102]) is insufficient to judge exactly how the temperature distributions along the walls of cavities of liquid-bath blackbodies were measured.

References

1. ASTM, *Manual on the Use of Thermocouples in Temperature Measurement*. ASTM manual series: MNL 12, 4th edn. (ASTM International, 1993)
2. ASTM E1137/E1137M-08, *Standard Specification for Industrial Platinum Resistance Thermometers* (ASTM International, West Conshohocken, PA, 2008)
3. ASTM E230/E230M-12, *Standard Specification and Temperature-Electromotive Force (emf) Tables for Standardized Thermocouples* (ASTM International, West Conshohocken, PA, 2012)
4. ASTM E2877-12$^{\varepsilon 1}$, *Standard Guide for Digital Contact Thermometers* (ASTM International, West Conshohocken, PA, 2012)
5. ASTM E879-20, *Standard Specification for Thermistor Sensors for General Purpose and Laboratory Temperature Measurements* (ASTM International, West Conshohocken, PA, 2020)
6. M.J. Ballico, Thermal modelling, in *Handbook of Temperature Measurement, Vol. 1. Temperature and Humidity Measurement*, ed. by R.E. Bentley (Springer, 1998), pp. 185–201

7. R.E. Bentley (ed.), *Handbook of Temperature Measurement. Vol. 1: Temperature and Humidity Measurement* (Springer, Singapore, 1998)
8. R.E. Bentley (ed.), *Handbook of Temperature Measurement. Vol. 2: Resistance and Liquid-in-Glass Thermometry* (Springer, Singapore, 1998)
9. R.E. Bentley (ed.), *Handbook of Temperature Measurement. Vol. 3: The Theory and Practice of Thermoelectric Thermometry* (Springer, Singapore, 1998)
10. F. Bernhard (Heraus.), *Handbuch der Technischen Temperaturmessung. 2. Auflage*—in German. *Handbook of Technical Temperature Measurement*, ed. by F. Bernhard, 2nd edn. (Springer, Berlin, 2014)
11. F.A. Best, H. E. Revercomb, Knuteson, et al. The Geosynchronous Imaging Fourier Transform Spectrometer (GIFTS) on-board blackbody calibration system. Proc. SPIE **5655**, 77–88 (2005)
12. BIPM, *Techniques for Approximating the International Temperature Scale of 1990.* Reprinting of the 1990 first ed. Organisation intergouvernementale de la Convention du Mètre, Sèvres, France (1997), https://www.bipm.org/utils/common/pdf/ITS-90/ITS-90-Techniques-for-Approximating.pdf. Accessed 17 Feb 2020
13. BIPM, *Guide to the Realization of the ITS-90. Platinum Resistance Thermometry* (2018), https://www.bipm.org/utils/common/pdf/ITS-90/Guide_ITS-90_5_SPRT_2018.pdf. Accessed 17 Feb 2020
14. W.R. Blevin, W.J. Brown, A precise measurement of the Stefan-Boltzmann constant. Metrologia **7**, 15–29 (1971)
15. G.W. Burns, D.C. Ripple, M. Battuello, Platinum versus palladium thermocouples: an emf—temperature reference function for the range 0 °C to 1500 °C. Metrologia **35**, 761–780 (1998)
16. R. Carvell, E. Usadi, N. Fox et al., Innovative black body for on-board calibration. Proc. SPIE **5234**, 411–422 (2004)
17. P. Castro, G. Machin, M.A. Villamañan et al., Calculation of the temperature drop for high-temperature fixed points for different furnace conditions. Int. J. Thermophys. **32**, 1773–1785 (2011)
18. P. Castro, G. Machin, P. Bloembergen et al., Thermodynamic temperatures of high-temperature fixed points: uncertainties due to temperature drop and emissivity. Int. J. Thermophys. **35**, 1341–1352 (2014)
19. D. Cárdenas-García, E. Méndez-Lango, Blackbody for metrological control of ear thermometers. AIP Conf. Proc. **1552**, 976–980 (2013). https://doi.org/10.1063/1.4821415
20. CCT, Mise en pratique for the definition of the kelvin in the SI. SI Brochure, 9th edn. Appendix 2 (2019), https://www.bipm.org/utils/en/pdf/si-mep/SI-App2-kelvin.pdf. Accessed 1 Feb 2020
21. K. Chahine, M. Ballico, J. Reizes et al., Temperature profile measurement of a graphite tube furnace using optical fibre and platinum thermocouples, in *Proceedings of 6th Biennial Conference of Metrology Society of Australia*, October 19–21, 2005 (Australian National University, Canberra, 2005), pp. 19–21, https://opus.lib.uts.edu.au/bitstream/10453/7262/1/2005002546.pdf. Accessed 25 Mar 2019
22. K. Chahine, M. Ballico, J. Reizes et al., Optimization of a graphite tube blackbody heater for a Thermogage furnace. Int. J. Thermophys. **29**, 386–394 (2008)
23. T. Chaowasakoo, T.H. Ng, J. Songninluck et al., Indium solder as a thermal interface material using fluxless bonding technology, in *25th Annual IEEE Semiconductor Thermal Measurement and Management Symposium* (San Jose, CA, 2009). https://doi.org/10.1109/stherm.2009.4810761
24. C. Chen, Evaluation of resistance-temperature calibration equations for NTC thermistors. Measurement **42**, 1103–1111 (2009)
25. B. Chu, G. Machin, The upgraded NPL blackbody calibration facility, in *The 7th International Symposium on Temperature and Thermal Measurements in Industry and Science (Tempmeko'99)*, ed. by J.F. Dubbeldam, M.J. de Groot, vol. 2 (NMi Van Swinden Laboratorium, Delft, Netherlands, 1999), pp. 543–548
26. B. Chu, H.C. McEvoy, J.W. Andrews, The NPL reference sources of blackbody radiation. Meas. Sci. Technol. **5**, 12–19 (1994)

27. J. Connolly, Resistance thermometer measurement, in *Handbook of Temperature Measurement, vol. 2. Resistance and Liquid-in-Glass Thermometry*, ed. by R.E. Bently (Springer, Singapore, 1998), pp. 55–81
28. C.J. Donlon, T. Nightingale, L. Fiedler et al., The calibration and intercalibration of sea-going infrared radiometer systems using a low cost blackbody cavity. J. Atm. Oceanic Technol. **16**, 1183–1197 (1999)
29. C.J. Donlon, W. Wimmer, I. Robinson et al., A second-generation blackbody system for the calibration and verification of seagoing infrared radiometers. J. Atm. Oceanic Technol. **31**, 1104–1127 (2014)
30. C.J. Donlon, P.J. Minnett, N. Fox et al., Strategies for the laboratory and field deployment of ship-borne fiducial reference thermal infrared radiometers in support of satellite-derived sea surface temperature climate data records, in *Optical Radiometry for Ocean Climate Measurements*, ed. by G. Zibordi, C.J. Donlon, A.C. Parr (Acad. Press, Amsterdam, Netherlands, 2014), pp. 557–603
31. F. Edler, Y.G. Kim, G. Machin et al., *Guide on Secondary Thermometry. Specialized Fixed Points above 0 °C.* (BIPM, 2018), https://www.bipm.org/utils/common/pdf/ITS-90/Specialized-FPs-above-0C.pdf. Accessed 19 Feb 2020
32. J. Felba, Thermally conductive adhesives in electronics, in *Advanced Adhesives in Electronics. Materials, Properties and Applications*, ed. by M.O. Alam, C. Bailey (Woodhead Publishing, Oxford, UK, 2011), pp. 15–52
33. S.L. Firebaugh, K.F. Jensen, M.A. Schmidt, Investigation of high-temperature degradation of platinum thin films with an *in situ* resistance measurement apparatus. J. Microelectromech. Syst. **7**, 128–135 (1998)
34. J. Fischer, Developments in infrared radiation thermometry, in *TEMPMEKO '99. The 7th International Symposium on Temperature and Thermal Measurements in Industry and Science*, ed. by J.F. Dubbeldam, M.J. de Groot, vol. 1 (NMi Van Swinden Laboratorium, Delft, Netherlands, 1999), pp. 27–34
35. J. Fischer, J. Seidel, B. Wende, The double-heatpipe black body: a radiance and irradiance standard for accurate infrared calibrations in remote sensing. Metrologia **35**, 441–445 (1998)
36. J. Fischer, M. de Podesta, Hill et al., Present estimates of the differences between thermodynamic temperatures and the ITS-90. Int. J. Thermophys. **32**, 12–25 (2011)
37. J.B. Forgione, D.H. Kemp, P.S. Grant et al., Modernization of blackbody temperature control for the Enhanced MODIS Airborne Simulator (eMAS), in *IEEE Aerospace Conference* (Big Sky, MT, 2018). https://doi.org/10.1109/AERO.2018.8396368
38. J.B. Fowler, A Third generation water bath based blackbody source. J. Res. Natl. Inst. Stand. Technol. **100**, 591–599 (1995)
39. J.B. Fowler, An oil-bath-based 293 K to 473 K blackbody source. J. Res. Natl. Inst. Stand. Technol. **101**, 629–637 (1996)
40. J. Geist, J.B. Fowler, *A water bath blackbody for the 5 to 60 °C temperature range: performance goal, design concept, and test results.* NBS Technical Note 1228 (Natl. Bureau of Standards, U.S. Department of Commerce, 1986)
41. J. Gröbner, Operation and investigation of a tilted bottom cavity for pyrgeometer characterizations. Appl. Opt. **47**, 4441–4447 (2008)
42. B. Gutschwager, E. Theocharous, C. Monte et al., Comparison of the radiation temperature scales of the PTB and the NPL in the temperature range from −57 °C to 50 °C. Meas. Sci. Technol. **24**, 065002 (2013)
43. L. Hanssen, V. Khromchenko, S. Mekhontsev, Validation of NIST's low temperature infrared spectral radiance scale, in *Presentation on CALCON Conference*. Logan, UT (2017), https://digitalcommons.usu.edu/cgi/viewcontent.cgi?article=1281&context=calcon. Accessed 19 Feb 2019
44. J. Hartmann, J. Fischer, Radiator standards for accurate IR calibrations in remote sensing based on heatpipe blackbodies. Proc. SPIE **3821**, 395–403 (1999)
45. J. Hartmann, D. Taubert, J. Fischer, Characterization of the double-heatpipe blackbody LABB for use at temperatures below 500 °C, in *The 7th International Symposium on Temperature and Thermal Measurements in Industry and Science (Tempmeko'99)*, ed. by J.F. Dubbeldam, M.J. de Groot, vol. 2 (NMi Van Swinden Laboratorium, Delft, Netherlands, 1999), pp. 511–516

46. J. Hartmann, J. Hollandt, B. Khlevnoy et al., Blackbody and other calibration sources, in *Radiometric Temperature Measurements. I. Fundamentals*, ed. by Z.M. Zhang, B.K. Tsai, G. Machin (Academic Press, Amsterdam, Netherlands, 2010), pp. 241–294
47. J. Hartmann, F. Bernhard, Strahlungstemperaturmessung, in *Handbuch der Technischen Temperaturmessung*, 2nd edn., ed. by F. Bernhard (Springer, Berlin, 2014), pp. 1161–1393
48. C.B. Herron, S.L. Steely, R.P. Young, Arnold Engineering Development Center low-background blackbody calibration. Technical Report AEDC-TR-93-17, AD-A273 671 (AEDC, US Air Force, 1993)
49. K.D. Hill, D.J. Woods, Characterizing the NRC blackbody sources for radiation thermometry from 150 °C to 962 °C. Int. J. Thermophys. **30**, 105–123 (2009)
50. J. Hollandt, J. Seidel, R. Klein et al., Primary sources for use in radiometry, in *Optical Radiometry*, ed. by A.C. Parr, R.U. Datla, J.L. Gardner (Academic Press, Amsterdam, 2005), pp. 213–290
51. J.P. Holman, *Heat Transfer*, 10th edn. (McGraw-Hill, Boston, MA, 2010)
52. C. Horrigan, Calibration enclosures, in *Handbook of Temperature Measurement, vol. 2. Resistance and Liquid-in-Glass Thermometry*, ed. by R.E. Bently (Springer, Singapore, 1998), pp. 145–160
53. IEC 60539-1, Edition 3.0. *Directly heated negative temperature coefficient thermistors—Part 1: Generic specification.* International Standard (International Electrotechnical Commission, Geneva, 2016)
54. IEC 60539-2, Edition 2.0. *Directly heated negative temperature coefficient thermistors—Part 2: Sectional specification—surface mount negative temperature coefficient thermistors.* International Standard (International Electrotechnical Commission, Geneva, 2019)
55. IEC 60584-1, Edition 3.0 2013-08. *International Standard. Thermocouples—Part 1: EMF specifications and tolerances* (International Electrotechnical Commission, 2013)
56. IEC 60751, Edition 2.0. *International Standard. Industrial platinum resistance thermometers and platinum temperature sensors.* (International Electrotechnical Commission, 2008)
57. F.P. Incropera, D.P. DeWitt, T.L. Bergman et al., *Fundamentals of Heat and Mass Transfer*, 6th edn. (Wiley, Hoboken, NJ, 2007)
58. V.S. Ivanov, B.E. Lisiansky, S.P. Morozova et al., Medium-background radiometric facility for calibration of sources or sensors. Metrologia **37**, 599–602 (2000)
59. K.B. Jarratt, Calibration of an AEDC low-temperature blackbody standard at NIST. Technical Report AEDC-TR-91-9, AD-A245 071 (AEDC, US Air Force, 1992)
60. JCGM 200:2012,*International Vocabulary of Metrology—Basic and General Concepts and Associated Terms (VIM)*, 3rd edn. 2008 Version with Minor Corrections (BIPM Joint Committee for Guides in Metrology, Paris, 2012)
61. P. Jimeno-Largo, Y. Yamada, P. Bloembergen et al., Numerical analysis of the temperature drop across the cavity bottom of high-temperature fixed points for radiation thermometry, in *TEMPMEKO 2004, Proceedings of 9th International Symposium on Temperature and Thermal Measurements in Industry and Science*, ed. by D. Zvizdić, vol. 1 (LPM/FSB, Zagreb, Croatia, 2005), pp. 335–340
62. T.M. Jung, A.C. Carter, D.R. Sears et al., LBIR fluid bath blackbody for cryogenic vacuum calibrations. Presentation in 2015 CALCON Technical Meeting, Utah State University, Space Dynamics Lab., Logan, UT (2015). https://digitalcommons.usu.edu/cgi/viewcontent.cgi?referer=https://www.google.com/&httpsredir=1&article=1186&context=calcon. Accessed 19 Feb. 2020
63. T.W. Kerlin, M. Johnson, *Practical Thermocouple Thermometry*, 2nd edn. (ISA—International Society of Automation, Research Triangle Park, NC, 2012)
64. H.-Y. Ko, B.-J. Wen, S.-F. Tsa et al., A high-emissivity blackbody with large aperture for radiometric calibration at low-temperature. Int. J. Thermophys. **30**, 98–104 (2009)
65. K.G. Kreider, F. DiMeo, Platinum/palladium thin-film thermocouples for temperature measurements on silicon wafers. Sens. Actuators A **69**, 46–52 (1998)
66. M. Kutz (ed.), *Heat-Transfer Calculation* (McGraw-Hill, New York, 2006)
67. B.G. Lipták (ed.), *Temperature Measurement* (CRC Press, New York, 1993)

68. M. Litorja, B.K. Tsai, Preliminary results in the characterization of a NIST variable temperature blackbody, in *Proceedings of 9th International Symposium on Temperature and Thermal Measurements in Industry and Science (TEMPMEKO 2004)*, ed. by D. Zvizdić, vol. 2 (LPM/FSB, Zagreb, Croatia, 2005), pp. 963–968
69. H.R. Little, J. Hiatt, K.A. Lienemann, Low Temperature Infrared Source Calibration and Traceability at Arnold Engineering Development Center (AEDC). Proc. SPIE **0416**, 136–144 (1983)
70. G. Machin, The kelvin redefined. Meas. Sci. Technol. **29**, 022001 (2018)
71. G. Machin, R. Simpson, M. Broussely, Calibration and validation of thermal imagers. Quant. InfraRed Thermogr. J. **6**, 133–147 (2009). https://doi.org/10.3166/qirt.6.133-147
72. C.V. Madhusudana, *Thermal Contact Conductance*, 2nd edn. (Springer, Cham, Switzerland, 2014)
73. L.C. Martin, J.D. Wrbanek, G.C. Fralick, *Thin Film Sensors for Surface Measurements*. NASA TM-2001-211149 (NASA Glenn Research Center, 2001)
74. M. Matus, Temperature measurement in dimensional metrology – Why the Steinhart-Hart equation works so well, in *Proceedings of MacroScale 2011 Conference "Recent developments in traceable dimensional measurements"* (Bern-Wabern, Switzerland, 04–06, October, 2011), https://oar.ptb.de/files/download/56d6a9edab9f3f76468b4643. Accessed 19 Feb 2020
75. L. Michalski, K. Eckersdorf, J. Kucharski, J. McGhee, *Temperature Measurement*, 2nd edn. (Wiley, Chichester, UK, 2001)
76. W.J. Minkowycz, E.M. Sparrow, J.Y. Murthy (eds.), *Handbook of Numerical Heat Transfer*, 2nd edn. (Wiley, Hoboken, NJ, 2006)
77. C. Monte, B. Gutschwager, S.P. Morozova et al., Radiation thermometry and emissivity measurements under vacuum at the PTB. Int. J. Thermophys. **30**, 203–219 (2009)
78. S.P. Morozova, N.A. Parfentiev, Lisiansky et al., Vacuum variable medium temperature blackbody. Int. J. Thermophys. **31**, 1809–1820 (2010)
79. S.P. Morozova, B.E. Lisyanskiy, A.A. Stakharny et al., Low-temperature blackbodies for temperature range from −60 °C to 90 °C. Int. J. Thermophys. **32**, 2544–2559 (2011)
80. J.V. Nicholas, D.R. White, *Traceable Temperatures: An Introduction to Temperature Measurement and Calibration*, 2nd edn. (Wiley, Chichester, UK, 2001)
81. M. Noorma, S. Mekhontsev, V. Khromchenko et al., Water heat pipe blackbody as a reference spectral radiance source between 50 °C and 250 °C. Proc. SPIE **6205**, 620502 (2006)
82. S.A. Ogarev, M.L. Samoylov, N.A. Parfentyev et al., Low-temperature blackbodies for IR calibrations in a medium-background environment. Int. J. Thermophys. **30**, 77–97 (2009)
83. S.A. Ogarev, S.P. Morozova, A.A. Katysheva et al., Blackbody radiation sources for the IR spectral range. AIP Conf. Proc. **1552**, 654–659 (2013)
84. C.-W. Park, Y.S. Yoo, B.-H. Kim et al., Construction and characterization of a large aperture blackbody for infrared radiometer calibration. Int. J. Thermophys. **32**, 1622–1631 (2011)
85. J.V. Pearce, V. Montag, D. Lowe et al., Melting temperature of high-temperature fixed points for thermocouple calibrations. Int. J. Thermophys. **32**, 463–470 (2011)
86. D.D. Pollock, *Thermocouples: Theory and Properties* (CRC Press, Boca Raton, FL, 1991)
87. R. Prasher, Thermal interface materials: historical perspective, status, and future directions. Proc. IEEE **94**, 1571–1586 (2006)
88. R. Prasher, C.-P. Chiu, Thermal interface materials, in *Materials for Advanced Packaging*, ed. by D. Lu, C.P. Wong (Springer, Cham, Switzerland, 2017)
89. H. Preston-Thomas, The international temperature scale of 1990 (ITS-90). Metrologia **27**, 3–10 (1990)
90. C.A. Pullins, T.E. Diller, Adaptation of the in-cavity calibration method for high temperature heat flux sensors. Int. J. Heat and Mass Transfer **54**, 3369–3380 (2011)
91. T.J. Quinn, *Temperature*, 2nd edn. (Academic Press, London, 1990)
92. B. Ramachandran, C.O. Justice, M.J. Abrams (eds.), Land Remote Sensing and Global Environmental Change. NASA's Earth Observing System and the Science of ASTER and MODIS (Springer, New York, 2011)

93. D.C. Ripple, G.W. Burns, *Standard Reference Material 1749: Au/Pt Thermocouple Thermometer*. NIST Special Publication 260–134 (NIST, U. S. Department of Commerce, 2002)
94. R. Rusby,*The Beginner's Guide to Temperature Measurement.* Measurement Good Practice Guide No. 127 (NPL, Teddington, UK, 2012)
95. J.S. Steinhart, S.R. Hart, Calibration curves for thermistors. Deep Sea Research and Oceanographic Abstracts **15**, 497–503 (1968)
96. M. Stock, J. Fischer, Friedrich et al., The double-heatpipe black body: a high-accuracy standard source of spectral irradiance for measurements of T-T_{90}. Metrologia **32**, 441–444 (1995/96)
97. M. Stock J. Fischer, R. Friedrich et al., Measurement of T-T_{90} in the range from 660 °C to 962 °C by absolute spectral radiometry employing a cryogenic radiometer and a double heatpipe black body, in *Proceedings of TEMPMEKO '96, 6th International Symposium on Temperature and Thermal Measurements in Industry and Science*, ed. by P. Marcarino (Levrotto & Bella, Torino, 1996), pp. 19–24
98. Surrey NanoSystems: Vantablack (Surrey NanoSystems Ltd, Newhaven, UK, 2016), https://www.surreynanosystems.com/assets/media/vantablack-vb-a4-data-brochure-2016-009-download.pdf. Accessed 19 Feb 2020
99. I.M. Tougas, M. Amani, O.J. Gregory, Metallic and ceramic thin film thermocouples for gas turbine engines. Sensors **13**, 15324–15347 (2013)
100. M.A. Vadivelu, C.R. Kumar, G.M. Joshi, Polymer composites for thermal management: a review. Compos. Interfaces (2016). https://doi.org/10.1080/09276440.2016.1176853
101. M.S. Van Dusen, Platinum-resistance thermometry at low temperatures. J. Am. Chem. Soc. **47**, 326–332 (1925)
102. J. Wang, Z. Yuan, X. Hao et al., A −30 °C to 80 °C stirred-liquid-bath-based blackbody source. Int. J. Thermophys. **36**, 1766–1774 (2015)
103. T. Wescott, PID Without a PhD (2016), http://www.wescottdesign.com/articles/pid/pidWithoutAPhd.pdf. Accessed 19 Feb 2020
104. D.R. White, M. Ballico, D. del Campo et al., Uncertainties in the realization of the SPRT sub-ranges of the ITS-90. Int. J. Thermophys. **28**, 1868–1881 (2007)
105. D.R. White, K. Hill, D. del Campo et al., *Guide on Secondary Thermometry. Thermistor Thermometry* (BIPM, 2014). http://www.bipm.org/utils/common/pdf/ITS-90/Guide-SecTh-Thermistor-Thermometry.pdf. Accessed 2 Feb 2020
106. X. Yang, C. Liang, T. Ma et al., A review on thermally conductive polymeric composites: classification, measurement, model and equations, mechanism and fabrication methods. Adv. Compos. Hybrid Mater. **1**, 207–230 (2018)
107. C. Yinghang, L. Yaping, L. Yongqian et al., A medium temperature radiation calibration facility using a new design of heatpipe blackbody as a standard source. Meas. Sci. Technol. **12**, 491–494 (2001)
108. G. Zeng, S. McDonald, K. Nogita, Development of high-temperature solders: review. Microelectron. Reliab. **52**, 1306–1322 (2012)
109. X. Zhang, C. Hongseok, A. Datta et al., Design, fabrication and characterization of metal embedded thin film thermocouples with various film thicknesses and junction sizes. J. Micromech. Microeng. **16**, 900–905 (2006)

Chapter 8
Radiation Thermometry of Blackbodies

Abstract Two aspects of radiation thermometry of blackbodies (with traceability to the ITS-90 and within framework of relative primary radiometric thermometry, with traceability to the kelvin, the base unit of the SI) are considered. The design features and common sources of systematic errors of narrowband radiation thermometers suitable for measuring blackbody temperatures above the freezing point of silver are discussed. Techniques of temperature extrapolation, interpolation, and least squares fitting based on the Planck and Sakuma-Hattori equations, as well as the methods for evaluating the measurement uncertainties are described. Finally, application of the radiation thermometry to the measurement of temperature nonuniformities over blackbody radiating surfaces is outlined.

Keywords Radiation thermometry · Relative primary radiometric thermometry · Sakuma-Hattori equation · Blackbody temperature · Uncertainty · Temperature nonuniformity

8.1 Introduction

Radiation thermometry (an older name is pyrometry) is the branch of measurement science and technique, which deals with the measurement of the temperature of remote objects using its thermal radiation. The measuring instrument implementing this technique is called radiation thermometer (RT). Radiation thermometry is a well-developed and extensive branch of measurement technique. We have neither the opportunity nor the need to consider the numerous aspects of radiation thermometry that are not directly related to our subject. The interested reader can consult with Coates and Lowe [29], who expound systematically main issues of radiation thermometry and design principles of modern RTs. A lot useful information is contained in a collective monograph edited by Zhang et al. [205, 206]; another collective monograph edited by DeWitt and Nutter [35] is a valuable source of theoretical and reference information, but it is outdated in some respects.

The most serious problem in radiation thermometry is unknown emissivity of a measured object. The thermal radiation of an opaque body comes from its surface. Therefore, the radiance or spectral radiance of radiation emitted by a real body

V. Sapritsky and A. Prokhorov, *Blackbody Radiometry*, Springer Series in Measurement Science and Technology, https://doi.org/10.1007/978-3-030-57789-6_8

are the products of corresponding quantities calculated for a perfect blackbody and the surface emissivity or spectral emissivity, respectively, which depend on many factors that unknown or difficult to control. Many design solutions and measurement methods have been developed for this problem. It is eliminated when working only with blackbodies. In most cases, their effective emissivities can be considered equal to one, and the deviations from the perfect blackbody can be either corrected for and/or taken into account in the uncertainty budget.

The ITS-90 uses an RT as a primary thermometer above the freezing point of silver (1234.93 K). Below the silver point, the uncertainties of contact thermometry are generally lower than those of radiation thermometry.

Among the various types of RTs, we consider only those that are suitable for measuring the temperature of blackbodies above the silver point. This narrows the range of suitable instruments to the RTs with a single operating waveband. The term "single-waveband" is used to differentiate these RTs from so-called two-channel, two-color, and ratio RTs, as well as multi-wavelength, multi-channel, and multi-color RTs, which were developed mainly to solve the problem of unknown emissivity on the base of its mathematical models of various level of complexity.

For instance, the ratio (two-color) radiation thermometry (see, e.g. [174]) employs two distinct wavebands to solve (at least partially) the emissivity problem. The ratio RT has never been considered as an instrument for precise measurement of blackbody temperatures. In radiation thermometry of blackbodies, the ratio technique plays rather an auxiliary role, for example, when it is necessary to eliminate the influence of an intervening medium or a window. Some modern RTs can operate in both single- and two-waveband mode, i.e. be used optionally as the ratio RTs.

Until the last decades of the 20th century, precision measurement of temperatures higher than the freezing point of silver was not much in demand. However, the cutting-edge research in material science, technological progress, and advances in space exploration led to the growing interest in high-temperature sources of thermal radiation and measuring their temperatures. At that time, commercially available blackbodies with graphite cavities capable of operating up to 2500 K appeared in the market. A little later, researchers at VNIIOFI (Moscow) presented blackbodies with radiating cavities made of pyrolytic graphite and operating up to 3500 K [142]. Their temperatures were measured using ITS-90, since the methods of thermodynamic temperature measurements were not yet sufficiently developed.

Discovery of the high-temperature fixed points (HTFPs) at the end of the 20th century [189, 190] and their further investigation led to crucial improvement of high-temperature measurements. The measurement techniques for thermodynamic temperatures based on the absolute primary radiometric thermometry (APRT, see [98]) allowed measuring the thermodynamic temperatures of the HTFPs bypassing the ITS-90, with traceability to the SI. The assignment of thermodynamic temperatures and associated uncertainties to the HTFPs [92, 188] created a new basis for high-temperature metrology. Although the methods for the determination of thermodynamic temperatures by means absolute radiometry developed rapidly at the beginning of the 21st century, it remained available only to a limited number of world-leading NMIs due to their complexity. Instead, the relative primary radiometric thermometry RPRT, see [99], which inherited the instruments and methods of high-temperature radiation thermometry, can be used if the HTFP blackbodies are available. Due to recent advances in HTFP, the importance of measuring the high temperatures traceable to the ITS-90 is steadily declining.

The major area of the radiation thermometry (both ITS-90-and SI-traceable) in the blackbody radiometry is the measurement of the temperature of VTBBs above the silver point.

In some cases, it is required to measure the temperature of VTBBs operating below the silver point using an RT. According to the ITS-90, the radiation thermometry is not the primary technique for temperature measurements below the silver point. Temperatures of VTBBs operating in this temperature range are commonly measured using contact thermometers. At the same time, there a few cases in the blackbody radiometry, where temperature measurements using an RT are preferable in comparison with contact measurements. For instance, if the temperature of a radiating surface of the VTBB differs significantly from the temperature measured with a contact sensor, measurements carried out with an RT can provide lower temperature uncertainty. Shimizu and Ishii [165] described the VTBB operating at temperature from 100 to 500 °C in a furnace heated by air circulated with fans to maintain the isothermal condition in the cavity consisting of a thin-walled graphite cylinder and a flat bottom also made of graphite, but with the vertically aligned carbon nanotube coating. Although the temperature of a cavity bottom is measured by two platinum resistance thermometers, their readings serve for temperature monitoring and control only. They are placed behind the rear surface of the cavity and, since the thermal resistance of the carbon nanotube coating is unknown, the temperature of the radiating surface of the cavity bottom can be measured only by an RT. In fact, the RT measures the radiance temperature; however, since the effective emissivity of the cavity bottom is very close to unity, the radiance temperature will be very close to the actual temperature.

The RT is often more suitable for measuring temperature distributions over a radiating surface of VTBBs, whose temperature fields can be distorted due to heat loss through lead wires of the contact sensor. In particular, the thin-walled cavities are prone to local disturbance in temperature uniformity near the positions of contact sensors. As a rule, too strict requirements are not imposed on these measurements; often it is enough to measure only temperature difference with respect to some reference temperature (e.g. at the center of the target area).

Finally, if the VTBB is calibrated in terms of radiance temperature, which is typical for the remote sensing, the use of RT often allows achieving lower uncertainty than the use of the contact temperature sensor coupled with the independent determination of the effective emissivity.

8.2 Design Consideration and Defining Parameters of Radiation Thermometers

8.2.1 Generalized Scheme and Measurement Equation

An RT is a radiometer calibrated to indicate the temperature of a perfect blackbody. The ultimate goal of radiation thermometry is to relate the signal S and temperature T of the target without measuring the spectral responsivity of the RT. A generalized, albeit simplified, diagram of an RT is shown in Fig. 8.1. Below we briefly discuss the most important components of the RTs commonly used to measure the temperature of blackbodies.

Radiation Detector The detector of optical radiation is a key component of an RT. Modern technologies provide a wide selection of photoelectric and thermal detectors of optical radiation. We recommend readers who are not familiar with the principles of operation and applications of various types of the optical radiation detectors to refer to classical books by Budde [19], Dereniak and Crowe [34], or Kingston [84]. A near state-of-the-art review of most aspects of photoelectric and thermal detectors can be found in Norton [116] and Wolfe and Kruse [184].

Distinctive features of photoelectric (photon) detectors are significant spectral selectivity and long-wavelength cut-off, high sensitivity, and a short response time, a parameter that determines the detector's ability to track rapid changes in the target temperature. Typical response time for photoelectric detectors ranges from a few to several tens of milliseconds, which is much lower than for the typical thermal detectors.

Thermal detectors have virtually flat but much lower sensitivity throughout a large spectral range. As a rule, photoelectric detectors (mainly, photodiodes and photoconductors) are used in narrowband RTs for high- and medium-temperature measurements, while thermal detectors (pyroelectric, thermoelectric, and bolometric) are most often employed in wideband RTs designed for measuring lower temperatures.

Only a relatively small number of radiation detectors are permanently used in the constructions of RTs.

The silicon (Si) photodiodes operating in the photovoltaic mode (without a bias) are the most commonly used photoelectric detectors for the narrow-band RTs of visible and NIR spectral ranges operating usually at wavelengths around 0.65 μm and 0.9 μm. The typical spectral responsivity curve of a Si photodiode is shown in Fig. 8.2a. The choice of an operational wavelength is dictated, among other condi-

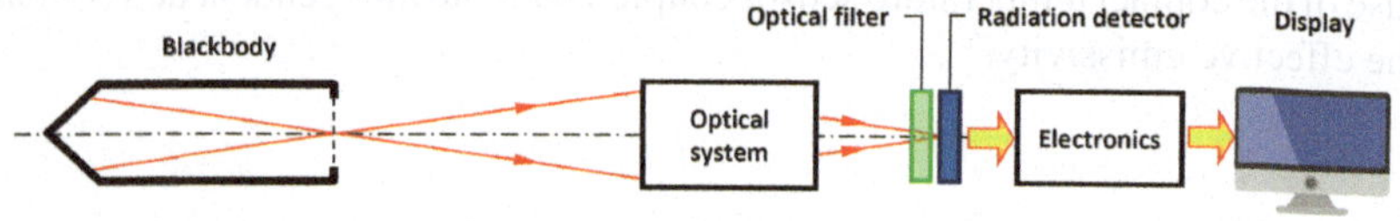

Fig. 8.1 A simplified scheme of a single-waveband RT

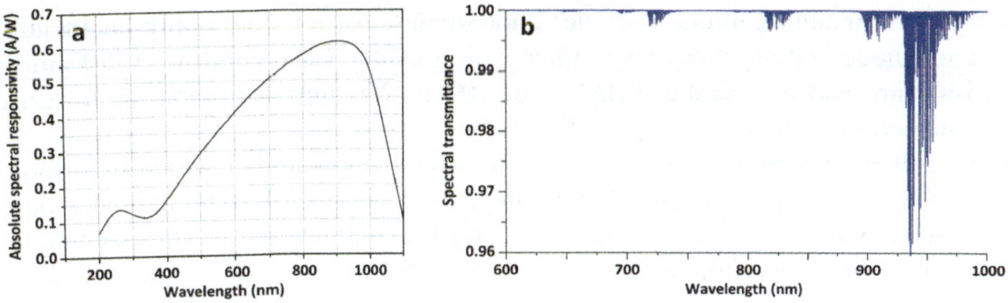

Fig. 8.2 Typical absolute spectral responsivity of **a** Si photodiode; **b** spectral transmittance of a 1 m horizontal path length in the standard atmosphere at sea level for the wavelength range from 0.6 μm to 1.0 μm

tions, by the spectral transmittance of the media, in which measurement is carried out. Since most RTs operate in laboratory conditions (in atmospheric air), it is advisable to work in the so-called atmospheric transparency windows, that is, in spectral ranges, in which absorption of radiation in air is absent or minimal. The spectral responsivity ranges of Si photodiodes, fortunately, coincide with the atmospheric transparency window shown in Fig. 8.2b. Spectral transmittances is given for a horizontal path 1 m long through a standard atmosphere at sea level and at 300 K.

Calculations with the spectral resolution of 0.25 nm were performed using the HITRAN2016 molecular database [52] and SpectralCalc online calculator (https://www.spectralcalc.com).

Although an RT is used as a relative measuring instrument (spectral comparator), the temperatures (and spectra) of the compared blackbodies can differ greatly, which leads to different absorption (mainly, in water vapor) even along identical optical paths. Choosing a suitable transparency window virtually eliminates the source of the corresponding systematic error.

A silicon photodiode with an interference filter having an FWHM of 10–20 nm and a central wavelength of about 650 nm makes it possible to measure the temperature from the freezing point of copper (1357.77 K) to 3000 K and even higher. The linearity range of the detector determines the measuring temperature range of the RT. As it was shown by Jung [74], Si photodiodes operating in the photovoltaic (unbiased) mode exhibit higher linearity and stability, lower drift, dark current, and noise than in the biased mode. For accurate measurements, the nonlinearity has to be accounted for (see Sect. 8.2.3). In order to obtain optimum short-term stability and resolution, the temperature the Si photodiodes should be stabilized. In some cases, it is necessary to take into account and correct measurements for the spatial uniformity of the responsivity of the Si photodiode and its dependence on the polarization of the incident radiation.

To measure temperatures below the Ag point, the RTs must employ detectors sensitive to IR radiation with wavelengths larger than 1 μm (i.e. above the cut-off wavelength of Si photodiodes) and filters having wider bandwidths. Precision RTs suitable for this task, may operate with the following detectors:

- InGaAs (indium gallium arsenide) photodiodes. An RT combining an InGaAs photodiode and an interference filter with a center wavelength of 1550 nm or 1600 nm and a typical FWHM from 100 to 200 nm is suitable to measure temperature from 420 K to 1300 K.
- So-called extended InGaAs (ex-InGaAs or EIGA) photodiodes, which are responsive up to 2.5 μm due to slightly increased amount of indium in the non-stoichiometric compound $In_xGa_{1-x}As$ compared with the regular InGaAs photodiodes. An RT with the ex-InGaAs photodiode cooled thermoelectrically to −85 °C in order to increase its shunt resistances has been used at NIST [40, 201]. The RT allows operation with ordinary glass optics in the atmospheric window between 2.0 and 2.5 μm for the measurement of blackbody temperatures from 20 to 50 °C with a thermal resolution of less than 3 mK at 50 °C.
- InSb photodiodes cooled with the liquid nitrogen (LN_2) to ~77 K have a high sensitivity from about 3 to 5 μm. Usually, the CaF_2 optics are used for these wavelengths. The bands of 3.8 ± 0.2 and 4.7 ± 0.3 μm can be used in the RTs either individually or together, with a notch filter blocking the band from 4.05 to 4.5 μm, where the atmospheric absorption is significant. Depending on filters and design features of the RT, the InSb-based RT allows measuring the temperature from −20 to 160 °C [70]; −30 to 430 °C [69]; from 100 to 420 °C [163]; from 150 to 962 °C [106].
- An alternative for deeply cooled IR photodetectors (e.g. InSb photodiodes, photoconductive lead salts PbS and PbSe, HgCdTe photodiodes and photoconductors—see Budzier and Gerlach [20], Rogalski [133], Kinch [83], Daniels [33], or Rogalski [134] for additional information) is uncooled thermal detectors, among which pyroelectric detectors are most widely used in radiation thermometry. The principle of operation of a pyroelectric detector is based on the ability of certain crystals to generate an electric charge on their surface when their temperature changes. If two faces of a crystal are electrically connected, an electric current will flow in the external circuit; otherwise, a modulated voltage difference will appear between the crystal faces. After modulating the radiation incident on the pyroelectric crystal, the generated output signal (alternating electric current or voltage) can be easily measured. In most widely used pyroelectric detectors, the crystals of triglycine sulfate (TGS), deuterated triglycine sulfate (DTGS), lithium tantalate ($LiTaO_3$), or pyroelectric ceramics such as zirconate titanate (PTZ) are used. For more information see Aggarwal et al. [2] or Batra and Aggarwal [8]. Like other thermal detectors, pyroelectric detectors are as spectrally selective as their absorbing coatings are spectrally selective. In order not to increase the thermal inertia of pyroelectric detectors, their receiving surfaces are usually coated with "metal blacks", microscopic particles of metals, such as gold, platinum, nickel, etc., deposited using the method of thermal evaporation. Most often, pyroelectric RTs are used in the spectral band from 8 μm to about 14 μm to measure the lowest temperatures, e.g., from −50 °C [171] or from 30 to 160 °C [70]. Using the filters with shorter wavelengths, the RT with the pyroelectric detectors are capable to cover a wide wavelength range from 273 to 1234 K [121, 196].

Optical Filter transmits radiation within a given spectral range and blocks radiation outside the nominal passband. The operational waveband of an RT is usually formed by a band-pass filter. A wider passband for some RTs can be formed by a combination of long- and short-pass filters. The bandwidth of an optical filter (and of an RT) can be characterized by the center (mean) wavelength λ_0 and an FWHM (full width at half maximum) $\Delta\lambda$; for wideband filters, it may be more convenient to use the short- and long-wave boundaries of the passband, λ_{min} and λ_{max}. Two types of filters are commonly used: absorptive filters and multilayer interference filters.

The principle of operation of absorptive filters is based on attenuation of optical radiation by absorption of certain wavelengths. Spectral absorptance depends on the thickness of the filter and the concentration of additives in a transparent matrix (substrate), which can be made of glass or other material, depending on the required spectral range. By controlling the thickness of the substrate, precise filtering across a large bandwidth can be achieved. Absorptive filters are used to block short- or long-wavelength radiation and to design RTs with a wide spectral bandpass. Although absorbing filters can be combined with each other, it is almost impossible to achieve the same narrow passband with them as with interference filters.

The principle of operation of interference filters is based on the phenomenon of interference in multiple thin layers of materials having different refractive indices [100]. The bandpass interference filters used in narrowband RTs must have high peak transmittances, narrow bandwidths, and low out-of-band transmittances. Radiation with wavelengths outside the passband in spectral regions, where the detector is still sensitive, should be blocked to a level of less than 10^{-4} to 10^{-6} of those inside the passband. If the interference filter does not sufficiently reduce radiation with undesired wavelengths, an auxiliary blocking filter can be added to enhance the out-of-band blocking. In most cases, measurements in radiation thermometry of blackbodies are performed in the ascending branch of the blackbody spectral radiance curve (at wavelengths shorter than the peak of the Planckian curve); therefore, suppressing the out-of-band radiation at longer wavelengths is the overriding task.

The spectral transmittance of interference filters varies with the filter temperature; therefore, it should be carefully controlled; filter heating by incident radiation is unacceptable. When an RT is used to measure temperatures of high-temperature blackbodies (HTBBs), it may be necessary to use additional absorption filter or some other means of reducing the intensity of incident radiation.

Any filters placed in the optical path should be oriented in such a way as to avoid reflections between them and to avoid passing through the interference filter at an angle to the axis. The tilt angles should be carefully chosen to eliminate unwanted reflections.

To reduce effects of ambient temperature and humidity on the long-term stability of the spectral transmittance, special coatings are applied to interference filters.

The spectral transmittance of interference filters depends also on the angle of radiation incidence; therefore, measurement of its transmittance should be carried out in authentic geometry of the incidence radiation.

The interference filters are sensitive to polarization of radiation. Although the radiation of blackbodies is considered as unpolarized because of multiple chaotic reflections inside the cavities, it can be not true for low-emissivity blackbodies formed by specularly reflecting surfaces. If an incandescent lamp is used as a transfer measuring device, partial polarization of its radiation should be taken into account. Finally, the optical system itself can polarize optical radiation, which subsequently falls at oblique angles on the interference filter. For some RTs designed to measure high temperatures, this may be a source of a noticeable systematic error [37].

In the wideband RTs with thermal detectors, the use of optical filters can sometimes be avoided. In these cases, the spectral responsivity of an RT is determined mainly by the spectral absorptance of the absorbing coating of the detector.

Optical System collects the radiant flux within a narrow solid angle and in a well-characterized spectral region at a distance from a radiation source. Optical system, which may consist of lens and/or mirrors and the stops (diaphragms), directs the radiation from the blackbody under measurement through the optical filter to the radiation detector. Since the spectral transmittance and spectral refractive index of lens materials depend upon wavelength, the lens made of certain material can be employed in the optical system only within a limited spectral range. Mirror optics is less spectrally selective and, therefore, can be used in RTs with a wider responsivity band. The stops (diaphragms) are used to define the field of view (FOV) of the RT and to prevent stray radiation from falling on the detector.

The f-number[1] of the optical systems of a typical RT is between f/10 and f/20. In a dioptric system, this is enough to correct the chromatic and spherical aberrations almost to diffraction-limited quality using an achromatic doublet composed of two lenses made of glasses with different refractive indices and often cemented together. The ability to manufacture lenses from ordinary glass simplifies visual focusing through a viewfinder, which is especially important for the RTs of the NIR (SWIR) spectral range.

A serious problem in the design of RTs is the suppression of scattered radiation (stray light), the source of which is the lens (or lenses) of the RT. All optical components should be of high optical quality and kept clean to minimize the scattering of radiation on imperfections and surface contamination. To minimize reflections, the anti-reflection coatings are applied to the lens surfaces. To suppress unwanted radiation from outside the target area propagating through the system by diffraction, reflection, and scattering from the mechanical or optical elements, the baffles and grooves are commonly used, in combination with low-reflectance coatings.

The detector output signal is fed into an electronic system, where it is conditioned, amplified, digitized, processed according to the prescribed algorithm, and displayed. The electronic system may include preamplifier, analog-to-digital converter (ADC), a specialized microprocessor (digital signal processor), and other components depending on the type of radiation detector and its output signal. The function of the electronics is to convert the output of the detector into a signal that can be processed. This conversion should be performed with minimal degradation in system performance, ensuring low noise level, high gain, high dynamic range and linearity. We

---

[1]The f-number (f/#) of an optical system is the ratio of the focal length to the diameter of the entrance pupil.

will not discuss electronic components because this goes far beyond the scope of our book. As an introduction to the problem we can recommend a book written by Teare [173].

Displaying of all necessary information can be shown on the liquid-crystal display of the RT or the PC monitor.

The general measurement equation relating the output signal S of the RT and the blackbody temperature T is given by

$$S(T) = \int_A \int_\Omega \int_{\lambda_a}^{\lambda_b} \tau_s(\lambda)\tau_f(\lambda)\tau_m(\lambda)s_d(\lambda)\varepsilon_{eff}(\lambda)L_{\lambda,bb}(\lambda, T)dA\, d\Omega\, d\lambda, \quad (8.1)$$

where A is the target area, the radiation of which reaches the detector through a solid angle Ω defining the FOV of the RT; λ is a wavelength belonging to a continuous wavelength range $[\lambda_a, \lambda_b]$ that define the waveband of the RT; τ_s, τ_f, and τ_m are spectral transmittances of the optical system, filter, and media, respectively, for the rays coming from the target to the detector; s_d is the absolute spectral responsivity of the detector (usually, in V/W or A/W); ε_{eff} is the effective emissivity of the blackbody under measurement; $L_{\lambda,bb}(\lambda, T)$ is the spectral radiance of the perfect blackbody at the wavelength λ and temperature T computed using Planck's law. For given viewing conditions (fixed values of A and Ω), Eq. 8.1 can be slightly simplified:

$$S(T) = \int_{\lambda_a}^{\lambda_b} s(\lambda)\varepsilon_{eff}(\lambda)L_{\lambda,bb}(\lambda, T)\, d\lambda, \quad (8.2)$$

where s is the absolute spectral responsivity of the RT as a whole.

8.2.2 Defining Parameters of Radiation Thermometers

In order to select an RT for a particular application, users should have a common set of parameters and characteristics of RTs. The International Electrotechnical Commission has issued an international standard, IEC/TS 62492-1 [65], which regulates the nomenclature of technical data for RTs. Another standard, IEC/TS 62492-2 [66], specified methods for their determination. However, these standards were developed more for RTs manufacturers than for their users. In contrast, the standard ASTM E2758-15a [6] provides guidance on the selection and use of RTs, but it covers only wideband low temperature IR RTs. The standard ASTM E1256-17 [5] describes the test methods for the six most important operational parameters of a single-waveband RT, namely the calibration accuracy, repeatability, field-of-view, response time, warm-up time, and long-term stability. Below, we discuss the parameters and characteristics of RTs used to measure the temperature of blackbodies, which require special consideration in connection with such use.

The most important characteristics of an RT are its measuring temperature and spectral ranges.

Measuring temperature range is the range of temperature, for which the RT is designed and within which its measurement uncertainties are guaranteed by its developers. In general, the RTs designed to measure the temperature of blackbodies in laboratory conditions should be able to measure temperatures from 200 to 3500 K. When working in cryovacuum chambers, the lower limit of the measured temperatures can reach 100 K and even lower. Zhang and Machin [204] classified single-waveband RTs according to their ranges of measured temperature (high-temperature RTs for temperature above 600 °C, mid-temperature RTs for temperatures from 100 to 600 °C, and low-temperature RTs for temperatures from below 0 °C to around a few hundred degrees Celsius). Due to the wide temperature range measured by RT, the linearity of the conversion of temperature or radiance of blackbodies into the output signal is of paramount importance. This linearity (or nonlinearity) of an RT is mainly determined by the features of the detector and therefore is not considered in the above standards. Because of importance of this characteristic, we devoted a separate section (Sect. 8.2.3) to it.

Spectral range is the parameter which gives the lower and upper limits of the wavelength range over which the RT operates. The spectral range should be determined taking into account the optical system, optical filter, and detector of the RT. The lower the temperature of the measured blackbody the lower the radiance of its thermal radiation and the wider spectral range is needed for the RT to achieve an acceptable level of the signal-to-noise ratio (SNR).

Conventional radiation thermometry does not require knowledge of the absolute spectral responsivity of an RT; the relative spectral responsivity is necessary only for realization of the ITS-90 above the silver point. In all other cases, it is enough to know a center (mean) wavelength λ_0, i.e. a wavelength usually near the middle of the spectral range over which an RT responds, and a full width at half maximum (FWHM) $\Delta\lambda$, at which the spectral responsivity has reached 50% of the maximum value, can be given.

It is customary to divide the single-waveband RTs into narrowband and wideband (relative bandwidths $\Delta\lambda/\lambda_0 < 0.05$ and $\Delta\lambda/\lambda_0 > 0.2$, respectively). Of course, the indicated numerical values are very conditional. It is common for narrow-band RTs to give the center wavelength of the spectral range and the FWHM, while for wideband RTs it is more convenient to specify lower and upper wavelength limits. The terms "monochromatic RT" and "total RT" are often used, although this is nothing more than idealization for extremely narrowband and extremely wideband RTs, respectively. Real RTs always have finite bandwidths.

Measuring distance is a distance or distance range between the RT and the target (measured object) for which the RT is designed. When a blackbody acts as a measured object, the distance is usually counted from the cavity aperture unless otherwise specified.

Field-of-view (FOV) is usually a circular, flat surface of a measured object, from which the RT receives thermal radiation. The FOV is sometimes referred to as the target area, target size, spot size, or measurement field. The FOV is determined by the optical components of the RT. IEC/TS 62492-1 [65] requires the relation between the FOV and the measuring distance as an equation or a graph, and this relation is most often given as a chart in users guides to commercial RTs. According IEC/TS 62492-1 [65], the dependence of FOV on measuring distance to be indicated, and this relation is most often given as a chart in user fixed-focus RTs. A simplified optical scheme of such an RT is presented in Fig. 8.3a. The optical system includes a lens, an aperture stop, and a field stop. The distance L_0 between the aperture stop and the image of the field stop (the focusing length—should not be confused with the focal length F of the objective lens) can be found using the thin lens formula:

$$L_0 = \frac{HF}{F - H}, \tag{8.3}$$

where H is the distance between the aperture and the field stop. At the focusing length L_0, the diameter D_0 of the FOV is equal to

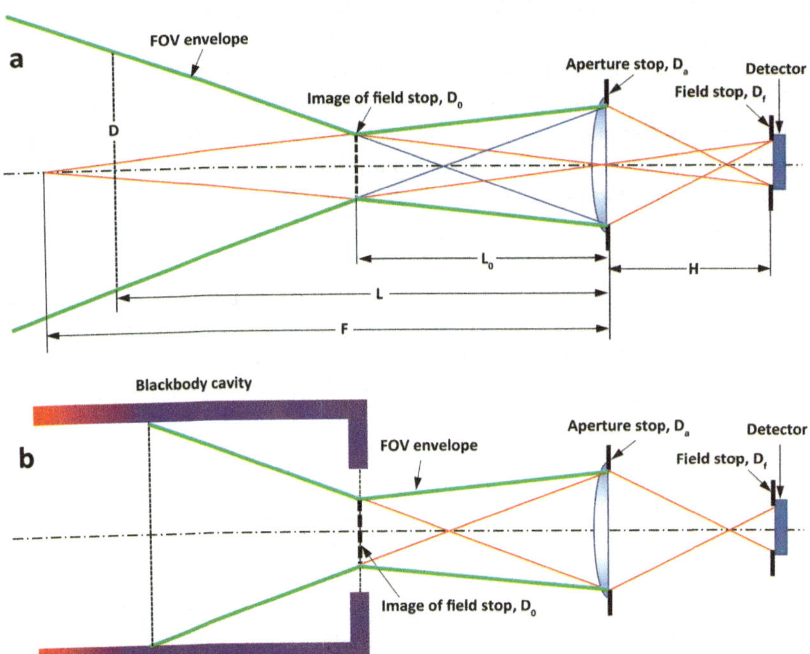

Fig. 8.3 The FOV of an RT (not to scale) for: **a** a flat uniform radiance target and **b** a blackbody cavity

$$D_0 = D_a\left(1 - \frac{H}{F-H}\right) \tag{8.4}$$

the circle of diameter D_0 is called the virtual target [117] or, more often, the target spot. At the distance L, the diameter of the FOV is expressed as

$$D = \begin{cases} \dfrac{(D_0 - D_a)L}{L_0} + D_a, & if\ L < L_0 \\ \dfrac{(D_0 + D_a)L}{L_0} - D_a, & if\ L > L_0 \end{cases}. \tag{8.5}$$

The envelope of the angular FOV is shown in thick polygonal line in Fig. 8.3a. Any uniformly radiating surface that completely fills the FOV (i.e. lying inside the FOV envelope) creates the same radiant flux incident on the detector, and therefore the same output signal RT, for any measurement distance L.

The situation is somewhat more complicated if the temperature of the blackbody cavity is measured (see Fig. 8.3b). In this case, the registered radiant flux, and therefore the output signal of the RT, depends on what part of the cavity falls into the FOV. Even if the cavity is isothermal, the effective emissivity and, therefore, the radiance of its walls both depend on the axial coordinate. Consequently, the RT output will depend on the location of the virtual target.

Usually, the RT is focused on the cavity aperture, when the virtual target is positioned at the center of the cavity aperture, so the focusing length L_0 is equal to the distance from the aperture stop of the RT to the cavity aperture. Of course, the diameter of the target spot should be less than the diameter of the cavity aperture.

If it is more convenient to focus the RT in another way, for example, on the bottom of the cavity, this should be specially reported; alternatively, the "visible" area of the cavity should be indicated.

The angular measure of the FOV is often used, when the FOV is expressed as a plane angle at the apex of the right circular cone, the height of which is the measuring distance and the base coincides with the flat circular area on the measured object. Therefore, it is necessary to clearly indicate what type of FOV is used, if this is not obvious from the context.

The FOV of a real RT has no sharp boundaries. The target spot is blurred due to lens aberrations, diffraction, scattering along the optical path, and interreflections inside the RT compared to the ideal target spot obtained by ray tracing. That means that the output signal of an RT depends to a greater or lesser extent on the size of the target and on the thermal radiation coming from outside of the nominal FOV. The quantitative measure of this dependence is the size-of-source effect (SSE). For precision measurement of the temperature of blackbodies, the SSE can bring one of the biggest contributions to the resulting uncertainty. We consider it in detail in Sect. 8.2.4.

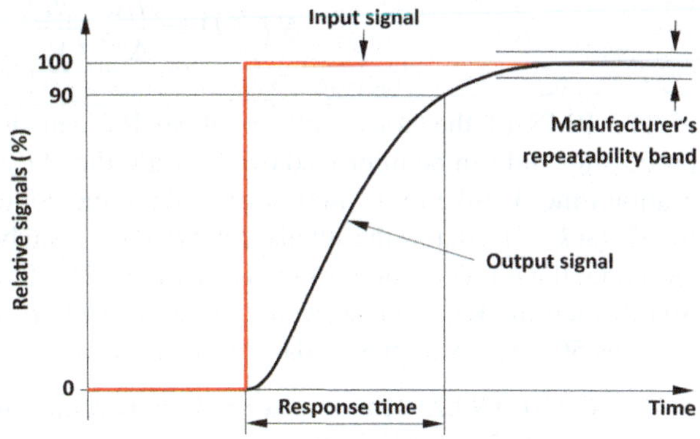

Fig. 8.4 Dependences of relative input and output signals of the RT illustrating the response time ($t_{R90\%}$ is shown)

Distance ratio (a.k.a. distance factor) is the ratio of the measuring distance to the diameter of the FOV when the target is in focus.

Warm-up time is a time period needed after switching on the RT so that the RT operates according to its specifications.

Response time is a time interval between the moment of an abrupt change in the value of the input parameter (object temperature or object radiation) and the moment, from which the measured value of the RT (output parameter) remains within specified limits of its final value.

The lower/upper temperature value for specifying the response time has to be a temperature value within, respectively, the lower/upper quartile of the measuring temperature range (see Fig. 8.4).

For an RT, the rise and fall times (response times for rising and falling temperature steps) may be different. If so, they should both be indicated. Examples of data on the response time are:

- $t_{R90\%} = 0.05$ s (25 °C, 100 °C), i.e. 0.05 s for 90% of the maximum value for the temperature step from 25 to 100 °C or
- $t_{R99\%} = 1$ s (20 °C, 1 000 °C), i.e. 1 s for 99% of the maximum value of the temperature step from 20 to 1 000 °C.

Noise equivalent temperature difference (NETD) is defined as the parameter that indicates the contribution of the measurement uncertainty due to instrument noise. The sources of noise are the detector and amplifier. The NETD is expressed in units of temperature (K or °C) and represents the temperature difference, which would produce a signal equal to the RT noise varying in time. Minkina and Dudzik [110] defines the NETD as the difference between the temperature of the observed object and the ambient temperature that generates a signal level equal to the noise level.

Alternatively, the NETD can be defined as the ratio of the root mean square (RMS) noise voltage V_n to the voltage increment ΔV_s generated by the difference in temperature between the measurement area of a blackbody T_{bb} and background temperature T_{bg}, divided by this difference:

$$NETD = \frac{T_{bb} - T_0}{\Delta V_s / V_n}. \tag{8.6}$$

The RMS of the noise voltage of an RT can depends on the quality signal processing and can be improved (to a certain limit) by increasing the response integration time. Unlike most metrological data, the confidence interval in this case is 68.3% (at $k = 1$), that is the standard deviation σ_n can be used instead of V_n. Roughly speaking, the NETD expresses the temperature resolution of the RT: if the NETD = 50 mK then the RT can resolve the target-to-background temperature difference as small as 50 mK. Examples of data on the NED are:

- 0.1 °C (20 °C/0.25 s) at a measured temperature of 20 °C and response time of $t_{R90\%} = 0.25$ s (for DC signal) or
- 0.1 °C (20 °C/100 Hz to 1 kHz) at a measured temperature of 20 °C and after the electric signal has passed through a bandpass filter from 100 Hz to 1 kHz (for AC signal).

The measuring temperature range, FOV, response time, and NETD are mutually dependent. The higher the temperature measured, the larger the field-of-view, and the longer the response time, the higher NETD can be achieved. On the contrary, the higher the measuring temperature and the larger the FOV, the faster detectors can be used to achieve the same value of the NETD.

Influence of the internal instrument or ambient temperature (temperature parameter) gives the additional uncertainty of the measured temperature depending on the deviation of the temperature of the RT from the value, for which the technical data is valid after warm-up time and under stable ambient conditions. An example of data for the temperature parameter: 0.2 °C/°C (25 °C, 600 °C), 0.02 °C/°C (25 °C, >700 °C), which means additional uncertainty of the measured temperature, where the internal temperature of the RT deviates from 25 °C for a target temperature of 600 °C and for target temperatures above 700 °C.

Influence of air humidity (humidity parameter) gives the additional uncertainty of the measured temperature value depending on the relative air humidity at a defined ambient temperature. Examples of data for the humidity parameter are:

- 0.2 °C/% (50%, 23 °C, 1 m, 600 °C), 0.1 °C/% (50%, 23 °C, 1 m, <500 °C), i.e. the additional uncertainty of the measured temperature where the relative humidity deviates from 50% at 23 °C for a measuring distance of 1 m for a target temperature of 600 °C and target temperatures below 500 °C;
- 0.02% of the measured value in °C/% (45%, 23 °C, 1 m), i.e. the additional uncertainty of the measured value where the relative humidity deviates from 45% at 23 °C for a measuring distance of 1 m for the complete measuring temperature range.

This list of influencing parameters given in IEC/TS 62492-1 [65] should be supplemented by at least one more—the polarization parameter. The spectral responsivity of any RT containing specularly reflecting surfaces and/or interference filter is polarization-sensitive. Its output signal in a varying degree depends on the state of polarization of radiation emitted by a target. Although the radiation of blackbodiest is usually assumed non-polarized due to multiple reflections inside a cavity, this is not true for radiation of some imperfect thermal sources, such as tungsten strip lamps, low-emissivity blackbodies with predominantly specular reflection of walls, and blackbodies with safety windows. In such cases, polarization dependence of the spectral responsivity should be corrected or accounted for in the uncertainty budget. Although the principles for taking into account the polarization parameter are understandable [104], this issue remains insufficiently studied.

8.2.3 Detector Nonlinearity

A characteristic feature of measuring the temperatures of blackbodies by their radiation is a wide dynamic range of the measured radiances. The spectral radiances of perfect blackbodies at temperatures of the ITS-90 fixed points (from the freezing point of silver to the freezing point of copper), HTFPs (Co–C, Pt–C, and Re–C), and at temperatures of 2500 K, 3000 K, and 3500 K are presented in Fig. 8.5.

As can be seen from this figure, the ratio of the spectral radiances with respect to the Ag fixed point at the wavelength of 0.65 μm typical for high-temperature radiation thermometry can reach five orders of magnitude.

According to the ITS-90 specification, the quasi-monochromatic RTs are the only measuring instruments that can be used for realization of temperature scale above the freezing point of silver. Since the temperature extrapolation is implemented using the ratio of spectral radiances, the linearity of the extrapolation instruments affects

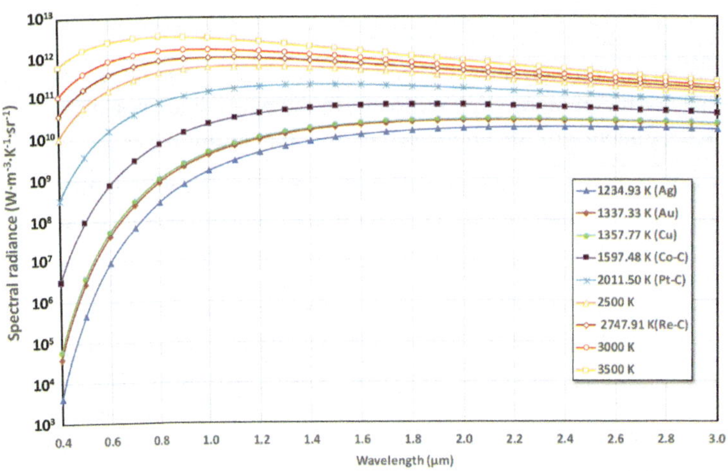

Fig. 8.5 Spectral radiances of the perfect blackbody at temperatures of the ITS-90 defining fixed points (from the freezing point of silver to the freezing point of copper), at temperatures of HTFPs (Co-C, Pt–C, and Re-C), and at temperatures of 2500 K, 3000 K, and 3500 K

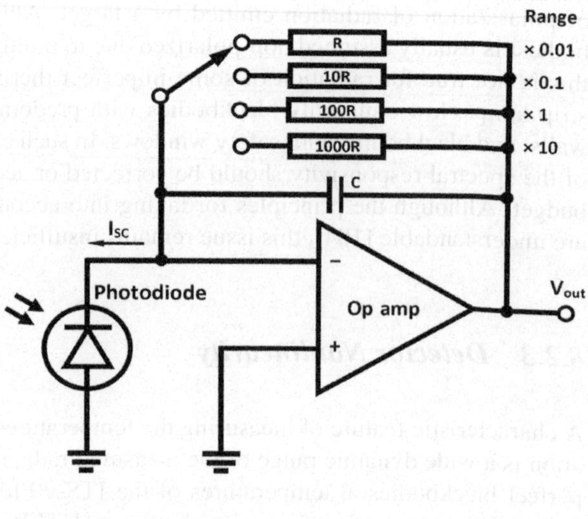

Fig. 8.6 A transimpedance amplifier for a photodiode operating in the photovoltaic mode. Four different feedback resistors allow four gain settings

crucially on the uncertainty in realization of the ITS-90 above the silver fixed point. The detectors of most common RTs used for high-temperature measurements are Si and InGaAs photodiodes operating in the photovoltaic mode (with zero bias voltage) for better linearity. The generated photocurrent is converted into a measured voltage using a transimpedance amplifier, which can be easily implemented [53, 78] with an operational amplifier (op amp) as shown in Fig. 8.6.

The photovoltaic mode is preferred when a photodiode is used in applications with low frequency (less than approximately 300 kHz) and low incident radiant flux. This mode is characterized by:

- the absence of dark current;
- the highest linearity of flux-to-current conversion;
- a low noise: the Johnson (thermal) noise generated by the shunt resistance (internal resistance of the photodiode) dominates in the photodiode noise;
- a weak dependence of the photocurrent on temperature.

The photodiode bias is maintained at zero volts by the virtual ground of the op amp, and the short-circuit current I_{SC} is converted into an output voltage V_{out} as

$$V_{out} = R \cdot I_{SC}, \tag{8.7}$$

where R is the resistance of the feedback resistor, which must be very large for the bias current of the amplifier to be very small. The higher the incident flux values, the lower the circuit gain should be set using a smaller feedback resistor.

As a rule, the detector makes the greatest contribution to the thermometer's nonlinearity, while the contribution of a properly designed amplifier (Eppeldauer [39, 42] can be minimized and accurately accounted for. To cover the measured temperature range for a particular RT, several different amplifier gains may be needed. If a multi-gain amplifier is used, the knowledge of the gain ratios is necessary to normalize all signals to a common gain value before interpolation or extrapolation using fixed points. Uncertainty may originate from the measurement of the gain ratios and possible temporal drift.

The actual gain ratios can be determined experimentally with a high-stability radiation source such as precision blackbodies, tungsten strip or halogen lamps, LEDs, or integrating sphere-based sources [77, 114, 138, 140, 200].

Nonlinearity in the conversion of the radiant flux into an electrical signal is not a problem specific only to radiation thermometry. Despite a large body of literature on this subject (see, e.g. [28, 30, 159, 169, 182, 193] and Sect. 3.3 in [19]), there is no unified approach to nonlinearity quantification. Various authors use different measures of nonlinearity depending on the method, which is applied to the nonlinearity measurement. In turn, the choice of such a method depends on the detector under measurement and the dynamic range of radiant fluxes measured by that detector.

In mathematics, a linear function $f(x)$ is a function that satisfies the superposition principle that includes properties of additivity and homogeneity:

$$f(\alpha x + \beta y) = \alpha f(x) + \beta f(y), \tag{8.8}$$

where $-\infty < \alpha, \beta < \infty$ are constants. The superposition principle is the base of experimental determination of nonlinearity. If the response of a measuring instrument is linear, its output signal S is proportional to the value of the measurand, and the responsivity s is constant.

On an intuitive level, we can define linearity of the response of a radiation detector as the extent to which the output signal of the detector is directly proportional to the incident radiant flux at a given wavelength, at constant geometry of irradiation, and a fixed polarization state.

The most appropriate for our purposes interpretation of nonlinearity was given by Saunders and White [153], who introduced the so-called nonlinearity function of an RT. It was assumed that nonlinearity in an RT's output signal is caused by nonlinearity of the detector's response, while the contribution from the amplifier or changes in its gain is negligible. The nonlinearity in the measured (raw) signal S' can be corrected by determining a continuous nonlinearity function $\eta(S')$ such that the linearized signal S is equal to

$$S = \eta(S')S'. \tag{8.9}$$

Two types of methods for determination the nonlinearity function η are best developed: the dual aperture and superposition methods. Both methods use the sampling of at η_j for $j = 1, \ldots, m$ discrete measured signal values, more or less uniformly distributed over the entire range of the RT signals. The values η_j are used to find (by interpolating or least squares fitting) the continuous function $\eta(S')$ in a prescribed form.

Fig. 8.7 Simplified arrangement for the dual aperture method

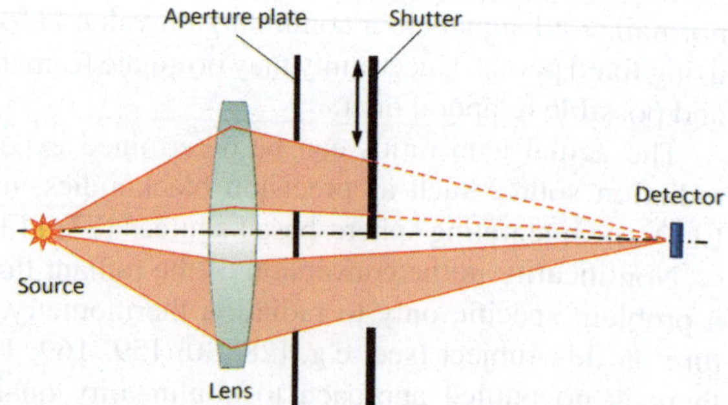

The dual aperture method implies obtaining two radiant fluxes (from the same source or from different sources), measuring signals S'_A and S'_B individually, and then comparing them to the signal S'_{A+B}, measured by optical combining of both fluxes [74, 108, 141]. A simple arrangement for measuring nonlinearity by the dual aperture method with the use of a single radiation source is shown in Fig. 8.7. The measured nonlinearity for such measurements is

$$\eta'_{A+B} = \left(S'_A + S'_B\right) / S'_{A+B}. \tag{8.10}$$

It should be noted that this is not true nonlinearity, since the signals S'_A and S'_B themselves are subject to nonlinearity. Let us assume, for simplicity, that two individual signals are arranged to be equal ($S'_A = S'_B \equiv S'_0$ and $S'_{A+B} \equiv S'_1$) so that the dual aperture method becomes the flux-doubling method.

The value of S'_0 is typically chosen to be close to the lowest operating signal, and is successively doubled up to the highest operating signal. The measured non-linearity at the first doubling is

$$\eta'_1 = \left(S'_0 + S'_0\right) / S'_1 = 2S_0 / \left(S'_1 \eta_0\right), \tag{8.11}$$

where S_0 is the linearized signal corresponding to S'_0 and η_0 is its true nonlinearity value, which is unknown.

The flux doubling proceeds by doubling the signal S'_1 to give S'_2, and so on. In general, after k doublings, the measured non-linearity is

$$\eta'_k = \frac{2^k S_0}{S'_k \eta_0 \prod\limits_{j=1}^{k-1} \eta'_j}. \tag{8.12}$$

Since $2^k S_0 = S_k$, we can obtain the expression for the true nonlinearity for the measured signal S'_k:

$$\eta_k = \eta_0 \prod_{j=1}^{k} \eta'_j. \tag{8.13}$$

Equation 8.13 still contains the unknown value of nonlinearity η_0, which can be found by the following way. The discrete data points η_j can be approximated by a predefined functional form, the adjustable parameters of which can be determined by the least square technique. Coslovi and Righini [30] suggested a simple equation for Si detectors:

$$\eta(S') = 1 - \left(a + bS'^c\right), \tag{8.14}$$

where a, b and c are adjustable parameters. If a flux-doubling method is used, then an equivalent form of Eq. 8.14 is

$$\eta(S') = 1 - a\left(1 - 2^{cx}\right), \tag{8.15}$$

where $x = \log_2\left(S'/S'_0\right)$ is the effective number of flux doublings from the initial signal S'_0. Equations 8.14 and 8.15 are directly related through $b = -a/S_0'^c$.

Since the dual aperture method (and the flux-doubling method as its special case) is based on the spatial subdivision of irradiating beam, it can provide only not so large diapason of nonlinearity measurements, about 3 orders of magnitude. This is not always enough for Si and InGaAs photodiodes, although is quite acceptable for IR detectors. A typical example of the embodiment of the dual aperture method is the developed at the NPL (UK) facility (Theocharous et al. [175], Theocharous [176], Theocharous [177], whose schematic diagram is presented in Fig. 8.8. The tungsten strip lamp with a sapphire window is used as the radiation source for wavelengths

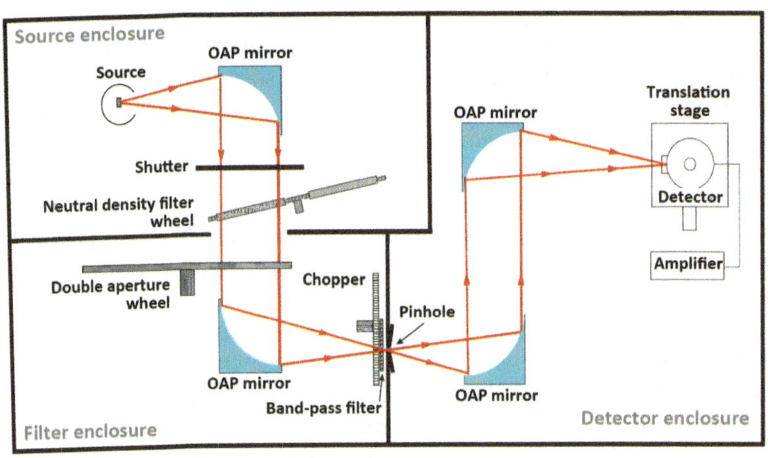

Fig. 8.8 Layout of the NPL facility for measuring detector linearity. Reproduced from [177] with permission of the author and NPL (UK)

Fig. 8.9 Photograph of the double aperture wheel. Reproduced from [177] with permission of the author and NPL (UK)

up to 5 μm, and a Nernst glower source is used for longer wavelengths. A fraction of the radiation emitted by the source is collimated using the off-axis parabolic (OAP) mirror. The collimated radiation, passed through the neutral density filter and the double-aperture wheel, is imaged onto a pinhole by the second identical OAP mirror. The filter wheel contains eight reflective neutral density filters deposited on a ZnSe substrate. The neutral density filters attenuate the beam in optical density[2] steps of about 0.3.

The filter wheel is positioned at an angle to the collimated beam to avoid interreflections between optical components. The double-aperture wheel (disk) shown in Fig. 8.9 includes the apertures A, B, and A + B. The surface of this disk is blackened with a paint that has low reflectance in the IR spectral range.

Both rotational disks are driven by computer-controlled stepper motors. Additionally, the radiant flux incident on the detector is varied using neutral density filters of different transmittance. Theocharous et al. [175] and Theocharous [176] described application of this technique to the linearity measurement of photovoltaic HgCdTe and pyroelectric $LiTaO_3$ detectors, respectively, in the IR spectral range. Another type of methods for nonlinearity measurement is presented by the superposition (or combinatorial) methods, which are based on the technique of beam conjoining to obtain integral and fractional multiples of a given radiance using beam splitters (Erminy [43]. This approach combined with the use a series of filters was employed by many authors (see, e.g. [30, 46, 56, 159, 195]). The measured values of signals

[2]Optical density is the logarithm to base 10 of the reciprocal value of the transmittance.

are used together with a model of the nonlinearity to write a series of equations, which should be solved for the unknown flux levels, the filter transmissions, and the parameters of the non-linearity model.

The non-linearity value at each signal level is equal to

$$\eta_j = K\Phi_j / S'_j, \tag{8.16}$$

where Φ_j is the radiant flux corresponding to the measured signal S'_j and K is an unknown proportionality factor between radiant fluxes and signals. Without additional information, the nonlinearity function given by Eq. 8.10 can be determined only up to a multiplication factor, similar to the flux-doubling method.

Let us consider as example the well-known facilities for measuring the nonlinearity of radiation detectors, so-called beam conjoiner, Beamcon III developed at NIST [179]. A simplified schematic diagram illustrating the technique of beam conjoining is shown in Fig. 8.10.

The input beam from the source is split into two optical paths by the first beam splitter BS1 and variably attenuated by filters on the filter wheels FW1 and FW2. After reflection off mirrors Ml and M2, the two beams are combined by the beam splitter BS2 and attenuated again by a third filter on the wheel FW3 before falling on the detector. All filters are metallized neutral density filters with quartz substrates. The filter wheels are tilted to avoid interreflections between filters. In addition, the beam splitters are wedged to prevent interference. Beamcon III allows using of both broadband (quartz halogen or arc lamp) and laser (He-Ne or Ar^+) sources. For narrowband operation with a broadband source, a monochromator can be used. The total number of filter combinations in the Beamcon III is 150, 30 of which are the dark signal measurements. The dark signal/filter combinations are randomized and inserted after every five measurements. Each dark signal is used to correct the succeeding five response signal measurements. The maximum throughput is 7% at a wavelength of 632.8 nm, and the total attenuation factor is about 3000 against a radiation source. The output beam is partially vertically polarized even for an unpolarized source since each tilted surface preferentially reflects the vertical component of the incident beam. The ratio of the polarization in the vertical direction to that in the horizontal direction is 1.17. Therefore, it is important to keep the polarization

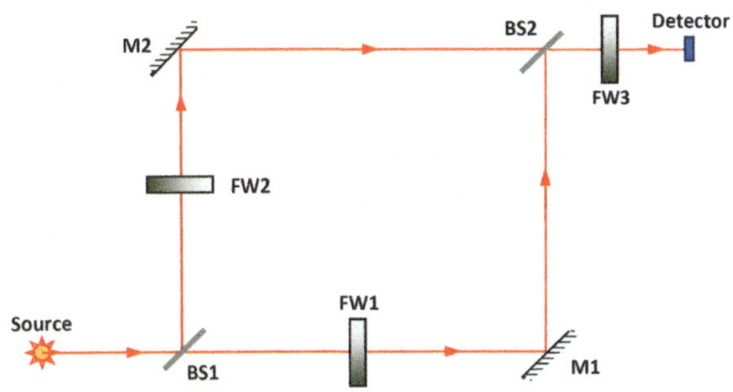

Fig. 8.10 Schematic optical diagram of Beamcon III showing the primary optical elements: BS1 and BS2 are beam splitters; M1 and M2 are mirrors, and FW1, FW2, and FW3 are filter wheels

direction of a source (e.g., laser) constant in order to eliminate this polarization preference. To suppress scattered radiation, the entire facility is subdivided into optically isolated chambers. The environmental humidity and temperature are monitored at the detector.

The principle of operation of the Beamcon III is the beam addition method, in which different filter wheel combinations lead in different radiant fluxes incident on the detector. This method assumes that the flux from the source remains constant during all measurements. The final flux Φ at the detector is the sum of the fluxes Φ_1 and Φ_2 from two paths. Indexing the fluxes by the filter combination,

$$\Phi(i, j, k) = \Phi_1(i, k) + \Phi_2(j, k), \qquad (8.17)$$

where $i = 0, 1, \ldots, 4$ designates the filter on the wheel FW1, $j = 0, 1, \ldots, 4$ the filter on FW2, and $k = 0, 1, \ldots, 5$ the filter on FW3. In each case, an index value of 0 corresponds to an opaque, shuttered position. For a single path, there are 30 possible filter combinations. Of these, no flux is transmitted for 10 combinations.

Thus, each path has 20 filter combinations that result in a response signal. Since the detector is assumed nearly linear, we can assume an n-th order's polynomial response function for the detector:

$$\Phi(i, j, k) = r_0 + r_1 S(i, j, k) + r_2 S^2(i, j, k) + \ldots + r_n S^n(i, j, k), \qquad (8.18)$$

where $S(i, j, k)$ is the measured response signal corrected by the dark signal and indexed by the filter combination, and $r_0, r_1, r_2, \ldots, r_n$ are the response function coefficients.

Since the detector is assumed to be nearly linear and in order to normalize the flux to the same units as the signal, we can set $r_1 = 1$ and obtain

$$\Phi_1(i, k) + \Phi_2(j, k) = r_0 + S(i, j, k) + r_2 S^2(i, j, k) + \ldots + r_n S^n(i, j, k), \qquad (8.19)$$

The system of linear equations written for all i, j, k should be solved for r_i, $i = 0, 2, \ldots, n$ and fluxes $\Phi_1(i, k)$ and $\Phi_2(j, k)$. The Beamcon III allows obtaining more equations than there are unknown variables, which makes it possible to find solutions using the least squares method.

The results obtained with the Beamcon III facility for nonlinearity of Si photodiode-amplifier system is illustrated by Fig. 8.11. For the signal model, a second-order polynomial was used. The linearity was determined for amplifier gains from 10^4 to 10^{10} but only the results for a gain of 10^6 is presented. The 1000 W tungsten-halogen lamp operated at a stabilized current of 7.6 A was used as a broadband radiation source. Over the three decades of signal measured at this gain, the detector system is linear to within the expanded (at $k = 2$) uncertainty of 0.027%.

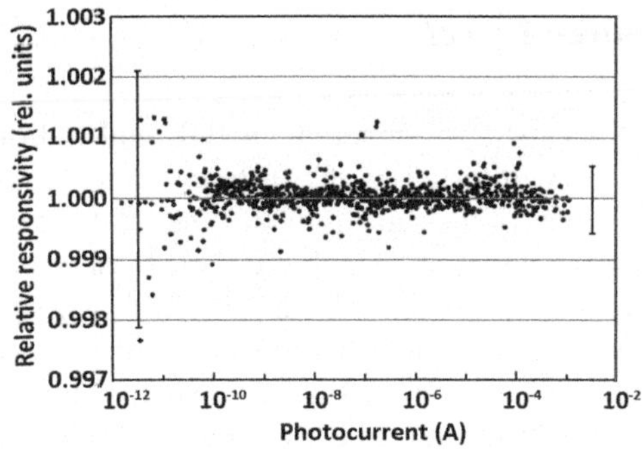

Fig. 8.11 Relative responsivity as a function of signal for a Si photodiode with amplifier over its entire signal range of nine decades. The left and right error bars indicate the expanded uncertainty estimated for the relative responsivity for signals less than 10^{-11} A and over the entire range of signal, respectively [179]. Reprinted courtesy of the National Institute of Standards and Technology, U.S. Department of Commerce. Not copyrightable in the United States

Modern instruments for nonlinearity measurement are increasingly use the light emitting diodes (LEDs) as radiation source. The researchers at KRISS[3] [166, 167] described a series of LED-based devices for assessing the linearity of optical radiation detectors. The instrument uses the principle of the flux addition of two high-brightness LEDs with individually controlled power supplies.

The output radiant fluxes are varied by adjusting the injection currents of LEDs. Each LED has an output radiant flux of 200 mW; the peak emission wavelength and the FWHM are 396 nm and 20 nm, respectively. The instrument is operated with a special algorithm that determines the sequence of LEDs switching and data reading to eliminate the measurement error caused by a drift of the LED radiant fluxes.

The linearity measurement with an uncertainty of as low as 0.01% in a dynamic range over six decades has been experimentally demonstrated for Si photodiodes. Later, the LED-based instrument was modified by arrangement of the LEDs with the center wavelengths of 390 nm, 670 nm, and 750 nm in an integrating sphere [167]. The non-linearity correction factors for Si photodiode responsivity were determined with the relative standard uncertainties of $(2\ldots6) \times 10^{-5}$ for photocurrent range from 10^{-8} to 10^{-3} A.

Saunders et al. [157] described a compact computer-controlled device for measuring the nonlinearity of Si photodiodes. It consists of the LED emitting at 625 nm, a set of beam-splitting cubes, and a set of neutral density filters. It was claimed that the device measures the nonlinearity with the combined uncertainty of order of 0.01% or better using a combinatorial technique.

[3] Korea Research Institute of Standards and Science (South Korea).

8.2.4 Size-of-Source Effect

The size-of-source effect (SSE) is inherent in RTs and radiometers operating in the radiance mode. It is manifested as the different detector output signals when comparing radiation sources of the same radiance, but of different size and/or shape. The SSE leads to instrumental errors, which is caused by shortcomings of the optical system (optical aberrations, reflections from lens surfaces, scattering on the dust particles and imperfections of optical components, diffraction on the diaphragm edges, etc.) As a result, instead of a well-defined, sharp-boundary, circular FOV of an idealized RT, a real RT has a blurred FOV. Thus, a certain amount of the radiant flux originated outside the nominal FOV reaches the detector and, conversely, a part of the radiation coming from regions within the target area does not. In the ideal case, corrections for the SSE should be made so that the corrected temperature matches the reading for an infinite (or sufficiently large) target with a uniform radiance. In such a case, the radiation losses from within the nominal FOV are compensated with those gained from outside the FOV and the reading corresponds to the correct target temperature.

The magnitude of the SSE is determined by the optical design and component quality of the RT, the sizes and temperatures of the sources being compared, and the wavelengths at which the comparison is made. Radiation thermometry of blackbodies deals with the sources that can have different sizes and shapes of radiating surfaces. The SSE correction is required to realize the radiation temperature scale regardless of the characteristics of the RT used for this purpose. When the fixed-point calibration is performed for the RT, the SSE correction is needed to normalize the fixed-point calibration to some reference diameter, which allows comparison of scales realized in different NMIs. The same applies to the comparison of FPBBs and VTBBs. The typical FPBBs have small, 3- to 6-mm diameter apertures, while the VTBBs usually have much larger openings. The SSE should be measured for every RT and, after introducing appropriate corrections, the uncertainty can be evaluated according the type B evaluation [71]. The uncertainty due to the SSE can be a dominant fraction of the total uncertainty in RT calibration and at comparison of blackbody radiation sources. Presently, this is an active area of research in the RT design and correction of the measurement results. It is necessary to have a rigorous definition for the quantitative measure of the SSE, especially since one can find in the literature the data for the SSE, both very close to one [55, 68, 91, 106, 171, 194, 203], and very close to zero [1, 11, 12, 67, 107, 115, 200]. Obviously, different definitions of the SSE, tightly connected with the different measurement methods, have been used in these references. There are two main techniques of measuring the SSE called direct method and indirect (differential) method. In recent decades, the method of the SSE measurement based on scanning an RT over a slit radiation source has been developed [7, 156, 181]. Liebmann and Kolat [90] proposed the "sweep" method, which uses scanning an RT over a circular radiation source of a fixed diameter. We will not discuss the scanning methods here due to their infrequent use. Besides, they require more complicated data processing than the ordinary direct and indirect methods.

Although there are many theoretical works attempting to harmonize definitions and relationships regarding the SSE [11, 58, 132, 150, 156], there are still no uniform terminology and notation in this area. Despite the importance of the SSE, not all of its aspects have received proper coverage in the literature. The SSE is not the only effect, which manifests itself in the RT signal changes due to changes in measurement geometry. Along with the SSE that is evaluated at one fixed measuring distance, the so-called distance effect (DE) and focus effect (FE) are inherent to RTs with adjustable measuring distance and focusing capabilities [131].

The DE can be defined as a change in the RT output due to a change in the measuring distance and is a potentially significant source of uncertainty for the IR RTs due to (but not only) the dependence of their signals on the atmospheric absorption in the optical path. It has been hypothesized that the RTs having a large SSE should have also a large distance effect [137] but generally the nature of the distance effect is not completely understood [61].

The focus effect (FE) is the change of the output signal of an RT with focusing capability when the measuring distance changes without changing the focus. The FE takes place always when the RT is aimed at a cavity and at least a part of its radiating surface is out of focus.

The literature on the DE and FE effects are extremely scarce; the experimental results remain mainly unconfirmed; there are no theoretical considerations regarding them. It can be assumed that for many practical cases, the DE and FE are masked by more influential SSE. Therefore, it is difficult to consider the SSE problem finally solved. The framework of this book does not allow us to delve into this problem, so we present only the most common definitions, explain the difference and reveal inter-connections between the direct and indirect methods. The applicability prerequisites for all SSE measurement methods are as follows:

- An RT is assumed linear, i.e. its signal S is proportional to the measured radiance.
- The target area (FOV) of the RT is circular, with the nominal radius R_0.
- A circular source of a uniform radiance L (spectral, total, or band-limited) is assumed coaxial and confocal with the FOV of the RT.
- A radius of the source, R_{max}, is much greater than the FOV of the RT: $R_{max} \gg R_0$.
- In further analytical calculations, the radiation source is implicitly assumed Lambertian.

Direct Method The direct method is applied mostly to the IR RTs. It employs the large-aperture blackbodies as the source of virtually uniform radiance. The direct method consists of measuring the RT signal when the radius r of the uniform radiance source varies from R_{min} to R_{max} (see Fig. 8.12a). This is easily accomplished using a set of cooled blackened diaphragms of variable radii r, $R_{min} \leq r \leq R_{max}$, similar to those shown in Fig. 8.13. We can define the SSE function $\sigma(r)$ for the direct measurement method as

$$\sigma(r) = S(r) \big/ S(R_{max}). \tag{8.20}$$

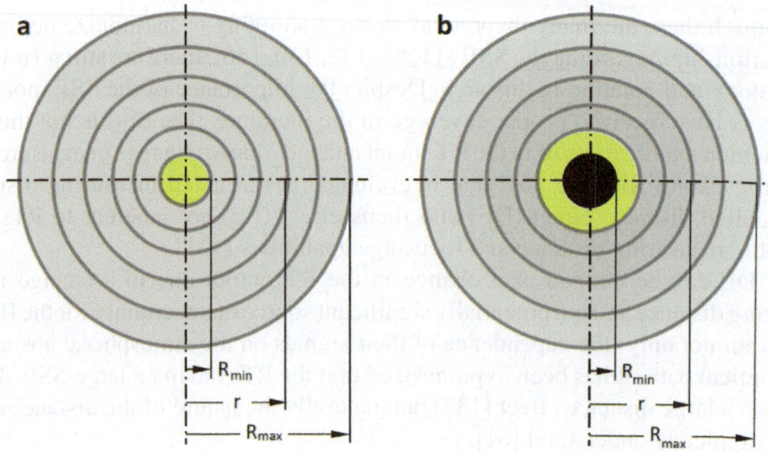

Fig. 8.12 Schematics for the SSE measurement methods: **a** direct and **b** indirect

Fig. 8.13 Set of exchangeable apertures for SSE measurements by the direct method. Reproduced from [112] according to Creative Commons Attribution-NoDerivatives 4.0 International Public License

The SSE $\sigma(r)$ represents the fraction of the radiation collected from the source of the radius r with respect to the radiation collected from a source of a larger radius $R_{\max}$. It is clear that $\lim_{r \to R_{\max}} \sigma(r) = 1$. Ideally, the radius $R_{\max}$ should be so large that its further increase does not change the output signal of the RT. However, this condition is not always satisfied in practice. It is often enough to use certain reference value of radius instead of $R_{\max}$.

We must associate definition (8.20) with the value of the correction factor, which takes into account the difference in the aperture radii when comparing different blackbodies. Let dp be the elementary probability that the radiation emitted by the elementary annular area dA (see Fig. 8.14) reaches the RT detector to contribute to its signal.

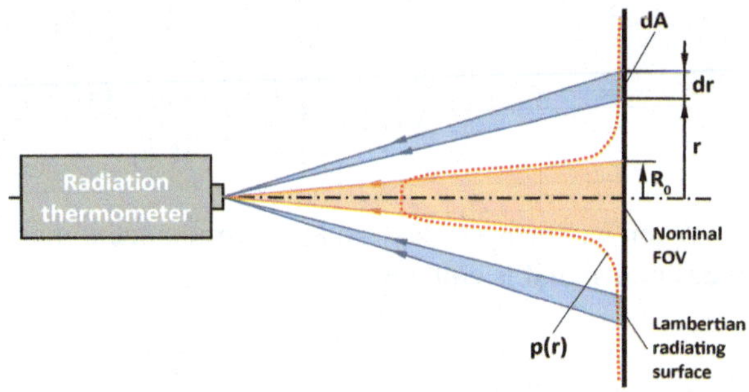

Fig. 8.14 Scheme for probabilistic interpretation of the SSE

If the optical system of the RT has axial symmetry and the radiating plane is a Lambertian radiator, the probability density function p can depend only on the radial coordinate r of the polar system of coordinates.

If R_{max} is sufficiently large, the normalization condition for p can be written in the form:

$$\int_{A_{max}} p(A)dA = 2\pi \int_0^{R_{max}} p(r)rdr = 1, \tag{8.21}$$

where A_{max} is the circular area of a radius R_{max}. The SSE $\sigma(r)$ can be expressed as the cumulative distribution function for $p(r)$:

$$\sigma(r) = \frac{\int_A p(A)dA}{\int_{A_{max}} p(A)dA} = \int_0^r p(r)rdr \tag{8.22}$$

and vice versa, the probability density can be expressed through $\sigma(r)$:

$$p(r) = \frac{d\sigma}{dA} = \frac{d\sigma}{2\pi rdr}. \tag{8.23}$$

A signal S of an ideal RT must be proportional to the measured radiance L:

$$S \propto L. \tag{8.24}$$

However, a signal of a real detector is proportional rather to the radiant flux that reaches the detector. Let us assume that the radiance of a measured object is constant within the circular area of a radius R_a and is zero outside it:

$$L(r) = \begin{cases} L = \text{const, if } 0 \le r < R_a \\ 0, \text{ if } R_a \le r \le R\text{max} \end{cases}. \tag{8.25}$$

When we measure the radiance or radiance temperature of using the real RT, its measured signal is equal to

$$S_{meas}(R_a) \propto \int_0^{\infty} L(r)p(r)rdr = L \cdot \int_0^{R_a} p(r)rdr = \frac{L\sigma(R_a)}{2\pi} \propto S(R_a)\sigma(R_a). \tag{8.26}$$

If we compare two blackbodies with the same radiance but different radii, R_a and R_b, the following equality holds:

$$\frac{S_{meas}(R_a)}{S_{meas}(R_b)} = \frac{\sigma(R_a)}{\sigma(R_b)}. \tag{8.27}$$

Let us suppose that we compare two blackbodies, the FPBB and VTBB with the aperture radii R_{FPBB} and R_{VTBB}, respectively. If we obtained $S_{meas,VTBB} = k_{meas} S_{meas,FPBB}$, we must correct k for the SSE as

$$k = k_{meas} \frac{\sigma(R_{VTBB})}{\sigma(R_{FPBB})}. \tag{8.28}$$

The precision of the direct method is limited by measuring small differences of large values, when the fluctuations in the source radiance can be comparable to these differences.

Indirect Method The indirect (differential) method has been developed to increase the accuracy of the SSE measurement. Its realization requires a uniform radiance source of a sufficiently large radius, at the center of which a circular opaque obstacle (non-radiating "black spot") of the constant radius is placed as shown in Fig. 8.12b. Although the "black spot" radius is usually slightly larger than the nominal FOV radius R_0, we will consider them equal for simplicity. The RT is focused onto the "black spot" center on the radiance source of a variable diameter. It is common for this purpose to use an integrating sphere having a sufficiently large opening covered with a translucent diffuser. The integrating sphere is made of material with a high diffuse reflectance (e.g. Spectralon, Halon, $BaSO_4$) and irradiated from inside. Multiple reflections from the internal walls of the sphere result in uniform irradiation of the sphere opening that has a radius $R_{\max}$. A translucent diffuser (e.g. roughened quartz plate) plays the role of a source of uniform radiance. The central circular zone of the diffuser has the diameter slightly greater than R_0 and made opaque and black. The diameter of the bright area can be varied using a set of interchangeable

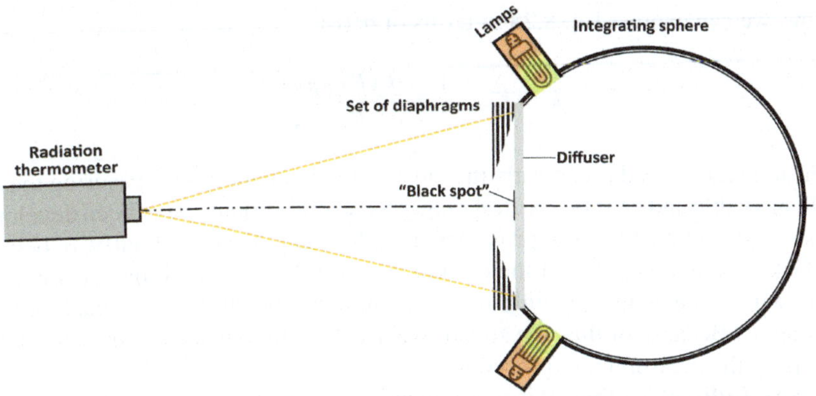

Fig. 8.15 A simplified scheme for indirect measurement of the SSE using an integrating sphere

diaphragms with radii $r \in [R_0, R_{max}]$. A simplified scheme for indirect measuring of the SSE using an integrating sphere is presented in Fig. 8.15.

Sakuma and Ma [139] described the integrating sphere source for the indirect measurement of the SSE developed at the NMIJ. The inner diameter of the sphere is 450 mm with the exit port diameter of 140 mm. The integrating sphere walls painted with $BaSO_4$ are illuminated by six halogen lamps arranged around the exit port inside the sphere. The constant voltage mode using DC stabilized power sources are used to feed the lamps. The exit port is covered with a quartz plate that has a black spot of 6 mm in diameter painted on it. The aperture size can be varied between 9 mm and 96 mm using the set of interchangeable diaphragms.

The indirect method with the use of the integrating sphere is applicable in the wavelength range from approximately 0.4–1.2 μm. Its main idea is to measure the signal $S(r, R_0)$ of the linear RT aimed at the "black spot" in the center of the bright field of a diameter r.

The SSE function for the indirect method is defined as the ratio of the signal measured for a given aperture radius with the black spot to the signal measured with the black spot removed:

$$\sigma'(r) = \frac{S(r, R_0)}{S(r)}. \quad (8.29)$$

It is easy to show that $\sigma(r)$ and $\sigma'(r)$ are interconnected:

$$\sigma'(r) = \frac{\int_{R_{min}}^{r} p(r)rdr}{\int_{0}^{r} p(r)rdr} = \frac{\int_{R_{min}}^{r} p(r)rdr}{\sigma(r)} = \frac{\int_{0}^{r} p(r)rdr - \int_{0}^{R_{min}} p(r)rdr}{\sigma(r)}$$

$$= 1 - \frac{\sigma(R_{min})}{\sigma(r)}. \quad (8.30)$$

Now, we can express Eq. 8.28 in terms of $\sigma'(r)$:

$$k = k_{meas} \frac{1 - \sigma'(R_{FPBB})}{1 - \sigma'(R_{VTBB})}. \tag{8.31}$$

The indirect method is currently the most common method for measuring the SSE of RTs of visible and near IR spectral range. Its several variants have been developed. If r is not too small, $S(r)$ is approximated by the signal measured halfway between the black spot and the edge of the current aperture. It was found that $S(r)$ changes slowly with changes in r, mainly due to additional reflections from back sides of diaphragms. Because of this, the signal $S(R\mathrm{max})$ is often measured instead of $S(r)$ to simplify the measurement procedure.

In Situ Indirect Method It is worth mentioning the in situ indirect measurement technique developed for the most responsible measurements with FPBBs. Its use is especially justified if several replaceable fixed points can be realized in the same furnace. The target area of the FPBB is usually clearly distinguishable against the background of a less bright annulus (diaphragm) made of polished metal with low emissivity. Nevertheless, the closest surroundings of the cavity aperture emit a non-zero radiance that distorts the readings of the RT. The essence of the in situ indirect method is to measure signals from the emitting cell and from its imitation, which eliminates radiation from the inside of the cavity. Unlike traditional methods, which employ the formalism of idealized measurement conditions, the in situ method allows one to determine the SSE correction for from the real background radiation falling into the FOV of an RT.

Jones and Tapping [73] proposed a method for estimating the SSE using a dummy crucible inserted in an FPBB instead of a gold fixed-point cell. A dummy crucible is designed so that the FOV of the RT freely passes through the hole in it. The furnace behind the crucible was modified so that no radiating parts of it were directly visible by the RT from the front of the FPBB. A cold black cavity (radiation trap) was placed behind the furnace not allowing the ambient radiation to enter the FOV of the RT. Envall et al. [38] implemented the idea of Jones and Tapping [73] for the Si photodiode-based RT with the center wavelength of 900 nm. The crucible, normally containing metal to be melted, is replaced in the furnace with a dummy cell having a radiating cavity without a conical bottom. An absorbing cylindro-conical trap is attached to the rear flange of the furnace. The schematic of the measuring setup is shown in Fig. 8.16.

Matveyev [105] proposed a method of “cooled target” and described a specially developed device applied to measurement of the SSE for FPBBs at freezing temperatures of silver, gold, and copper in the national primary measurement temperature standard of Russia. The target simulator cooled with the flow of cold nitrogen gas is placed instead of the radiating fixed-point cell in the furnace, at a place that is sighted by an RT.

The measurements are carried out in the furnace heated up to the reference temperature. The radiation from a sufficiently cooled target simulator is basically

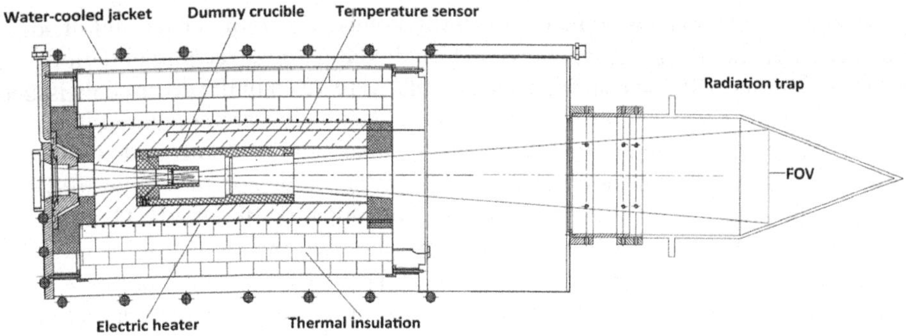

Fig. 8.16 Schematic of the measurement setup for using an FPBB in the SSE measurements. Reproduced from [38] with permission of Springer Nature

the reflected part of the radiation incident from the hot environment onto the simulator. If the geometry and surface reflectance of the simulator are known, this radiation can be estimated.

The techniques mentioned above require some modification of the structure of the FPBB. In contrast, Shimizu and Ishii [164] proposed a method for determining the in situ correction for the SSE, based on the integrating sphere-based measurements of σ' according to Ma and Sakuma [94], and a separate measurement of the radiance distribution around the cavity aperture of the FPBB. The radiance temperature distributions were obtained by scanning the front flange of the FPBB using a 1.6 μm InGaAs RT. The minimum target size of 3 mm in diameter at a distance of 400 mm made it possible to obtain the radial radiance temperature distributions over the cavity aperture and the metal annulus around it. The product of the radiance distribution and the SSE function forms the value of the correction due to the SSE at real background radiation.

The relative differences in the spectral radiance $\Delta L_\lambda / L_\lambda$ and in temperature $\Delta T / T$ are interconnected by expression

$$\frac{\Delta L_\lambda}{L_\lambda} = \frac{x}{1 - \exp(-x)} \frac{\Delta T}{T}, \tag{8.32}$$

where $x = c_2 / (\lambda T)$, λ is the central wavelength of the RT, and c_2 is the second radiation constant in the Planck law. Therefore, the RT raw signals have to be accessible to apply the correction for the SSE according to Eq. 8.32. The situation is more complicated with RTs, whose readings are temperatures (so-called direct reading RTs). They calculate temperatures by the prescribed algorithm using detector signals during calibration and measurement and, therefore, correction must be done in terms of temperature. This case, more typical for industrial rather than for metrological applications, was studied by Pušnik et al. [128, 129], Saunders [148, 151], Manoi and Saunders [101].

The uncertainty in measurement of the SSE contributes to the uncertainty budget for measurement of the temperature of blackbodies. Some recommendations on calculation of the SSE measurement uncertainty were given in the documents issued on behalf of the CCT [47, 49]. When different blackbodies are compared using an RT, the target areas of approximately uniform temperatures are surrounded by areas with significantly non-uniform radiances. For instance, the aperture of an FPBB is encircled by colder front walls of a crucible. To compare different blackbodies, for each of them, the effective source radius must be used in calculation of the SSE blackbody instead of the radius of its aperture. The effective source radius R_{eff} is the radius of a disk source with abrupt and uniform radiance distribution that gives the same SSE as the real source with distributed radiance. This approach suggested by Bloembergen [10] was developed in more details in [11, 12]. It can be expressed by the equation:

$$\sigma^*\left(R_{eff}\right) = \sigma^*(R_0) + \frac{1}{S_0}\sum_{i=1}^{N}\left[\sigma^*(R_i) - \sigma^*(R_{i-1})\right]S_i, \tag{8.33}$$

where σ^* is σ or σ', R_0 and S_0 are the radius and measured signal, respectively, of the central uniform zone of the source divided into N annular zones, each limited by radii (r_{i-1}, r_i) and having a uniform exitance corresponding to the signal S_i. It is assumed that the SSE is small (close to 1 for σ or close to 0 for σ').

Studies of the last two decades [161, 198–200, 203] have made it possible to develop some common recommendations that can reduce the SSE to less than $5{\cdot}10^{-5}$ (we provide examples of implementing these recommendations in specific design solutions in Sect. 8.3.2):

1. Reducing the number of reflecting surfaces in an optical system.
2. Preventing dust from entering the RT.
3. Use the optical schemes with the Lyot stop.
4. Placement of the filter between collimator lenses.
5. Formation of a radiation beam underfilling the photodetector.
6. Suppressing the radiation scattered inside the RT by blackening the internal surfaces and placing additional partitions and baffles.

8.3 Realization of the ITS-90 Above the Freezing Point of Silver

8.3.1 Solution of Measurement Equation

Above the freezing point of silver, the ITS-90 [126] is defined by the extrapolation based on the Planck equation:

$$\frac{L_\lambda(\lambda, T_{90})}{L_\lambda(\lambda, T_{90}(X))} = \frac{\exp\left(\frac{c_2}{\lambda T_{90}(X)}\right) - 1}{\exp\left(\frac{c_2}{\lambda T_{90}}\right) - 1}, \tag{8.34}$$

where T_{90} is the unknown temperature; $T_{90}(X)$ denotes the silver, gold, or copper freezing point (1234.93 K, 1337.33 K, or 1357.77 K, respectively); $L_\lambda(\lambda, T_{90})$ and $L_\lambda(\lambda, T_{90}(X))$ are the spectral radiance of a perfect blackbody at the wavelength λ in vacuo at temperatures T_{90} and $T_{90}(X)$, respectively; $c_2 = 0.014388$ m K is the second radiation constant. The values for the fixed points and for c_2 adopted in the original text of the ITS-90 should be used in Eq. 8.34, despite their possible later amendments or changes [202]. There are at least four points in this definition, to which we should pay attention.

1. The monochromatic approximation is used in (8.34), i.e. the waveband $\Delta\lambda \to 0$ of an RT is assumed. A real RT may have a very narrow but still finite waveband.
2. The radiance approach implies observation of a blackbody within an infinitesimal solid angle. A real RT collects radiation within a small but finite solid angle, that is, a real RT reacts rather to a radiant flux than to radiance.
3. It is assumed that signals S of the RT should be linearly proportional to the corresponding spectral radiances, i.e. the equality $S(\lambda, T_1)/S(\lambda, T_2) = L_{\lambda,bb}(\lambda, T_1)/L_{\lambda,bb}(\lambda, T_2)$ takes place.
4. Equation (8.34) is written for two perfect blackbodies, for which the effective emissivities are the same (for example, equal to unity). In practice, this condition is not be met.

These deviations of the actual measurement conditions from idealized ones lead to systematic errors, corrections of which must be made, and uncorrected residuals should be accounted for as uncertainties of type B. The experimentally measured ratio r of detector signals can be written as

$$r = \frac{S(T_{90})}{S(T_{90}(X))} = \frac{\int_0^\infty \varepsilon_e(\lambda, T_{90}) L_{\lambda,bb}(\lambda, T_{90}) s_r(\lambda) d\lambda}{\int_0^\infty \varepsilon_e(\lambda, T_{90}(X)) L_{\lambda,bb}(\lambda, T_{90}(X)) s_r(\lambda) d\lambda}, \tag{8.35}$$

where $S(T_{90})$ and $S(T_{90}(X))$ are the measured detector signals (voltage or electric current) when the RT is aimed at a target with unknown temperature T_{90} and at the ITS-90 FPBB at the temperature $T_{90}(X)$, respectively; $\varepsilon_e(\lambda, T_{90})$ and $\varepsilon_e(\lambda, T_{90}(X))$ are the spectral effective emissivities of corresponding blackbodies; $s_r(\lambda)$ is the relative spectral responsivity of the RT; $L_{\lambda,bb}(\lambda, T_{90})$ and $L_{\lambda,bb}(\lambda, T_{90}(X))$ are the spectral radiances calculated according the Planck law:

$$L_{\lambda,bb}(\lambda, T_{90}) = \frac{c_{1L}}{n^2(\lambda)\lambda^5}\left[\exp\left(\frac{c_2}{n(\lambda)\lambda T_{90}}\right) - 1\right]^{-1}, \tag{8.36}$$

where $n(\lambda)$ is the spectral refractive index of the medium in which the RT is immersed (usually, in air); λ is the wavelength in this medium; c_{1L} is the first radiation constant for the spectral radiance.

The integrands in Eq. 8.35 are near zero outside the passband of the RT, which allows reducing the limits of integration to a finite wavelength range, where $s_r(\lambda) \neq 0$.

The $n(\lambda)$ for air is a slowly varying function of wavelength (see Sect. 3.2.6); therefore, the n^2 factor can be taken outside the integral signs in (8.35) and cancelled. However, the refractive index inside the exponential term in the Planck law can be significant for high-accuracy measurements.

The spectral effective emissivities of an FPBB and of the blackbody with unknown temperature are usually considered constant over the bandwidth of the RT, so these two quantities can be taken outside the respective integrals in (8.35). We can rewrite Eq. 8.35 for the signal ratio r':

$$r' = \frac{S'_{SSE,NL}(T_{90})}{S'(T_{90}(X))} = \frac{\int_0^\infty L_{\lambda,bb}(\lambda, T_{90}) s_r(\lambda) d\lambda}{\int_0^\infty L_{\lambda,bb}(\lambda, T_{90}(X)) s_r(\lambda) d\lambda}, \tag{8.37}$$

where $S'_{SSE,NL}$ is the corrected for the effective emissivity, nonlinearity, and SSE signal of the RT aimed to a blackbody, whose temperature to be measured; S' is the emissivity-corrected signal of the RT aimed to a reference FPBB. We can rewrite (8.37) in the compact form:

$$F(T_{90}) = 0, \tag{8.38}$$

where

$$F(T_{90}) = \frac{\int_0^\infty L_{\lambda,bb}(\lambda, T_{90}) s_r(\lambda) d\lambda}{\int_0^\infty L_{\lambda,bb}(\lambda, T_{90}(X)) s_r(\lambda) d\lambda} - r'. \tag{8.39}$$

Equation 8.39 can be solved for T_{90} using conventional numerical methods for nonlinear equations such as Newton's iteration method:

$$T_{90,k+1} = T_{90,k} - F(T_{90,k}) / F'(T_{90,k}), \quad k = 0, 1, 2, \ldots, \tag{8.40}$$

where k is the number of iteration, $F'(T) = dF/dT$, and T_0 is a zero approximation (initial guess). The derivative in (8.40) can be easily computed analytically (see Sect. 4.2.4); then

$$T_{90,k+1} = T_{90,k} + \frac{r'G - \int\limits_0^\infty s_r(\lambda) L_{\lambda,bb}\left(\lambda, T_{90,k}\right) d\lambda}{\frac{c_2}{T_{90,k}^2} \int\limits_0^\infty \frac{s_r(\lambda) L_{\lambda,bb}\left(\lambda, T_{90,k}\right) d\lambda}{n(\lambda)\lambda\left[1 - \exp\left(-c_2 / \left(n(\lambda)\lambda T_{90,k}\right)\right)\right]}}, \quad k = 0, 1, 2, \ldots \tag{8.41}$$

where $G = \int\limits_0^\infty s_r(\lambda) L_{\lambda,bb}(\lambda, T_{90}(X))\, d\lambda$ is a constant value, which can be computed before iterations. The convergence of Newton's algorithm expressed by (8.41) is achieved usually after 5–10 iterations with the error less than 1 μK. Modern PCs do this almost instantly.

Alternatively, extrapolation using the Sakuma-Hattori (SH) equation can be applied to solve the Eq. 8.34. We will consider it in Sect. 8.4.4.

8.3.2 Primary Standard Radiation Thermometers

A primary standard radiation thermometer is an RT that is used to realize the ITS-90 above the silver point. The most suitable RTs for this purpose use silicon photodiodes and narrowband interference filters with the center wavelength λ_0 from 650 to 900 nm. Such RTs must have low SSE, high linearity within 5–6 decades, small target size (of order of 1 mm), and low NETD. Currently, the RTs suitable for precision measurements of high-temperature blackbodies can be found in the relevant market segment. Commercial RTs are used to realize the vast majority of national high-temperature scales. In this section, we briefly discuss the features of RTs from two manufacturers, which are currently most often used for national realizations of the ITS-90.

The RTs known under the abbreviation "LP" (from German "linearpyrometer") from KE Technologie GmbH (Stuttgart, Germany, http://ke-gmbh.de/linearpyrometer) are very popular in radiation thermometric community. All "Linearpyrometers" (LP2 and LP3, which are no longer manufactured but still in use in some laboratories, as well as newer LP4 and LP5) have a highly linear relation between the detector output signal and the spectral radiance of a target, exceptional performance and metrological characteristics. They employ the same design principles: a low-scattering objective lens, a collimation system for the filter, the Lyot stop, and underfilling irradiation of the photodiode. The main differences between these RTs are in interference filters and the electronic systems.

A good example of the primary standard RT from KE Technologie GmbH is the LP4 [76]. Its scheme is presented in Fig. 8.17. LP4 can operate with Si or InGaAs photodiodes in the spectral range from 500 to 1700 nm and measure temperatures from 680 to 3500 °C, depending on the detector and filter selection.

For operating wavelengths up to 900 nm, the silicon photodiode Hamamatsu S1336-18BK [57] is used. The relative spectral responsivity curves of the LP4 with two interference filters (center wavelengths of 650 nm with the FWHM of 15 nm and

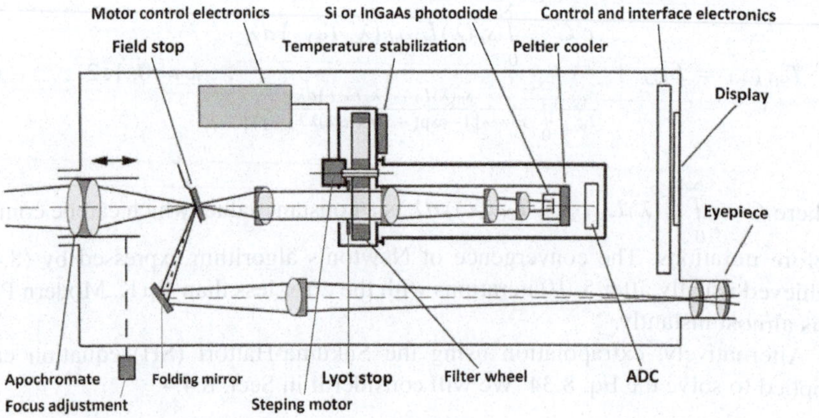

Fig. 8.17 A schematic of the LP4 RT. Reproduced from [3] with permission of IOP Publishing

900 nm with the FWHM of 21 nm) thoroughly measured using the monochromator-based setup with the supercontinuum radiation source (see Woodward et al. [185]) are shown in Fig. 8.18.

The LP4 apochromatic objective lens with a focal length of 143 mm and the aperture diameter of 40 mm forms the image of the target in the field stop, which is made in the inclined mirror. The circular aperture of the field stop has a diameter of 0.25 mm, which provides the target size of 0.8 mm at the target distance of 600 mm from the objective lens and increases to 3.4 mm as the target distance increases to 2 m. The image of the aperture is reflected by a flat inclined mirror to the eyepiece used to aim at the center of the aperture of a blackbody.

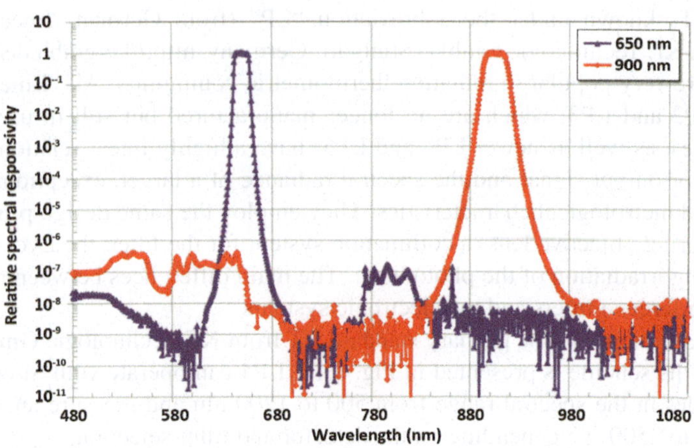

Fig. 8.18 Relative spectral responsivities of the LP4 with silicone photodiode for two interference filters with the center wavelength of 650 nm and 900 nm. Reproduced from [197] with permission of IOP Publishing

After passing the field stop, the radiation passes through the collimator and underfills the receiving surface of the silicon photodiode. The bandpass filter is placed in a collimated beam between the collimator lenses. The LP4 has six places for filters in a filter wheel operated with a stepping motor. A high reproducibility of the filter position is achieved using a special snap mechanism for the filter wheel. A Lyot stop installed directly in front of the filter prevents scattered radiation from reaching the detector.

A silicon photodiode and a high-gain transimpedance amplifier including an ADC are incorporated in thermostatically controlled chamber. The photocurrent amplifier and the ADC handle the range from 10^{-12} to $8{\cdot}10^{-7}$ A with two gain settings. In order to increase stability and minimize the noise, the photodiode is cooled and temperature-stabilized by a Peltier element.

The LCD displays at the rear faceplate of the measuring head indicates the photocurrent expressed as a floating-point number with 5-digit mantissa, temperature calculated from the measured photocurrent, position of the filter wheel, and some other information. Additionally, the LP4 can communicate with a PC via the RS-232 serial or USB interface: receive commands (e.g. to perform measurements or to switch the filter wheel to another position) and send measurement information (e.g. calculate temperature using the previously obtained calibration data or plot the dependences of measured temperature on time). With a 650 nm filter, the LP4 measures temperature from 1000 K and 3000 K with the standard uncertainty from 1 to 6 K and the repeatability of 0.2–1.0 K. The NETD is 0.1 K at 1200 K and 0.02 K at 1700 K.

With a 900 nm filter, the LP4 allows measuring temperatures up to 2000 K. It should be noted that all the data for LP4 presented above can only be considered as indicative. LP4 is a highly customizable instrument; it is often assembled according to customer requirements. Besides, customers sometimes take their own measures aimed at improving the operational and metrological characteristics of the LP4 (see, e.g. [197]).

Another well-established manufacturer of RTs (and other measurement equipment, including blackbodies), CHINO Corp. (http://www.chino.co.jp/english/) offers presently two Si photodiode-based RTs of the highest accuracy: IR-RST65H with the center wavelength λ_0 of 650 nm and IR-RST90H with $\lambda_0 = 900$ nm [27]. These instruments have a similar optical design presented in Fig. 8.19. These RTs have been developed under collaboration of CHINO Corp. and the NMIJ (Japan). The relative spectral responsivities of the Chino IR-RST90H measured by NMIs of Thailand (NIMT) and Japan (NMIJ) are presented in Fig. 8.20.

The objective lens of the IR-RST is composed of three elements. The first two lenses form a doublet, which is separated by 64 mm (more than 70% of the objective focal length) from the third lens. The interference filter is located between the lenses of the collimator, in a beam of parallel rays. The radiation beam is focused on the receiving surface of the Si or InGaAs photodiode.

The temperatures of the detector and the interference filter are controlled at about 30 °C. At the distance of 400 mm, the target size is 0.6 mm for IR-RST65H and 3.0 mm IR-RST90H.

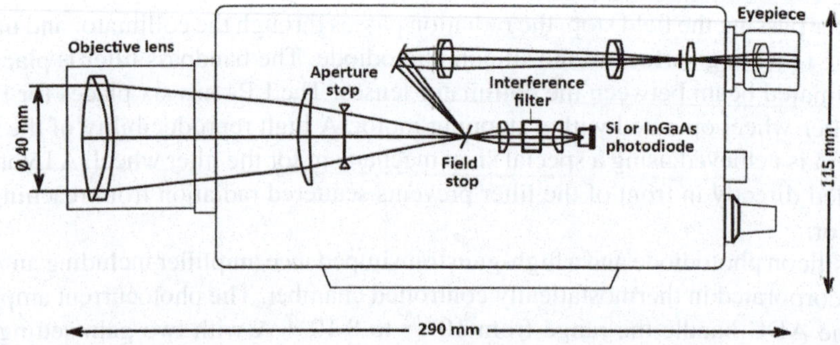

Fig. 8.19 Optical scheme of CHINO IR-RST series RT. Reproduced from [192] with permission of Springer Nature

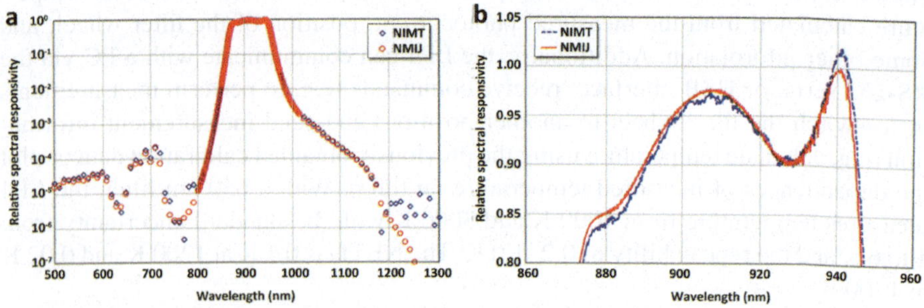

Fig. 8.20 Relative spectral responsivities of the Chino IR-RST90H measured by NIMT (Thailand) and NMIJ (Japan): **a** full logarithmic scale and **b** top of the curve in linear scale. Reproduced from [77] with permission of Springer Nature

The Si photodiode-based IR-RST65H and IR-RST90H allow measuring temperatures from 1000 to 3000 °C and from 400 to 2000 °C, respectively. The NETD for both RTs at 1000 °C is 0.1 °C. The IR-RST65H is more applicable for realization of the ITS-90, since its FOV is compatible with the apertures of FPBBs (usually, 3 mm in diameter). Normally, the IR-RST65H is calibrated at the copper fixed point.

8.3.3 *Temperature Uncertainty of the Planckian Extrapolation*

Comprehensive analysis of uncertainties in measurement of blackbody temperatures by means of radiation thermometry is a complex and time-consuming problem. There is obsolete in some respects but remaining in effect to date the CCT/03-03 document "Uncertainty Budgets for Realisation of Scales by Radiation Thermometry" elaborated by the CCT-WG5[4] on radiation thermometry [47, 48].

The CCT/03-03 document was developed a long time ago. During the past decade, some issues thoroughly developed in the document are no longer relevant, while the others, mainly related to the modern measurement instrumentation and novel measurement methods, came to the fore. The CCT/03-03 document employs the mean effective wavelength approach to evaluate the combined standard uncertainty. This approach was very popular in the precomputer and early computer era, when the problem of the finite bandwidth of RTs was solved by approximate analytical methods. The so-called mean effective wavelength λ_m is defined by the following equation:

$$\frac{\exp\left(\frac{c_2}{\lambda_m T_{90}(X)}\right) - 1}{\exp\left(\frac{c_2}{\lambda_m T_{90}}\right) - 1} = \frac{\int_0^\infty s_r(\lambda) L_{\lambda,bb}(\lambda, T_{90}) d\lambda}{\int_0^\infty s_r(\lambda) L_{\lambda,bb}(\lambda, T_{90}(X)) d\lambda}, \tag{8.42}$$

where the right-hand part is the ratio of signals of an RT with a finite bandwidth. For simplicity, the effective emissivities ratio was assumed equal to unity in Eq. 8.42.

There is no exact solution of this equation with respect to λ_m for any T_{90} and $T_{90}(X)$; the solution λ_m depends on both temperatures T_{90} and $T_{90}(X)$. Nevertheless, it is possible to find the value of λ_m, which satisfies Eq. 8.42 approximately for sufficiently different temperatures T_{90} and $T_{90}(X)$. The detailed derivation of the analytical expression for the mean effective wavelength can be found in Nutter [118]. We restrict ourselves to writing the simple equation:

$$\lambda_m = 2 \Big/ \left(\lambda_{T_{90}}^{-1} + \lambda_{T_{90}(X)}^{-1}\right), \tag{8.43}$$

where $\lambda_{T_{90}}$ and $\lambda_{T_{90}(X)}$ are the so-called limiting effective wavelengths computed for the blackbody temperatures T_{90} and $T_{90}(X)$ using the definition given by Kostkowski and Lee [85]:

$$\lambda_T = \frac{\int_0^\infty s_r(\lambda) L_{\lambda,bb}(\lambda, T) d\lambda}{\int_0^\infty \frac{s_r(\lambda)}{\lambda} L_{\lambda,bb}(\lambda, T) d\lambda}. \tag{8.44}$$

Saunders [146] analyzed uncertainties of the mean effective wavelength approximation applied to calculation of the ITS-90 temperature measured by an RT above the silver point. For the RTs with the center wavelengths from 650 nm to 900 nm, the

---

[4] Now, the CCT Working Group for Non-Contact Thermometry (CCT-WG-NCTh) Task Group on Radiation Thermometry (https://www.bipm.org/en/committees/cc/wg/cct-wg-ncth.html).

error arising due to effective wavelength approximation was found negligibly small up to 2000 °C if the RT bandwidth is narrower than 50 nm and up to 3000 °C if the bandwidth does not exceed 20 nm.

With the widespread use of personal computers, applying the concept of the effective wavelength to solve nonlinear equations of radiation thermometry has become an anachronism.

However, this approach is used until now for uncertainty analysis, since it allows analytical differentiation of the measurement equation. With the mean effective wavelength λ_m calculated via Eqs. 8.43 and 8.44, Eq. 8.42 can be rewritten as

$$r' = \frac{S'(T_{90})}{S'(T_{90}(X))} = \frac{\exp\left(\frac{c_2}{\lambda_m T_{90}(X)}\right) - 1}{\exp\left(\frac{c_2}{\lambda_m T_{90}}\right) - 1}, \tag{8.45}$$

where $S'_{SSE,NL}$ is the corrected for the effective emissivity, nonlinearity, and SSE signal of the RT aimed to a blackbody, whose temperature to be measured; S' is the emissivity-corrected signal of the RT aimed to a reference FPBB (cf. Eq. 8.37). Solution of Eq. 8.45 can be easily found:

$$T_{90} = \frac{c_2}{\lambda_m \ln\left\{\frac{1}{r'}\left[\exp\left(\frac{c_2}{\lambda_m T_{90}(X)}\right) - 1\right] + 1\right\}}. \tag{8.46}$$

To avoid overcomplicated expressions for the sensitivity coefficients, the Wien approximation was used in the CCT/03-03 document for analytical differentiation. Fortunately, the Wien radiation law satisfactorily approximates the Planck law for temperatures above the silver point and wavelengths in the visible and the near-IR spectral ranges; for such conditions, derivatives of these distributions with respect to wavelength and temperature also differ negligibly small (see Sect. 3.2.3).

For instance, in order to evaluate uncertainty in determination temperature above the silver point, a simplified relationship between uncertainties in determination of temperatures $T_{90}(X)$ and T_{90} can be derived assuming validity of the Wien approximation:

$$\frac{L_{\lambda,\mathrm{bb}}^{\mathrm{Wien}}(\lambda, T_{90}(X))}{L_{\lambda,\mathrm{bb}}^{\mathrm{Wien}}(\lambda, T_{90})} = \frac{\exp\left(\frac{c_2}{\lambda T_{90}}\right)}{\exp\left(\frac{c_2}{\lambda T_{90}(X)}\right)}, \tag{8.47}$$

from which

$$u(T_{90}) = \frac{\lambda T_{90}^2}{c_2} \frac{u\left(L_{\lambda,\mathrm{bb}}^{\mathrm{Wien}}\right)}{L_{\lambda,\mathrm{bb}}^{\mathrm{Wien}}} u(T_{90}(X)). \tag{8.48}$$

If to assume that the measurement uncertainty of the spectral radiance does not depend on the spectral radiance level, one can write:

$$u(T_{90}) = \left(T_{90}/\,T_{90}(X)\right)^2 u(T_{90}(X)). \tag{8.49}$$

This widely known expression contains no dependence on wavelength, which reflects a weak influence of the center wavelength of a narrow-band RT on the uncertainty of temperature measurement if λT is well within the domain of the Wien approximation validity.

The CCT/03-03 describes three possible schemes for realization of the ITS-90 by means of radiation thermometry. Two first schemes imply the use of a tungsten strip lamp (TSL) as a transfer measurement device, by which spectral radiance of the blackbody with unknown temperature is compared with the FPBB at the silver, gold, or copper freezing temperature.

In recent years, highly fragile TSLs as transfer standards are being supplanted by radiation thermometers because of their better stability and durability (the latter one is especially important for transportation during international comparisons). Currently, only the third scheme described in the CCT/03-03 remains relevant. It consists in calibration of an RT against a FPBB and then converting the signal of the calibrated RT aimed at the blackbody under measurement into the temperature T_{90}. According to GUM [71], the square of the combined standard uncertainty is presented as

$$u_c^2(\lambda, T_{90}) = \left(\frac{T_{90}}{T_{90}(X)}\right)^4 \sum_i u_i^2 + \left(\frac{T_{90}}{\lambda}\right)^2 \left(\frac{T_{90}}{T_{90}(X)} - 1\right)^2 \sum_j u_j^2 + \left(\frac{\lambda T_{90}^2}{c_2}\right)^2 \left(N^2 u_{NL}^2 + \sum_k u_k^2\right). \tag{8.50}$$

In Eq. 8.50, the summation over i refers to the components associated with the calibration of RT against the FPBB and expressed in units of temperature (due to the impurity of metal, temperature drop across the cavity bottom, freezing plateau identification, and repeatability); summation over j is for components related to the measurements of the relative spectral responsivity of the RT (in terms of the mean effective wavelength); summation over k is made for components expressed via relative uncertainties (due to the effective emissivity of the FPBB, SSE, nonlinearity, drift, gain ratio, repeatability of the RT signals, and variability of its signals due to ambient conditions); u_{NL} is the uncertainty due to nonlinearity of the RT, and N is the number of doubling steps in the nonlinearity measurement

Although, from a formal point of view, the CCT/03-03 remains valid to date, the approach to uncertainty budgeting presented in it does not seem quite consistent. In the latter edition of the guide to the realization of the ITS-90 by radiation thermometry [202], an alternative approach based on the SH equation is recommended.

8.4 Temperature Interpolation and Extrapolation Using Sakuma-Hattori Equation

8.4.1 Sakuma-Hattori Equation

The problem of temperature interpolation using an RT has been intensively developed since 1970s to provide a unified framework for non-contact measurement of medium and low temperatures. To interpolate between reference temperatures, which can be embodied in the FPBB at the fixed points of the defined temperature scale or in precision VTBBs, whose temperatures can be determined by contact methods, it is necessary to construct a suitable interpolant, a function used to generate interpolated temperature values. This interpolant should:

- take into account the finite bandwidth of the RT in the dependence of its signal on the blackbody temperature;
- have the necessary and sufficient number of free parameters, providing the appropriate adjustment for various RTs with different relative spectral responsivities and different sets of reference temperatures;
- allow deriving the adjustable parameters from the results of the RT calibration against the reference temperatures.

Since the 1980s, a lot of interpolants with several adjustable parameters were proposed to provide a sufficient flexibility, e.g. in choice of an RT and a set of fixed points.

The general form for an interpolation equation is

$$S = S(T, a_1, a_2, \ldots, a_n), \tag{8.51}$$

where S is the signal measured by the RT, T is the unknown (current) temperature, and a_i, $i = 1, 2, \ldots, n$ is a set of n parameters. If Eq. 8.51 has a single exact solution for T, the function S passes through each pair (T_i, S_i), $i = 1, 2, \ldots, n$. Then each adjusting parameter is a function of n measured pairs of reference temperatures and corresponding signals of the RT:

$$a_i = a_i(T_1, S_1; T_2, S_2; \ldots; T_n, S_n), \quad i = 1, 2, \ldots, n. \tag{8.52}$$

Thus, Eq. 8.51 can be rewritten as

$$S = S(T; T_1, T_2, \ldots, T_n; S_1, S_2, \ldots, S_n). \tag{8.53}$$

he most influential interpolant has been proposed by Sakuma and Hattori [135] in the form that indicates its origin from the Wien approximation of the Planck law:

$$S(T) = C \exp\left(-\frac{c_2}{AT + B}\right), \tag{8.54}$$

where c_2 is the second radiation constant; A, B, and C are adjustable parameters. It is easy to see that Eq. 8.54 uses an approximate expression for the mean effective wavelength in the denominator of the exponent:

$$\lambda_m = A + B/T. \tag{8.55}$$

Sakuma and Kobayashi [136] compared a number of various interpolants, including Eq. 8.54, and proposed the so-called Planck version of the SH equation:

$$S(T) = \frac{C}{\exp\left(\frac{c_2}{AT+B}\right) - 1}. \tag{8.56}$$

If temperature measurements of blackbodies using RTs are carried out in atmospheric air, it is enough to employ $c_2' = c_2/n$, where n is the refractive index of the air in the optical path, instead of c_2 in Eq. 8.56.

Despite the fact that Eq. 8.56 was introduced by Sakuma and Kobayashi [136], it continues to be called the SH equation because of the long history of the use of its Wien's form (8.54). Sometimes, the refined name "SH equation in the Planck form" is used for Eq. 8.56. This equation has gained great popularity in radiation thermometry below the silver point, as reflected in the relevant CCT guide [49, 155].

The inverse SH equation can be obtained by solving (8.56) for the temperature T:

$$T(S) = \frac{c_2}{A\ln(C/S+1)} - \frac{B}{A}. \tag{8.57}$$

The parameters A, B, and C should be found from the RT calibration at the available FPBBs. Saunders [145] proposed a systematic approach to derivation and analysis of the SH equation in the Wien and Planck forms, as well as their relationships to the mean effective wavelength concept. This analysis led to a better understanding of physical basis of the SH equation and allowed Saunders and White [152] to found a simple physical interpretation of the parameters A, B, and C. They showed that, for sufficiently narrow bandwidths, these parameters depend only slightly on the shape of the spectral responsivity curve. If to consider the spectral responsivity $s(\lambda)$ as a distribution function (non-normalized), these parameters can be expressed via its first moment μ_1 (the mean value λ_0) and the second central moment μ_2 (the variance σ_λ^2) of the distribution $s(\lambda)$:

$$\mu_1 = \lambda_0 = \int_0^\infty \lambda s(\lambda)d\lambda \bigg/ \int_0^\infty s(\lambda)d\lambda = \int_0^\infty \lambda s_r(\lambda)d\lambda, \tag{8.58}$$

$$\mu_2 = \sigma_\lambda^2 = \int_0^\infty (\lambda - \lambda_0)^2 s(\lambda)d\lambda \bigg/ \int_0^\infty s(\lambda)d\lambda = \int_0^\infty (\lambda - \lambda_0)^2 s_r(\lambda)d\lambda, \tag{8.59}$$

where s_r is the relative spectral responsivity of the RT. It is known that the distribution function can be defined uniquely by the infinite set of its moments, if the function $s_r(\lambda)$ is nonzero only on a finite segment (this is so-called moment problem—see, e.g. [168]). If only a finite number of moments is known, the function $s_r(\lambda)$ can be recovered only approximately [172]. Saunders and White [152] showed that the condition $\sigma_\lambda / \lambda_0 \ll 1$ is sufficient to achieve an acceptable accuracy in the majority of practical cases, if only information on the first two moments is used instead of actual relative spectral responsivity $s_r(\lambda)$. For practical purpose, it is convenient to relate variances σ_λ^2 to the FWHM for most common models of spectral responsivity. Three symmetrical distribution shown in Fig. 8.21 have the same variance. The mean wavelength λ_0 of a symmetrical spectral responsivity curve coincides with its center wavelength (median) and the central moments of odd orders are equal to zero.

The FWHM $\Delta\lambda$ for distributions presented in Fig. 8.21 can be easily expressed via the standard deviation σ_λ:

$$\Delta\lambda = \begin{cases} 2\sigma_\lambda\sqrt{3} \approx 3.464\sigma_\lambda & \text{(rectangular)} \\ \sigma_\lambda\sqrt{6} \approx 2.449\sigma_\lambda & \text{(triangular)}. \\ 2\sigma_\lambda\sqrt{2\ln 2} \approx 2.355\sigma_\lambda & \text{(Gaussian)} \end{cases} \tag{8.60}$$

By expanding parameters A, B, and C in the series in powers of the relative bandwidth $\sigma_\lambda / \lambda_0$ and neglecting the terms of power greater than two, Saunders and White [152] obtained the simple approximate expressions for the constants A, B, and C:

$$A = \lambda_0\left[1 - 6(\sigma_\lambda / \lambda_0)^2\right], \tag{8.61}$$

$$B = c_2(\sigma_\lambda / \lambda_0)^2 / 2. \tag{8.62}$$

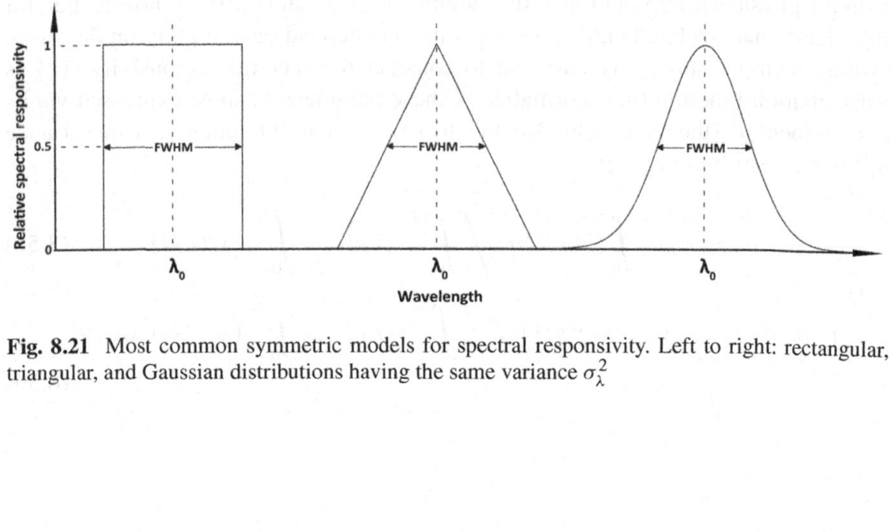

Fig. 8.21 Most common symmetric models for spectral responsivity. Left to right: rectangular, triangular, and Gaussian distributions having the same variance σ_λ^2

We omit the expression for C obtained by Saunders and White [152], because of its more complicated expression compared with equation for A and B. Parameters A, and B computed via (8.61) and (8.62) are too crude estimates to be used directly in the calibration equation, but they can serve as initial guesses for the nonlinear fitting software. In practice, there is no need to use the omitted expression for C to calculate the initial guess for the parameter C. It can be found using values of initial guesses $A^{(0)}$ and $B^{(0)}$, and the measured signals S_i corresponding to the reference temperatures T_i, $i = 1, 2, \ldots, n$ as

$$C^{(0)} = \frac{1}{n}\sum_{i=1}^{n} S_i \cdot \left[\exp\left(\frac{c_2}{A^{(0)}T_i + B^{(0)}}\right) - 1\right]. \tag{8.63}$$

Equations 8.56 through 8.59, 8.61, and 8.62 comprise the mathematical framework for solution and uncertainty analysis for typical schemes of temperature measurements in radiation thermometry.

The SH equation is widely utilized in the calibration and use of IR RTs measuring the temperatures below the silver point, where the ITS-90 provides a fairly dense set of fixed points.

The SH technique is described in detail in the document issued on behalf of the CCT[5] [49] and in the accompanied article by Saunders et al. [155]. The RTs of the medium- and low-temperature ranges are used to solve various tasks in science, industry, and medicine. However, these RTs are applied to measuring the temperature of VTBBs only sporadically, in very rare situations where the use of contact sensors is impossible or undesirable.

Until recently, the SH equation was widely used only in radiation thermometry below the silver point, where the sufficiently dense set of fixed points exist. For temperatures above the silver point, the SH equation was almost useless until the discovery of high-temperature fixed points (HTFPs) and their introducing in the practice of thermal metrology through the *Mise en Pratique*[6] document on the RPRT released by the CCT [99]. The HTFPs provides temperature references above the freezing point of copper (1357.77 K), which makes possible to use temperature interpolation instead of extrapolation prescribed by the ITS-90 when high temperatures are measured. Although the RPRT deals with thermodynamic temperatures, it uses the same mathematical apparatus, including the SH equation, as in the guideline for calibration of RTs below the silver point [49, 155].

Since the original text of ITS-90 provides the only way to measure temperature above the silver point—single-point extrapolation from the silver, gold, or copper point, the range of applicability of the SH equation was limited mainly by low-temperature radiation thermometry, where interpolation between fixed points is possible. As we already noted, the situation changed dramatically after discovery

[5]Consultative Committee for Thermometry of the BIPM (International Bureau of Weights and Measures; French: Bureau international des poids et mesures).

[6]Mise en pratique (French)—practical realization. A collective name for a series of updatable regulatory documents.

of HTFPs on the base of metal-carbon eutectics [189, 190], metal carbide-carbon eutectics [144] and peritectics [191]. The question of possible incorporation of some HTFPs in the future temperature scale arose almost simultaneously with their discovery. Currently, the thermodynamic temperatures of solid-to-liquid phase transitions of Co–C, Pt–C, and Re–C eutectic alloys were recommended by the BIPM as the fixed points for measuring temperature up to ~2748 K [98, 99]. In this way, in connection with the recent revision of the SI, redefinition of the kelvin [95], and the release of the document "*Mise en Pratique* for the Definition of the Kelvin in the SI" [24], the scope of the SH equation has expanded. It introduces the RPRT as the indirect method of measuring the thermodynamic temperature using a specified equation of state, with one or more key-parameter values determined from n temperature fixed points, for which values for the thermodynamic temperature T and their uncertainties are known from previous APRT or RPRT.

The unified approach to the analysis of uncertainty propagation using GUM for thermodynamic temperature dissemination above the silver point was most consistently and completely described by Saunders [149]. This approach is also presented in the CCT document CCT/10-12/rev1 [97].

8.4.2 Uncertainty Components Related to Fixed Points

As we already mentioned, the RPRT and conventional radiation thermometry employ similar measuring instruments and measurement methods. The main difference in these techniques is the temperature scale used. The radiation thermometry deals with temperature T_{90}, while the RPRT operates with thermodynamic temperature T. The values $T_{90}(X)$ of fixed points in the ITS-90 has zero uncertainties by definition; uncertainties arise only due to imperfect conditions of their practical realization. The same fixed points have finite uncertainties in their thermodynamic temperatures obtained from the APRT ($n = 0$). To these uncertainties, should be added the uncertainties due to fixed point realization, components of which are the same as for the ITS-90 with conventional radiation thermometry. The SH equation can be used both in radiation thermometry and in RPRT and they use the same values of uncertainty components, with the exception of the uncertainties associated with fixed points.

A variety of implementation schemes in RPRT is conveniently classified according to the number n of reference temperatures (fixed point) used. The case of $n = 0$ corresponds naturally to the APRT, when the complete determination of the absolute spectral responsivity and no measurements at fixed points are required. Recently, the CCT released the guide on the uncertainty evaluation in APRT [158].

If $n = 1$, we deal with the single-point extrapolation. In most practical cases, extrapolation is started from the Ag, Au, or Cu fixed points, similar to that in the ITS-90.

Presently, when at least three HTFPs (melting temperature of Co–C, Pt–C, and Re–C eutectic alloys—see Appendix B) received assigned thermodynamic temperatures, corresponding uncertainties, and legal status in the RPRT [99], we can expect that the importance of extrapolation approach will decrease as the HTFP cells and high-temperature furnaces become easily accessible to measurement laboratories.

The SH equation throws the bridge between radiation thermometry in its classical version (with traceability to the ITS-90) and primary radiometric thermometry (with traceability to the SI). For extrapolation ($n = 1$) and interpolation ($n \geq 2$), the SH equation offers a unified approach to measuring both temperatures T_{90} and thermodynamic temperature T. It should be noted that the importance of measuring T_{90} above the silver point is steadily declining, especially in high-temperature blackbody radiometry, where the use of T_{90} instead of T may lead to unacceptably high errors. However, the measurement (SH) equation is the same for both temperatures; therefore, the sensitivity coefficients are the same, and the difference is only in the values of the uncertainty components. Hereafter, we will not distinguish T and T_{90}, except where necessary.

In contrast to the ITS-90 Planck extrapolation, which operates with relative spectral responsivity, for the SH equation, it is enough to know λ_0 and σ_λ, which can be determined from the first two moments of the relative spectral responsibility (see Eqs. 8.58 and 8.59).

Application of the SH equation consists of two phases. The first phase, often referred to as "realization" is determination of SH parameters A, B, and C, which can be considered as calibration of the RT. The second, "in-use", phase is calculation (by extrapolation or interpolation) of unknown temperature using Eq. 8.57.

A part of "realization" uncertainties is related to the reference temperatures (fixed points). They are the products of the uncertainties in $u(T_i)$, $i = 1, \ldots, n$ and the corresponding sensitivity coefficients $\partial T / \partial T_i$, $i = 1, \ldots, n$. Each $u(T_i)$ consists of similar components listed in Table 8.1 for the case of thermodynamic temperature.

Please note that the numerical data in Table 8.1 correspond to the best accuracy level achievable for the thermodynamic temperatures to the date of publication of [149]. The uncertainty set for the thermodynamic temperature of the Au point and HTFPs corresponds to the value of the radiance temperature derived from the

Table 8.1 Standard uncertainty components (in mK) related to thermodynamic temperatures of fixed points used for calculations of combined standard uncertainties (after [149])

Uncertainty component due to	Fixed point (°C)		
	1064.22 (Au)	1738 (Pt–C)	2749 (WC–C)
APRT	33	90	200
Impurities	5	10	20
Cavity bottom temperature drop	2	3	14
Plateau identification	1	20	10
Furnace effect	0	45	100
Repeatability	5	5	5
Total	34	100	230

APRT with the corrections associated with the impurities in the fixed-point material, temperature drop across the cavity wall, and plateau identification, which are needed to convert the radiance temperature to the true temperature. Below, we briefly comment the selected uncertainty components listed in Table 8.1.

Impurities The processes of melting and freezing of metals and alloys are well studied, at least at the phenomenological level. For definiteness, we consider extrapolation from a metal fixed point, which corresponds to the most common practical case. Phase transitions between the solid and liquid states of pure metals proceed at constant temperatures. This phenomenon constitutes the thermodynamic basis for the defining fixed points of the ITS-90.

The radiating fixed-point cells typically contain from 0.5 to 1.5 kg to provide a sufficient freezing time [107].

The presence of impurities leads to decrease of freezing temperature with respect to the freezing-point temperature of the pure material and distortion of the freezing curve (dependence of temperature on time). The ingots of Ag, Au, and Cu usually contain metals with a purity between 99.999% (5 N) and 99.9999% (6 N); there is no experimental evidence of fixed-point temperature changes when using a purer metal.

Typical freezing curves for metal at three purity levels (say, 5 N or 99.999% or 10 μg/g, 5 N+ or 99.9995%, and 6 N or 99.9999%) are presented in Fig. 8.22. The purer the metal, the flatter the freezing plateau, the higher the freezing curve, and the deeper the supercooling (cooling the metal below the freezing temperature without solidification due to the lack of solidification centers). A calculation of the depression of the freezing-point temperature is extremely complicated problem because it depends on the type of impurities and their arrangement in the crystal lattice of the metal, their concentration, distribution inside the ingot, and many other factors. The CCT has published a list of common impurities for metallic fixed-point materials of the ITS-90 [130]. A simplest technique for assessment of the impurity effect consists in representation of a slightly impure metal as a dilute alloy and analysis of an equilibrium binary phase diagram for a dominant impurity at a low concentration. A detailed analysis of the effects of impurities can be found in [44] and in the references therein.

Currently, thermodynamic and thermochemical databases and modeling software such as MTDATA developed by the NPL [31, 123, 124] or Thermo-Calc

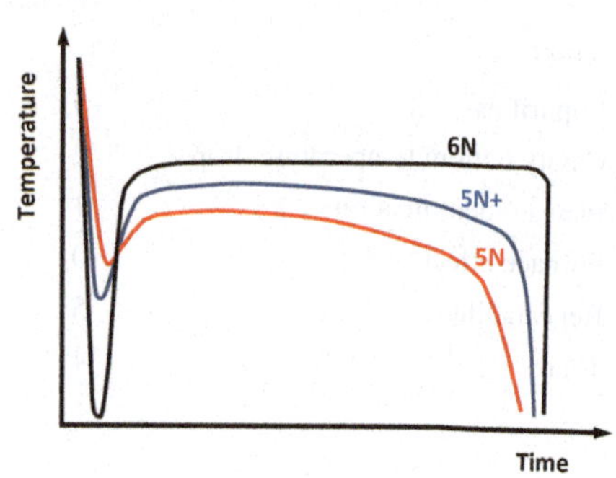

Fig. 8.22 Typical freezing curves of metals of three different levels of purity

from Thermo-Calc Software AB (Solna, Sweden, https://www.thermocalc.com/) are increasingly used for calculation of phase diagrams. According to assessment adopted in [47], the standard uncertainty (of type B) due to impurity of the fixed-point materials is less than 10 mK for Ag, Au, and Cu.

Temperature Drop This effect is similar to that described in Sect. 7.3.2. Even if the outer surface of the FPBB radiating cavity has a constant temperature equal to the freezing temperature of the metal, but there is heat transfer from the inner surface, its temperature will differ from the temperature of the outer surface due to the finite thermal conductivity k of the cavity wall.

At temperatures above the freezing point of Ag, the dominant mechanism of heat transfer is radiation. The loss of radiant flux through the aperture (it depends on the distance to the cavity aperture) produces a difference between the average temperature of the target area of the cavity and temperature of its outer surface. To assess the temperature drop, Fischer and Jung [45] proposed the following formula:

$$\Delta T \approx \varepsilon(T)\sigma T^4 \frac{\delta}{k}\left(\frac{r_a}{L}\right)^2, \tag{8.64}$$

where ΔT is the temperature drop, $\varepsilon(T)$ is the total emissivity of graphite, σ is the Stefan-Boltzmann constant, T is the temperature in Kelvin, δ is the thickness of the cavity bottom, k is the thermal conductivity of graphite, r_a is the aperture radius, and L is the cavity length.

Equation 8.64 is written for the radiative heat loss from a flat surface through the cavity opening.

However, Eq. 8.64 is very approximate because the actual temperature field in the crucible is two-dimensional, while δ / k is the thermal resistance for the one-dimensional heat flux through a flat wall. A factor $(r_a / L)^2$ is used instead of a diffuse view factor between a surface element on the cavity bottom and the cavity opening. Besides, the total emissivity $\varepsilon(T)$ takes into account only direct radiative heat loss through the cavity aperture but does not consider indirect heat loss after multiple reflection.

The solution of the steady-state heat equation (Laplace's equation) for the 2D temperature field together with the constant temperature boundary condition on the outer surface of the cavity and the radiative boundary conditions on its internal surface. An example of such a solution was given by Fuksov et al. [51]. They modeled a crucible of the copper-point FPBB using the finite element analysis software. One half of the longitudinal section of the axisymmetrical crucible with the calculated temperature field superimposed is shown in Fig. 8.23. The authors of this work have solved about 80 variants of this problem with various geometrical and thermophysical parameters. The results presented in Fig. 8.23 correspond to the conditions close to the real experiment: the front ring-shaped insert had a temperature $T_{Cu} - 0.9$ K, the back solid insert had a temperature $T_{Cu} + 0.1$ K, the emissivity was $\varepsilon = 0.88$, and the thermal conductivity of graphite $k = 46$ W m^{-1} K^{-1}. Calculations allow obtaining the temperature drop across the cavity bottom as 7 mK and the decrease in temperature towards the cavity aperture, excluding the front 10 mm, as 10 mK. The relative

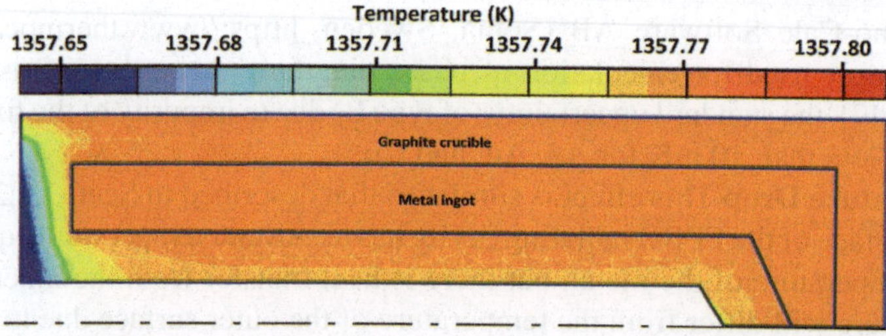

Fig. 8.23 Results of thermophysical modeling of the temperature field in the crucible at the copper freezing temperature. Reproduced from [51] with permission of Springer Nature

uncertainty of the numerical calculation of the temperature field is estimated as 35% and depends mainly on the uncertainty of the thermal conductivity of graphite.

Similar studies have been undertaken in recent decades in connection with studies of high-temperature fixed points [15, 16, 23, 96]. It is also worth noting an attempt to experimentally determine the temperature drop in the cavity of the FPBB undertaken by Ma [93].

Taking into account the contribution due to inaccurate knowledge of the parameters included in Eq. 8.64, Fischer et al. [47] concluded that the contribution of temperature drop to the uncertainty budget is usually limited to a few millikelvins. This estimate is in good agreement with data for Au presented in Table 8.1.

Plateau Identification The uncertainty due to identification of the freezing plateau is closely related to the freezing curve shape. The modern high-temperature furnaces with multiple zone temperature control and heat-pipe isothermal liners allow identification of the freezing temperature from the freezing curve with the uncertainties of 1 millikelvin or less. For HTFP, the uncertainty due to identification of point of inflection (PoI) on the melting curve is used instead.

Furnace Effect It manifests itself in different melting curves obtained for the same fixed-point cell in different furnaces, even if the same temperature schedule is used. The effect of the furnace was observed both for metals and for HTFPs, but for HTFPs, it is more pronounced [18, 36, 67]. The reason for this effect has not yet been explained.

Repeatability The repeatability of freezing temperature is also determined mainly by the controllability and uniformity of the temperature field in the furnace used for the fixed points realization. For Ag, Au, and Cu fixed points, the typical repeatability of 5 mK is achievable with modern furnaces, with the typical duration of freezing plateaux from 30 to 100 min.

8.4.3 Uncertainty Components Related to Measured Signals

Another category of uncertainties that always should be accounted for in the uncertainty budget at the SH extrapolation and interpolation consists of the *n* "realization" components $u(S_i)$, $i = 1, \ldots, n$, where S_i is the signal of the RT aimed at the FPBB having the temperature T_i. The "in-use" uncertainty $u(S)$ of the signal of the RT, aimed at the blackbody with unknown temperature, stands apart: its components differs from those of the "realization" uncertainties. The relative uncertainties $u_r(S_i) = u_r(S_i)/S_i$ and $u_r(S) = u_r(S)/S$ for all these components are given in Table 8.2. Below, we will briefly comment on data in Table 8.2.

Emissivity The effective emissivities are specified only for FPBBs, which are used as temperature references at the SH extrapolation and interpolation. The temperature of the blackbody under measurement is determined as the radiance temperature. It the effective emissivity of the blackbody under measurement is known, the radiance temperature can be easily converted in the true temperature.

The effective emissivity of the radiating cavity of an FPBB is determined by the following factors: radiation characteristics of the cavity walls, geometric characteristics of the cavity, temperature distribution over the radiating surface, geometrical conditions of cavity observation with the RT. These factors (together with that due to imperfectness of cavity manufacturing) were analyzed in the CCT/03-03 [47] but this was done on the base of a very primitive expression for the effective emissivity of a diffuse isothermal cylindro-conical cavity.

Its useful to note here that the methodology for estimating the corresponding error has not yet been developed, despite the fact that the capabilities of computer modeling of effective emissivities have grown immeasurably over the past two decades.

Now, we are able to model nonisothermal cavities with internal walls, reflection from which is described by the BRDF experimentally determined for graphite samples at the wavelength close to the center wavelength of the RT. Patrick et al. [122] presented the BRDF of graphite samples measured at room temperature for the wavelength of 658 nm, which is very close to typical center wavelengths of RTs used for realization of the ITS-90 above the silver point (see Fig. 6.37b in Chap. 6).

Table 8.2 Uncertainty components related to measured signals used for calculations of combined standard uncertainties (after [149])

Uncertainty source	Uncertainty subcomponent	Relative uncertainty (%)
"Realization" signals S_i	Effective emissivity of FPBB	0.01
	Ambient conditions	0.002
	Repeatability	0.003
	Total S_{ref}	0.011
"In-use" signal S	SSE	0.01
	Non-linearity	0.005
	Gain ratios	0.01
	Ambient conditions	0.002
	Drift	0.01
	Repeatability	0.003
	Total in-use S	0.018

However, there are no similar studies conducted at high temperatures; in fact, we do not know how heating to high temperatures affects the reflective properties of graphite.

The least information we have regarding the temperature distribution over the cavity radiating surface. Currently, there is no experimental technique that allows measuring temperature distributions inside the radiating fixed-point cells, which usually have very small apertures (3–6 mm in diameters). In principle, the temperature distribution can be calculated on the base of a steady-state 2D thermophysical model and commercial finite element software. This problem has a direct relation to the calculation of the temperature drop between the radiating surface of a cavity and its outer surface interfaced with a liquid metal. Moreover, from mathematical point of view, this is the same problem.

The CCT/03-03 [47] provides the value $u(\varepsilon_{eff}) = 0.002\%$ for the effective emissivity$\varepsilon_{eff} = 0.99997$ for the "best" accuracy level. However, in view of the foregoing, the values $\varepsilon_{eff} = 0.9999$ and $u(\varepsilon_{eff}) = 0.01\%$, specified in Saunders [149], seems to be more realistic.

Ambient Conditions The ambient temperature affects the output signal of a 0.65 μm RT, while the effect of relative humidity is negligible. If the change in the detector temperature is about 1 °C, the residual error in the corrected value of the RT signal has the order of 0.002%. The same uncertainties should be applied to both signals, S_{ref} and S.

Repeatability There are three principal sources of uncertainties in calibration of RTs in terms of relative spectral responsivity:

- Imperfect reproducibility of geometrical conditions at calibration and measurement. The inclination of the incident beam shifts the center wavelength of the interference filters; the real detectors have not perfect angular and spatial uniformity of the responsivity. Therefore, the RT must be calibrated in the same geometry as it is used.
- The center wavelength of the RT is sensitive to ambient temperature and, therefore, should be held at a constant temperature for both calibration and measurement. Generally, the peak transmission of the filter shifts to longer wavelengths with increasing temperature. Uncontrolled temperature variations may lead to the corresponding subcomponent of uncertainty in the mean effective wavelength.
- Temperature-induced changes in the detector spectral responsivity. For the RT operating at 900 nm, changes in the output signal with temperature are minimal since the photodiode works near the flat portion of its spectral responsivity. At shorter wavelengths, the signal decreases with increasing ambient temperature since the center wavelength of the filter shifts to the shorter wavelength. Major effects occur at wavelengths beyond the peak sensitivity, where detectors are very sensitive to changes in ambient temperature.

Size-of-Source Effect Correction for the SSE is required in order to make the ITS-90 realization independent of the characteristics of the specific RT used and allow scale comparisons among different NMIs. For more detail, see Sect. 8.2.4. Uncertainty due to the SSE consists of three components:

- uncertainty due to the measurement of the SSE for the RT, which includes subcomponents related to (i) the imperfectness of uniformity, stability, and other characteristics of the radiation source used in the SSE measurements, (ii) particular technique of the SSE measurement (direct, indirect, etc.), and (iii) mathematical procedures of introducing the SSE correction;
- drift of the SSE characteristics that is possible mainly due to optical elements contamination;
- radial temperature distribution in the surroundings of the fixed-point cavity.

Uncertainty due to the SSE is applied only to the signal S, which reflects the relative character of the SSE.

Nonlinearity Deviation from linearity caused by the non-ideal performance of the detector and electronics, were considered in Sect. 8.2.3. The numerical values (0.005%) given in Table 8.1 corresponds to the flux doubling method common for the RTs of visible and NIR spectral ranges. Uncertainty due to the nonlinearity is applied only to the signal S.

Gain Ratio To cover the temperature range from the fixed points to a highest temperature measured, different amplification gains are often required. Since the actual value of gain ratio at two adjacent ranges may differ from its nominal value. The corresponding uncertainty originates from the uncertainty in the measurement of gain ratios and in its possible drift. Uncertainty due to the gain ratio is applied only to the signal S.

8.4.4 *Uncertainty of Single-Point Sakuma-Hattori Extrapolation*

The single-point extrapolation scheme ($n = 1$) employed by ITS-90 requires knowledge of the relative spectral responsivity of the RT. Using the SH equation allows one to avoid numerical integration using only two first moments (Eqs. 8.58 and 8.59) or even not using information about the relative spectral responsivity. The measurements with the single reference point T_1 are modeled as the ratio of the SH expressions written for temperatures T and T_1:

$$\frac{S}{S_1} = \frac{S(T)}{S(T_1)} = \frac{\exp(AT_1 + B) - 1}{\exp(AT + B) - 1}. \tag{8.65}$$

Equation 8.65 can be easily solved for T:

$$T = \frac{\ln\left[\frac{S_1}{S}\frac{1}{\exp(AT_1+B)-1} + 1\right] - B}{A}, \tag{8.66}$$

where A and B are calculated using Eqs. 8.61 and 8.62; λ_0 and σ_λ are determined from the relative spectral responsivity measurements.

To calculate the sensitivity coefficients, Saunders [149] rewrote Eq. 8.65 in the Wien form, employed the approximate Eqs. 8.61 and 8.62, found the partial derivatives, and, after neglecting the terms of the order of smallness more than one for the relative bandwidth $\sigma_\lambda / \lambda_0$, obtained:

$$\frac{\partial T}{\partial \lambda_0} = \frac{\partial S}{\partial \lambda_0} \Big/ \frac{\partial S}{\partial T} \approx \frac{T}{\lambda_0}\left(1 - \frac{T}{T_1}\right), \tag{8.67}$$

$$\frac{\partial T}{\partial \sigma_\lambda} = \frac{\partial S}{\partial \sigma_\lambda} \Big/ \frac{\partial S}{\partial T} \approx \left(\frac{1}{T_1} - \frac{1}{T}\right)\left[12 - \frac{c_2}{\lambda_0}\left(\frac{1}{T_1} + \frac{1}{T}\right)\right]\frac{T^2 \sigma_\lambda}{\lambda_0^2}, \tag{8.68}$$

$$\frac{\partial T}{\partial T_1} = \frac{\partial S}{\partial T_1} \Big/ \frac{\partial S}{\partial T} \approx -\frac{T^2}{T_1^2}, \tag{8.69}$$

$$\frac{\partial T}{\partial S_1} = \frac{\partial S}{\partial S_1} \Big/ \frac{\partial S}{\partial T} \approx \frac{\lambda_0 T^2}{c_2 S_1}, \tag{8.70}$$

$$\frac{\partial T}{\partial S} = 1 \Big/ \frac{\partial S}{\partial T} \approx \frac{\lambda_0 T^2}{c_2 S}, \tag{8.71}$$

For signals, the relative uncertainties are specified in Table 8.2. Therefore, the expressions for the sensitivity coefficients in (8.70) and 8.71) should be transformed:

$$\frac{\partial T}{\partial S_1} u(S_1) \approx \frac{\lambda_0 T^2}{c_2} u_r(S_1), \tag{8.72}$$

$$\frac{\partial T}{\partial S} u(S) \approx \frac{\lambda_0 T^2}{c_2} u_r(S). \tag{8.73}$$

The combined standard uncertainty is calculated as the GUM [71] prescribes (no correlations between the components is assumed):

$$u_c(T) = \left\{\left[\frac{\partial T}{\partial \lambda} u(\lambda_0)\right]^2 + \left[\frac{\partial T}{\partial \sigma_\lambda} u(\sigma_\lambda)\right]^2 + \left[\frac{\partial T}{\partial T_1} u(T_1)\right]^2 + \left[\frac{\partial T}{\partial S_1} S_1 \frac{u(S_1)}{S_1}\right]^2 + \left[\frac{\partial T}{\partial S} S \frac{u(S)}{S}\right]^2\right\}^{\frac{1}{2}}. \tag{8.74}$$

To calculate the combined standard uncertainty, we need to know uncertainties in determination of variables λ_0 and σ_λ derived from the relative spectral responsivity.

The relative spectral responsivity of an RT depends on the relative spectral responsivity of the detector, the spectral transmittance of the interference filter and other filters (if any), and the spectral transmittance of lenses and any other optical components. The dominant role in the determination of the spectral responsivity of the RT plays the spectral transmittance of the interference filter because the spectral responsivity of the photodiode and the spectral transmittance of optical components vary slowly with the wavelength.

To account for multiple reflections inside the RT, it should be calibrated as a whole. This can be done using a monochromator, a tunable laser and optical parametric oscillator source (Schuster et al. [162], Woodward et al. [186], or a monochromator equipped with a supercontinuum laser source (Hong et al. [63], Yoo et al. [195], Yoo et al. [197], Woodward et al. [186]. Using a monochromator with an incandescent lamp is the simplest, least expensive, and most common method (see Fig. 8.24), although the least accurate. The measurements are carried out by substitution method [88], when the RT under measurement and the reference detector are alternately irradiated with quasi-monochromatic radiation from a monochromator at each wavelength. The reference detector does not have to be calibrated, but must have a spectrally invariant or at least predictable relative spectral responsivity. Typically, blackened pyroelectric detector or thermopile is used as a reference detector. Better results can be achieved with a photodiode previously characterized in term of relative spectral responsivity due to its better SNR. Even better is to use a trap detector (TD) consisting of several silicon photodiodes, due to its highly predictable spectral responsivity. In order to eliminate the effect of radiant flux fluctuations, a beamsplitter and monitor detector, which are assumed to be highly stable over the measurement time, are used. The monitor detector watches the radiant flux fluctuations, so the ratio of the RT to monitor detector signals will be free of fluctuations.

The specific uncertainties are caused by the spectral resolution of the monochromator. The measured profile of the spectral responsivity is the convolution of the true profile and the apparatus function of the monochromator. In practice, this effect leads to broadening of the spectral responsivity profile and loss of its fine structure. Effect of bandwidth in spectral responsivity measurements using a monochromator was discussed theoretically by Kostkowski [86] and studied experimentally by Boivin [17]. In particular, the spectral resolution affects strongly on the account for the out-of-band transmittance of the interference filter.

The spectral responsivity of the RT outside the nominal passband should be zero, otherwise a systematic error appears. The grating monochromator with the incandescent lamp as a radiation source allows measuring the relative spectral responsivity within the dynamic range of 5 decades. This range can be obtained as a compromise between the level of the radiant flux at the exit slit and the slit width, which determines the spectral resolution and, therefore, the smoothing of the fine structure in

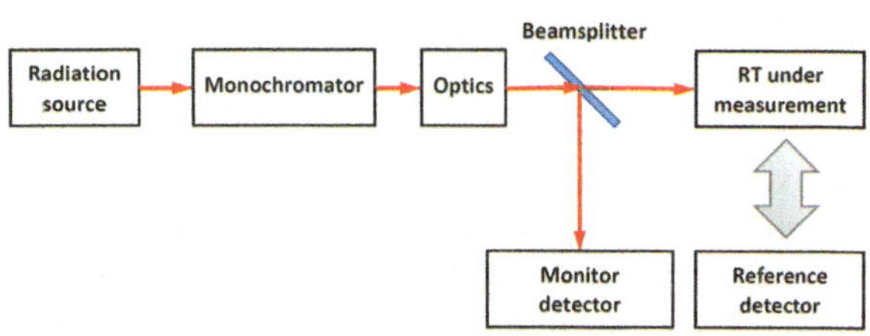

Fig. 8.24 Schematic for measuring relative spectral sensitivity

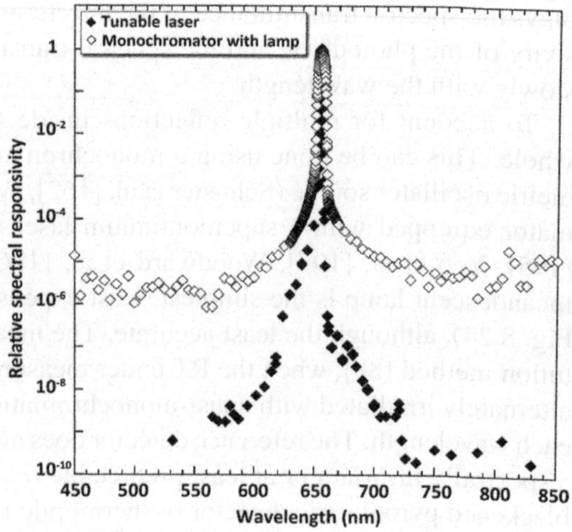

Fig. 8.25 Relative spectral responsivity of the RT measured on the laser-based and the monochromator-based facilities at NIST. Reproduced from [41] with permission of Elsevier

the spectrum. A wider dynamic range can be achieved with tunable lasers, which provide a narrower bandwidth and a higher level of radiant flux than a lamp. An example of relative spectral responsivity measured by both techniques is given in Fig. 8.25. Measurements of the relative spectral responsivity of the PEP (PhotoElectric Pyrometer) were carried out using the monochromator-based facility SCF (spectral comparator facility) and laser-based facility for spectral irradiance and radiance responsivity calibrations using uniform sources (SIRCUS).

Measurements of the out-of-band relative spectral responsivity within such a large dynamic range are complicated, time-consuming, and available only to a few NMIs. However, the improvement of the out-of-band blocking from 10^{-5} to 10^{-7} decreases the corresponding temperature error to 0.2 mK at the Ag point. A larger dynamic range in measurements of out-of-band blocking may be required for much higher temperatures.

Scattering inside the monochromator can lead to the appearance of radiation with different wavelengths at the exit slit. This radiation may be detected by a wideband reference detector, but not by the narrowband RT. When a single monochromator is used, the stray radiation allows measuring the out-of-band blocking of the interference filter to 10^{-5}, while using a double monochromator this level can be reduced down to about 10^{-7}. It is highly desirable to minimize polarization of the source and polarization sensitivity of the RT. As we already mentioned, blackbody radiation is unpolarized; however, transfer standard lamps and calibration sources can exhibit some degree of polarization. The polarization sensitivity of the detector can be reduced by including as few optical elements as possible, ensuring that those elements are in a stress-free mounting, and illuminating all optical components normally and on axis. The polarization of calibration sources can be minimized by introducing radiation from the exit slit of the monochromator or laser into an integrating sphere.

The spectral responsivity of an RT changes over time due to ageing of the interference filter and, to a lesser extent, of the detector. This drift may vary over time, and can occasionally exhibit sudden jumps of up to 1% in the spectral responsivity. Filter ageing is difficult to predict since it is very dependent on the ambient conditions at the use and storage. As a rule, the center wavelength of an interference filter increases by 0.06 nm per year, but changes are also observed up to 0.2 nm per year. Therefore, the RTs should be calibrated regularly. It is highly desirable to measure the blackbody temperature with more than one RT or with an RT that has more than one independently characterized wavelength.

Saunders [149] assessed the total uncertainties in λ_0 and σ_λ at the laser-based measurement of the relative spectral responsivity of the RT with silicon photodiode, $\lambda_0 = 650$ nm, and $\sigma_\lambda / \lambda_0 = 0.01$. His estimates are $u(\lambda_0) = 0.014$ nm and $u(\sigma_\lambda) = 0.00014$ nm.

Figure 8.31 presents the dependences of each uncertainty component (the product of the standard uncertainty and absolute value of the corresponding sensitivity coefficient) and the combined standard uncertainty for T computed using Eq. 8.74. The most influencing component is due to the reference temperature T_1. The uncertainty component due to σ_λ is negligibly small on the scale of the graph in Fig. 8.26.

Saunders [149] performed also calculations for T_{90}. Because of much higher importance of the thermodynamic temperature for the blackbody radiometry, we will not discuss these results but only recall that the difference between T and T_{90} increases proportional to T^2, as the sensitivity coefficient (8.69) predicts. According to Fischer et al. [50], the best estimates for the differences $T - T_{90}$ at the Ag, Au, and Cu points are 46.2 mK (14 mK), 39.9 mK (20 mK), and 52.1 mK (20 mK), respectively; the standard uncertainties $u(T - T_{90})$ are indicated in brackets. If one of these points is used as the reference for extrapolation in ITS-90, then at a temperature of, say, 3000 °C, according to Eq. 8.69 $T - T_{90}$ is equal to 325 mK, 239 mK, or 303 mK, with standard uncertainties of 100 mK, 120 mK, or 120 mK, respectively, depending on the fixed point used.

Recently, Manoi et al. [102] studied propagation of both $u(T_{90})$ and $u(T)$ using Eq. 8.68. The experimental confirmation of its validity was obtained using the RPRT by extrapolation from Ag and Cu fixed points to the point of inflection (PoI) of the melting curve of the Co-C eutectic HTFP (Woolliams et al. [188], see also Table B.1

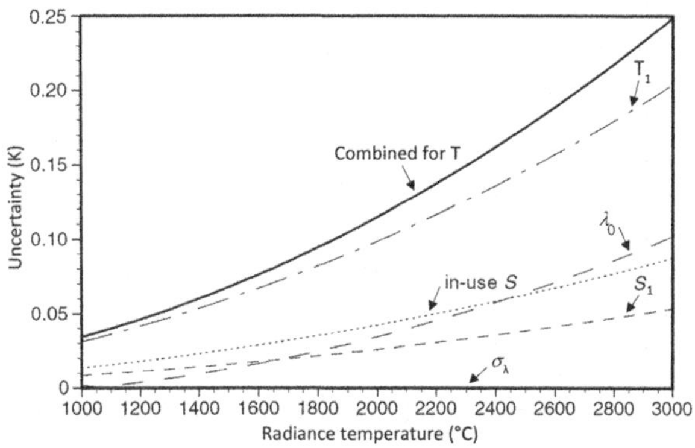

Fig. 8.26 Uncertainty components and combined standard uncertainties for the $n = 1$ scheme for the thermodynamic temperature T. Extrapolation is performed from the gold point as the reference. Reproduced from [149] with permission of Springer Nature

of Appendix B). Thermodynamic temperature was preliminary obtained using the APRT.

It is also worth mentioning that that the latter edition of the ITS-90 guide on radiation thermometry [202] recommends the approximate expression for the extrapolated value of T_{90} for small values of λ_0/σ_λ, i.e. for narrowband RTs:

$$T_{90} \approx \frac{c_2}{n\lambda_0\left(1-\frac{6\sigma_\lambda^2}{\lambda_0^2}\right)\ln\left[\exp\left(\frac{c_2}{n\lambda_0\left(1-\frac{6\sigma_\lambda^2}{\lambda_0^2}\right)T_{90}(X)+\frac{c_2\sigma_\lambda^2}{2\lambda_0^2}}\right)+r'-1\right]}-\frac{\sigma_\lambda^2}{2\lambda_0^2}, \quad (8.75)$$

where

$$r' = S'_{SSE,NL}(T_{90})/S'(T_{90}(X)) \quad (8.76)$$

is the ratio of signals, corrected for the effective emissivity, SSE, and nonlinearity; λ_0 and σ_λ are the mean wavelength and bandwidth of the spectral responsivity expressed by Eqs. 8.58 and 8.59, respectively; n is the refractive index of air at the wavelength λ_0.

8.4.5 *Uncertainty of Two-Point Sakuma-Hattori Interpolation*

This scheme uses two fixed points ($n = 2$), either one metal fixed point and one HTFP, or two HTFPs. Since the SH equation has three parameters to be determined, while only two measurements can be carried out at the calibration phase, it is necessary to determine one parameter independently. A key idea to solve this problem was presented by Saunders and White [152] and applied to the two-point interpolation by Saunders et al. [154]. This approach consists of the following. Since the sensitivity coefficient for the bandwidth σ_λ is small, we can use a rough estimate of σ_λ as a known value. Then, using expressions (8.61) and (8.62) are used to form the system of two nonlinear equations for λ_0 and C:

$$\begin{cases} S_1 = \dfrac{C}{\exp\left\{c_2\left[\lambda_0\left(1-6(\sigma_\lambda/\lambda_0)^2\right)T_1+c_2/\left(2(\sigma_\lambda/\lambda_0)^2\right)\right]^{-1}\right\}-1} \\ S_2 = \dfrac{C}{\exp\left\{c_2\left[\lambda_0\left(1-6(\sigma_\lambda/\lambda_0)^2\right)T_2+c_2/\left(2(\sigma_\lambda/\lambda_0)^2\right)\right]^{-1}\right\}-1} \end{cases}. \quad (8.77)$$

We can form the ratio S_1/S_2 to cancel the parameter C and obtain the nonlinear equation for λ_0:

$$\frac{S_1}{S_2} = \frac{\exp\left\{c_2\left[\lambda_0\left(1 - 6(\sigma_\lambda/\lambda_0)^2\right)T_2 + c_2\Big/\left(2(\sigma_\lambda/\lambda_0)^2\right)\right]^{-1}\right\} - 1}{\exp\left\{c_2\left[\lambda_0\left(1 - 6(\sigma_\lambda/\lambda_0)^2\right)T_1 + c_2\Big/\left(2(\sigma_\lambda/\lambda_0)^2\right)\right]^{-1}\right\} - 1}. \tag{8.78}$$

Solution of Eq. 8.78 for λ_0 can be found by standard numerical technique. Substituting the found value of λ_0 into any of the equations of the system (8.77), we can find the value of the parameter C and, using Eqs. 8.61 and 8.62, we find parameters A and B.

To analyze the uncertainties of the two-point interpolation, Saunders [149] derived five "realization" sensitivity coefficients using the same approximations as in the case of $n = 1$:

$$\frac{\partial T}{\partial \sigma_\lambda} \approx -\frac{(T - T_1)(T - T_2)c_2\sigma_\lambda}{T_1 T_2 \lambda_0^3}, \tag{8.79}$$

$$\frac{\partial T}{\partial T_1} \approx \frac{T(T - T_2)}{T_1(T_1 - T_2)}, \tag{8.80}$$

$$\frac{\partial T}{\partial T_2} \approx \frac{T(T - T_1)}{T_2(T_2 - T_1)}, \tag{8.81}$$

$$\frac{\partial T}{\partial S_1} \approx \frac{\lambda_0 T T_1(T - T_2)}{c_2(T_1 - T_2)S_1}, \tag{8.82}$$

$$\frac{\partial T}{\partial S_2} \approx \frac{\lambda_0 T T_2(T - T_1)}{c_2(T_2 - T_1)S_2}. \tag{8.83}$$

The "in-use" sensitivity coefficient $\partial T/\partial S$ is expressed by Eq. 8.71. Transforming the sensitivity coefficients for signals S_1, S_2, and S to be used with relative uncertainties and assuming no correlation between the uncertainty components, we can write the combined standard uncertainty of unknown temperature as

$$u_c(T) = \left\{\left[\frac{\partial T}{\partial \sigma_\lambda}u(\sigma_\lambda)\right]^2 + \left[\frac{\partial T}{\partial T_1}u(T_1)\right]^2 + \left[\frac{\partial T}{\partial T_2}u(T_2)\right]^2 + \left[\frac{\partial T}{\partial S_1}S_1 u_r(S_1)\right]^2 + \left[\frac{\partial T}{\partial S_2}S_2 u_r(S_2)\right]^2 + \left[\frac{\partial T}{\partial S}S u_r(S)\right]^2\right\}^{\frac{1}{2}}. \tag{8.84}$$

Figure 8.27 presents the dependences of each uncertainty component and the combined standard uncertainty for T predicted using Eq. 8.84.

The freezing point of gold and the melting point of peritectic alloy WC–C were used as the reference points. The WC–C peritectic is still not officially adopted as a HTFP. Saunders [149] assumed that $T_2 = 2749$ °C for WC–C, which is very

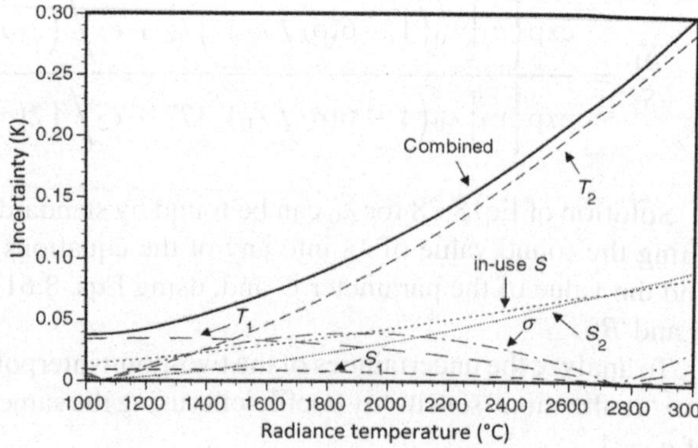

Fig. 8.27 Uncertainty components and combined standard uncertainty in thermodynamic temperature for the HTFP two-point interpolation scheme ($n = 2$) for the case where the Au point and WC–C peritectic are used as the interpolation nodes. Reproduced from [149] with permission of Springer Nature

close to later determinations (Khlevnoy et al. [82], Khlevnoy and Grigoryeva [79], Grigoryeva et al. [54].

The uncertainty in determination of the higher temperature, $u(T_2) = 230$ mK, dominates in the combined standard uncertainty $u_c(T)$. The uncertainty $u(T_2)$ consists of the components determined from APRT (200 mK), due to the furnace effect (100 mK), impurities (20 mK), temperature drop (14 mK), identification of the inflexion point (10 mK), and repeatability (5 mK).

According to Saunders and White [153], correlations between uncertainty components of some input values (in particular, between S_1, S_2, and S due to nonlinearity) reduce the total uncertainty compared with that plotted in Fig. 8.27. Such a correlation entirely cancels the corresponding components of the uncertainty in each fixed point, with partial cancelation between the fixed points. To obtain more realistic estimate of $u(T)$, Saunders [149] assumed that only "in-use" measurement of S is affected by nonlinearity, while the signals S_1 and S_2 are not affected. The effect of correlation on the uncertainties at two-point interpolation has been recently studied by Manoi et al. [103].

The two-point interpolation scheme is a minimalist scheme for calculating any temperature between two calibration points. Bloembergen et al. [14] were the first to draw attention to the convenience of using two-point interpolation for the continuation of the ITS-90 to the high temperatures using HTFPs. The method used by Bloembergen et al. [14] had no relation to the SH equation and required to know the exact shape of the spectral responsivity measured relative to some center wavelength, and the center wavelength is subsequently determined by fitting the integral of the product of the measured relative spectral responsivity and the Planckian radiance to the measured signal at the two fixed points. Later, Bloembergen and Yamada [13] further developed the non-SH approach and considered both interpolation between reference temperatures and extrapolation beyond them. They also stressed that the simplicity in realization two-point scheme may largely "outweigh" the need in realization of additional HTFPs in the schemes $n \geq 3$.

8.4.6 Uncertainty of Sakuma-Hattori Interpolation Using Three or More Fixed Points

For three reference temperatures ($n = 3$), the curve expressing SH equation passes exactly through three reference points and the problem is reduced to solution of the system of three nonlinear equations:

$$S_i = \frac{C}{\exp\left(\frac{c_2}{AT_i + B}\right) - 1}, \quad i = 1, 2, 3. \tag{8.85}$$

It is common to reduce the problem of solution of such systems to the equivalent problem of minimization of the objective functions of three variables (for Eq. 8.56 and 8.57, respectively):

$$F_S(A, B, C) = \sum_{i=1}^{3} [S_i - S(T_i, A, B, C)]^2 \tag{8.86}$$

or

$$F_T(A, B, C) = \sum_{i=1}^{3} [T_i - T(S_i, A, B, C)]^2. \tag{8.87}$$

Since functions F_S and F_T are the sums of non-negative terms, the have absolute minima $F_S = 0$ and $F_T = 0$ when $S_i = S(T_i, A, B, C)$ and $T_i = T(S_i, A, B, C)$ for $i = 1, 2, 3$, respectively.

If more than three reference points ($n > 3$) are involved in the interpolation, the least square technique should be applied so the curve will be drawn in such a way that the sums of squared curve-to-point distances (the objective function)

$$F_S(A, B, C) = \sum_{i=1}^{n} [S_i - S(T_i, A, B, C)]^2 \tag{8.88}$$

or

$$F_T(A, B, C) = \sum_{i=1}^{n} [T_i - T(S_i, A, B, C)]^2. \tag{8.89}$$

will be minimal. As can be seen, the case of $n = 3$ can be considered as a special case of more general case $n \geq 3$. The only difference is that the functional to be minimized reaches zero for $n = 3$, but is equal to a small positive value for the least square fitting. This allows the use of the same numerical technique for any $n \geq 3$.

Equations 8.88 and 8.89 have the same solutions, although, in general, $\min(F_S) \neq \min(F_T)$. For better physical plausibility of solutions, each term in the right-hand sides in Eqs. 8.88 and 8.89 can be multiplied by a weight inversely proportional to the corresponding variance [146].

Minimization of the objective functions expressed by (8.86) to (8.89) is a typical minimization problem for non-linear functions of several variables commonly encountered in the least squares curve fitting. Among many numerical techniques developed to this end, the Levenberg-Marquardt (LM) algorithm has become a de facto standard technique of multidimensional nonlinear regression (curve fitting) and has been included in various software packages for numerical analysis, statistics, and graph plotting. It is very flexible (has many versions and can be easily modified to accelerate convergence for particular tasks) and robust (i.e. provides convergence for very large range of initial guesses and prevents accumulation of rounding errors). A detailed description of the LM algorithm is given by Nash [113], Press et al. [125]; it can be found in many textbooks on numerical methods, data analysis, and applied and engineering optimization.

For illustration purpose, we have used the LM procedure [21] to compute parameters A, B, and C for two examples presented in Table 8.3. The example **a** involves the silver point (1234.93 K) and two HTFPs: Co–C (1597.48 K) and Pt–C (2011.50 K). In the example **b**, the gold point (1337.33 K) and Re–C (2747.91 K) are added. The equilibrium liquidus thermodynamic temperatures were used for the eutectic transitions of HTFPs (see Appendix B, Table B.2). The surfaces of $\log_{10}[F_T(A, B)]$ plotted for optimized values of C are shown in Fig. 8.28. In both examples, measurements with the RT that has the rectangular spectral responsivity with the center wavelength $\lambda_0 = 0.65$ μm and the FWHM $\Delta\lambda = 10$ nm were modeled. As can be seen from calculation results provided in Table 8.3, the parameters A and C found for examples **a** and **b** differ in sixth significant digits only, while the difference is in fourth significant digit is observed for the parameter B. Therefore, the parameter B is more sensitive to the set of fixed point than the parameter A, which remain almost the same as λ_0 and $\Delta\lambda$ are the same in both examples.

The signals S_i were computed numerically as integrals of Planck's distribution (up to a constant multiplier) over the wavelength interval $\left[\lambda_0 - \frac{\Delta\lambda}{2}, \lambda_0 + \frac{\Delta\lambda}{2}\right]$:

$$S_i = \int_{\lambda_0 - \frac{\Delta\lambda}{2}}^{\lambda_0 + \frac{\Delta\lambda}{2}} \lambda^{-5}\left[\exp\left(\frac{c_2}{\lambda T_i}\right) - 1\right]^{-1} d\lambda. \tag{8.90}$$

In all cases, initial guesses for A, B, and C were calculated according to (8.61) to (8.63). The surface shown in Fig. 8.28a has a very narrow and deep minimum ($F_T = 0$, $\log_{10}(F_T) = -\infty$), which lies at the bottom of a narrowing down valley. The surface shown in Fig. 8.28b has a deeper valley with a less deep minimum, which corresponds to the finite residuals of the least square interpolation. The criterion for stopping the iterative LM process was the achievement of relative changes for each

Table 8.3 Two examples (**a** and **b**) of calculating parameters A, B, and C for the SH equations

	T_i (K)	λ_0 (μm)	$\Delta\lambda$ (nm)	A (μm)	B (μm·K)	C(arb. units)	Direct		Inverse	
							F_S	N_S	F_T	N_T
a	1234.93 1597.48 2011.50	0.65	10	0.649963	0.140175	9.94998×10^{-3}	7.3946×10^{-51}	20	1.4791×10^{-31}	14
b	1234.93 1597.48 2011.50 2747.91	0.65	10	0.649962	0.140385	9.95002×10^{-3}	3.1805×10^{-28}	19	2.7168×10^{-10}	13

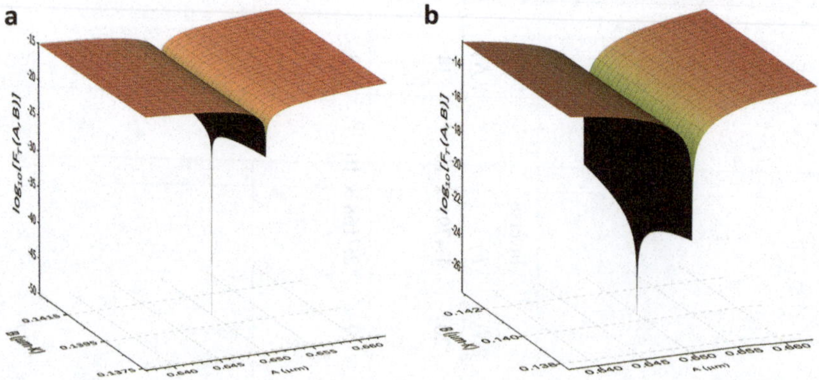

Fig. 8.28 3D plots of the objective functions $F_T(A, B)$ at the optimized value of parameter C for examples **a** and **b** of Table 8.3

of three parameters less than 10^{-9} (for compactness, obtained values for A, B, and C are presented in Table 8.3 with only 6 significant digits). The N_S and N_T are the number of iterations spent for achieving the convergence for the direct and inverse SH equations, respectively. Convergence for the inverse equation is slightly faster, but this is not critical because all calculations take very little time (less than one second per example).

It was found that too large deviation of initial guess from that computed using (8.61) to (8.63) may require more (up to 1000) iterations, but we could not detect areas of reasonable initial guesses where the LM algorithm diverges.

Among other things, the undoubted advantage of the SH technique is that it does not require accurate measurement of the RT spectral responsivity. All the necessary information for this is contained in the measured signals S_i and reference temperatures T_i; statistical confidence is provided with information on uncertainties $u(S_i)$, $u(T_i)$, and correlations between different variables (if any). A common disadvantage of all interpolation techniques is that the uncertainty tends to oscillate between the interpolation nodes.

Uncertainty propagation according to GUM [71] for the SH equation has been studied by Saunders [149] in relation to the thermodynamic temperature measurement above the silver point. The same approach was recommended in the *MeP*-K on RPRT. Alternatively, the Monte Carlo method for propagation of distributions according to [72] can be used to model uncertainties at the SH interpolation [190].

Evaluation of uncertainty in determination of temperature using the SH equation for an arbitrary $n \geq 3$ is non-trivial problem. The desired temperature cannot be written as an explicit function of signals and reference temperatures; it can be obtained only via nonlinear least squares fitting. This makes it impossible to directly apply the GUM approach. At present, there is no standard procedure for estimating the uncertainty of quantity values resulting from the application of the least squares method. While the JCGM (Joint Committee on the Guidance on Metrology) is developing the guide "JCGM 107. Evaluation of measurement data. Least Squares Application", which is only planned to be released [9], researchers are forced to develop their own methods for this purpose.

For $n = 3$, Saunders [149] used the Wien approximation and the second-order Lagrange polynomials in temperature to obtain the following approximate expressions for sensitivity coefficients with respect to T_1, T_2, T_3, S_1, S_2, and S_3:

$$\frac{\partial T}{\partial T_1} \approx \frac{(T - T_2)(T - T_3)}{(T_1 - T_2)(T_1 - T_3)}, \tag{8.91}$$

$$\frac{\partial T}{\partial T_2} \approx \frac{(T - T_1)(T - T_3)}{(T_2 - T_1)(T_1 - T_3)}, \tag{8.92}$$

$$\frac{\partial T}{\partial T_3} \approx \frac{(T - T_1)(T - T_2)}{(T_3 - T_1)(T_3 - T_2)}, \tag{8.93}$$

$$\frac{\partial T}{\partial S_1} \approx \frac{\lambda_0 T_1^2}{c_2 S_1} \frac{(T - T_2)(T - T_3)}{(T_1 - T_2)(T_1 - T_3)}, \tag{8.94}$$

$$\frac{\partial T}{\partial S_2} \approx \frac{\lambda_0 T_2^2}{c_2 S_2} \frac{(T - T_1)(T - T_3)}{(T_2 - T_1)(T_2 - T_3)}, \tag{8.95}$$

$$\frac{\partial T}{\partial S_3} \approx \frac{\lambda_0 T_3^2}{c_2 S_3} \frac{(T - T_1)(T - T_2)}{(T_3 - T_1)(T_3 - T_2)}. \tag{8.96}$$

The "in-use" sensitivity coefficient $\partial T / \partial S$ is expressed by Eq. 8.71. Assuming no correlations between the components, Saunders [149] obtained the following expression for the combined standard uncertainty $u(T)$:

$$u_c(T) = \left[\left(\frac{\partial T}{\partial T_1} u(T_1) \right)^2 + \left(\frac{\partial T}{\partial T_2} u(T_2) \right)^2 + \left(\frac{\partial T}{\partial T_3} u(T_3) \right)^2 + \right.$$
$$\left. \left(\frac{\partial T}{\partial S_1} u(S_1) \right)^2 + \left(\frac{\partial T}{\partial S_2} u(S_2) \right)^2 + \left(\frac{\partial T}{\partial S_3} u(S_3) \right)^2 + \left(\frac{\partial T}{\partial S} u(S) \right)^2 \right]^{\frac{1}{2}}. \tag{8.97}$$

The uncertainty of the reference temperature of T_1, T_2, and T_3 consists of several components, depending on which fixed point (metal or HTFP) is used as a reference temperature (see Table 8.1). The *MeP*-K on RPRT [99] recommends to use one metal fixed point and two HTFPs or, alternatively, three HTFPs. Extrapolation beyond the end points is not recommended due to the rapid growth of the uncertainty. Saunders [149] considered the three-point scheme with Au fixed point, Pt–C eutectic, and

WC–C peritectic HTFPs used as the reference points. The results of calculation of uncertainty components and the combined standard uncertainty for this case are shown in Fig. 8.29. The uncertainty component due to S_1 is too low to be visible on the scale of this graph.

A comparison of the combined standard uncertainties for the schemes with $n =$ 0, 1, 2, and 3 is shown in Fig. 8.30. As expected, the APRT scheme ($n = 0$) has the lowest combined standard uncertainty. All other schemes are traceable to the thermodynamic temperature via, at least, one $n = 0$ measurement. They require

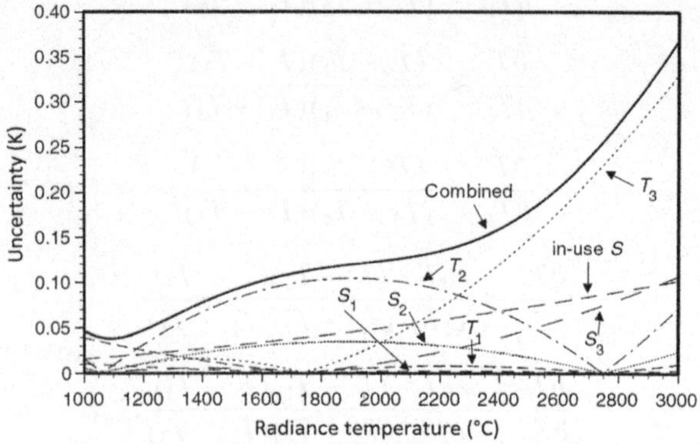

Fig. 8.29 Uncertainty components and combined standard uncertainty in thermodynamic temperature for the three-point SH interpolation scheme, where the Au fixed point, Pt–C eutectic, and WC–C peritectic HTFPs are used as the reference points. Reproduced from [149] with permission of Springer Nature

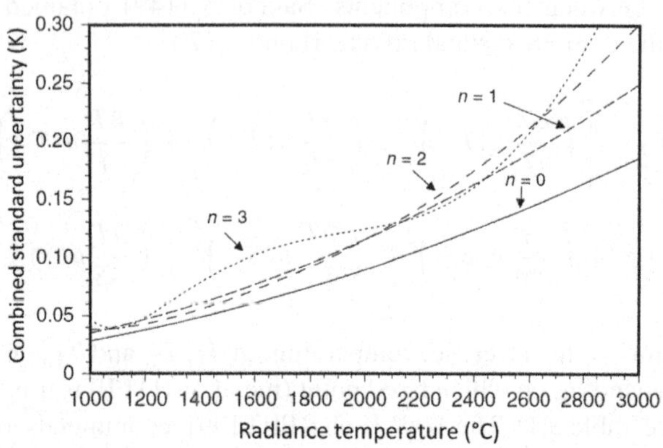

Fig. 8.30 Comparison of combined standard uncertainties in radiance temperature for different scheme in the SH implementation of the RPRT. The curve for $n = 0$ represents contemporary results for the APRT. The curve for $n = 1$ corresponds to the single-point extrapolation from the gold point. For $n = 2$, the Au and WC–C fixed points are used as the interpolation nodes. The curve $n = 3$ corresponds to the interpolation using Au, Pt–C, and WC–C fixed point. Reproduced from [97] with permission of Springer Nature

Table 8.4 Combined standard uncertainties for SH interpolation with different number of fixed points (after [147])

Interpolated temperature (°C)	Combined standard uncertainty for different number n of fixed points (mK)			
	$n = 3$	$n = 4$	$n = 5$	$n = 12$
1500	140	125	115	90
2000	190	170	160	140
2500	230	220	210	190
3000	420	405	390	320

much less technical capability than the APRT and capable to achieve reasonable uncertainties, only about two times higher than the APRT. A comparison between the RPRT methods with $n \geq 1$ shows that they are very close in their combined standard uncertainties. In particular, we do not observe a cardinal decrease in the combined standard uncertainty for $n = 3$ against $n = 2$.

If more than three fixed points are used, the least squares technique can be applied to the measured pairs (T_i, S_i), $i = 1, 2, \ldots, n$, where $n > 3$. A general methodology for calculating the sensitivity coefficients was proposed by Saunders [147]. Unfortunately, we could not find any evidence of its use in any work published to date, with the exception of some examples presented by Saunders [147] himself. Although the initial data for these examples do not quite correspond to the real ones, they allow us to make some important conclusions. Saunders considered the cases of 3, 4, 5, and 12 reference temperature, including Ag (961.78 °C) and Cu (1084.62 °C) metal fixed points, as well as some prospective metal-carbon and metal carbide-carbon HTFPs, namely[7]: Fe–C (1153 °C), Ni–C (1329 °C), Pd–C (1492 °C), Rh–C (1657 °C), Pt–C (1738 °C), Ru–C (1953 °C), Ir–C (2290 °C), Re–C (2474 °C), TiC–C (2776 °C), ZrC–C (2927 °C). It was assumed that the uncertainty in each fixed point (except for Ag and Cu) is equal to 0.1 °C. The uncertainty of each simulated signal was supposed to be 0.05%.

Modeling was performed in the temperature interval between 1000 °C and 3000 °C for a realistic relative spectral responsivity curve with a center wavelength of 650 nm and a bandwidth of 20 nm. Some results of this modeling are presented in Table 8.4. According to these modeling results, we can talk about a significant reduction in the combined standard uncertainty compared to three-point interpolation only for the case of 12 interpolation nodes, which is difficult and expensive to put into practice.

The *MeP*-K on RPRT [99] stated that the uncertainty components associated with the calibration of the RT are generally reduced approximately by a factor of $\sqrt{n/3}$; therefore, the total uncertainty decreases as the number of reference temperatures increases. However, since the "in-use" uncertainty is the same for any number of fixed

[7]The temperatures of phase transition are indicated as they were be determined to the date of publication.

points, the combined standard uncertainty cannot be significantly lower than that the three-point interpolation. An advantage of the multi-point least-squares interpolation is rather in additional reliability, as the residuals from the least-squares fit can serve as a measure of the quality of the interpolation, than in lower total uncertainty.

8.5 Measuring Blackbody Temperature Distributions Using Radiation Thermometry

8.5.1 *Radiance Temperature Scanning*

When contact methods are difficult or impossible to use for measuring temperature distributions over the radiating surfaces of blackbodies, these distributions can be evaluated via measuring the radiance temperatures. Although this method has obvious drawbacks (to derive temperature from the radiance temperature, the emissivity should be known; distribution of radiance temperature can be measured from outside the cavity only over directly viewable area), in some cases, scanning of radiance temperature is the only way to obtain experimental information on temperature uniformity of a radiator. The most straightforward way to measure the radiance temperature distribution is to perform the angular scanning of the internal surface of the cavity through its opening using the RT having a narrow angular FOV as it is shown in Fig. 8.31 for on-axis and off-axis positioning of the RT. If the cavity under measurement has no baffle and other obstacles that hinder the scanning, the radiance temperature can be measured over almost entire cavity surface.

Otherwise, only a part of the cavity internal surface marked by the dashed line in Fig. 8.31 is available for measurements.

Miklavec et al. [109] described the custom-made automated bench that allows angular scanning the distribution of the radiance temperature over the cavity internal surface using a commercial RT. The bench has five degrees of freedom: three of them allow translations of the radiation thermometer in orthogonal directions and two enable rotation of the RT relative to the main axis of the blackbody in the two mutually perpendicular planes. Movement and rotation of the RT is performed using the stepping motor connected to a PC via a RS-232 interface. The RT is and positioned

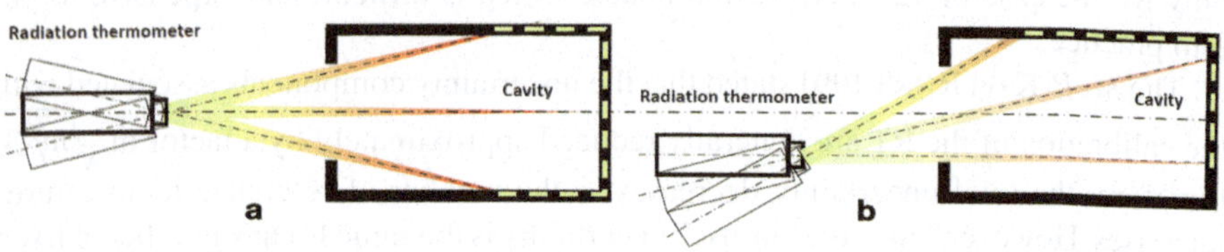

Fig. 8.31 Schemes of the angular scanning of a cavity internal surface by **a** on-axis and **b** off-axis radiation thermometer

at 1300 mm from the bottom of the cavity, where the FOV of the RT is of 6 mm in diameter. Such an approach is sufficient for qualitative evaluation of laboratory blackbodies used for routine calibrations of RTs in terms of radiance temperature. However (this is inherent drawback of all methods based on the radiance temperature scanning), it does not allow measuring the actual temperature distribution, which is necessary for calculation of deviation from Planck's law. Angular scanning is most suitable for lightweight RTs because it is hard to provide precise angular positioning for instrument of larger size having greater mass and, correspondingly, inertia.

The scheme with linear scanning (see Fig. 8.32) is more common in the precision blackbody radiometry since the linear movement of the RT can be easily implemented with sufficient accuracy than rotation. Besides, alignment of the system is simplified and the position of a scanned point on the cavity internal surface can be easily referenced to the translation of an RT. As for the angular scanning, the greatest inconvenience faced by the researchers is that the elements of blackbody construction or cavity baffle may obstruct some areas of interest on the cavity internal surface. This issue is less critical for large-aperture and shallow blackbodies but becomes the main problem for deep cavities with relatively small apertures. It is highly desirable to employ the RT that has the angular FOV as narrow as possible; in other words, the RT must have a large f-number (the ratio of the focal length of the objective lens to its aperture diameter). Usually, RTs with f/10–f/15 are easily available for visual and near-IR spectral range.

Sperfeld et al. [170] and Woolliams et al. [187] applied the scheme similar to that depicted in Fig. 8.32a to investigation of the pyrolytic graphite blackbodies, whose radiating cavities are formed by internal surfaces of the stacked pyrolytic graphite rings of the blackbody heater [142]. Due to modular design, it is possible to use the various cavity bottoms and aperture diaphragm with internal diameter from 22 to 26 mm. The base model has the cavity with the facing inwardly conical bottom (the apex angle is 110°) made of graphite (see Fig. 8.33). Both groups of researchers—Sperfeld et al. [170] and Woolliams et al. [187]—used similar filter radiometers with 800 nm center wavelength and 20 nm FWHM to scan pyrolytic

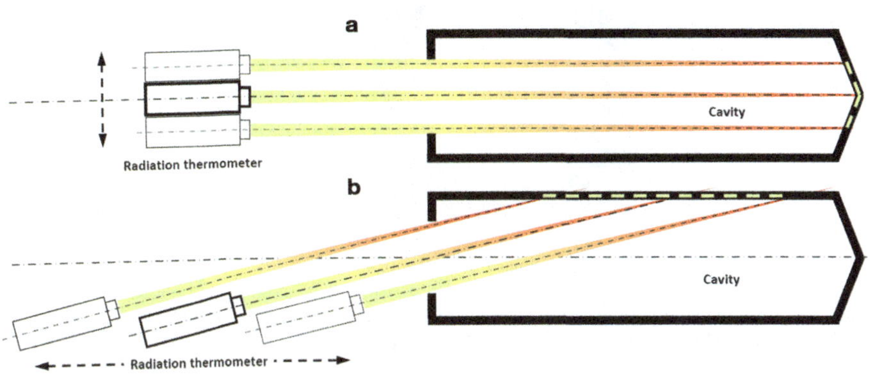

Fig. 8.32 Schemes of the linear scanning by the RT of **a** bottom and **b** lateral walls of a cavity

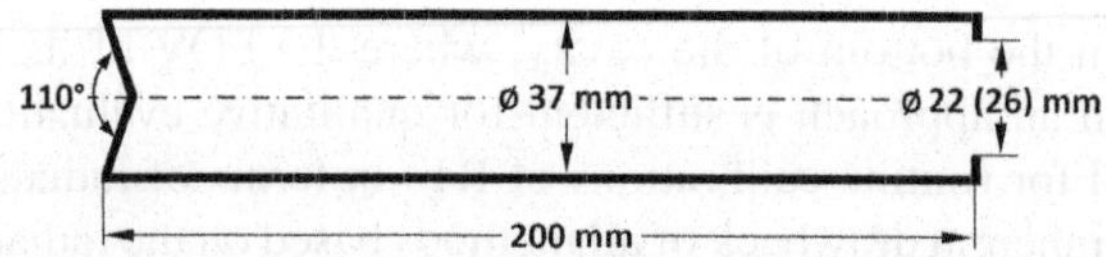

Fig. 8.33 Schematic of the BB3200pg radiating cavity. Due to modular design, the bottom of other shapes can be used; aperture diameter can also be varied

graphite cavities in irradiance mode. The scanning area was limited by a precision water-cooled aperture placed in front of the blackbody to provide measurement in the irradiance mode, which were the main objectives of both investigations.

The radiance from inside the blackbody was imaged by a lens on to the radiometer aperture with a diameter of 1 mm. The lens and filter radiometer form together the RT able to scan using a translation stage almost entire blackbody aperture across a plane perpendicular to the cavity axis. Sperfeld et al. [170] performed scans along the diameter of 15-mm precision aperture and found that temperature uniformity of the BB3200pg is about 0.1% at temperatures higher than 2800 K (i.e. non-uniformity does not exceed ± 3 K). Woolliams et al. [187] presented the 2D map of radiance temperature across the part of the cavity bottom viewable through the precision aperture of 12 mm in diameter. The gray-scale map of one of such scans showing radiance temperature variations from 3104 K to 3107 K is presented in Fig. 8.34.

Hartmann et al. [59] reported the results of measuring the radiance temperature distributions along the lateral wall of the BB3200pg radiating cavity with the opening diameter of 26 mm. Measurements were carried out using the LP3 RT with the 650 nm interference filter and the angular FOV of about 0.1°. At the distance of 1230 mm, the spot diameter was 1.5 mm. The LP3 was tilted by 5° with respect to the cavity axis and moved linearly with the step of 5 mm keeping the cavity wall in the focus, which corresponds to the scheme shown in Fig. 8.32b.

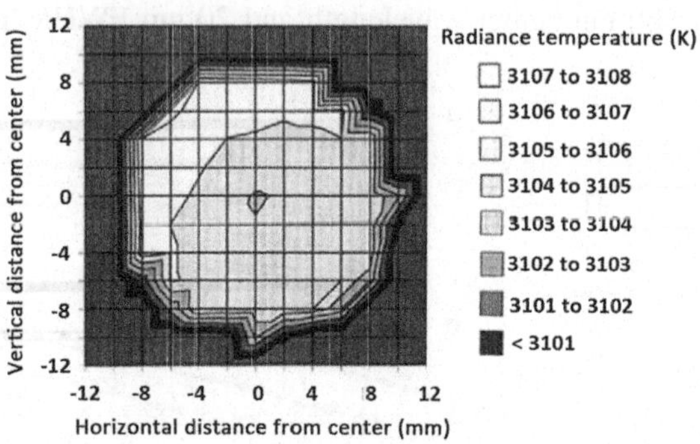

Fig. 8.34 Radiance temperature map for the directly viewable part of the BB3200pg cavity bottom. Reproduced from [187] with permission of IOP Publishing

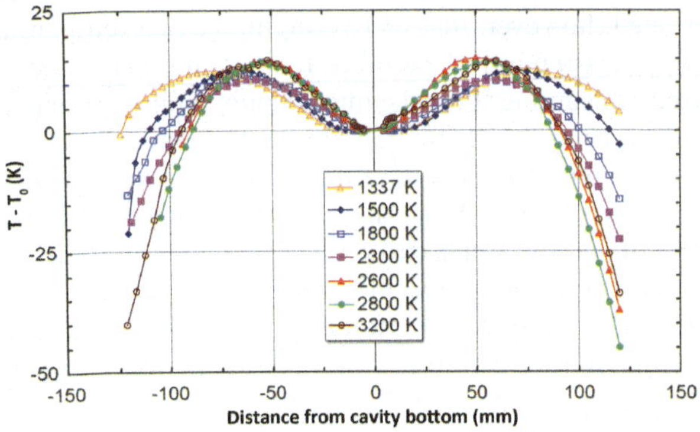

Fig. 8.35 The radiance temperature distribution at 650 nm measured by LP3 radiation thermometer along the lateral walls of the BB3200pg cavity. Positive and negative values of the abscissa correspond to the cavity opposite walls; $T_0 = T(0)$ is the radiance temperature at the joint with the cavity bottom. Reproduced from [60] with permission of Elsevier

Only a part of the cavity wall of 120 mm in length adjacent to the bottom could be scanned. Measured distributions of the radiance temperature at 650 nm are presented in Fig. 8.35 for seven cavity temperatures from 1337 K up to 3200 K.

All temperature distributions plotted in Fig. 8.35 show temperature rises by 10 K to15 K over temperature T_0 near the cavity bottom followed by a temperature drop towards the cavity aperture. Such a behavior cannot be explained only by the effective emissivity variation along the isothermal cylindrical walls but undoubtedly related to real temperature gradient along it.

Hartmann et al. [59] suggested that the actual temperatures could be iteratively derived from the distribution of the radiance temperature. Computational procedure begins from the initial guess for temperature distribution followed by the Monte Carlo modeling of the radiance temperature distribution along the cavity walls. After necessary alterations in the distribution of actual temperatures, process is repeated until its convergence. The temperature of the cavity bottom that is assumed uniform and serves as the reference temperature. In order to accelerate calculations, the measured distributions of radiance temperatures were approximated by a polynomial of the third order.

A similar iterative procedure was applied recently by De Lucas and Segovia [32] to extract the temperature distribution from the measured distribution of radiance temperature along the sidewall of the cylindro-conical cavity of the custom-made VTBB operating in the temperature range from 400 to 1000 °C. The radiating cavity made of SiC is heated in the commercial three-zone furnace from Carbolite[8], having an independent temperature control for each zone.

---

[8]Carbolite (UK) and Carbolite Gero (Germany) joined in 2016 forming one company under the name of CARBOLITE GERO. The currently available universal tube furnace can be found at https://www.carbolite-gero.com/products/tube-furnace-range/universal-tube-furnaces/.

It should be noted, however, that recovering the actual temperature distribution from the measured distribution of radiance temperature can be significantly more time- and resource-consuming if to take into account temperature non-uniformity of the cavity bottom and more realistic than perfectly diffuse model of reflection for the cavity wall. Since the tilting angle of the radiation thermometer with respect to the cavity axis is very small (5°), the major part of registered radiation leaves the cylindrical surface of the cavity at large (about 85°) angles with the normals. One can expect noticeable rise of specular reflection at such grazing angles and corresponding decrease of directional emissivity. Measurements of the BRDF at grazing angles are very unreliable if possible. Therefore, results of temperature distribution recovery should be considered as approximate.

The non-contact measurement methods are often the only way to evaluate temperature distributions for high-temperature blackbodies and becomes widely used in the last decade due to introducing of HTFPs into the practice of radiation thermometry and blackbody radiometry.

The pyrolytic graphite blackbodies developed at VNIIOFI [81, 143], which have designs similar to that of the BB3200pg, are currently the only blackbodies capable to operate at temperatures as high as 2500–3500 K and thus can serve as the furnaces for realization of HTFPs. For the better quality of the phase transition plateau, temperature uniformity of a fixed-point cell plays the critical role. The HTFP cell is placed inside the blackbody in the cell holder. It is necessary to choose the zone inside this holder, where temperature distribution has minimal nonuniformity. For such an optimization, absolute values of temperature are not important, so the selection of the uniform zone can be done using the measured distribution of the radiance temperature. The problem of obstruction the FOV of the RT remains and becomes dominant due to mounting of baffles, which prevent radiative heat exchange inside the furnace. To overcome this problem for temperatures around 2500 °C, Khlevnoy et al. [80] used a dummy blackbody cavity made of graphite and able to move inside the C/C (carbon-carbon composite) cell holder (see Fig. 8.36). Two RTs were involved in measurements. The narrow-beam optical fiber RT sighting through the furnace rear channel at the graphite back cap of the cell holder was used for the furnace temperature control; the LP3 RT was aimed through the front opening at the movable target that is a graphite cylindrical cavity with a soot-blackened bottom and aperture diameter of 5 mm. Initially, the target is placed inside the C/C tube near its front end, the furnace is heated up to the temperature of 2500 °C, and the target temperature

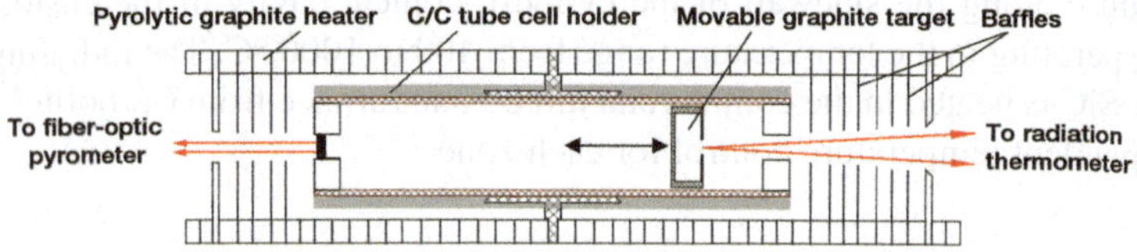

Fig. 8.36 Schematic for measuring the temperature distribution along the fixed-point cell holder mounted inside the pyrolytic graphite furnace. Reproduced from [80] with permission of Springer Nature

is measured using the LP3 RT. Then the furnace temperature is set to 1500 °C, the target is pushed using a rode made of alumina (Al_2O_3, the melting temperature is 2072 °C) and moved by 10 to 20 mm. The furnace is heated again to 2500 °C, and the second measurement of the temperature is taken. This sequence is repeated until the backside of the target reaches the holder's edge. To validate this technique, measurements performed using the LP3 for temperature of 1500 °C were compared with those obtained using the type R thermocouple. Results obtained using two methods were in agreement within 1 °C.

For heavily non-isothermal blackbodies, the measurement of radiance temperature distribution remains the only way to compute the effective emissivity. The fee for using radiance temperatures instead of thermodynamic temperatures is the increase of the uncertainty in determining the effective emissivity.

At the LNE-Cham (France), where the blackbody from Thermo Gauge Instruments, Inc. (http://thermogauge.com/) is used as a comparison source to calibrate customers' instruments against a reference RT in terms of the ITS-90, a complex investigation of temperature distributions in the radiating cavity of the Thermo Gauge blackbody HT-9500 was described by Kozlova et al. [87]. The HT-9500 has the operating temperature range from 500 to 3000 °C, a 25.4 mm diameter of the aperture, and a 142 mm length cylindrical cavity. The design of this type of dual-cavity blackbodies was shown in Fig. 5.13a; a schematic of the Thermo Gauge graphite HTBB was presented in Fig. 7.34. This blackbody employs the principle of direct resistance heating of a tubular graphite heater using a large AC current and low voltage to bring a high electric power to a poorly insulated heater.

Poor insulation of the graphite tube results in a large temperature nonuniformity, which may lead to a noticeable uncertainty because the effective emissivity of the cavity depends on the temperature distribution, which, in turn, depends on the set temperature. Radial distribution of temperature was measured using the LP3 RT focused on the cavity bottom (tube septum). The scanning of the bottom was performed along the vertical and horizontal axes according to the scheme shown in Fig. 8.32a. The results of measurement were interpolated to produce a 2D maps presented in Fig. 8.37. For all distributions, the rotational symmetry is broken,

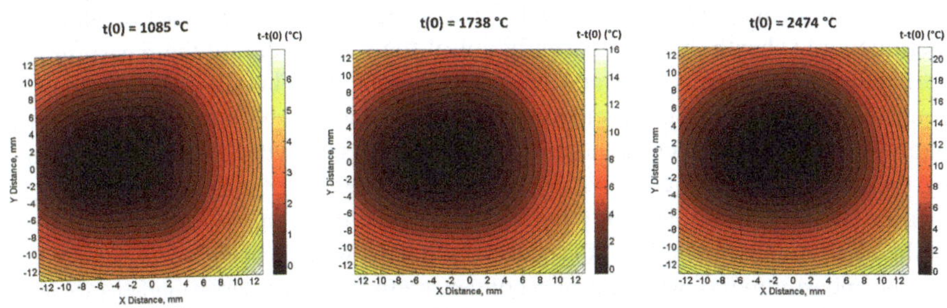

Fig. 8.37 2D color maps for the temperature distributions over the radiating cavity bottom obtained at three temperature $t(0)$ of the cavity bottom center. Reproduced from [87] with permission of Springer Nature

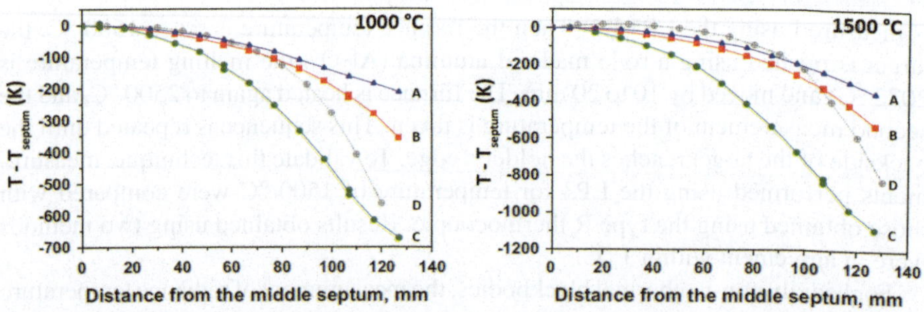

Fig. 8.38 Comparison of temperature distributions along the sidewalls of the Thermo Gauge HT-9500 blackbody measured by techniques A to C (see in the text) for temperatures of 1000 and 1500 °C. Reproduced from [87] with permission of Springer Nature

perhaps, due to flow of purge argon.

For temperature distributions along the lateral walls of the cylindrical cavity, Kozlova et al. [87] compared different contact and non-contact measurement techniques:

A. Measurement of the lateral wall temperature with the titled RT (see Fig. 8.32b).
B. Measurement of the radiance temperature of the movable target inside the cylindrical cavity using the RT. A 24 mm diameter graphite disk served as the target for the RT. This disk was attached to a 5 mm diameter graphite rod in order to have the possibility to move it.
C. Contact measurement of the temperature distribution using a movable Type S thermocouple in an alumina sheath. This thermocouple touches a graphite disk inserted in the radiating cavity and served as the target for the method B.
D. Measurement performed with the Type R thermocouple forming a loop [26]. We considered this technique in Sect. 7.5.1 (see also Fig. 7.35).

The results of comparison presented in Fig. 8.38 for two temperatures, 1000 °C and 1500 °C, show a noticeable discrepancy between distributions obtained by different methods. The contact technique C shows a larger gradient compared to the measurements made using the RT. This is explained by the heat loss along the thermocouple sheath that was partially out of the furnace during the measurement. A smaller gradient was obtained by the technique A. This can be explained by contribution of the grazing incidence. The technique B with a displacing target is likely to be the most accurate among the methods employed. The results obtained by the technique D have shown fewer discrepancies with non-contact method of measurements. During the study performed by Kozlova et al. [87], the dependence of the non-contact methods on the wavelength used was checked.

The measurements with the non-contact methods were performed using three different filters of the LP3 RT (650 nm, 750 nm, and 900 nm) and no significant differences in the temperature distribution were found.

8.5.2 Radiation Thermometry with Optical Fibers

In this section, we consider the optical fiber RTs, which employ fiber probes inserted into the radiating cavity to transport its thermal radiation to the measuring head arranged outside the hot zone and operating at near-room temperature. This idea is not as new as it may seem. In 1956, Sutcliffe and Carroll [160] described application of a single-crystal sapphire rode to transmit radiation from the furnace with the molten silicon to the RT placed outside the furnace. Practical advantage of such an arrangement over the sighting of the molten silicon through an opening in the carbon crucible has been demonstrated.

Two materials are suitable for manufacturing optical fibers, which can be used at temperatures above 1000 °C. These are silica (SiO_2, quartz) with the melting temperature of about 1600 °C and sapphire (α-Al_2O_3) with the melting temperature of 2030 °C to 2050 °C. The silica fibers have minimal losses at wavelengths around 1.5 μm, sapphire fibers—at about 2.5 μm, but at distances of order of 1 m, losses of both materials are small enough within a wide wavelength range spanning the visible and near-IR spectral ranges employed in radiation thermometry.

The use of optical fiber thermometer (OFT) from Luxtron (now LumaSense Technologies Inc.) to evaluate axial temperature distribution inside the Thermogage graphite blackbody operating at temperatures around 1100 °C is described by Horn and Abdelmessih [64]. The OFT includes a single-crystal sapphire lightpipe bent at the angle $\beta = 90°$ at the tip, a fiber optic cable, and a receiving unit that consists of an optical detector and signal conditioning electronics. The optical fiber (lightpipe) is the only part of the system, which enters the blackbody cavity. The thermal radiation is transmitted from the lightpipe to the receiving unit through the fiber optic cable. The receiving unit converts the incoming IR radiation into a DC voltage, which is routed to a data acquisition system. The OFT can measure temperatures from 400 to 1900 °C with a resolution of 0.4 °C. Response time of the OFT is 0.04 s. The OFT is mounted on the slide track and crossbar assembly. Both the control and reference RTs measure the blackbody temperature. These are identical commercially available RTs capable of measuring temperatures from 800 to 3100 °C. Their FOV covers approximately one-quarter of the diameter of the bottom of the cylindrical blackbody. The control RT is used in the control feedback system to maintain stable temperatures and it is calibrated to be accurate within ± 1 °C at 1100 °C. The reference RT was calibrated at NIST prior to measurement with the expanded (at $k = 2$) uncertainty of 0.7 °C at 1100 °C. The OFT mount was designed to allow the sensor measuring the axial temperature distribution at various locations around the interior circumference of the blackbody cavity using rotation of the sapphire fiber probe with the angle increment of 45°. An acceptance angle θ of the OFT is about 52°. The mount allows approach the tip of the sensor to a distance $x_{\min}$ between 3 mm and 4 mm to the cavity bottom (Fig. 8.39). The axial temperature profiles were obtained by fixing the OFT at the desired angular location and then slowly inserting the OFT into the blackbody. The OFT was stopped at several axial locations as it was moved into the

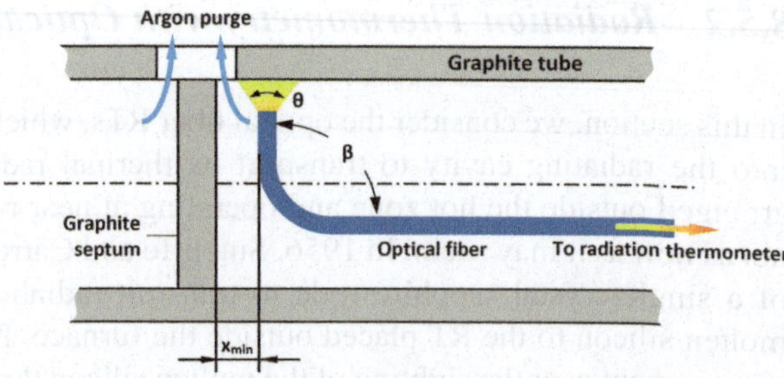

Fig. 8.39 The measurement scheme for temperature distribution in the cylindrical cavity of the Thermogage blackbody using an optical fiber (after [64])

blackbody in order to obtain steady-state readings. The axial location of the sensing tip was measured using a deflection potentiometer connected to the crossbar.

During testing, it was noticed that a significant amount of energy was entering the OFT through the bend in the probe. This effect was obvious when the indicated temperature rose above the minimum reading of 400 °C by placing the OFT outside the cavity opening and in line with the hot blackbody with the probe tip aimed at ambient surroundings and disappeared when the OFT probe was moved away from the axis of the hot blackbody while still viewing the ambient surface.

A rough estimate of the energy traveling down the axis of the probe, which entered through the bend instead of the probe tip, was performed using Monte Carlo ray tracing. The analysis included a model of the blackbody radiating surface and demonstrated that the radiation entering the probe at the bend may reach about 10% of the total energy traveling along the axis of the OFT. The analysis showed that this percentage remained essentially constant at all axial locations in the portion of the blackbody between the copper electrodes.

A part of the radiant flux entering the bend was likely to be present, and thereby accounted for, during the manufacturer's calibration process. Therefore, it was inappropriate to apply a 10-percent reduction to the radiant power measured by the sensor.

A correction scheme was developed, which utilizes the known temperature (from the reference optical pyrometer) at the center partition to correct the OFT measurements. The measurement uncertainty for the corrected temperature is found to be ± 5 °C.

Chahine et al. [25] described application of similar technique to the Thermogage blackbody. Its radiator is a 300 mm long and 25 mm in diameter graphite tube, which has in the middle a septum subdividing the tube onto two cavities. One cavity is working, the second serves for measuring and control the blackbody temperature by an RT included in the temperature feedback system.

A silica optical fiber bent at $\beta = 45°$ was used to collect thermal radiation from the graphite tube internal surface. Scanning of the cylindrical cavity surface is carried out by linear movement of the silica fiber probe that transmits radiant energy to the RT. The bent silica fiber probe is connected to a multimode optical fiber, whose end is viewed by the radiation thermometer with a usual f/10 optics and a 1 mm target spot.

The bending of the silica fiber was done as follows. The cladding of the silica fiber was removed and it was heated in a hydrogen flame until it began to bend under its own weight. The heating of the fiber was ceased when the bend angle β reached around 45°. Removal of the cladding and bending result in the radiation leakage at the bend of the fiber. Unlike the previous case, when the RT was calibrated together with the light guide, the leakage of the OFT in this work has been determined experimentally at the operating wavelength of 850 nm of the RT using the measurement setup with the integrating sphere. Radiation from the QTH (quartz tungsten halogen) lamp was collimated, filtered by the interference filter with the center wavelength of 850 nm, and focused onto a 1-mm multimode fiber via a fiber coupler. Two measurement has to be made in order to evaluate the leakage.

The first measurement is made when the fiber is entirely inside the integrating sphere of 100 mm in diameter, so its wall is irradiated by both radiation exiting the fiber end and the radiation leaking the fiber at its bend. The second measurement is made when the tip of the fiber is brought out through a small hole in the integrating sphere wall. A digital multimeter measured the photocurrent of a silicon photodiode mounted in the sphere wall. The total radiant flux Φ, launched into the fiber was assessed by placing it entirely inside the integrating sphere. The flux leakage $\Delta\Phi$ from the bent area was measured by the thoroughly aligning fiber tip, so that the ratio $\Delta\Phi/\Phi = 7.3\%$ was measured. Although the measured leakage is relatively large, its actual contribution to the total temperature uncertainty is lower by the order of magnitude because the leakage area is surrounded by the radiating surface at nearly the same temperature as that measured, which partially compensates the radiant flux losses at the bent. The leakage to other areas of the tube and the cavity aperture was assessed as about 10%, so the resulting error due to the leakage is about 0.73%, which leads to the uncertainty in temperature of about 0.7 °C at 1000 °C. Computational account of radiative heat losses from the external surface of the fiber probe through the cavity opening allows evaluating the total uncertainty in measuring temperature distribution as a function of the distance from the graphite tube center. Measured temperatures were confirmed by measurements with the Pt/Pt-Rh thermocouple within ± 1.5 °C accuracy.

Fiber optics thermometry looks very attractive method for measuring temperature distributions but, unfortunately, it has relatively low upper limit due to lack of materials of optical fiber more heat-resistant than sapphire that has the melting temperature of 2030–2050 °C.

8.5.3 Camera-Based Technique

Camera-based technique is the visualization of temperature distribution over a part of the blackbody (usually, the bottom of the radiating cavity) by means of imaging optics. There are many books covering various topics of thermal imaging (see, e.g. Kaplan [75], Williams [183], Vollmer and Möllmann [180]). A measuring camera renders the image captured using false colors assigned to each pixel according to the output signal of a detector (in the scanning devices) or each detector of a detector array (in the non-scanning devices) so that it is possible to derive the radiance or radiance temperature corresponding to each pixel of the output image. In the thermal IR spectral range, we deal with thermographic or thermovision system depending on whether the camera captures an instant image or a sequence of images separated by short time intervals. In the visible and near-IR spectral ranges, we deal with systems, whose principle is similar to digital photography or television. Application of imaging technique to evaluation of temperature nonuniformities of the blackbody radiating surfaces has a relatively short history although the reverse case—the calibration of a thermal imager against a blackbody—has been used for a long time. Modern measuring cameras are, as a rule, non-scanning thermovision system. Due to high cost, it is hardly expedient to purchase a high-quality thermovision measuring system exclusively for evaluation of blackbody temperature distributions. Therefore, when these measurements are carried out, researchers are forced to use any available camera, often with an outdated design, sometimes those, whose production has long been discontinued. For such measuring systems, we do not give their characteristics, except for the most necessary.

The early thermovision measuring devices had the LN_2-cooled IR detectors conjugated with optical scanners that form the camera's FOV. Currently, the scanning devices are superseded by digital cameras with detector arrays; cooling with liquid nitrogen has been substituted by miniature Stirling cycle refrigerators and solid-state Peltier elements. The newest thermal imagers use focal point arrays (FPAs) of uncooled microbolometers. Nevertheless, many old thermal imagers, including scanning ones, are still in use in some laboratories and capable to interact with modern PC via custom-made drivers and software.

An example of application of the scanning thermovision device to investigation of the low-temperature blackbody BB100-V1 is described by Ogarev et al. [119]. This blackbody was designed at the VNIIOFI (Russia) as a reference source of IR radiation in the wavelength range from 1.5 to 15 μm for ground calibration of space-borne radiometers. The BB100-V1 (its design was shown schematically in Fig. 7.14) is the large-aperture (100 mm in diameter), liquid-circulation blackbody capable to operate at temperatures from −45 to 200 °C in the vacuum chamber using the external thermostat with the heat-transfer liquid KRYO-51 (polydimethylphenylsiloxane). Miniature Pt100 sensors are embedded into the V-grooved bottom and cylindrical walls of the cavity; one of them is included in the temperature stabilization feedback loop. Numerical modeling using FEA software performed at the design stage predicted excellent thermal uniformity of the radiating cavity [127].

For experimental verification of sufficient homogeneity of temperature field over the cavity bottom, the BB100-V1 was placed inside the vacuum LN_2-cooled chamber so that the temperature distribution across the cavity bottom could be viewable through the chamber's IR window. The thermal imager AGA Thermovision 780 placed outside the chamber was used to scan the cavity bottom. The AGA-780 is a scanning instrument that employs the LN_2-cooled HgCdTe (8–14 μm spectral band) or InSb (3–7.6 μm spectral band) detector. Scanning performed using two refractive elements rotating about mutually perpendicular axes provides two-dimensional pseudo-color thermal imaging with 280 lines per frame and 100 elements per line. Two sample scans (for cavity reference temperatures of ± 30 °C) across the BB100-V1 aperture are shown in Fig. 8.40. The upper screenshots present the color maps of the scanned two-dimensional temperature distributions; the lower ones depict one-dimensional horizontal distributions and their central parts.

Palchetti et al. [120] reported investigation of temperature non-uniformity over the bottoms of blackbodies designed for in-flight calibration of the polarizing Fourier-transform spectrometer for measuring the Earth's upwelling spectral radiance during REFIR (Radiation Explorer in the Far InfraRed) space mission planned and funded by European Union. To the date of publication, the REFIR prototype passed a field campaign on the board of a stratospheric balloon. Three on-board blackbodies operating at temperatures from −10 to 120 °C are used for the Fourier-transform spectrometer calibration. To maintain temperature of each blackbody, a Peltier element and a resistive heater are used. The larger blackbody, RBB, is used as reference for the second input port operating with the nearly collimated beam. Two smaller blackbodies serve as the hot and cold calibration sources (HBB and CBB, respectively) and operate with the focused beam. Sectional views of these two types of blackbodies are shown in Fig. 8.41. Three Pt100 sensors are embedded in the bottom and sidewall of each blackbody for accurate temperature measurement. The NTC thermistor having electrical resistance of 10 kΩ at 25 °C is used for active stabilization of the source temperature.

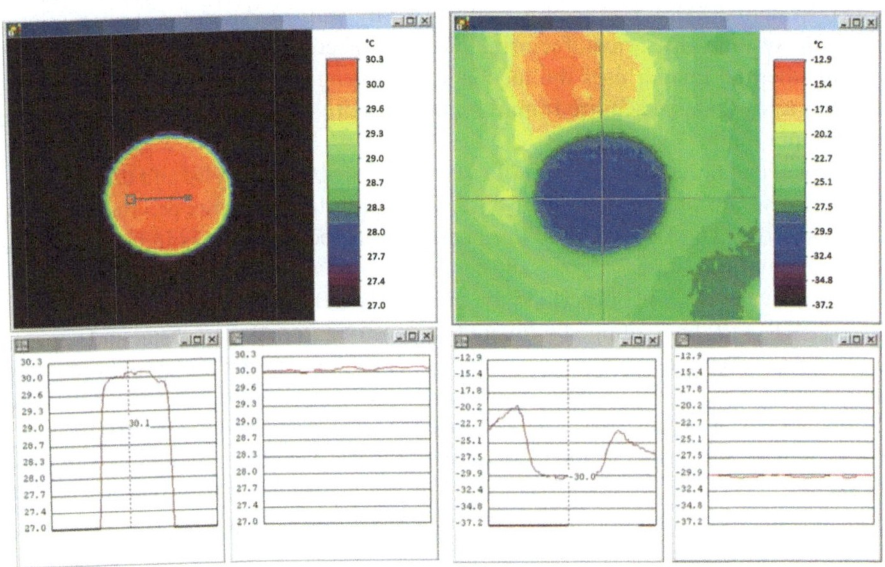

a b

Fig. 8.40 Temperature maps and the aperture scans of the BB100-V1 at temperature settings **a** $T = 30$ °C and **b** $T = -30$ °C made by a thermal imager placed outside the cryo-vacuum chamber near the IR window. Reproduced from [119] with permission of Springer Nature

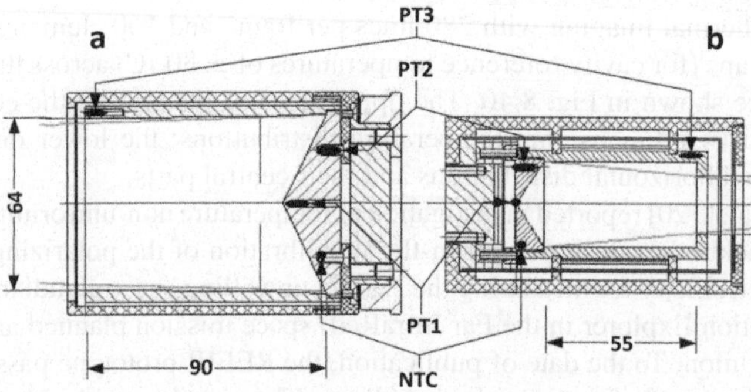

Fig. 8.41 Sectional views of the REFIR calibration blackbodies: **a** RBB, **b** HBB and CBB, PT1, PT2, and PT3 are the Pt100 temperature sensors, and NTC is a 10 kΩ NTC thermistor. All dimensions are in mm. Reproduced from [120] with permission of Elsevier

The three Pt100 sensors indicate radial temperature gradient across the bottom and axial gradient along the lateral wall. The Pt100 sensors measure temperature under the 27.5 μm layer of Xylan high-emissivity coating with low thermal conductivity (it was supposed that the thermal conductivity of Xylan is close to the typical value of 0.25 W·K^{-1}m^{-1} for PTFE). To make correction for the temperature drop, additional measurements of the radiance temperature distributions in the 3–7.4 μm spectral range across the cavity bottoms have been performed using the Probeye® Thermal Video System 2000. It has been manufactured by Hughes Aircraft Company; later, by FLIR Systems, Inc.; manufacturing has been discontinued many years ago.

This thermal imager uses the mechanical Sterling refrigerator to cool down the detector array and can acquire frames on the hard drive, process them using PC, and then output on the TV screen.

Vertical and horizontal sections of measured distributions (solid and dashed lines, respectively) are presented in Fig. 8.42. Operating temperatures are measured by Pt100 sensors. The FOV of the Fourier transform spectrometer is indicated by the vertical grey lines.

Morozova et al. [111] described the blackbody BB100K1 designed based on the BB100-V1 [119] but equipped with another type of external thermostat and capable to operate in vacuum, in inert gas or dry air, or in open air. Its maximum operating temperature is 90 °C; minimum operating temperature is −60 °C in vacuum, −40 °C in inert gas or dry air, and −20 °C in open air. The destination of the BB100K1 is practically the same as of its predecessor, BB100-V1. The ThermaCAM SC640 from FLIR Systems, Inc. was used for thermal imagery. This hand-held thermal imager uses the uncooled microbolometer FPA with resolution of 640 × 480 pixels and the spectral sensitivity range from 7.5–13 μm.

Temperature distributions over the cavity bottom at 0 °C indicated by the central Pt100 sensor were measured by this thermal imager under ambient air conditions. The range of temperature change across the 100-mm aperture is about 0.5 °C and 1 °C in horizontal and vertical directions, respectively, is shown in Fig. 8.43. The

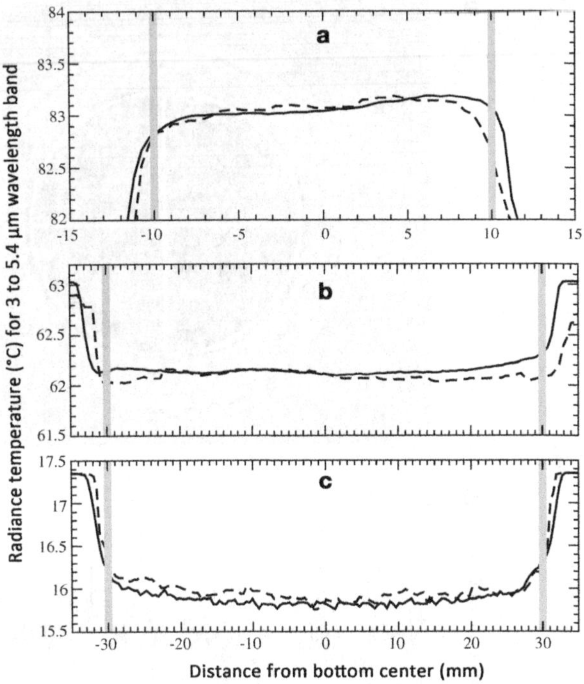

Fig. 8.42 Profiles of radiance temperature corresponding to the wavelength band from 3 μm to 7.4 μm across the bottoms of the REFIR calibration blackbodies: **a** HBB at the operating temperature about 83 °C, **b** RBB at the operating temperature of about 63 °C, and **c** RBB at the operating temperature of about 14 °C. Reproduced from [120] with permission of Elsevier

larger magnitude of temperature change in the vertical direction caused, most likely, by free convection because measurements were carried out in the open air.

The applicability of camera-based technique is not limited by low-temperature blackbodies and the thermal IR spectral range. Anhalt et al. [4] described application of a radiance camera of visible spectral range to investigate the time-resolved temperature inhomogeneity within high-temperature fixed-point cells at temperatures up to 3000 K. According to the Wien displacement law, the Planck distribution's peak corresponds to the wavelength of around 1 μm; therefore, it is desirable to have an imaging sensor with the sensitivity bandwidth in the red part of visible and the near-IR wavelength ranges.

The main aim of the experiment was to watch evolution of the temperature distribution over the bottom of the fixed-point cell cavity. For this purpose, the radiometric camera should have a high spatial resolution and ability to acquire a sufficient number of images as the phase transition proceeds, i.e. the frame rate must be of several tenth per second.

The luminance and color measuring camera LMK 98-3 Color from TechnoTeam Bildverarbeitung GmbH (Germany) was chosen to solve this problem. The camera LMK 98-3 is the imaging radiometer that uses the CCD (charge-coupled device) array as the imaging sensor. Image of a scene under measurement is projected onto the CCD matrix by the camera lens. A wheel with six filters is arranged between the lens and the CCD matrix, which converts the incident radiant flux into electric charges

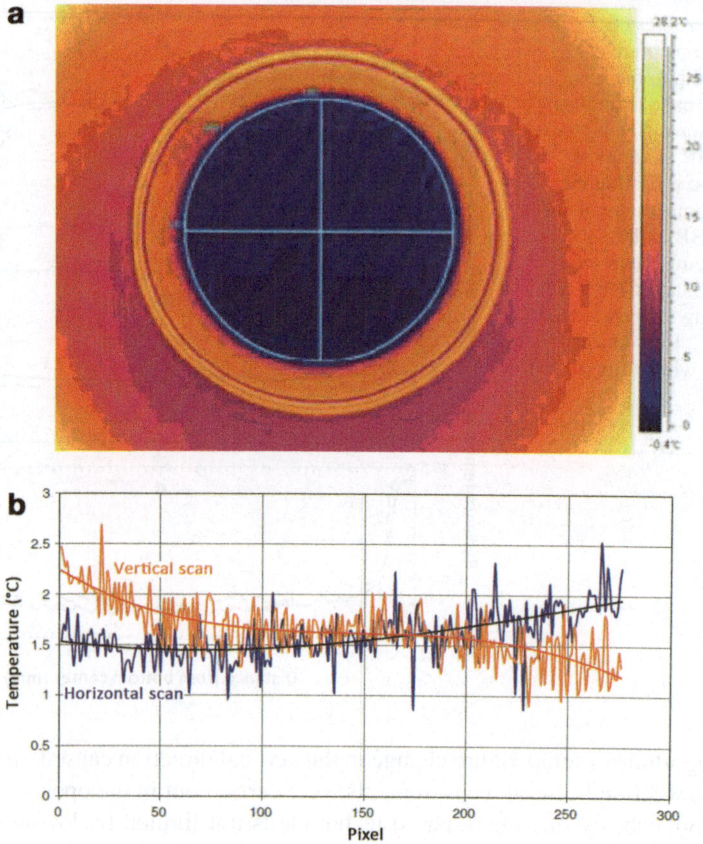

Fig. 8.43 Spatial distribution of radiation temperature across 100 mm aperture of BB100K1: **a** color map at $T = 0$ °C and **b** vertical and horizontal temperature profiles with added trendlines. Reproduced from [111] with permission of Springer Nature

according to the response of each pixel. The spectral responsivity of the CCD array can be modified using optical filters. Such effects as the dark signal, non-linearity and spatial non-uniformity of the response, as well as the CCD-specific phenomena are subjected to hardware correction. Readers can familiarize themselves with the basic principles and the state-of-the-art of the CCD imaging in the brief review written by Lesser [89]. For an in-depth study of the scientific CCD imaging, we can recommend the monographs by Holst [62] or Theuwissen [178].

The CCD imaging is a rapidly developing branch of electronics, so the LMK 98-3 resolution of 1.8 MP (megapixels) may seem modest compared to modern professional digital cameras, for which the resolution of 28 MP at the performance of 10 frames per second are not something extraordinary. However, characteristics of the LMK 98-3 camera were quite enough for solution of the problems stated.

For realization of phase transitions in the fixed-point cells of copper (1084.62 °C) and high-temperature eutectics Pt–C (1738 °C) and ZrC–C (2882 °C), Anhalt et al. [4] used the pyrolytic graphite blackbody BB3200pg [142] as a high-temperature furnace.

Instead of reversed cone, the flat V-grooved bottom made of ordinary graphite was mounted in the middle part of the heater composed from pyrolytic graphite ring. The rear cavity formed in such a way served as a radiation source for the radiation thermometer included in the feedback loop of the temperature stabilization system. The fixed-point cell was placed in the most isothermal part of the frontal cavity.

Prior to the application of the LMK 98-3 camera to investigation of the phase transition proceeding, its usability for the measurement of the radiance distributions in high-temperature blackbodies has been tested on the BB3200pg without the fixed-point cell at temperature of about 2500 °C. The camera without a photometric filter was focused onto the cavity bottom plane. The measured radiance distribution is presented in Fig. 8.44. The concentric pattern is formed by V-grooved graphite bottom and caused signal variations of around 0.5%, which corresponds to the temperature variations of about 2 K between the peaks and the valleys of concentric V-grooves. The signal decreases by about 2.5% toward the bottom center that corresponds to the temperature decrease of about 10 K.

The main drawbacks of the camera-based technique are the same as for measuring temperature distributions using radiation thermometers with high f-number optics. Both these techniques allow measuring only directly viewable part of the cavity. For both techniques, the dependence of the effective emissivity on the viewing angle may introduce noticeable distortion in measured temperatures. In fact, without knowing of the effective emissivity, only distributions of radiance temperature can be obtained.

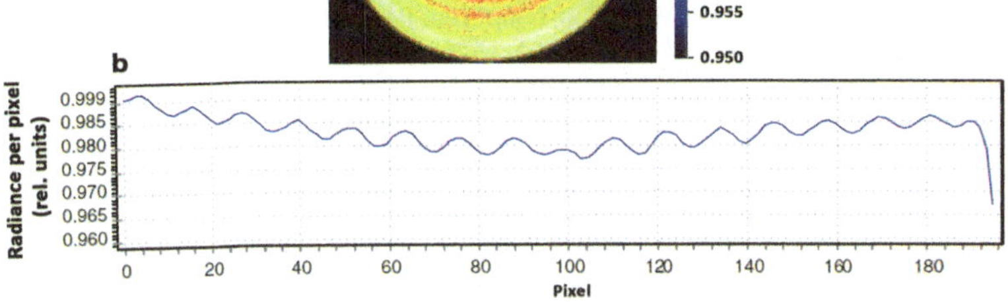

Fig. 8.44 Relative radiance distribution over the BB3200pg V-grooved bottom at about 2500 °C: **a** color map and **b** its horizontal section along the dashed white line. Reproduced from [4] with permission of IoP Publishing

The CCD FPA allows obtaining instantaneous temperature distribution, which is important for solving the main problem—study of melting dynamics for metals and carbonaceous eutectic alloys.

Bünger et al. [22] described the calibration of the similar camera LMK 98-4 Color from the same manufacturer. The camera, equipped with the narrowband interference filters in the visible spectral range for temperature measurements above 1200 K, was characterized with respect to its temperature response traceable to ITS-90 and with respect to the absolute spectral radiance responsivity. The calibration traceable to ITS-90 was performed at a high-temperature blackbody source using a radiation thermometer as a transfer standard. The use of Planck's law and the absolute spectral radiance responsivity of the camera system allows the determination of the thermodynamic (radiance) temperature. For the determination of the absolute spectral radiance responsivity, a monochromator-based setup with a supercontinuum white-light laser source was developed.

The CCD-camera was also characterized with respect to the dark signal-non-uniformity, the non-uniformity of the photoresponse, the nonlinearity, and the SSE. The results of two different calibration schemes in the temperature range from 1200 to 1800 K gave the expanded (at $k = 2$) uncertainty for the absolute spectral responsivity of the camera of 0.56%.

References

1. A. Achmadi, E. Juliastuti, A. Handojo et al., Light source for scanning method of size-of-source effect measurement. Int. J. Thermophys. **36**, 3330–3340 (2015)
2. M.D. Aggarwal, A.K. Batra, P. Guggilla, P., et al., *Pyroelectric Materials for Uncooled Infrared Detectors: Processing, Properties, and Applications.* NASA/TM–2010–216373 (NASA, Marshall Space Flight Center, AL, 2010), https://ntrs.nasa.gov/archive/nasa/casi.ntrs.nasa.gov/20110008068.pdf. Accessed 15 Mar 2020
3. M.G. Ahmed, K. Ali, F. Bourson et al., Comparison of the copper blackbody fixed-point cavities between NIS and LNE-Cnam. Meas. Sci. Technol. **24**, 095902 (2013)
4. K. Anhalt, D. Lindner, U. Krüger et al., Camera-based investigation of high-temperature fixed points for radiometric application. Metrologia **46**, S260–S265 (2009)
5. ASTM E1256-17. *Standard Test Methods for Radiation Thermometers (Single Waveband Type)* (ASTM International, West Conshohocken, PA, 2017)
6. ASTM E2758-15a. *Standard Guide for Selection and Use of Wideband, Low Temperature Infrared Thermometers* (ASTM International, West Conshohocken, PA, 2015)
7. M. Bart, E.W.M. van der Ham, P. Saunders, A new method to determine the size-of-source effect. Int. J. Thermophys. **28**, 2111–2117 (2007)
8. A.K. Batra, M.D. Aggarwal, *Pyroelectric Materials: Infrared Detectors. Particle Accelerators and Energy Harvesters* (SPIE Press, Bellingham, WA, 2013)
9. W. Bich, M.G. Cox, P.M. Harris, Evolution of the '*Guide to the Expression of Uncertainty in Measurement*'. Metrologia **43**, S161–S166 (2006)
10. P. Bloembergen, On the correction for the size-of-source effect corrupted by background radiation, in *Proceedings of TEMPMEKO '99, the 7th International Symposium on Temperature and Thermal Measurement in Industry and Science*, ed. by J.F. Dubbeldam, M.J. de Groot, vol. 2 (NMi Van Swinden Laboratorium, Delft, Netherlands, 1999), pp. 607–612
11. P. Bloembergen, On the uncertainty in the correction for the size-of-source effect. Metrologia **46**, 544–553 (2009)

12. P. Bloembergen, Analytical representations of the size-of-source effect. Metrologia **46**, 534–543 (2009)
13. P. Bloembergen, Y. Yamada, Measurement of thermodynamic temperature above the silver point on the basis of the scheme n = 2. Int. J. Thermophys. **32**, 45–67 (2011)
14. P. Bloembergen, Y. Yamada, N. Yamamoto et al., Realizing the high-temperature part of a future ITS with the aid of eutectic metal-carbon fixed points. AIP Conf. Proc. **684**, 291–296 (2003)
15. P. Bloembergen, B.B. Khlevnoy, P. Jimeno Largo et al., Spectral and total effective emissivity of a high-temperature fixed-point radiator considered in relation to the temperature drop across its back wall. Int. J. Thermophys. **29**, 370–385 (2008)
16. P. Bloembergen, L.M. Hanssen, S.N. Mekhontsev et al., A determination study of the cavity emissivity of the eutectic fixed points Co–C, Pt–C, and Re–C. Int. J. Thermophys. **32**, 2623–2632 (2011)
17. L.-P. Boivin, Study of bandwidth effects. Appl. Opt. **41**, 1929–1935 (2002)
18. F. Bourson, S. Briaudeau, B. Rougié et al., Determination of the furnace effect of two high-temperature furnaces on metal-carbon eutectic points. AIP Conf. Proc. **1552**, 380–385 (2013)
19. W. Budde, *Physical Detectors of Optical Radiation* (Optical Radiation Measurements, vol. 4) (Academic Press, New York, 1983)
20. H. Budzier, G. Gerlach, *Thermal Infrared Sensors. Theory, Optimisation and Practice* (Wiley, Chichester, UK, 2011)
21. J.D. Buys, K. van Gadow, A PASCAL program for fitting non linear regression models on a micro-computer. EDV in Medizin und Biologie **18**, 105–107 (1987), http://www.forestgrowth.wzw.tum.de/fileadmin/publications/239.pdf. Accessed July 17, 2019
22. L. Bnger, K. Anhalt, R.D. Taubert, Traceability of a CCD-camera system for high-temperature measurements. Int. J. Thermophys. **36**, 1784–1802 (2015)
23. P. Castro, G. Machin, P. Bloembergen et al., Thermodynamic temperatures of high-temperature fixed points: uncertainties due to temperature drop and emissivity. Int. J. Thermophys. **35**, 1341–1352 (2014)
24. CCT: *Mise en pratique* for the definition of the kelvin in the SI. SI Brochure, 9th ed. Appendix 2 (2019), https://www.bipm.org/utils/en/pdf/si-mep/SI-App2-kelvin.pdf. Accessed 21 Feb 2020
25. K. Chahine, M. Ballico, J. Reizes et al., Temperature profile measurement of a graphite tube furnace using optical fibre and platinum thermocouples, in *Proceedings of the Sixth Biennial Conference of Metrology Society of Australia* (Australian National University, Canberra, 2005), pp. 19–21. https://opus.lib.uts.edu.au/bitstream/10453/7262/1/2005002546.pdf. Accessed 16 July 2019
26. K. Chahine, M. Ballico, J. Reizes et al., Optimization of a graphite tube blackbody heater for a Thermogage furnace. Int. J. Thermophys. **29**, 386–394 (2008)
27. CHINO: Standard Radiation Thermometer. Model IR RST Series. (CHINO Corp., 2018), https://www.chino.co.jp/english/wp/wp-content/themes/chino_en/pdf/IR-RST_PSE-600.pdf. Accessed 12 Oct. 2019
28. J.F. Clare, Correction for nonlinearity in the measurement of luminous flux and radiant power. Meas. Sci. Technol. **13**, N38–N41 (2002)
29. P. Coates, D. Lowe, *The Fundamentals of Radiation Thermometers* (CRC Press, Boca Raton, FL, 2017)
30. L. Coslovi, F. Righini, Fast determination of the nonlinearity of photodetectors. Appl. Opt. **19**, 3200–3203 (1980)
31. R.H. Davies, A.T. Dinsdale, J.A. Gisby et al., MTDATA—thermodynamic and phase equilibrium software from the National Physical Laboratory. Calphad **26**, 229–271 (2002)
32. J. De Lucas, J.J. Segovia, Measurement and analysis of the temperature gradient of blackbody cavities, for use in radiation thermometry. Int. J. Thermophys. **39**, 57 (2018)
33. A. Daniels, *Field Guide to Infrared Systems, Detectors, and FPAs*, 3rd edn. (SPIE Press, Bellingham, WA, 2018)
34. E.L. Dereniak, D.G. Crowe, *Optical Radiation Detectors* (Wiley, New York, 1984)

35. D.P. DeWitt, G.D. Nutter (eds.), *Theory and Practice of Radiation Thermometry* (Wiley, New York, 1988)
36. W. Dong, D. Lowe, G. Machin et al., Investigation of the furnace effect in cobalt-carbon high-temperature fixed-point cells. Measurement **106**, 88–94 (2017)
37. E.K. Ejigu, Simulating radiation thermometer temperature measurement error from the performance change of an interference filter due to polarization effect. Measurement **114**, 471–477 (2018)
38. J. Envall, S.N. Mekhontsev, Y. Zong, Spatial scatter effects in the calibration of IR pyrometers and imagers. Int. J. Thermophys. **30**, 167–178 (2009)
39. G.P. Eppeldauer, Traceability of photocurrent measurements to electrical standards. MĀPAN – J. Metrology Soc. India **24**, 193–202 (2009)
40. G.P. Eppeldauer, H.W. Yoon, AC-mode short-wavelength IR radiation thermometers for measurement of ambient temperatures. Int. J. Thermophys. **29**, 1041–1051 (2008)
41. G.P. Eppeldauer, S.W. Brown, K.R. Lykke, Transfer standard filter radiometers: applications to fundamental scales, in *Optical Radiometry*, ed. by A. Parr, R.U. Datla, J.L. Gardner (Academic Press, Amsterdam, Netherlands, 2005), pp. 155–211
42. G.P. Eppeldauer, H.W. Yoon, D.G. Jarrett et al., Development of an *in situ* calibration method for current-to-voltage converters for high-accuracy SI-traceable low dc current measurements. Metrologia **50**, 509–517 (2013)
43. D.E. Erminy, Scheme for obtaining integral and fractional multiples of a given radiance. J. Opt. Soc. Am. **53**, 1448–1449 (1963)
44. B. Fellmuth, K.D. Hill, J.V. Pearce et al., *Guide to the Realization of the ITS-90. Fixed Points: Influence of Impurities* (Consultative Committee for Thermometry under the auspices of the International Committee for Weights and Measures, 2015), https://www.bipm.org/utils/common/pdf/ITS-90/Guide-ITS-90-Impurities-2015.pdf. Accessed 30 Mar 2020
45. J. Fischer, H.J. Jung, Determination of the thermodynamic temperatures of the freezing points of silver and gold by near-infrared pyrometry. Metrologia **26**, 245–252 (1989)
46. J. Fischer, L. Fu, Photodiode nonlinearity measurement with an intensity stabilized laser as a radiation source. Appl. Opt. **32**, 4187–4194 (1993)
47. J. Fischer, M. Battuello, M. Sadli et al., CCT/03-03, in *Uncertainty Budgets for Realisation of Scales by Radiation Thermometry* (BIPM, 2003), http://www.bipm.org/cc/CCT/Allowed/22/CCT03-03.pdf. Accessed 30 Mar 2020
48. J. Fischer, M. Battuello, M. Sadli et al., Uncertainty budgets for realization of ITS-90 by radiation thermometry. AIP Conf. Proc. **684**, 631–638 (2003)
49. J. Fischer, P. Saunders, M. Sadli et al., *Uncertainty Budgets for Calibration of Radiation Thermometers Below the Silver Point*. Version 1.71 (CCT-WG5 on Radiation Thermometry, 2008), http://www.bipm.org/wg/CCT/CCT-WG5/Allowed/Miscellaneous/Low_T_Uncertainty_Paper_Version_1.71.pdf. Accessed 3 Mar 2020
50. J. Fischer, M. de Podesta, K.D. Hill et al., Present estimates of the differences between thermodynamic temperatures and the ITS-90. Int. J. Thermophys. **32**, 12–25 (2011)
51. V.M. Fuksov, A.I. Pohodun, M.S. Matveyev, Experimental and numerical investigation of the temperature field of a fixed-point cavity. Int. J. Thermophys. **32**, 337–347 (2011)
52. I.E. Gordon, L.S. Rothman, C. Hill et al., The HITRAN2016 molecular spectroscopic database. J. Quant. Spectrosc. Radiat. Transfer **203**, 3–69 (2017)
53. J.G. Graeme, *Photodiode Amplifiers: OP AMP Solutions* (McGraw-Hill, Boston, MA, 1996)
54. I.A. Grigoryeva, B.B. Khlevnoy, M.V. Solodilov, Reproducibility of WC–C high-temperature fixed point. Int. J. Thermophys. **38**, 69 (2017)
55. B. Gutschwager, R. Gärtner, J. Hollandt, An Infrared Precision Radiation Thermometer for the Calibration of Remote Sensing Instrumentations under Vacuum. Proc. SPIE **7477**, 747719 (2009)
56. A. Haapalinna, T. Kübarsepp, P. Kärhä et al., Measurement of absolute linearity of photodetectors with a diode laser. Meas. Sci. Technol. **10**, 1075–1078 (1999)
57. Hamamatsu: Si photodiodes S1336 series (Hamamatsu Photonics K.K., Solid State Division, 2015), https://www.hamamatsu.com/resources/pdf/ssd/s1336_series_kspd1022e.pdf. Accessed 39 Mar 2020

58. X. Hao, Z. Yuan, X. Lu et al., Size-of-source effect difference between direct and indirect methods of radiation thermometers. Int. J. Thermophys. **32**, 1655–1663 (2011)
59. J. Hartmann, S. Schiller, R. Friedrich et al., Non-isothermal temperature distribution and resulting emissivity corrections for the high temperature blackbody BB3200, in *Proceedings of TEMPMEKO 2001, 8th International Symposium on Temperature and Thermal Measurements in Industry and Science*, vol. 1, ed. by B. Fellmuth, J. Seidel, G. Scholz (VDE Verlag, Berlin, 2002), pp. 227–232
60. J. Hartmann, J. Hollandt, B. Khlevnoy et al., Blackbody and other calibration sources, in *Radiometric Temperature Measurements. I. Fundamentals*, ed. by Z.M. Zhang, B.K. Tsai, G. Machin (Academic Press, Amsterdam, Netherlands, 2010), pp. 241–295
61. K.D. Hill, D.J. Woods, Exploring the size-of-source and distance effects of radiation thermometers, in *Proceedings of TEMPMEKO 2004, 9th International Symposium on Temperature and Thermal Measurements in Industry and Science*, vol. 1, ed. by D. Zvizdić (FSB/LPM, Zagreb, Croatia, 2005), pp. 599–604
62. G.C. Holst, *CCD Arrays, Cameras, and Displays*, 2nd edn. (SPIE Press, Bellingham, WA, 1998)
63. K.S. Hong, D.-H. Lee, S. Park et al., A new method for measuring the relative spectral responsivity of detectors based on a pulsed OPO tunable from 210 to 2000 nm, in *Proceedings of the Conference on Precision Electromagnetic Measurements* (Washington, DC, 2012), https://doi.org/10.1109/cpem.2012.6251106
64. T.J. Horn, A.N. Abdelmessih, *Experimental and Numerical Characterization of a Steady-State Cylindrical Blackbody Cavity at 1100 Degrees Celsius*. NASA/TM-2000-209022 (NASA Dryden Flight Research Center, Edwards, CA, 2000), https://www.nasa.gov/centers/dryden/pdf/88687main_H-2403.pdf. Accessed 17 July 2019
65. IEC/TS 62492-1. *Industrial Process Control Devices—Radiation Thermometers—Part 1: Technical Data for Radiation Thermometers* (International Electrotechnical Commission, Geneva, Switzerland, 2008)
66. IEC/TS 62492-2. *Industrial Process Control Devices—Radiation Thermometers—Part 2: Determination of the Technical Data for Radiation Thermometers* (International Electrotechnical Commission, Geneva, Switzerland, 2013)
67. M. Imbe, Y. Yamada, Experimental study of the furnace effect with the copper point blackbody. Meas. Sci. Technol. **27**, 125020 (2016)
68. J. Ishii, M. Kobayashi, F. Sakuma et al., DC-operated InSb radiation thermometer for precision measurement of near room temperatures, in *Proceedings of TEMPMEKO '99—The 7th International Symposium on Temperature and Thermal Measurements in Industry and Science*, vol. 2, ed. by J.F. Dubbeldam, M.J. de Groot (NMi Van Swinden Laboratorium, Delft, Netherlands, 1999), pp. 625–630
69. J. Ishii, A. Ono, Low-temperature infrared radiation thermometry at NMIJ, in:*Temperature: Its Measurement and Control in Science and Industry*, vol. 7, ed. by D.C. Ripple, part 2. AIP Conf. Proc. **684**, 657–662 (2003)
70. J. Ishii, Y. Yamada, N. Sasajima et al., Radiation Thermometry standards at NMIJ from −30 °C to 2800 °C. AIP Conf. Proc. **1552**, 666–671 (2013)
71. JCGM 100:2008. *GUM. Evaluation of Measurement Data. Guide to the Expression of Uncertainty in Measurement. (GUM 1995 with Minor Corrections)* (BIPM Joint Committee for Guides in Metrology, Paris, 2008)
72. JCGM 101:2008. *GUM. Evaluation of Measurement Data—Supplement 1 to the "Guide to the Expression of Uncertainty in Measurement" —Propagation of Distributions Using a Monte Carlo Method* (BIPM Joint Committee for Guides in Metrology, Paris, 2008)
73. T.P. Jones, J. Tapping, A precision photoelectric pyrometer for the realization of the IPTS-68 above 1064.43 °C. Metrologia **18**, 23–32 (1982)
74. H.J. Jung, Spectral nonlinearity characteristics of low noise silicon detectors and their application to accurate measurements of radiant flux ratios. Metrologia **15**, 173–181 (1979)
75. H. Kaplan, *Practical Applications of Infrared Thermal Sensing and Imaging Equipment*, 3rd edn. (SPIE Press, Bellingham, WA, 2007)

76. KE: Linearpyrometer LP4. Technical Documentation KE 256-6.2007 (KE-Technologie GmbH, Stuttgart, Germany, 2010), http://www.apsmyx.com/uploadfile/2016/1027/20161027055515539.pdf. Accessed 29 Mar 2020
77. T. Keawprasert, Y. Yamada, J. Ishii, Pilot comparison of radiance temperature scale realization between NIMT and NMIJ. Int. J. Thermophys. **36**, 315–326 (2015)
78. W. Kester, S. Wurcer, Ch. Kitchin, High impedance sensors, in *Op Amp Applications Handbook*, ed. by W. Jung (Newnes, Amsterdam, Netherlands, 2005), pp. 257–305
79. B.B. Khlevnoy, I.A. Grigoryeva, Long-term stability of WC-C peritectic fixed point. Int. J. Thermophys. **36**, 367–373 (2015)
80. B. Khlevnoy, M. Sakharov, S. Ogarev et al., Investigation of furnace uniformity and its effect on high-temperature fixed-point performance. Int. J. Thermophys. **29**, 271–284 (2008)
81. B.B. Khlevnoy, M.L. Samoylov, I.A. Grigoryeva et al., Development of high-temperature blackbodies and furnaces for radiation thermometry. Int. J. Thermophys. **32**, 1686–1696 (2011)
82. B.B. Khlevnoy, I.A. Grigoryeva, D.A. Otryaskin, Development and investigation of WC–C fixed-point cells. Metrologia **49**, S59–S67 (2012)
83. M.A. Kinch, *State-of-the-Art Infrared Detector Technology* (SPIE Press, Bellingham, WA, 2014)
84. R.H. Kingston, *Optical Sources, Detectors, and Systems. Fundamentals and Applications* (Academic Press, San Diego, CA, 1995)
85. H.J. Kostkowski, R.D. Lee, *Theory and Methods of Optical Pyrometry*. NBS Monograph 41 (National Bureau of Standards, U.S. Department of Commerce, 1962)
86. H.J. Kostkowski, The relative spectral responsivity and slit-scattering function of a spectroradiometer, in *Self-Study Manual on Optical Radiation Measurements*, ed. by F.E. Nicodemus, Part I—Concepts, Chapters 7, 8, and 9 (National Bureau of Standards, U.S. Department of Commerce, Washington, DC, 1979)
87. O. Kozlova, S. Briaudeau, L. Rongione et al., Calibration of radiation thermometers up to 3000 °C: effective emissivity of the source. Int. J. Thermophys. **36**, 1726–1742 (2015)
88. T.C. Larason, S.S. Bruce, A.C. Parr, *Spectroradiometric Detector Measurements: Part I — Ultraviolet Detectors and Part II —Visible to Near-Infrared Detectors*. NIST Spec. Publ. 250-41 (National Institute of Standards and Technology, U.S. Department of Commerce, Washington, DC, 1988)
89. M. Lesser, Charge coupled device (CCD) image sensors, in *High Performance Silicon Imaging. Fundamentals and Applications of CMOS and CCD Sensors*, ed. by D. Durini (Elsevier–Woodhead Publ., Amsterdam, Netherlands, 2014), pp. 78–97
90. F. Liebmann, T. Kolat, Radiation thermometer size-of-source effect testing using aperture. AIP Conf. Proc. **1552**, 613–618 (2013)
91. D. Lowe, H.C. McEvoy, M. Owen, A Mid-IR pyrometer calibrated with high-temperature fixed points for improved scale realization to 2,500 °C. Int. J. Thermophys. **28**, 2059–2066 (2007)
92. D.H. Lowe, A.D.W. Todd, R. Van den Bossche et al., The equilibrium liquidus temperatures of rhenium–carbon, platinum–carbon and cobalt–carbon eutectic alloys. Metrologia **54**, 390–398 (2017)
93. C.K. Ma, Experimental investigation of the temperature drop across the wall of a copper-freezing-point blackbody. AIP Conf. Proc. **684**, 651–656 (2003)
94. L. Ma, F. Sakuma, Improvement of size of source effect measurement of standard radiation thermometers, in *Proceedings of the SICE Annual Conference* (Kagawa University, Japan, 2007), pp. 1743–1748. https://doi.org/10.1109/sice.2007.4421265
95. G. Machin, The kelvin redefined. Meas. Sci. Technol. **29**, 022001 (2018)
96. G. Machin, L. Wright, D. Lowe et al., Optimizing the implementation of high-temperature fixed points through the use of thermal modeling. Int. J. Thermophys. **29**, 261–270 (2008)
97. G. Machin, K. Anhalt, P. Bloembergen et al., Realisation and dissemination of thermodynamic temperature above the silver point (1234.93 K). CCT/10-12/rev1, version 8 (2010), https://www.bipm.org/cc/CCT/Allowed/25/D12r_MeP-HT_v8.pdf. Accessed 28 Apr 2020

98. G. Machin, K. Anhalt, P. Bloembergen et al., *MeP-K Absolute Primary Radiometric Thermometry* (BIPM, 2017), https://www.bipm.org/utils/en/pdf/si-mep/MeP-K-2018_Absolute_Primary_Radiometry.pdf. Accessed Apr 30, 2020
99. G. Machin, K. Anhalt, P. Bloembergen et al., *MeP*-K Relative Primary Radiometric Thermometry (2017), https://www.bipm.org/utils/en/pdf/si-mep/MeP-K-2018_Relative_Primary_Radiometry.pdf. Accessed 24 Feb 2020
100. H.A. Macleod, *Thin-Film Optical Filters*, 5th edn. (CRC Press, Boca Raton, FL, 2018)
101. A. Manoi, P. Saunders, Size-of-source effect in infrared thermometers with direct reading of temperature. Int. J. Thermophys. **38**, 101 (2017)
102. A. Manoi, P. Wongnut, X. Lu et al., $T - T_{90}$ for radiation thermometry realization above the copper point. Int. J. Thermophys. **41**, 28 (2020)
103. A. Manoi, P. Wongnut, X. Lu et al., Calibration of standard radiation thermometers using two fixed points. Metrologia **57**, 014002 (2020)
104. M.S. Matveev, A.Yu. Makarenko, V.G. Tsorin, The influence of the polarization of the radiation on the accuracy of the transmission of the dimensions of the unit of temperature in radiation thermometry. Meas. Techniques **47**, 908–914 (2004)
105. M.S. Matveyev, New method for measuring the size-of-source effect in standard radiation thermometry, in *Proceedings of TEMPMEKO 2001, 8th International Symposium on Temperature and Thermal Measurements in Industry and Science*, vol. 1, ed. by B. Fellmuth, J. Seidel, G. Scholz (VDE Verlag, Berlin, 2002), pp. 167–171
106. H.C. McEvoy, D.H. Lowe, M. Owen, The NPL InSb-based radiation thermometer for the medium temperature range (< 962 °C). Int. J. Thermophys. **28**, 2067–2075 (2007)
107. H.C. McEvoy, M. Sadli, F. Bourson et al., A comparison of the NPL and LNE-Cnam silver and copper fixed-point blackbody sources, and measurement of the silver/copper temperature interval. Metrologia **50**, 559–571 (2013)
108. K.D. Mielenz, K.L. Eckerle, Spectrophotometer linearity testing using the double-aperture method. Appl. Opt. **11**, 2294–2303 (1972)
109. A. Miklavec, J. Drnovšek, I. Pušnik, Procedure for automated evaluation of a blackbody and a surface calibrator with a radiation thermometer. Int. J. Thermophys. **32**, 1674–1685 (2011)
110. W. Minkina, S. Dudzik, *Infrared Thermography. Errors and Uncertainties* (Wiley, Chichester, UK, 2009)
111. S.P. Morozova, B.E. Lisyanskiy, A.A. Stakharny et al., Low-temperature blackbodies for temperature range from −60 °C to 90 °C. Int. J. Thermophys. **32**, 2544–2559 (2011)
112. I. Müller, A. Adibekyan, B. Gutschwager et al.: Calibration capabilities at PTB for radiation thermometry, quantitative thermography and emissivity, in *Proceedings of the 14th Quantitative InfraRed Thermography Conference (QIRT 2018)*, pp. 329–333 (2018). https://doi.org/10.21611/qirt.2018.015
113. J.C. Nash, *Compact Numerical Methods for Computers. Linear Algebra and Function Minimization* (Adam Hilger, Bristol and New York, 1990)
114. H. Nasibov, A. Diril, O. Pehlivan, Comparative study of two InGaAs-based reference radiation thermometers. Int. J. Thermophys. **38**, 112 (2017)
115. M. Noorma, S. Mekhontsev, V. Khromchenko et al.: Design and characterization of Si and InGaAs pyrometers for radiance temperature scale realization between 232 °C and 962 °C. Proc SPIE **6205**, 620501 (2006)
116. P.R. Norton, Photodetectors, in *Handbook of Optics*, ed. by M. Bass, vol. II. Design, Fabrication, and Testing; Sources and Detectors; Radiometry and Photometry, 3rd ed. (McGraw-Hill, New York, 2010, pp. 24.3–24.102
117. G.D. Nutter, Radiation thermometers: design principles and operating characteristics, in *Theory and Practice of Radiation Thermometry*, ed. by D.P. DeWitt, G.D. Nutter (Wiley, New York, 1988), pp. 231–337
118. G.D. Nutter, Spectral-band radiation thermometers, in *Theory and Practice of Radiation Thermometry*, ed. by D.P. DeWitt, G.D. Nutter (Wiley, New York, 1988), pp. 341–458
119. S.A. Ogarev, M.L. Samoylov, N.A. Parfentyev et al., Low-temperature blackbodies for IR calibrations in a medium-background environment. Int. J. Thermophys. **30**, 77–97 (2009)

120. L. Palchetti, G. Bianchini, F. Castagnoli, Design and characterisation of black-body sources for infrared wide-band Fourier transform spectroscopy. Infrared Phys. Technol. **51**, 207–215 (2008)
121. S.-N. Park, B.-H. Kim, C.-W. Park et al., Realization of radiance temperature scale from 500 K to 1,250 K by a radiation thermometer with a thermal detector. Int. J. Thermophys. **29**, 301–311 (2008)
122. H.J. Patrick, L.M. Hanssen, J. Zeng et al., BRDF measurements of graphite used in high-temperature fixed point blackbody radiators: a multi-angle study at 405 nm and 658 nm. Metrologia **49**, S81–S92 (2012)
123. J.V. Pearce, Distribution coefficients of impurities in metals. Int. J. Thermophys. **35**, 628–635 (2014)
124. J.V. Pearce, J.A. Gisby, P.P.M. Steur, Liquidus slopes of impurities in ITS-90 fixed points from the mercury point to the copper point in the low concentration limit. Metrologia **53**, 1101–1114 (2016)
125. W.H. Press, S.A. Teukolsky, W.T. Vetterling et al., *Numerical Recipes. The Art of Scientific Computing*, 3rd ed. (Cambridge University Press, Cambridge, UK, 2007)
126. H. Preston-Thomas, The International Temperature Scale of 1990 (ITS-90). Metrologia **27**, 3–10 (1990)
127. A.V. Prokhorov, V.B. Khromchenko, *Precision Large-Area Low Temperature Blackbody BB-100V1 for IR Calibration. Conceptual Design and Computer Modeling*. (Virial, Inc., Gaithersburg, MD, 2005). www.virial.com/pdf/BB100V1.pdf. Accessed Mar 29, 2019
128. I. Pušnik, G. Grgić, J. Drnovšek, System for the determination of the size-of-source effect of radiation thermometers with the direct reading of temperature. Meas. Sci. Technol. **17**, 1330–1336 (2006)
129. I. Pušnik, G. Grgić, J. Drnovšek, Calculated uncertainty of temperature due to the size-of-source effect in commercial radiation thermometers. Int. J. Thermophys. **29**, 322–329 (2008)
130. D. Ripple, A. Pokhodun, P.P.M. Steur et al., *Guide to the Realization of the ITS-90. Fixed Points: Influence of Impurities. Appendix 4: Recommended List of Common Impurities for Metallic Fixed-point Materials of the ITS-90* (Consultative Committee for Thermometry under the auspices of the International Committee for Weights and Measures, 2015), https://www.bipm.org/utils/common/pdf/ITS-90/Guide-ITS-90-Impurities-2015-Appendix-4.pdf. Accessed 1 Apr. 2020
131. H. Rodriguez, E. Mendez-Lango, The influence of distance and focus on the calibration of infrared radiation thermometers. Int. J. Thermophys. **30**, 179–191 (2009)
132. H. Rodríguez-Arteaga, D. Cárdenas-García, An attempt to simplify the determination or the size-of-source effect in radiation thermometers. Int. J. Thermophys. **38**, 104 (2017)
133. A. Rogalski, *Infrared Detectors*, 2nd edn. (CRC Press, Boca Raton, FL, 2011)
134. A. Rogalski, *Infrared and Terahertz Detectors*, 3rd edn. (CRC Press, Boca Raton, FL, 2019)
135. F. Sakuma, S. Hattori, Establishing a practical temperature standard by using a narrow-band radiation thermometer with a silicon detector, in *Temperature: Its Measurement and Control in Science and Industry*, vol. 5, part 1, ed. by J.F. Schooley (American Institute of Physics, New York, 1982), pp. 421–427
136. F. Sakuma, M. Kobayashi, Interpolation equations of scales of radiation thermometers, in *Proceedings of TEMPMEKO '96, 6th International Symposium on Temperature and Thermal Measurements in Industry and Science*, ed. by P. Marcarino (Levrotto and Bella, Torino, Italy, 1997), pp. 305–310
137. F. Sakuma, L. Ma, Z. Yuan, Distance effect and size-of-source effect of radiation thermometers, in *Proceedings of TEMPMEKO 2001, 8th International Symposium on Temperature and Thermal Measurements in Industry and Science*, vol. 1, ed. by B. Fellmuth, J. Seidel, G. Scholz (VDE Verlag GmbH, Berlin, 2002), pp. 161–166
138. F. Sakuma, L. Ma, Calibration and characterization of the transfer standard radiation thermometer for an APMP intercomparison, in *Temperature: Its Measurement and Control in Science and Industry*, vol. 7, ed. by D.C. Ripple, part 2. AIP Conf. Proc. **684**, pp. 589–594 (2003)

139. F. Sakuma, L. Ma, Calibration facilities for high temperature radiation thermometers at NMIJ, in *2006 SICE-ICASE International Joint Conference* (Busan, South Korea, 2006), pp. 3336–3341, https://doi.org/10.1109/sice.2006.315082
140. F. Sakuma, L. Ma, Range ratio of Topcon radiation thermometers, in *SICE Annual Conference* (Kagawa, Japan, 2007). https://doi.org/10.1109/sice.2007.4421264
141. C.L. Sanders, Accurate measurements of and corrections for nonlinearities in radiometers. J. Res. Natl. Bureau Stand. **76A**, 437–453 (1972)
142. V.I. Sapritsky, B.B. Khlevnoy, V.B. Khromchenko et al., Precision blackbody sources for radiometric standards. Appl. Opt. **36**, 5403–5408 (1997)
143. V.I. Sapritsky, B.B. Khlevnoy, V. B. Khromchenko et al., Blackbody sources for the range 100 K to 3500 K for precision measurements in radiometry and radiation thermometry, in *Temperature: Its Measurement and Control in Science and Industry, ed. by D.C. Ripple,* vol. 7, part 2. AIP Conf. Proc. **684**, 619–624, 2003)
144. N. Sasajima, Y. Yamada, F. Sakuma, Investigation of fixed points exceeding 2500 °C using metal carbide-carbon eutectics. AIP Conf. Proc. **684**, 279–282 (2003)
145. P. Saunders, General interpolation equations for the calibration of radiation thermometers. Metrologia **34**, 201–210 (1997)
146. P. Saunders, Uncertainty arising from the use of the mean effective wavelength in realizing ITS-90, in *Temperature: Its Measurement and Control in Science and Industry*, ed. by D.C. Ripple, vol. 7, part 2. AIP Conf. Proc. **684**, pp. 639–644 (2003)
147. P. Saunders, Propagation of uncertainty for non-linear calibration equations with an application in radiation thermometry. Metrologia **40**, 93–101 (2003)
148. P. Saunders, Calibration and use of low-temperature direct-reading radiation thermometers. Meas. Sci. Technol. **20**, 025104 (2009)
149. P. Saunders, Uncertainties in the realization of thermodynamic temperature above the silver point. Int. J. Thermophys. **32**, 26–44 (2011)
150. P. Saunders, Correcting radiation thermometry measurements for the size-of-source effect. Int. J. Thermophys. **32**, 1633–1654 (2011)
151. P. Saunders, Dealing with the size-of-source effect in the calibration of direct-reading radiation thermometers. AIP Conf. Proc. **1552**, 619–624 (2013)
152. P. Saunders, D.R. White, Physical basis of interpolation equations for radiation thermometry. Metrologia **40**, 195–203 (2003)
153. P. Saunders, D.R. White, Propagation of uncertainty due to non-linearity in radiation thermometers. Int. J. Thermophys. **28**, 2098–2110 (2007)
154. P. Saunders, P. Bloembergen, D.R. White, Uncertainty in temperatures realised by radiation thermometry using two fixed points, in *Proceedings of TEMPMEKO 2004: 9th International Symposium on Temperature and Thermal Measurements in Industry and Science*, vol. 2, ed. by D. Zvizdić (FSB/LPM, Zagreb, Croatia, 2005), pp. 1149–1154
155. P. Saunders, J. Fischer, M. Sadli et al., Uncertainty budgets for calibration of radiation thermometers below the silver point. Int. J. Thermophys. **29**, 1066–1083 (2008)
156. P. Saunders, H. Edgar, On the characterization and correction of the size-of-source effect in radiation thermometers. Metrologia **46**, 62–74 (2009)
157. P. Saunders, D.R. White, H. Edgar, A compact combinatorial device for measurement of nonlinearity of radiation detectors. Int. J. Thermophys. **36**, 290–302 (2015)
158. P. Saunders, E. Woolliams, H. Yoon et al., *Uncertainty Estimation in Primary Radiometric Temperature Measurement* (Under the auspices of the Consultative Committee for Thermometry, BIPM, 2018), https://www.bipm.org/utils/en/pdf/si-mep/MeP-K-2018_Absolute_Primary_Radiometry_Uncertainty.pdf. Accessed 15 Apr 2020
159. R.D. Saunders, J.B. Shumaker, Automated radiometric linearity tester. Appl. Opt. **23**, 3504–3506 (1984)
160. C.H. Sutcliffe, T.J. Carroll, Temperature control system for a silicon pulling furnace. Rev. Sci. Instrum. **27**, 656 (1956)
161. E. Schreiber, G. Neuer, Reduction of the size-of-source effect and distance dependency of a high precision infrared radiation thermometer, in *Proceedings of TEMPMEKO 2004: 9th*

International Symposium on Temperature and Thermal Measurements in Industry and Science, vol. 1, ed. by D. Zvizdić (FSB/LPM, Zagreb, Croatia, 2005), pp. 527–532
162. M. Schuster, S. Nevas, A. Sperling et al., Spectral calibration of radiometric detectors using tunable laser sources. Appl. Opt. **51**, 1950–1961 (2012)
163. Y. Shimizu, J. Ishii, Middle temperature scale for infrared radiation thermometer calibrated against multiple fixed points. Int. J. Thermophys. **29**, 1014–1025 (2008)
164. Y. Shimizu, J. Ishii, Calibration of radiation thermometers and blackbodies in the temperature range from 160 °C to 420 °C. Int. J. Thermophys. **32**, 2560–2578 (2011)
165. Y. Shimizu, J. Ishii, Blackbody thermal radiator with vertically aligned carbon nanotube coating. Jap. J. Appl. Phys. **53**, 068004 (2014)
166. D.-J. Shin, D.-H. Lee, C.-W. Park et al., A novel linearity tester for optical detectors using high-brightness light emitting diodes. Metrologia **42**, 154–158 (2005)
167. D.-J. Shin, S. Park, K.-L. Jeong et al., High-accuracy measurement of linearity of optical detectors based on flux addition of LEDs in an integrating sphere. Metrologia **51**, 25–32 (2014)
168. J.A. Shohat, J.D. Tamarkin, *The Problem of Moments* (Am. Math. Soc, Providence, RI, 1943)
169. J.B. Shumaker, Linearity considerations and calibrations, in *Self-study Manual on Optical Radiation Measurements: Part 1—Concepts*, ed. by F.E. Nicodemus, Chapter 11. NBS Tech. Note 910-7 (NBS, U.S. Dept. of Commerce, Washington, DC, 1984)
170. P. Sperfeld, J. Metzdorf, N.J. Harrison et al., Investigation of high-temperature black body BB3200. Metrologia **35**, 419–422 (1998)
171. O. Struß, Transfer radiation thermometer covering the temperature range from −50 °C to 1000 °C, in *Temperature: Its Measurement and Control in Science and Industry*, vol. 7, part 2. AIP Conf. Proc., ed. by D.C. Ripple **684** (American Institute of Physics, Melville, NY, 2003), pp. 565–570
172. G. Talenti, Recovering a function from a finite number of moments. Inverse Prob. **3**, 501–517 (1987)
173. S.W. Teare, *Practical Electronics for Optical Design and Engineering* (SPIE Press, Bellingham, WA, 2016)
174. A.S. Tenney, Radiation ratio thermometry, in *Theory and Practice of Radiation Thermometry*, ed. by D.P. DeWitt, G.D. Nutter (Wiley, New York, 1988), pp. 459–494
175. E. Theocharous, J. Ishii, N.P. Fox, Absolute linearity measurements on HgCdTe detectors in the infrared region. Appl. Opt. **43**, 4182–4188 (2004)
176. E. Theocharous, Absolute linearity measurements on $LiTaO_3$ pyroelectric detectors. Appl. Opt. **47**, 3397–3405 (2008)
177. E. Theocharous, Absolute linearity measurements on PV HgCdTe detectors in the infrared. Presented on NEWRAD 2011 Conference (Maui, HW, 2011), https://physics.nist.gov/newrad2011/presentations/day3/AM/6_Theo_PV_CMT_Presentation_11_09_20.pdf. Accessed 10 Oct. 2019
178. A.J.P. Theuwissen, *Solid-State Imaging with Charge-Coupled Devices* (Kluwer Academic Publishers, New York, 2002)
179. A. Thompson, H.-M. Chen, Beamcon III, a linearity measurement instrument for optical detectors. J. Res. Natl. Inst. Stand. Technol. **99**, 751–755 (1994)
180. M. Vollmer, K.-P. Möllmann, *Infrared Thermal Imaging. Fundamentals, Research and Applications*, 2nd ed. (Wiley, Weinheim, Germany, 2018)
181. E.M. Vuelban, Determining size-of-source effect using the scanning slit method. Meas. Sci. Technol. **24**, 105007 (2013)
182. D.R. White, M.T. Clarkson, P. Saunders et al., A general technique for calibrating indicating instruments. Metrologia **45**, 199–210 (2008)
183. T.L. Williams, *Thermal Imaging Cameras: Characteristics and Performance* (CRC Press, Boca Raton, FL, 2009)
184. W.L. Wolfe, P.W. Kruse, Thermal detectors, in *Handbook of Optics, Vol. II. Design, Fabrication, and Testing; Sources and Detectors; Radiometry and Photometry*, ed. by M. Bass, 3rd ed. (McGraw-Hill, New York, 2010), pp. 28.1–28.14

185. J.T. Woodward, A.W. Smith, C.A. Jenkins et al., Supercontinuum sources for metrology. Metrologia **46**, S277–S282 (2009)
186. J.T. Woodward, P.-S. Shaw, H.W. Yoon et al., Advances in tunable laser-based radiometric calibration applications at the National Institute of Standards and Technology, USA. Rev. Sci. Instrum. **89**, 091301 (2018)
187. E.R. Woolliams, N.J. Harrison, N.P. Fox, Preliminary results of the investigation of a 3500 K black body. Metrologia **37**, 501–504 (2000)
188. E.R. Woolliams, K. Anhalt, M. Ballico et al., Thermodynamic temperature assignment to the point of inflection of the melting curve of high-temperature fixed points. Phil. Trans. R. Soc. A **374**, 20150044 (2016)
189. Y. Yamada, H. Sakate, F. Sakuma et al., Radiometric observation of melting and freezing plateaus for a series of metal-carbon eutectic points in the range 1330 °C to 1950 °C. Metrologia **36**, 207–209 (1999)
190. Y. Yamada, H. Sakate, F. Sakuma, A. Ono, A possibility of practical high temperature fixed points above the copper point, in *Proceedings of TEMPMEKO '99: the 7th International Symposium on Temperature and Thermal Measurements in Industry and Science*, vol. 2, ed. by J.F. Dubbeldam, M.J. de Groot (NMi Van Swinden Laboratorium, Delft, Netherlands, 1999), pp. 535–540
191. Y. Yamada, Y. Wang, N. Sasajima, Metal carbide-carbon peritectic systems as high-temperature fixed points in thermometry. Metrologia **43**, L23–L27 (2006)
192. Y. Yamaguchi, Y. Yamada, Thermodynamic temperature measurement to the indium point based on radiance comparison. Int. J. Thermophys. **38**, 49 (2017)
193. S. Yang, I. Vayshenker, X. Li et al., *Optical Detector Nonlinearity: Simulation.* NIST Tech. Note 1376 (NIST, United States Department of Commerce, Washington, DC, 1995)
194. Y.S. Yoo, B.-H. Kim, S.D. Lim et al., Realization of a radiation temperature scale from 0 °C to 232 °C by a thermal infrared thermometer based on a multiple-fixed-point technique. Metrologia **50**, 409–416 (2013)
195. Y.S. Yoo, D.H. Lee, C.W. Park et al., High dynamic range measurement of spectral responsivity and linearity of a radiation thermometer using a super-continuum laser and LEDs. AIP Conf. Proc. **1552**, 693–697 (2013)
196. Y.S. Yoo, C.W. Park, B.H. Kim et al., Radiance temperature scales of KRISS realized by two radiation thermometers using pyroelectric detector and silicon photodiode from 273 K to 2900 K. AIP Conf. Proc. **1552**, 698–703 (2013)
197. Y.S. Yoo, G.J. Kim, S. Park et al., Spectral responsivity calibration of the reference radiation thermometer at KRISS by using a super-continuum laser-based high-accuracy monochromatic source. Metrologia **53**, 1354–1364 (2016)
198. H.W. Yoon, D.W. Allen, R.D. Saunders, Methods to reduce the size-of-source effect in radiometers. Metrologia **42**, 89–96 (2005)
199. H.W. Yoon, D.W. Allen, R.D. Saunders, Methods to reduce the size-of-source effect in radiation thermometers, in *Proceedings TEMPMEKO 2004, 9th International Symposium on Temperature and Thermal Measurements in Industry and Science*, vol. 1, ed. by D. Zvizdić (FSB/LPM, Zagreb, Croatia, 2005), pp. 521–526
200. H.W. Yoon, C.E. Gibbson, V. Khromchenko et al., SSE- and noise-optimized InGaAs radiation thermometer. Int. J. Thermophys. **28**, 2076–2086 (2007)
201. H.W. Yoon, G.P. Eppeldauer, Measurement of thermal radiation using regular glass optics and short-wave infrared detectors. Opt. Express **16**, 937–949 (2008)
202. H. Yoon, P. Saunders, G. Machin, *Guide to the Realization of the ITS-90. Radiation Thermometry* . CCT under the auspices of the International Committee for Weights and Measures (2017), https://www.bipm.org/utils/common/pdf/ITS-90/Guide_ITS-90_6_RadiationThermometry_2018.pdf. Accessed 20 Feb 2020
203. H.W. Yoon, V. Khromchenko, G.P. Eppeldauer, Improvements in the design of thermal infrared radiation thermometers and sensors. Opt. Express **27**, 14246–14259 (2019)
204. Z.M. Zhang, G. Machin, Overview of radiation thermometry, in *Radiometric Temperature Measurements, I. Fundamentals*, ed. by Z.M. Zhang, B.K. Tsai, G. Machin (Academic Press , Amsterdam, Netherlands, 2010), pp. 1–28

205. Z.M. Zhang, B.K. Tsai, G. Machin (eds.), *Radiometric Temperature Measurements, I. Fundamentals* (Academic Press, Amsterdam, Netherlands, 2010)
206. Z.M. Zhang, B.K. Tsai, G. Machin (eds.), *Radiometric Temperature Measurements, II. Applications* (Academic Press, Amsterdam, Netherlands, 2010)

Chapter 9
Absolute Primary Radiometric Thermometry

Abstract Absolute primary radiometric thermometry is a recently introduced term for an approach to measuring the thermodynamic temperature of a blackbody using the accurate determination of the radiant flux that it emits in a known spectral band and known solid angle. The instrumental and metrological infrastructure for measuring thermodynamic temperature of high-temperature blackbodies are described. Four schemes of the filter radiometer calibration in terms of the spectral power, irradiance, and radiance and corresponding sources of measurement uncertainties are discussed. Calibration facilities on the base of tunable lasers and monochromators are described and compared. The state-of-the-art uncertainties in measuring temperature of the high-temperature blackbodies are discussed.

Keywords Thermodynamic temperature · Cryogenic radiometer · Trap detector · Filter radiometer · Calibration · Uncertainty · Supercontinuum source

9.1 Principles of Absolute Primary Radiometric Thermometry

The absolute primary radiometric thermometry (APRT) is the term introduced in the *MeP*-K [140] for the methods of measuring the temperature of a blackbody, based on the determination of the optical power, emitted over a known spectral band and known solid angle by a blackbody of known effective emissivity.

The main idea of application of the absolute radiometry technique to measurement of thermodynamic temperature of blackbodies is straightforward: if to perform absolute measurement of a total, spectral (actually, in a narrow spectral interval), or band-limited radiometric quantity of the radiation emitted by a blackbody, its temperature can be derived from the Stefan-Boltzmann or Planck law, bypassing the ITS-90. Initially, this approach has been implemented for the total radiometric quantities and the Stefan-Boltzmann law. Attempts of applying the absolute total radiometry for determination of thermodynamic temperature (if the Stefan-Boltzmann constant is assumed known) or, on the contrary, for experimental determination of the Stefan-Boltzmann constant (if the blackbody thermodynamic temperature is assumed known) were undertaken since the end of 19th century; an incomplete list of such

V. Sapritsky and A. Prokhorov, *Blackbody Radiometry*, Springer Series in Measurement Science and Technology, https://doi.org/10.1007/978-3-030-57789-6_9

experiments is provided by Macoy [139]. The most influential measuring experiments in the absolute total radiometric thermometry were performed by Blevin and Brown [15] and Quinn and Martin [167]. Presently, these works are rather of historical and methodological interest. Blevin and Brown [15] described measurements the Stefan-Boltzmann constant σ and the freezing temperature T_{Au} of gold; Quinn and Martin [165–167] and Martin et al. [143] described measurements of σ and thermodynamic temperatures between -130 and $+100$ °C. Blevin and Brown employed the spectrally non-selective thermal detector of radiation with electrical substitution, i.e. electrical-substitution radiometers (ESR). A lot of useful information concerning the ESRs can be found in the book edited by Hengstberger [106]; we consider briefly the electrical substitution principle for the thermal detectors of optical radiation in Sect. 9.2.1. Instead of the room-temperature ESR, Quinn and Martin used an absolute cryogenic radiometer (ACR). Up to date, the ACR (see Sect. 9.2.4 and references therein) is considered the most accurate radiometric instrument that allows measuring optical power with the relative uncertainty as low as 10^{-4}. However, such a low uncertainty is attainable only for monochromatic radiation of CW lasers with a definite level of output optical power. Currently, the total radiation thermometry is practically out of use due to revolutionary changes that have taken place in the optical radiometry in the last quarter of the 20th century. They can be briefly characterized by introducing of three innovative measuring devices into the practice of optical radiometry: absolute cryogenic radiometer (ACR), filter radiometer (FR), and trap detector (TD). These three measuring instruments [69, 70] became the basis for absolute primary spectral-band radiation thermometry traceable to two SI units, the watt and the meter.

The development of the modern APRT owes much to the inception of so-called high-temperature fixed points (HTFPs) at the very end of the previous century. Yamada et al. [222, 223] reported successful realization at NMIJ[1] (Japan) of sufficiently reproducible HTFPs based on the binary metal-carbon eutectic alloys, whose melting temperatures greatly exceed the freezing temperature of copper, the upper fixed point of the ITS-90. Later, the HTFPs based on metal carbide-carbon eutectics [178] and peritectics [224] were realized in the radiating cells. To date, thermodynamic temperatures with associated uncertainties have been assigned to three metal-carbon eutectic HTFPs (see Appendix B).

The corresponding *MeP*-K documents [140, 180] give practical guidelines for the realization of methods and techniques of APRT, as well as typical uncertainty estimates. Modern level of standard uncertainties in APRT is of order of 0.1 mK at 2800 K.

Measurement of the optical power is carried out with a filter radiometer (FR), which consists of spectral filter and a detector, with known absolute spectral responsivity. The FR optical system typically includes two coaxial circular apertures separated by a known distance to determine the solid angle, and may further include lens(es). The refractive index of the medium, in which the measurement is made, must also be known.

---

[1]National Metrology Institute of Japan.

There are several implementations of the APRT using FRs, which differ in the type of absolute spectral responsivity for which they are calibrated, in measured quantities, optical system designs, and geometric conditions of measurement and calibration [10]. Regardless of what quantities are involved in calibrating the FR and measuring with it, they must be traceable to the appropriate SI units, in particular, the meter and electrical units.

Despite the fact that the name APRT was coined recently, various methods for measuring the thermodynamic temperature of blackbodies have been developed since the last quarter of the 20th century. They were often referred to as the detector-based temperature measurements, in contrast to conventional radiation thermometry traceable to the ITS-90, which was considered as source-based due to the calibration of radiation thermometers (RTs) using FPBBs.

All modern variations of the APRT require two essential elements of a calibration infrastructure, at which we look in more detail in the next Sections:

1. A primary standard of the optical power. At present, this is the absolute cryogenic radiometer (ACR), the most accurate measuring instrument for the optical power (radiant flux). It measures the power of the optical radiation at several discrete wavelengths by substituting the optical power with the power of electric current.
2. A transfer standard detector (transfer measurement device) with a predictable relative spectral responsivity. To date, the best choice for the transfer standard detector is the trap detector (TD) composed of several silicon detectors with a high quantum efficiency. The TD is calibrated in terms of absolute spectral responsivity at several discrete wavelengths against the ACR. The predictable relative spectral responsivity of the TD allows calculating its absolute spectral responsivity over a spectral range sufficient to calibrate the FR in terms of the absolute spectral responsivity.

9.2 Absolute Cryogenic Radiometer

9.2.1 Electrical Substitution Principle

The absolute radiometers have long been considered as the most suitable, convenient, and reliable, if not the only, absolute radiometric instruments.

The method of substitution of the radiant power absorbed in the detector's receiving element with the power of the electric direct current, proposed at the very end of 19th century, is laid in the base of the electrical substitution radiometer (ESR). The radiant flux is measured using the ESR during cyclic operations with alternating phases of radiation heating and electrical substitution. A temperature sensor, which can be uncalibrated, serves as an indicator of temperature equality in both heating phases. For the most general case, when the receiving element of the ESR is preheated by an electric current to some operating temperature, we can write the following measurement equation:

$$\Phi = K \cdot (P_s - P_r), \tag{9.1}$$

where Φ is the measured radiant flux (optical power); P_r and P_s are electric powers released in the heating element in the phases of the radiative heating and electric substitution, respectively; K is the correction factor accounting for incomplete absorption of the incident radiation by the receiving element and for imperfect substitution, i.e. for various deviations of a real-world ESR from the measurement method in its pure (idealized) form, when the equality of optical and electric powers is deduced from the equality of temperatures registered by a temperature sensor.

The measurement principle (see [114]) of the ESR is the transformation of the optical radiation energy absorbed by the receiving element into the heat. The Joule law is the underlying fundamental physical law for the ESR. Schematics of the ESRs used for the radiant flux (optical power) and irradiance measurements are shown in Fig. 9.1.

In the power mode, the incident beam of the optical radiation underfills the receiving area. In the irradiance mode, the overfilling beam uniformly distributed over its cross-section is passed through the aperture diaphragm of a known area A. The measurement equation for the irradiance E can be written as follows:

$$E = K \cdot (P_s - P_r)/A. \tag{9.2}$$

In the simplest case of the DC use, electric powers P_r and P_s can be expressed through the Joule law using any two of the three easily measurable quantities: the electric resistance R of the electric heater, electric current I flowing through the electric heater, and voltage V applied to the heater.

Measurement of electric quantities (resistance, current, and voltage) can be done with sufficiently high accuracy, so the overall accuracy of measurements performed using the ESRs depends largely by accuracy in determination of the correction

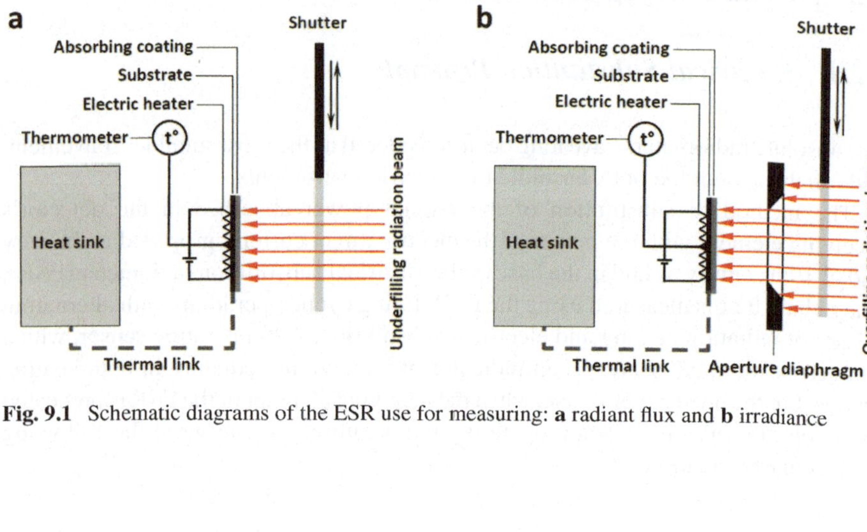

Fig. 9.1 Schematic diagrams of the ESR use for measuring: **a** radiant flux and **b** irradiance

factors K. In the case of irradiance measurements, uncertainty in measurement of the aperture area also brings a large contribution into the uncertainty budget.

9.2.2 Room-Temperature Absolute Radiometers

The ESR is an absolute instrument that can serve as a primary standard if and only if the correction factor K is determined experimentally or computed on the base of measured physical parameters of the components of the ESR.

As a thermal detector, the ESR has an indisputable advantage over optical radiation detectors of other types, primarily, the photoelectric ones: the wider operational spectral range, which is determined only by absorptive characteristics of the receiving element.

The ESRs are the oldest detectors of optical radiation; their earliest use dates back to the second half of the 19th century. For a long time, ESRs operating at room temperature were the only absolute measuring instrument for the radiant flux and the irradiance. A number of historical examples can be found in Hengstberger [106] and reference therein. Since the mid-1960s, the ESRs are used for absolute measurements of the power of laser radiation in the range from a few tenths of milliwatt to several kilowatts. All the time after invention of the substitution method, the ESRs have been employed for measuring the total solar irradiance [206].

Extensive experience in using the ESRs has revealed their shortcomings, including those that can be eliminated using relatively simple design solutions and measurement techniques, as well as others that require fundamentally new approaches to overcome them.

Among avoidable shortcomings of the ambient-temperature ESRs, the main place belongs to the problem of determining the absorptance of the receiving element. Although thermal detectors are often considered as spectrally non-selective devices, their actual spectral responsivity depends upon the spectral absorptance of the receiving surface. Measurement of the spectral absorptance can be carried out presently with the relative uncertainty of about 0.1%, which brings one of the greatest contributions to the uncertainty budget for measurements using ESRs. Additionally, absorptance of the receiving element are influenced by environmental characteristics (temperature, humidity, etc.), which decreases long-term repeatability of measurements. Fabrication of a receiving element in the form of a cavity allows the uncertainties in determination and maintaining the long-term stability of the ESR absorptance to be significantly reduced due to so-called "cavity effect," which we discussed in Sect. 4.1.2. However, there are some principally unavoidable disadvantages of the ambienttemperature ESRs:

1. The first and major inherent shortcoming is incomplete equivalence of the measured optical power and the Joule heat released in the electric heater. Among the many causes of this non-equivalence, the main one in the difference in the spatial distributions of radiant and electric heat fluxes. For instance, since the

electric heater lies necessarily beneath the absorbing coating, the different paths, along which the heat fluxes flow at the radiative and electrical heating, lead to a significant systematic error similar to that due to the temperature drop effect, which we considered in Sect. 7.3.2.

2, 3. The second and third drawbacks of an ambient-temperature ESR, which should be considered together, are a small sensitivity and large thermal inertia. The sensitivity of the ESR is typically much lower and the response time is much greater than those for photoelectric detectors. Sensitivity of a thermal detector is inversely proportional to the heat capacity of and heat losses (due to thermal conduction, convection, and thermal radiation) from the receiving element. Since the decrease of heat losses leads to increase of detector's thermal inertia, a reasonable trade-off between the sensitivity and the thermal time constant should be chosen. Regardless of what attempts were made to minimize the heat capacity of the receiving element, the time constant of a room-temperature ESR can hardly be reduced below one minute without a significant decrease in sensitivity. Additionally, a significant thermal inertia of room-temperature ESRs, especially of those with the cavity receiving elements, makes problems at measuring the time-varying radiant fluxes; a large (up to tens of minutes) time needed for the complete cycle of measurements entails an excessive heating of the instrument with accompanying drift of operating characteristics of the electronic circuits.

4. The fourth drawback of a room-temperature ESR consists in relatively large thermal noises. Thermal noises in the heater and temperature sensor limit the signal-to-noise ratio (SNR), when low-level radiant fluxes are measured. As a result, sensitivities of absolute ambient-temperature ESRs are usually not sufficient for precise measurement of *spectral* characteristics of conventional thermal radiation sources, such as Sun, incandescent lamps, blackbodies, etc. With some exceptions, the ESRs were used to measure the total (integrated over entire spectrum) radiant flux.

Perhaps, the most known exception is the use of a room-temperature ESR for the absolute spectroradiometric measurements of the thermodynamic temperature of the gold freezing point [149]. This unique ESR operating in vacuum had the responsivity of about 0.27 V/W. The duration of the substitution phase was 3 min, so the full measurement cycle was about twice as long. Overall correction factor for the systematic effects was assessed as 1.0010 with the standard uncertainty of about 0.06%. This experiment allowed determination of the gold freezing point with the resulting uncertainty of 0.23 K. The corresponding relative uncertainty expressed in terms of spectral radiance can be roughly assessed as 0.3%. However, participation in this experiment was the swan song of the room-temperature absolute ESRs: practically all resources for improving their accuracy were exhausted. Limitations imposed on the accuracy of the absolute ESRs are caused by thermal properties of materials at ambient temperatures and by uncertainties in determination of physical quantities used at computational correction for the substitution non-equivalence.

9.2.3 Physical Properties of Materials at Cryogenic Temperatures

Although the idea of overcoming the disadvantages of a room-temperature ESR temperature—deep cooling of the receiving element—was almost obvious, it could only be realized by the beginning of the 1980s. A principal possibility of significant increase of the ESR accuracy by cooling down to the liquid helium (LHe) temperatures was proved by Ginnings and Reilly [86].

Their cryogenic radiometer was developed to measure the thermodynamic temperatures of certain fixed points relative to the triple point of water using the Stefan-Boltzmann law. This goal was not accomplished (presumably, due to inaccurate account for the diffraction losses in the complicated system of diaphragms separating the radiometer and blackbody radiator), so the work by Ginnings and Reilly [86] had no appreciable effect on the progress in absolute radiometry.

The rise of absolute cryogenic radiometers is associated with the experiment on the determination of the Stefan-Boltzmann constant conducted by Quinn and Martin [165–167], who showed the possibility of measuring the power of optical radiation with precision of about 100 times better than is possible with conventional room-temperature ESRs. We will not consider the design of the apparatus built for this experiment, since modern ACRs are significantly different from ACRs developed by Quinn and Martin. We will focus on ACRs, which are typically at the top of the traceability chain for measuring the thermodynamic temperature of blackbodies, and those that are used to establish spectral responsivity scales for detectors in the Vis and NIR spectral ranges. Such ACRs are cooled by non-superfluid LHe with the boiling temperature of about 4 K (it depends on the proportion of the common isotope ^{4}He and rarer isotope ^{3}He), while the Quinn-Martin's ACR was cooled with superfluid (the He II phase) isotope ^{4}He at about 2 K, just below the ^{4}He lambda point (~2.17 K). Besides, Quinn and Martin built their apparatus for the total absolute radiation thermometry applications, whilst the spectral (or, more exactly, band-limited) absolute radiation thermometry holds presently a dominant position in the APRT. Unlike a blackbody source in the Quinn and Martin experiment, laser sources of highly collimated and stable quasi-monochromatic optical radiation are more often used together with modern ACRs to achieve the highest radiometric accuracy.

For better understanding basic principles of the ACRs thermal design, we should make a brief excursion into the temperature behavior of constructional material typically used in cryogenic applications. Temperature dependences of the thermal conductivity k and specific heat capacity at constant pressure c_p are of greatest importance for the ACR design. For isotropic bodies, these characteristics are defined as

$$k = -q \cdot \left(\frac{dT}{dx}\right)^{-1}, \tag{9.3}$$

$$c_p = \frac{1}{m}\frac{dQ}{dT}\bigg|_{p=Const}, \tag{9.4}$$

where T is the temperature, m is the mass, q [W m^{-2}] is the heat flux density, x is the coordinate (the definition for k is given for the one-dimensional heat flow), Q is the heat flux, and p is the pressure.

There are many texts, in which the theories and mechanisms explaining thermal behavior of these characteristics at low temperature are considered (see, e.g. [12, 52, 199, 205]).

The thermal conductivity is very sensitive to the material purity, structural defects, and heat treatment. Because of this and due to difficulty of accurate experimental determination of thermal conductivity in cryogenic conditions, the literature data on temperature dependences of thermal conductivity should be considered as averaged, with large relative uncertainties (10–50%) assigned.

The theory of thermal conductivity of solids is complex since many separate mechanisms act simultaneously to transmit energy through the material (for detail, see [204]). To show the variety in thermal behavior of metals and non-metals, we presented temperature dependences of thermal conductivity for some engineering materials for temperature below 300 K in Fig. 9.2a.

The low-temperature thermal conductivity values reported in the literature may differ significantly. For relatively pure metals such as copper, these differences can be attributed to tiny amounts of impurities that have large effects on thermal conductivity at cryogenic temperatures.

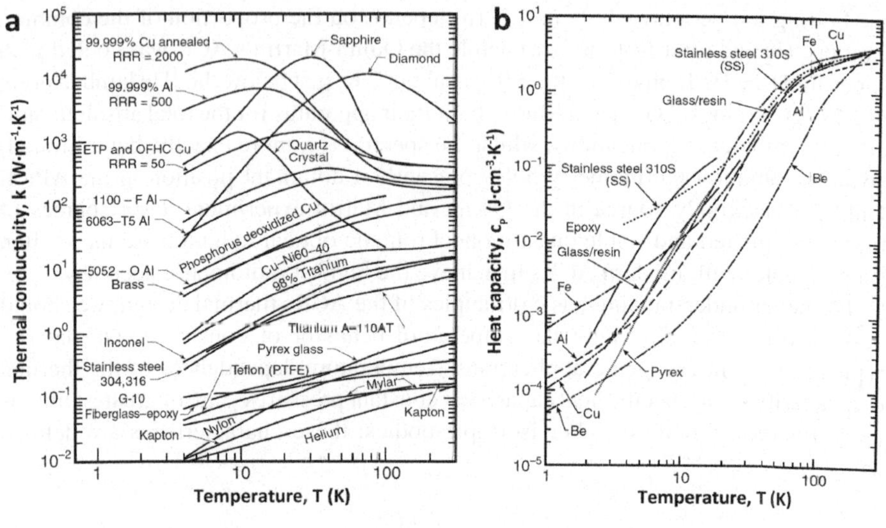

Fig. 9.2 Temperature dependences of **a** thermal conductivity k and **b** specific heat at constant pressure c_p for some materials typically used in cryogenic applications. Reproduced from [50] with permission of Oxford Publishing Limited

For example, the thermal conductivity of two copper samples at 10 K can differ by an order of magnitude, while the same samples at room temperature can vary by only a few percent.

As can be seen in Fig. 9.2a, this effect manifests itself differently for copper of various purities expressed in RRR, the residual resistivity ratio, defined as

$$\mathrm{RRR} = r(273\ \mathrm{K})\big/ r(4\ \mathrm{K}), \tag{9.5}$$

where r is the electrical resistivity [Ω m] of a sample[2].

To specify the RRR range we can deal with, we indicate that the RRR of the copper wire used for telephone lines is less than 50; commercial oxygen-free high-conductivity (OFHC) copper has an RRR from 50 to 500; for polycrystalline copper with a purity of 99.999% (5N) annealed in high vacuum at temperatures from 600 to 700 °C, the RRR reaches a value of 2000 and higher. For heavily alloyed copper, the RRR may drop down to 2 or less.

It is important to note that unconditional increase in the thermal conductivity takes place only for copper of sufficient purity (starting from RRR $\approx$ 100), but even in this case, the ratio of thermal conductivities $k(T)\big/ k(300\ \mathrm{K})$ for temperature T strongly depend on RRR.

Measurement of specific heat by various relative techniques is quite easy, while absolute measurements are difficult. The data reported in literature usually have uncertainty of about 1% below 20 K. Above this temperature, measurements can be carried out with better accuracy, so often extrapolation toward the lower temperature based on the Debye theory (see, e.g. [90] or [199]) can be used. The general trend in the specific heat presented in Fig. 9.2b is universal for all solids and demonstrates reducing by several orders of magnitude as temperature decreases from ambient down to cryogenic [12, 199].

9.2.4 *Modern Absolute Cryogenic Radiometers*

After pioneering work performed by T. Quinn and J. Martin at the NPL in the mid-1980s, the ACR began the triumphant march around the world. Radiometrists quickly recognized the potential of the new measuring instrument to improve radiometric measurements in a wide range of applications. Almost immediately after the first ACRs appeared, the tendency towards their specialization manifested itself in the development of task-oriented ACRs, the design of which was focused on solving problems that differ in ranges of the measured power, the geometric conditions of irradiation, the spectral composition of the measured radiation, etc. Over the past decades, specialized ACRs were developed to measure the radiant flux (optical

[2]Sometimes, the RRR is defined via electrical resistance [Ω] and called residual resistance ratio. The upper and lower temperatures in the RRR definition may be slightly different, e.g. 298 and 4.2 K. Usually, difference in these two definitions is not critical.

power) for such specific tasks as measurement of faint sources of thermal radiation in space-like environments [33, 34, 37], measurement of the optical power delivered through optical fibers (see, e.g. [137]), or measurement of irradiance by the direct solar radiation [144, 208].

However, the greatest impact on the optical radiometry had the ACRs designed to measure the optical power of radiation generated by continuous-wave (CW) and quasi-CW lasers. The highest accuracy (0.01% or better) of these ACRs is achievable at the power level of order of 1 mW. Measuring systems with such ACRs serve as national primary standards for realization of the optical power unit at fixed wavelengths, so all radiometric measurements have to be traceable to these primary standards. Dissemination of the optical watt is carried out over a continuous spectral range (approximately from 400 to 900 nm) using silicon photodiodes-based TDs that play the role of transfer standard detectors (transfer measurement devices). The responsivity of TDs (in A/W or V/W) are determined at fixed wavelengths and then extends on the continuous spectral interval by interpolation that takes into account physical properties of semiconductor structures that form silicon photodiodes.

It is difficult to overestimate the impact of ACRs on the blackbody radiometry. The FRs calibrated in terms of absolute spectral responsivity and traceable to the ACR make possible measurement of radiant flux emitted by a blackbody within a narrow spectral band. This, in turn, allows deriving the thermodynamic temperature of a blackbody from Planck's radiation law and eliminate the linkage of blackbody radiometry to the currently accepted (defined) temperature scale, which, by definition, is only an approximation of the scale of thermodynamic temperature.

As a rule, absorbing coatings with the near-specular reflection are employed since they have higher density as compared with looser diffusely reflecting ones and hence lower thermal resistance. This decreases correction for the substitution non-equivalence and, correspondingly, reduces uncertainty of this correction. A simplified schematic of the ACR is shown in Fig. 9.3.

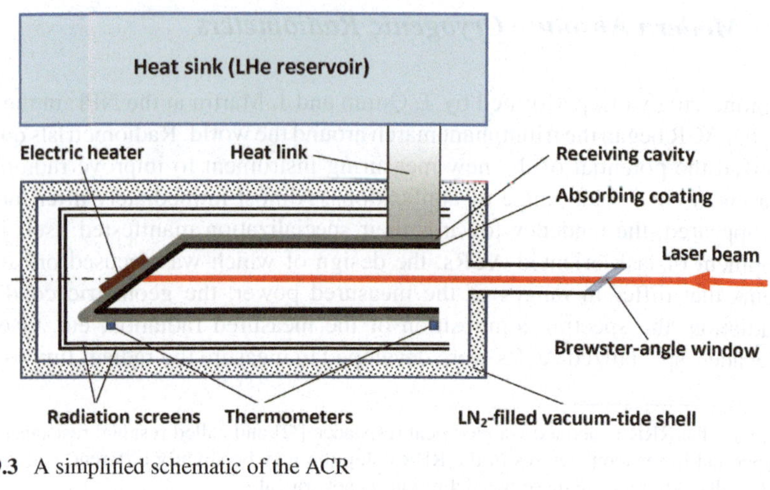

Fig. 9.3 A simplified schematic of the ACR

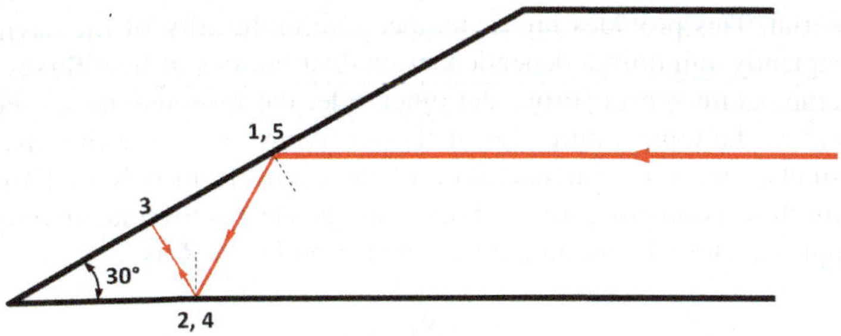

Fig. 9.4 Tracing of a central ray in the specularly-reflecting cylindrical cavity with a flat bottom inclined at 30° to the axis of the cylindrical part. Figures indicate the sequence of reflections

The near-unity effective absorptance of the cavity is achieved by the choice of its shape. Most frequently, the cylindrical cavity with a flat bottom, inclined at 30° relative to the cylinder's axis is used. A ray entering the cavity and lying in the plane of its symmetry undergoes the three successive reflections until reversing and then two more until escaping the cavity in the opposite direction (see Fig. 9.4). If, say, the absorptance of the coating $\alpha = 0.9$ (and, correspondingly, the reflectance $\rho = 1 - \alpha = 0.1$, the effective absorptance of the cavity must be $\alpha_{eff} = 1 - \rho^5 = 0.99999$. Due to imperfect specular reflection from the real-world coatings, the effective absorptance of such a cavity takes lower values, say, 0.9999. This is in accordance with the measurements carried out using integrating sphere (IS) reflectometers within the spectral range from 325 to 1550 nm (see, e.g. [138, 150]) with the relative standard uncertainties of order of several ppm (parts per million).

A cylindrical cavity with a flat inclined bottom has additional advantage in easier manufacturing and possibility to reliably mount the electric heater (wired, thick-film, or of chip-type) on the flat bottom with guaranteed low contact thermal resistance. Annealed OFHC copper is the most appropriate material for the cavity fabrication.

A radiant flux entering the receiving cavity is absorbed in the black coating applied to its internal surface. The radiant power cannot be absorbed uniformly over the entire cavity, but once absorbed it must flow along the only path through the heat link towards the LHe reservoir. The same occurs with the power released in the electric heater as the Joule heat.

Considering the ACR as a lumped-parameter thermal system (see, e.g. [115]), we can write out the simple equation for the average temperature T_c of the receiving cavity using the electro-thermal analogy:

$$T_c = T_s + P_{r,e} \big/ R_T, \tag{9.6}$$

where $T_s \approx 4$ K is the temperature of the heat sink, R_T [W K^{-1}] is the thermal resistance of the heat link (see Fig. 9.3), and $P_{r,e}$ [W] is the power of incident radiation or electric power released in the heater.

Material and geometry of a heat link is chosen so that the average temperature of the cavity corresponds to the maximal thermal conductivity of copper used as the cavity material. This provides higher temperature uniformity of the cavity surface and, consequently, minimizes dependences on distributions of heat fluxes and positions of temperature sensor. From the other side, the responsivity of the ACR is proportional to the temperature rise of the receiving cavity for the given optical power. Usually, the average temperature of the cavity from 6 K to 10 K is good trade-off for these competing trends. Following the electro-thermal analogy, we can express approximately the thermal time constant of the ACR as

$$\tau = R_T \cdot C, \tag{9.7}$$

where $C = c \cdot m$, m is the mass of the cavity receiver.

Since the specific heat c of pure copper at these temperatures is about three orders of magnitude lower than that at room temperature (see Fig. 9.2b), one can expect a lower value of τ for the ACR as compared with the analogous room-temperature ESR temperature, even at higher value of R_T and other things being equal.

The systematic error due to substitution non-equivalence of the ACR is defined primarily by the thermal resistance of the layered back wall of the cavity and the heat losses from both its sides. Since the receiving cavity of the ACR is in the vacuum environment by perforce, the convective heat loss is absent. It can participate only in radiative heat exchange. To minimize radiative heat losses from the cavity external surface, it is surrounded by low-emissivity radiation screens with controllable temperature. The inner screen has temperature close to that of the cavity, the outer one has temperature close to that of the vacuum-tide casing filled with the liquid nitrogen (LN_2), and intermediate screens have temperatures gradually changing between the first and the last ones. Minor heat losses from the cavity internal surface occur due to radiative heat exchange with the cavity surroundings through its opening. The bottom of the cavity can "see" the room-temperature environment only within a narrow solid angle. The greater the cavity length the less radiative heat losses from the cavity bottom.

Below, we consider some typical ACRs and measuring systems on their base that are either used as primary standard facilities at world-leading NMIs, or employed for calibration of optical radiation detectors for visible and near IR spectral ranges in terms of absolute spectral responsivity and typically head the traceability chains when blackbody thermodynamic temperatures are measured.

Depending on whether narrow-band radiation from lasers or monochromatized radiation of broadband sources is measured by an ACR, measuring systems with ACRs are divided into laser-based and monochromator-based.

Lasers of various types are used in the laser-based systems serving as highly collimated and almost monochromatic sources of optical radiation. Presently, a variety of laser types suitable for this purpose is commercially available. These include gas (He–Ne and CO_2), metal-vapor (He–Cd), and ion (Ar^+ and Kr^+) CW lasers, solid-state (Nd:YAG and Ti:sapphire) quasi-CW lasers, semiconductor quantum cascade lasers, dye lasers, and fiber-coupled laser diodes. Some of these lasers allow a continuous tuning over a relatively wide wavelength range. Additionally, shorter wavelengths can be obtained using optical frequency multipliers; larger wavelengths can be obtained

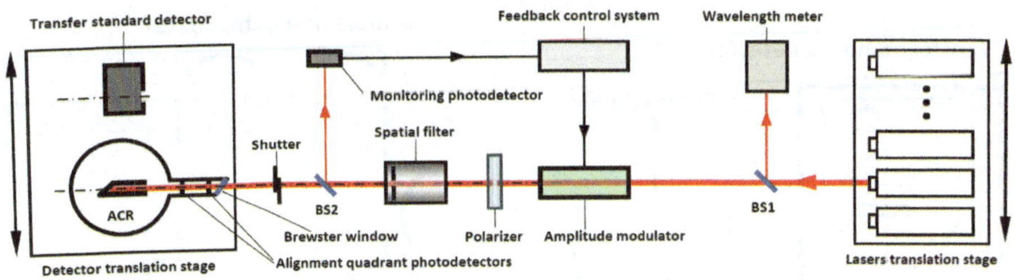

Fig. 9.5 A schematic of a laser-based facility with ACR; BS1 and BS2 denote beamsplitters

using optical parametric oscillators that convert an input laser beam into two output beams with the wavelengths larger than that of the input (pumping) beam. Suchwise, lasers of different types provide a coverage of the spectral range roughly from 0.3 to 12 μm with more closely spaced emission lines approximately between 400 and 900 mm, i.e. just in the range of the best performance of silicon photodiodes and TDs on their base.

A generalized scheme of a laser-based facility with the ACR intended for realization of the optical power unit and calibration of a transfer standard detector is presented in Fig. 9.5. A set of lasers is installed on a translation stage that allows choosing the laser that generates the wavelength required. If a tunable laser is used, a small part of its power is deflected by the beam-splitter BS1 to measure the actual wavelength of the radiation generated.

External stabilization of the laser intensity is carried out using electro-optic or acousto-optic amplitude modulators via a feedback control system receiving a signal from a photodetector monitoring the measured power of the laser beam via the beamsplitter BS2. A spatial filter serves to remove diffraction pattern and produce a near-Gaussian beam profile with sharp boundaries. A polarizer allows adjustment of the beam polarization to minimize reflection from the vacuum-tight Brewster-angled window of the ACR. The Brewster angle is adjusted before measurements at each wavelength. The total correction for the window-related radiation losses due to reflectance and transmittance depends on cleanliness of the window itself and on the optical quality of the laser beam. A typical value of the corresponding correction factor equals a few parts in 10^4, with the standard uncertainty of a few parts in 10^5.

Alternatively, optical radiation detectors, including transfer standards, can be calibrated against an ACR within a continuous spectral range using monochromator-based measuring systems. A simplified scheme of a monochromator-based system with an ACR is presented in Fig. 9.6. A monochromator selects a narrow band of wavelength from a wider spectrum of a source (e.g., tungsten-halogen lamp). Typical values of optical power that has to be measured by the ACR in monochromator-based systems are between 1 and 50 μW (cf. 0.5–1 mW typical for laser-based measurements). Therefore, an ACR operating with a monochromator must be more sensitive and have a lower level of thermal noise than those operating in laser-based systems. As a rule, this requirement can be fulfilled by deeper cooling of a receiving cavity.

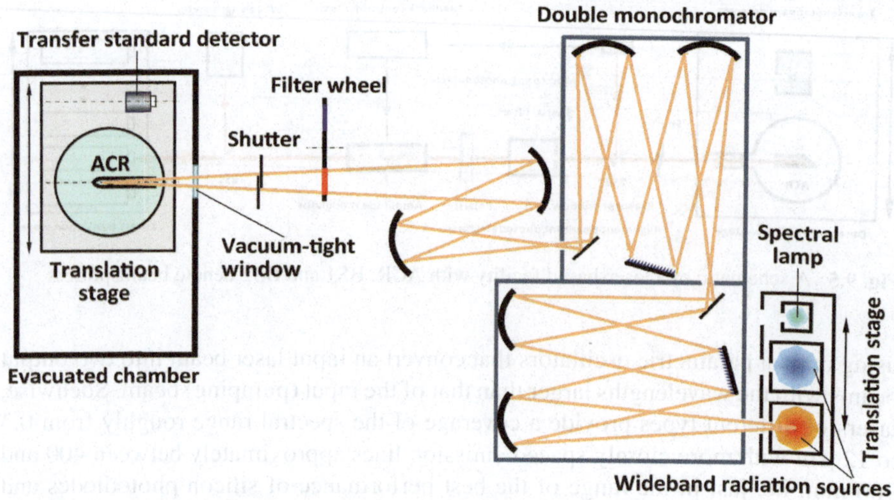

Fig. 9.6 A general schematic of a monochromator-based facility

It is also desirable that such an ACR has the shortest possible response time to allow a fast scanning over a continuous wavelength range. Unlike highly collimated and linearly polarized laser beams, output beams of monochromators are polarized only partially and may have noticeable divergences (up to several degrees) at the ACR's cavity entrance.

Brewster-angled windows become ineffective for such beams, so both the ACR and the detector under calibration must be placed in a common evacuated chamber to eliminate the uncertainty associated with the window's transmittance measurement as, e.g. it is shown in Fig. 9.6. The whole complex of the mentioned problems leads to the fact that monochromator-based facilities are usually less accurate and often more complicated than the laser-based systems.

However, monochromator-based system is usually less expensive than that equipped with a set of laser sources having quality and performance sufficient for achieving the highest accuracies in calibration of transfer standard detectors within a required spectral range. Some of the disadvantages of laser-based systems make cryogenic radiometry based on monochromators very attractive today with the advent of high-intensity radiation sources that can replace incandescent lamps.

To illustrate the general concepts of the ACR design, we overview constructions of some particular instruments.

POWR (NIST, USA) The LHe-coled ACR called HACR (High-accuracy Absolute Cryogeric Radiometer) was the primary radiometric standard at the NIST (USA) since the mid-1990s. Its traditionalist design is described in many works (see, e.g. [109, 160]). Circa 2005, the HACR has been replaced by the POWR (Primary Optical Watt Radiometer), more versatile instrument with the modular design [110]. The POWR can operate at 4.2 K to measure optical power at milliwatt level and at 2 K at microwatt level, with exchangeable receiving cavity. A schematic and a photograph of the POWR are presented in Fig. 9.7. The standard operational temperature of 4.2 K is achieved with the LHe reservoir filled with the mix of isotopes ^{4}He and ^{3}He; the optional temperature mode (2 K) with the reservoir filled with liquid ^{4}He is used for measurements, which require a greater sensitivity.

The replaceable detector module is mounted under the cold plate (the bottom of the LHe reservoir). The front optics section is attached to the high-vacuum sealed flange of the cryostat, in which the LN_2 reservoir surrounds the LHe reservoir. A laser beam is introduced into the system through a window mounted at the at the Brewster angle to the beam's plane of polarization to minimize reflection losses. The window's transmittance measured in situ at 633 nm is 0.99992 with the standard uncertainty of order of 0.00001.

The input beam can be directed either into the receiving cavity of the POWR or to the detector under calibration that is installed in the vacuum-tight side port.

The detector module consists of the cold block and the baffle section. The receiving cavity is placed inside the cold block and attached to the heat sink through the heat link (the ring made of polyimide film). Both the heat sink and the cold block are made of OFHC copper that is plated first with nickel and then with gold. All surfaces are polished to reduce their emissivity and thus radiative heat losses. The baffle section attached to the cold block contains the off-axis parabolic mirror with a central hole allowing the laser beam to pass through and additional baffles along the optical

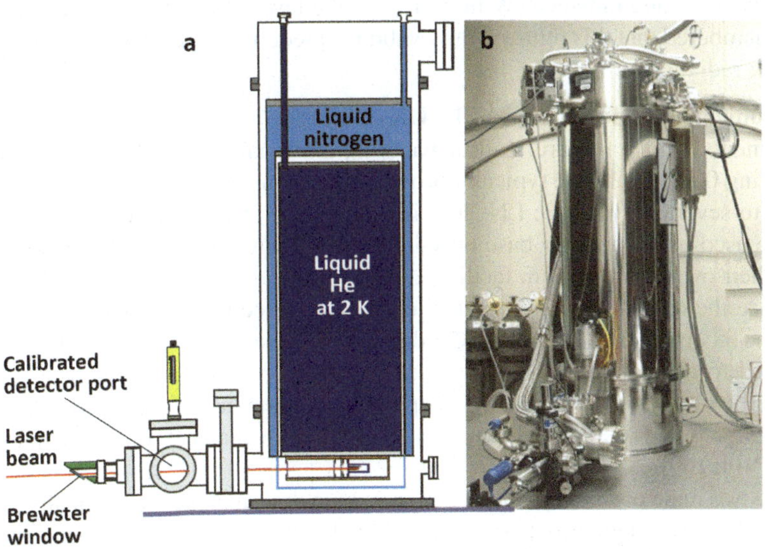

Fig. 9.7 The NIST Primary Optical Watt Radiometer (POWR), the National standard for optical power responsivity: **a** schematic and **b** photograph (after [152]). Reprinted courtesy of the National Institute of Standards and Technology, U.S. Department of Commerce. Not copyrightable in the United States

path. To measure scattered radiation, the mirror collects the scattered radiation on a calibrated silicon photodiode.

The receiving cavity is an electroformed copper cylinder, 20 mm in diameter and 150 mm long, with a flat inclined bottom that forms the angle of 30° with the axis of the cylinder. The cavity internal surface is coated with a specular black paint providing the cavity effective absorptance of 0.999995 (as it was measured at 633 nm). Its uncertainty brings a negligibly small contribution to the POWR uncertainty budget.

One heater is bifilarly wrapped near the cavity bottom; two chip heaters are attached to the bottom. Two germanium resistance thermometers are mounted on the cavity's external cylindrical wall. The POWR operates with the feedforward active control, periodic shuttering, and phase-sensitive detection.

The POWR is a national primary standard of the USA reproducing the optical power unit with the standard uncertainty not exceeding 0.01% for the power ranged from 1 μW to 1 mW. Dissemination of the optical watt is carried out via transfer standards calibrated in terms of absolute responsivity against the POWR. The TDs composed of silicon photodiodes are used as transfer standards in the Vis and NIR spectral ranges. Besides, the POWR is used to maintain and disseminate two of the seven SI base units, the kelvin above the freezing point of silver (1234.93 K) with the standard uncertainty of about 25 mK and the candela, the SI base unit of the luminous intensity

Most optical radiation measurements at the NIST are traceable to the POWR. Presently, the POWR-based facilities provide the measurement accuracy for irradiance (W m^{-2}) and radiance (W m^{-2} sr^{-1}) at the level of 0.01%. However, it should be remembered that this figure is only valid for predefined spectral and power ranges of laser radiation.

Mechanically-Cooled ACR (NPL, UK) The LHe-cooled ACRs are complicated instruments, which require skillful handling and qualified staff. The Dewar flack accepting 0.5–3 L of LHe typically provides operating time of an ACR from several hours to several days if the LN_2 loss is replenished. Therefore, LHe-cooled ACRs can be used on the regular base only in large laboratories (as a rule, at NMIs) that have their own liquid helium facilities.

Since the early 1980s, mechanical cryocoolers (see [207] for basics) capable to cool the receiving element of an ACR down to 10 K become commercially available. This led to development of less expensive and more compact helium-free ACRs. Presently, mechanically cooled ACRs provide performance comparable with that of most LHe-cooled ACRs. A helium-free ACR with a mechanical cooler was developed at the NPL [69].

The two main problems had to be overcome in design of the mechanically cooling ACR. These are the thermal noise caused by the vibration of the mechanical cooler and the increase in specific heat of materials at higher temperatures (e.g. 15 K) compared with 4 K. Figure 9.8 show a schematic representation of the mechanically cooled ACR developed at the NPL. This ACR uses a two-stage cryocooler based upon the Stirling cycle. The first stage cools the outer radiation shield down to about 30 K. The second-stage cold head cools the inner radiation shield and the

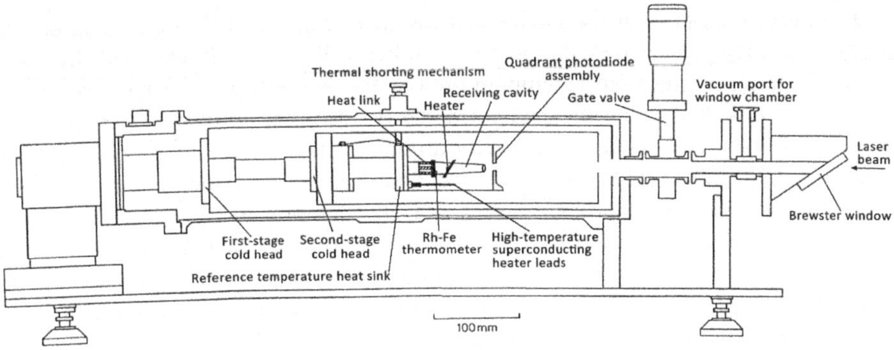

Fig. 9.8 A schematic of the NPL mechanically cooled cryogenic radiometer. Reproduced from [68] with permission of Elsevier

reference-temperature heat sink. A fast cool-down time is achieved by reducing the thermal resistance of the stainless-steel heat link between the cold-head block and the reference-temperature heat sink using the thermal shorting mechanism. Cooling the ACR from room temperature down to 13 K requires about 8 h.

The reference-temperature heat sink made of copper is maintained at temperature of about 15 K monitored using a thin-film Rh–Fe resistance thermometer and a resistance bridge.

Heat is supplied to the heat sink using a 1 MΩ thin-film heater connected to a high-resolution source of electric current.

A commercial control software is used to maintain the constant temperature of the heat sink via the IEEE interface, which is connected also with the resistance bridge and the electric current source. The measuring cycle frequency is optimized to reduce the impact of the cryocooler engine thermal noise and the relative temperature instability of the heat sink down to 10^{-6}. Since the receiving cavity operates at temperatures of 12–13 K, the specific heat capacity of copper is higher than that for the LHe-cooled ACRs operating at lower temperatures. In order to decrease the thermal constant of the ACR, the mass of the cavity and hence its dimensions should be reduced.

The receiving cavity made of 0.1 mm thick electroformed copper has the shape of a truncated cone, with an average diameter of 10.5 mm, a length of about 40 mm, and a flat inclined bottom. The latter is coated with a glossy Ni-P black [122], while the sidewall of the cavity is coated with a diffusely reflecting platinum black. The effective absorptance of the cavity has been evaluated experimentally via measurement of its effective reflectance at a 647 nm wavelength of the power-stabilized laser using an IS and found to be 0.99998 ± 0.00005.

The electrical heater is a 1 kΩ surface-mounted resistor firmly attached to the rear side of the cavity bottom. The cavity temperature is measured by a thin-film Rh–Fe resistance thermometer mounted behind the cavity assembly.

The heat link between the cavity and the heat sink at a reference temperature consists of three thin-walled stainless-steel tubes, whose length and wall thickness are selected to provide a 0.6 K temperature rise for an input power of 1 mW.

The heat link determines the cavity sensitivity: the noise equivalent power (NEP) of the ACR at 1 mW of measured power is around 10 nW. Unlike deep-cooled ACRs, a mechanically cooled ACR cannot be used in monochromator-based application but it provides a sufficient accuracy if laser radiation is employed.

A 6 mm thick silica window is set at the Brewster angle to the plane of polarization of the incident laser beam. The measured transmittance of the window is 0.9997 in the Vis spectral range. The window unit can be isolated from the main vacuum chamber of the radiometer by means of a gate valve.

The window chamber has its own vacuum port; therefore, the Brewster window can be detached for its transmittance measurement without warming up the radiometer. A unit with a quadrant photodiode mounted in front of the cavity aperture is used for the laser beam alignment; besides, being roughly calibrated at 15 K, it is used to determine a residual scattering from the Brewster window. This ACR is employed at the NPL as the primary standard to realize the optical radiation scales and the scale of luminous intensity. The accuracy of mechanically cooled ACR at measurement of the power of an intensity-stabilized laser beam is comparable with that of the LHe-cooled ACRs.

CryoRad III (L-1 Standards and Technology, Inc., USA) After mid-1990s, advances in cryocoolers' technology have made it possible to create very compact, closed-cycle, LHe-free refrigerators capable of cooling down to temperatures of 4 K or even lower a small thermal mass, in particular, a receiving cavity of an ACR. This, in turn, made possible the use of mechanical refrigeration for not only laser-based ACRs but also ACRs for measuring systems with monochromators. A brief overview of modern cryocoolers is given by Radebaugh [168]; a more extensive article by De Waele [38] discusses the basics of various mechanical refrigerators and contains their simplified thermodynamic analysis. Presently, L-1 Standards and Technology, Inc. (USA) offers CryoRad III, an LHe-free multi-purpose ACR designed for detector calibrations in both laser- and monochromator-based systems. Specification of the CryoRad III is given in Table 9.1.

A photograph of the CryoRad III with the Brewster window mount for operations with the laser radiation is presented in Fig. 9.9a. Figure 9.9b shows the modification of the CryoRad III that can be employed in both laser-based and monochromator-based systems by virtue of the flexible Y-shaped bellows and the transfer detector chamber that forms the common evacuated volume with the ACR. Both the CryoRad III and the transfer detector chamber are mounted on the rotary stage.

Such a configuration is similar to those described by Boivin and Gibb [21] and Schrama et al. [185] and eliminates the necessity to measure the window transmittance. The rotary stage moves either the CryoRad III or the transfer detector chamber into the laser or monochromator beam path to measure its power alternately by the CryoRad III and transfer detector. The measurement accuracy is claimed to be of 0.005% at the laser beam measurements with a Brewster-angled window.

Table 9.1 Specification of the CryoRad III ACR (after [127])

Characteristic	Value
Spectral range (μm)	0.2–30
Receiver effective thermal time constant (s)	<8
Servo settling time (to 0.01% of signal) (s)	40
Receiver responsivity, K/mW	2 ± 0.25
Minimum dynamic range, μW	1–1000
Noise (rms for 30 samples over), nW (varies with operating power)	<2
Absolute accuracy (%)	
For collimated beam For divergent f/8 beam	<0.005 ~0.01
Cryostat dimensions, Depth (mm) × Height (mm)	~300 × 700
Cavity aperture diameter (mm)	6–8
Cavity operating temperature (K)	4–9
Cavity effective absorptance at 632.8 nm	>0.9999
Initial cool down time (hrs)	12
Radiation shields temperature (K)	45 and 4
Thermal instability of heat sink assembly, rms (μK)	<3

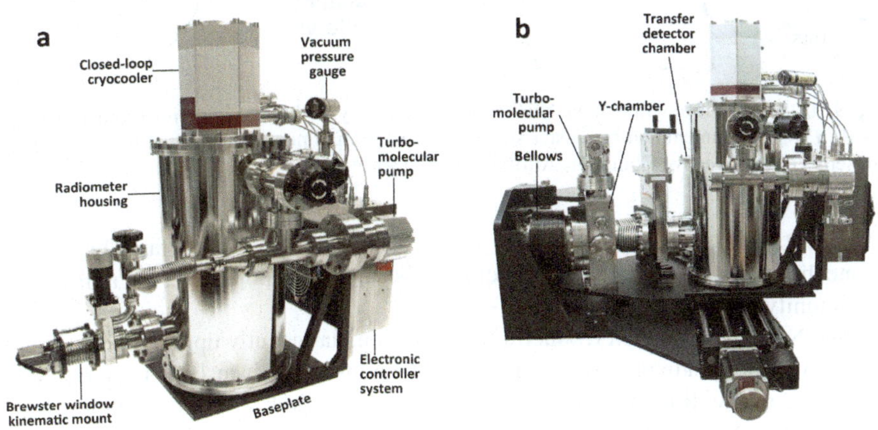

Fig. 9.9 Cryogenic radiometer CryoRad III: **a** with the Brewster window for laser-based measurement and **b** with the Detector Comparator Rotary System (*source* http://www.l-1.biz/). Reproduced with permission of L-1 Standards and Technology

At operations with the f/8 monochromator's beam, the accuracy of the CryoRad III is of order of 0.01%. The noise threshold is below 2 nW even when receiving elements "view" the ambient-temperature window. Due to the fast response, background drifts are minimized.

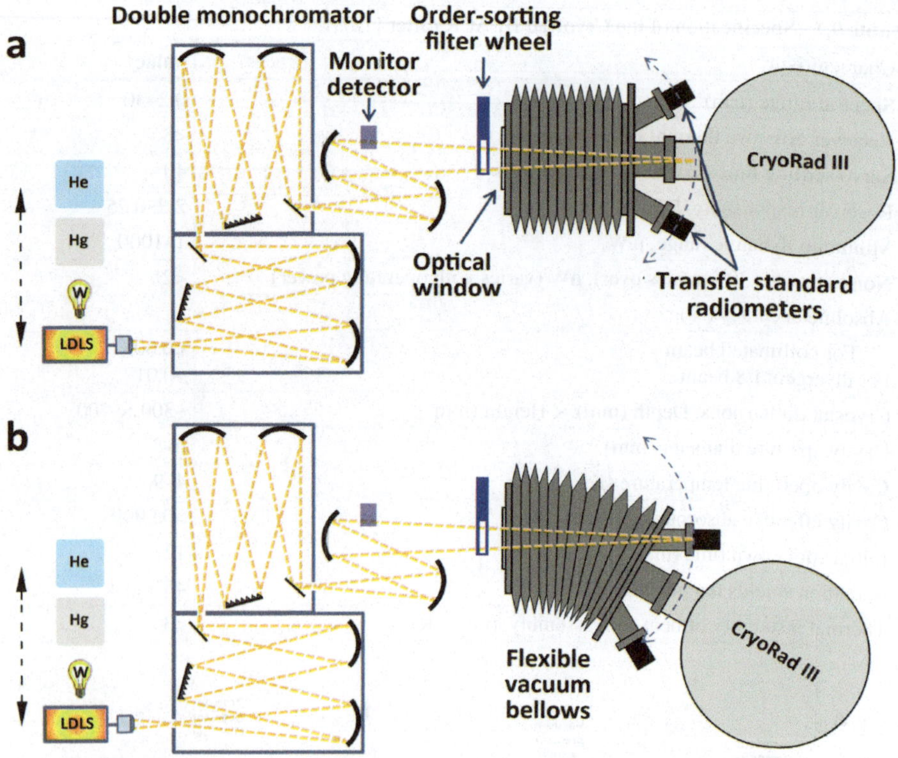

Fig. 9.10 Schematic diagram of the NRC calibration facility on the base of monochromator and the ACR: **a** ACR alignment and **b** transfer standard detector alignment. Reproduced from [73] under the terms of the Creative Commons Attribution 9.0 license

CryoRad III, due to possibility of incorporating into both laser-based and monochromator-based calibration facilities, became very popular in the radiometric community around the world.

The National Research Council (NRC) of Canada recently upgraded its primary scales in optical radiant power and spectral responsivity. Since 1994 [21], the NRC employed the ACR manufactured by Cambridge Research & Instrumentation, Inc.[3] (Cambridge, MA) in the monochromator-based facility. Presently, this ACR is at the end of its service life and is being replaced with the CryoRad III. Gamouras et al. [73] described the new facility, whose structure is presented in Fig. 9.10. This facility employs a laser-driven light source (LDLS) from Energetiq Technology, Inc. (https://www.energetiq.com/ldls-laser-driven-light-source-duv-broadband) and a tungsten lamp to cover the spectral range of 0.25–2.7 μm. These sources are mounted on a linear translation stage at the monochromator entrance allowing for rapid and repeatable alignment to the monochromator entrance slit. Two concave

[3]Now, a Division of Perkin-Elmer Inc. re-oriented on biomedical imaging systems.

spherical mirrors focus radiation from the exit slit of a double-subtractive monochromator through the order-sorting filter wheel to a tilted optical window mounted on a flexible vacuum bellows assembly. This large vacuum assembly was mounted on a linear stage with a pin and pivot system to rotate the cryogenic radiometer and transfer standard detectors for alignment to the common optical path. The LDLS delivers an optical power to the ACR of 2.3 μW at 300 nm and 5.3 μW at 400 nm.

When the tungsten lamp is aligned to the monochromator input, the optical radiant power at the ACR is 1.2 μW at 450 nm and 2.9 μW at 1000 nm. It was stated that the new ACR provides a significant improvements in uncertainty for the realization of the optical power scale, reducing the overall ACR uncertainty contribution from the 0.017 to 0.011%.

The ACRs of various types play the role of primary standards heading up the scales of optical power and spectral responsivity of detectors throughout the world. Many modern radiometric and pyrometric measurements are traceable to the SI unit of power, the watt, and more specifically, to the ACR.

9.3 Trap Detector as a Transfer Standard

9.3.1 Design Features of Trap Detectors

The FRs measure different physical quantities (radiant flux, irradiance, or radiance), while the ACR measures only the optical power (radiant flux). Besides, the FR and the ACR operate with radiation sources with different energy, spatial, angular, and spectral characteristics. Therefore, the FR cannot be calibrated against the ACR directly, but only using an intermediate detector playing the role of the transfer standard (transfer measuring device). Such a transfer detector should have a high SNR, a small nonlinearity, short time constant, low polarization sensitivity, good temporal stability and spatial uniformity over a receiving area large enough. Its relative spectral responsivity should be sufficiently smooth to provide easy and accurate interpolation between wavelengths, at which the absolute spectral responsivity can be determined using comparison with the ACR.

The best transfer detector available to date is the so-called trap detector (TD) [63], which consists of three or more silicon photodiodes connected optically in series and electrically in parallel. At present, the most widespread design of TD, shown schematically in Fig. 9.11, includes three photodiodes with the windows removed. The photodiodes are arranged each to lie in a different plane, to reduce any sensitivity to polarization of the incident radiation. In this arrangement, the incident radiation undergoes five specular reflections before emerging from the trap detector and is thus nearly totally absorbed. Due to the five reflections, the overall reflectance of the TD is much lower than the reflectance of a single photodiode as can be seen in Fig. 9.12.

A photograph of a TD with a front wall removed and large-area silicon photodiodes forming the TD in the foreground is given in Fig. 9.13.

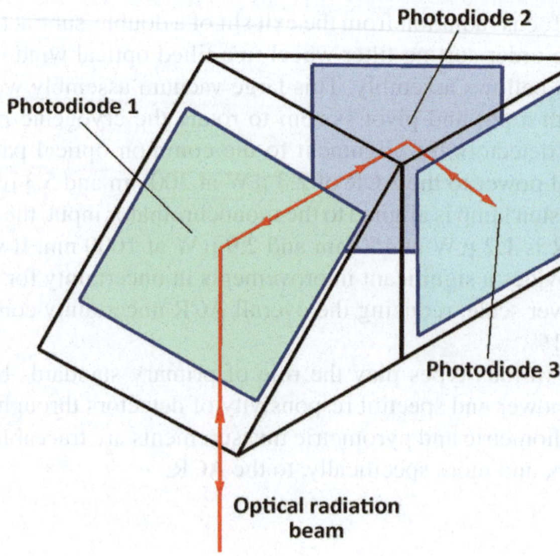

Fig. 9.11 A schematic of a TD composed of three photodiodes

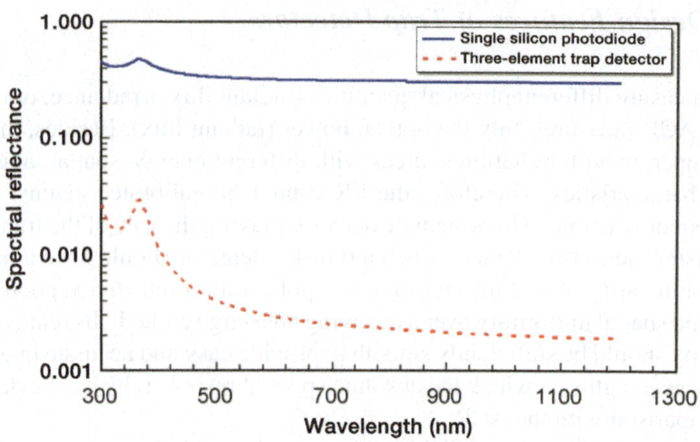

Fig. 9.12 Spectral reflectance of a silicon photodiode and spectral effective reflectance of a three-element trap detector. Reproduced from [210] with permission of Elsevier

The TDs, for which the incident and reflected beams are directed in the same half-space, belong to the reflection-type TDs. Otherwise, TDs are referred to as transmission or tunnel TDs [74, 75, 125, 126, 131]. They are free of back reflectance, therefore separately calibrated filters or other optical components could be attached to their inputs without introducing additional uncertainties.

Fig. 9.13 The interior of a three-element TD. Large-area photodiodes forming the TD are shown in the foreground. Reproduce from [203] with permission of Laboratoire national de métrologie et d'essais (France)

Eppeldauer and Lynch [53] described the tunnel-TD developed at NIST. The tunnel-TD consists of two medium size (10 mm by 10 mm) Hamamatsu 1337 and four large size (18 mm by 18 mm) Hamamatsu 6337 silicon photodiodes connected electrically in parallel. The photodiodes are oriented such that the incident radiation reflected from one detector is incident on the next detector and so forth as shown in Fig. 9.14.

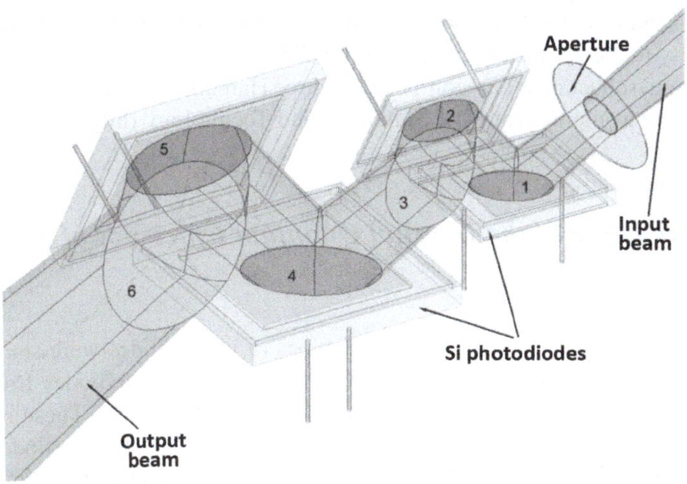

Fig. 9.14 Beam propagation through the six-element transmission tunnel-TD. Detectors 3 and 6 are not shown in the diagram [53]. Reprinted courtesy of the National Institute of Standards and Technology, U.S. Department of Commerce. Not copyrightable in the United States

The spectral external quantum efficiency (spectral EQE) of the TD, $\eta_e(\lambda)$, is the ratio of the number of photoelectrons forming the photocurrent to the number of incident photons. It is related to the spectral responsivity $s(\lambda)$ by the following simple relation

$$\eta_e(\lambda) = \frac{hc}{n_{\text{air}} e \lambda} s(\lambda), \tag{9.8}$$

where λ is the wavelength of the incident radiation in air, n_{air} is the refractive index of air, h is the Planck constant, c is the speed of light in vacuum, and e is the elementary charge. Using values of physical constants provided in Appendix A and $n_{air} \approx 1.00029$, we obtain:

$$\eta_e(\lambda) \approx 1239.48 \frac{s(\lambda)}{\lambda}. \tag{9.9}$$

Assuming that the spectral internal quantum efficiency (spectral IQE), $\eta_i(\lambda)$, is the same for each photodiode in the TD, we obtain the expression for the spectral EQE:

$$\eta_e(\lambda) = [1 - \rho_e(\lambda)]\eta_i(\lambda), \tag{9.10}$$

where ρ_e is the effective reflectance of the TD.

Therefore, if we know the effective reflectance of the TD and the spectral IQE of the individual photodiode, we can calculate the spectral EQE of the TD and its relative spectral responsivity, which both depend on several constant parameters. These constants can be determined by fitting the expression for the relative spectral responsivity to the results of calibrating the TD against the ACR.

At present, reflectance TDs composed of three Hamamatsu photodiodes S1337 are commercially available from L-1 Standards and Technology (http://www.l-1.biz/).

9.3.2 *Modeling of the Internal Quantum Efficiency of Silicon Photodiodes*

Attempts to calculate the IQE for special types of photodiodes were started at the very end of 1970s, when mass production of large-area photodiodes from silicon with a small amount of impurities and defects became possible. These researches led to the development of the so-called "self-calibrated" silicon photodiodes [23, 77, 78, 120, 236] and TDs [237] on their base. Although it was anticipated that the TD would be used as the primary radiometric standard, it did not have time to become widespread as they could not withstand competition with the ACR, developed in the mid-1980, which provided a wider spectral range and lower measurement uncertainty. Nevertheless, the developed technique for modeling the IQE of photodiodes

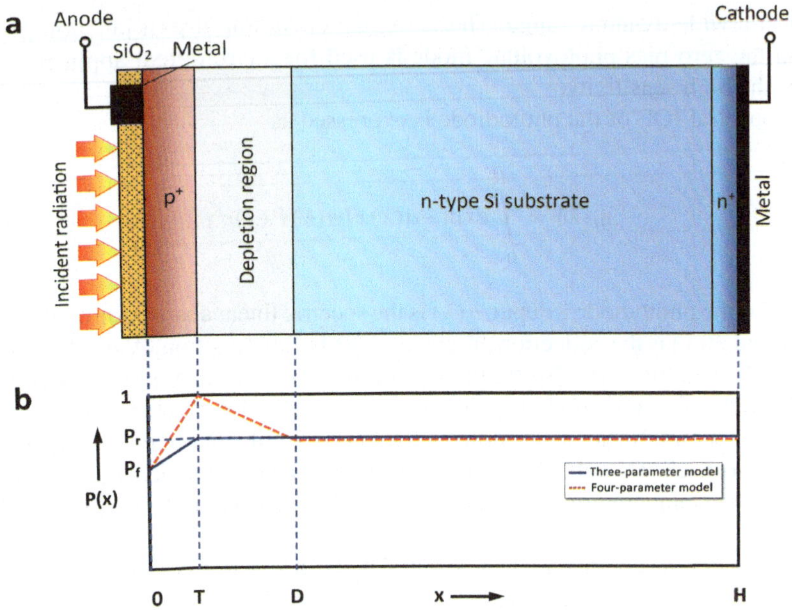

Fig. 9.15 **a** A Simplified cross-section of shallow junction p^+–p–n silicon photodiode and **b** two model distributions of collection efficiency $P(x)$ with depth x into the photodiode

and the results obtained proved to be very useful for modeling the relative spectral responsivity of TDs used as transfer standards.

Typical photodiodes used in the design of modern TDs are large area silicon photodiodes of S1337 series from Hamamatsu Photonics K. K. [95], with the windows removed. The Hamamatsu S1337 is a p^+–n–n^+ photodiode with a structure shown in Fig. 9.15a.

It is fabricated from the 0.3 mm thick silicon substrate lightly doped to provide the n type of conductivity. The front surface of the photodiode has a thermally grown passivating layer of SiO_2 with a nominal thickness of 25 nm. The p^+ layer may be formed by thermal diffusion or ion implantation to depths of about 1 μm.

The interface between the p-doped Si and the n-doped Si of the substrate comprises a shallow p–n junction, so that the depletion region is formed in the n-type region. The heavily doped n^+-zone of about 1 μm thick at the rear side of the photodiode has a high conductivity to ensure ohmic electrical contact between the semiconductor and the metal conductor.

The absorption of optical radiation creates electron-hole pairs that are subsequently separated by the electric field of the p–n junction. The electrons move toward the n-layer and holes move toward the p-layer. The open-circuit voltage generated by the photoinduced charge is depends linearly on the incident radiant flux. If the anode and cathode of the photodiode are connected via a resistive load, the migration of internal charge manifests itself as a photocurrent that is also linear with the radiant flux over a wide dynamic range. Therefore, the operation of p–n junction photodiodes in the zero-bias photovoltaic mode is used for metrological applications that require the high sensitivity.

The spectral IQE of the photodiode is expressed as

$$\eta_i(\lambda) = \int_0^H \exp[-a(\lambda)x]\, a(\lambda) P(x) dx, \tag{9.11}$$

where H is the photodiode's depth, $a(\lambda)$ is the spectral linear absorption coefficient of silicon, and $P(x)$ is the collection efficiency distribution function. A similar equation was first derived by Geist [76] for the IQE of the inversion layer photodiode. Therefore, the IQE of a photodiode is completely determined by two functions: $a(\lambda)$ and $P(x)$. The spectral linear absorption coefficients $a(\lambda)$ of silicon are usually adopted from the most reliable published data. Figure 9.16 presents the dependences $a(\lambda)$ at 300 K and the temperature coefficient $C_{a,T}(\lambda)$ plotted using data provided by Green [92].

The temperature coefficient of the spectral linear absorption coefficient a is defined as

$$C_{a,T} = \frac{1}{a}\frac{da}{dT} = \frac{d(\ln a)}{dT}. \tag{9.12}$$

The simplest way to use $C_{a,T}(\lambda)$ assumes a linear dependence of $a(\lambda)$ on a small temperature difference ΔT:

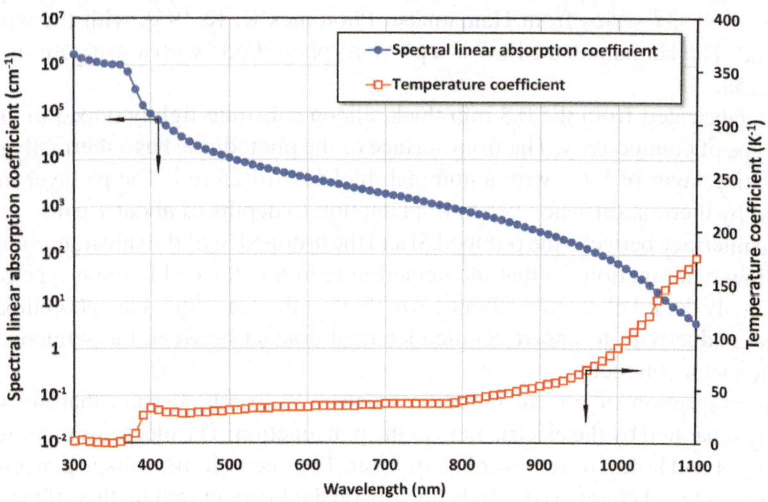

Fig. 9.16 Spectral linear absorption coefficient $a(\lambda)$ of silicon at 300 K and temperature coefficient $C_{a,T}(\lambda)$ (after [92])

$$a(\lambda, T_0 + \Delta T) = a(\lambda, T_0)[1 + C_{a.T}(\lambda)\Delta T], \tag{9.13}$$

where T_0 is a reference temperature (300 K for $a(\lambda)$ presented in Fig. 9.16).

Geist et al. [78] derived an equation for $\eta_i(\lambda)$ with the assumption of the collection efficiency independent on the depth x of penetration into the photodiode. More accurate models for the collection efficiency must take into account its variation with x. Two piecewise-linear models for $P(x)$ are shown in Fig. 9.15b. The simplest model implies that $P(x)$ starts from P_f at the air-SiO_2 interface and increases linearly with x reaching the bulk value P_b of silicon at the p–n junction depth T. The value P_b is maintained up to the rear surface of the diode at the depth H. The internal quantum efficiency for this model referred to as three-parameter model is expressed by the following equation:

$$\eta_i(\lambda) = P_f + \frac{P_b - P_f}{a(\lambda)T}\left[1 - e^{-a(\lambda)}T\right], \tag{9.14}$$

where T, P_f, and P_b can be considered as fitting parameters; if measurements are carried out at three wavelength of laser radiation, the exact values of these parameters can be found uniquely; if number of laser wavelengths is greater than 3, the least square approach can be applied to find optimal values for T, P_f, and P_b.

Gentile et al. [83] employed for $P(x)$ a piecewise-linear approximation, according to which $P(0) = P_f$ at the SiO_2–Si interface (the constant P_f is affected mainly by the trapped surface charges and the lattice defects caused by mismatch of the lattice constants for SiO_2 and Si). Then $P(x)$ increases linearly and reaches unity at the depth T; further inwards, a linear decrease follows until the bulk value P_b of silicon is reached at the depth D. The following expression for $\eta_i(\lambda)$ was obtained:

$$\eta_i(\lambda) = P_f + \frac{1 - P_f}{a(\lambda)T}\left[1 - e^{-a(\lambda)T}\right] - \frac{1 - P_b}{a(\lambda)(D - T)}\left[e^{-a(\lambda)T} - e^{-a(\lambda)D}\right] - P_b e^{-a(\lambda)h}, \tag{9.15}$$

where h is the known thickness of Si in a photodiode; the four fitting parameters (P_f, T, P_b, and D) indicated in Fig. 9.15b, require measurements at least at four wavelengths.

Werner and Hartmann [210] described an extension of this model for NIR spectral range accounted for the fact that $a(\lambda)$ for silicon is very small for wavelengths longer than about 950 nm, so the penetration depth of photons becomes larger than H.

The photons, partly reflected from the rear surface of a photodiode, pass through it once again. This leads to correction in the last term in Eq. 9.15:

$$\eta_i(\lambda) = P_f + \frac{1 - P_f}{a(\lambda)T}\left[1 - e^{-a(\lambda)T}\right] - \frac{1 - P_r}{a(\lambda)(D - T)}\left[e^{-a(\lambda)T} - e^{-a(\lambda)D}\right] - (1 + \rho_b)P_r e^{-a(\lambda)H}, \tag{9.16}$$

where ρ_b is the reflectance of the back side of the photodiode assumed to be wavelength-independent.

Extensive experience in using TDs for establishing national radiometric scales in NMIs has shown that the IQE of a photodiode is more sensitive to the spectral linear absorption coefficient of silicon than to the distribution of the collection efficiency. Eppeldauer and Lynch [53] also stated that $\eta_i(\lambda)$ does not depend on typical variations in the doping profile of individual photodiodes.

The above-described approach to modeling η_i is not the only possible. Ferrero et al. [55] proposed for the quantum efficiency of a p–n photodiode a more complicated physical model capable to explain photodiode's non-linearity and spatial non-uniformity. It is worth mentioning about computer programs that allow simulating the transfer of charge carriers in semiconductor structures to obtain various characteristics of photodiodes. For instance, Jon Geist of NIST with coauthors [79–81] described the application of the semiconductor device one-dimensional finite element modeling software to high-accuracy simulation of silicon photodiodes. Dönsberg et al. [41] used the 3D model and the commercial semiconductor simulation software to precise calculation of IQE of a prospective candidate for a primary radiometric standard on the base of large-area silicon photodiodes [39]. However, the development of more accurate model for photodiode's IQE was not chosen as the most promising way for the spectral interpolation using the TDs. Below, we will see why.

9.3.3 Spectral Interpolation Using Trap Detectors

Equation 9.8 can be rewritten taking into account of (9.10) as

$$s(\lambda) = n_{air}\eta_i(\lambda)\frac{\lambda e}{hc}[1 - \rho_e(\lambda)]. \tag{9.17}$$

Substituting one of Eqs. 9.14, 9.15, or 9.16 into Eq. 9.17, we obtain the spectral responsivity of the TD with three or four adjustable parameters. However, before calculating η_i, we must determine the effective reflectance ρ_e of the TD. In principle, it can be done by computational methods, on the base of Fresnel's equation, taking into account different angles of the radiation incidence on each photodiode, polarization state of the incident beam, and multiple reflections in the thin film of SiO_2 at the interface with air and silicon. For such calculations, the complex refractive indices for Si and SiO_2 are usually taken from the reference books (e.g. [47] or [1]).

The attempts of calculating ρ_e were undertaken multiple times [19, 32, 45, 94, 146]; however, the experimental determination of ρ_e (or at least measuring of spectral, angular, and polarization dependences of reflectance of individual photodiodes) is required in any case to confirm the calculated values.

Typical calibration results of the trap detector composed of the three Hamamatsu S6337 silicon photodiodes are presented in Fig. 9.17. The circles denote spectral responsivities measured at laser wavelengths with an uncertainty below 10^{-4} against the Radiation Thermometry Cryogenic Radiometer [194] at the PTB[4]. The solid line

[4]Physikalisch-Technische Bundesanstalt, the NMI of the Federal Republic of Germany.

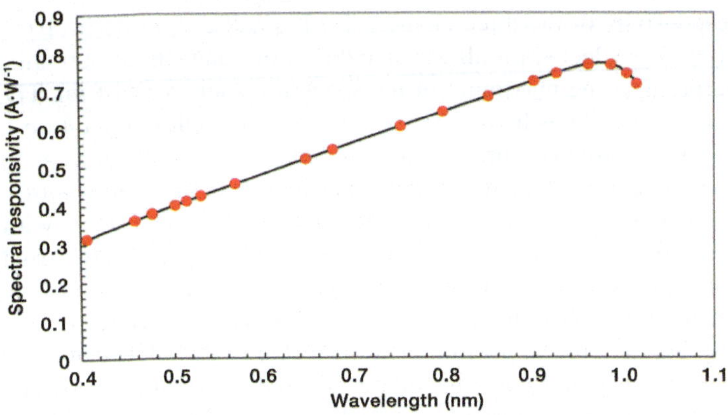

Fig. 9.17 Spectral responsivity of a silicon trap detector measured at the laser lines (circles) and obtained by interpolation using a physical model (line). Reproduced from [210] with permission of Elsevier

shows the spectral responsivity model with parameters fitted to the calibration results at laser wavelength.

The use of a trap detector allows establishing at the PTB a continuous scale of the spectral responsivity from 400 nm to 1020 nm with an overall relative standard uncertainty of 2×10^{-4} in the wavelength range from 500 nm up to 900 nm and a somewhat higher uncertainty of up to 14×10^{-4} outside this range. The main uncertainty sources in calibration of the TD and measurement with it are the TD non-linearity, spatial, temperature, and polarization variations of its responsivity (see details in [20, 209]).

The temperature dependence of the spectral responsivity of a TD is the most significant source of uncertainty of transferring the units of spectral radiometric quantities from the ACR to secondary radiometers (e.g. FRs). This effect was studied for photodiodes by Lei and Fischer [132] and for TDs by Werner et al. [209], who measured the signals of the TD irradiated by power-stabilized laser radiation while the detector temperature is changed using a temperature-controlled water jacket.

The greatest contribution to the total temperature coefficient of the spectral responsivity (i.e. the relative changes of the spectral responsivity $\Delta s/s$ per temperature change ΔT) brings the temperature-dependent spectral linear absorption coefficient, which, in turn, originates in the temperature dependence of the bandgap of silicon and deeper penetration of radiation at larger wavelengths.

Hartmann et al. [99] discussed various analytical models for the temperature dependence of the spectral responsivity of silicon and applied it to the three-photodiodes TD. The reflectance of a TD is the product of the reflection coefficients for the five reflections inside it. Assuming that the temperature dependence of these five reflections is the same for all incidence angles, the relative temperature dependence of the reflectance is five times larger for a TD than for a single photodiode. Werner and Hartmann [210] showed that the temperature coefficient of spectral responsivity of the three-element TD is small enough (5×10^{-5} K^{-1} at $\lambda = 650$ nm) to be neglected for all wavelength shorter than about 900 nm. However, it rises sharply near the maximum of the spectral responsivity of the TD, where it can reach values nearly as high as 0.5% per 1 K. The high temperature coefficient around 1000 nm requires accurate control and measurement of the detector temperature during operations. If the uncertainty contribution of the temperature correction is to be kept below 6×10^{-5} at 1 μm, the temperature has to be stable within 0.02 K [209]. If to keep the TD temperature constant within ±0.3 K, which is easily achievable using temperature stabilization of an outer jacket by flowing water, a remaining relative uncertainty can reach 4×10^{-5} after temperature corrections.

Presently neither precise measurement traceable to the ACR can be carried out without the use of TDs and, correspondingly, without determining its spectral responsivity. However, the calculations of the effective reflectance cannot take into account such factors as the divergence of the radiation beam and imperfect specular reflection (finite width of the specularly reflected lobe, diffuse reflection, volumetric scattering, etc.) There is an alternative approach to determination of $\rho_e(\lambda)$ using its suitable approximation by a function with several fitting parameters. Hartmann [97] described the use of an analytical approximation, sufficiently accurate for the wavelength range from 400 to 1015 nm

$$\rho_e(\lambda) = a \exp(b/\lambda) + c\lambda + d, \tag{9.18}$$

which adds four fitting parameters a, b, c, and d, but requires the use of at least four additional wavelengths at the TD calibration against the ACR.

One can conclude that the physical-based modeling of the IQE seems to be a best way to build the interpolant if the number of laser wavelengths, at which the spectral responsivity of a TD is measured using the ACR, is not so large (say, less than 20). The modern calibration facilities with ACRs typically employ a set of tuneable lasers and optical parametric oscillators, emission wavelengths of which cover the spectral range between 400 and 1100 nm very densely. The use of supercontinuum light sources [213] in combination with a monochromator makes it possible to operate at hundreds of wavelengths with bandwidths and spectral radiant fluxes comparable to those of lasers.

Although the physically-based models of the quantum efficiencies are typically employed for interpolation of the TD spectral power responsivity, the use of such models are not necessary when calibration is carried out at many wavelengths. More suitable for this purpose can be a model developed at NIST [84] that is not physically-based. The nine-parameter expression for the external quantum efficiency $\eta_e(\lambda)$ valid for wavelength λ from 350 to 1010 nm consists of an exponentially decaying function with fitting parameters a_1, a_2, and a_3 combined with the two Lorentzian functions with fitting parameters b_1, b_2, b_3 and c_1, c_2, c_3:

$$\eta_e(\lambda) = \frac{a_1}{1 + \exp[-(\lambda - a_2)/a_3]} - \frac{b_1}{1 + [(\lambda - b_2)/b_3]^2} - \frac{c_1}{1 + [(\lambda - c_2)/c_3]^2}. \quad (9.19)$$

Comparison of results obtained with three- and four-parameter models for IQE and calculated values of $\rho_e(\lambda)$, on the one hand, and with nine-parameter model for EQE (9.19), on the other hand, shows that results are consistent within 0.005% over almost all of the 400–900 nm range but differ sharply near endpoints of this interval. The difference between $s(\lambda)$ computed according to the three-parameter and nine-parameter models reaches 0.02% at 900 nm [84].

9.3.4 Uncertainties in Calibration of Trap Detectors

Due to the difficulty in directly comparing the FR and the ACR, the TD is an indispensable link between them, when the FR is used as an absolute primary radiometric thermometer. The spectral power responsivity $s_{\Phi,TD}(\lambda)$ of the TD is the ratio of its photocurrent I to the incident optical power P causing the photocurrent and measured by the ACR:

$$s_{\Phi,TD}(\lambda) = I/P. \quad (9.20)$$

The photocurrent measurement is performed usually by a current-to-voltage converter and a digital voltmeter. The measured spectral responsivity of the TD can be interpolated over the entire wavelength range of interest using an empirical or physical model for $s_{\Phi,TD}(\lambda)$. When the TD is applied to calibrate an FR, the TD spectral responsivity has to be corrected for the experimental conditions (e.g., radiant power level, temperature, geometry of measurements, polarization state, etc.) during the calibration of the FR, which may differ significantly from those during the calibration of the TD against the ACR.

Table 9.2 gives an example of the uncertainty budget for the TD power responsivity determination. The uncertainty components in Table 9.2 are subdivided into four groups:

I. Uncertainty of the radiant power measurement by the ACR. These contributions mainly depend on the type of ACR used and the effort spent to characterize the device.
II. Uncertainty contributions related to the TD calibration against the ACR. These contributions mainly depend on the calibration principle, the radiation source, and the facility used to calibrate a TD against an ACR.
III. Uncertainty components arising from the interpolation of the TD spectral responsivity.

Table 9.2 A typical uncertainty budget for the spectral power responsivity of a TD (after [180])

Source of uncertainty	Relative standard uncertainty (%)
Electrical power measurements	0.0005
Non-equivalence of electrical/optical power	0.0010
Cavity absorptance	0.0010
Window transmittance	0.0018
Sensitivity of the radiometer	0.0010
Repeatability	0.0015
Stray radiation	0.0020
Photocurrent	0.0010
Wavelength	0.0002
Distance and diameter effects	0.0020
Interpolation/fit of the spectral responsivity	0.0200
Spatial non-uniformity	0.0040
Linearity correction	0.0020
Temperature correction	0.0010
Beam polarization orientation	0.0010
Temporal stability	0.0020
Total	0.0211

IV. Uncertainty contributions related to the use of the calibrated TD under experimental conditions different from those during its calibration. These contributions depend on the properties of the FR to be calibrated and the degree of mismatch of the experimental conditions of the calibration and the use of the TD.

The combined relative standard uncertainty of determining $s_{\Phi,TD}(\lambda)$ is equal (in average) to 0.02%, although numerical values for components listed in Table 9.2 may vary depending on the design of the TD and ACR and on thoroughness of their characterization, as well as on features of the calibration process.

9.4 Methods of Absolute Primary Radiometric Thermometry

9.4.1 *The Basic Measurement Equations*

When the absolute spectral responsivity of a TD is determined at several wavelength using an ACR, the filter radiometer (FR) can be calibrated against the TD and then be used to measure the spectral radiant flux emitted by a blackbody in a particular waveband and in a defined solid angle.

The general principles of the APRT are summarized in the *MeP*-K [140] and are detailed in a report elaborated by the CCT[5] [180], which describes the different methods for implementing the APRT and provides a unified framework for assessing the uncertainty in measured thermodynamic temperature of a blackbody. Within this framework, any FR measure the radiance of the blackbody (integrated within the FR's bandwidth), while four following methods of the FR calibration can be used.

1. The spectral power method. An FR, calibrated in terms of power responsivity, is used in combination with the pair of apertures.
2. The irradiance method. An FR, calibrated for irradiance responsivity, is used in combination with a single aperture.
3. The hybrid method. An FR, calibrated for irradiance responsivity, is used in combination with an aperture and a simple single objective lens.
4. The radiance method. An FR is incorporated within an optical system consisting of several lenses and appropriate baffling. The resulting measuring device (RT) is calibrated for radiance responsivity.

The methods 1 and 2 are non-imaging, while the methods 3 and 4 use optics to facilitate the measurement of small sources. Whichever scheme is used, the basic measurement equation can be written as

$$i = \int_0^\infty s_L(\lambda) L_{\lambda,bb}(\lambda, T) d\lambda, \tag{9.21}$$

where i is the measured photocurrent, $s_L(\lambda)$ is the absolute spectral radiance responsivity [A W^{-1} m^2 sr] of the measuring instrument, $L_{\lambda,bb}(\lambda)$ is the spectral radiance of the perfect blackbody, λ is the wavelength in air, and T is the thermodynamic temperature. Equation 9.20 should be somewhat modified for the different schemes. The spectral radiance of a perfect blackbody is given by Planck's law. For measurements where the detector is in air,

[5]Consultative Committee for Thermometry of the BIPM (International Bureau of Weights and Measures; French: Bureau international des poids et mesures).

$$L_{\lambda,bb}(\lambda,T) = \frac{c_{1L}}{n_{air}^2\lambda^5}\frac{1}{\exp\left[c_2/(n_{air}\lambda T)\right]-1}, \quad (9.22)$$

where c_{1L} is the first radiation constant for the radiance, c_2 is the second radiation constant in the Planck law (their values are given in Appendix A).

Below, we consider in detail the four above-mentioned methods, noting the advantages and disadvantages of each of them.

9.4.2 The Spectral Power Method

For the spectral power method, the FR is calibrated for flux responsivity [A/W] with the radiation beam underfilling the FR's aperture. The second aperture is added for measuring the blackbody temperature (Fig. 9.18). Equation 9.21 is implemented using the spectral power responsivity:

$$s_L(\lambda) = \pi g s_\Phi(\lambda), \quad (9.23)$$

where $s_\Phi(\lambda)$ [A W^{-1}] is the spectral power responsivity of the FR, π [sr] has the dimension of the solid angle, and g [m^2] is the so-called geometric factor for the double-aperture system.

The geometric factor g was introduced as an alternative to the diffuse view factor $F_{A_1-A_2}$(see Sect. 3.4.2). According Eq. 3.143, for two coaxial discs lying in the parallel planes:

$$g = \frac{2\pi r_1^2}{(r_1^2+r_1^2+d^2)+\sqrt{(r_1^2+r_1^2+d^2)^2-4r_1^2r_1^2}}, \quad (9.24)$$

where r_1 and r_2 are the radii of two aperture and d is the distance between them (see Fig. 9.18b).

In principle, calibration of an FR in the power mode can be done directly against the ACR; however, in such a case, the FR must have a uniform response across the

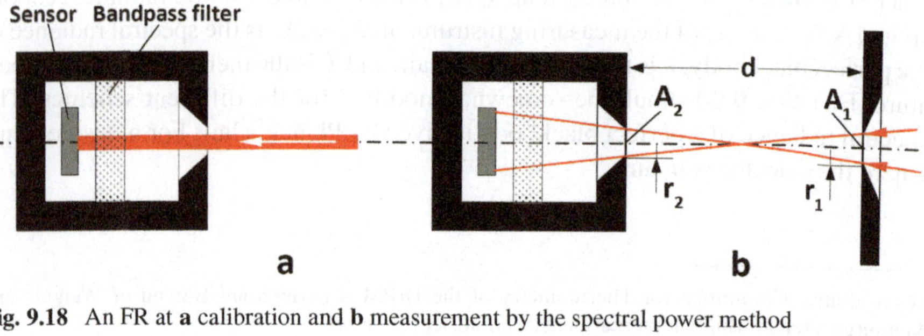

Fig. 9.18 An FR at **a** calibration and **b** measurement by the spectral power method

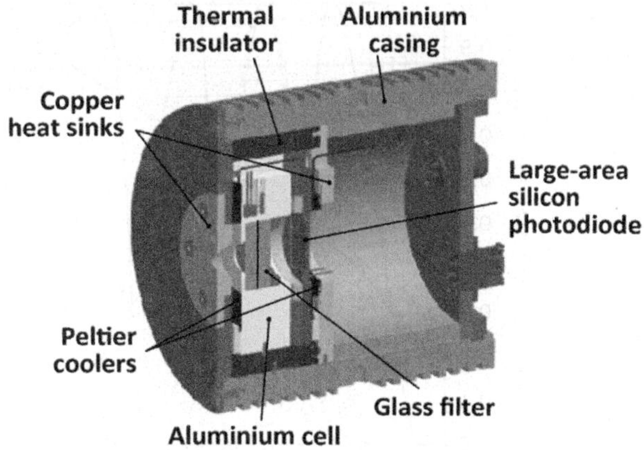

Fig. 9.19 Three-dimensional sectional view of the NRC's FR. Reproduced from [22] with permission of Taylor & Francis

aperture, which cannot be guaranteed in practice. Therefore, an intermediate detector should be used to transfer the unit of optical watt from the ACR to the FR. More often, the power responsivity is converted to spectral irradiance responsivity before measurements (see, e.g. [228]), so the power method, in its pure form, is used rarely. We are aware of the only such a case described by Boivin et al. [22] and applied to the measurement of thermodynamic temperature of a VTBB operating in the temperature range from 1800 to 3500 K at the NRC[6] (Canada). Note that this case is in some ways atypical. First, the filter used in the FR is not narrowband interference, but made of glass and rather broadband. Second, instead of TD, a working standard silicon photodiode was used as transfer detector. However, this work is an example of a thorough study of all elements of the measuring system in order to determine its metrological characteristics. Later, the FRs calibrated in terms of power responsivity were used to calibrate the RT LP3 applied to measuring temperatures of HTFP cells.

Since a narrow spectral bandpass was not necessary, combinations of filter glasses were used to obtain an FWHM of about 100 nm.

The design of an FR developed at the NRC is shown in Fig. 9.19.

Spectral responsivity was measured by direct comparison with the Si photodiode working standard using the facility comprising a double grating monochromator with a quartz tungsten halogen (QTH) lamp used as a source. At the output slit of the monochromator, a quasi-monochromatic radiation with a bandwidth of 4 nm FWHM was obtained. Relative spectral responsivities of the five wide-band FRs are presented in Fig. 9.20 together with the relative spectral responsivity of Hamamatsu S1337 Si photodiode used as a working standard.

For better stability, the thin shims are used instead of cement to hold together components of composite filters separated by air space. Multiple interreflections occurred at air-glass interfaces are suppressed by chamfering and blackening the internal surfaces of the FRs. The FR aperture diameters of 6 mm closely match the diameters of the apertures of the working standard detectors used for the FRs

[6] National Research Council of Canada (https://nrc.canada.ca/en).

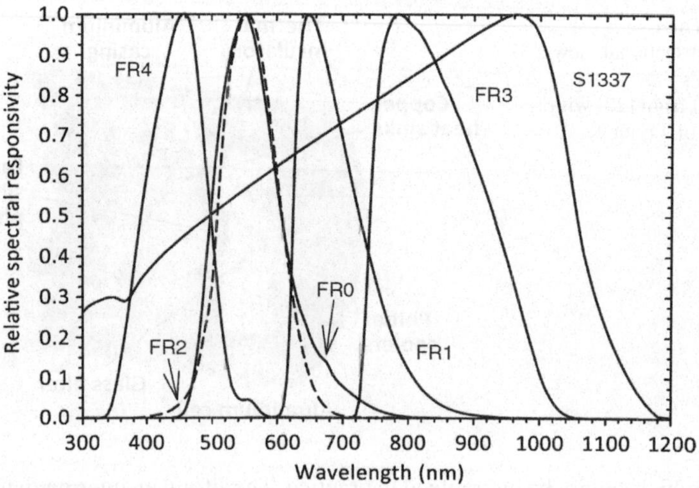

Fig. 9.20 Relative spectral responsivities of the five NRC FRs (from FR0 to FR4) and of the working standards (Hamamatsu S1337 Si photodiode) used for their calibration. Reproduced from [22] with permission of Taylor & Francis

calibration. To meet strong requirements to the spatial uniformity of a sensor that is critical for the power mode, the large-area (18 mm × 18 mm) Hamamatsu S6337 silicon photodiode with uniformity better than 0.1% in the central 10 mm region throughout the spectral range from approximately 400–1100 nm was chosen.

Each FR was characterized in terms of its spatial uniformity. Distributions of the relative spectral responsivity along diameters of the five FRs were obtained by scanning of the FR apertures by a narrow beam of broadband radiation for azimuth angles of 0, 45, 90, and 135°. While four FRs demonstrated quite uniform spatial responsivity (better than 0.1% in any direction, one FR (with a 0.48 μm center wavelength) showed larger values (up to 0.5%).

The optical filter and the sensor are mounted in an aluminum cell thermally insulated by a Delrin cylinder. The temperature of the cell is controlled using two annular Peltier coolers mounted on the front and rear of the aluminum cell. The output of an integrated circuit temperature transducer imbedded in the aluminum cell is used to control the cooling power of the Peltier elements. The three resistance temperature detectors measure independently temperatures of the aluminum cell, the back of the front flange, and the rear side of the detector. The normal operating temperature of the FR is 30.0 °C. The temperature stability of the FRs was checked out by monitoring the temperatures of the external heat sink and the internal temperature-controlled cell with the FR placed 40 cm from the heat load from a 1 kW incandescent lamp. Measurements indicated that the internal cell temperature variation was less than 0.04 °C even during abrupt changes in the external heat load. The Peltier coolers allow the FR temperature to be varied from 20 to 40 °C for measuring the FR temperature coefficient of responsivity. Figure 9.21 presents the spectral responsivity of the FR1

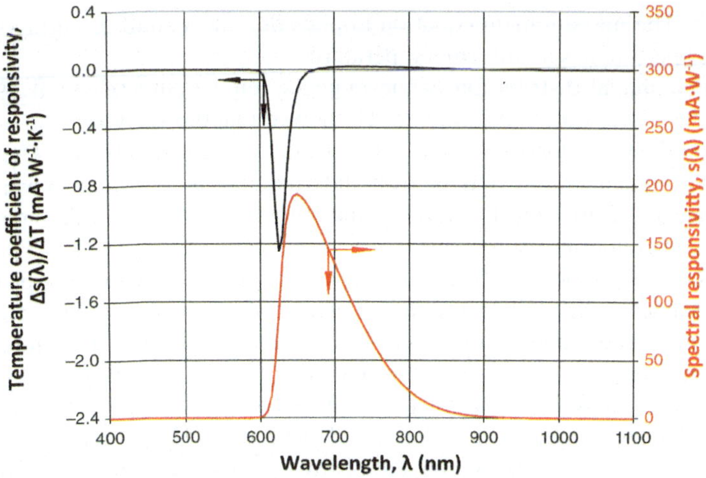

Fig. 9.21 Spectral responsivity and temperature sensitivity of spectral responsivity for filter radiometer FR1. Reproduced from [22] with permission of Taylor & Francis

having the largest temperature sensitivity, together with its temperature coefficient of responsivity that allows estimating the effect of the FR temperature variation on the blackbody temperature measurement.

The FR circular apertures of 9.35 mm diameter were fabricated by swaging the aperture of slightly larger diameter onto steel rods by immersion in liquid nitrogen. This results in smooth round apertures whose area can be accurately measured. Measurements were made at every 10° around the circumference using a coordinate measuring machine. The root-mean-square of the 36 measured diameters was used to determine the final diameter.

Temperature measurements were carried out for the pyrolytic graphite BB3500-M [156, 177] developed and built at the VNIIOFI (Russia). The temperature of this high-temperature blackbody (HTBB) was obtained by measuring the spectral radiant flux in a beam defined by the apertures at the HTBB (the source aperture of radius r_1) and the FRs (the detector aperture of radius r_2). The apertures were aligned to be centered on the cavity optical axis and to lie in the planes perpendicular to it. The source and the detector apertures are separated by the distance of about $d = 165$ mm, which is set using a gauge bar.

This geometry ensured that the radiant flux measured by the FRs is originated from the radiating cavity bottom only. The five FRs and the RT LP3 were mounted on the optical rail that allows their repeatable positioning in front of the HTBB with the uncertainty less than 1 mm.

For each measured temperature, the HTBB temperature was ramped up and the desired setpoint was maintained using the feedback loop. Once the HTBB had reached a sufficiently stable temperature, an initial measurement using LP3 was followed by two series of five FR measurements, after which the RT measurement is repeated. The measurement equation implies that the virtual radiating surface at the source aperture has a uniform temperature.

Measured spatial distribution of the radiance temperature of about 2900 K by scanning the source aperture using the RT showed that the variations do not exceed 0.1 and 0.6 K in the horizontal and vertical directions, respectively. The spatially weighted mean temperature derived from the irradiance measurements was found to be less than 0.1 K different from the temperature at the center of the aperture (for radiance temperature measurements).

The FRs were calibrated using a slightly divergent (f/20) beam of approximately 2 mm in diameter underfilling the radiometer aperture, while the HTBB radiation overfilled the 6 mm diameter aperture. The uncertainties due to this difference were estimated from the measurement of the spatial uniformity of the FR response.

In order to check independently the FR readings, the radiation temperature of the HTBB was measured using a RT LP3 from KE Technologie GmbH (Stuttgart, Germany, http://ke-gmbh.de/linearpyrometer) calibrated according to the CCT recommendations [58].

To measure the radiance temperature, the LP3 was positioned at 77.5 cm from the HTBB and focused on a 1.3 mm diameter spot at the center of the HTBB aperture (8.6 mm diameter). The standard uncertainties for the LP3 were 1.9 K at 2500 K, 2.2 K at 2700 K, and 2.6 K at 2900 K. A good agreement (within the limits of the estimated uncertainties) was found between the HTBB temperatures determined by each of the five filter radiometers, from one side, and with radiance temperature measurement made using the LP3, from the other side.

Filter radiometers from FR0 to FR3 agree with each other to better than 1 K, while the FR4 differs from the average of the other four radiometers by approximately 2 K, which is larger than the associated uncertainty. The reason for the discrepancy between the FR4 and the others was, as assumed, the poor spatial uniformity of the FR4.

The main contributor to the uncertainty budget of the FRs is the aging of the working standards (a single silicone photodiode). The main contribution to the uncertainty budget for the operation of the HTBB is the drift and stability of the HTBB, even though its temperature is controlled using feedback loop with the temperature-stabilized monitor detector.

It was also observed that the temperature of the laboratory was drifting by several degrees during the measurements.

Generally, the measurements presented by Boivin et al. [22] demonstrate that the NRC FRs can be used to perform the HTBB temperature measurements using the power method with the combined standard uncertainties assessed as 0.54–0.75 K for the HTBB temperature of 2500 K, 0.60–0.94 K for 2700 K, and 0.67–0.94 K at 2900 K.

9.4.3 The Irradiance Method

For the irradiance method, the FR is calibrated for irradiance responsivity s_E [A W^{-1} m^2] with an overfilled aperture. Then it is used with a second aperture added for the blackbody measurement (see Fig. 9.22). Equation 9.21 is implemented using

$$s_L(\lambda) = \frac{\pi g s_E(\lambda)}{A_2}, \tag{9.25}$$

where A_2 is the FR's aperture area.

To calibrate the FR against the TD, the latter one should be equipped with a precision aperture of known diameter. The incident beam should be expanded to overfill the aperture with the largest diameter. Prior to calibration, it is necessary to make sure that the irradiance is uniformly distributed over the beam cross section.

Measurement of blackbody thermodynamic temperature is carried out according to the same scheme as for spectral power method. It is clear from Fig. 9.22 that the detector areas irradiated at calibration and measurement do not coincide. If the detector response is spatially inhomogeneous, a non-uniform irradiance of the detector aperture may result in significant uncertainty. To improve irradiance uniformity, the distance d should be large enough. However, increase in distance between apertures leads to decrease of radiant flux transferred between them. If the radius of the detector aperture r_2 is fixed (it must be less than the area of the detector's receiving surface), one can increase the radiant flux registered by an FR by increasing the radius r_1 of the source aperture.

However, first, its upper bound is limited by the radius of the blackbody opening; second, this again makes worse the irradiance uniformity. Therefore, the choice of geometrical parameters for the irradiance method should be done as a trade-off between two following requirements: (i) it is necessary to provide a sufficient spectral response of an FR for the blackbody at a given temperature and (ii) the uncertainty due to non-uniformity of the FR irradiation at that blackbody temperature must be minimized.

A typical aperture diameter of the silicon photodiode-based FR is 5 mm, while the most suitable for the irradiance mode diameter of the source aperture is about

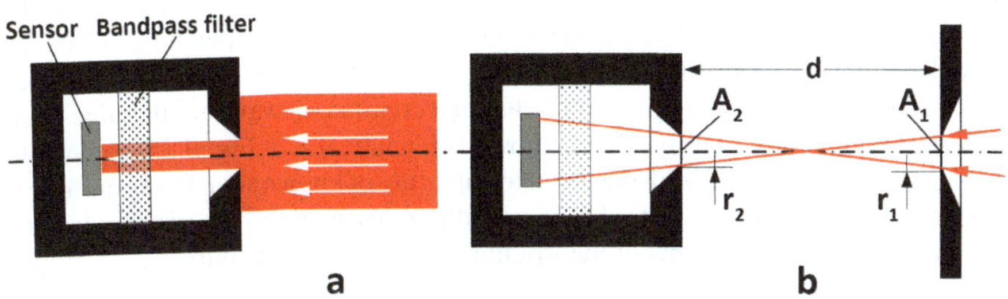

Fig. 9.22 An FR at **a** calibration and **b** measurement using in the irradiance method

20 mm. To ensure uniformity of the FR aperture irradiation, the distance d between apertures is usually from 400 to 2000 mm and must be measured with high accuracy, e.g. using an interferometer (see, e.g. [51]) providing an uncertainty of less than 1 μm. The geometric factor is approximately proportional to the inverse square of the aperture separation; therefore, the smaller the distance, the higher the required accuracy. With a blackbody at the freezing temperature of silver in a distance $d =$ 1 m, the FR with an interference filter of 800 nm wavelength and 20 nm FWHM receives a radiant power of approximately 20 nW. This imposes some geometrical restriction on the arrangement of the aperture pair for each combination of the FR and blackbody temperature.

The irradiance method of APRT was intensively developed at the PTB[7] since early 1990s for the determination of difference $T - T_{90}$ between the thermodynamic and the ITS-90 scales. The objective of these studies was to eliminate a partial inconsistency (up to 30 mK) of the results obtained with the high-temperature gas thermometers from about 500 K to the freezing point of silver (1234.93 K). The APRT could resolve the problem with the uniqueness of the ITS-90 determination in this temperature range (more details can be found in [59]). This problem has rather fundamental than practical significance, because its effect is nearly negligible for IR blackbody radiometry. However, the measuring devices and measurement methodology developed at the PTB for radiometric determination of thermodynamic temperatures below the silver point [56, 100, 155, 195, 200] have later been applied to determination of thermodynamic temperature of the HTBBs, particularly, with HTFP cells [6, 7, 202]. Investigations performed and the results obtained at the PTB in the APRT with the FRs that operate according to the irradiance method were systematized by Hartmann [97] and Hartmann et al. [103].

A variety of FRs has been developed at the PTB to measure the thermodynamic temperature of blackbodies, which are the primary standard radiators for the realization and dissemination of both the spectral radiance [71] and the spectral irradiance [191] scales. Most of them use narrowband interference filters. For the visible and the NIR spectral ranges, the FRs with the silicon photodiodes are used. Figure 9.23 presents the schematic design of the PTB's FRs with the Si and InGaAs photodiodes, the spectral irradiance responsivities of which are shown in Fig. 9.24.

The FRs are calibrated in a two-step procedure against the primary detector standard, the Radiation Thermometry Cryogenic Radiometer (RTCR) with the relative uncertainty of several hundredths of a percent [70, 200]. At the first stage, a transfer detector (silicon photodiode-based TD or InGaAs photodiode, depending on the FR's photodetector) is calibrated against the ACR at several discrete laser emission lines.

At the second stage, the FR is calibrated in a continuous wavelength range using a spectral comparator facility (SCF), shown schematically in Fig. 9.25, against the transfer detector equipped with the precision aperture of known area. Being uniformly irradiated by the monochromator beam overfilling this aperture, the transfer detector becomes recalibrated in terms of the irradiance spectral responsivity.

---

[7] Physikalisch-Technische Bundesanstalt, the NMI of Federal Republic of Germany.

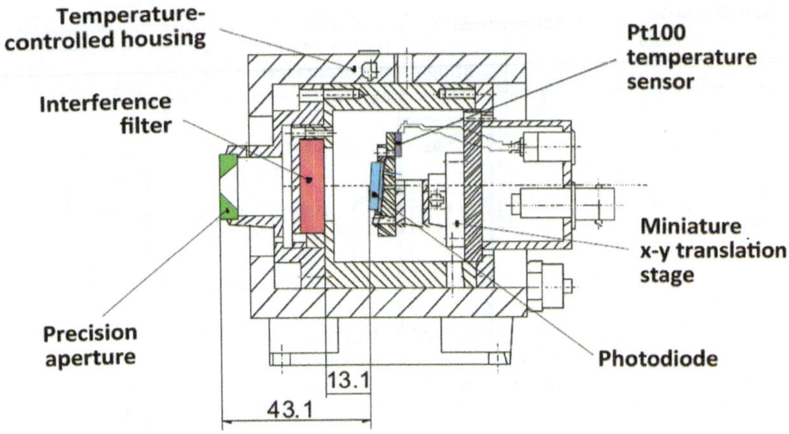

Fig. 9.23 Schematic diagram of an FR used at PTB for thermodynamic temperature determinations. Reproduced from [102] with permission of Springer Nature

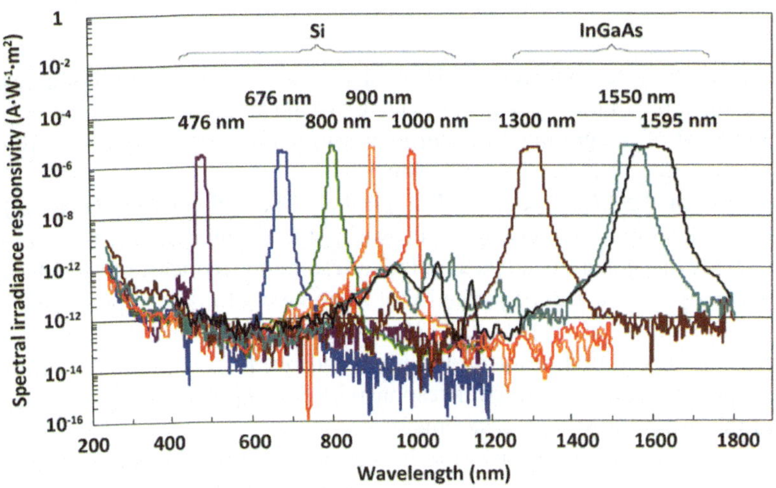

Fig. 9.24 The spectral irradiance responsivities of the PTB narrow-band FRs with Si and InGaAs photodiodes. Reproduced from [102] with permission of Springer Nature

The calibration of narrow-band FRs against broadband transfer detectors requires suppression of the out-of-band stray light to provide the SNR of about 10^9. The typical value for the stray light contribution for a single grating monochromator is 10^{-5}. To reduce the stray light level and, at the same time, provide sufficiently high optical output, a prism predisperser is used.

Unlike a double-grating monochromator, such a combination does not require additional filters suppressing higher-order spectral components.

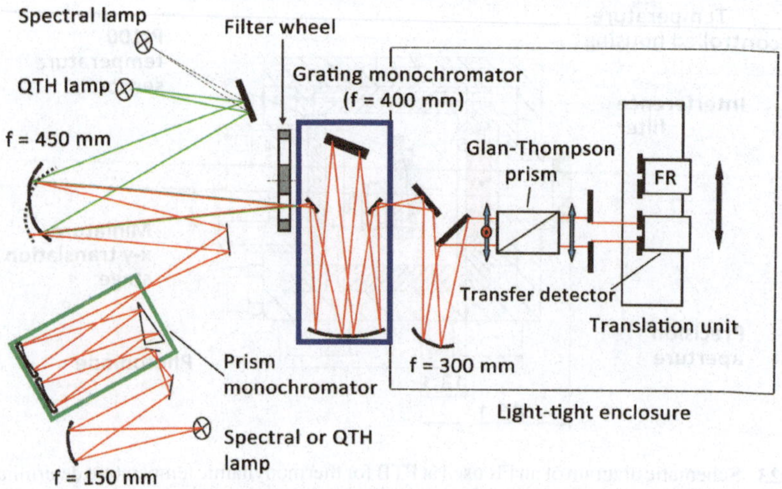

Fig. 9.25 The PTB spectral comparator facility (SCF) for the calibration of narrow-band FRs against a broadband transfer detector. Reproduced from [200] with permission of IOP Publishing

The quartz tungsten-halogen (QTH) lamps are used as the sources of continuous spectra at the input of the predisperser. The use of the prism monochromator (predisperser) in front of the grating monochromator limits the spectral range of the QTH and minimizes the stray light contributions.

Radiation of the QTH, passing through the SCF, is polarized. This cardinally differs the FR calibration phase from measuring the thermodynamic temperature of the blackbody that generally emits unpolarized radiation. To measure the polarization dependence of the spectral responsivity of radiometers to be compared, the Glan–Thompson prism polarizer (see e.g. [35, 89]) is placed behind the exit slit of the monochromator. Calibration of an FR is carried out for two mutually orthogonal states of polarization. The spectral responsivity for unpolarized light is the mean value of the two orthogonal measurements.

The TD is used with the radiation in the same polarization state as calibrated at the RTCR. This avoids a possible polarization effect of spectral responsivity for nonideal TDs as described by Goebel [88]. As a result, the calibration of the FR is independent of the polarization at the exit slit of the monochromator.

The accuracy of the thermodynamic temperature measurements using the irradiance method is illustrated by Fig. 9.26 [202], where the difference of the radiance thermodynamic temperature $T_{S,i}$ measured with i-th FR and the weighted mean value

$$T_S = \sum_i \frac{1}{u_i^2} T_{S,i} \Big/ \sum_i \frac{1}{u_i^2}, \tag{9.26}$$

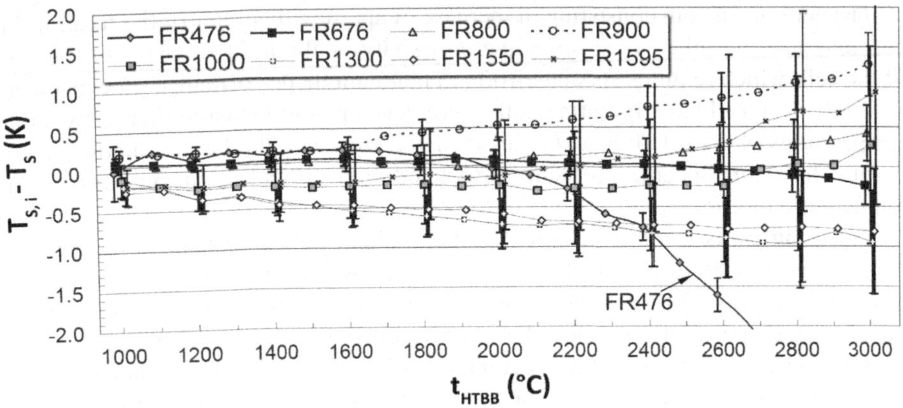

Fig. 9.26 Results for the thermodynamic radiance temperature determination of the HTBB in the temperature range from 1000 to 3000 °C with Si and InGaAs photodiode-based FRs. The uncertainty bars denote the standard uncertainties. Reproduced from [202] with permission of American Institute of Physics

where u_i is the standard uncertainty in measuring temperature with i-th FR. The x-axis represents the setpoint temperature t_{HTBB} of the directly heated pyrolytic graphite cavity [176]. As can be seen in Fig. 9.26, there is a good agreement between results obtained using four Si photodiode-based FRs (FR676, FR800, FR900, and FR1000) and three InGaAs photodiode-based FRs (FR1300, FR1550, and FR1595), although the use of InGaAs-based FRs is not typical for high-temperature measurements.

The results obtained with Si photodiode-based FR476 ($\lambda_0 = 476$ nm) have the largest deviation from the mean value, as was found, due to absorption bands of molecular carbon (C_2) that sublimated from the surface of the pyrolytic graphite cavity at temperatures above 2000 K [192]. The error bars denote the corresponding standard uncertainties.

In general, the results obtained demonstrate independence of the thermodynamic radiance temperature of the HTBB upon the wavelength within a standard uncertainty of 0.1% (in terms of irradiance) over the investigated temperature range. This confirms applicability of the irradiance mode measurements to determination of the thermodynamic temperatures of the HTBBs with sufficiently large openings.

The irradiance technique, being initially developed for measuring the thermodynamic temperatures of a large-aperture VTBBs on the base of heat pipes with alkali metal working substance, encounters serious difficulties when it is applied to the FPBBs, which typically has cavity opening diameters of 2–3 mm. Because of lower radiant flux emitted by such cavities, to keep the SNR of an FR at an acceptable level, the distance between apertures has to be reduced.

However, this (i) increases the irradiance non-uniformity across the FR aperture and (ii) leads to increased diffraction loss.

The natural way out consisting in increase of aperture diameter of the fixed-point cell may lead to either lower effective emissivity or the total increase in the size of the cell, which requires special efforts to maintain its performance. For instance, specially designed large radiating cells were developed to measure thermodynamic temperatures of the HTFP blackbodies [8, 101]. However, the large radiating fixed-point cells create specific problems associated with difficulties in providing their temperature uniformity.

Concluding this section, we note that the irradiance mode of measuring the thermodynamic temperature of blackbodies is, very likely, the easiest and straightforward in implementation. As a non-imaging mode, it is free of various sources of uncertainty related to the transmittance of and scattering in the objective lens. At the same time, the inherent drawback of this technique is obvious: to provide a sufficiently high level of the FR signal and low diffraction losses, the source aperture must be sufficiently large. Serious difficulties arise with the measurement of the thermodynamic temperatures of the FPBBs, whose radiating cells have small openings.

9.4.4 The Hybrid Method

For the hybrid method, the setup is similar to the irradiance method, but an additional lens is introduced to enable the measurement of smaller sources (see Fig. 9.27). Equation 9.21 is implemented with

$$s_L(\lambda) = \frac{\pi g s_E(\lambda)\tau(\lambda)}{A_2}, \tag{9.27}$$

where g is the geometric factor for the double aperture system given by (9.23), after the second aperture is added, A_2 is the FR's aperture area, $s_E(\lambda)$ is the irradiance responsivity of the FR (see Sect. 9.4.3), and $\tau(\lambda)$ is the spectral transmittance of the lens.

An FR operating by the hybrid method is equipped with a precision aperture and calibrated with the overfilling radiation beam similar to the irradiance method, as it is shown in Fig. 9.27a, but the measurement of the blackbody temperature is carried

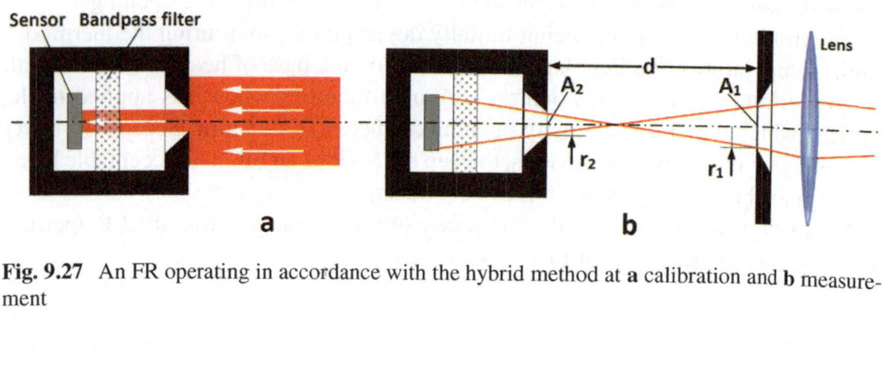

Fig. 9.27 An FR operating in accordance with the hybrid method at **a** calibration and **b** measurement

out with the lens typically mounted together with the source aperture as it is shown in Fig. 9.27b. The lens allows collecting a greater fraction of radiation emitted by a blackbody on the precision aperture of the FR.

The system should be calibrated by parts: the FR is calibrated in terms of the irradiance responsivity; the lens transmittance and the geometric factor are determined separately. The hybrid approach was first applied to the radiometric determination of thermodynamic temperatures of the FPBBs at the freezing temperatures of gold, silver, and aluminum at the NPL[8] although the term "hybrid" was coined much later. Fox et al. [65–67] and Fox [64] described in detail the pioneering work of the NPL researchers, who became the principal and the most consistent advocates of the hybrid method.

The FR, originally developed at the NPL to measure the thermodynamic temperatures of the gold, silver, and aluminum fixed points, had a simple structure. The detector, a 20 mm diameter silicon photodiode, is placed in a casing hermetically sealed with a wedged at a 0.1° window.

Four interference filters with the spectral bandwidths of about 20 nm FWHM were prepared to provide the FR peak responsivity wavelengths at 468, 568, 647, and 800 nm, close to emission lines of the krypton-ion laser. To avoid multiple reflections among filters, the interference layers were deposited directly on the 25 mm in diameter wedged glass filters that block the out-of-band radiation. Most measurements were performed for the spectral responsivities centered at 647 and 800 nm. All optical surfaces of the FRs had anti-reflection coating. The radiometer apertures of 4- or 5-mm internal diameter are diamond-machined from 30 mm diameter and 6 mm thick brass discs, by cutting a cone from one side. They were manufactured with lands of approximately 100 μm on the inner edge.

The FR is mounted inside a water jacket, which is maintained at a constant temperature of 20 ± 0.1 °C.

The FRs exhibited the linearity of better than 0.01% for the power levels below 100 μW and the wavelengths shorter than 900 nm. The amplified output photocurrent from the FR is converted to a voltage using a high-performance operational amplifier.

Calibration of the FRs and measurements of the lens spectral transmittance were performed using the integrating sphere (IS) source described by Anderson et al. [5] and Fox [64] and shown schematically in Fig. 9.28.

The core of the IS source is an IS with an internal diameter of 25 mm and is coated on the inside with $BaSO_4$ paint having the near-Lambertian reflectance higher than 0.97 throughout the visible spectral range. A continuously tunable CW dye laser with a typical spectral bandwidth of 0.002 nm pumped by the krypton-ion laser is used as a radiation source. The dye laser is coupled into a step-index, plastic-clad silica multimode optical fiber with a core diameter of 200 μm. The optical radiation from the fiber enters the IS at right angles to the sphere exit port.

The spatial coherence of laser radiation causes the formation of speckle patterns on the wall of the sphere. To eliminate this non-uniformity, approximately 1 m of the optical fiber was immersed in the vibration ultrasonic bath with the nominal

[8] National Physical laboratory, the NMI of the UK (http://www.npl.co.uk).

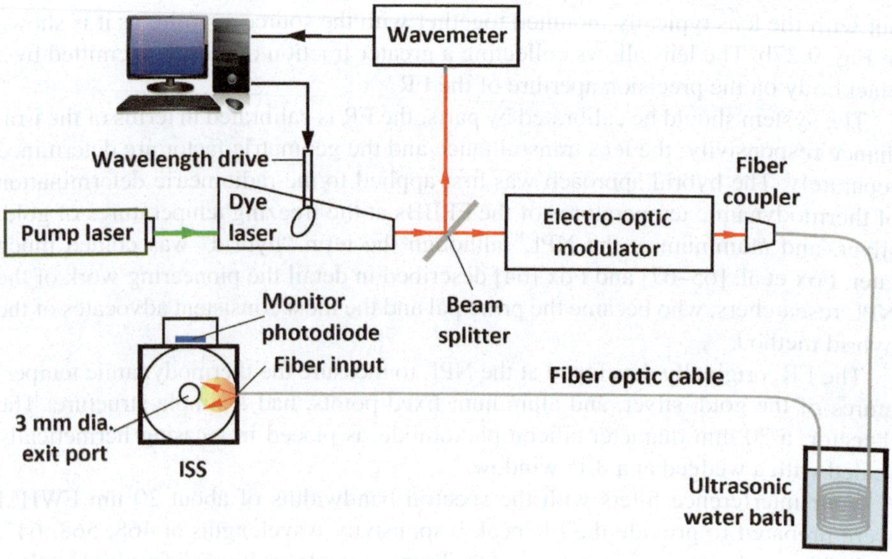

Fig. 9.28 Schematic representation of the NPL IS source (after [64])

frequency of vibrations of 40 kHz and the ultrasonic power dissipated into the water of 150 W. If a detector, observing the exit port of the sphere, has a time constant longer than the reciprocal of the vibration frequency, it will register a time-averaged response, the speckle pattern disappears, and this time-averaged response should be uniform across the sphere exit port. To stabilize the radiant power of the IS source, a monitor photodiode is mounted in the third port arranged at right angles to the other two ports in the sphere wall. The signal of this detector is used to stabilize the radiant power of the dye laser by passing the linearly polarized beam through an electro-optic modulator that is controlled by the output of the monitor photodiode. This feedback allows the output radiant flux to be stabilized up to 10^{-5} over periods of a few minutes.

The wavelength of the IS source is measured by attenuation of a small part of the dye laser beam using a beam splitter to a commercial interferometer-based wavemeter with a helium-neon laser line at 632.8 nm as a reference. By tuning the dye laser, the source can be set to any wavelength available from the spectral coverage of the particular dye used; the wavelength setting is determined by the resolution (typically 0.02 nm) of the of the dye laser tuning element. The dye laser selectivity allows radiation at other wavelengths to be suppressed below the detectable level.

The IS source has been characterized with respect to the radiance uniformity.

In the near field, the radiance uniformity was obtained by scanning a magnified image of the sphere exit port by a photodiode with a 1 mm diameter pinhole aperture. The reproducibility of the results of measuring the radiance uniformity was of ±0.01%. The sphere exit port was found uniform to within ±0.05%. This non-uniformity is partly caused by aberrations and diffraction at the imaging lens.

Some asymmetry of the distribution was attributed to the imperfectness of the IS, particularly, to deviation of the reflection behavior of the sphere wall from Lambert's law.

The uniformity of the irradiance in the far field, at a distance of 670 mm, scanned by a photodiode with a 4 mm diameter aperture was assessed as ±0.01% from the measurement's reproducibility. It was concluded that the far-field uniformity is sufficient for the characterization of most detectors and optical components.

The main feature that distinguishes the hybrid method from all the others is that the FR is calibrated without a lens, while the measurements are performed with a lens. Two correction factors (for the SSE and lens transmittance) with the corresponding uncertainty components can be determined either separately or in the same measuring experiment. Woolliams et al. [217] described a high-accuracy facility for measuring transmittance of the lenses used in the FRs of the NPL's design. The facility employs monochromatic radiation of a tunable laser to carry out measurements of the lens transmittance at discrete wavelengths. The spectral transmittance from 10^{-4} to 10^{-3}, depending on wavelength, can be measured for lenses with focal lengths from 300 to 750 mm. The experimental configuration is shown in Fig. 9.29. All components are mounted on computer-controlled stages. A laser is coupled to a multimode optical fiber that passes through an ultrasonic bath and then enters an IS that was described above. The fiber is immersed into the ultrasonic bath to removes the speckle effect caused by the coherence of the laser beam and creates a spatially uniform field over the internal surface of the IS. The signal from a silicon photodiode attached to an additional port of the IS is used to stabilize the laser by controlling an electro-optic modulator in the laser beam. The two lenses shown in Fig. 9.29 are nominally identical in both size and focal length. A high-precision thin-film aperture on the first lens defines the beam geometry. The lens under test is placed in the second position on a stage perpendicular to the main optical axis, so that this lens can be inserted

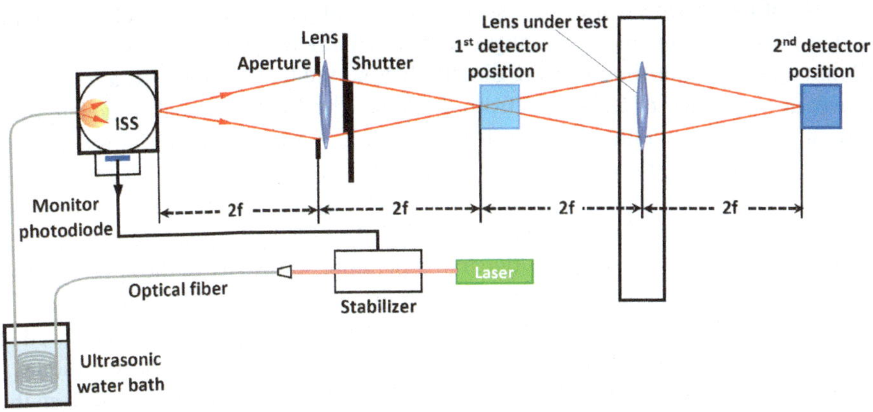

Fig. 9.29 A configuration of the experimental facility for measuring spectral transmittance of a lens. The source, first lens and detector are on stages allowing them to move along the optical axis to the positions. The lens under test is on a stage that moves perpendicular to the optical axis (after [217])

into the optical path and removed from it allowing the detector (the TD is used to minimize back-reflections and hence on-axis stray light) to be moved from the 1st to the 2nd position. The lenses are used to image the IS exit port onto the two detector positions without magnification since the distances from the source to the lens and from the lens to the image are equal $2f$ (twice the focal length). Because of limited FOV of the TD, there is the limit in the f-number of the lens that can be measured. The shutter allows measuring the dark signal in both detector positions to account for out-of-axis stray light.

In most applications, the size-of-source effect (SSE, see Sect. 8.2.4) is a relative effect: the instrument is sensitive to the difference in the size and radiance distribution of the measured and reference sources. For the hybrid method, the SSE has the absolute character. At calibration, the SSE can be caused only by a lens that is used to create the uniform irradiance used to compare the FR (without a lens) and the TD. However, the use of the same radiation source cancels the SSE at the FR calibration. However, at measurement, when the lens is added to the FR, there is no reference source size, so the absolute SSE has to be measured as the difference when viewing the test source compared to an infinite uniform source. This effect is difficult to measure because due to simplicity of design, the FR operating by the hybrid method usually exhibits a large SSE when measuring the blackbody temperature. So, an extended source of uniform radiance may be required to trace the SSE curve until it reaches asymptotically a constant value. Even more difficult is to evaluate the corresponding uncertainty.

Winkler et al. [212] proposed the solution that allows avoiding the absolute SSE measurements. To do this, the lens transmittance and the SSE should be measured simultaneously by performing measurements of a large-aperture radiance source both with and without the lens.

The method proposed was applied to the absolute radiation thermometer (ART) designed for accurate determinations of thermodynamic temperature at the NPL on the base of pioneering works published by Fox et al. [65, 66]. The ART is a simple device consisting of a linearly arranged a 75 mm diameter plano-convex silica lens, a 40 mm diameter thin-film aperture directly behind the lens, and an FR, as described above, at a distance of 2100 mm from the lens. The FR has a peak wavelength of 800 nm and a bandwidth of 20 nm. It is fitted with a 5 mm diameter brass aperture and a black-anodized cover. All optical elements are embedded in an Invar frame and covered by a common enclosure. To reduce stray light, all mechanical elements are blackened and two baffles have been added. The FR is temperature-stabilized to ±0.02 °C to minimize any effects due to temperature sensitivity of the spectral filter.

The setup for simultaneous measurement the SSE and the lens transmittance is shown in Fig. 9.30. In the "radiance mode," a 5.3 mm diameter aperture is placed in the object plane of ART and the signal V_{rad} is obtained from large aperture the IS.

In the "irradiance mode," the lens and the 5.3 mm aperture are removed from the ART, and the signal V_{irrad} is measured. V_{irrad} was corrected for diffraction effects. This corrected signal would be measured if the lens transmits radiation perfectly and if there is no SSE associated with the lens. The diffraction correction K_{diffr} was

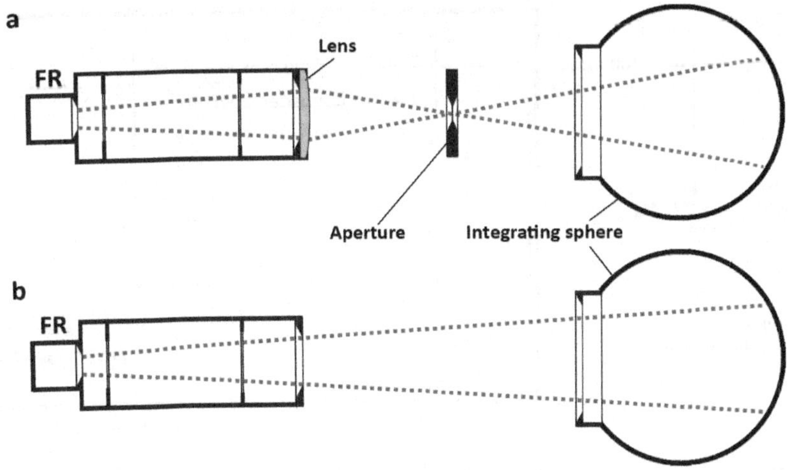

Fig. 9.30 Combined lens transmittance/SSE experiment: **a** "radiance mode" and **b** "irradiance mode." Reproduced from [212] with permission of Springer Nature

calculated using the method of Edwards and McCall [48]. Both methods gave the same value of $K_{\mathrm{diffr}} = 0.99978$ accurate to five significant digits.

The same IS that was also used for the direct SSE measurement was placed such that the FR with and without lens were irradiated by radiation originating from the same part of the sphere, which was calculated to be approximately 1600 mm from the ART's lens aperture.

The sensitivity of the measurement to the positioning of the sphere was evaluated by moving the sphere over a distance of 1600 ± 200 mm along the optical axis, separately for the "radiance" and "irradiance mode," and was repeated several times. Both signals decreased in a linear fashion and at the same rate of approximately 0.02% per 100 mm.

Their ratio was therefore independent of the exact positioning of the sphere. The ratio between the radiance mode signal and the diffraction-corrected irradiance mode signal is the correction factor necessary to calibrate the ART as a whole against a source of known radiance and of known finite size. For a source diameter of 5.3 mm, the value of the correction factor $K_{cal} \approx 0.93034$ was found using the equation:

$$K_{cal} = \frac{1}{K_{\mathrm{diffr}}} \cdot \frac{V_{\mathrm{rad}}}{V_{\mathrm{irrad}}}. \tag{9.28}$$

Figure 9.31 presents the setup for measuring the thermodynamic temperature of the FPBBs. A blackbody cavity surrounded by a metal ingot (Au, Ag, or Al) heated in an electrical furnace was imaged onto an FR using a simple plano-convex lens (glass or fused silica, depending on the interference filter selected) of 75 mm diameter and 500 mm focal length. A shutter placed between the cavity and the lens is used for the

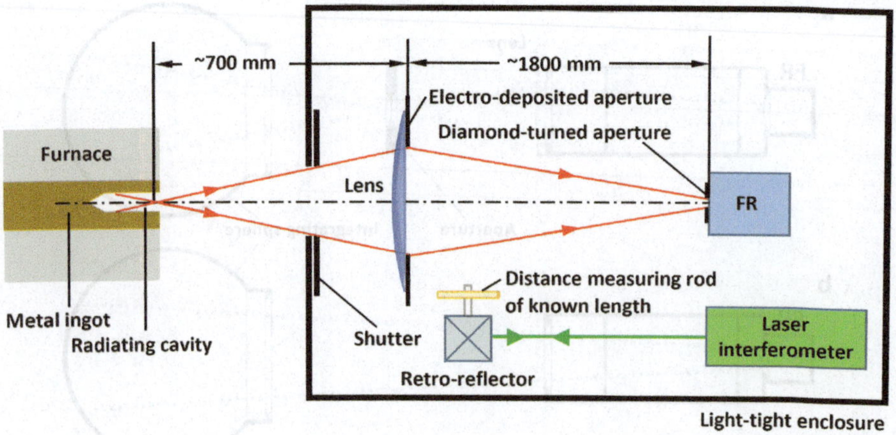

Fig. 9.31 A schematic of the optical system used to measure the spectral radiance of an FPBB (after [65, 66])

radiometric zero measurements. An electro-formed nickel aperture is placed against the plane face of the lens, which faced the FR.

Although reversing the lens would slightly improve the image quality, it would make it difficult to mount the lens aperture. To minimize scattered radiation, the lens aperture was made extremely thin (less than 25 μm) at its edge. All optical components were mounted on an optical bench within a light-tight enclosure.

The optical system was designed with a magnification of ~2.5, which set the distance between lens and the FR at approximately 1.8 m. Measurements of FPBBs were carried out over a range of solid angles and magnifications.

Fox et al. [66] obtained the thermodynamic temperatures for Al (933.509 K), Ag (1235.009 K), and Au (1337.330) with the standard uncertainties of 27 mK, 44 mK, and 49 mK, respectively. These results are in good agreement with the values adopted in the ITS-90.

The hybrid method was further developed mainly at the NPL and later has been applied to determination of thermodynamic temperatures of HTFPs by the research teams of NPL [173, 218] and NIM[9] [96, 235]. An exhaustive analysis of the uncertainties of the hybrid method was performed by Fox et al. [65, 66] and updated by Woolliams et al. [219]. They reviewed and summarized the hybrid method and evaluated the uncertainty associated with measurements of HTFP blackbodies using the hybrid radiometric technique. In particular, it was shown that, if the SSE and lens transmittance are determined separately, they have typical standard uncertainties of 0.06% and 0.04%, respectively. If they are measured together (as described above), their overall standard uncertainty can be as small as 0.01%. Within the framework of the HTFP research program, the ART was used to determine the thermodynamic

[9]National Institute of Metrology, Beijing, China.

temperatures of Re–C, Pt–C, and Co–C cells with the standard uncertainties of 549 mK, 292 mK, and 168 mK, respectively.

The hybrid approach has been used for about 25 years and played a significant role in the development of the technique for thermodynamic temperature measurements in the high-temperature range. Designed specifically for measuring temperatures of the small-aperture FPBB, it has been widely used in the study of HTFPs. Uncertainties that can be achieved with the hybrid mode are comparable to those of other modes. The only drawback of the hybrid method is the complexity in characterization of the FR, namely the accurate determination of the absolute SSE and lens transmittance.

9.4.5 The Radiance Method

The scheme for calibration and use of the FR operating according to the radiance method (see Fig. 9.32) is similar to that for the conventional RT. For the radiance method, Eq. 9.21 can be used directly, since the instrument is calibrated in terms of radiance responsivity.

Design of an FR operating by the radiance method does not differ from that of a typical RT considered in Chap. 8. Moreover, every RT or imaging radiometer can be calibrated as an FR operating by the radiance method. For brief, we will call the absolute RT any FR designed to be used according the radiance method. Conventional RT is a measuring device that is employed for interpolation between reference points of the temperature scale or for extrapolation above the reference point with the highest available temperature. To measure with a usual RT, it is enough to know its relative spectral responsivity. To measure with the absolute RT, its absolute radiance responsivity has to be known but reference temperature points are not required. The thermodynamic temperature of a blackbody is derived from Eqs. 9.21 and 9.22.

The possibility of calibrating the conventional RTs in terms of absolute radiance responsivity made the radiance method especially attractive. Among methods of the thermodynamic temperature determination based on absolute radiometry, this one is the oldest and most thoroughly researched. The basics of the radiance method have been laid by Mielenz et al. [149], who employed the spectroradiometric technique

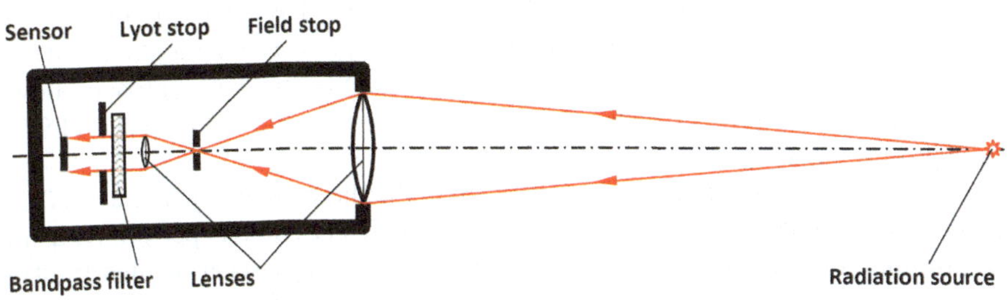

Fig. 9.32 An FR operating by the radiance method at calibration and measurement

and the room-temperature electrical substitution radiometer (see Sect. 9.2.1) as a primary standard of optical watt to determine the freezing temperature of gold.

However, the intensive development of the FRs operating by the radiance method and applicable to measuring the blackbody thermodynamic temperatures began only after the advent of the ACRs (see Sect. 9.2.4). The world-leading NMIs such as the NIST (USA) and LNE-CNAM (France) have done a lot for the remarkable progress in this area.

Recently, other NMIs, attracted by the simplicity of design and the ability to use experience in characterizing conventional RTs, have developed their own absolute RTs or have calibrated well-established RTs in terms of absolute radiance responsivity.

At the NIST, the FR operating by the radiance method (so-called *Advanced* or *Absolute Pyrometer*), AP1 has been developed to measure blackbody thermodynamic temperatures by means of absolute detector-based radiometry [4, 229–232]. The spectral selection in the AP1 is carried out by the narrowband interference filters with the center wavelengths at 650 nm. An additional IR blocking filter is used to suppress radiation beyond 800 nm. The filters are temperature-stabilized near the room temperature. A hermetically sealed Si photodiode with the NEP of 10 fW (10^{-14} W) has a two-stage Peltier cooler for operation down to −15 °C.

The successor of the AP1, the AP2 [231] have been constructed at NIST to achieve the small FOV, low SSE and long-term stability. The latter one was achieved by temperature stabilization of the filter and detector. The filter is manufactured using ion-assisted deposition to form a hard, humidity-resistant coating. The substrate is made wedged to reduce interference. The AP2 employs a low-scatter objective lens with a Lyot stop to reduce the SSE.

Yoon et al. [233] described the construction, calibration, and use of the NIST Thermodynamic Radiation Thermometer (TRT) designed to measure the temperatures from 800 to 2700 °C. The Si photodiode with enhanced responsivity in the NIR spectral range can be combined with five interference filters to provide narrow bands with the center wavelengths from 400 nm to 1000 nm. The calibration of the NIST's absolute RTs is performed in the NIST Spectral Irradiance and Radiance responsivity Calibration Facility (SIRCUS), which is discussed in Sect. 9.5.2. The radiance responsivity of the 650-nm band of the TRT is shown in Fig. 9.33. The characteristics of the TRT are listed in Table 9.3.

A special feature of the precision RTs designed at NIST is the ability to calibrate them both against the FPBBs with the traceability to the ITS-90 and against the ACR (via the TD), providing the traceability to the SI. Such an approach allows recalibration of the precision RTs developed earlier to operate within the ITS-90 paradigm for the use as absolute RTs.

The basics of the modern NIST's approach to the thermodynamic radiation thermometry is given by Yoon et al. [234]. Measurements at NIST is performed using the absolute RT calibrated for radiance responsivity using laser-irradiated IS. The radiance of the IS source is determined with the help of the silicon photodiode-based TD, whose spectral power responsivity is traceable to the ACR. A diagram of the calibration steps is depicted in Fig. 9.34.

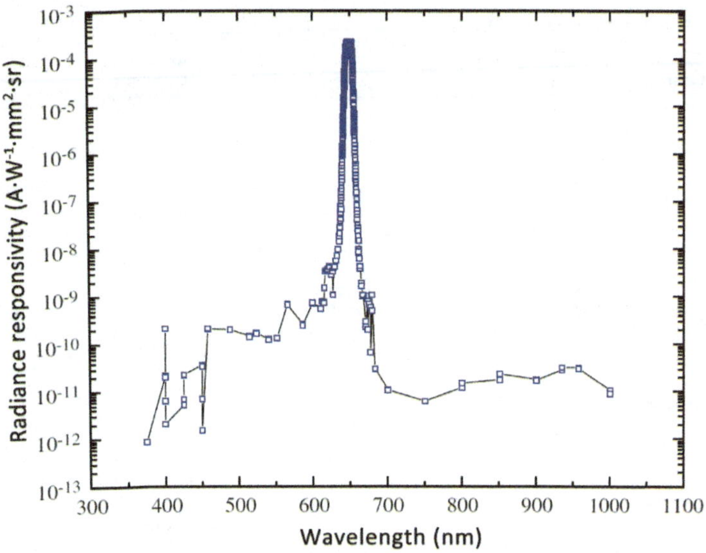

Fig. 9.33 The radiance responsivity of the TRT measured in the NIST SIRCUS facility. Reproduced from [233] with permission of Springer Nature

Table 9.3 Parameters and design specifications of the NIST TRT (after [233])

Parameter	Value
Object distance	50 cm
Image distance	33.3 cm
Objective focal length	20 cm
Field stop diameter	0.6 mm
Target diameter	0.9 mm
f/#	12
Lyot stop diameter	5 mm
Detector size	5.8 × 5.8 mm
Detector temperature	−25°C
Filter wheel temperature	30°C
Collimator lens focal length	70 mm
Image diameter on detector	5.0 mm
TRT diameter	26 cm
TRT length	89 cm

At the first stage, the optical power of a laser source is determined using the ACR. The laser wavelength is measured by an interferometric wavemeter. The spectral power responsivity of the ACR in the spectral range from 400 to 920 nm is transferred to the silicon photodiode-based TD with spatially and angularly uniform

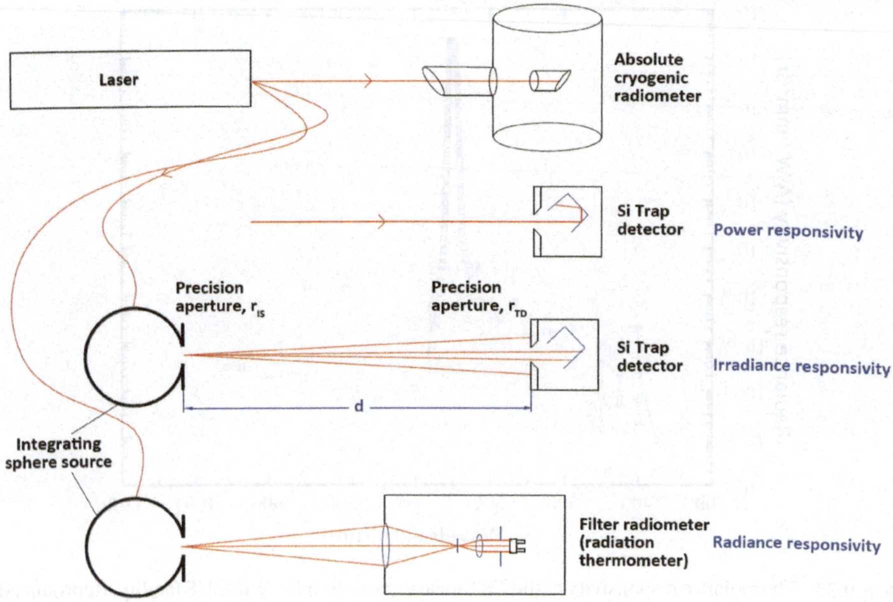

Fig. 9.34 The successive steps in the FR calibration by the radiance method. Reproduced from [234] with permission of Springer Nature

responsivity. The third step consists in determination of the FR radiance responsivity by comparison with the TD, equipped with a precision aperture, using the IS source.

Researchers of the NPL (UK), who traditionally (since 1990s) developed another imaging technique (namely, the FRs operating in the hybrid mode, see Sect. 9.4.4), in addition to the NPL's Absolute Radiation Thermometer (ART) operating by the hybrid method, developed also an FR operating by the radiance method to precisely measure the thermodynamic temperatures of the HTFPs [46]. The new instrument called THOR (THermodynamic Optical Radiometer) is portable and compact, with a much lower SSE than that of the ART and the improved temperature sensitivity. Unlike the ART that requires calibration of individual components, the THOR is calibrated in the radiance mode as a whole instrument, which results in lower uncertainties. A schematic of the THOR and its photograph are shown in Fig. 9.35.

The calibration facility that was used to calibrate the ART for spectral irradiance responsivity was modified to make possible the radiance responsivity calibration required for the THOR (see Fig. 9.36).

A Ti:sapphire laser pumped by the Nd:YAG laser was used as a tunable source with a resolution of 0.1 nm across the 750–850 nm spectral range that covers the 800 nm peak transmittance of the THOR interference filter. The out-of-band responsivity measurements were carried out over the full responsivity range of the THOR silicon photodiode sensor using the NPL designated monochromator-based facility.

The intensity of the Ti:sapphire laser was stabilized using a Pockels cell controlled by the feedback the signal of the monitor detector mounted on the IS. A wavelength

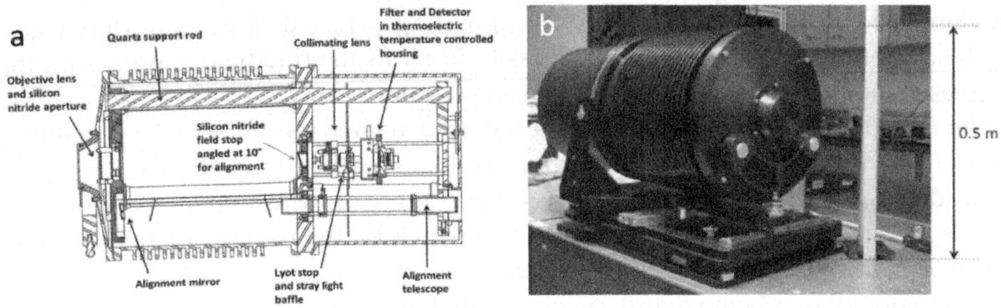

Fig. 9.35 The THOR instrument: **a** schematic diagram and **b** photograph. Reproduced from [46] with permission of American Institute of Physics

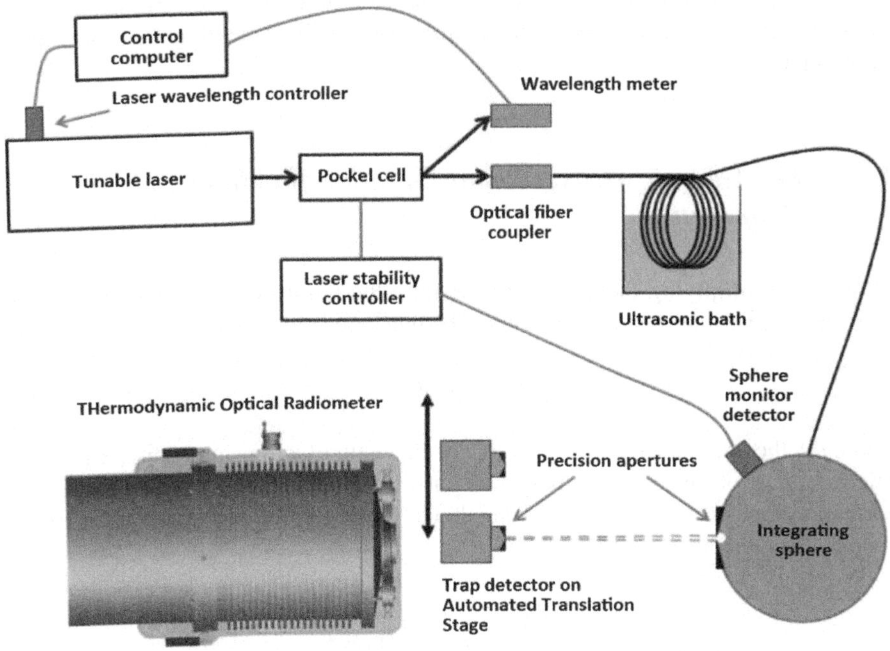

Fig. 9.36 Schematic diagram of the facility used for calibration of the THOR for spectral radiance responsivity. Reproduced from [46] with permission of American Institute of Physics

meter provides the feedback to a PC that varies the laser wavelength. The laser radiation irradiates the IS via an optical fiber passing through an ultrasonic bath to reduce the speckle effect. The 300 mm diameter IS internally coated with $BaSO_4$ has several symmetrically arranged inputs for a quadfurcated optical fiber and a 5 mm diameter output port. Each fiber is attached to the sphere using a lens system allowing the beam diameter to be controlled. The TDs traceable to the NPL ACR are mounted on a computer-controlled stage for automated measurements of the IS source radiance. Apertures attached to the IS and the TDs define the geometry of the system. The radiance method becomes the most widespread despite some drawbacks such susceptibility to the SSE and necessity in an IS intermediate source with the increased requirements to the spatial and angular uniformity.

Perhaps, the growing popularity of the radiance method is related to the possibility of dealing with a usual RT in a familiar way to obtain the thermodynamic temperature instead of T_{90}. The principal restrictions on the ranges of measured temperature and the target size remain the same as for conventional RTs. In this case, no fundamental changes in the design of the RT are required.

The same device being appropriately calibrated can serve as a conventional RT measuring blackbody temperatures according to the ITS-90 and as an FR operating in the radiance mode and measuring the thermodynamic temperature. The calibration results for the spectral radiance responsivity can be easily converted into the calibration equation indicated the temperature as a function of the FR signal. With minor alterations in electronics, each RT can serve as a direct-reading thermodynamic thermometer.

9.4.6 *The Generic Measurement Equation for Blackbody Temperature*

Additional multiplying factors should be introduced into Eq. 9.21 to account for correction of various systematic errors. The measured signal is, as a rule, a voltage

$$V = i \cdot G, \tag{9.29}$$

where i is the photocurrent and G [V A^{-1}] is the gain of a transimpedance amplifier. The right part of this equation has to be corrected for systematic effects. A part of correction factors is common to all methods, other correction factors are method-specific. In general,

$$V = \frac{\pi g}{A_{FR}} \varepsilon_{eff} G \prod_i K_i \int_0^\infty s_E(\lambda) \tau(\lambda) L_{\lambda,bb}(\lambda, T) d\lambda, \tag{9.30}$$

where K_i is the correction factor for an individual systematic effect. For the purposes of uncertainty analysis, it is helpful to write the measurement equation in the most general way to maintain the uniformity of analytical expressions. Thus, the generic version of the measurement equation is:

$$S(T) = K \int_0^\infty s(\lambda)\, L_{\lambda,bb}(\lambda, T) d\lambda, \tag{9.31}$$

where $S(T)$ is the measured temperature-dependent signal, which may be in volts, amperes, or digital numbers, depending on the device; $s(\lambda)$ is the radiance, irradiance, or power responsivity of the FR, as appropriate (for the hybrid method, it also includes the lens transmittance); $K = \prod_j K_j$ is the total correction factor; $L_{\lambda,bb}(\lambda, T)$ is the spectral radiance of the perfect blackbody. Equation 9.31 can be solved numerically by varying iteratively T until convergence. The Newton algorithm starting from an initial estimate T_0, recommended by Saunders et al. [180], proceeds by forming successive approximations T_i for $i = 1, 2, 3, \ldots$ using the iterative formula

$$T_{i+1} = T_i + \frac{S - K\int_0^\infty s(\lambda) L_{\lambda,bb}(\lambda, T_i) d\lambda}{\frac{c_2}{T_i^2} K \int_0^\infty s(\lambda) \frac{L_{\lambda,bb}(\lambda, T_i)}{n\lambda\left[1-\exp(-c_2/(n\lambda T_i))\right]} d\lambda}. \quad (9.32)$$

Convergence to better than 0.1 mK is usually achieved in fewer than 5–10 iterations, depending on how close the initial guess T_0 to the true temperature. Equation 9.32 gives a base for the uniform approach to evaluation of the uncertainty in determination of the thermodynamic temperature by any of the APRT method.

9.5 Modern Facilities for Calibrating Filter Radiometers

9.5.1 Preliminary Considerations

The modern schemes of the FR calibration are traceable to the ACR. calibration of the FR is carried out using a transfer detector (a TD on the base of silicon photodiodes or, less commonly, a single photodiode). In early measuring facilities, the CW lasers (e.g., with Kr^+ working medium) generating radiation at widely separated lines were used as the radiation sources. Modern systems employ tunable lasers or a monochromator in conjunction with a broadband radiation source.

Further, we will assume that the TD is already calibrated against the ACR, i.e. the absolute spectral (power) responsivity values are known for a desired wavelength range. The FR can now be calibrated in terms of power, irradiance, or radiance spectral responsivity over that wavelength range (wavelength-by-wavelength) by comparison with the TD. Depending on the technique of obtaining the monochromatic radiation, the methods of the FR calibration are subdivided onto two groups. The earlier of them is based on the use of a monochromator, which isolates a narrow band in the spectrum of a broadband source using prisms or diffraction gratings. The advent of tunable lasers and optical parametric oscillators (OPOs, see [42, 161, 175]) provided intense and stable radiation sources in the visible and NIR ranges and ensured greater SNR and, therefore, higher accuracy and reliability of the FR calibrations. The traditional monochromator-based technique retained its significance mainly for IR calibrations. Currently, this situation is changing due to the introduction of intensive broadband radiation sources, such as laser driven light sources [108] and, especially, supercontinuum light sources [213], into metrological practice.

To calibrate the FR, the spatial and angular distributions of the radiation from the laser or exit slit of a monochromator should be modified.

For the irradiance responsivity calibration, a uniform irradiance in the reference plane must be provided. To expand the radiation beam of a laser or after the exit slit of monochromator, the reflective beam expanders are used. The central, most spatially uniform part of the expanded beam is cut to provide a near-uniform irradiance in the plane perpendicular to the beam axis. This allows alternately irradiating the TD and FR, each of which is equipped with a precise aperture.

The IS source allows calibrating the FR against the TD in terms of both irradiance and radiance responsivity. The IS is a simple and versatile measuring instrument widely used in solution of various tasks of radiometry, photometry, reflectometry, and illuminating engineering, including the design of the uniform radiance sources. Initial information on the IS can be found in Technical Guides issued by Labsphere, Inc. [128, 129] and available online.

The IS is a hollow sphere, that is fabricated monolithically from PTFE[10]-based materials or use a white diffuse coating such as $BaSO_4$ applied to the sphere internal surface. Radiation can be injected through the entrance ports into the sphere, or come from sources (e.g. lamps or LEDs) arranged inside it. Even if the sources of the primary radiation irradiate the internal surface of the IS nonuniformly, each consecutive reflection from Lambertian walls creates the irradiance uniformly distributed over the entire internal surface of the IS. If the directly irradiated sphere regions cannot be viewed through the exit port (this can be achieved by proper arrangement of baffles), the observed exitance will be spatially uniform. Since the internal surface of the IS is Lambertian, the exit port of the sphere will behave like a Lambertian disk.

To perform calibration of the FR against the TD by means of the IS source, the TD must be equipped with the precision aperture of known area A_2.

For the irradiance responsivity calibration, the IS source must create a uniform irradiance in the aperture plane of the detectors compared. In the far-field zone, the irradiance at the detector aperture plane is equal to

$$E = \Phi / A_2, \tag{9.33}$$

where Φ is the power measured by the TD calibrated against the ACR.

To calibrate an FR against a TD in terms of irradiance responsivity, the IS must be also equipped with the precision aperture of known area A_1.

The radiance of the IS source is determined by the known irradiance E measure by the TD. The spectral radiance of the IS source is given by

[10]Polytetrafluoroethylene, a synthetic fluoropolymer of tetrafluoroethylene; the most known trade mark of PTFE-based material is Spectralon developed by Labsphere, Inc. (https://www.labsphere.com/).

$$L = E/\left(\pi F_{A_1-A_2}\right), \tag{9.34}$$

where $F_{A_1-A_2}$ is the diffuse view factor from the TD aperture to the IS aperture. The assigned spectral radiance of the IS is used to determine the radiance responsivity s_L of the FR.

The radiation characteristics of real-world IS sources differ from ideal ones due to the presence of internal baffles, spatial inhomogeneity of the wall reflectance, and deviation of wall reflection from Lambertian behavior.

The exit port of the IS has a finite size; therefore, its Lambertian behavior can be checked in the far-field zone only. The angular radiance uniformity of the IS source can be measured using a TD with the spectral power responsivity proportional to the cosine of the angle between the direction of the incidence and the normal to the TD aperture. Instead of angular scanning, the linear scanning that is easier to implement can be carried out in the plane perpendicular to the source. The angular uniformity of the IS radiance can be evaluated using the well-known relationship (see, e.g. [157]) for the irradiance created by an elementary Lambertian source in the plane perpendicular to its axis:

$$E(\theta) = E(0)\cos^4\theta. \tag{9.35}$$

Figure 9.37 presents the $\cos^4\theta$ curve compared with the measured distribution of the relative irradiance created at the distance of 500 mm by the IS with a diameter of

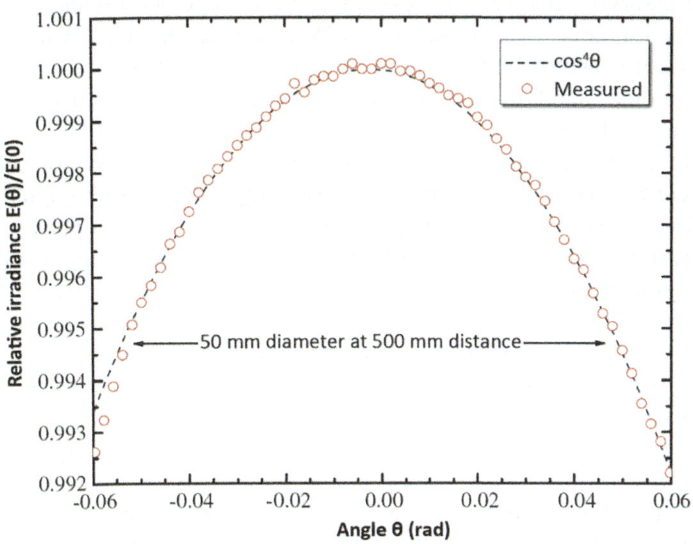

Fig. 9.37 Measurement result for the 308 mm diameter IS source. Reproduced from [234] with permission of Springer Nature

300 mm and the exit port with a diameter of 75 mm. This plot shows that deviations from the Lambertian output of the IS is negligible within ±0.05 rad (±2.9°).

Since the end of 20th century, a numerous setups, installations, and facilities were designed at world leading NMI in order to calibrate the FR in terms of irradiance and radiance spectral responsivity. Below, we discuss the practical implementations of the most common calibration techniques, remembering, however, that they may vary significantly depending on the availability of the necessary measurement equipment and adopted methods of the radiometric scale realization in individual laboratories. We also discuss some of the most influential developments in this area, without claiming their comprehensive consideration. Readers who need in additional information can refer to the bibliography contained in *MeP*-K documents [140, 180]. Since the objects of our primary interests are the FRs applicable to measurements of thermodynamic temperature of high-temperature blackbodies, we focus on the calibration facilities for visual and near IR spectral ranges.

9.5.2 Calibrations with Tunable Lasers

For a long time, the FR calibrations carried out using laser sources were considered more accurate than calibrations made on monochromator-based facilities. Calibration of an FR in a narrow spectral range cannot be ensured only by continuous-wave (CW) lasers producing monochromatic radiation of constant power.

It is necessary to involve lasers providing pulsed operation with some duration at some repetition rate, such as the quasi-CW, mode-locked lasers, which, as it was shown by Gentile and Cromer [82], can be used for calibrations of radiometers with silicon photodiodes and do not introduce additional uncertainties due to the pulsed nature of their radiation.

The continuously tunable solid-state and dye lasers can be used alone or to pump the OPOs converting an input frequency into two output waves (OPO signal and OPO idler) of lower frequencies by means of second-order nonlinear optical interaction (see details in [238]). In such a way, laser sources allow generating a wide variety of wavelengths from the UV to IR. In order to outline the current state of the FR calibrations using tunable lasers, we consider below two facilities developed in the world's leading NMIs: the SIRCUS facility at the NIST (USA) and the TULIP facility at the PTB (Germany). Both of these facilities are multifunctional, so we confine ourselves to discussing only their features that are directly related to the calibration of FRs designed to measure the thermodynamic temperature of high-temperature blackbodies.

One of the most known, investigated, and described in detail [27–29] calibration facility with laser sources is the SIRCUS (Spectral Irradiance and Radiance Responsivity Calibrations using Uniform Sources) facility developed and built at the NIST (USA), the scheme of which is presented in Fig. 9.38. The SIRCUS uses the broadly tunable narrow-band lasers as radiation sources (see details in [214]) in two subsequent radiometric calibrations. First, the tunable lasers are used to calibrate

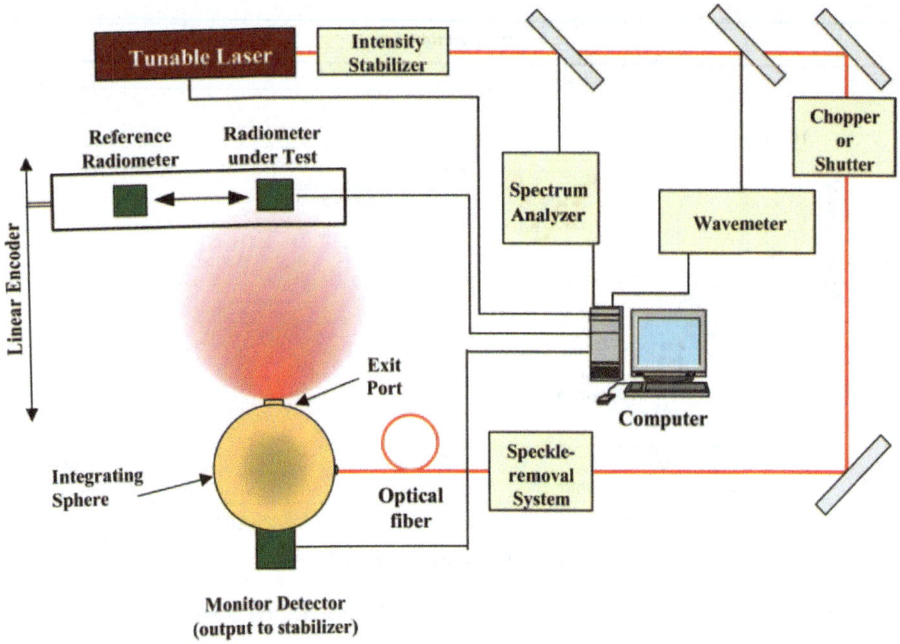

Fig. 9.38 Schematic diagram of the SIRCUS facility. Reproduced from [29] with permission of Optical Society of America

the spectral power responsivities of primary standard detectors against an absolute cryogenic radiometer (ACR).

The second function is calibration in terms of the absolute radiance and irradiance responsivities of detectors using uniform radiation sources, which are the ISs coupled with the lasers.

The radiant flux from the uniform sources is determined by the ACR-calibrated primary standard detector. Together they are used to transfer radiometric scales to a variety of optical instruments with low uncertainties. To cover the spectral range from about 0.2 to 1 μm, various lasers emitted at discrete wavelengths, as well as continuously tunable Ti:sapphire and dye lasers are used as radiation sources.

The output power and spectral coverage of the laser sources in the SIRCUS facility is shown in Fig. 9.39. Laser radiation is passed through an intensity stabilizer that holds the beam power within ±0.1% of the set value. The first beam splitter sends a fraction of the laser beam power onto the spectrum analyzer on the base of a Fabry-Perot interferometer to control the laser mode stability and measure the radiation bandwidth. Using the second beam splitter, another fraction of the laser beam power is directed onto the Michelson interferometer to measure the wavelength with the absolute accuracy of 0.001 nm. Laser radiation is introduced using an optical fiber into the IS, whose size may vary depending on the task solved.

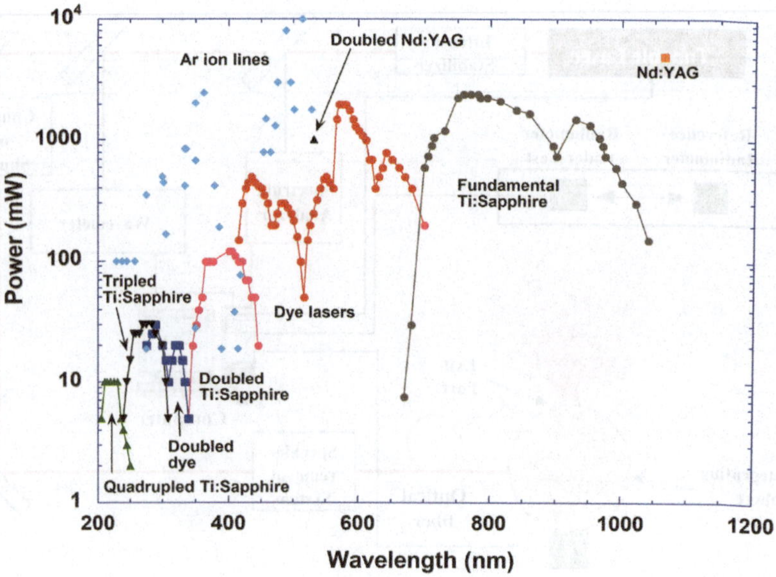

Fig. 9.39 Output power and spectral coverage of the SIRCUS facility laser sources. Reproduced from [29] with permission of Optical Society of America

Due to interference of the coherent laser radiation, the image of the IS opening contains a speckle pattern, which can be almost completely removed by immersing a short part of the optical fiber in an ultrasonic bath [5, 153]. The photographs presented in Fig. 9.40 illustrate the result of the ultrasonic bath action of the on the speckle

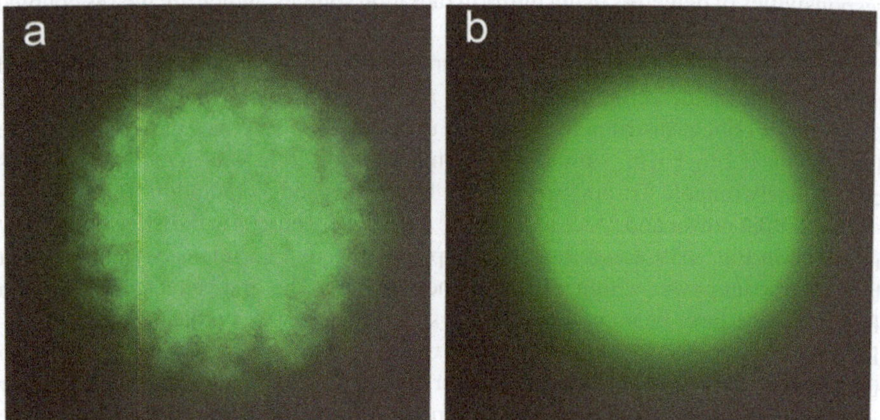

Fig. 9.40 Photographs of the output from a multimode optical fiber with injected 532 nm laser radiation: **a** with the ultrasonic bath turned off and **b** turned on. Reproduced from [214] with permission of AIP Publishing

pattern at the output of a multimode optical fiber with the injected laser radiation of a 532 nm wavelength. The sources and detectors are placed in a light-tight box blackened from inside to suppress the stray radiation.

The transfer detectors of the highest accuracy level are the Si tunnel-TDs (see Fig. 9.14), configured to operate according both power and irradiance method in the spectral range from 300 nm to 960 nm. The deviation of the relative responsivity of the tunnel-TD equipped with a circular aperture of 5 mm in diameter from the cosine function is less than 0.02% within a 6° FOV. For an 8° FOV, the deviation is approximately 0.05%. The reflectance-type TD composed of three Si photodiodes is designed especially for shortwave calibrations, from 210 nm to 350 nm. For the FR calibration, it is equipped with the precision aperture.

Parameters of the continuous empirical model expressed by Eq. 9.19 were found by the fitting the model to measurements at various laser wavelengths. Residuals of the fitted spectral power responsivity of the TD are typically within $\pm 5 \times 10^{-4}$ for wavelengths from about 450 to 1000 nm.

The SIRCUS facility employs different ISs for different types of calibration. Small ISs, with diameters from 25.4 to 50.8 mm, equipped with precision apertures with diameters ranging from 3 to 8 mm are typically used for irradiance responsivity calibrations.

Larger ISs, of 300 mm in diameter and with exit port diameters from 50 to 100 mm, are used for radiance responsivity calibration. The areas of the exit apertures are measured at the NIST facility for aperture area measurement (see [62]). The ISs has the sintered PTFE-based coating with the diffuse reflectance of higher than 0.99 between about 0.4 and 1.0 μm. Typical irradiance levels at 1 m using a 25.4 mm diameter IS with a 5 mm diameter aperture range about from 1 to 10 $\mu W\ cm^{-2}$. Radiance levels between 1 and 5 $mW\ cm^{-2}\ sr^{-1}$ are typical for a 300 mm diameter IS with a 75 mm diameter of the exit port. The uncertainty components due to stability of the wavelength and radiant flux coming out of the IS are of greatest importance. A laser intensity stabilizer keeps the laser power constant to within 0.1%.

The lasers used as excitation sources are equipped with Fabry-Pérot etalons to reduce the wavelength uncertainty from mode hopping. The absolute uncertainty of wavelength determination is about 10^{-3} nm with a typical laser bandwidth of 10^{-3} nm.

The wavelength uncertainty contributes to the uncertainty budget mainly by the dependence of the refractive index of air on temperature, humidity, and pressure. The wavemeter can be programmed to correct for these dependences. A monitor detector is mounted in the IS auxiliary port to correct for fluctuations in the IS radiant flux during calibrations.

For the irradiance calibration, the irradiance created by the IS source in the reference plane should be uniform to minimize the correction for different aperture of the detectors compared. Measurements were made in the vertical and horizontal directions in the reference plane at approximately 1 m from the exit port of a 25.4 mm diameter IS with a 5 mm aperture diameter. The irradiance in the reference plane is uniform to within 0.1% over the central zone of ±20 mm in both the horizontal and vertical directions.

At the radiance responsivity calibration, the TD measures the average radiance over the entire IS exit port, while the FR typically measures the central area of the exit port that corresponds to the FR's FOV. Therefore, a spatial nonuniformity of the IS exit port radiance may lead to systematic errors when the TD and the FR are compared. The ISs used for the radiance responsivity calibrations in the SIRCUS facility were designed in such a way that the laser radiation is introduced in the sphere wall near the exit port. The spatial distribution of the radiance for the 300 mm diameter PTFE-coated IS with a 75 mm diameter aperture was mapped using an FR with an FOV of about 1°. Fixed at a distance of 500 mm from the sphere, the target spot diameter of the FR at the sphere exit port was approximately 12.5 mm. The radiometer was translated in 6 mm increments and the radiance recorded at each position. The measured radiance nonuniformity does not exceed 0.1% over the central 90% of the IS exit aperture. The uncertainty budget for calibrations of the FRs against the transfer TDs is presented in Table 9.4.

The uncertainties due to the aperture areas refer to both the TD and the IS apertures. The distance d between them was measured mechanically and optically, using an original method described by Brown et al. [29] and Yoon et al. [234]. Its essence is as follows.

The irradiance E created by a point source is proportional to d^{-2} (the inverse square law). However, the exit port of the IS has a finite radius, so the dependence of $E(d)$ is more complicated. The approximate equation

$$E = E_0 / (d - d_0)^2 \tag{9.36}$$

was used to describe the dependence of the irradiance E on the distance d roughly measured by an electronic ruler (linear encoder). The parameters E_0 and d_0 are found by the least squares fitting the model (9.36) to experimentally determined ratios S_{TD}/S_{TM}, where S_{TD} and S_{TM} are signals of the TD and the monitor detector of the IS source, respectively. The uncertainties in determination of the distance d can

Table 9.4 The uncertainty budget of the spectral radiance responsivity calibrations of the FRs at the SIRCUS facility (after [234])

SIRCUS uncertainty components	Type	Relative uncertainty value (%)
TD responsivity	A	0.025
Aperture areas	A	0.004
Distance between apertures	B	0.01
Sphere spatial and angular uniformity	B	0.025
Amplifier gain	A	0.005
Temperature coefficient of TD	B	0.002
Temporal stability of TD	B	0.013
Combined standard uncertainty		0.040
Expanded uncertainty ($k = 2$)		0.080

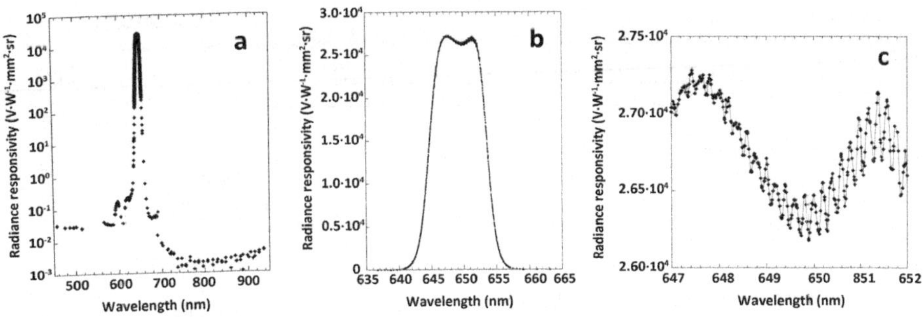

Fig. 9.41 Absolute spectral radiance responsivity of an AP1 pyrometer (FR) shown on **a** a logarithmic scale, **b** a linear scale and c expanded view of the responsivity showing interference fringes near the responsivity peak. Reproduced from [29] with permission of Optical Society of America

be evaluated from the residuals of the fit. It was claimed that the lower uncertainty can be achieve using this approach than with theoretical formula.

Yoon et al. [234] provided the uncertainty budget for measurement of the thermodynamic temperature by means of the FR calibrated in terms of the radiance responsivity using the SIRCUS facility. At the freezing point of gold, the expanded (at $k = 2$) temperature uncertainty is equal to 0.116 K. This is about two times lower than the uncertainty achieved at NIST by Mielenz et al. [149], who determined the freezing temperature of gold on the eve of the adoption of the ITS-90.

According to Yoon et al. [234], one of the dominant components in the uncertainty budget is the uncertainty in determination of the FR's spectral radiance responsivity (~0.04%). It seems that this component cannot be reduced significantly due to interference of coherent radiation in the FR, when it is calibrated in terms of the spectral radiance responsivity using the laser radiation.

Figure 9.41 presents the spectral radiance responsivity of an AP1 pyrometer (the FR operating by radiance method, see [232]). Figures 9.41a and 9.42b depict the same curve in the logarithmic and linear scales, respectively, obtained using the SIRCUS facility.

The peak value of the spectral responsivity lies between 647 and 652 nm; the FWHM is about 10 nm; the out-of-band leakage is less that 10^{-7}. Figure 9.41c shows a zoomed area at the top of the spectral responsivity curve. The high-frequency oscillations observed in Fig. 9.41c can be undoubtedly attributed to the optical interference. The high-frequency oscillations (interference fringes) are reproducible and temporally stable. They were observed by many independent researchers (see, e.g. [10, 25, 145]). Briaudeau et al. [25] reported development the laser-based setup at the LNE-Cnam[11] for calibration of the LP-5 RT (KE Technologie GmbH, Stuttgart, Germany, http://ke-gmbh.de/linearpyrometer) and subsequent measurements of thermodynamic temperatures of the HTFP blackbodies. The LNE setup uses the same principles as the NIST facility although differs from it in some details. The TD is equipped with two apertures, each of 6 mm in diameter.

[11] The NMI of France.

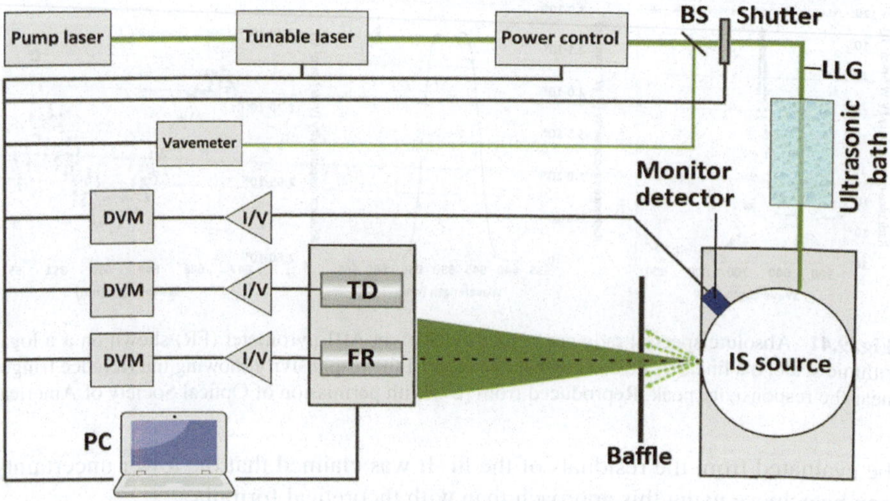

Fig. 9.42 A simplified diagram of the TULIP facility. Reproduced from [182] with permission of Optical Society of America

One diaphragm is attached to the IS output port and the other one is attached to the TD. The distance of 600 mm between apertures holds constant by means of three metal rods. This allows calibrate the TD directly in terms of radiance responsivity. The LP5 is calibrated against the TD by means of the IS source. An interference filter of about 25 nm spectral width defines the spectral responsivity of the LP5 around the center wavelength of 800 nm.

Despite numerous precautions against stray radiation (wedging of the filter, baffling, application of antireflection coatings, etc.), due to residual interferences of coherent radiation, the recorded spectral responsivity of the LP5 exhibits an amplitude modulation with a contrast of about 3% and a period of 0.2 nm.

If the period of modulation is much narrower than the spectral width of a bandpass interference filter, the detected amplitude modulation caused by interferences can be removed using a digital filtering. It was found that the relative uncertainty in the thermodynamic temperature associated with the optical interference might reach 0.05%, which is equivalent to 1.5 K at 3000 K.

A diffraction pattern was also reported by Anhalt and Machin [10] for the LP3 RT (KE Technologie GmbH) calibrated in terms of the spectral radiance responsivity using a tuneable laser. The interference fringes of about 2% difference in spectral responsivity between neighboring maximum and minimum and an interference period of 0.15 nm are observed. The observation of these fringes requires a high spectral resolution of the measurement system, less than 0.02 nm, used for the spectral radiance responsivity calibration.

Interference fringes arise from multiple reflections of the coherent quasi-monochromatic radiation at surfaces of lens, windows, and other optical elements, if they are not wedged or an antireflection coating is not applied. The interference fringes either the increase the difficulty in the calibration (if they must be measured during the calibration) or increase the calibration uncertainty. The analogous problem was encountered if an FR operating by the irradiance mode is calibrated in terms of irradiance responsivity using coherent radiation.

At the PTB, the tunable laser system TULIP (TUnable Lasers In Photometry), covering mainly the visible spectral range, is used, in particular, to determine the irradiance and radiance responsivity of the FRs designed for measuring thermodynamic temperatures of high-temperature blackbodies. A simplified diagram of the TULIP [182] is presented in Fig. 9.42. The following acronyms are used: DVM is for digital voltmeter, I/V denotes current-to-voltage converter, BS is for beam splitter, LLG is for liquid light guide.

The FR and the reference TD are mounted on a translation stage located in a light-tight measurement chamber. Temperature-stabilized three-element reflection-type silicon TD with precision aperture is used as references, where the spectral power responsivities of the TD is traceable to the ACR, the primary radiometric standard of PTB [209]. The TULIP facility includes two tunable laser systems.

A CW laser system covers the spectral range from 360 to 960 nm, except for a gap from 460 nm to 565 nm. The optical output power of the laser systems accessible for the FR calibrations is shown in Fig. 9.43. The wavelength-tunable lasers are pumped at 532 nm using a frequency-doubled Nd:YVO_4 laser. Two dye lasers are used to cover the spectral ranges from 565 to 700 nm. A Ti:Sapphire (Ti:Sa) laser and additional intracavity frequency doubling allow measurements from 360 to 460 nm and from 690 to 960 nm. The gap between 460 and 565 nm is partly covered by the single lines of an Ar^+-laser and the pump laser itself (not shown in Fig. 9.37). A quasi-CW laser system with pulse duration between 130 and 200 femtoseconds and a repetition rate of 80 MHz is used to close the gap in the wavelength range of the CW laser setup. It also extends the spectral range for the spectral calibration and characterization of FRs throughout the spectral range from 230 to 3000 nm. The laser tuning over spectral ranges from 680 to 1080 nm, 345 to 540 nm, and 230 to 346 nm

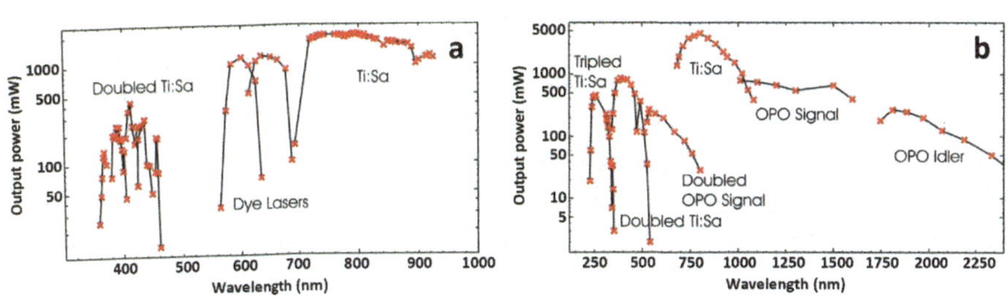

Fig. 9.43 Output power of various laser sources used in the TULIP facility: **a** tunable CW lasers and **b** quasi-CW laser system. Reproduced from [182] with permission of Optical Society of America

is successively facilitated by a quasi-CW mode-locked Ti:Sa laser with frequency doubling and tripling.

The spectral range above 1000 nm is covered by an OPO with a tunable signal range from 1000 to 1600 nm. The idler radiation is available from 1700 to 3000 nm. Finally, the spectral range from 525 to 720 nm is covered by frequency doubling of the OPO signal. The bandwidth of the laser radiation lies between 2 nm at the 400 nm wavelength and 9 nm at the 900 nm wavelength but can be reduced down to 0.1 nm by a diffraction grating if necessary.

The wavelength and the bandpass of the laser radiation are monitored continuously during the measurements. A calibrated wavemeter is used to measure the wavelength of the CW lasers. A high-resolution Echelle spectrograph is applied to record the spectral distribution of the quasi-CW laser radiation. If needed, additional diffractive parts in the optical path are used to reduce the spectral bandwidth of the femtosecond pulses. A uniform irradiance field, necessary for the measurement of large-area detectors operating by the irradiance method, is achieved by feeding the laser beams into the IS with a $BaSO_4$-coated IS.

Since the $BaSO_4$ coating fluoresces at shortwave irradiation, the IS source is not used for wavelengths shorter than 500 nm.

The laser radiation is guided into the IS source by a liquid light guide (LLG), which has a higher throughput than the ordinary light guides made of silica fibers.

In order to suppress the speckle patterns at the measurement plane of the IS source, the LLG is partially immersed in the ultrasonic bath that scrambles the propagation modes and hence reduces the speckle pattern[12].

A laser power stabilizer is used for the active compensation of the drift and the fluctuation of the laser power. In addition, monitor technique is used to account for source-dependent residual fluctuations and changes the amplitude of the radiation field at the measurement plane during the measurements. The typical variation of the spectral irradiance at the wavelength of 570 nm in the measurement plane at a distance of 700 mm from the IS exit port is found to be ±0.05% over the central area with a diameter of 30 mm. The vertical and horizontal scans at other wavelengths showed no wavelength dependences of the irradiance nonuniformity.

The laser-based measurements allow resolving the interference oscillations occurred due to multiple reflections of the highly monochromatic coherent laser radiation between parallel surfaces of the optical components in the radiometers. Since a TD has no such surfaces, it is insensitive with respect to internal interference effects. An FR is affected by interference fringes (periodical variations of the measured responsivity values with respect to the wavelength of the applied radiation) due to the presence of specularly reflected parallel surfaces. These interference effects are less pronounced if short-pulse lasers are used instead of CW lasers, since the fringes are spectrally smoothed due to the broader line width of this type of lasers.

[12]It should be noted that the speckle pattern is observed by imaging radiometers operating in the radiance mode. It is not clear whether this pattern will be observed in the measuring plane and whether it affects the measurements in the irradiance mode.

Schuster et al. [182] reported a strong dependence of the fringes on the measurement geometry, which is difficult to reproduce with an appropriate accuracy. The appearance of the fringes differs slightly with every measurement, which makes it difficult the evaluation of uncertainty due to impossibility of realizing the normal operating condition. Therefore, it is highly desired to remove interference fringes from measurement data. Schuster et al. [182] proposed to treat periodic oscillations appeared due to interference using digital filtering, which suppresses high-frequency components.

We will not delve into the issues of choosing a digital filter and its proper application.

This is a routine task of the digital signal processing, the discipline, foundations of which can be found in many textbooks, entry level books (e.g. [159]) and monographs (see, e.g. [2, 11, 183]). Algorithmically, the sequence of operations is as follows. The set of experimental points (spectral responsivity data) is subjected to the discrete Fourier transform. The Fourier image was multiplied by the window function that suppress high frequencies in the transformed data. Schuster et al. [182] employed the lowpass Chebyshev type II filter. The inverse Fourier transform returns the spectral responsivity data "cleared" of high-frequency oscillations. An example of the action of the digital filtering is presented in Fig. 9.44. The spectral irradiance responsivity of a Si photodiode Hamamatsu S1337 with a protective window and a limiting aperture was measured using a laser-based setup. In Fig. 9.44a, this curve with interference fringes are plotted together with the interference-corrected data and the calibration data obtained using a monochromator-based setup. Figure 9.44b presents the amplitude of the Fourier spectrum of the initial data and the data with suppressed bursts on both side of the main peak, which are responsible for interference oscillation in the initial data. As the result, the fringes were damped to the noise level in Fig. 9.44b. The algorithm of the digital filtering, direct and inverse fast Fourier transform algorithm [169] was realized using the commercial software

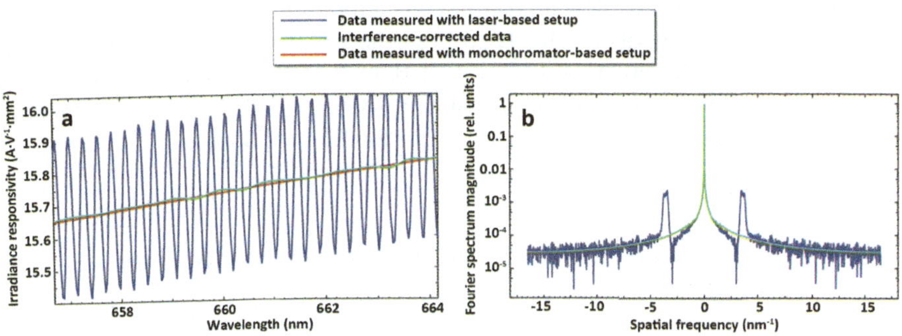

Fig. 9.44 **a** Spectral irradiance responsivity of a Si photodiode with a protective window and a limiting aperture measured using a laser-based setup plotted together with the interference-corrected data and the calibration data carried out on a monochromator-based setup; **b** the magnitude of the Fourier transform of measurement data before and after digital filtering. Reproduced from [182] with permission of Optical Society of America

package MATLAB® (https://www.mathworks.com). The common problem in the digital filtering applications is the possible unwanted distortion of the signal restored, especially in bands, where it varies sharply.

The uncertainty introduced by the digital filtering algorithm was evaluated by computer simulations of the incoherent-based calibration data, addition of the interference oscillations, Gaussian noise, and wavelength instability. The resulting deviation between the filtered curve and the initial data was found to be about 0.05% for the pessimistic values of affected parameters (e.g., the relative amplitude of the interference fringes was set to 5%).

The presence of the fringes does not create problems if the radiance responsivity calibrations can adequately measure all peaks and valleys of the fringes, but the number of calibrations at the different wavelengths could be very time-consuming. Schuster et al. [182] preferred to solve radically the interference problem by introducing a diffuse scattering element (diffuser) into the FR structure and to limit themselves to an uncertainty of the spectral irradiance responsivity of 0.2%.

The laser-based facilities for calibration of the FRs in terms of radiance and irradiance responsivity were considered until recently as providing the lowest calibration uncertainties. However, these facilities have some obvious drawbacks. They are complicated, expensive, and need highly qualified staff. Only the largest NMIs can afford such facilities for the most responsible calibrations and establishing their own spectral responsivity scales.

For a long time, it was believed that the main prospect of increasing the accuracy of calibrations with tunable lasers is to increase their output power, which will allow the use of sufficiently large ISs having a good spatial uniformity over the output apertures.

However, the high-resolution spectral responsivity calibrations with the laser-based facilities bring to the fore the problem of the interference patterns in measured spectra. Researchers faced a choice: to increase the amount of experimental data (and the time spent on measurements) in order to track all interference fringes and digitally filter them, or to allow an increase in calibration error.

The interference effects are likely to limit the uncertainty of laser-based spectral responsivity calibration at the level of 0.08–0.15%. On the other hand, the emergence of high-power incoherent radiation sources revived the development of monochromator-based calibration facilities for the FRs applicable to the APRT.

9.5.3 Monochromator-Based Calibrations for the Radiance Method

Since the radiation at the output of the monochromator is incoherent, measurements using monochromator-based systems are free of interference effects. The monochromator-based facilities have proven themselves in calibrating FRs in terms of the spectral power responsivity [22] and irradiance responsivity [200–202].

However, when trying to use a monochromator-based system to calibrate FRs for the radiance method, serious difficulties arise. The radiant flux at the exit slit of the monochromator is much lower than the output power of CW lasers or the mean output power of quasi-CW lasers. The calibration of an FR in terms of the radiance responsivity is performed using an IS, so the bottleneck point of monochromator-based calibration of FRs operating by the radiance method is the insufficient radiance of the IS source, which makes it impossible to accurately determine the out-of-band responsivity of the FR (RT).

To date, several techniques have been developed to solve this problem.

Anhalt et al. [9] reported the calibration of the RT LP3 using the assembled at the PTB experimental setup, which can be qualified as a transition form between approaches based on a monochromator and a laser.

Researchers at the PTB fruitfully developed the irradiance technique of the blackbody thermodynamic temperature measurement since 1990s (overviews of the PTB achievements in this field are given by Hollandt et al. [107] and Hartmann [97]). However, motivated by necessity of measuring the thermodynamic temperatures of the HTFP blackbodies with the radiating cells, whose apertures are restricted to 3 mm in diameter, they developed a monochromator-based setup for the absolute calibration of RTs in terms of the spectral radiance responsivity.

The concept of this setup is to measure the relative spectral responsivity of the RT using the conventional monochromator-based facility for the FRs calibration with a spectral bandwidth of about 0.5 nm. The absolute spectral radiance responsivity is measured at a single wavelength of the radiation of a laser diode, which is guided with an optical fiber into an IS source equipped with a precision aperture of known area. The spectral radiance of the IS is determined via the spectral irradiance measured by a TD calibrated against the ACR. A schematic of the monochromator-based spectral comparator for measuring the relative spectral responsivity is presented in Fig. 9.45 (see also [200]).

To measure the relative spectral responsivity of the LP3, its imaging optics was removed to ensure overfilling of the field stop that defines the FOV of the instrument. With the imaging optics, the image of the monochromator exit slit overfills the aperture only partly, which results in an unstable measured signal. To avoid this, the exit slit of the grating monochromator was imaged with a magnification ratio of ~2 onto a variable aperture that was set to a diameter of about 0.25 mm. The overfilled aperture was then imaged with a magnification of ~1.8 alternately onto the LP3 aperture and the TD. Measurements of the LP3 relative spectral responsivity were performed with a 4 nm/mm reciprocal dispersion. Within the bandpass of the interference filter, i.e., from 620 to 680 nm, the monochromator spectral bandwidth was 0.5 nm and the wavelength step of 0.5 nm was used for calibrations. For the measurement of the out-of-band regions of the interference filter (420–620 nm and 680–1200 nm), the spectral bandwidth of 6 nm and the wavelength step of 5 nm were used for calibration.

The absolute measurement of the LP3 radiance responsivity was performed using an IS source irradiated by a laser diode emitting about 5 mW of optical power around a wavelength of 650.1 nm. The laser diode is operated using a low-noise power supply,

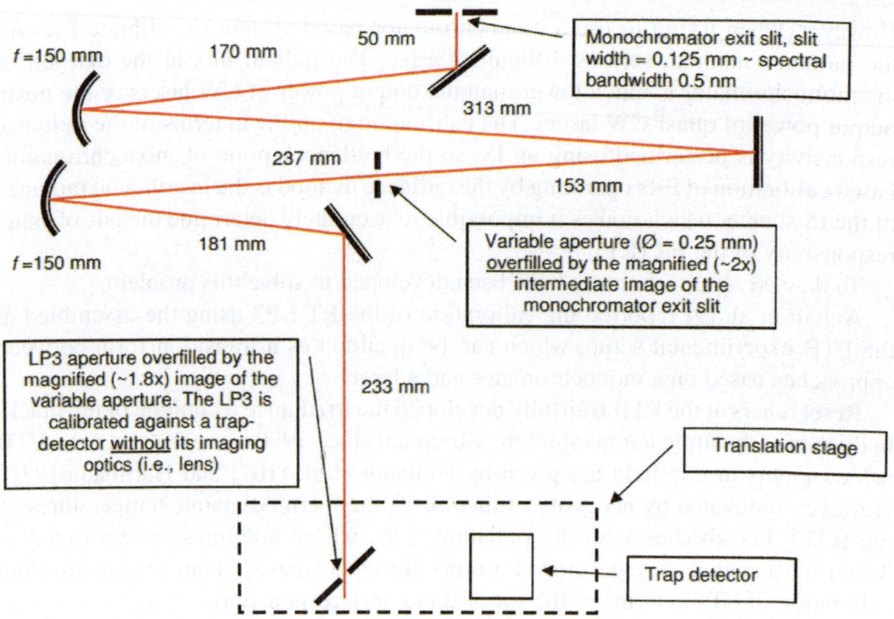

Fig. 9.45 A schematic of the setup for the LP3 relative spectral responsivity measurement. Reproduced from [9] with permission of Springer Nature

typically with 70 mA at 2.4 V, and a temperature controller. Using an optical spectrum analyzer, the stability of the laser diode was found to be heavily dependent upon the electrical power and temperature settings. Even minor changes of the settings lead to alterations of the generated radiation modes, ranging from a single mode to up to three modes within a 1.5 nm bandwidth. Only after a sufficiently long settling time, the optical power becomes stable within a few parts per thousand. The laser radiation was introduced into the IS using a 200 cm long optical fiber. To reduce the speckle pattern in the IS wall, a 100 cm portion of the optical fiber with removed the jacket lining and buffer was immersed in an ultrasonic bath.

The photographs of the opening of the IS obtained with the LP3 optics are shown in Fig. 9.46. The central black spot corresponds to the FOV limiting aperture of the LP3. When the laser radiation is directly injected into the IS, the speckle pattern can clearly be seen in Fig. 9.46a. This pattern is highly unstable, changing with thermal and mechanical noise, and leads to variations in the spectral radiance at the output aperture of the IS, which increases the overall noise in the measured signal. Ultrasonic vibrations of the optical fiber disrupt the coherence of the laser radiation and remove the speckle pattern as it is shown in Fig. 9.46b.

After removing the speckle pattern, the temporal stability of the spectral radiance emitted by the IS was measured by the TD.

The results of these measurements presented in Fig. 9.47 show that the temporal variations of the spectral radiance emitted by the IS source are within 0.3% over

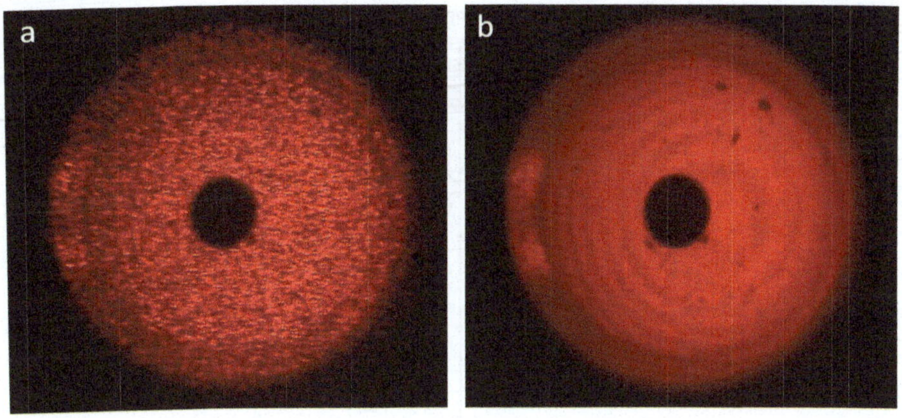

Fig. 9.46 Photograph of the opening of the IS source (as it is seen through the LP3 optics) irradiated by laser radiation via the glass fiber: **a** directly and **b** after passing through an ultrasonic bath. Reproduced from [9] with permission of Springer Nature

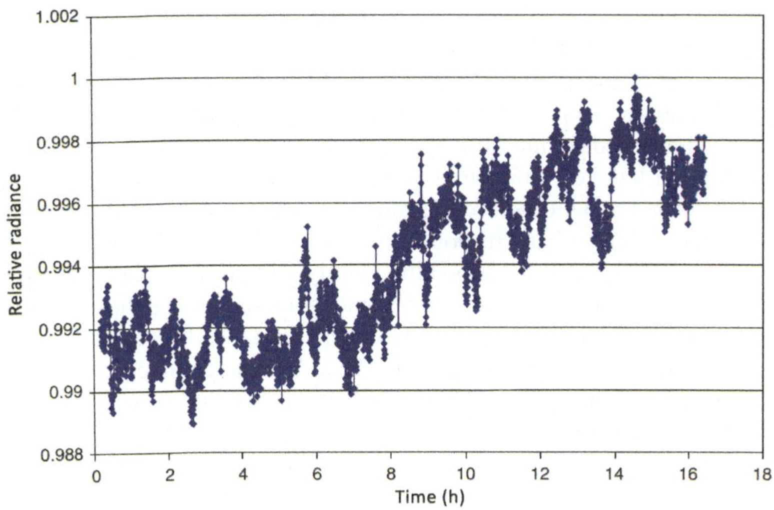

Fig. 9.47 Stability of the spectral radiance emitted by the IS source measured by the TD. Reproduced from [9] with permission of Springer Nature

5 h. The spatial radiance uniformity of the IS was measured by scanning with the measuring spot of the LP3 across the IS output port and recording the obtained photocurrents.

These measurements showed that the relative variations of the IS's spectral radiance have the order of 0.5%. The overall standard uncertainty of the setup is about 0.26% of the determined spectral responsivity of the RT. It was also stated that there

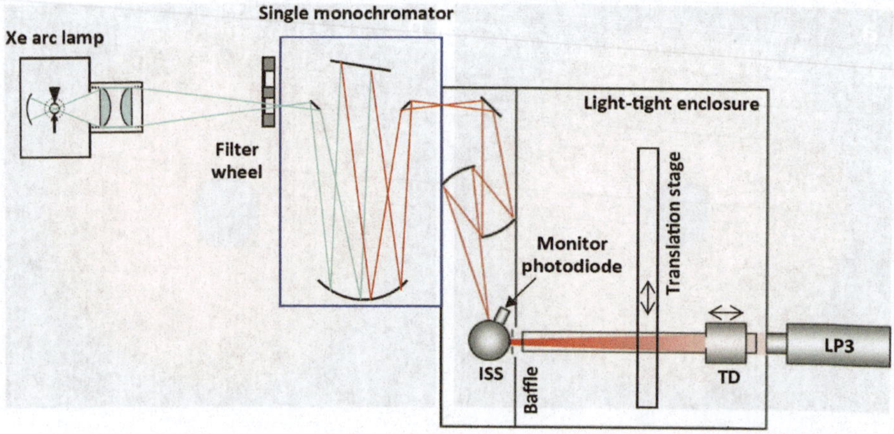

Fig. 9.48 Schematic diagram of the monochromator-based calibration facility. Reproduced from [118] with permission of Springer Nature

is possibility to reduce this value down to 0.08%, primarily, by improving the IS's homogeneity to a value of 0.05%.

The principal factor limiting the accuracy of the monochromator-based calibrations is the small optical power at the exit slit of a monochromator.

Since the first decade of the 21th century, there were numerous attempts to substitute the tungsten halogen lamps traditionally used in monochromators with more intensive wideband radiation sources, primarily, to increase the SNR and reduce the uncertainty due to out-of-band responsivity.

Keawprasert et al. [118] described the facility with a 75 W Xe-arc lamp with a concave mirror and condenser lens, a single monochromator, and the IS source. This facility, whose scheme is shown in Fig. 9.48, was used to calibrate the LP3 in terms of the absolute spectral radiance responsivity. Calibrations are traceable to the PTB's ACR via the three-element TD calibrated in terms of spectral power responsivity at several laser wavelengths.

The radiation of the lamp was imaged onto the entrance slit of the monochromator using a pair of achromatic lenses. Since the Xe-arc lamp also produces intensive UV radiation, a filter wheel with a set of bandpass filters was used. To provide a higher throughput, a single grating monochromator was employed instead of a double monochromator. Wavelength calibration was performed at 50 spectral lines of different spectral lamps in the wavelength range from 370 to 1100 nm. Since a bandwidth of at most 1 nm is required for the calibration of a narrow-band RT, the entrance and exit slit widths of the monochromator were set to 0.5 mm. With such a slit width, the optical power at the exit slit ranges from 0.1 μW/nm at 400 nm to more than 1 μW/nm from 500 to 700 nm.

The radiation behind the exit slit was focused using a mirror system onto an entrance port of an IS with a diameter of 5.1 cm. A precision aperture of a nominal diameter of 5 mm was used to define the radiating area for calculation of the spectral radiance for absolute measurements. For the relative spectral responsivity measurements, the IS was replaced by a mirror with an adjustable aperture to match the f-number of the LP3. To minimize the effect of stray radiation, a blackened variable-diameter aperture was placed between the IS source and the detectors. A monitor photodiode was attached to the IS for correction of short-term radiance variations during the measurements.

Measurements were performed by comparison the signals of the TD and the LP3 at each wavelength using a translation stage that moves the TD relative to a fixed position at a certain distance in front of the LP3 in a light-tight box.

The TD, in combination with two diamond-turned precision apertures, is employed to calculate the absolute spectral radiance via the measured spectral irradiance.

The geometric factor g is calculated using known areas of the apertures and the distance between them. The spectral radiance responsivity is calculated as

$$s_L(\lambda) = \frac{\pi g\, s_{E,TD}(\lambda)}{A_2} \cdot \frac{i_{LP3}}{i_{TD}}, \tag{9.37}$$

where $s_{E,TD}(\lambda)$ is the spectral irradiance responsivity of the TD, A_2 is area of the TD aperture, and the i_{LP3} and i_{TD} are the photocurrents of the LP3 and the TD, respectively.

Within the nominal waveband of the LP3, the relative measurements were performed with spectral bandwidths of 0.2 nm to reduce the measurement duration. It was found that the stray radiation level for the single monochromator is suppressed to less than 10^{-5} even for a monochromator bandwidth of 1 nm. As a result, the out-of-band responsivity of the LP3 on the setup described was limited by this level.

The absolute spectral radiance responsivity of the LP3 is plotted together with the relative results obtained with a bandwidth of 0.2 nm in the logarithmic and in the linear scales in Fig. 9.49a and b, respectively. It can be seen that the out-of-band leakage of the LP3 is less than 10^{-5} in the wavelength range from 400 to 1000 nm.

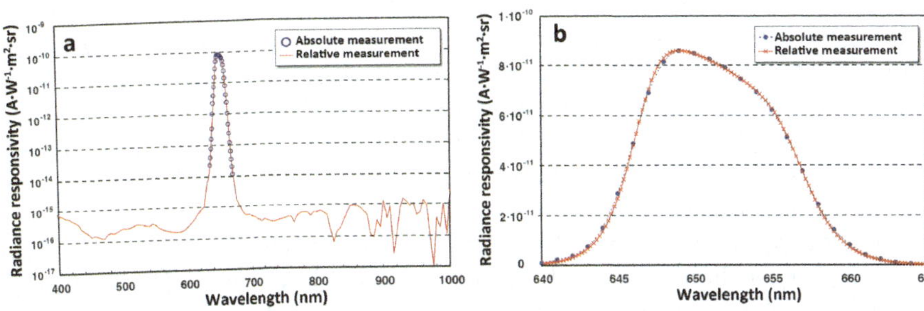

Fig. 9.49 Absolute spectral radiance responsivity of the LP3 measured with the monochromator-based facility: **a** in a logarithmic scale and **b** around the center wavelength in a linear scale. Reproduced from [118] with permission of Springer Nature

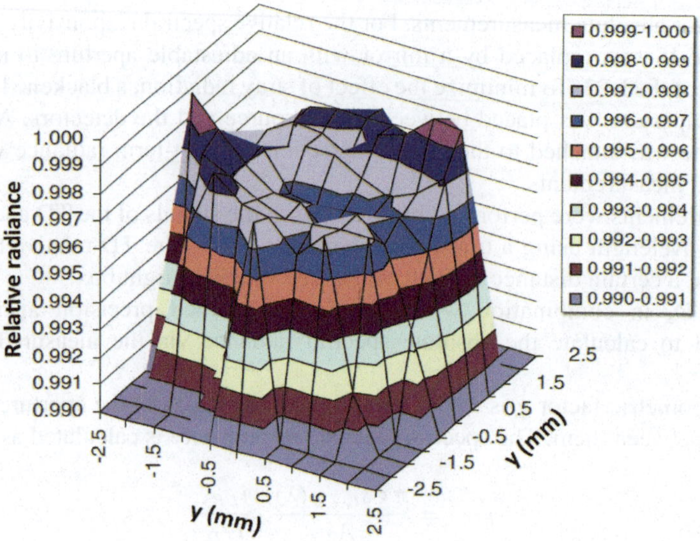

Fig. 9.50 Measured spatial uniformity of the spectral radiance across the radiating opening of the IS at 650 nm with a bandwidth of 9 nm. Reproduced from [118] with permission of Springer Nature

It was found that the greatest contribution in the uncertainty budget brings the spatial uniformity of the spectral radiance across the IS exit port. It was measured by scanning the aperture with the LP3 under test at 650 nm center wavelength with a bandwidth of 9 nm. The relative distribution of the spectral radiance at 650 nm across the 5 mm aperture for the 5.1 cm diameter IS is shown in Fig. 9.50. The relative standard uncertainty $u(s_L)/s_L$ in determining s_L was assessed as 0.169%.

This value can be approximately converted into the temperature uncertainty using the following expression:

$$u(T) \approx \frac{\lambda T^2}{c_2} \cdot \frac{u(s_L)}{s_L}. \tag{9.38}$$

Using the LP3 calibrated for the absolute spectral radiance responsivity, the thermodynamic temperatures of the Au and Cu FPBBs were determined as 1064.232 ± 0.154 °C and 1084.570 ± 0.158 °C (uncertainties are indicated for $k = 1$), which are in good agreement with the ITS-90 values (1064.18 °C and 1084.62 °C, respectively). Uncertainty analysis showed that the uncertainty due to the IS spatial non-uniformity remains the dominant component.

Later, Keawprasert et al. [119] reported the comparison of the results obtained using this monochromator-based facility and the laser-based TULIP facility described in the previous section. Due to the limited output power at the monochromator-based setup, the absolute calibration was performed with the exit slit spectral width of 1 nm for the wavelength range from 635 to 665 nm.

In the out-of-band region, the monochromator-based calibrations were performed using relative measurements and then adjusted to the absolute calibration results in the overlapping wavelength range around the peak wavelength. The spectral radiance responsivities measured in both facilities exhibit very good agreement, except for some wavelength ranges outside the nominal bandwidth. However, since these results were obtained with different bandwidths, their direct comparison is not quite correct.

The detailed uncertainty budgets provided by Keawprasert et al. [119] indicate that the combined relative standard uncertainties in the radiance responsivity of the LP3 was 0.17% for the calibration using the monochromator-based facility and 0.145% using the TULIP facility.

Verification of the results was carried out by measuring the thermodynamic temperature of the Au FPBB by the LP3. Measurements gave the thermodynamic temperature of 1337.37 $\pm$ 0.32 K for the monochromator-based calibration and 1337.58 $\pm$ 0.27 K for laser-based calibration (expanded uncertainties at $k = 2$ are specified), which are agreed well with the ITS-90 value of 1337.33 K.

The Xe arc lamp is not the only radiation source that can be used at the input of monochromator-based calibration facilities. Mantilla et al. [141] and Martín et al. [145] described the use of a high power Xe–Hg discharge lamp for the absolute calibration of the RTs.

However, the traditional lamp sources can be considered only as a palliative solution. They give relatively small gain in brightness of the IS source and not for all wavelengths. Besides, the stability of these sources is often insufficient and require further investigation and often hardware improvement.

9.5.4 The Use of Supercontinuum Light Sources and Modern Monochromatizing Devices

The generation of supercontinuum (SC) radiation, a phenomenon known since the early 1970s, manifests itself in a drastic spectral broadening of ultrashort laser pulses propagating through a strongly nonlinear medium. The optical output of an SC source consists of very short pulses with a high repetition rate. However, most detectors used in optical radiometry have sufficiently large time constants, which allow considering the SC sources as an equivalent replacement for CW sources.

The physical processes that contribute to the supercontinuum generation is the combination of nonlinear effects and group velocity dispersion, the discussion of which goes far beyond the scope of this book. The interested readers can find many useful information in the relevant literature, e.g. in the collective monograph edited by Alfano [3].

The advent of photonic-crystal (microstructured) fibers [43, 163, 172], capable of generating the SC radiation under laser pulse pumping has resulted in development of emergence of commercially available fiber-based SC sources, often referred to as SC or white lasers, for a variety of applications [91]. The commercial SC lasers typically span the spectral range from 400 to 2500 nm at the spectral power density of up to tens of mW/nm, which result in output optical powers of up to tens of watt, providing ultra-short pulses at up to 100 MHz repetition rate. The characteristics of the SC radiation generated in photonic-crystal fibers can be easily tailored by varying the fiber material, length, and microstructure, as well as parameters of the pumping laser radiation [44].

The temporal and spectral coherence characteristics of SC radiation strongly depend on the pump pulse energy and dispersion regime in the photonic-crystal fiber [85]. The SC light sources, combining the power and focusing properties of a laser with the continuous spectrum of an incandescent lamp, are well suited for spectroradiometric purposes in combination with a monochromator or other means of spectral selection.

Woodward et al. [213] drew the attention of the radiometric community to the prospects of using SC sources for measurements, which require high input power and high output spectral resolution, including spectral radiance responsivity calibrations.

Eppeldauer et al. [54] published the preliminary results of comparisons between measurements of the radiance responsivity of an RT using the NIST laser-based SIRCUS facility and the monochromator-based SCF (Spectral Comparator Facility), the major facility for dissemination of the NIST radiance temperature scale to industry. In order to provide a sufficient level of the optical power at the output of the IS, an SC source was used as the input to the prism-grating double monochromator. The SC source provided about 1 mW/nm output throughout the spectral range from 400 to 2400 nm; therefore, more than 1 mW of optical power could be delivered within a bandpass of 4 nm to the IS input port. The SC source was pumped by a laser at a frequency of 40 MHz. This high repetition rate has no effect on the measurements using the relatively slowly responding silicon photodiodes. To minimize the effects due to the line profiles and width differences of the laser- and the monochromator-based sources, the RT with a spectrally broad photopic-response filter (so-called $V(\lambda)$ filter, see Fig. 2.5 in Chap. 2) was used. The $V(\lambda)$ filter has a greater thickness than an interference filter, which reduces interference fringes that are unavoidable at the laser-based calibrations. The tunnel-TD traceable to the ACR was used as a transfer detector. The radiance responsivities obtained by the laser-based and the monochromator-based calibrations were compared via calculation of the illuminance responsivities by integrating over the visible wavelength range. It was found that the agreement of the illuminance responsivities is within 0.04%. The verification was performed via comparison of the thermodynamic temperature of the VTBB determined using the radiance responsivity measured with the SCF and the ITS-90 temperature measured against the Au FPBB. In the latter case, the wide-band FR was used as a conventional RT. Temperature differences were ranged from 0.09 °C at 1300 °C to 1.17 °C at about 2500 °C. It was shown that monochromators fed by a SC sources can be employed for spectral radiance responsivity calibrations of RTs to realize thermodynamic temperature scales.

Yamaguchi et al. [225, 226] described a monochromator-based radiometric facility with an SC source developed at NMIJ (Japan) and applied to absolute calibration of the LP5 RT. A schematic diagram of the facility is presented in Fig. 9.51.

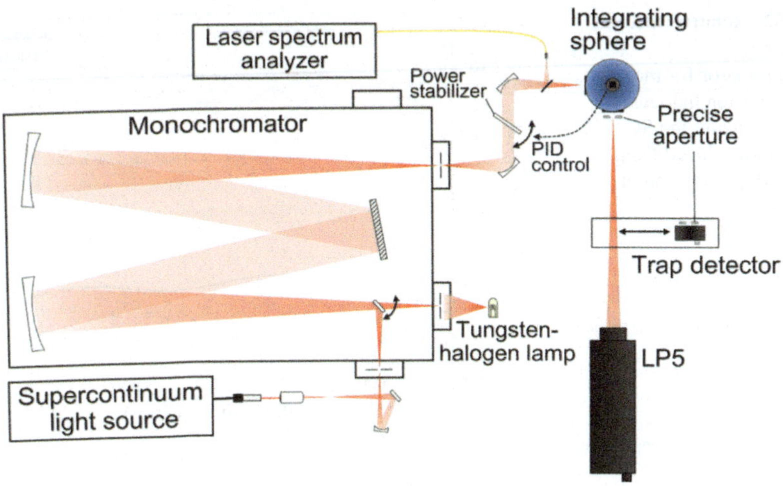

Fig. 9.51 Schematic diagram of the SC-based calibration facility. Reproduced from [226] with permission of Springer Nature

A commercial SC source with a wavelength range of 480–2400 nm and a total power of 6 W is coupled with a single monochromator, whose source can be switched between the SC source and a QTH lamp. The monochromator has a focal length of 1 m and a reciprocal linear dispersion of 0.68 nm/mm at 800 nm. The output optical radiation goes through an order-sorting filter and is collimated and focused with gold-coated parabolic mirrors through a power stabilizer controlled by a PID feedback loop. The center wavelength and the bandwidth of the output spectrum are continuously monitored with a laser spectrum analyzer. The focused optical radiation enters the IS with an 84 mm diameter made of sintered PTFE.

The position and the angle of the radiation first incidence onto the sphere internal surface were adjusted to optimize the radiance homogeneity at the exit port. The detector mounted in the additional port monitors the power stability and provides the feedback to the stabilizer. A pair of knife-edged precise apertures with a diameter of 4.5 mm is placed in front of the IS and the TD. The distance between apertures is 400 mm. The TD can be moved with an automatic positioning stage into the path of the RT to conduct the comparison with the LP5 RT with the center wavelength of 793 nm and a fixed focal length of 700 mm. The relative spectra of the SC source and the QTH lamp passing through the monochromator and directly measured by the silicon TD are shown in Fig. 9.52.

The QTH lamp operated with 12 V and 8.6 A is imaged on the entrance slit with a 0.75 mm width. The monochromator's narrow slit width and the low numerical aperture result in a lower throughput. However, the SC optical radiation emitted from the single-mode fiber coupled to the entrance slit provides a much higher radiant flux that the QTH lamp does.

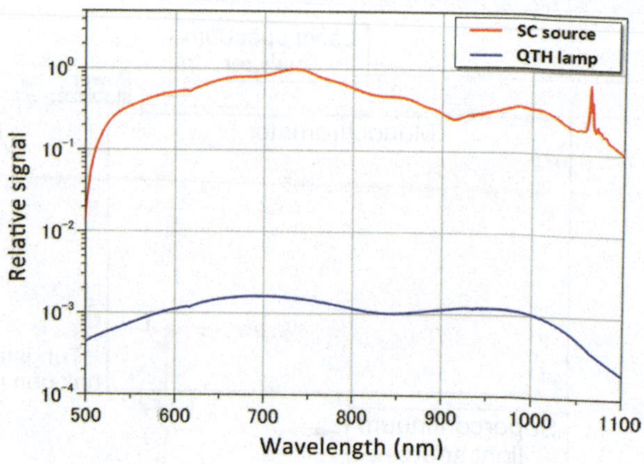

Fig. 9.52 Relative spectral outputs of the monochromator for the quartz tungsten halogen (QTH) lamp and the SC source. Reproduced from [226] with permission of Springer Nature

Around 800 nm, the output power of the SC source is more than 600 times greater than that of the QTH lamp, so the comparison between the TD and the RT using the IS can be carried out with a bandwidth of less than 0.5 nm. The continuous spectrum laser sources are currently available with a mean spectral power of up to 5 mW/nm. Typical beam diameters are about 1 mm in the visible region and 2 mm in the near IR region.

If the grating with the linear dispersion of 2 nm/mm is used, an output spectral power from the monochromator of at least 1 mW/nm would be achieved.

However, instability of the SCs due to noise is from 0.4 to 2% for the averaging time of 30 s and depends on wavelength. Although the instability of the SC source total power is less than ±0.5%, its spectral distribution varies due to nonlinear optical effects.

To maintain the output monochromatized radiation stable, a stabilizer consisting of a beam sampler and an automatic rotation stage is placed between the monochromator and the IS.

The beam sampler reflects about 5% of the optical radiation at 45°, and its Fresnelian reflectance depends on the incident angle, which is controlled by a PID feedback loop with the signal of the monitor detector at the IS's wall. The radiance fluctuations are of order of 1% without the stabilizer and within ±0.03% with the stabilizer. A typical time required for the comparison between the TD and the RT at one wavelength is about 2 min. Within this time interval, the standard deviation of the stabilized radiance is less than 0.015%.

Unlike the IS irradiated by conventional lasers, the use of an SC source leads to the less pronounced speckle pattern and hence the lower amplitude of the interference fringes in the spectral responsivity curve because the bandwidth broadening reduces the coherence length. The relative responsivities of LP5 measured with different bandwidths of the SC source-based system are shown in Fig. 9.53. The amplitude of fringes is greater than 1% with a bandwidth of 0.01 nm, but they become almost indistinguishable with a bandwidth of 0.5 nm.

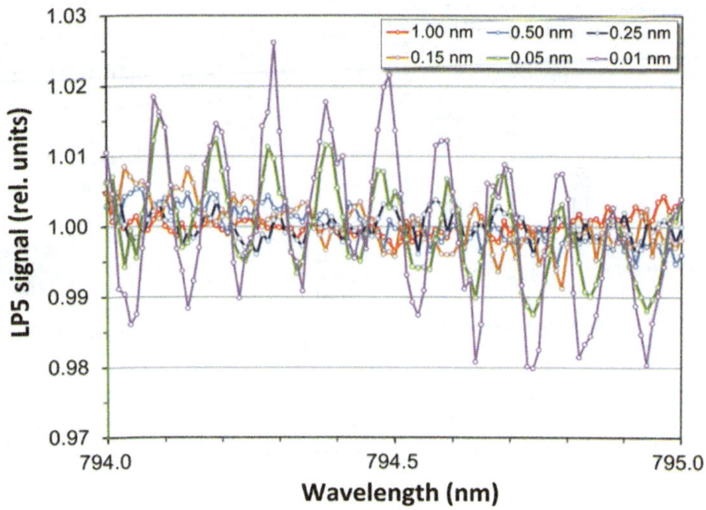

Fig. 9.53 Interference effect with the SC source for different slit width. Reproduced from [226] with permission of Springer Nature

Non-uniformity of the IS radiance is one of the major components of uncertainty at the calibrations in the radiance mode. The radiance relative distribution over the IS port scanned by LP5 with a target size of 1.3 mm is about 0.15%. The angular non-uniformity of the IS has been measured with the TD. The angular dependence of the radiant flux at the TD moving in the plane perpendicular to the optical axis is described by the fourth power of cosine law. The measured signal is in agreement with the $\cos^4\theta$ dependence within ± 0.036 rad from normal, which corresponds to the lens diameter of LP5 at the focal length.

The uncertainty due to angular non-uniformity is estimated to be less than 0.02%. The expanded (at $k = 2$) uncertainty in the radiance responsivity was assessed as 0.24%.

This corresponds to the expanded (at $k = 2$) temperature uncertainty of 239 mK. Since the calibrations of the TD responsivity and the aperture areas have not been completed, the preliminary values for the uncertainties were given. Yamaguchi et al. [226] concluded that the combined uncertainty of the determination of thermodynamic temperature could be sufficiently reduced if the TD is adequately calibrated and the apertures areas are thoroughly measured.

A conventional monochromator is not the only instrument that can be used for the spectral selection of the wideband sources, including SC radiation sources.

Kozlova et al. [123] described a spectroradiometric device on the base of an acousto-optical tunable filter (AOTF). The instrument was developed for high-temperature metrology. Its major design feature is the attenuation of the diffraction side lobes using a double-pass configuration. The radiofrequency tuning of the AOTF allows continuous, fine, and rapid wavelength control over the spectral range from

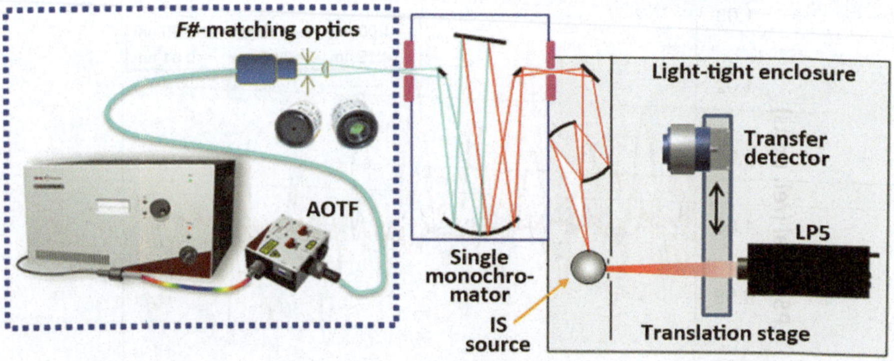

Fig. 9.54 Schematic of the PTB spectral comparator facility with a supercontinuum radiation source and a single monochromator. The calibration is based on an irradiance/radiance responsivity comparison between the transfer detector and the RT. Reproduced from [31] with permission of Springer Nature

0.65 to 1 μm. The wavelength scale of the instrument is calibrated with an Ar spectral lamp with the reproducibility of about 10^{-5} over one week. The stability over several hours of the signal is of order of 10^{-4} and of the wavelength is 10^{-5}. The spectral scanning is performed by varying the radio frequency of the AOTF.

At the PTB (Germany), the existing monochromator-based calibration setup was upgraded with an SC source [10, 30, 31]. The calibration scheme for the NIR RT in terms of radiance responsivity is shown in Fig. 9.54.

The transfer detector is an InGaAs photodiode placed in the housing, temperature-stabilized at 25 °C. The absolute spectral power responsivity of the transfer detector was determined using the PTB ACR [155]. The SC source gives a significant gain in the output optical power as compared to the QTH lamp for the same spectral interval.

For example, the ratio of the spectral power of the SC source the QTL lamp is about 100 and even higher for shorter wavelengths [31].

In the PTB's spectral comparator facility, the SC source emits a broadband radiation, which is preselected spectrally with an acousto-optic tunable filter (AOTF) at the desired wavelength. Behind the AOTF, the collimated beam is focused on the entrance slit of the monochromator by a dedicated optical element, which matches the acceptance angle of the monochromator. After the monochromator, the radiation is focused into the entrance port of the IS with an inner diameter of about 50 mm.

The IS exit port is equipped with a precision aperture of known diameter. A transfer detector in a known distance allows determining the spectral irradiance created by the IS source.

A research team from the NPL (UK) presented a novel approach to design of quasi-monochromatic source on the base of SC laser. Levick et al. [133] described the spectrally tunable absolute irradiance and radiance source (STAIRS) developed and characterized for calibrating the radiometric detectors in terms of radiance, irradiance, and spectral power responsivity. The setup consists of an SC laser, wavelength-tunable bandpass filter, power-stabilizing feedback control scheme, and

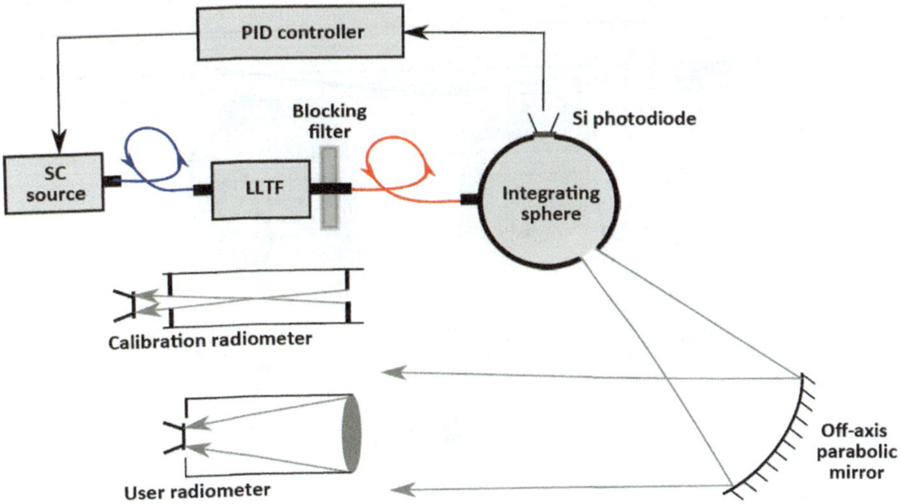

Fig. 9.55 Schematic diagram of a typical setup for STAIRS. Reproduced from [133] with permission of Optical Society of America

output coupling optics (see Fig. 9.55). The near-IR RT LP5 with the center wavelength of at about 1.57 μm was characterized with respect to its nonlinearity, SSE, distance effect, and the consistency of its temperature measuring ranges. A relative standard uncertainty of 0.11% for the spectral radiance responsivity of the LP5 was achieved in the near IR spectral region. The validation of radiometric calibration was carried out by measuring thermodynamic temperatures of a heat-pipe VTBB (from 773 K to 1235 K) and a gold FPBB (1337.33 K) and comparison of them with their ITS-90 values.

The radiation of the SC source with a wavelength range from 400 nm and 2500 nm and a total power of 6 W is monochromatized using the laser line tunable filter (LLTF from Photon etc. Inc., http://www.photonetc.com/) tunable between 400 and 1000 nm and has a bandwidth of 1 or 2 nm depending on the wavelength selected.

The LLTF is a non-dispersive tunable bandpass filter based on volume holographic gratings [13, 72, 87, 104]. The photograph of the LLTF CONTRAST™ is presented in Fig. 9.56. This is a compact device; its dimensions are 23 × 16 × 17 cm (L × W × H). A single LLTF can be tuned from 400 to 1000 nm (VIS) or 1000 nm to 2500 nm (SWIR), with the bandwidths (FWHM) of less than 2.5 nm and less than 5 nm, respectively. A power-stabilizing feedback control scheme was incorporated to stabilize the source to better than 0.01% for averaging times longer than 100 s. The out-of-band transmission of the LLTF filter is estimated to be less than 3×10^{-7}. Figure 9.57 compares the spectra of the STAIRS source at 633 nm and the He–Ne laser at 633 nm both measured using Si array spectrometer.

The quality of filtration suggests that an SC in combination with the LLTF can serve as an adequate replacement for a SC source in combination with a monochromator and for a set of tunable lasers.

Fig. 9.56 The photograph of the Laser Line Tunable Filter LLTF CONTRAST™ (*Source* http://www.photonetc.com/api/images/5418979f781045b7470013f7?maxWidth=747) Reproduced with kind permission of Photon etc., Inc. (Montreal, Canada)

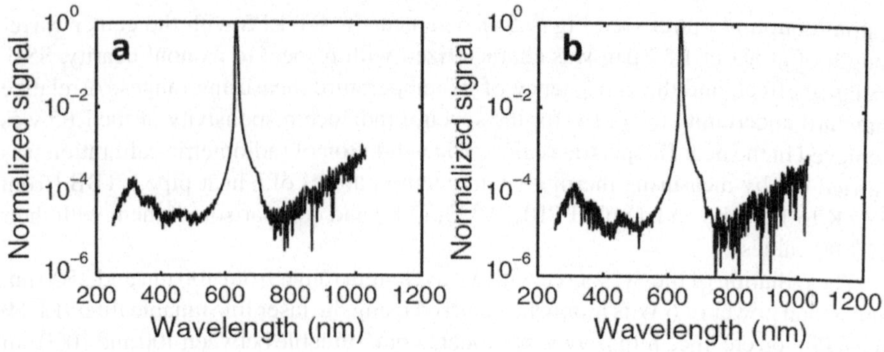

Fig. 9.57 Measured output from Si array spectrometer with **a** the STAIRS source at 633 nm and **b** HeNe laser at 633 nm. Reproduced from [133] with permission of Optical Society of America

Currently, the SC sources are introduced into calibration facility of many NMIs and research institutions around the world [142, 145, 221].

9.6 Uncertainty in Calibrations of Filter Radiometers

9.6.1 *Overview of Uncertainty Components*

Methods for determining the uncertainties associated with the FR calibrating using various methods of the APRT are summarized in the document prepared under the auspices of the CCT [180] and references therein. The classification of the components of the uncertainty in the calibration of FRs developed in this document is shown in Fig. 9.58.

It should be remembered that this document, as part of *MeP*-K, is updatable and does not pretend to be the only correct solution. In particular, some components of uncertainty include several different effects that can be rearranged.

Some sources of uncertainty for the FR calibration are common to all calibration methods (the uncertainties in the wavelength scale, spectral responsivity of the TD, geometric factor, and the amplifier gain, due to the stray light, and out-of-band responsivity), while others are method-specific.

Since the CCT document [180] is freely available, there is no need to retell the contents of its seventy pages. We restrict ourselves to the presenting only several tables, which list the uncertainty components and provide numerical values for the

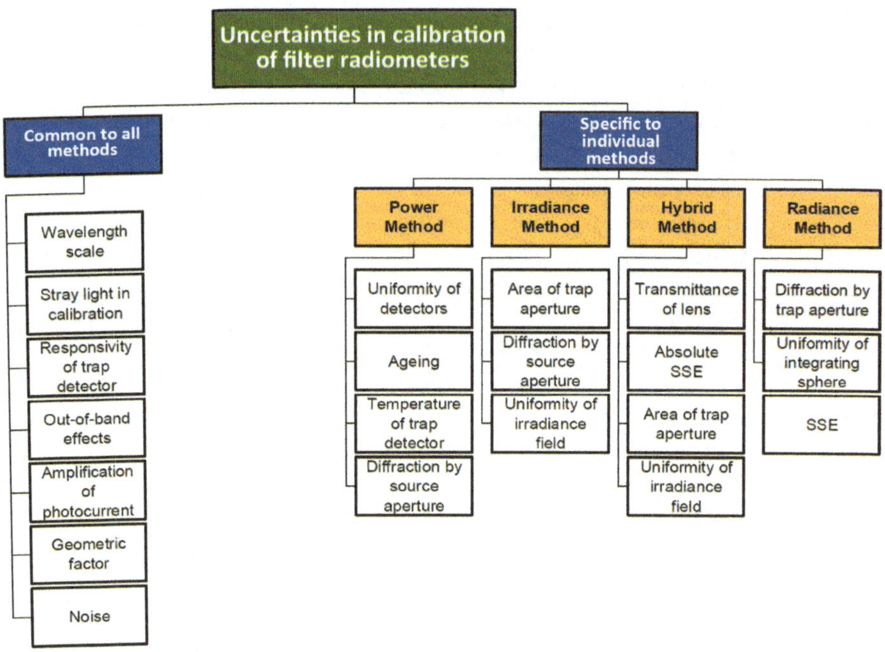

Fig. 9.58 A diagram for the classification of uncertainty components in calibration of FRs (after [180])

"normal" and "best" levels for each uncertainty component for calculation of the combined standard uncertainties in temperature. "Normal" uncertainty values are easily obtainable at the NMIs, while the "best" uncertainty values can be obtained with considerable effort by a small number of world-leading NMIs. We provide the extended comments to the components that, in the opinion of the authors, not fully addressed in the CCT document [180].

On the base of the CCT document [180], we have composed Table 9.5, where the "normal" and "best" values of the uncertainty components (both common for all methods and method-specific) are quoted as the relative standard uncertainties. The letters P, I, H, and R in the second column of Table 9.5 indicate the participation of the given component in the power, irradiance, hybrid, and radiance method, respectively. Below we briefly comment on some of the components.

Table 9.5 Typical contributions to the uncertainty budget for the calibration of FRs

Uncertainty component		Methods	"Normal" relative standard uncertainty (%)	"Best" relative standard uncertainty (%)
Wavelength scale		PIHR	0.02 (for monochromator-based systems)	0.00002 (for tunable laser-based systems)
Stray light in calibration		PIHR	0.01	0.0003
Power responsivity of a transfer detector		PIHR	0.07	0.013
Out-of-band response		PIHR	0.025	0.01
Photocurrent amplification		PIHR	0.0026	0.001
Geometric factor	Distance	PIHR	0.1	0.004
	FR aperture radius	P	0.02	0.01
	Source aperture radius	PR	0.02	0.01
	TD aperture radius	IHR	0.02	0.01
	Aperture co-alignment	PIHR	0.01	0.0001
	Aperture non-roundness	PIHR	0.004	0.001
	Aperture change	PIHR	0.02	0
Noise		PIHR	0.05	0.01
FR spatial uniformity		P	0.02	0.02
TD ageing and repeatability		P	0.03	0.01
Temperature of TD		P	0.01	0.003
Diffraction		PIR	0.07	0.001
Lens transmittance		H	0.02	0.02
Irradiance uniformity		IH	0.03	0.013
Absolute SSE		H	0.01	0.002
SSE		R	0.01	0.002

9.6.2 *Wavelength Scale*

The FR is calibrated against the transfer detector (typically, the TD) at several wavelengths using a quasi-monochromatic source (a monochromator with a broadband source or tuneable lasers).

In the monochromator-based methods, the wavelength scale of a prism or grating monochromator is calibrated using a set of low-pressure gas-discharge (usually, Hg, Ne, Ar, Kr, and D_2) lamps, which are used to generate a sufficient number of spectral peaks at known wavelengths. These wavelengths can be found in in atomic-transition databases, the largest and most famous of which is the Atomic Spectra Database (ASD) maintained by NIST and containing critically evaluated data on atomic energy levels, spectral lines, and transition probabilities [124, 174, 211]. Its latest version is freely available at https://www.nist.gov/pml/atomic-spectra-database. The uncertainty in the wavelengths of atomic transition assignments is about 10^{-5} nm.

The spectral lamp is commonly placed in the IS to provide the radiant flux uniformly distributed in the angular and spatial domains at the input of the monochromator. This allows to perform calibration of the monochromator's wavelength scale with the uncertainty from 0.02 to 0.1 nm. As a rule, uncertainties in the wavelength calibration of a monochromator are at least partially correlated.

In the laser-based methods, the laser radiation wavelengths are measured by wavemeters, which use interferometers with an internal laser reference source. This provides the lowest uncertainties in the wavelength determination (0.003–0.0001 nm), but can be applied to coherent radiation only. The uncertainty correlation can appear due to interpolation.

9.6.3 *Power Responsivity of a Transfer Detector*

The spectral power responsivity of a transfer detector (a single photodiode or, more commonly, photodiode-based TD) is measured at the set of discrete wavelengths by comparison with the ACR. The components of uncertainty in the spectral power responsivity of a transfer detector are divided into four groups:

1. Uncertainty of the radiant power measurement by the ACR. These contributions mainly depend on the ACR design and thoroughness of its characterization.
2. Uncertainty components related to the calibration of the transfer detector against the ACR. They mainly depend on the calibration principle, the radiation source stability, and the calibration facility used.
3. Uncertainties arising from the interpolation of the transfer detector spectral responsivity.
4. Uncertainty contributions related to the use of the calibrated transfer detector under calibration conditions and those during its use for the FR calibration.

These components depend on the properties of the FR to be calibrated and the difference in the conditions of the calibration and the use of the transfer detector.

A typical relative combined uncertainty in the power responsivity of a TD is of order of 0.02%; in Table 9.5, for the value of 0.07% is specified for "normal" uncertainty (the worst case in calibration of a single photodiode) and 0.013% for the "best" uncertainty (the most accurate calibration of the TD).

9.6.4 Out-of-Band Response

The response of every FR consists of two unequal parts. The largest part is the response of the detector to incident radiation, the wavelengths of which fall into the nominal passband $[\lambda_1, \lambda_2]$ of the filter. The remaining part corresponds to the wavelengths $\lambda < \lambda_1$ and $\lambda > \lambda_2$ that are out of the nominal passband. If the suppression (blocking) of the out-of-band (OOB) radiation is insufficient, the systematic error occurs. When the FR is calibrated using a quasi-monochromatic beam, the OOB signal is nearly zero. When the FR measures the thermodynamic temperature of the blackbody, the measured signal, unaccounted-for the OOB transmittance will be higher than expected by the factor

$$K_{OOB} = 1 + \frac{\int_0^{\lambda_1} s(\lambda) L_{\lambda,bb}(\lambda, T) d\lambda + \int_{\lambda_2}^{\infty} s(\lambda) L_{\lambda,bb}(\lambda, T) d\lambda}{\int_{\lambda_1}^{\lambda_2} s(\lambda) L_{\lambda,bb}(\lambda, T)\, d\lambda} = \frac{\int_0^{\infty} s(\lambda) L_{\lambda,bb}(\lambda, T)\, d\lambda}{\int_{\lambda_1}^{\lambda_2} s(\lambda) L_{\lambda,bb}(\lambda, T) d\lambda}, \tag{9.39}$$

where $L_{\lambda,bb}(\lambda, T)$ is the spectral radiance of a perfect blackbody at the temperature T.

The temperature error, ΔT, resulting from uncorrected OOB radiation can be calculated as

$$\Delta T \approx (K_{OOB} - 1) \frac{n \lambda_0 T^2}{c_2}, \tag{9.40}$$

where λ_0 is the mean wavelength of the FR spectral responsivity. The monochromatic Wien approximation was used to derive Eq. 9.40.

An experimental method to determine the OOB correction relies on the use of a notch (band-stop) filter to cut out the in-band radiation falling on the FR. The value of K_{OOB} is determined by measuring the FR signals with and without the notch filter in place:

$$K_{OOB} \approx 1 + \frac{S_{\text{with notch filter}}}{S_{\text{without notch filter}}}. \tag{9.41}$$

For a real notch filter, the blocking of in-band radiation and transmission of OOB radiation cannot be complete; therefore, the appropriate corrections have to be made in Eq. 9.41.

9.6.5 Geometric Factor

The geometric factor is expressed by Eq. 9.24. It is defined by two apertures of known areas separated by a known distance. By introducing auxiliary variables α and β as

$$\beta = \sqrt{\left(r_1^2 + r_2^2 + d^2\right)^2 - 4r_1^2 r_2^2} \tag{9.42}$$

and

$$\alpha = r_1^2 + r_2^2 + d^2 + \beta, \tag{9.43}$$

where r_1 and r_2 are the radii of the source and detector apertures (the order does not matter), d is the distance between them. Now we can rewrite Eq. 9.24 in the form

$$g = \frac{2\pi r_1^2 r_2^2}{\alpha}. \tag{9.44}$$

The sensitivity coefficients are given by:

$$\frac{\partial g}{\partial r_1} = \frac{4\pi r_1 r_2^2}{\alpha}\left[1 - \frac{r_1^2\left(\alpha - 2r_2^2\right)}{\alpha\beta}\right], \tag{9.45}$$

$$\frac{\partial g}{\partial r_2} = \frac{4\pi r_1^2 r_2}{\alpha}\left[1 - \frac{r_2^2\left(\alpha - 2r_1^2\right)}{\alpha\beta}\right], \tag{9.46}$$

$$\frac{\partial g}{\partial d} = -\frac{4\pi r_1^2 r_2^2 d}{\alpha\beta}, \tag{9.47}$$

The combined standard uncertainty in g is

$$u(g) = \sqrt{\left[\frac{\partial g}{\partial r_1}u(r_1)\right]^2 + \left[\frac{\partial g}{\partial r_2}u(r_2)\right]^2 + \left[\frac{\partial g}{\partial d}u(d)\right]^2} \tag{9.48}$$

and the corresponding relative uncertainty equal to

$$u_r(g) = u(g)/g. \tag{9.49}$$

The aperture radius is never measured directly. It is derived from the measured diameter or the area of the aperture. It should be remembered that the relative uncertainty in determining the radius is half the relative uncertainty in measuring the area.

The distance between apertures can be measured using a gauge bar of known length or, for lower uncertainty, by interferometric methods.

The calculation of the geometric factor assumes that the apertures lie in the parallel planes and the optical axis passes through the aperture centers. Any possible deviation from this ideal configuration can be easily evaluated using the view factor algebra (see Sect. 3.4.2 and reference therein). In the next section, we consider the quality of the aperture manufacturing and methods for measuring their areas.

9.6.6 *Precision Apertures*

In order to achieve an acceptable temperature uncertainty in the APRT, the measurement geometry defined by the aperture pair, must be known very accurately. A schematic of a precision aperture with the detailed structure of the edges in a zoomed area is shown in Fig. 9.59. Typically, precise apertures are made of aluminum, copper, or their alloys by diamond turning. An ideal aperture should be knife-edged to reduce scattering on uneven edges. The required edge thickness is 10–20 μm. Hartmann [97] presented the electron microscope photographs illustrated the difference between high-quality and low-quality aperture edges of diamond-turned copper diaphragms (Fig. 9.60). Thin high-quality apertures can also be made by electrochemical etching.

Once a nearly ideal aperture is manufactured, its area should be precisely measured. To be measured using contact methods, the land area (cylindrical zone) should be at least 100 μm thick; otherwise, non-contact methods have to be used.

The conventional mechanical methods for measuring the aperture areas employ the diameter-measuring machines. A moving ball-ended probe contacts with the aperture edges, and the corresponding contact force is detected.

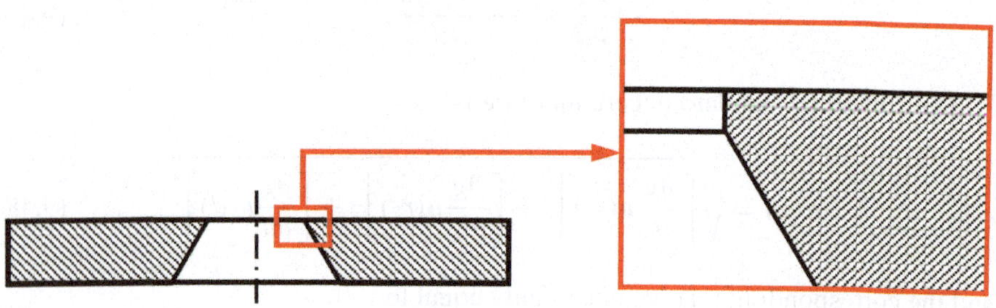

Fig. 9.59 Schematic of a precision aperture showing the detailed structure of the edges in a zoomed area on the right

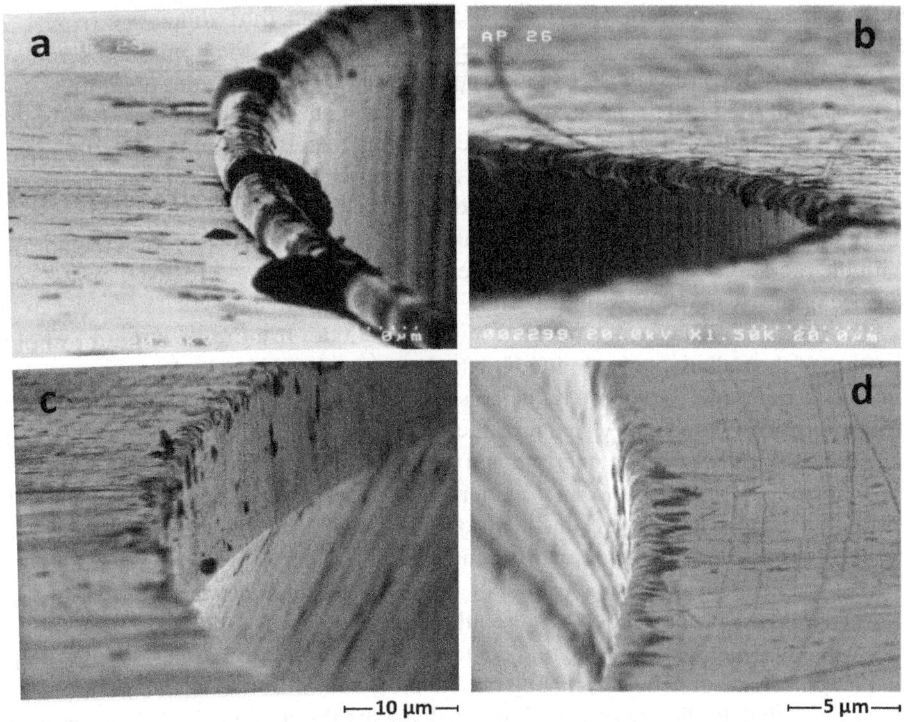

Fig. 9.60 Electron microphotographs of diamond-turned apertures: **a**, **b** non-ideal edges and **c**, **d** nearly ideal edges. Reproduced from [97] with permission of Elsevier

The probe movements are measured by a laser interferometer. The contact methods allow accurate calibration of the aperture diameter with the standard uncertainty of 0.15 μm. Due to fragility of the aperture lands, a ball-ended probe cannot be applied directly to measurement of diameters of precision aperture. However, the use of an optical fiber as an edge probe with the low-force optical sensors [151] still allows contact measurement of the aperture areas with the relative combined standard uncertainty of 5×10^{-5}.

Modern optical methods of absolute measurement of apertures are divided into two fundamentally different types. The first one implies measuring the diameters of apertures by detecting the edge point locations. This approach was implemented in the NIST aperture measurement facility [61, 62], which is a machine vision-based system, consisting of a high-accuracy XY stage that carries the aperture and has a laser interferometer for measurement of displacements and a microscope mounted on a high-resolution vertical stage for edge detection.

An automated system operates according to the measurement algorithm to find the coordinates of the aperture edges.

The edge points are used by a circle or ellipse least squares fitting algorithm [116] to determine the aperture area. The combined standard uncertainty of area measurements varies from 4×10^{-4} to 2.5×10^{-5} for aperture diameters from 1 to 50 mm. For the aperture with the diameter of about 5.2 mm, the combined standard uncertainty of the area measurement is equal to 0.0025%. A similar technique was developed at the NMI[13] [135]. With the translation stage available, the combined standard uncertainty of 0.034% was achieved for the aperture diameter of 6 mm.

At the PTB, the absolute non-contact measurement method has been developed [57, 98] also on the base of aperture edge detecting. However, instead of transmitted, the reflected radiation was used. Hartmann [97] described its recent implementation with the radiation beam of a laser diode emitting at 780 nm and focused on the top surface of the diaphragm, which aperture has to be measured.

The reflected radiation is collected by a photodiode. If more than the threshold fraction of the radiant power emitted by the laser is reflected back, the laser focus is outside the aperture. When the reflected signal drops down the threshold value, the aperture edge is detected. A signal of the photodiode indicates when the reflected signal decreases below. The aperture is fixed on a rotation stage, which is mounted on a horizontal stage with sub-micrometer resolution. An interferometer with spatial resolution of 0.1 μm measures the position. Rotating the aperture enables measurements of the aperture diameters at different angular positions. A correction is applied for the diameter of the laser beam spot with the known Gaussian distribution of power across the beam. The system is calibrated using the chromium-on-glass structure with known dimensions and reflectance. An additional correction is required for diaphragms made of materials having optical properties different from chromium. It was proved that, for the aperture of 5 mm in diameter, the combined relative standard uncertainty achievable with this technique is 2×10^{-5} in diameter and is 4×10^{-5} in the area.

Littler et al. [136] reported the comparison results for the reflection-based technique developed at the PTB (Germany) and the transmission-based technique of the NMI (Australia). It was concluded that the measurements performed at both NMIs are in excellent agreement in the values measured, with the measurements agreeing well within the 95% confidence intervals.

A principally different approach to the absolute optical measurement of the aperture areas was proposed by Lassila et al. [130] and Ikonen et al. [111], studied by Stock and Goebel [196], and recently adapted to modern photometry tasks by Dönsberg et al. [40]. The essence of this method is as follows. A spatially uniform, known irradiance over the measured aperture is formed by the overlapping identical and parallel laser beams centered at the nodes of a constant-step orthogonal grid. The ratio of the throughput power and irradiance gives the area of the aperture. The relative standard uncertainty of the area measurement was estimated as 1.6×10^{-4} for a circular aperture of 3 mm in diameter [196].

In addition to the absolute, the relative techniques were developed for less responsible measurements of aperture areas by comparison of radiant fluxes transmitted by the standard and the studied apertures. Fowler and Dezsi [60] achieved the relative

[13] National Measurement Institute, the NMI of Australia (https://www.industry.gov.au/strategies-for-the-future/national-measurement-institute).

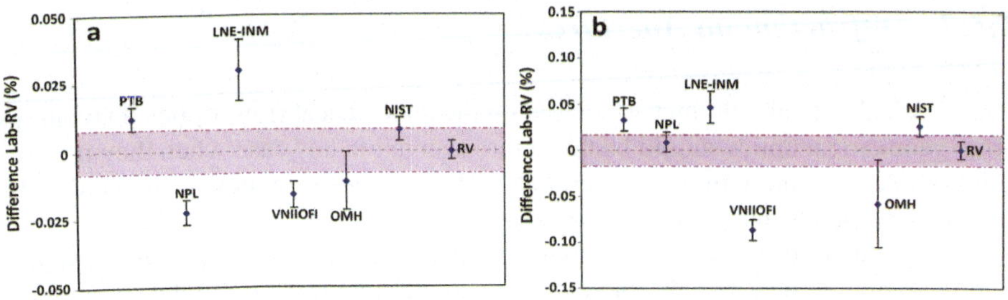

Fig. 9.61 Differences of each laboratory's measurement and the reference value (RV) for the diamond-turned copper aperture with diameters: **a** 25 mm and b **5** mm. The horizontal band denotes expanded uncertainty of the RV at $k = 2$. Reproduced from [134] with permission of IOP Publishing

standard uncertainty of the 0.04%, including the uncertainty of the standard aperture area of 0.022%.

The BIPM CCPR-S2 Supplementary Comparison of Area Measurements of Apertures for Radiometry [134] revealed significant disagreement in measurements carried out by different NMI. Figure 9.61 illustrates the of measurement results obtained by the participating NMIs for the areas of two apertures (diameters of 25 mm and 5 mm).

Error bars of individual results indicate their standard uncertainties; RV is for the reference (mean) value; the acronym OMH is for Office of Measures of Hungaria (existed until 2007); the shadowed bar bracketing the RV represents the standard deviation of the mean of the relative differences. As can be seen in Fig. 9.61, the results did not agree within the uncertainties.

The following conclusions were made on the base of CCPR-S2 comparison.

1. The transfer artifacts (eight circular apertures of different sizes and profiles, made of different materials) were measured by contact and non-contact methods at the participating NMIs, when were possible. The measurement results from the contact methods showed more consistency than the non-contact methods.
2. It is apparent that most NMIs underestimated uncertainty of their measurements.
3. The apertures of smaller areas showed larger relative deviations from the RV regardless of the material. However, the absolute uncertainties in area were comparable for all aperture sizes.
4. Scanning electron micrographs of the edges showed many defects, even for the diamond-turned lands, which are the scattering centers. The contributions to the measurement uncertainty due to scattering may vary with the technique of each laboratory.
5. Some laboratories showed a consistent bias in the measurements.

Saunders et al. [180] recommended increasing all declared uncertainties by at least 3 times. In any case, this issue cannot be considered closed and requires additional investigations.

9.6.7 Diffraction on Apertures

The classical optical radiometry is based on the geometrical (ray) optics. Deviations from geometrical optics should be taken into account in any case when they become comparable with the required uncertainty. This primarily relates to diffraction by apertures, which is manifested in two ways.

Firstly, due to diffraction by an aperture diaphragm located between the source and the radiation detector, the radiant flux actually incident on the detector differs from the flux calculated on the basis of geometric optics. Secondly, a part of the diffracted flux contributes to the scattered radiation, which can reach the detector only after intermediate reflections.

The diffraction effects occur on the different apertures for the different calibration methods and can be interpreted in different ways.

1. The spectral power method. There is no diffraction correction when the FR is calibrated with an underfilling incident beam (see Fig. 9.18a). When the calibrated FR is used to measure the blackbody temperature, the diffraction occurs at both source and detector apertures shown in Fig. 9.62.
2. Irradiance method. When the FR is calibrated against the transfer detector using the near-collimated beam of monochromator (see, e.g. Fig. 9.25), the diffraction loss at a single aperture (Fig. 9.63) should be considered. At the measurement of the blackbody temperature, the two-aperture diffraction must be taken into account (Fig. 9.62).
3. Hybrid method. For the hybrid method, diffraction effects are considered as the part of the absolute SSE determined experimentally.
4. Radiance method. For the radiance method, diffraction has to be accounted for the apertures used with the TD for its calibration. Diffraction at the FR aperture is considered as the part of the relative SSE and is determined experimentally.

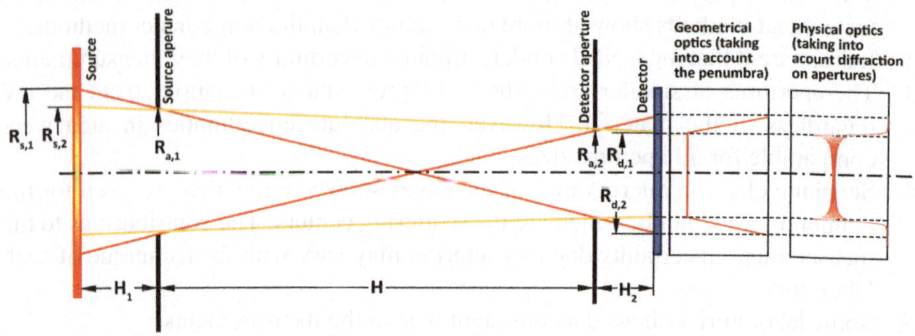

Fig. 9.62 Typical geometrical configuration for calculation of diffraction loss at the aperture pair. Irradiance distributions calculated in the frameworks of geometric and wave optics are shown on the right

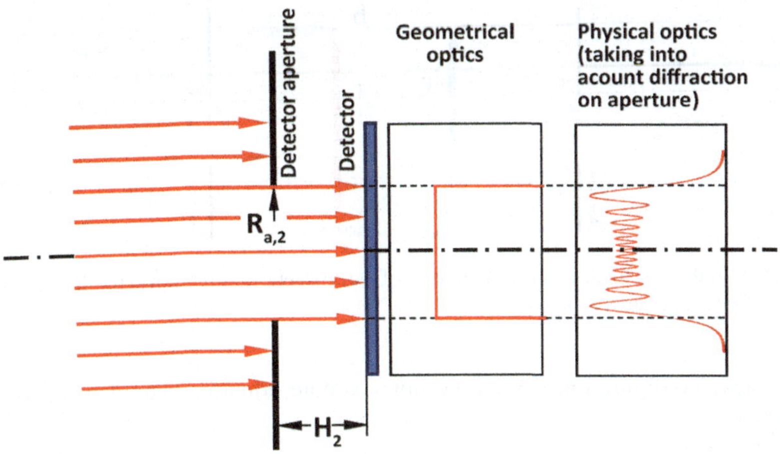

Fig. 9.63 Geometrical configuration for calculation of diffraction loss at a single aperture. Irradiance distributions calculated in the frameworks of geometrical and wave optics are shown on the right

As can be seen, evaluation of deviation (due to diffraction) of the measured radiant flux from that derived from the geometric optics is required in each calibration method. This makes diffraction calculation very important in the APRT, as, however, it is important for optical radiometry as a whole.

Diffraction at a single aperture results in a difference δ between the measured radiation flux, Φ_{rad}, and the flux calculated by means of geometrical optics, Φ_{geom}:

$$\Phi_{rad} = (1 + \delta)\Phi_{\text{geom}}. \tag{9.50}$$

Generally, the deviation δ is small for a large ratio of aperture diameter to wavelength, but can have a positive or negative sign. For typical measurement configurations in optical radiometry, radiation overfills the aperture and $0 < \delta \ll 1$, i.e. we deal with diffraction loss of the radiant flux.

By the end of the 1960s, the accuracy of radiometric measurements reached a level, at least in the NMI, when diffraction effects can no longer be ignored. A survey of the early works on evaluation of diffraction losses was made by Shirley [188].

Steel et al. [193] presented the approximate analytical expression for diffraction correction for a coaxial system consisting of a uniform circular source

Despite some inaccuracies later corrected by Boivin [16–18], this work is still one of the most influential in optical radiometry. Figure 9.64 presents two common configurations that are mutually equivalent due to the Helmholtz reciprocity.

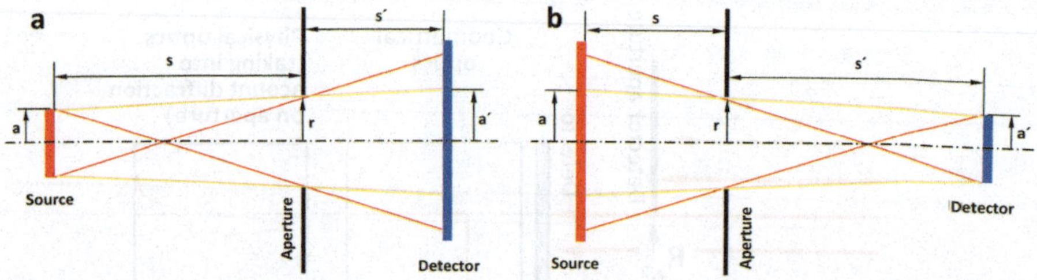

Fig. 9.64 Simplified schemes for the calculation of the diffraction correction at the detector aperture (top) and source aperture (bottom)

For these configurations, a simple approximate equation was derived by Steel et al. [193]:

$$\delta(u, w, w') \approx -\frac{1}{2\pi w} \ln \frac{(w' + w)^2 - u^2}{(w' - w)^2 - u^2}, \tag{9.51}$$

where $u = \frac{2\pi r^2}{\lambda}\left(\frac{1}{s} + \frac{1}{s'}\right)$, $w = \frac{2\pi ra}{\lambda s}$, and $w' = \frac{2\pi ra'}{\lambda s'}$.

This expression was used by Anhalt [6], Noulkhow [154], and Hartmann et al. [103] for calculation of the diffraction correction in the calibration of the FR operating in the irradiance mode. For calculation of the diffraction correction in the blackbody measurement, when diffraction occurs at both of the two apertures defining the measurement geometry, the following heuristic approximate equation was adopted:

$$\delta_{\text{aperture},1+2} = \delta_{\text{aperture},1} + \delta_{\text{aperture},\,2}, \tag{9.52}$$

where two summands in the right-hand parts of Eq. 9.52 correspond to reciprocal configurations shown in Fig. 9.64.

Calculation of diffraction losses with an accuracy sufficient for the vast majority of problems in optical radiometry can be performed using the Kirchhoff scalar theory of diffraction. It is described in detail in many monographs and textbooks on optics, e.g. Guenther [93], Born and Wolf [24], Brooker [26], Römer [170], Hecht [105].

The electromagnetic field coming from a point source P_0 and diffracted at the circular aperture A, in the point P (see Fig. 9.65) is expressed by the Fresnel-Kirchhoff diffraction formula:

$$U(P) = -\frac{iA}{2\lambda} \iint_A \frac{e^{ik(r+s)}}{rs} [\cos(n, r) - \cos(n, s)]\, dS, \tag{9.53}$$

where $U(P)$ is the complex function, $k = 2\pi/\lambda$, λ is the wavelength, and $i^2 = -1$.

Despite apparent simplicity of Eq. 9.53, the calculation of a wave field $U(P)$ are extremely time-consuming due to oscillatory behavior of the integrand. The shorter the wavelength, the higher the oscillation frequency and the greater the number of

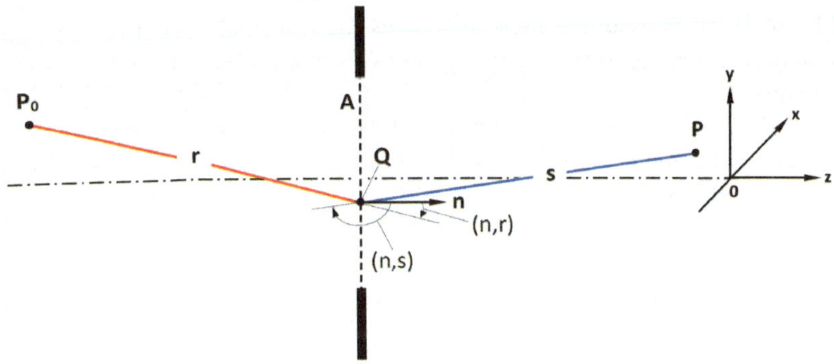

Fig. 9.65 Geometrical scheme for the Fresnel-Kirchhoff diffraction formula

nodes on the aperture is necessary achieve acceptable accuracy in the numerical calculation of the integral in (9.53). Once the integral is calculated, we can find the irradiance E produced by a point P_0 at a point P that is equal to the squared wave field (up to a scaling factor):

$$E(P) \sim |U(P)|^2 = U(P)U^*(P), \tag{9.54}$$

where the U^* is the complex value conjugated to U.

In order to obtain a diffracted field produced by an extended source, it is necessary to integrate $U(P)$ over the entire area of the source. Finally, in order to find the radiant flux received by the detector, the distribution of irradiance in the plane of the detector must be integrated over the detector area. Therefore, in order to find the correction for the diffraction effect at one wavelength, it is necessary to calculate the triple surface integral. Fortunately, in the APRT, we deal with interference filters having a very narrow bandpass. Therefore, there is no need to carry out additional integration over the spectral range of the source radiation. It is enough to compute the diffraction loss for the mean effective wavelength or even the center wavelength of the filter [14]. This can be not true for the broadband radiation (see, e.g. [16, 187, 190]).

A straightforward approach based on the multidimensional numerical integration is practically unrealizable not only because of unacceptable calculation time, but also because of the inevitable accumulation of rounding errors.

To solve this problem, a number of approximate methods have been proposed. The most common is the Fresnel approximation, which allows one to fairly accurately calculate diffraction losses for the vast majority of problems in optical radiometry.

In the Fresnel approximation, the finite distances P_0Q and QP are large compared to the wavelength and the diameter of the aperture. Therefore, we should assume that

$$P_0Q, QP \gg 1/k. \tag{9.55}$$

Additionally, it is assumed these distances their normal derivatives do not vary appreciably over the aperture, so that a part of integrand can be taken outside the integral sign.

Fresnel diffraction theory is the paraxial form of Kirchhoff theory in compliance with the above geometric conditions. The Fresnel-Kirchhoff diffraction formula is expressed as follows:

$$U(x, y) = \frac{\exp[ik(z_P - z_Q)]}{i\lambda(z_P - z_Q)} \iint_A U(x_Q, y_Q) \exp\left[\frac{ik}{2ik(z_P - z_Q)}(x_P - x_Q)^2 + (y_P - y_Q)^2\right] dS, \tag{9.56}$$

where (x_P, y_P, z_P) and (x_Q, y_Q, z_Q) are the Cartesian coordinates of points P and Q, respectively.

The most comprehensive study of the Fresnel diffraction in a simple radiometric system consisting of coaxial circular source, aperture, and detector was undertaken by Edwards and McCall [48] and Edwards et al. [49]. Figure 9.66 shows the geometrical configurations considered. They differ from each other by the position of the aperture. Most previous studies have examined the case F1; the case F2 (the case of non-limiting aperture) mainly was ignored; Edwards and McCall [48] examined also F2 and F3, the most difficult cases to analyze. The two configurations in the top row of Fig. 9.66 are equivalent (configuration F1) because the source and detector are interchangeable according to the Helmholtz reciprocity principle. In these cases, either the source cannot be "visible" entirely from any point of the detector, or any

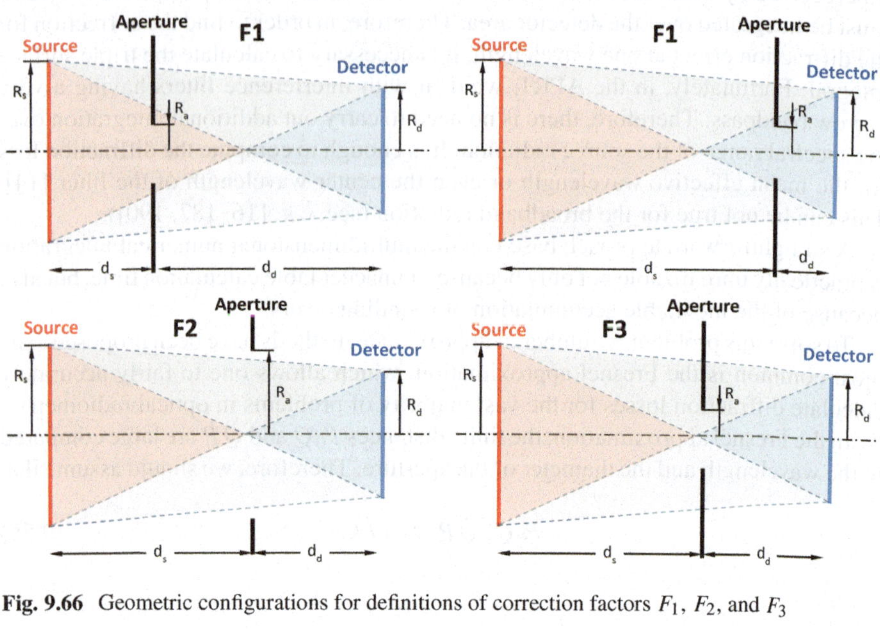

Fig. 9.66 Geometric configurations for definitions of correction factors F_1, F_2, and F_3

point on the source can irradiate only a part of the detector surface. The configuration F2 corresponds to non-limiting aperture, when each point on the detector can "see" entire source and vice versa. In this case, the geometric irradiation of the detector is unaffected by the aperture, while the diffracted radiation alternates the detector's irradiance. Finally, the configuration F3 corresponds to the case, in which the detector is partially filled by the penumbra. To describe the configurations shown in Fig. 9.66, it is convenient to define the parameters u, v_s, and v_d:

$$u = \frac{2\pi r_a^2}{\lambda}\left(\frac{1}{d_s} + \frac{1}{d_d}\right), \tag{9.57}$$

$$v_s = \frac{2\pi r_a r_s}{\lambda d_s}, \tag{9.58}$$

$$v_d = \frac{2\pi r_a r_d}{\lambda d_d}, \tag{9.59}$$

where λ is the wavelength in *vacuo*, the parameters r_s, r_a, r_d, d_s, and d_d are also defined in Fig. 9.66. For ease of geometry classification and total flux calculation, it is useful to introduce the following four parameters:

$$v_0 = \max(v_s, v_d), \tag{9.60}$$

$$\omega = \frac{\min(v_s, v_d)}{\max(v_s, v_d)}, \tag{9.61}$$

$$\phi = \max(u, v_0), \tag{9.62}$$

$$\xi = \frac{\min(u, v_0)}{\max(u, v_0)}. \tag{9.63}$$

Any geometric configuration can be completely specified by three angular parameters (u,v_0, and ω) that incorporate wavelength. The parameter u is interpreted as the angular size of the aperture, v_0 as the angular size of the larger of source and detector, and ω is the ratio of the angular size of the smaller of source and detector to the larger of them. The areas of change of defining angular parameters are $0 \le \omega \le 1$, $0 \le u < \infty, 0 \le v_0 < \infty$, although, in practice, the paraxial scalar diffraction theory limits the upper values for u and v_0. For configurations F1 and F2, the radiant flux Φ_{geom} can be easily expressed via the angular parameters:

$$\text{F1},\ v_0(1-\omega) \ge u:\quad \Phi_{\text{geom}} = L \cdot \left(\frac{u}{v_0}\frac{\pi r_s r_d}{d_s + d_d}\right)^2 = L \cdot \left(\frac{\lambda v_0^2 \omega}{2\phi}\right)^2, \tag{9.64}$$

$$\text{F2},\ v_0(1+\omega) \ge u:\quad \Phi_{\text{geom}} = L \cdot \left(\frac{\pi r_s r_d}{d_s + d_d}\right)^2 = L \cdot \left(\frac{\lambda v_0^2 \omega}{2\phi}\right)^2. \tag{9.65}$$

A more complicated expression for taking into account the penumbra effect was obtained for the configuration F3:

$$\text{F3, } \nu_0(1-\omega) < u < \nu_0(1+\omega): \quad \Phi_{geom} = \left(\frac{\lambda}{2}\right)^2 \left[2u^2 \int_{\frac{\phi(1-\xi)}{u}}^{\frac{\omega\nu_0}{u}} A(u_0/u, x)x dx + \nu_0^2(1-\xi)^2 \right], \tag{9.66}$$

where

$$A(a,b) = \frac{1}{\pi}\left\{ cos^{-1}\left(\frac{1+b^2-a^2}{2b}\right) + a^2 cos^{-1}\left(\frac{b^2-1+a^2}{2ba}\right) - b\left[1 - \frac{\left(1+b^2-a^2\right)^2}{4b^2}\right]^{1/2} \right\}. \tag{9.67}$$

Edwards and McCall [48] derived expressions for the radiant fluxes Φ_1, Φ_2, and Φ_3 (accounted for diffraction) that reach the detector in the case of each geometrical configuration. We will save the reader from the need to deal with sophisticated mathematical expressions for the radiant fluxes containing infinite series and special functions, especially since some programming efforts are still required to obtain numerical results. Instead, we provide the contour maps (Fig. 9.67) plotted for four values of ω showing the diffraction correction values for a range of u and ν_0. The behavior of the diffraction correction as a function of physical parameters can be deduced from these maps. For small values of ν_0, the diffraction loss is highly structured and changes greatly in magnitude, whereas for the case in which u is small, the change is smooth and gradual. Diffraction loss trends in the F1 and F2 regions are approximately independent of the value of ω. The exception is with the F1 region that is no longer present when $\omega = 1$. The contour maps corresponding to small values of ω give insight into why geometries that fall into the F3 classification are generally avoided by practitioners of radiometry. The diffraction loss fluctuations in the F3 region are complex and large in magnitude; only small changes in u and ν_0 are required for the diffraction loss to change of the order of 1%. The situation is reversed when the value of ω approaches 1 and the angular sizes of u and ν_0 are large. Now a large portion of the F3 region has highly oscillatory fluctuations, but significantly these fluctuations are small in magnitude and the diffraction loss parameter remains close to unity. Generally, the value of diffraction loss is approximately equal to 1, when $u \approx \nu_0$.

The effect on diffraction loss that is due to a change in a physical parameter of the geometry can also be investigated by use of the contour maps. For lack of a more suitable example, we give an example from Edwards and McCall [48] for the hypothetic case of $\lambda = 5.8$ μm, $r_s = 1$ mm, $r_d = 20$ mm, $d_s = d_d = 1$ m. In Fig. 9.68a, the aperture radius is increased (all other dimensions being fixed) along the dotted curve. The values of the correction factor calculated along this curve is shown in Fig. 9.68b. The curve crosses the zero diffraction loss regime when $u \approx \nu_0$ and

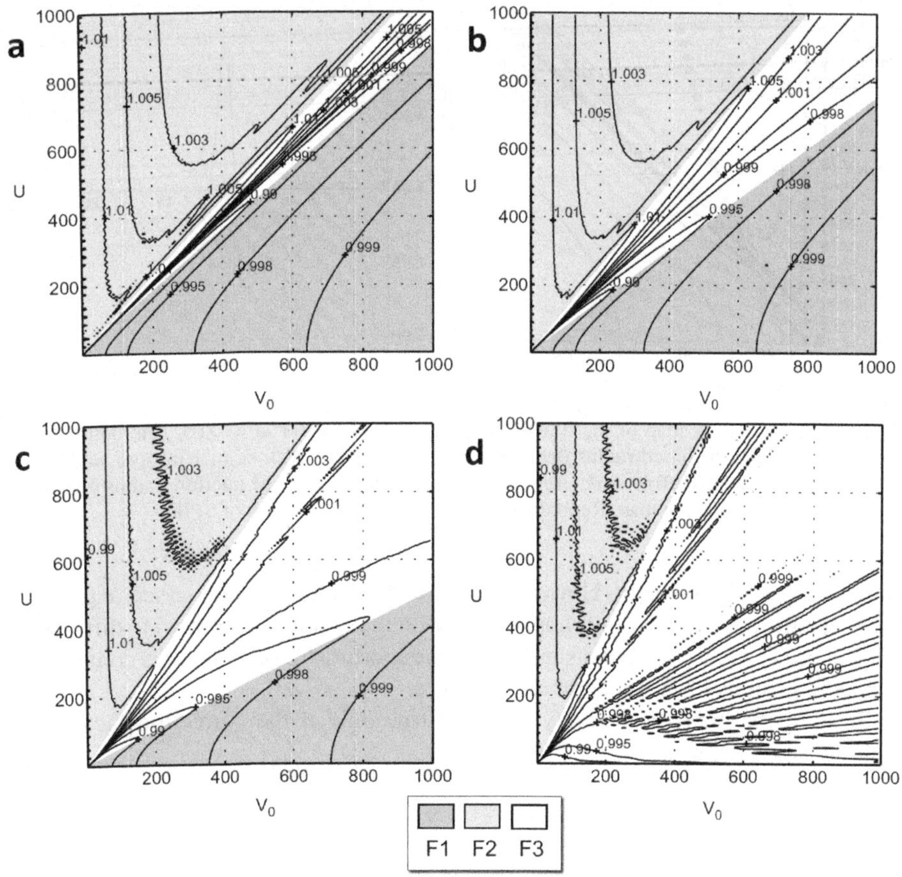

Fig. 9.67 Contour maps of the diffraction correction factor for $0 \le u, v_0 \le 1000$ and $\omega = 0.1$, 0.25, 0.5, and 1.0 (**a** to **d**, respectively). The grayscale coding is applied to geometry configuration F1, F2, and F3. Reproduced from [49] with permission of SPIE

clearly shows the large changes in the value of diffraction correction factor through the F3 region when $\omega \ll 1$.

The diffraction correction can bring a substantial contribution to the uncertainty budget of the FR calibration and/or measurement. For instance, Hartmann [97] provided the diffraction correction values (expressed in temperatures) for three FRs with the center wavelengths of 676, 800, and 900 nm (the spectral irradiance responsivities of these FRs were shown in Fig. 9.24), operating in the irradiance mode. For measuring the temperature from 2748 to 3155 K, the diffraction corrections are from 0.65 to 1.14 K.

According to the *MeP*-K document [180], the uncertainty in the diffraction correction is typically of order 0.02%. We believe this estimate is too optimistic. It can be correct only if we assume that the adequate mathematical model of diffraction is

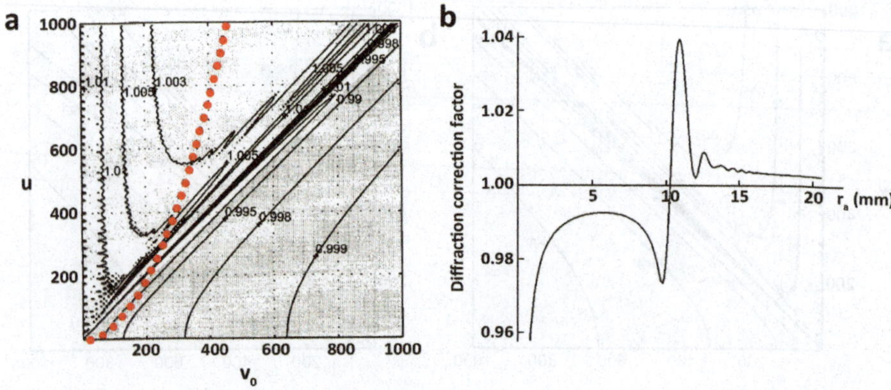

Fig. 9.68 **a** A contour map of the diffraction correction factor for $\omega = 0.05$. The dotted curve corresponds to the increased radius of the aperture for $\lambda = 5.8$ μm, $r_s = 1$ mm, $r_d = 20$ mm, $d_s = d_d = 1$ m and **b** Diffraction correction factor as a function of the aperture radius computed along the dotted line in the map. Reproduced from [49] with permission of SPIE

built and applied correctly. However, there are reasonable doubts (see, e.g. [147, 148]) that the paraxial theory is applicable (with the required accuracy) to diffraction at the aperture of an FR if the distance from the aperture to the detector is comparable to the dimensions of the detector and aperture.

Besides, the heuristic equality (9.52) is true only if the correction factors are multiplicative, i.e. if

$$1 + \delta_{1,2} = (1 + \delta_1)(1 + \delta_2). \tag{9.68}$$

However, this hypothetical property of the correction factors has never been proven. In recent decades, a number of studies have been published on calculating the diffraction effect in a coaxial system containing more than one aperture [171, 186, 189, 197, 198]. Equation 9.66 does not follow from any of these works.

The experimental verification mentioned in Saunders et al. [180]—varying the distance between the two apertures—can confirm the calculated diffraction loss only up to a scaling factor. A direct experimental check for the calculated diffraction correction is extremely difficult; to our knowledge, no such experimental studies were published. We believe that the problem of estimating diffraction losses cannot yet be considered complete.

9.7 Uncertainty Estimation in Measuring Thermodynamic Temperatures of Blackbodies

9.7.1 Introductory Notes

Any method of the APRT include two stages:

- calibration of the FR, at which its absolute spectral responsivity $s(\lambda)$ and associated uncertainties are determined;
- measuring the thermodynamic temperature of the blackbody when the resulting temperature uncertainty has to be evaluated.

Analyzing the generic measurement equation (9.31), we see that the uncertainties in the determination of $s(\lambda)$ do not directly affect the measured signal S, but are integrated with the Planck function. Assuming that the spectral responsivity $s(\lambda)$ differs from zero only in the wavelength interval $\lambda_1 \leq \lambda \leq \lambda_N$, Eq. 9.31 can be rewritten as

$$S(T) = K \int_{\lambda_1}^{\lambda_N} s(\lambda)\, L_{\lambda,bb}(\lambda, T) d\lambda, \tag{9.69}$$

where K is the correction factor for systematic effects and $L_{\lambda,bb}(\lambda, T)$ is the spectral radiance of the perfect blackbody at the temperature T.

The generic measurement equations (9.69) relate the measured signal S to the blackbody temperature T. This equation contains some peculiarities, that are not covered by GUM [112].

- This expression cannot be written in the explicit form $T = f(x_1, x_2, \ldots)$; therefore, to apply the GUM approach, the implicit differentiation should be used.
- The temperature T must be extracted from the function, which, multiplied by the spectral sensitivity $s(\lambda)$ of the FR, is placed under the integral sign.
- The spectral responsivity $s(\lambda)$ is the function of wavelengths, in which both λ and $s(\lambda)$ are subjects to random and systematic errors.
- The spectral responsivity $s(\lambda)$ is known on the discrete set of wavelengths λ_i, $i = 1, 2, \ldots, N$. The distribution of nodal wavelengths in the interval $[\lambda_i, \lambda_N]$ is important here, since it determines the choice of the quadrature formula for numerical integration.

The straightforward solution to all these problems is the Monte Carlo simulation as recommended in the Supplement 1 to the GUM [113]. However, since the overall uncertainty of thermodynamic temperature measurements is calculated using the traditional GUM approach [112], the authors of MeP-K [180], in order to ensure overall compatibility, were forced to remain within the GUM approach, accepting the uncertainty propagation paradigm and analytical calculation of sensitivity coefficients also for integrated spectral quantities. They described two purely GUM-based methods.

The first method is based on the generic measurement equation (9.69). For a given measured value of S, the value of T can be found from (9.69) using a suitable numerical technique, e.g. the iterative Newton-Raphson method [164]:

$$T_{j+1} = T_j + \frac{S - K\int_{\lambda_1}^{\lambda_N} s(\lambda) L_{\lambda,bb}(\lambda, T_j)\, d\lambda}{\frac{c_2 K}{T_j^2}\int_{\lambda_1}^{\lambda_N} \frac{n(\lambda)s(\lambda)L_{\lambda,bb}(\lambda,T_j)}{\lambda\left[1-\exp\left(-\frac{c_2}{n(\lambda)\lambda T_j}\right)\right]} d\lambda}, \quad j = 0, 1, 2, \ldots, \tag{9.70}$$

where T_0 is a physically plausible zero approximation (initial guess) and n is the refractive index of air. To evaluate the temperature uncertainty, the equivalent closed-form implicit equation (9.69) is used.

An alternative method proposed by Saunders [179] uses the Planck version of the inverse Sakuma-Hattori (SH) equation instead of Eq. 9.69:

$$T = \frac{c_2}{A \ln\left(\frac{C}{S} + 1\right)} - \frac{B}{A}, \tag{9.71}$$

where A, B, and C are the key spectral parameters determined at the FR calibration. Equation 9.71 allows expressing T via the explicit closed-form equation. However, as we will see below, evaluation of uncertainty in integrated spectral quantities is an inevitable stage in the uncertainty propagation through Eq. 9.71.

Therefore, whichever approach of these two (based on the generic measurement equation or SH equation) is used, we must be able to calculate uncertainty propagation through the integrals of spectral quantities. In this regard, we devoted the next Section to this problem.

In both approaches, the same elementary components of the calibration uncertainty, even if grouped in different ways, contribute to the combined standard uncertainty $u_c(T)$. Measuring the thermodynamic temperature of a blackbody leads to additional components of uncertainty due to the FR (linearity and temperature sensitivity) and due to the blackbody (determination of the phase transition for FPBBs, stability and uniformity for VTBBs, and effective emissivity for both types). The "normal" and "best" values for these components are presented in Table 9.6.

9.7.2 *Uncertainty of the Integrated Spectral Quantity*

The problem of determining the uncertainty associated with the integral of the spectral quantity is formulated as follows. We have a spectral quantity, whose values q are measured at any wavelength $\lambda \in [\lambda_1, \lambda_N]$. Both λ and $q(\lambda)$ are subject of random and systematic errors. We need to find the uncertainty of the quantity, whose value

Table 9.6 Typical contributions to the uncertainty budget for the measuring the blackbody thermodynamic temperature. All components quoted as relative standard uncertainties (after [180])

Uncertainty Component	"Normal" standard uncertainty (%)	"Best" standard uncertainty (%)	Notes
Linearity	0.01	0.002	
FR temperature stability	0.01	0.002	
Effective emissivity of blackbody	0.008	0.008	
Source uniformity	0.02	0.01	For VTBB
Source stability	0.12	0.03	For VTBB
Fixed-point transition definition	See [58, 220]		For FPBB

is expressed as

$$\bar{q} = \int_{\lambda_1}^{\lambda_2} q(\lambda) d\lambda. \tag{9.72}$$

The quantification of uncertainty of integrated spectral quantity is a very common problem in optical radiometry, photometry, colorimetry, and radiation thermometry [36, 117, 162, 184]. In most cases, the solution of this problem is found by the Monte Carlo method, according to the Supplement 1 to the GUM [113].

Woolliams [215, 216] and Saunders [179] proposed similar GUM-based techniques to evaluate of the uncertainty propagation through Eq. 9.69. The integral in (9.69) is substituted with a finite sum using the trapezoidal quadrature rule. The wavelength and responsivity measurements are modelled as:

$$\lambda_i = \lambda_{\text{true},i} + \lambda_{\text{rand},i} + \lambda_{\text{sys}}, \tag{9.73}$$

$$s_i = s_{true,i}\left(1 + s_{rand,i} + s_{sys}\right), \tag{9.74}$$

where $\lambda_{true,i}$ and $\lambda_{rand,i}$, are the unknown true value and random error of the i-th wavelength, respectively; λ_{sys} is the unknown systematic error common to all wavelength; $s_{true,i}$ and $s_{rand,i}$ are the unknown true value and relative random error of i-th spectral responsivity, respectively; s_{sys} unknown systematic error common to all spectral responsivity measurements.

Each of these errors is characterized by zero expectation and uncertainties $u\left(\lambda_{rand,i}\right)$, $u\left(\lambda_{sys}\right)$, $u_{rel}\left(s_{rand,i}\right)$, and $u_{rel}\left(s_{sys}\right)$. The expectations of $\lambda_{true,i}$ is λ_i and the expectation of $s_{true,i}$ is s_i.

Further exposition follows Saunders [179], who described the approximation of the integral in (9.69) using trapezoidal rule for arbitrarily distributed nodal wavelength λ_i on the interval $[\lambda_1, \lambda_N]$. Equation 9.69 can be rewritten as

$$S = \frac{1}{2} K \sum_{i=1}^{N} s_i L_{\lambda,bb}(\lambda_i, T) \Delta\lambda_i , \tag{9.75}$$

where

$$\Delta\lambda_i = \begin{cases} \lambda_2 - \lambda_1 & \text{if } i = 1 \\ \lambda_{i+1} - \lambda_{i-1} & \text{if } 1 < i < N \\ \lambda_N - \lambda_{N-1} & \text{if } i = N \end{cases} . \tag{9.76}$$

Substituting Eqs. 9.73 and 9.74 into Eq. 9.75 gives:

$$\begin{aligned} S = \tfrac{K}{2} [& s_{\text{true},1} (1 + s_{\text{ran},1} + s_{sys}) L_{\lambda,bb}(\lambda_{\text{true},1} + \lambda_{\text{ran},1} + \lambda_{sys}, T)(\lambda_{\text{true},2} + \lambda_{\text{ran},2} - \lambda_{\text{true},1} - \lambda_{\text{ran},1}) \\ & + \sum_{i=2}^{N-1} s_{\text{true},i} (1 + s_{\text{ran},i} + s_{sys}) L_{\lambda,bb}(\lambda_{\text{true},i} + \lambda_{\text{ran},i} + \lambda_{sys}, T)(\lambda_{\text{true},i+1} + \lambda_{\text{ran},i+1} - \lambda_{\text{true},i-1} - \lambda_{\text{ran},i-1}) \cdot \\ & + s_{\text{true},N} (1 + s_{\text{ran},N} + s_{sys}) L_{\lambda,bb}(\lambda_{\text{true},N} + \lambda_{\text{ran},N} + \lambda_{sys}, T)(\lambda_{\text{true},N} + \lambda_{\text{ran},N} - \lambda_{\text{true},N-1} - \lambda_{\text{ran},N-1})] \end{aligned} \tag{9.77}$$

Equation 9.77 allows analytical differentiating with respect to each of $\lambda_{\text{ran},i}$, λ_{sys}, $s_{\text{ran},i}$, s_{sys}, K, and T:

$$\frac{\partial S}{\partial \lambda_{\text{ran},i}} = \frac{K}{2} \times \begin{cases} s_1 \left. \frac{\partial L_{\lambda,bb}(\lambda,T)}{\partial \lambda} \right|_{\lambda=\lambda_1} \Delta\lambda_1 - s_1 L_{\lambda,bb}(\lambda_1, T) - s_2 L_{\lambda,bb}(\lambda_2, T) & \text{if } i = 1 \\ s_i \left. \frac{\partial L_{\lambda,bb}(\lambda,T)}{\partial \lambda} \right|_{\lambda=\lambda_i} \Delta\lambda_i - s_{i-1} L_{\lambda,bb}(\lambda_{i-1}, T) - s_{i+1} L_{\lambda,bb}(\lambda_{i+1}, T) & \text{if } 1 < i < N \\ s_N \left. \frac{\partial L_{\lambda,bb}(\lambda,T)}{\partial \lambda} \right|_{\lambda=\lambda_N} \Delta\lambda_N - s_{N-1} L_{\lambda,bb}(\lambda_{N-1}, T) - s_N L_{\lambda,bb}(\lambda_N, T) & \text{if } i = N \end{cases} , \tag{9.78}$$

$$\frac{\partial S}{\partial \lambda_{sys}} = \frac{K}{2} \sum_{i=1}^{N} s_i \Delta\lambda_i \left. \frac{\partial L_{\lambda,bb}(\lambda, T)}{\partial \lambda} \right|_{\lambda=\lambda_i} , \tag{9.79}$$

$$\frac{\partial S}{\partial s_{\text{ran},i}} = \frac{K}{2} s_i \Delta\lambda_i L_{\lambda,bb}(\lambda_i, T), \tag{9.80}$$

$$\frac{\partial S}{\partial s_{sys}} = S, \tag{9.81}$$

$$\frac{\partial S}{\partial K} = \frac{S}{K}, \tag{9.82}$$

$$\frac{\partial S}{\partial T} = \frac{K}{2} \sum_{i=1}^{N} s_i \Delta\lambda_i \left. \frac{\partial L_{\lambda,bb}(\lambda, T)}{\partial T} \right|_{\lambda=\lambda_i} , \tag{9.83}$$

where

$$\left.\frac{\partial L_{\lambda,bb}(\lambda,T)}{\partial\lambda}\right|_{\lambda=\lambda_i} = \frac{L_{\lambda,bb}(\lambda_i,T)}{\lambda_i}\left\{\frac{c_2}{n\lambda_i T\left[1-\exp\left(-\frac{c_2}{n\lambda_i T}\right)\right]}-5\right\}, \approx \frac{L_{\lambda,bb}(\lambda_i,T)}{\lambda_i}\left(\frac{c_2}{n\lambda_i T}-5\right) \tag{9.84}$$

$$\left.\frac{\partial L_{\lambda,bb}(\lambda,T)}{\partial T}\right|_{\lambda=\lambda_i} = \frac{c_2 L_{\lambda,bb}(\lambda_i,T)}{n\lambda_i T^2\left[1-\exp\left(-\frac{c_2}{n\lambda_i T}\right)\right]} \approx \frac{c_2 L_{\lambda,bb}(\lambda_i,T)}{n\lambda_i T^2} \tag{9.85}$$

n is the refractive index of air assumed to be wavelength-independent. The sign $\approx$ denotes the Wien approximation of the Planck radiation law.

9.7.3 Uncertainty Propagation Using Generic Measurement Equation

The combined standard uncertainty associated with the temperature T is calculated in accordance with the law of propagation of uncertainties [112]:

$$u_c(T) = \left\{\sum_i\left[\frac{\partial T}{\partial x_i}u(x_i)\right]^2 + \sum_i\sum_{j,j\neq i}\frac{\partial T}{\partial x_i}\frac{\partial T}{\partial x_j}r(x_i,x_j)\,u(x_i)u(x_j)\right\}^{\frac{1}{2}}, \tag{9.86}$$

where x_i denote one of $\lambda_{\mathrm{ran},i}$, λ_{sys}, $s_{\mathrm{ran},i}$, s_{sys}, K, and S; $r(x_i, x_j)$ is the correlation coefficient between the uncertainties $u(x_i)$ and $u(x_j)$. The sensitivity coefficients for temperature are obtained by applying the rule for implicit differentiation:

$$\frac{\partial T}{\partial x_i} = -\frac{\partial S}{\partial x_i}\Big/\frac{\partial S}{\partial T} \tag{9.87}$$

with x_i formally substituted with one of variables listed above. We omit very cumbersome intermediate expressions, the more especially as the interested reader can always refer to the original CCT document.

Saunders [179] assumed that the uncertainties in wavelength and spectral responsivity are either fully correlated or fully uncorrelated, i.e. the correlation coefficients are either 1 or 0:

$$r(\lambda_i, \lambda_j) = \delta_{ij};\ r(\lambda_{\mathrm{ran},i}, \lambda_{sys}) = 0;\ r(\lambda_{sys}, \lambda_{sys}) = 1, \tag{9.88}$$

$$r(s_i, s_j) = \delta_{ij};\ r(s_{\mathrm{ran},i}, s_{sys}) = 0;\ r(s_{sys}, s_{sys}) = 1, \tag{9.89}$$

where δ_{ij} is the Kronecker delta ($\delta_{ij} = 0$ if $i \neq j$ and $\delta_{ij} = 1$ if $i = j$). Then (9.86) is reduced to

Table 9.7 Combined standard uncertainties (in mK) of measuring high-temperature VTBBs for four calibration schemes, propagated from the "normal" and "best" values given in Tables 9.5 and 9.6 for three temperatures of a VTBB (after [180])

FR calibration method	VTBB temperature (°C)					
	1000		2000		3000	
	Normal	Best	Normal	Best	Normal	Best
Power	0.28	0.04	0.58	0.11	1.09	0.225
Irradiance	0.28	0.03	0.56	0.097	1.08	0.200
Hybrid	0.28	0.04	0.58	0.11	1.03	0.223
Radiance	0.28	0.03	0.56	0.095	1.07	0.196

$$u_c(T) = \left\{ \sum_{i=1}^{N} \left[\frac{\partial T}{\partial \lambda_{\mathrm{ran},i}} u\left(\lambda_{\mathrm{ran},i}\right) \right]^2 + \sum_{i=1}^{N} \left[\frac{\partial T}{\partial s_{\mathrm{ran},i}} u_{rel}\left(s_{\mathrm{ran},i}\right) \right]^2 + \left[\frac{\partial T}{\partial \lambda_{sys}} u\left(\lambda_{sys}\right) \right]^2 \right. \\ \left. + \left[\frac{\partial T}{\partial s_{sys}} u_{rel}\left(s_{sys}\right) \right]^2 + \left[\frac{\partial T}{\partial K} u(K) \right]^2 + \left[\frac{\partial T}{\partial S} u(S) \right]^2 \right. . \tag{9.90}$$

Equations 9.77 through 9.87 (including omitted ones) and (9.90) are not suitable for manual calculations; however, they can be easily programmed or implemented in a spreadsheet application.

Saunders et al. [180] referred to the GUM Tree Calculator (GTC), a freeware developed by MSL[14]. Currently, it is available for download at https://measurement.govt.nz/download/46. Working with this program requires certain skills. To handle non-standard situations, at least basic knowledge of Python programming language is needed.

The *MeP*-K document [180] gives the detailed instructions on calculations of the combined standard uncertainty $u_c(T)$, as well as the typical "normal" and "best" values of all contributing components. Table 9.7 presents the calculated uncertainties for measurements of a VTBB at temperature 1000 °C, 2000 °C, and 3000 °C for each of four calibration methods, for the "normal" and "best" uncertainty values, respectively, based on the values given in Tables 9.5 and 9.6. These uncertainties were calculated for uncorrelated uncertainty components and demonstrated a little difference in accuracy of four methods.

In the last years, the most accurate results were obtained using FRs calibrated in terms of radiance responsivity. An additional argument in favor of the radiance method is the ability to use conventional radiation thermometers, recalibrated in terms of absolute radiance responsivity, for measurement of the thermodynamic temperature of blackbodies. Moreover, the radiance approach has the largest potential in further reducing measurement accuracy.

We would give the second place to the irradiance method that, at least, for the VTBB, allows achieving comparable uncertainties. Recently developed large-aperture radiating cells for FPBBs [121, 158, 227] allow up to ten times increasing the

---

[14] Measurement Standards Laboratory, the NMI of New Zealand.

radiant flux emitted by FPBBs, including HTFPs. This, in turn, noticeably increases the SNR, which allows reduce the measurement accuracy. Currently, the greatest component (0.07%) in the uncertainty budgets for calibrating the FR in terms of irradiance responsivity is associated with the diffraction loss on the pair of apertures. Refining the value of the diffraction correction should lead to reducing the uncertainty in measurement of thermodynamic temperature down to the value comparable to that for the radiance method. Since the uncertainty components in the irradiance method differ from those in the radiance method, the independence use of both of them could detect possible unrecognized systematic errors and, as a result, increase confidence of the thermodynamic temperature measurements.

9.7.4 Uncertainty Propagation Using Sakuma-Hattori Equation

An alternative method of propagating the uncertainties in the APRT is to employ the Planck version of the inverse SH Eq. (9.71) instead of Planck's equation to redefine the integral in Eq. 9.69 in terms of key spectral parameters A, B, and C of the FR. Saunders [179] showed that the SH equation introduces the error less than 2 mK for measuring temperatures between 1000 and 3000 °C using an FR typical for the APRT (the central wavelength $\lambda_0 = 0.65$ μm and the standard deviation $\sigma_\lambda = 10$ nm of the spectral responsivity bandwidth).

According to Saunders and White [181], the key spectral parameters can be expressed as

$$A = n\lambda_0\left(1 - 6r^2\right), \tag{9.91}$$

$$B = \frac{c_2 r^2}{2}, \tag{9.92}$$

$$C = K\frac{c_1}{n^2\lambda_0^5}\left(1 + 15r^2\right)\int\limits_{\lambda_1}^{\lambda_N} s(\lambda)d\lambda = \frac{H}{n^2\lambda_0^5}\left(1 + 15r^2\right), \tag{9.93}$$

with the relative bandwidth

$$r = \sigma_\lambda / \lambda_0. \tag{9.94}$$

The values of H, λ_0, and σ can be calculated from $s(\lambda)$ through the following integrals:

$$H = Kc_1 \int_{\lambda_1}^{\lambda_N} s(\lambda) d\lambda, \tag{9.95}$$

$$\lambda_0 = \left[\int_{\lambda_1}^{\lambda_N} \lambda s(\lambda)\, d\lambda \right] \Big/ \int_{\lambda_1}^{\lambda_N} s(\lambda)\, d\lambda, \tag{9.96}$$

$$\sigma_\lambda^2 = \left[\int_{\lambda_1}^{\lambda_N} (\lambda - \lambda_0)^2 s(\lambda)\, d\lambda \right] \Big/ \int_{\lambda_1}^{\lambda_N} s(\lambda)\, d\lambda, \tag{9.97}$$

Without taking into account the correlations, the combined standard uncertainty in T evaluated according to the GUM [112] and the SH equation is equal to

$$u_c(T) = \left\{ \left[\frac{\partial T}{\partial \lambda_0} u(\lambda_0) \right]^2 + \left[\frac{\partial T}{\partial \sigma} u(\sigma_\lambda) \right]^2 + \left[\frac{\partial T}{\partial H} u(H) \right]^2 + \left[\frac{\partial T}{\partial n} u(n) \right]^2 + \left[\frac{\partial T}{\partial S} u(S) \right]^2 \right\}^{\frac{1}{2}}. \tag{9.98}$$

The standard uncertainties $u(\lambda_0)$, $u(\sigma_\lambda)$, $u(H)$, $u(n)$, and $u(S)$, as well as the corresponding sensitivity coefficients in (9.97) are calculated in the same manner as in Sect. 9.7.3. Since the uncertainty $u(n)$ is small, the fourth term in Eq. 9.97 can be neglected. Each of the integrals in Eqs. 9.93 and 9.95 to 9.97 are approximated by the finite sums similar to (9.77). Saunders [179] used the narrowband approximation, i.e. the components of the order of r^2 and higher were neglected. The uncertainties $u(H)$, $u(\lambda_0)$, and $u(\sigma_\lambda)$ are expressed through the uncertainties $u(\lambda_{\text{rand},i})$, $u(\lambda_{sys})$, $u_{rel}(s_{\text{rand},i})$, $u_{rel}(s_{sys})$, and $u(K)$ also in a manner similar to that was used in the previous Section. We again omit bulky analytical transformations, noting only that the resulting ready-to-use formulae are no less cumbersome than those obtained for Eq. 9.69 and also not suitable for manual calculation. Therefore, from the point of view of computational efficiency, the SH equation hardly has an advantage with respect to the generic measurement equation (9.69).

Saunders [179] showed that uncertainties computed using Eqs. 9.90 and 9.97 differ by no more than 0.1 mK in the range from 1000 to 3000 °C for the FR with $\lambda_0 = 0.65$ μm and $\sigma_\lambda = 10$ nm.

References

1. S. Adachi, *Optical Constants of Crystalline and Amorphous Semiconductors. Numerical Data and Graphical Information* (Springer, NY, 1999)
2. S.M. Alessio, *Digital Signal Processing and Spectral Analysis for Scientists. Concepts and Applications* (Springer, Cham, Switzerland, 2016)
3. R.R. Alfano (ed.), *The Supercontinuum Laser Source. The Ultimate White Light*, 3rd edn. (Springer, New York, 2016)

4. D.W. Allen, R.D. Saunders, B.C. Johnson et al., The development and characterization of an absolute pyrometer calibrated for radiance responsivity. AIP Conf. Proc. **684**, 577–582 (2003)
5. V.E. Anderson, N.P. Fox, D.H. Nettleton, Highly stable, monochromatic and tunable optical radiation source and its application to high accuracy spectrophotometry. Appl. Opt. **31**, 536–545 (1992)
6. K. Anhalt, *Radiometric Measurement of Thermodynamic Temperatures During the Phase Transformation of Metal-Carbon Eutectic Alloys for a New High-Temperature Scale Above 1000 °C*. Ph.D. thesis (Technischen Universität Berlin, 2008), https://depositonce.tu-berlin.de/bitstream/11303/2310/1/Dokument_17.pdf. Accessed 30 May 2020
7. K. Anhalt, J. Hartmann, D. Lowe et al., Thermodynamic temperature determinations of Co–C, Pd–C, Pt–C and Ru–C eutectic fixed-point cells. Metrologia **43**, S78–S83 (2006)
8. K. Anhalt, Y. Wang, Y. Yamada, J. Hartmann, Large- and small-aperture fixed-point cells of Cu, Pt-C, and Re-C. Int. J. Thermophys. **29**, 969–983 (2008)
9. K. Anhalt, A. Zelenjuk, D.R. Taubert et al., New PTB setup for the absolute calibration of the spectral responsivity of radiation thermometers. Int. J. Thermophys. **30**, 192–202 (2009)
10. K. Anhalt, G. Machin, Thermodynamic temperature by primary radiometry. Philos. Trans. R. Soc. A **374**, 20150041 (2016)
11. A. Antoniou, *Digital Signal Processing. Signals, Systems and Filters* (McGraw-Hill, New York, 2006)
12. T.H.K. Barron, G.K. White, *Heat Capacity and Thermal Expansion at Low Temperatures* (Springer, New York, 1999)
13. G. Bianco, M.A. Ferrara, F. Borbone et al., Volume holographic gratings: fabrication and characterization. Proc. SPIE **9508**, 950807 (2015)
14. W.R. Blevin, Diffraction losses in radiometry and photometry. Metrologia **6**, 39–44 (1970)
15. W.R. Blevin, W.J. Brown, A precise measurement of the Stefan-Boltzmann constant. Metrologia **7**, 15–29 (1971)
16. L.P. Boivin, Diffraction losses associated with tungsten lamps in in absolute radiometry. Appl. Opt. **14**, 197–200 (1975)
17. L.P. Boivin, Diffraction corrections in radiometry: comparison of two different methods of calculation. Appl. Opt. **14**, 2002–2009 (1975)
18. L.P. Boivin, Diffraction corrections in the radiometry of extended sources. Appl. Opt. **15**, 1204–1209 (1976)
19. L.-P. Boivin, Spectral responsivity of various types of silicon photodiode at oblique incidence: comparison of measured and calculated values. Appl. Opt. **40**, 485–491 (2001)
20. L.P. Boivin, Realization of spectral responsivity scales, in *Optical Radiometry*, ed. by A.C. Parr, R.U. Datla, J.L. Gardner (Academic Press, Amsterdam, 2005), pp. 97–154
21. L.P. Boivin, K. Gibb, Monochromator-based cryogenic radiometry at the NRC. Metrologia **32**, 565–570 (1995/96)
22. L.P. Boivin, C. Bamber, A.A. Gaertner et al., Wideband filter radiometers for blackbody temperature measurements. J. Mod. Opt. **57**, 1648–1660 (2010)
23. R.L. Booker, J. Geist, Induced junction (inversion layer) photodiode self-calibration. Appl. Opt. **23**, 1940–1945 (1984)
24. M. Born, E. Wolf, *Principles of Optics. Electromagnetic Theory of Propagation, Interference and Diffraction of Light*, 7th (expanded) edn. (Cambridge University Press, Cambridge, UK, 1999)
25. S. Briaudeau, M. Sadli, F. Bourson et al., Primary radiometry for the *mise-en-pratique*: the laser-based radiance method applied to a pyrometer. Int. J. Thermophys. **32**, 2183–2196 (2011)
26. G. Brooker, *Modern Classical Optics* (Oxford University Press, Oxford, UK, 2003)
27. S.W. Brown, G.P. Eppeldauer, K.R. Lykke, NIST facility for spectral irradiance and radiance responsivity calibrations with uniform sources. Metrologia **37**, 579–582 (2000)
28. S.W. Brown, G.P. Eppeldauer, J.P. Rice et al., Spectral irradiance and radiance responsivity calibrations using uniform sources (SIRCUS) facility at NIST. Proc. SPIE **5542**, 363–374 (2004)

29. S.W. Brown, G.P. Eppeldauer, K.R. Lykke, Facility for spectral irradiance and radiance responsivity calibrations using uniform sources. Appl. Opt. **45**, 8218–8237 (2006)
30. L. Bünger, R.D. Taubert, K. Anhalt, Spectrally resolved radiometric characterization and calibration using a supercontinuum laser in the NIR, in *AMA Conferences 2015—SENSOR 2015 and IRS^2 2015* (2015), pp. 895–898. https://doi.org/10.5162/irs2015/1.2
31. L. Bünger, R.D. Taubert, B. Gutschwager et al., Absolute radiation thermometry in the NIR. Int. J. Thermophys. **38**, 61 (2017)
32. J. Campos, P. Corredera, A. Pons et al., Reflectance dependencies of silicon trap detectors. Metrologia **35**, 455–460 (1998)
33. A.C. Carter, S.R. Lorentz, T.M. Jung et al., ACR II: improved absolute cryogenic radiometer for low background infrared calibrations. Appl. Opt. **44**, 871–875 (2005)
34. A.C. Carter, S.I. Woods, S.M. Carr et al., Absolute cryogenic radiometer and solid-state trap detectors for IR power scales down to 1 pW with 0.1% uncertainty. Metrologia **46**, S146–S150 (2009)
35. R.A. Chipman, W.-S.T. Lam, G. Young, *Polarized Light and Optical Systems* (CRC Press, Boca Raton, FL, 2019)
36. R.R. Cordero, G. Seckmeyer, D. Pissulla et al., Uncertainty of experimental integrals: application to the UV index calculation. Metrologia **45**, 1–10 (2008)
37. R.U. Datla, K. Stock, A.C. Parr et al., Characterization of an absolute cryogenic radiometer as a standard detector for radiant-power measurements. Appl. Opt. **31**, 7219–7225 (1992)
38. A.T.A.M. de Waele, Basic operation of cryocoolers and related thermal machines. J. Low Temp. Phys. **164**, 179–236 (2011)
39. T. Dönsberg, M. Sildoja, F. Manoocheri et al., A primary standard of optical power based on induced-junction silicon photodiodes operated at room temperature. Metrologia **51**, 197–202 (2014)
40. T. Dönsberg, H. Mäntynen, E. Ikonen, Optical aperture area determination for accurate illuminance and luminous efficacy measurements of LED lamps. Opt. Rev. **23**, 510–521 (2016)
41. T. Dönsberg, F. Manoocheri, M. Sildoja et al., Predictable quantum efficient detector based on n-type silicon photodiodes. Metrologia **54**, 821–836 (2017)
42. F.J. Duarte (ed.), *Tunable Lasers Handbook* (Academic Press, San Diego, CA, 1995)
43. J.M. Dudley, G. Genty, S. Coen, Supercontinuum generation in photonic crystal fiber. Rev. Mod. Phys. **78**, 1135–1184 (2006)
44. J.M. Dudley, J.R. Taylor (eds.), *Supercontinuum Generation in Optical Fibers* (Cambridge University Press, Cambridge, UK, 2010)
45. M. Durak, M.H. Aslan, Optical characterization of the silicon photodiodes for the establishment of national radiometric standards. Opt. Laser Technol. **36**, 223–227 (2004)
46. M.R. Dury, T.M. Goodman, D.H. Lowe et al., Development of a new radiometer for the thermodynamic measurement of high temperature fixed points, in *Temperature: Its Measurement and Control in Science and Industry*, vol. 8, AIP Conf. Proc. **1552**, 65–70 (2013)
47. D.F. Edwards, Silicon (Si), in *Handbook of Optical Constants of Solids I*, ed. by E.D. Palik (Academic Press, San Diego, CA, 1998), pp. 547–569
48. P. Edwards, M. McCall, Diffraction loss in radiometry. Appl. Opt. **42**, 5024–5032 (2003)
49. P. Edwards, M. McCall, E. Usadi, Use of the single wavelength approximation in radiometric diffraction loss calculations. Proc. SPIE **5189**, 36–44 (2003)
50. J.W. Ekin, *Experimental Techniques for Low-Temperature Measurements. Cryostat Design, Material Properties, and Superconductor Critical-Current Testing* (Oxford University Press, Oxford, UK, 2006)
51. J.D. Ellis, *Field guide to displacement measuring interferometry* (SPIE Press, Bellingham, WA, 2014)
52. C. Enss, S. Hunklinger, *Low-Temperature Physics* (Springer, Berlin, 2005)
53. G.P. Eppeldauer, D.C. Lynch, Opto-mechanical and electronic design of a tunnel-trap Si radiometer. J. Res. Natl. Inst. Stand. Technol. **105**, 813–828 (2000)

54. G.P. Eppeldauer, H.W. Yoon, J. Neira et al., Comparison of laser-based and monochromator-based thermodynamic temperature measurements. AIP Conf. Proc. **1552**, 60–64 (2013)
55. A. Ferrero, J. Campos, A. Pons et al., New model for the internal quantum efficiency of photodiodes based on photocurrent analysis. Appl. Opt. **44**, 208–216 (2005)
56. J. Fischer, F. Lei, M. Stock, Present state of the determination of thermodynamic temperatures near the freezing point of silver by absolute cryoradiometry. Metrologia **28**, 243–246 (1991)
57. J. Fischer, M. Stock, A non-contact measurement of radiometric apertures with an optical microtopography sensor. Meas. Sci. Technol. **3**, 693–698 (1992)
58. J. Fischer, M. Battuello, M. Sadli et al., CCT/03-03. Uncertainty Budgets for Realisation of Scales by Radiation Thermometry (2003), http://www.bipm.org/cc/CCT/Allowed/22/CCT03-03.pdf. Accessed 9 Oct 2019
59. J. Fischer, M. de Podesta, K.D. Hill et al., Present estimates of the differences between thermodynamic temperatures and the ITS-90. Int. J. Thermophys. **32**, 12–25 (2011)
60. J.B. Fowler, G. Dezsi, High accuracy measurement of aperture area relative to a standard known aperture. J. Res. Natl. Inst. Stand. Technol. **100**, 277–283 (1995)
61. J.B. Fowler, R.D. Saunders, A.C. Parr, Summary of high-accuracy aperture-area measurement capabilities at the NIST. Metrologia **37**, 621–623 (2000)
62. J. Fowler, M. Litorja, Geometric area measurements of circular apertures for radiometry at NIST. Metrologia **40**, S9–S12 (2003)
63. N.P. Fox, Trap detectors and their properties. Metrologia **28**, 197–202 (1991)
64. N.P. Fox, Advances in radiometric methods for the determination of thermodynamic temperature and its impact on the future of ITS-90, in *Proceedings of the TEMPMEKO 2001, 8th International Symposium on Temperature and Thermal Measurements in Industry and Science* ed. by B. Fellmuth, J. Seidel, G. Scholz, vol. 1 (2002), pp. 27–35
65. N.P. Fox, J.E. Martin, D.H. Nettleton, Radiometric aspects of an experiment to determine the melting/freezing temperature of gold. Metrologia **28**, 221–227 (1991)
66. N.P. Fox, J.E. Martin, D.H. Nettleton, Absolute spectral radiometric determination of the thermodynamic temperatures of the melting/freezing points of gold, silver and aluminium. Metrologia **28**, 357–374 (1991)
67. N.P. Fox, J.E. Martin, D.H. Nettleton, Absolute spectral radiation thermometry at the National Physical Laboratory, in *Temperature: Its Measurement and Control in Science and Industry* ed. by J.F. Schooley, vol. 6, part 1 (American Institute of Physics, New York, 1992), pp. 41–45
68. N.P. Fox, J.P. Rice, Absolute radiometers, in *Optical Radiometry*, ed. by A.C. Parr, R.U. Datla, J.L. Gardner (Academic Press, Amsterdam, 2005), pp. 35–96
69. N.P. Fox, P.R. Haycocks, J.E. Martin et al., A mechanically cooled portable cryogenic radiometer. Metrologia **32**, 581–584 (1995/96)
70. R. Friedrich, J. Fischer, M. Stock, Accurate calibration of filter radiometers against a cryogenic radiometer using a trap detector. Metrologia **32**, 509–513 (1995/96)
71. R. Friedrich, J. Fischer, New spectral radiance scale from 220 nm to 2500 nm. Metrologia **37**, 539–542 (2000)
72. D. Gagnon, L.-I. Dion-Bertrand, *Widely Tunable Filter. Technology & Measurement of Critical Specifications* (Photon etc., Inc., 2015), http://photonetc.ekomobi.com/api/files/5556025262d46ee4300003bd-Photon-etc-White-Paper-Widely-Tunable-Filter.pdf. Accessed 9 Oct 2019
73. A. Gamouras, A.D.W. Todd, É. Côté et al., The development of the advanced cryogenic radiometer facility at NRC. J. Phys.: Conf. Ser. **972**, 012014 (2018)
74. J.L. Gardner, Transmission trap detectors. Appl. Opt. **33**, 5914–5918 (1994)
75. J.L. Gardner, A four-element transmission trap detector. Metrologia **32**, 469–472 (1995/96)
76. J. Geist, Quantum efficiency of the p-n junction in silicon as an absolute radiometric standard. Appl. Opt. **18**, 760–762 (1979)
77. J. Geist, Current status of, and future directions in, silicon photodiode self-calibration. Proc. SPIE **1109**, 246–256 (1989)
78. J. Geist, E.F. Zalewski, A.R. Schaefer, Spectral response self-calibration and interpolation of silicon photodiodes. Appl. Opt. **19**, 3795–3799 (1980)

79. J. Geist, D. Chandler-Horowitz, A.M. Robinson et al., Numerical modeling of silicon photodiodes for high-accuracy applications. Part I. Simulation programs. J. Res. Natl. Inst. Stand. Technol. **96**, 463–469 (1991)
80. J. Geist, R. Köhler, R. Goebel et al., Numerical modeling of silicon photodiodes for high-accuracy applications. Part II. Interpreting oxide-bias experiments. J. Res. Natl. Inst. Stand. Technol. **96**, 471–479 (1991)
81. J. Geist, A.M. Robinson, C.R. James, Numerical modeling of silicon photodiodes for high-accuracy applications. Part III. Interpolating and extrapolating internal quantum-efficiency calibrations. J. Res. Natl. Inst. Stand. Technol. **96**, 481–492 (1991)
82. T.R. Gentile, C.L. Cromer, Mode-locked lasers for high-accuracy radiometry. Metrologia **32**, 585–587 (1996)
83. T.R. Gentile, J.M. Houston, C.L. Cromer, Realization of a scale of absolute spectral response using the National Institute of Standards and Technology high-accuracy cryogenic radiometer. Appl. Opt. **35**, 4392–4403 (1996)
84. T.R. Gentile, S.W. Brown, K.R. Lykke et al., Internal quantum efficiency modeling of silicon photodiodes. Appl. Opt. **49**, 1859–1864 (2010)
85. G. Genty, A.T. Friberg, J. Turunen, Coherence of supercontinuum light. Prog. Opt. **61**, 71–112 (2016)
86. D.C. Ginnings, M.L. Reilly, Calorimetric measurement of thermodynamic temperatures above 0 °C using total blackbody radiation, in *Temperature, Its Measurement and Control in Science and Industry* ed. by H.H. Plumb, vol. 4, Part 1 (Instrument Society of America, Pittsburgh, PA, 1972), pp. 339–348
87. A.L. Glebov, O. Mokhun, A. Rapaport et al., Volume Bragg gratings as ultra-narrow and multiband optical filters. Proc. SPIE **8428**, 84280C (2012)
88. R. Goebel, S. Yilmaz, R. Pello, Polarization dependence of trap detectors. Metrologia **33**, 207–213 (1996)
89. D. Goldstein, *Polarized Light*, 3rd edn. (CRC Press, Boca Raton, FL, 2011)
90. E.S.R. Gopal, *Specific Heats at Low Temperatures* (Plenum Press, New York, 1966)
91. N. Granzow, Supercontinuum white light lasers: a review on technology and applications. Proc. SPIE **11144**, 1114408 (2019)
92. M.A. Green, Self-consistent optical parameters of intrinsic silicon at 300 K including temperature coefficients. Sol. Energy Mater. Sol. Cells **92**, 1305–1310 (2008)
93. R.D. Guenther, *Modern Optics* (Wiley, New York, 1990)
94. A. Haapalinna, P. Kärhä, E. Ikonen, Spectral reflectance of silicon photodiodes. Appl. Opt. **37**, 729–732 (1998)
95. Hamamatsu, Si photodiodes. Selection guide. Hamamatsu Photonics K.K. (2020), https://www.hamamatsu.com/resources/pdf/ssd/si_pd_kspd0001e.pdf. Accessed 9 May 2020
96. X. Hao, Z. Yuan, X. Lu, Lens transmission measurement for an absolute radiation thermometer, in *Temperature: Its Measurement and Control in Science and Industry*, vol. 8, AIP Conf. Proc. **1552**, 649–653 (2013)
97. J. Hartmann, High-temperature measurement techniques for the application in photometry, radiometry and thermometry. Phys. Rep. **469**, 205–269 (2009)
98. J. Hartmann, J. Fischer, J. Seidel, A non-contact technique providing improved accuracy in area measurements of radiometric apertures. Metrologia **37**, 637–640 (2000)
99. J. Hartmann, J. Fischer, U. Johannsen et al., Analytical model for the temperature dependence of the spectral responsivity of silicon. J. Opt. Soc. Am. B **18**, 942–947 (2001)
100. J. Hartmann, D.R. Taubert, J. Fischer, Measurement of $T-T_{90}$ down to zinc point temperatures with absolute filter radiometry, in *Proceedings of the TEMPMEKO 2001, 8th International Symposium on Temperature and Thermal Measurements in Industry and Science*, ed. by B. Fellmuth, J. Seidel, G. Scholz, vol. 1 (VDE Verlag GmbH, Berlin, 2002), pp. 337–382
101. J. Hartmann, K. Anhalt, P. Sperfeld et al., Thermodynamic temperature measurement of the melting curves of Re–C, TiC–C and ZrC–C eutectics, in *Proceedings of TEMPMEKO 2004, 9th International Symposium on Temperature and Thermal Measurements in Industry and Science*, ed. by D. Zvizdić (LPM/FSB, Zagreb, Croatia, 2005), pp. 189–194

102. J. Hartmann, J. Hollandt, P. Meindl et al., Traceable radiometric calibration of semiconductor detectors and their application for thermodynamic temperature measurement. MĀPAN J. Metrol. Soc. India **25**, 3–10 (2010)
103. J. Hartmann, K. Anhalt, R.D. Taubert et al., Absolute radiometry for the *MeP*-K: the irradiance measurement method. Int. J. Thermophys. **32**, 1707–1718 (2011)
104. F. Havermeyer, W. Liu, C. Moser et al. Volume holographic grating-based continuously tunable optical filter. Opt. Eng. **43**, 2017–2021 (2004)
105. E. Hecht, *Optics*, 5th edn. (Pearson, Boston, MA, 2017)
106. F. Hengstberger (ed.), *Absolute Radiometry. Electrically Calibrated Thermal Detectors of Optical Radiation* (Academic Press, Boston, MA, 1989)
107. J. Hollandt, R. Friedrich, B. Gutschwager, D.R. Taubert, J. Hartmann, High-accuracy radiation thermometry at the National Metrology Institute of Germany, the PTB. High Temp.–High Press. **35/36**, 379–415 (2003/2004)
108. S. Horne, D. Smith, M. Besen et al., A novel high-brightness, broadband light-source technology from the VUV to the IR. Proc. SPIE **7680**, 76800L (2010)
109. J.M. Houston, C.L. Cromer, J.E. Hardis et al., Comparison of the NIST high accuracy cryogenic radiometer and the NIST scale of detector spectral response. Metrologia **30**, 285–290 (1993)
110. J.M. Houston, J.P. Rice, NIST reference cryogenic radiometer designed for versatile performance. Metrologia **43**, S31–S35 (2006)
111. E. Ikonen, P. Toivanen, A. Lassila, A new optical method for high-accuracy determination of aperture area. Metrologia **35**, 369–372 (1998)
112. JCGM 100:2008, *GUM. Evaluation of Measurement Data. Guide to the Expression of Uncertainty in Measurement* (GUM 1995 with Minor Corrections) (BIPM Joint Committee for Guides in Metrology, Paris, 2008)
113. JCGM 101:2008, *GUM. Evaluation of Measurement Data—Supplement 1 to the "Guide to the Expression of Uncertainty in Measurement"—Propagation of Distributions Using a Monte Carlo Method* (BIPM Joint Committee for Guides in Metrology, Paris, 2008)
114. JCGM 200:2012, *International Vocabulary of Metrology—Basic and General Concepts and Associated Terms (VIM)*, 3rd edn.—2008 version with minor corrections (BIPM Joint Committee for Guides in Metrology, Paris, 2012)
115. R. Karwa, *Heat and Mass Transfer* (Springer, Singapore, 2017)
116. I. Kàsa, A circle fitting procedure and its error analysis. IEEE Trans. Instrum. Meas. **IM-25**, 8–14 (1976). https://doi.org/10.1109/TIM.1976.6312298
117. P. Kärhä, A. Vaskuri, H. Mäntynen et al., Method for estimating effects of unknown correlations in spectral irradiance data on uncertainties of spectrally integrated colorimetric quantities. Metrologia **54**, 524–534 (2017)
118. T. Keawprasert, K. Anhalt, D.R. Taubert et al., Monochromator-based absolute calibration of radiation thermometers. Int. J. Thermophys. **32**, 1697–1706 (2011)
119. T. Keawprasert, K. Anhalt, D.R. Taubert et al., A comparison of absolute calibrations of a radiation thermometer based on a monochromator and a tunable source, in *Temperature: Its Measurement and Control in Science and Industry*, vol. 8. AIP Conf. Proc. **1552**, 682–687 (2013)
120. P.J. Key, N.P. Fox, M.L. Rastello, Oxide-bias measurements in the silicon photodiode self-calibration technique. Metrologia **21**, 81–87 (1985)
121. B. Khlevnoy, I. Grigoryeva, K. Anhalt et al., Development of large-area high-temperature fixed-point blackbodies for photometry and radiometry. Metrologia **55**, S43–S51 (2018)
122. S. Kodama, M. Horiuchi, K. Kuroda et al., Ultra-black nickel-phosphorus alloy optical absorber, in *6th IEEE Record, Instrumentation and Measurement Technology Conference* (Washington, DC, 1989), pp. 495–497. https://doi.org/10.1109/IMTC.1989.36909
123. O. Kozlova, A. Sadouni, D. Truong et al., Tunable transportable spectroradiometer based on an acousto-optical tunable filter: Development and optical performance. Rev. Sci. Instrum. **87**, 125101 (2016)

124. A. Kramida, Recent developments in the NIST atomic databases. AIP Conf. Proc. **1344**, 81–95 (2011)
125. T. Kübarsepp, P. Kärhä, E. Ikonen, Characterization of a polarization-independent transmission trap detector. Appl. Opt. **36**, 2807–2812 (1997)
126. T. Kübarsepp, M. White, Ten-element photodetector for optical power and attenuation measurements. Appl. Opt. **49**, 3774–3779 (2010)
127. L-1 Standards and Technology, Inc., http://www.l-1.biz/cryorad3.html. Accessed 8 June 2020
128. Labsphere, *Integrating Sphere Uniform Light Source Applications. Technical Guide* (Labsphere, Inc., 2008), https://www.labsphere.com/site/assets/files/2552/a-guide-to-integrating-sphere-uniform-light-source-applications.pdf. Accessed 9 Oct 2019
129. Labsphere, *Integrating Sphere Radiometry and Photometry. Technical Guide* (Labsphere, Inc., 2017), https://www.labsphere.com/site/assets/files/2550/radiometry_and_photometry_tech_guide-1.pdf. Accessed 9 Oct 2019
130. A. Lassila, P. Toivanen, E. Ikonen, An optical method for direct determination of the radiometric aperture area at high accuracy. Meas. Sci. Technol. **8**, 973–977 (1997)
131. J.H. Lehman, C.L. Cromer, Optical tunnel-trap detector for radiometric measurements. Metrologia **37**, 477–480 (2000)
132. F. Lei, J. Fischer, Characterization of photodiodes in the UV and visible spectral region based on cryogenic radiometry. Metrologia **30**, 297–303 (1993)
133. A.P. Levick, C.L. Greenwell, J. Ireland et al., Spectral radiance source based on supercontinuum laser and wavelength tunable bandpass filter: the spectrally tunable absolute irradiance and radiance source. Appl. Opt. **53**, 3508–3519 (2014)
134. M. Litorja, J. Fowler, J. Hartmann et al., Report on the CCPR-S2 supplementary comparison of area measurements of apertures for radiometry. Metrologia **44**, 02002 (2007)
135. I.C.M. Littler, E.G. Atkinson, P.J. Manson, Non-contact aperture area measurement by occlusion of a laser beam. Metrologia **50**, 596–611 (2013)
136. I.C.M. Littler, E.G. Atkinson, P.J. Manson, Aperture area measurement using two different traceability routes. Meas. Sci. Technol. **26**, 125201 (2015)
137. D.J. Livigni, N.A. Tomlin, C.L. Cromer et al., Optical fibre-coupled cryogenic radiometer with carbon nanotube absorber. Metrologia **49**, S93–S98 (2012)
138. M. López, H. Hofer, S. Kück, Measurement of the absorptance of a cryogenic radiometer cavity in the visible and near infrared. Metrologia **42**, 400–405 (2005)
139. N.H. Macoy, Total radiant exitance measurements. Proc. SPIE **1311**, 16–40 (1990)
140. G. Machin, K. Anhalt, P. Bloembergen et al., *MeP*-K Absolute Primary Radiometric Thermometry (2012), https://www.bipm.org/utils/en/pdf/si-mep/MeP-K-2018_Absolute_Primary_Radiometry.pdf. Accessed 30 Apr 2020
141. J.M. Mantilla, M.L. Hernanz, J. Campos et al., Monochromator-based absolute calibration of a standard radiation thermometer. Int. J. Thermophys. **35**, 493–503 (2014)
142. J.M. Mantilla, M.J. Martin, J. Campos et al., Testing irradiance and radiance methods for absolute radiation thermometry based on InGaAs detectors in the NIR at CEM/CSIC. J. Phys.: Conf. Ser. **1065**, 122005 (2018)
143. J.E. Martin, T.J. Quinn, B. Chu, Further measurements of thermodynamic temperature using a total radiation thermometer: the range −130 °C to +60 °C. Metrologia **25**, 107–112 (1988)
144. J.E. Martin, N.P. Fox, Cryogenic Solar Absolute Radiometer (CSAR). Metrologia **30**, 305–308 (1993)
145. M.J. Martín, J.M. Mantilla, D. del Campo et al., Performance of different light sources for the absolute calibration of radiation thermometers. Int. J. Thermophys. **38**, 138 (2017)
146. T. Menegotto, M.S. Lima, G.B. Almeida et al., Direct modeling of external quantum efficiency of silicon trap detectors. Proc. SPIE **8083**, 808313 (2011)
147. S. Mezouari, A.R. Harvey, Validity of Fresnel and Fraunhofer approximations in scalar diffraction. J. Opt. A: Pure Appl. Opt. **5**, S86–S91 (2003)
148. K.D. Mielenz, Optical diffraction in close proximity to plane apertures. I. Boundary-value solutions for circular apertures and slits. J. Res. Natl. Inst. Stand. Technol. **107**, 355–362 (2002)

149. K.D. Mielenz, R.D. Saunders Jr., J.B. Shumaker, Spectroradiometric determination of the freezing temperature of gold. J. Res. Natl. Inst. Stand. Technol. **95**, 49–67 (1990)
150. J.C. Molina, J.J.S. Bernal, H.A. Castillo et al., Electrical substitution radiometer cavity absorptance measurement. Measurement **64**, 89–93 (2015)
151. B. Muralikrishnan, J.A. Stone, J.R. Stoup, Area measurement of knife edge and cylindrical apertures using ultra low force contact fiber probe on a CMM. Metrologia **45**, 281–298 (2008)
152. NIST, Strategic Plan 2010. U.S. Department of Commerce. National Institute of Standards and Technology. Physical Measurement Laboratory. Optical Technology Division (2010). https://tsapps.nist.gov/publication/get_pdf.cfm?pub_id=905286. Accessed 11 Oct 2019
153. M. Noorma, P. Toivanen, F. Manoocheri et al., Characterization of filter radiometers with a wavelength-tunable laser source. Metrologia **40**, S220–S223 (2003)
154. N. Noulkhow, *Radiometric Measurement of Thermodynamic Temperatures in the Temperature Range from 419 °C to 660 °C for an Improved Temperature Scale*. Ph.D. thesis (Technischen Universität Berlin, 2009), https://depositonce.tu-berlin.de/bitstream/11303/2615/2/Dokument_27.pdf. Accessed 30 May 2020
155. N. Noulkow, R.D. Taubert, P. Meindl et al., Infrared filter radiometers for thermodynamic temperature determination below 660 °C. Int. J. Thermophys. **30**, 131–143 (2009)
156. S.A. Ogarev, B.B. Khlevnoy, M.L. Samoylov et al., New high-temperature pyrolitic graphite blackbody sources for precision measurements in radiation thermometry, in *TEMPMEKO 2004, Proceedings of the 9th International Symposium on Temperature and Thermal Measurements in Industry and Science*, ed. by D. Zvizdić (LPM/FSB, Zagreb, Croatia, 2005), pp. 569–574
157. J.M. Palmer, B.G. Grant, *Art of radiometry* (SPIE Press, Bellingham, WA, 2010)
158. U. Pant, H. Meena, G. Gupta et al., Development and realization of Fe–C and Co–C eutectic fixed-point blackbodies for radiation thermometry at CSIR-NPL. Int. J. Thermophys. **41**, 101 (2020)
159. M. Parker, *Digital Signal Processing 101. Everything you Need to Know to Get Started*, 2nd edn. (Newnes, Oxford, UK, 2017)
160. A.C. Parr, *A National Measurement System for Radiometry, Photometry, and Pyrometry Based Upon Absolute Detectors*. NIST Technical Note 1421. National Institute of Standards and Technology, U.S. Department of Commerce (1996)
161. R. Paschotta, *Field Guide to Lasers* (SPIE Press, Bellingham, WA, 2008)
162. D. Pavanello, R. Galleano, R.P. Kenny, Uncertainty propagation of spectral matching ratios measured using a calibrated spectroradiometer. Appl. Sci. **8**, 186 (2018)
163. F. Poli, A. Cucinotta, S. Selleri, *Photonic Crystal Fibers. Properties and Applications* (Springer, Dordrecht, The Netherlands, 2007)
164. W.H. Press, S.A. Teukolsky, H.A. Bethe et al., *Numerical Recipes. The Art of Scientific Computing*, 3rd edn. (Cambridge University Press, Cambridge, UK, 2007)
165. T.J. Quinn, J.E. Martin, Radiometric measurement of thermodynamic temperature between 327 and 365 K, in *Temperature: Its Measurement and Control in Science and Industry*, ed. by J.F. Schooley, vol. 5, part I (American Institute of Physics, New York, 1982), pp. 103–107
166. T.J. Quinn, J.E. Martin, Radiometric determination of the Stefan-Boltzmann constant, in *Proceedings of the 2nd International Conference on "Precision Measurement and Fundamental Constants II"*, ed. by B.N. Taylor, W.D. Phillips (National Bureau of Standards, Gaithersburg, MD, 8–12 June 1981) NBS Spec. Publ. **617**, pp. 291–297 (National Bureau of Standards, US Department of Commerce, 1984)
167. T.J. Quinn, J.E. Martin, A radiometric determination of the Stefan-Boltzmann constant and thermodynamic temperatures between −40 °C and +100 °C. Philos. Trans. R. Soc. Lond. A **316**, 85–189 (1985)
168. R. Radebaugh, Cryocoolers: the state of the art and recent developments. J. Phys.: Condens. Matter **21**, 164219 (2009)
169. K.R. Rao, D.N. Kim, J.J. Hwang, *Fast Fourier Transform: Algorithms and Applications* (Springer, Dordrecht, Netherlands, 2010)
170. H. Römer, *Theoretical Optics. An Introduction* (Wiley, Weinheim, Germany, 2005)

171. J.S. Rubin, E.L. Shirley, Z.H. Levine, Acceleration of diffraction calculations in cylindrically symmetrical optics. Appl. Opt. **57**, 788–793 (2018)
172. P. Russell, Photonic crystal fibers. Science **299**, 358–362 (2003)
173. S.G.R. Salim, E.R. Woolliams, M. Dury et al., Furnace uniformity effects on Re–C fixed-point melting plateaux. Metrologia **46**, 33–42 (2009)
174. J.E. Sansonetti, W.C. Martin, Handbook of basic spectroscopic data. J. Phys. Chem. Ref. Data **34**, 1559–2259 (2005)
175. W. Silfvast, *Laser Fundamentals*, 2nd edn. (Cambridge University Press, Cambridge, UK, 2004)
176. V.I. Sapritsky, B.B. Khlevnoi, V.B. Khromchenko et al., High-temperature blackbody sources for precision radiometry. Proc. SPIE **2815**, 2–10 (1996)
177. V.I. Sapritsky, B.B. Khlevnoy, V.B. Khromchenko et al., Precision blackbody sources for radiometric standards. Appl. Opt. **36**, 5403–5408 (1997)
178. N. Sasajima, Y. Yamada, F. Sakuma, Investigation of fixed points exceeding 2500 °C using metal carbide-carbon eutectics, in *Temperature: Its Measurement and Control in Science and Industry*, ed. by D. C. Ripple, vol. 7, part 1. AIP Conf. Proc. **684**, 279–285 (American Institute of Physics, Melville, NY, 2003)
179. P. Saunders, Uncertainty propagation through integrated quantities for radiation thermometry. Metrologia **55**, 863–871 (2018)
180. P. Saunders, E. Woolliams, H. Yoon et al., Uncertainty Estimation in Primary Radiometric Temperature Measurement (2018), https://www.bipm.org/utils/en/pdf/si-mep/MeP-K-2018_Absolute_Primary_Radiometry_Uncertainty.pdf. Accessed 9 Oct 2019
181. P. Saunders, D.R. White, Physical basis of interpolation equations for radiation thermometry. Metrologia **40**, 195–203 (2003)
182. M. Schuster, S. Nevas, A. Sperling et al., Spectral calibration of radiometric detectors using tunable laser sources. Appl. Opt. **51**, 1950–1961 (2012)
183. D. Schlichthärle, *Digital Filters. Basics and Design*, 2nd edn. (Springer, Heidelberg, Germany, 2011)
184. F. Schmähling, G. Wübbeler, U. Krüger et al., Uncertainty evaluation and propagation for spectral measurements. Color Res. Appl. **4**, 6–16 (2018)
185. C.A. Schrama, R. Bosma, K. Gibb et al., Comparison of monochromator-based and laser-based cryogenic radiometry. Metrologia **35**, 431–435 (1998)
186. E.L. Shirley, Intuitive diffraction model for multistaged optical systems. Appl. Opt. **43**, 735–743 (2004)
187. E.L. Shirley, Diffraction effects on broadband radiation: formulation for computing total irradiance. Appl. Opt. **43**, 2609–2620 (2004)
188. E.L. Shirley, Diffraction effects in radiometry, in *Optical Radiometry*, ed. by A.C. Parr, R.U. Datla, J.L. Gardner (Academic Press, Amsterdam, 2005), pp. 409–451
189. E.L. Shirley, Higher-order boundary-diffraction-wave formulation. J. Mod. Opt. **54**, 515–527 (2007)
190. E.L. Shirley, Refined treatment of single-edge diffraction effects in radiometry. J. Opt. Soc. Am. A **33**, 1509–1522 (2016)
191. P. Sperfeld, K.-H. Raatz, B. Nawo, W. Möller, J. Metzdorf, Spectral-irradiance scale based on radiometric black-body temperature measurements. Metrologia **32**, 435–439 (1995/96)
192. P. Sperfeld, S. Pape, B. Khlevnoy et al., Performance limitations of carbon-cavity blackbodies due to absorption bands at the highest temperatures. Metrologia **46**, S170–S173 (2009)
193. W.H. Steel, M. De, J.A. Bell, Diffraction corrections in radiometry. J. Opt. Soc. Am. **62**, 1099–1103 (1972)
194. M. Stock, J. Fischer, R. Friedrich et al., Present state of the comparison between radiometric scales based on three primary standards. Metrologia **30**, 439–449 (1993)
195. M. Stock, J. Fischer, R. Friedrich, H.-J. Jung, B. Wende, Measurement of $T-T_{90}$ in the range from 500 °C to 962 °C by absolute spectral radiometry employing a cryogenic radiometer and a double heatpipe black body, in *Proceedings of the TEMPMEKO '96, 6th International Symposium on Temperature and Thermal Measurements in Industry and Science*, ed. by P. Marcarino (Levrotto & Bella, Torino, Italy, 1997), pp. 19–24

196. M. Stock, R. Goebel, Practical aspects of aperture-area measurements by superposition of Gaussian laser beams. Metrologia **37**, 633–636 (2000)
197. J.G. Suárez-Romero, E. Tepichín-Rodríguez, K.D. Mielenz, Cross-spectral density propagated through a circular aperture. Metrologia **38**, 379–384 (2001)
198. J.G. Suárez-Romero, E. Tepichín-Rodríguez, Irradiance measurements without explicit diffraction corrections. Metrologia **40**, S189–S191 (2003)
199. A. Tari, *The Specific Heat of Matter at Low Temperatures* (Imperial College Press, London, 2003)
200. D.R. Taubert, R. Friedrich, J. Hartmann et al., Improved calibration of the spectral responsivity of interference filter radiometers in the visible and near infrared spectral range at PTB. Metrologia **40**, S35–S38 (2003)
201. D.R. Taubert, R. Friedrich, J. Hartmann et al., Long term stability of the spectral responsivity of filter radiometers at the PTB, in *Proceedings of TEMPMEKO 2004, 9th International Symposium on Temperature and Thermal Measurements in Industry and Science*, ed. by D. Zvizdić (LPM/FSB, Zagreb, Croatia, 2005), pp. 977–982
202. R.D. Taubert, J. Hollandt, Radiometric characterization of a high temperature blackbody in the visible and near infrared. AIP Conf. Proc. **1552**, 607–612 (2013)
203. F. Tayeb-Chandoul, J.-M. Coutin, J. Bastie, Méthode de mesure de la sensibilité spectrale des détecteurs pièges de référence. Revue française de métrologie, n° 25, vol. 2011-1, 45–54 (2011). https://doi.org/10.1051/rfm/2011001
204. T.M. Tritt (ed.), *Thermal Conductivity. Theory, Properties, and Applications* (Kluwer Academic/Plenum Publishers, New York, 2004)
205. G. Ventura, M. Perfetti, *Thermal Properties of Solids at Room and Cryogenic Temperatures* (Springer, Dordrecht, Netherlands, 2014)
206. F. Vignola, J. Michalsky, T. Stoffel, *Solar and Infrared Radiation Measurements* (CRC Press, Boca Raton, FL, 2012)
207. G. Walker, *Cryocoolers. Part 1: Fundamentals* (Springer, New York, 1983)
208. B. Walter, R. Winkler, F. Graber et al., Direct solar irradiance measurements with a cryogenic solar absolute radiometer, in *Radiation Processes in the Atmosphere and Ocean (IRS2016)*. AIP Conf. Proc. **1810**, 080007 (2017)
209. L. Werner, J. Fischer, U. Johannsen et al., Accurate determination of the spectral responsivity of silicon trap detectors between 238 nm and 1015 nm using a laser-based cryogenic radiometer. Metrologia **37**, 279–284 (2000)
210. L. Werner, J. Hartmann, Calibration and interpolation of the spectral responsivity of silicon photodiode-based detectors. Sens. Actuators A **156**, 185–190 (2009)
211. W.L. Wiese, The expanding NIST atomic spectra database. Phys. Scr. **T105**, 85–89 (2003)
212. R. Winkler, E.R. Woolliams, W.S. Hartree et al., Calibration of an absolute radiation thermometer for accurate determination of fixed-point temperatures. Int. J. Thermophys. **28**, 2087–2097 (2007)
213. J.T. Woodward, A.W. Smith, C.A. Jenkins et al., Supercontinuum sources for metrology. Metrologia **46**, S277–S282 (2009)
214. J.T. Woodward, P.-S. Shaw, H.W. Yoon et al., Invited article: advances in tunable laser-based radiometric calibration applications at the National Institute of Standards and Technology, USA. Rev. Sci. Instrum. **89**, 091301 (2018)
215. E.R. Woolliams, Determining the uncertainty associated with integrals of spectral quantities. A report of the EMRP Joint Research. EMRP-ENG05-1.3.1, Version 1.0. NPL, Teddington, UK (2013), http://eprintspublications.npl.co.uk/5826/1/PDID6709.pdf. Accessed 3 June 2020
216. E.R. Woolliams, Uncertainty analysis for filter radiometry based on the uncertainty associated with integrated quantities. Int. J. Thermophys. **35**, 1353–1365 (2014)
217. E.R. Woolliams, D.F. Pollard, N.J. Harrison et al., New facility for the high-accuracy measurement of lens transmission. Metrologia **37**, 603–605 (2000)
218. E.R. Woolliams, B.B. Khlevnoy, N.J. Harrison et al., Measurements of the melting and freezing plateaus of Re-C eutectic fixed-point blackbodies, in *TEMPMEKO 2004, Proceedings of the 9th International Symposium on Temperature and Thermal Measurements in Industry and Science*, ed. by D. Zvizdić (LPM/FSB, Zagreb, Croatia, 2005), pp. 183–188

219. E.R. Woolliams, M.R. Dury, T.A. Burnitt et al., Primary radiometry for the *Mise-en-Pratique* for the definition of the kelvin: the hybrid method. Int. J. Thermophys. **32**, 1–11 (2011)
220. E.R. Woolliams, K. Anhalt, M. Ballico et al., Thermodynamic temperature assignment to the point of inflection of the melting curve of high-temperature fixed points. Philos. Trans. R. Soc. A **374**, 20150044 (2016)
221. N. Xu, Y. Lin, H. Gan et al., Spectral responsivity calibration of silicon photodetectors using monochromator-based cryogenic radiometer. Proc. SPIE **10155**, 1015513 (2016)
222. Y. Yamada, H. Sakate, F. Sakuma, A. Ono, Radiometric observation of melting and freezing plateaus for a series of metal-carbon eutectic points in the range 1330 °C to 1950 °C. Metrologia **36**, 207–209 (1999)
223. Y. Yamada, H. Sakate, F. Sakuma, A. Ono, A possibility of practical high temperature fixed points above the copper point, in *Proceedings of the TEMPMEKO '99. The 7th International Symposium on Temperature and Thermal Measurements in Industry and Science*, ed. by J.F. Dubbeldam, M.J. de Groot, vol. 2 (1999), pp. 535–540
224. Y. Yamada, Y. Wang, N. Sasajima, Metal carbide-carbon peritectic systems as high-temperature fixed points in thermometry. Metrologia **43**, L23–L27 (2006)
225. Y. Yamaguchi, Y. Yamada, J. Ishii, New setup for the absolute spectral responsivity of radiation thermometers with a supercontinuum source, in *XX IMEKO World Congress "Metrology for Green Growth,"* (Busan, Republic of Korea, 2012), https://www.imeko.org/publications/wc-2012/IMEKO-WC-2012-TC12-P1.pdf. Accessed 9 Oct 2019
226. Y. Yamaguchi, Y. Yamada, J. Ishii, Supercontinuum-source-based facility for absolute calibration of radiation thermometers. Int. J. Thermophys. **36**, 1825–1833 (2015)
227. Y. Yamaguchi, Y. Yamada, Temperature standard for radiometric calibration of earth-observation sensor, in *Proceedings of the SICE Annual Conference 2017*, (Kanazawa, Japan, 2017), pp. 424–426. https://doi.org/10.23919/SICE.2017.8105490
228. H.W. Yoon, C.E. Gibson, P.Y. Barnes, Realization of the National Institute of Standards and Technology detector-based spectral irradiance scale. Appl. Opt. **41**, 5879–5890 (2002)
229. H.W. Yoon, C.E. Gibson, D.W. Allen et al., The realization and the dissemination of the detector-based kelvin, in *TEMPMEKO 2004, Proceedings of the 9th International Symposium on Temperature and Thermal Measurements in Industry and Science*, ed. by D. Zvizdić (LPM/FSB, Zagreb, Croatia, 2005), pp. 55–70
230. H.W. Yoon, D.W. Allen, C.E. Gibson et al., Temperature determination of the Ag- and Au-freezing points using a detector-based radiation thermometer, in *TEMPMEKO 2004, Proceedings of the 9th International Symposium on Temperature and Thermal Measurements in Industry and Science*, ed. by D. Zvizdić (LPM/FSB, Zagreb, Croatia, 2005), pp. 113–118
231. H.W. Yoon, The realization and the dissemination of thermodynamic temperature scales. Metrologia **43**, S22–S26 (2006)
232. H.W. Yoon, D.W. Allen, C.E. Gibson, M. Litorja, R.D. Saunders, S.W. Brown, G.P. Eppeldauer, K.R. Lykke, Thermodynamic-temperature determinations of the Ag and Au freezing temperatures using a detector-based radiation thermometer. Appl. Opt. **46**, 2870–2880 (2007)
233. H.W. Yoon, C.E. Gibson, V. Khromchenko et al., Thermodynamic radiation thermometry for the next SI. Int. J. Thermophys. **29**, 285–300 (2008)
234. H.W. Yoon, C.E. Gibson, G.P. Eppeldauer et al., Thermodynamic radiation thermometry using radiometers calibrated for radiance responsivity. Int. J. Thermophys. **32**, 2217–2229 (2011)
235. Z. Yuan, X. Lu, X. Hao et al., Thermodynamic temperature measurements of silver freezing point and HTFPs, in *Temperature: Its Measurement and Control in Science and Industry*, vol. 8. AIP Conf. Proc. **1552**, 56–59 (2013)
236. E.F. Zalewski, J. Geist, Silicon photodiode absolute spectral response self-calibration. Appl. Opt. **19**, 1214–1216 (1980)
237. E.F. Zalewski, C.R. Duda, Silicon photodiode device with 100% external quantum efficiency. Appl. Opt. **22**, 2867–2873 (1983)
238. Y. Zong, S.W. Brown, G.P. Eppeldauer et al., A new method for spectral irradiance and radiance responsivity calibrations using kilohertz pulsed tunable optical parametric oscillators. Metrologia **49**, S124–S129 (2012)

Correction to: Theoretical Basis of Blackbody Radiometry

Correction to:
Chapter 3 in: V. Sapritsky and A. Prokhorov, *Blackbody Radiometry*, Springer Series in Measurement Science and Technology, https://doi.org/10.1007/978-3-030-57789-6_3

The original version of this book was inadvertently published with a wrong equation 3.143 in the chapter 'Theoretical Basis of Blackbody Radiometry'. The equation has now been corrected. The book and the chapter have been updated with the changes.

The updated version of this chapter can be found at
https://doi.org/10.1007/978-3-030-57789-6_3

V. Sapritsky and A. Prokhorov, *Blackbody Radiometry*, Springer Series in Measurement Science and Technology, https://doi.org/10.1007/978-3-030-57789-6_10

Appendix A
Values of Some Fundamental Physical Constants

Table A.1 Exact values of defining fundamental physical constants of the SI (after BIPM,The International System of Units (SI), 9th ed. (2019) https://www.bipm.org/utils/common/pdf/si-brochure/SI-Brochure-9-EN.pdf Accessed 5 Jan. 2020)

Constant	Symbol	Value	Unit
Unperturbed ground state hyperfine transition frequency of the cesium 133 atom	$\Delta\nu_{Cs}$	9,192,631,770	Hz
Speed of light in vacuum	c	299,792,458	$m\ s^{-1}$
Planck constant	h	$6.62607015 \times 10^{-34}$	J s
Elementary charge	e	$1.602176634 \times 10^{-19}$	C
Boltzmann constant	k	1.380649×10^{-23}	$J\ K^{-1}$
Avogadro constant	N_A	$6.02214076 \times 10^{23}$	mol^{-1}
Luminous efficacy of monochromatic radiation of frequency 540×10^{12} Hz	K_{cd}	683	lm/W

V. Sapritsky and A. Prokhorov, *Blackbody Radiometry*, Springer Series in Measurement Science and Technology, https://doi.org/10.1007/978-3-030-57789-6

Table A.2 Exact values of some derived fundamental physical constants associated with radiation of a perfect blackbody. The ellipsis (...) in the value means that further digits can be obtained if necessary

Constant	Symbol/expression	Value	Unit
Molar gas constant	$R = N_A k$	8.314462618...	J mol^{-1} K^{-1}
First radiation constant	$c_1 = 2\pi hc^2$	3.741771852... × 10^{-16}	W m^2
First radiation constant for spectral radiance	$c_{1L} = c_1 / \pi$	1.191042972... × 10^{-16}	W m^2 sr^{-1}
Second radiation constant	$c_2 = hc / k$	1.438776877... × 10^{-2}	m K
Wien wavelength displacement law constant	$b = \frac{hc/k}{4.965114231}$	2.897771955... × 10^{-3}	m K
Wien frequency displacement law constant	$b' = 2.821439372 \cdot k / h$	5.878925757... × 10^{10}	Hz K^{-1}
Stefan-Boltzmann constant	$\sigma = \frac{2\pi^5 k^4}{15h^3 c^2}$	1.191042972... × 10^{-16}	W m^2 sr^{-1}

Appendix B
Fixed Points

Table B.1 Defining fixed points of the ITS-90. Symbols V, G, T, M, and F in the last column are for vapor pressure point, gas thermometer point, triple point, melting point, and freezing point, respectively. Adopted from H. Preston-Thomas, The International Temperature Scale of 1990 (ITS-90). Metrologia **27**, 3–10 (1990)

Number	Temperature		Substance	State
	T_{90}(K)	t_{90}(°C)		
1	3 to 5	−270.15 to −268.15	He	V
2	13.8033	−259.3467	e–H_2	T
3	≈17	≈−256.15	e–H_2 (or He)	V (or G)
4	≈20.3	≈−252.85	e–H_2 (or He)	V (or G)
5	24.5561	−248.5939	Ne	T
6	54.3584	−218.7916	O_2	T
7	83.8058	−189.3442	Ar	T
8	234.3156	−38.8344	Hg	T
9	273.16	0.01	H_2O	T
10	302.9146	29.7646	Ga	M
11	429.7485	156.5985	In	F
12	505.078	231.928	Sn	F
13	692.677	419.527	Zn	F
14	933.473	660.323	Al	F
15	1234.93	961.78	Ag	F
16	1337.33	1064.18	Au	F
17	1357.77	1084.62	Cu	F

V. Sapritsky and A. Prokhorov, *Blackbody Radiometry*, Springer Series in Measurement Science and Technology, https://doi.org/10.1007/978-3-030-57789-6

Table B.2 High-temperature fixed points (HTFPs). Adopted from: G. Machin, K. Anhalt, P. Bloembergen et al., MeP-K Relative Primary Radiometric Thermometry (2017) https://www.bipm.org/utils/en/pdf/si-mep/MeP-K-2018_Relative_Primary_Radiometry.pdf Accessed June 30, 2019

HTFP	Thermodynamic temperature (PoI) (K)	Expanded uncertainty of PoI at $k = 2$ (K)	Thermodynamic temperature (ELq) (K)	Expanded uncertainty of ELq at $k = 2$ (K)
Co–C	1597.39	0.13	1597.48	0.14
Pd–C	1764.85	0.7		
Pt–C	2011.43	0.18	2011.50	0.22
Re–C	2747.84	0.35	2747.91	0.44

PoI denote the point of inflection in the melting curve; ELq is for equilibrium liquidus thermodynamic temperatures for the eutectic transitions.

Appendix C
Glossary of Metrology Terms

The definitions of key metrology terms are taken from:

JCGM 200:2012. International vocabulary of metrology—Basic and general concepts and associated terms (VIM), 3rd ed.—2008 version with minor corrections (BIPM Joint Committee for Guides in Metrology, Paris, 2012).

Calibration	The operation that, under specified conditions, in a first step, establishes a relation between the quantity values with measurement uncertainties provided by measurement standards and corresponding indications with associated measurement uncertainties and, in a second step, uses this information to establish a relation for obtaining a measurement result from an indication.
Calibration hierarchy	A sequence of calibrations from a reference to the final measuring system, where the outcome of each calibration depends on the outcome of the previous calibration.
Calibrator	Measurement standard used in calibration.
Certified reference material	Reference material, accompanied by documentation issued by an authoritative body and providing one or more specified property values with associated uncertainties and traceabilities, using valid procedures.
Combined standard measurement uncertainty	Standard measurement uncertainty that is obtained using the individual standard measurement uncertainties associated with the input quantities in a measurement model.
Correction	The compensation for an estimated systematic effect that causes a systematic error to occur. The

V. Sapritsky and A. Prokhorov, *Blackbody Radiometry*, Springer Series in Measurement Science and Technology, https://doi.org/10.1007/978-3-030-57789-6

	compensation can take a form of an addend (additive correction) or a factor (multiplicative correction), or can be deduced from a table, graph, or algorithm.
Coverage factor	A number larger than one by which a combined standard measurement uncertainty is multiplied to obtain an expanded measurement uncertainty.
Coverage interval	The interval containing the set of true quantity values of a measurand with a stated probability, based on the information available. A coverage interval does not need to be symmetrical about the chosen measured quantity value.
Coverage probability	The probability that the set of true quantity values of a measurand is contained within a specified coverage interval. The coverage probability is also termed "level of confidence" in the GUM.
Expanded measurement uncertainty	The product of a combined standard measurement uncertainty and a factor (termed coverage factor) larger than the number one.
Indicating measuring instrument	A measuring instrument providing an output signal carrying information about the value of the quantity being measured and, perhaps, recording of indication or transmission to one or more other devices.
Influence quantity	The quantity that, in a direct measurement, does not affect the quantity that is actually measured, but affects the relation between the indication and the measurement result.
Input quantity	A measurement model is the quantity that must be measured, or a quantity, the value of which can be otherwise obtained, in order to calculate a measured quantity value of a measurand. Indications, corrections and influence quantities can be input quantities in a measurement model.
Intermediate measurement precision	Measurement precision under a set of intermediate precision conditions of measurement.
Intermediate precision condition of measurement	The condition of measurement, out of a set of conditions that includes the same measurement procedure, same location, and replicate measurements on the same or similar objects over an extended period of time, but may include other conditions involving changes.
International measurement standard	Measurement standard recognized by signatories to an international agreement and intended to serve worldwide.

Material measure A measuring instrument reproducing or supplying, in a permanent manner during its use, quantities of one or more given kinds, each with an assigned quantity value.

Measurand The quantity intended to be measured.

Measurement (of a physical quantity) The process of experimentally obtaining one or more quantity values that can reasonably be attributed to that quantity. Measurement is a set of operations for determining the value of a quantity by comparison of homonymous quantities (for continuously varying quantity values) or counting of entities (for discrete quantity values). Measurement implies a description of the quantity commensurate with the intended use of a measurement result, a measurement procedure, and a calibrated measuring system operating according to the specified measurement procedure, including the measurement conditions.

Measurement accuracy A closeness of agreement between a measured quantity value and a true quantity value of a measurand. Measurement accuracy is sometimes understood as closeness of agreement between measured quantity values that are being attributed to the measurand.

Measurement bias The estimate of a systematic measurement error.

Measurement error A measured quantity value minus a reference quantity value.

Measurement function A function of quantities, the value of which, when calculated using known quantity values for the input quantities in a measurement model, is a measured quantity value of the output quantity in the measurement model. Measurement function is considered as a special case of a measurement model, when Eq. 1.56 can be explicitly resolved for the measurand Y:

$$Y = f(X_1, \ldots, X_n). \qquad \text{(C.1)}$$

Measurement method A generic description of a logical organization of operations used in a measurement.

Measurement model The mathematical relation among all quantities known to be involved in a measurement. A general form of a measurement model is the equation

$$h(Y, X_1, \ldots, X_n) = 0, \qquad \text{(C.2)}$$

where Y is the measurand (output quantity) of the measurement model, for which the quantity value has to be inferred from information about input quantities $(X_1, \ldots, X_n)$ of the measurement model. In the cases of simultaneous measurements of more than one quantity of the same or different kind, the measurement model contains more than one equation.

Measurement precision The closeness of agreement between indications or measured quantity values obtained by replicate measurements on the same or similar objects under specified conditions. Measurement precision can be expressed numerically by the standard deviation, variance, or coefficient of variation under the specified conditions of measurement (repeatability conditions, intermediate precision conditions, or reproducibility conditions).

Measurement principle A phenomenon serving as a basis of a measurement.

Measurement procedure A detailed description of a measurement in accordance with one or more measurement principles and to a given measurement method, based on a measurement model and including any calculation to obtain a measurement result.

Measurement result Set of quantity values being attributed to a measurand together with any other available relevant information

Measurement repeatability Measurement precision under a set of repeatability conditions of measurement.

Measurement reproducibility Measurement precision under reproducibility conditions of measurement.

Measurement result The set of quantity values being attributed to a measurand together with any other available relevant information.

Measurement standard Realization of the definition of a given quantity, with stated quantity value and associated measurement uncertainty, used as a reference.

Measurement trueness A closeness of agreement between the average of an infinite number of replicate measured quantity values and a reference quantity value. Measurement trueness is inversely related to systematic measurement error, but is not related to random measurement error.

Measuring instrument The device used for making measurements, alone or in conjunction with one or more supplementary devices.

Measuring system A set of one or more measuring instruments and often other devices, assembled and adapted to give information used to generate measured quantity values within specified intervals for quantities of specified kinds.

Metrological traceability A property of a measurement result whereby the result can be related to a reference through a documented unbroken chain of calibrations, each contributing to the measurement uncertainty.

Metrological traceability chain A sequence of measurement standards and calibrations that is used to relate a measurement result to a reference.

Metrological traceability to a measurement unit The metrological traceability where the reference is the definition of a measurement unit through its practical realization.

National measurement standard Measurement standard recognized by national authority to serve in a state or economy as the basis for assigning quantity values to other measurement standards for the kind of quantity concerned.

Output quantity A measurement model is the quantity, the measured value of which is calculated using the values of input quantities in a measurement model.

Primary measurement standard Measurement standard established using a primary reference measurement procedure, or created as an artifact, chosen by convention.

Primary reference measurement procedure Reference measurement procedure used to obtain a measurement result without relation to a measurement standard for a quantity of the same kind.

Quantity The property of a phenomenon, body, or substance, where the property has a magnitude that can be expressed as a number and a reference. A reference can be a measurement unit, a measurement procedure, a reference material, or their combination.

Quantity value The number and reference together expressing magnitude of a quantity.

Quantity-value scale	Ordered set of quantity values of quantities of a given kind of quantity used in ranking, according to magnitude, quantities of that kind.
Random measurement error	A component of measurement error that in replicate measurements varies in an unpredictable manner.
Reference material	Material, sufficiently homogeneous and stable with reference to specified properties, which has been established to be fit for its intended use in measurement or in examination of nominal properties.
Reference measurement procedure	Measurement procedure accepted as providing measurement results fit for their intended use in assessing measurement trueness of measured quantity values obtained from other measurement procedures for quantities of the same kind, in calibration, or in characterizing reference materials.
Reference measurement standard	Measurement standard designated for the calibration of other measurement standards for quantities of a given kind in a given organization or at a given location.
Reference quantity value	A quantity value used as a basis for comparison with values of quantities of the same kind.
Relative standard measurement uncertainty	The standard measurement uncertainty divided by the absolute value of the measured quantity value.
Repeatability condition of measurement	The condition of measurement, out of a set of conditions that includes the same measurement procedure, same operators, same measuring system, same operating conditions and same location, and replicate measurements on the same or similar objects over a short period of time.
Reproducibility condition of measurement	The condition of measurement, out of a set of conditions that includes different locations, operators, measuring systems, and replicate measurements on the same or similar objects.
Secondary measurement standard	Measurement standard established through calibration with respect to a primary measurement standard for a quantity of the same kind.
Sensor	An element of a measuring system that is directly affected by a phenomenon, body, or substance carrying a quantity to be measured.

Standard measurement uncertainty	A measurement uncertainty expressed as a standard deviation.
Systematic measurement error	The component of measurement error that in replicate measurements remains constant or varies in a predictable manner.
Transfer measurement device	Device used as an intermediary to compare measurement standards.
Travelling measurement standard	Measurement standard, sometimes of special construction, intended for transport between different locations.
True quantity value	A quantity value consistent with the definition of a quantity.
Type A evaluation of measurement uncertainty	Evaluation of a component of measurement uncertainty by a statistical analysis of measured quantity values obtained under defined measurement conditions.
Type B evaluation of measurement uncertainty	Evaluation of a component of measurement uncertainty determined by means other than a Type A evaluation of measurement uncertainty. Evaluation of the Type B can be based on information associated with authoritative published quantity values, associated with the quantity value of a certified reference material, obtained from a calibration certificate, from the accuracy class of a measuring instrument, etc.
Uncertainty budget	A statement of a measurement uncertainty, of the components of that measurement uncertainty, and of their calculation and combination. An uncertainty budget should include the measurement model, estimates, and measurement uncertainties associated with the quantities in the measurement model, covariances, type of applied probability density functions, degrees of freedom, type of evaluation of measurement uncertainty, etc.
Verification	Provision of objective evidence that a given item fulfils specified requirements.
Validation	Verification, where the specified requirements are adequate for an intended use.
Working measurement standard	Measurement standard that is used routinely to calibrate or verify measuring instruments or measuring systems.

Appendix D
Cumulative Fractional Blackbody Function *F*

λT (μm K)	λ / λ_{max}	*F*
600	0.2071	9.29E-08
800	0.2761	0.0000164
1000	0.3451	0.000321
1200	0.4141	0.00213
1400	0.4831	0.00779
1600	0.5521	0.01972
1800	0.6212	0.03934
2000	0.6902	0.06673
2200	0.7592	0.10089
2400	0.8282	0.14026
2500	0.8627	0.16136
2600	0.8972	0.18312
2700	0.9317	0.20536
2800	0.9663	0.22789
2900	1.0008	0.25056
3000	1.0353	0.27323
3100	1.0698	0.29578
3200	1.1043	0.31809
3300	1.1388	0.34011
3400	1.1733	0.36173
3500	1.2078	0.38291
3600	1.2423	0.4036
3800	1.3113	0.44337
4000	1.3804	0.48087
4200	1.4494	0.51599

(continued)

V. Sapritsky and A. Prokhorov, *Blackbody Radiometry*, Springer Series in Measurement Science and Technology, https://doi.org/10.1007/978-3-030-57789-6

(continued)

λT(μm K)	λ/λ_{max}	F
4400	1.5184	0.54878
4600	1.5874	0.57926
4800	1.6564	0.60754
5000	1.7255	0.63373
5200	1.7945	0.65795
5400	1.8635	0.68034
5600	1.9325	0.70102
5800	2.0015	0.72013
6000	2.0706	0.73779
6400	2.2086	0.7692
6800	2.3466	0.7961
7200	2.4847	0.81918
7600	2.6227	0.83907
8000	2.7607	0.85625
8400	2.8988	0.87116
8800	3.0368	0.88413
9200	3.1748	0.89547
9600	3.3129	0.90541
10,000	3.4509	0.91416
11,000	3.7960	0.93185
12,000	4.1411	0.94505
13,000	4.4862	0.95509
14,000	4.8313	0.96285
15,000	5.1764	0.96893
16,000	5.5215	0.97377
18,000	6.2117	0.98081
20,000	6.9018	0.98555
25,000	8.6273	0.99217
30,000	10.3528	0.99529
35,000	12.0782	0.99695
40,000	13.8037	0.99792
45,000	15.5291	0.99852
50,000	17.2546	0.9989
60,000	20.7055	0.99935
70,000	24.1564	0.99959
75,000	25.8819	0.99967
100,000	34.5092	0.99986

Index

V. Sapritsky and A. Prokhorov, *Blackbody Radiometry*, Springer Series in Measurement Science and Technology, https://doi.org/10.1007/978-3-030-57789-6

K

L

M

N

O

P

The manufacturer's authorised representative in the EU is Springer Nature Customer Service Centre GmbH, Europaplatz 3, 69115 Heidelberg, Germany. If you have any concerns regarding our products, please contact ProductSafety@springernature.com

Printed and bound by CPI Group (UK) Ltd, Croydon, CR0 4YY
28/09/2026
02224067-0011